Waste Production and Utilization in the Metal Extraction Industry

Waste Production and Utilization in the Metal Extraction Industry

Sehliselo Ndlovu, Geoffrey S. Simate,
and Elias Matinde

CRC Press
Taylor & Francis Group
Boca Raton London New York

CRC Press is an imprint of the
Taylor & Francis Group, an **informa** business

CRC Press
Taylor & Francis Group
6000 Broken Sound Parkway NW, Suite 300
Boca Raton, FL 33487-2742

First issued in paperback 2020

ISBN 13: 978-0-367-57350-8 (pbk)
ISBN 13: 978-1-4987-6729-3 (hbk)

Library of Congress Cataloging-in-Publication Data

Names: Ndlovu, Sehliselo, author. | Simate, Geoffrey S., author. | Matinde, Elias, author.
Title: Waste production and utilization in the metal extraction industry / Sehliselo Ndlovu, Geoffrey S. Simate, Elias Matinde.
Description: Boca Raton : Taylor & Francis, CRC Press, 2017. | Includes bibliographical references and index.
Identifiers: LCCN 2016051880| ISBN 9781498767293 (hardback : alk. paper) | ISBN 9781498767309 (ebook)
Subjects: LCSH: Metallurgical plants--Waste minimization. | Metallurgy--Waste minimization. | Mineral industries--Environmental aspects. | Extraction (Chemistry) | Metal wastes.
Classification: LCC TN675.5 .N39 2017 | DDC 669.028/6--dc23
LC record available at https://lccn.loc.gov/2016051880

Visit the Taylor & Francis Web site at
http://www.taylorandfrancis.com

and the CRC Press Web site at
http://www.crcpress.com

Contents

Preface

The depletion of high-grade and easy-to-process metallic ores and demand for finished metal products is posing a serious challenge to the metals industry. The challenge of resource depletion, coupled with the difficulties in finding new metal resources, could have serious implications for global economic growth. In the meantime, significant amounts of mineral and metal waste with potential economic value continue to pile up as waste on the earth's surface. If these wastes can be recycled and reused efficiently, not only will it result in a reduction in the exploitation of virgin primary mineral resources, but it will also lead to the utilization of abandoned minerals and metallic waste, which would otherwise be an environmental nuisance. In recent years, there have been many developments in the field of recycling and reuse of minerals and metallic wastes; therefore, the prime aim of this book is to provide a useful reference for engineers and researchers in industry and academia on the current status and future trends in the recycling and reuse of minerals and metallic wastes. The book details the current trends in technology, policy and legislation regarding waste management in the mining, beneficiation, metal extraction and manufacturing processes and includes some aspects on metal recovery from post-consumer waste.

The book is divided into 10 chapters, each providing comprehensive reviews of major research accomplishments and up-to-date methods, tools and techniques for recycling and reusing of minerals and metallic wastes. The first chapter is of an introductory nature. Chapter 2 considers the impact of the waste on the environment and health and the associated policies and legislation that have been created to regulate and manage waste production. Chapter 3 gives a general background on the mining and beneficiation processes and the subsequent production of mining and beneficiation wastes. This is followed by Chapters 4 through 6, which cover the pyrometallurgical processes and related waste. These chapters deal specifically with the types and/or categories of the waste produced in the selected pyrometallurgical unit processes and cover the emerging trends and innovations in terms of the reduction, recycling and reuse of the mining and metallurgical waste in the selected processing units. Chapter 7 focuses on the generation, recycling and reuse of solid and liquid waste from hydrometallurgical unit processes. Chapter 8 addresses the production of and the reuse and/or recycling of manufacturing and metal finishing waste. The recycling of post-consumer waste materials, which include electronic waste, spent catalysts, automobile and household equipment is covered in Chapter 9. The final chapter integrates the various issues covered in the whole book. The chapter also explores the existing and emerging challenges and discusses future prospects.

This book does not oblige the readers to proceed in a systematic manner from one chapter to the next. By the nature of the subject, there are many parallel developments and cross-linkages between the chapters. Therefore, the readers can start at any point in the book and work backward or forward, and we have provided adequate cross-references within the different chapters.

No undertaking of this nature can be accomplished in isolation; therefore, we greatly acknowledge the enormous contribution to our knowledge and understanding by sources which are listed in the references at the end of each chapter. Finally, we thank the University of the Witwatersrand, Johannesburg, South Africa, for the support given during the compilation of this book.

Authors

Sehliselo Ndlovu is an associate professor in the School of Chemical and Metallurgical Engineering at the University of the Witwatersrand in Johannesburg, South Africa. Her specialization is in extractive metallurgy and in particular, mineral processing, hydrometallurgy, biohydrometallurgy and the treatment of industrial and mining effluents. She holds a Diploma of Imperial College (DIC) in hydrometallurgy and a PhD in minerals engineering from Imperial College, London, in the United Kingdom. She has more than 12 years of experience in the metallurgical engineering field. Sehliselo currently holds a DST/NRF SARChI professorial chair in hydrometallurgy and sustainable development at the University of the Witwatersrand.

Geoffrey S. Simate is an associate professor in the School of Chemical and Metallurgical Engineering at the University of the Witwatersrand, Johannesburg. He has more than 9 years of hydrometallurgical industrial experience and has held various senior metallurgical engineering positions. He holds a PhD in chemical engineering from the University of the Witwatersrand, and has published several scientific journal articles, seven book chapters and one edited book.

Elias Matinde is a senior lecturer in the School of Chemical and Metallurgical Engineering at the University of the Witwatersrand, Johannesburg. He has extensive experience in the scientific and applied research and development of metallurgical processes and in technical consultancy and metallurgical advisory in diverse areas of metallurgical processes. He holds a PhD in metallurgical engineering from Tohoku University in Japan and is a registered professional engineer with the Engineering Council of South Africa.

1

Overview

1.1 Introduction

Mining is a complex process whose activities range from exploration, mine development, mineral extraction, mineral processing (mineral beneficiation), refining and reclamation and remediation (for mine closure). It also covers a wide range of mineral commodities, including copper, iron, gold, nickel, lead, zinc and diamonds. These metals and minerals are the primary raw materials used in many industrial applications and thus are essential to the world economy. Copper, for example, is essential to the electronics and construction industries, while iron ore provides the base material for the steel, automotive and transportation industries. However, the extraction and beneficiation of these minerals from raw ore lead to the generation of large quantities and different types of waste. The effect of some of these wastes on the environment is dire and may also threaten the health and safety of humans. This book discusses the different types of mine waste generated and recovery thereof during the mining process right from the mineral extraction to the beneficiation and recovery processes to the manufacturing of finished products. This introductory chapter is a preamble to that discussion.

1.2 Legislation and Impacts of Mining and Metallurgical Waste

Mining and metallurgical wastes come from the survey or prospecting or mining and processing of solid, liquid and gaseous deposits, which due to their high content of heavy metals and other hazardous substances constitute a potential hazard (to human health, plant life and aquatic species) and may cause irreversible environmental damage. Some of the potential environmental problems associated with mining are listed in Table 1.1 (UNEP, 2000). In view of this, a wide range of regulations have been introduced to mitigate the environmental and potential health impacts of mining and metallurgical waste.

1.2.1 Legislation and Policies

The problems described in Table 1.1 are not insurmountable, but do present significant challenges. To begin with, the environmental agenda has now become too complex for ad hoc approaches. Therefore, several waste management regulations and laws have been

TABLE 1.1

Some Potential Environmental Impacts of Mining

Environmental Impacts	Pollution Impacts	Occupational Health Impacts
Destruction of natural habitat at the mining site and at waste disposal sites Destruction of adjacent habitats as a result of emissions and discharges Destruction of adjacent habitats arising from the influx of settlers Changes in river regime and ecology due to siltation and flow modification Alteration in water tables Change in landform Land degradation due to inadequate rehabilitation after closure Land instability Danger from failure of structures and dams Abandoned equipment, plant and buildings	Drainage from mining sites, including acid mine drainage and pumped mine water Sediment run-off from mining sites Pollution from mining operations in riverbeds Effluent from minerals processing operations Sewage effluent from the site Oil and fuel spills Soil contamination from treatment residues and spillage of chemicals Leaching of pollutants from tailings and disposal areas and contaminated soils Air emissions from minerals processing operations Dust emissions from sites close to living areas or habitats Release of methane from mines	Handling of chemicals, residues and products Dust inhalation Fugitive emissions within the plant Air emissions in confined spaces from transport, blasting, combustion Exposure to asbestos, cyanide, mercury or other toxic materials used on-site Exposure to heat, noise and vibration Physical risks at the plant or at the site Unsanitary living conditions

Source: UNEP, *Ind. Environ.*, 23, 1, 2000.

enacted and promulgated over the years. In general, waste management laws govern the transport, treatment, storage and disposal of waste (Qasioun, 2015; Shantgiri, 2015).

Generally, the mining and metallurgical industries produce solid, liquid and gaseous wastes. It is the management of these wastes that is a target of several international, national and regional laws. At the moment, the ISO 14000 family of standards is the only (or most) recognized framework that may help an organization to identify those aspects of its business that have a significant impact on the environment, to set objectives and targets to minimize these impacts and to develop programmes to achieve targets and implement operational control measures to ensure compliance with environmental policies (Beejadhur et al., 2007; ISO, 2009). Nevertheless, various countries have developed different approaches for managing waste which differ both in the legislative scope and effectiveness (Kumar and Singh, 2013). Therefore, in the absence of a single agreed-upon legislation and/or policy governing mining wastes, this book will only discuss the mining regulations from a selected number of countries. The book also discusses the best available practices and/or technologies for managing waste rock, tailings, slags, sludge and gaseous emissions.

1.2.2 Environmental and Health Impacts of Mining and Metallurgical Waste

Environmental and human health protection is the number one priority of any piece of legislation in the mining and metallurgical industry. In addition to the legislation and policies governing mining waste, Chapter 2 of this book also analyses the environmental and health impacts of various mining wastes, some of which are summarized in Table 1.2.

TABLE 1.2

Pollutants in Wastewater Streams and Their Impacts

Type of Pollutant	Impact	References
Heavy metals	The toxicity or poisoning of heavy metals results from the disruption of metabolic functions. Heavy metals disrupt the metabolic functions in two ways: (1) they accumulate in vital organs and glands such as the heart, brain, kidneys, bone and liver where they disrupt their important functions, and (2) they inhibit the absorption, interfere with or displace the vital nutritional minerals from their original place, thereby, hindering their biological functions. Plants experience oxidative stress upon exposure to heavy metals that leads to cellular damage and disturbance of cellular ionic homeostasis, thus disrupting the physiology and morphology of plants.	Simate and Ndlovu (2014)
Acidity	There is an indirect, but devastating effect of high acidity. For example, acidification increases bioconversion of mercury to methylmercury, which accumulates in fish, increasing the risk to toxicity in people who eat fish. When the soil pH is low, nitrogen, phosphorus and potassium are tied up in the soil and not available to plants. Calcium and magnesium, which are essential plant nutrients, may also be absent or deficient in low-pH soils. At low pH, toxic elements such as aluminium, iron and manganese are also released from soil particles, thus increasing their toxicity. Furthermore, if soil pH is low, the activity of soil organisms that break down organic matter is reduced.	Goyer et al. (1985); Simate and Ndlovu (2014)
Residual organic chemicals	Adverse health effects of exposure to high levels of organic chemicals, particularly persistent organic pollutants, include cancer, damage to the nervous system, reproductive disorders, disruption of the immune system, allergies and hypersensitivity, sensory system, endocrine system and death. During the decomposition process of organic compounds, the dissolved oxygen in the receiving water may be used up at a greater rate than it can be replenished, causing oxygen depletion and having severe consequences on the biota. Organic effluents also frequently contain large quantities of suspended solids which reduce the light available to photosynthetic organisms and, on settling out, alter the characteristics of the river bed, rendering it an unsuitable habitat for many invertebrates.	WHO (2010); Lenntech (2016)
Organometallic compounds	Organometallic compounds are metal complexes containing at least one direct, covalent metal-carbon bond. Organification (e.g. methylation) of metals generally increases their toxicity by rendering them more lipid soluble and facilitating their crossing of lipid barriers such as the cell membrane or blood-tissue (e.g. blood–brain) barriers. For example, tin in its inorganic form is considered non-toxic, but is toxic in organotin compounds. The toxicity of organometallic compounds is related to their accumulation degree in cells of living organisms, and their actions include blocking metabolic processes, disorders of barrier functions of intestine epithelial cells and the initiation of the apoptosis of nerve cells.	WHO (2006); Engel et al. (2009)

(Continued)

TABLE 1.2 (*Continued*)

Pollutants in Wastewater Streams and Their Impacts

Type of Pollutant	Impact	References
Sulphates	There is no definite sulphate dose that would induce adverse health effects in adults, and thus there is no set a health-based standard for sulphate in drinking water. However, dehydration and diarrhoea are common side effects following the ingestion of large amounts of sulphate-containing compounds.	Fawell et al. (2004a)
Hydrogen sulphide	There is scantily available data on the oral toxicity of hydrogen sulphide. However, alkali sulphides irritate mucous membranes and can cause nausea, vomiting and epigastric pain following ingestion. The oral dose of sodium sulphide fatal to humans has been estimated at 10–15 g.	Fawell et al. (2003)
Nitrates	The toxicity of nitrate to humans is mainly attributable to its reduction to nitrite. The major biological effect of nitrite in humans is its involvement in the oxidation of normal haemoglobin (Hb) to methaemoglobinaemia (metHb), which is unable to transport oxygen to the tissues. Methaemoglobinaemia, causes cyanosis and, at higher concentrations, asphyxia. Other suggested effects include congenital malformations, goitre and cancer.	Cotruvo et al. (2011)
Cyanide	The health effects of large amounts of cyanide are similar, whether through breathing air, eating food or drinking water that contains it. The effects of acute cyanide exposure are dominated by central nervous system and cardiovascular disturbances. Typical signs of acute cyanide poisoning include tachypnoea, headache, vertigo, lack of motor coordination, weak pulse, cardiac arrhythmias, vomiting, stupor, convulsions and coma. Other symptoms include gastrointestinal tract disturbances, trauma and enlarged thyroid. Pathological findings may include tracheal congestion with haemorrhage, cerebral and pulmonary oedema, gastric erosions and petechiae of the brain meninges and pericardium.	WHO (2004); Taylor et al. (2006)
Fluoride	The most obvious health effect of excess fluoride exposure is dental fluorosis, which when mild includes white streaks, and when severe can include brown stains, pits and broken enamel. Elevated fluoride intakes can also have more serious effects on skeletal tissues.	Fawell et al. (2004b)

1.3 Mining and Beneficiation Waste Production and Utilization

As already stated in the introductory section of this chapter, mining is a complicated process involving several different steps. Chapter 3 will only give a general background on mining and beneficiation processes and the subsequent production and utilization of mining and beneficiation wastes. Some of these mining and beneficiations wastes are briefly introduced in the succeeding text. Acid mine drainage (AMD), a product of the oxidative dissolution of sulphide minerals in mining and beneficiation wastes, is also discussed in Chapter 3.

1.3.1 Mine Overburden and Waste Rock

Mining operations generate two types of wastes: overburden and waste rock or mine development rock. Overburden results from the development of surface mines (or open pits), while waste rock is a by-product of mineral extraction in both surface and underground mines. Overburden includes all soil above the bedrock, whereas waste rock consists of rock which is extracted from the mine, but has less than the cut-off grade for the valuable minerals. Both the overburden and waste rock can be used in a number of industrial applications (e.g. aggregate for asphalt roads, road fill, commercial sandblasting and anti-skid surfacing of bridges).

1.3.2 Mineral Beneficiation Waste

Mineral beneficiation is a major component and is the first step after mining in the value chain for the production of final metal products. Mineral beneficiation consists of the manipulation of the physical properties of the ore to produce the desired particle sizes so as to liberate and concentrate the valuable mineral from the gangue material. The process consists of size reduction (through crushing and grinding) followed by the concentration or enrichment step which is achieved by the separation of valuable and gangue minerals using various techniques that are based on the physical and chemical properties of the minerals such as gravity concentration, dense media separation, magnetic separation and flotation. The concentration or enrichment steps result in the production of a valuable stream with a high concentration of the target metals and the waste stream composed mostly of gangue minerals and a small amount of the valuable metals.

In the past, the waste streams were disposed of into tailings dams. However, in recent years there has been a trend towards the reprocessing of some of these old tailings in order to recover valuable metals. These tailings have also been used as raw material in the construction and building industry, such as in the production of concrete mix, bricks, tiles and ceramics.

1.3.3 Acid Mine Drainage

AMD is considered to be one of the devastating environmental pollution problems associated with the mining industry (Udayabhanu and Prasad, 2010). Ideally, the AMD may be generated from mine waste rocks or tailings or mine structures such as active or closed or abandoned pits and underground workings. However, the causes of AMD are not only limited to the mining industry, but can also occur where sulphide materials are exposed, for example, in highway and tunnel construction and other deep excavations (Skousen, 1995; Skousen et al., 1998; ZDun, 2001). Chapter 3 describes AMD generation and its associated technical issues. The removal and recycling of metal contaminants and other by-products are also covered in Chapter 3.

1.4 Ferrous Metals Waste Production and Utilization

Ferrous metals constitute one of the most ubiquitous metal commodities driving the growth of the world economy. Basically, ferrous metals refer to metals and alloys containing iron, and the term is generically used to refer to iron and steel products. The production of iron and steel products has been the key driver of the global economy, mainly due to their irreplaceable utilizations in specialized applications such as in construction, mining, transport, manufacturing and energy sectors. In addition, iron and steel products are inextricably linked to economic growth and prosperity and are a major driver of economic growth for most manufacturing and resource-producing countries (World Steel Association, 2015a,b). According to World Steel Association (2015a) estimates for 2014, the steel industry contributed about 954 billion US$ to society directly and indirectly, including about 120 billion US$ in tax contributions.

The primary production of iron and steel products produced from ores is dominated by the integrated blast furnace process, which technically accounts for about 70% of global crude steel production. The electric furnace process, mostly utilizing steel scrap and other metallic charges, currently contributes about 30% of the global crude steel production (World Steel Association, 2012). The crude steel from the blast furnace and/or electric arc melting process is further refined to remove deleterious impurities and is alloyed to produce value-added steel products such as stainless steels. Depending on the desired quality and properties of the final products, the typical processes used in the refining of steel products include unit processes such as basic oxygen steelmaking, argon-oxygen decarburization, vacuum and ladle refining (Fruehan, 1998). Finally, the refined steel is cast into semi-finished products for manufacturing of end-use products.

Despite the ubiquity of ferrous metals in the development of mankind, the pyrometallurgical production of these metal products is associated with the production of waste materials which are detrimental to the environment. In general, there is a strong expectation on the part of producers to proactively adopt environmentally sustainable business models in their operations in order to reduce the anthropogenic effects from the production of iron and steel products (World Steel Association, 2015b). According to the World Steel Association (2015b), the steel industry has adopted a circular economy model which promotes zero waste through the reuse and recycling of by-product materials. In this regard, the focus is on finding new markets and applications for steel by-products and wastes such as process slags, particulate emissions, spent refractory materials and process gases and sludges (Atkinson and Kolarik, 2001; World Steel Association, 2015b).

Chapter 4 provides a technical overview of the unit processes in the smelting and refining of iron and steel, and highlights the different types and categories of metallurgical wastes produced in the respective unit processes. Furthermore, the chapter provides an overview of the recycling and valorization opportunities and challenges of the various metallurgical wastes from the perspective of the unit processes producing them and discusses the current practices and trends in terms of waste reduction, recycling and utilization in the iron and steel sector.

1.5 Ferroalloys Waste Production and Utilization

Ferroalloys refer to the various master alloys of iron with a high proportion of one or more other elements, such as chromium, manganese, nickel, molybdenum and silicon (Eric, 2014). With the exception of silicon which has unique industrial and commercial applications, such as photovoltaic and microelectronics applications, most of ferroalloys are consumed in the production of iron and steel products (Holappa, 2010; Gasik, 2013; Tangstad, 2013; Holappa 2013; Eric, 2014). In fact, the growth of the ferroalloys industry has largely been driven by advances in the steel industry and the growing demand for high-performance materials in various applications (Holappa, 2010).

In general, ferroalloys are mainly used as alloying agents in steelmaking in order to obtain an appropriate chemical composition of liquid steel and impact distinct mechanical and chemical properties of solidified steels (Holappa, 2010; Gasik, 2013). Furthermore, the various grades of ferroalloys also play an important role in the metallurgical refining and solidification of various grades of steels, and these include, inter alia, in the refining, deoxidation and control of inclusions and precipitates (Pande et al., 2010; Gasik, 2013; Holappa and Louhenkilpi, 2013). In fact, the advent of ferroalloy metallurgy has been credited for the development of various ubiquitous grades of stainless and specialty steels (Gasik, 2013).

Naturally, the mining, beneficiation and smelting of ferroalloy ores are associated with the generation of toxic and hazardous wastes in the form of furnace slags, particulate dusts, beneficiation residues and leach residues (Richard and Bourg, 1991; EPA, 1995; Barnhart, 1997a,b; Ma and Garbers-Craig, 2006). Traditionally, the waste materials were disposed of in landfills, but the anthropogenic effects associated with the disposal of such wastes in landfills, particularly the potential leachability of toxic heavy metal species such as chromium, silicon, nickel, manganese, lead and zinc compounds, have resulted in a protracted shift towards holistic management practices incorporating the recycling and/or reuse and solidification and/or stabilization, as alternatives to disposal in landfills (Richard and Bourg, 1991; Barnhart, 1997a,b; Tanskanen and Makkonen, 2006; Olsen et al., 2007; Durinck et al., 2008; Kumar et al., 2013; Tangstad, 2013).

The processes used in the production of various grades of bulk ferroalloys are reviewed in Chapter 5. Furthermore, Chapter 5 discusses the characteristics and anthropogenic effects of the waste streams produced in the typical bulk ferroalloy unit processes. The potential and applicable waste mitigation, valorization and inertization techniques and processes are also addressed. In terms of waste production and utilization, the various considerations taken into account in Chapter 5 include the raw materials effects, equipment design effects, process chemistry effects and valorization potential of the different waste streams.

1.6 Base Metals Waste Production and Utilization

Base metals have irreplaceable industrial and commercial applications, and their consumption is directly correlated with global economic growth (IMF, 2015). Basically, base metals broadly refer to a category of non-ferrous metal commodities typically consisting of copper, nickel, cobalt, lead, zinc and mercury (Craig and Vaughan, 1990; Jones, 1999; Cole and Ferron, 2002; Crundwell et al., 2011; Schlesinger et al., 2011). In general, the primary pyrometallurgical production of base metals involves the following steps (Jones, 1999; Cole and Ferron, 2002; Crundwell et al., 2011; Schlesinger et al., 2011): (1) smelting of base metal concentrates in flash or electric furnaces to produce base metal–rich matte, (2) matte converting under hydrodynamic oxidizing conditions to produce base metal–enriched matte (significantly higher than the smelter furnace matte) and (3) the refining and purification process to produce high-purity metals.

The production of base metals is not immune to the typical environmental challenges associated with the generic production of metals. In general, the production processes of these metals inherently produce by-products and waste streams that can potentially cause significant health effects and environmental pollution (European Commission, 2014). In terms of green economy, the base metals industry is focusing on not only in meeting the demand of base metal products, but is also focusing on mitigating the environmental effects of waste streams and by-products from base metals production (European Commission, 2014).

Chapter 6 discusses the primary pyrometallurgical production processes of base metals (nickel, copper and cobalt) and platinum group metals from sulphide ores. Furthermore, the chapter highlights the production of waste streams, particularly process slags and particulate dusts, from the respective unit processes, and discusses the relevant processes and techniques applicable in the recycling and valorization of base metals. The chapter also discusses the platinum group metals (PGMs) since the primary production of PGMs is closely associated with the production processes of base metals.

1.7 Hydrometallurgical Waste Production and Utilization

Hydrometallurgy has been widely used for the production of an array of metals. This branch of extractive metallurgy is concerned with the production of metals or metallic compounds from ores, concentrates or other intermediate materials by a sequence of operations carried out in aqueous solutions. Its popularity when compared to pyrometallurgical extraction has been mostly related to its associated lower capital and operational costs, lower energy consumption, and its ability to treat an extensive range of feed materials such as complex, primary, high- and low-grade ores and secondary waste materials. It is also been regarded as much more environmentally friendly with the ability to use dilute and also natural inorganic and organic reagents which can be regenerated and recycled in the process flowsheets.

The hydrometallurgical circuit is divided into different stages which include metal extraction, solution purification and metal recovery. A generalized flow diagram illustrating many of the key hydrometallurgical processes as they occur in sequence is shown in Figure 1.1.

Chapter 7 will look at some of the hydrometallurgical production processes for particular metals, such as copper, zinc, cobalt, nickel, aluminium, PGMs, gold and uranium. The focus will be on the general flowsheet for the production of these metals, the waste streams generated and the potential value and resource recovery that can be attained from each particular stream. Some of the highlights of Chapter 7 are given in the following sections.

1.7.1 Production of Copper

The production of copper can be achieved using two approaches: (1) a full hydrometallurgical approach which is commonly used for both low- and high-grade oxide and secondary sulphide ores and (2) a pyrometallurgical route which is applied to high-grade sulphide ores. In the hydrometallurgical route known as the leach, solvent extraction and electrowinning (L/SX/EW) route, the feed material first undergoes leaching, followed by solvent extraction for solution purification and a concentration upgrade, then lastly electrowinning to obtain the required solid metallic copper cathodes. In the copper production by L/SX/EW, the potential residues that are generated include crud from the solvent extraction circuit and cell sludge from the electrowinning circuit. Crud treatment is usually

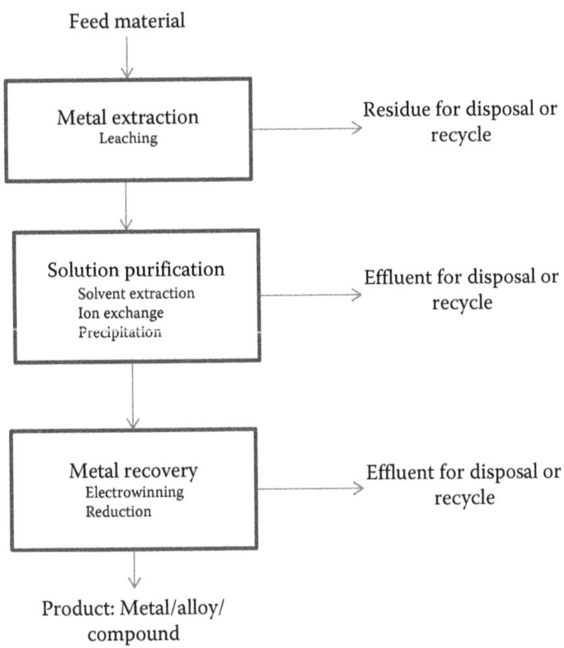

FIGURE 1.1
A generalized flow diagram illustrating key hydrometallurgical processes.

undertaken in order to recover some of the organic matter and thus minimize losses which leads to cost savings as the organic matter can be recycled into the extraction circuit.

In the pyrometallurgical process, the impure copper which is generated is purified using electrorefining. The electrorefining process results in the production of anode slime/sludge which constitutes significant levels of valuable metals such as PGMs. The recovery of these valuable metals by treating the anode slime is of great importance in the copper production industry as it significantly adds to the company's revenues stream. A number of processes are currently being used in industry and some of these are discussed in detail in Chapter 7.

1.7.2 Production of Zinc

The roast–leach–electrowin process has been widely used for the production of zinc from sulphide ores. The process involves the roasting of the zinc sulphide concentrates to generate an oxide concentrate which is then leached to get a zinc pregnant leach solution and followed by electrowinning to obtain the zinc metal. A number of variations to this basic zinc hydrometallurgical production process flowsheet have been adopted with the key focus being on the minimization of the amount of iron in the pregnant leach solution going to the electrowinning circuit. It is imperative to reduce the amount of iron in the solution that goes to the electrowinning circuit since the electrowinning efficiency of zinc from aqueous solutions can be negatively affected by the presence of even a trace amount of iron. Current efficiency and product purity is usually lowered. The efficient removal of iron from the zinc leach liquors have been achieved through a number of processes, for example the jarosite, the goethite and the hematite processes. These processes are named after the iron product that is generated during the iron removal stage, and recovery of valuable metallic iron from these waste products has been significantly considered through numerous research and development activities. In addition to iron recovery, the jarosite and goethite produced have found ready use in a variety of industrial applications, such as the production of construction and building material (e.g. as concrete, bricks, cement and tiles), catalysts, pigments, gypsum and refractories, composites, glass and ceramics (Pelino, 2000; Pappu et al., 2006a,b).

1.7.3 Production of Alumina

Alumina (or aluminium hydroxide) has been traditionally produced from bauxite using the Bayer process. In the Bayer process, bauxite is leached in a hot solution of sodium hydroxide; the resulting hydroxide solution is then cooled and the dissolved aluminium hydroxide precipitates out as a solid. The hydroxide is then calcined decomposing the aluminium hydroxide to alumina (Al_2O_3) and giving off water vapour in the process. During the leaching process some components of bauxite do not dissolve and are removed from the solution in a thickener during the clarification stage, resulting in a mixture of solid impurities known as red mud or bauxite residue.

The various uses of red mud include the production of iron and steel, titanium and alumina and the recovery of minor metallic constituents such as rare earths (Liu and Li, 2015). It has also been used as a raw material in the building and construction industry (e.g. bricks, light weight aggregates, roofing and flooring tiles, cements, etc.) and in catalysis, ceramics, tiles, glasses and glazes (Klauber et al., 2009, 2011; Samal et al., 2013; Liu and Naidu, 2014). Other alternative uses include the treatment of industrial wastewater and as a fertilizer in agricultural industries (Klauber et al., 2009, 2011).

1.7.4 Production of Precious Metals

The production of precious metals, such as gold and PGMs, has been well established for a number of years. Precious metals tend to be much nobler and, as a result, require very strong leaching reagents in order to get them into solution. Usually a complexing agent in the presence of an oxidant is used in the dissolution process.

For the production of gold, cyanide is used as a complexing agent in the presence of oxygen as the oxidant. The gold is then recovered from the solution using carbon adsorption technology. In order to ensure good leaching kinetics and high overall gold recovery, it is always necessary to add more cyanide during leaching process. This excess cyanide goes to the tailings as uncomplexed or free cyanide. However, rising costs associated with cyanide consumption and strict environmental regulations have led to an increased interest in technologies that can be used to recover and recycle some of the cyanide. Technologies such as the sulphidization–acidification–recycle–thickening process (the Hannah process); the acidification, volatilization and reneutralization process and application of membranes have been used to detoxify and recover some cyanide from the solution stream before discharge to the environment (Fleming, 2003). The advantage of these processes is the potential recovery of saleable zinc and copper metal products and cyanide which can be recycled to the leaching process (Fleming, 2003).

In the production of PGMs, chloride solutions are usually used as complexing agents in the presence of an oxidant such as hydrogen peroxide. The leaching process is followed by solvent extraction or ion exchange to purify and concentrate the solution. Unfortunately, some traces of PGM remain in the leach solution at levels which are not economic to purify and are thus lost to the effluent stream.

Over time, effluent which contains traces of PGMs, eventually, leads to the accumulation of metal losses into the environment which represents a cost to the company in terms of revenue lost. Therefore, a wide range of techniques, such as precipitation, cementation, ion exchange and adsorption, have been studied for the recovery and separation of PGMs from the effluent streams (Guibal et al., 1999; Amos et al., 2000; Saktas and Morcali, 2011; Izatt et al., 2012). Furthermore, some of these techniques have been used to develop innovative industrial technologies such as the Smopex® and Molecular Recognition Technology (MRT) for the recovery of the PGMs from the effluent streams (Amos et al., 2000; Izatt et al., 2012).

1.7.5 Production of Strategic and Critical Metals

The section on strategic metals looks at the hydrometallurgical processes for the production of uranium oxide and rare earth elements (REE). The hydrometallurgical process for the production of the uranium oxide involves a number of steps, such as leaching, ion exchange, solvent extraction and precipitation. In this process, a major portion of the radionuclides present in the ore remains in the tailings after uranium has been extracted and they can have a long-term impact on the environment due to the long half-lives and the ready availability of some of the toxic radionuclides (Weil, 2012). Current literature suggests that there is not much value or resource recovery from these tailings as they are generally dumped in special ponds or piles and covered by clay and topsoil with enough rock to resist erosion.

The production of REE is mostly achieved from monazite and bastnasite ores. The REEs are extracted from these ores by sulfuric and hydrochloric acid digestion. Thereafter, processes such as fractional crystallization and precipitation, solvent extraction, ion exchange and reduction are used to recover REE from the leach solutions. Once all processing has been undertaken, the resulting waste generally has no commercial value and is, therefore, disposed of (US Department of Energy, 2009, 2011). Although the waste is also considered hazardous, containing slightly radioactive thorium and uranium and other toxic substances, it is usually either dumped into a tailings pond in liquid form where it is able to leach into groundwater or in the form of a cemented paste laid out at a disposal site.

1.8 Metal Manufacturing and Finishing Waste Production and Utilization

Metal manufacturing and finishing refer to processes used to transform the metal intermediate products from the solidification processes into value-added manufactured and finished goods and components such as machine components, machinery, instruments and tools which can directly be used by end users (Kaushish, 2010). Some of the typical metal manufacturing processes include metal forming (either hot or cold forming), rolling and extrusion, machining and forging, and these, essentially, involve using an externally applied force to plastically deform a metal product into various desired shapes and sizes (Rentz et al., 1999; Boljanovic, 2010; Kaushish, 2010; Klocke, 2014).

In addition to the generic metal manufacturing processes, foundry processes also play an integral role in the production of near-net finished products by casted and solidified ferrous and non-ferrous metals (Davies, 1996; Oman, 1988; Bawa, 2004; European Commission, 2005; Jezierski and Janerka, 2011).

Generally, metal finishing processes refer to a generic group of surface coating techniques adopted to enhance the surface finish of steel products and impart properties such as corrosion and chemical resistance, biocompatibility, reflectivity and appearance, hardness, electrical conductivity, wear resistance and other engineering or mechanical properties (Roto, 1998; Lindsay, 2007; Regel-Rosocka, 2009; Rögener et al., 2012). Technically, the metal finishing industry involves two distinct operations, that is, plating (i.e. electroplating and electroless plating) and surface treatment (i.e. chemical and electrochemical conversion, case hardening, metallic coating and chemical coating) processes (EPA, 1997).

Despite the criticality of the metal forming, manufacturing and finishing processes, these processes are associated with the production of toxic and hazardous by-products (EPA, 1997, 2002; Rentz et al., 1999; Atkinson and Kolarik, 2001; Twidwell and Dahnke, 2001; Regel-Rosocka, 2009; Rögener et al., 2012). Chapter 8 discusses some of the technical aspects of the typical metal forming, manufacturing and finishing processes, the different waste products produced and the possible techniques and opportunities in reducing, recycling and utilizing the waste materials. Chapter 8 also discusses the production, recycling and utilization of foundry wastes.

1.9 Post-Consumer Waste Production and Utilization

Metals significantly contribute to the vibrancy of the modern-day economy. Unfortunately, rapidly growing demand for metals has led to the diminishing of high-grade metallic ores. Therefore, alternatives to the primary supply of metals from mined ores are being developed to bring relief to the metal market. Chapter 9 discusses some of the metallic recyclables from post-consumer materials (e.g. steel products, electronic waste, spent catalysts, automobile, household resources) including their reuse potential, recycling technologies, problems and solutions, and various international legislation and policies governing a selected number of post-consumer wastes.

As alluded to, recycling saves resources, as it reduces the primary metal supply from primary mining by using the resources temporarily locked up in post-consumer materials (UNEP, 2013). For example, 'critical' elements, not abundant in nature, must be preserved and reused as much as possible. A significant advantage of metal recycling is that the energy consumption of metal recovery from recycled sources is usually less than that of primary production, as recycling often 'only' involves the re-melting of metals.

1.10 Conclusions and Future Outlook

Waste materials from the various metal value chain processes have the potential to act as a secondary value resource and/or utilized in a number of alternative industrial applications. However, a number of challenges such as lack of adequate regulatory frameworks and policies, the complexity of the residue material versus available technology, lack of a comprehensive cost–benefit analysis, market availability and acceptance of the secondary products add to the myriad of problems faced by secondary resource or waste utilization industries. Despite the challenges mentioned, a number of countries are focusing on the development of diverse options for residue treatment, management and utilization as a means of industrial sustainability. Waste management options such as a zero waste process approach are shown to be more preferred from an economic and environmental point of view. However, these are generally difficult to accomplish in the mineral and metal extraction sector. A synergistic approach among residue producers and also between residue producers and potential downstream users is seen as having more potential for the future since this would promote the development of technologies that can lead to more complementary and comprehensive exploitation of the different waste materials being generated in the industry today.

References

Amos, G., Hopkins, W., Izatt, S.R., Bruening, R.L., Dale, J.B. and Krakowiak, K.E. (2000). Extraction, recovery, and recycling of metals from effluents, electrolytes, and product streams using molecular recognition technology. Plenary lecture, V International Conference on Clean Technologies for the Mining Industry, Santiago, Chile, May 9–13, 2000. In: Sánchez, M.A., Vergara, F., Castro, S.H. (eds.), *Environmental Improvements in Mineral Processing and Extractive Metallurgy,* Vol. 2, University of Concepción, Concepción, Chile, 2000, pp. 1–16.

Atkinson, M. and Kolarik, R. (2001). *Steel Industry Technology Roadmap.* The American Iron and Steel Institute, Washington, DC.

Barnhart, J. (1997a). Occurrences, uses, and properties of chromium. *Regulatory Toxicology and Pharmacology* 26: S3–S7.

Barnhart, J. (1997b). Chromium chemistry and implications for the environmental fate and toxicity. *Journal of Soil Contamination* 6(6): 561–568.

Bawa, H.S. (2004). Foundry. *Manufacturing Processes II.* Tata McGraw Hill, New Delhi, India, pp. 14–16.

Beejadhur, Y., Gujadhur, K. and Ghizzoni, L. (2007). An introduction to ISO 14000. Available at http://www.intracen.org/uploadedFiles/intracenorg/Content/Exporters/Exporting_Better/Quality_Management/Redesign/EQB78%20En%2015.09.2010.pdf. Accessed 11 May 2016.

Boljanovic, V. (2010). *Metal Shaping Processes: Casting and Molding, Particulate Processing, Deformation Processes and Metal Removal.* Industrial Press Inc., New York.

Cole, S. and Ferron, C.J. (2002). A review of the beneficiation and extractive metallurgy of the platinum group elements, highlighting recent process innovations. Technical Paper 2002–02. SGS Minerals Services, pp. 1–43.

Cotruvo, J., Fawell, J.K., Giddings, M., Jackson, P., Magara, Y., Ngowi, A.V.F. and Ohanian, E. (2011). Nitrate and nitrite in drinking-water. Available at http://www.who.int/water_sanitation_health/dwq/chemicals/nitratenitrite2ndadd.pdf. Accessed 4 March 2016.

Craig, J.R. and Vaughan, D.J. (1990). Compositional and textural variations of the major iron and base metal sulphide minerals. In: Gray, P.M.J., Bowyer, G.J., Castle, J.F., Vaughan, D.J. and Warner, N.A. (eds.), *Sulphide Deposits – Their Origin and Processing.* The Institution of Mining and Metallurgy, London, UK, p. 1.

Crundwell, F.K., Moats, M.S., Ramachandran, V., Robinson, T.G. and Davenport, W.G. (2011). *Extractive Metallurgy of Nickel, Cobalt and Platinum-Group Metals.* Elsevier, Oxford, UK.

Davies, J.R. (1996). *Cast Irons. ASM Specialty Handbook.* ASM International, Materials Park, OH. Available at www.asminternational.org, Accessed 2 June 2016, p. 133.

Durinck, D., Engstrom, F., Arnout, S., Huelens, J., Jones, P.T., Bjorkman, B., Blanpain, B. and Wollants, P. (2008). Hot stage processing of metallurgical slags: Review. *Resources, Conservation and Recycling* 52: 1121–1131.

Engel, G., Podolak, M. and Man, D. (2009). The effect of selected organometallic compounds on membrane voltage. *Polish Journal of Environmental Studies* 18(2): 207–211.

EPA (1995). Compilation of air pollutant emission factors. Volume I: Stationary point and area sources. Available at https://www3.epa.gov/ttn/chief/ap42/ch12/final/c12s03.pdf. Accessed 23 May, 2016.

Environmental Protection Agency (EPA) (1997). Pollution prevention for the metal finishing industry: A manual for pollution prevention technical assistance providers. EPA/742/B-97/005. U.S. Environmental Protection Agency, Washington, DC.

Environmental Protection Agency (EPA) (2002). Development document for final effluent limitations guidelines and standards for the iron and steel manufacturing point source category. EPA-821-R-02-004. U.S. Environmental Protection Agency, Washington, DC.

Eric, R.H. (2014). Production of ferroalloys. In: Seetharaman, S., McLean, A., Guthrie, R. and Seetharaman, S. (eds.), *Treatise on Process Metallurgy, Volume 3: Industrial Processes, Part A.* Elsevier, Oxford, UK, pp. 477–532.

European Commission (2005). Integrated pollution prevention and control: Reference document on the best available techniques in the smitheries and foundries industry. Available at http://eippcb.jrc.ec.europa.eu/reference/BREF/sf_bref_0505.pdf. Accessed 6 June 2016.

European Commission (2014). Processes to produce copper and its alloys from primary and secondary raw materials. Best Available Techniques (BAT) Reference Document for the Non-Ferrous Metal Industries. Final Draft. Industrial Emissions Directive 2010/75/EU (Integrated Pollution Prevention and Control).

Fawell, J.K., Lund, U. and Mintz, B. (2003). Hydrogen sulfide in drinking-water. Available at http://www.who.int/water_sanitation_health/dwq/chemicals/en/hydrogensulfide.pdf. Accessed 7 March 2016.

Fawell, J.K., Ohanian, E., Giddings, M., Toft, P., Magara, Y. and Jackson, P. (2004a). Sulfate in drinking-water. Available at http://www.who.int/water_sanitation_health/dwq/chemicals/sulfate.pdf. Accessed 24 March 2016.

Fawell, J.K., Ohanian, E., Giddings, M., Toft, P., Magara, Y. and Jackson, P. (2004b). Fluoride in drinking-water. Available at http://www.who.int/water_sanitation_health/dwq/chemicals/fluoride.pdf. Accessed 11 March 2016.

Fleming, C.A. (2003). The economic and environmental case for recovering cyanide from gold plant tailings. SGS Minerals Services Technical Paper 2003-02.

Fruehan, R.J. (1998). Overview of steelmaking processes and their development. In: Fruehan, R.J. (ed.), *The Making, Shaping and Treatment of Steels: Steelmaking and Refining*, Vol. 2, 11th edn. The AISE Steel Foundation, Pittsburgh, PA, p. 1.

Gasik, M.M. (2013). Introduction to ferroalloys. In: Gasik, M.M. (ed.), *Handbook of Ferroalloys: Theory and Technology*. Butterworth-Heinemann, Elsevier, Oxford, UK, pp. 3–7.

Goyer, R.A. et al. (1985). Potential human health effects of acid rain: Report of a workshop. *Environmental Health Perspectives* 60: 355–368.

Guibal, E., Vincent, T., Larkin, A. and Tobin, J.M. (1999). Chitosan sorbents for platinum recovery from dilute solutions. *Journal of Industrial & Engineering Chemistry Research* 38(10): 4011–4022.

Holappa, L. (2010). Towards sustainability in ferroalloy production. *Journal of the Southern African Institute of Mining and Metallurgy* 110: 703–710.

Holappa, L. (2013). Basics of ferroalloys. In: Gasik, M.M. (ed.), *Handbook of Ferroalloys: Theory and Technology*. Butterworth-Heinemann. Elsevier, Oxford, U.K., pp 9–27.

Holappa, L. and Louhenkilpi, S. (2013). On the role of ferroalloys in steelmaking. *13th International Ferroalloys Congress: Efficient Technologies in Ferroalloys Industry*, Almaty, Kazakhstan, pp. 1083–1090.

International Monetary Fund (IMF) (2015). Commodity market developments and forecasts, with a focus on metals in the world economy. Commodity special feature. Available at https://www.imf.org/external/np/res/commod/pdf/WEOSpecialOCT15.pdf. Accessed 17 June 2016.

ISO (2009). Environmental management: The ISO 14000 family of international standards. Available at http://www.iso.org/iso/theiso14000family_2009.pdf. Accessed 10 May 2016.

Izatt, S.R., Bruening, R.L. and Izatt, N.E. (2012). Some applications of molecular recognition technology (MRT) to the mining industry. In: Wang, S., Dutrizac, J.E., Free, M.L., Hwang, J.Y. and Kim, D. (eds.), *T.T. Chen Honorary Symposium on Hydrometallurgy, Electrometallurgy and Materials Characterisation*. TMS, Warrendale, PA, pp. 51–63.

Jezierski, J. and Janerka, K. (2011). Waste utilization in foundries and metallurgical plants. In: Kumar, S. (ed.), *Integrated Waste Management*, Vol. 1. INTECH, Rijeka, Croatia. Available at http://www.intechopen.com/books/integrated-waste-management-volume-i/solid-waste-utilization-in-foundries-and-metallurgical-plants. Accessed 9 June 2016.

Jones, R.T. (1999). Platinum smelting in South Africa. *South African Journal of Science* 95: 525–534.

Kaushish, J.P. (2010). *Manufacturing Processes*, 2nd edn. PHI Learning (Pvt) Ltd, New Delhi, India, p. 624.

Klauber, C., Gräfe, M. and Power, G. (2009). Bauxite residue issues: II. Options for residue utilization, CSIRO Document 3609, pp. 1–79.

Klauber, C., Gräfe, M. and Power, G. (2011). Bauxite residue issues: II. Options for residue utilization. *Hydrometallurgy* 108: 11–32.

Klocke, F. (2014). *Manufacturing Processes 4: Forming*, RWTH edn. Springer, Aachen, Germany, pp. 12–15.

Kumar, S., Garcia-Trinanes, P., Teixeira-Pinto, A. and Bao, M. (2013). Development of alkali activated cement from mechanically activated silico-manganese slag. *Cement and Concrete Composites* 40: 7–13.

Kumar, U. and Singh, D.N. (2013). E-waste management through regulations. *International Journal of Engineering Inventions* 3(2): 6–14.

Lenntech (2016). Effects of organic pollution on freshwater ecosystems. Available at http://www.lenntech.com/aquatic/organic-pollution.htm. Accessed March 2016.

Lindsay, T.C. (2007). Metal finishing and electroplating. In: Kutz, M. (ed.), *Environmentally Conscious Manufacturing*. Wiley Series in Environmentally Conscious Engineering, John Wiley & Sons, Inc., Hoboken, NJ, p. 123.

Liu, Y. and Naidu, R. (2014). Hidden values in bauxite residue (red mud): Recovery of metals. *Waste Management* 34: 2662–2673.

Liu, Z. and Li, H. (2015). Metallurgical process for valuable elements recovery from red mud—A review. *Hydrometallurgy* 155: 29–43.

Ma, G. and Garbers-Craig, A.M. (2006). A review on the characteristics, formation mechanism and treatment processes of Cr (IV)-containing pyrometallurgical wastes. *Journal of the Southern African Institute of Mining and Metallurgy* 106: 753–763.

Olsen, S.E., Tangstad, M. and Lindstad, T. (2007). *Production of Manganese Ferroalloys*. SINTEF and Tapir Akademisk Forlag, Trondheim, Norway.

Oman, D.E. (1988). Hazardous waste minimization: Part IV Waste minimization in the foundry industry. *Journal of the Air Pollution Control Association* 38(7): 932–940.

Pande, M.M., Guo, M., Guo, X., Geysen, D., Devisscher, S., Blainpain, B. and Wollants, P. (2010). Ferroalloy quality and steel cleanliness. *Ironmaking and Steelmaking* 37(7): 502–511.

Pappu, A., Saxena, M. and Asolekar, S. (2006a). Hazardous jarosite use in developing nonhazardous product for engineering application. *Journal of Hazardous Materials* 137(3): 1589–1599.

Pappu, A., Saxena, M. and Asolekar, S.R. (2006b). Jarosite characteristics and its utilisation potentials. *Science of the Total Environment* 359(1): 232–243.

Pelino, M. (2000). Recycling of zinc-hydrometallurgy wastes in glass and glass ceramic materials. *Waste Management* 20: 561–568.

Qasioun (2015). Biggest illegal waste dump in Europe. Available at http://qasioun.net/en/article/Biggest-illegal-waste-dump-in-Europe/1816/. Accessed 3 November 2015.

Regel-Rosocka, M. (2009). A review on methods of regeneration of spent pickling solutions from steel industry. *Journal of Hazardous Materials* 177: 57–69

Rentz, O., Jochum, R. and Schultmann, F. (1999). Report on best available techniques (BAT) in the German Ferrous Metals Processing Industry. Final Draft. Duetsch-Französiches Institut für Umweltforschung. Available at https://www.umweltbundesamt.de/sites/default/files/medien/publikation/long/2490.pdf. Accessed 2 June 2016.

Richard, F.C. and Bourg, A.C.M. (1991). Aqueous chemistry of chromium: A review. *Water Research* 25(7): 807–816.

Rögener, F., Sartor, M., Bàn, A., Buchloh, D. and Reichardt, T. (2012). Metal recovery from spent stainless steel pickling solutions. *Resources, Conservation and Recycling* 60: 72–77.

Roto, P. (1998). Smelting and refining. In: Stellman, J.M. (ed.), *Encyclopedia of Occupational Health and Safety: Chemical, Industries and Occupations*, Vol. III, 4th edn. International Labour Office, Geneva, Switzerland, 82.2.

Saktas, S. and Morcali, M.H. (2011). Platinum recovery from dilute platinum solutions using activated carbon. *Transactions of Nonferrous Metals Society of China* 21: 2554–2558.

Samal, S., Ray, A.K. and Bandopadhyay, A. (2013). Proposal for resources, utilization and processes of red mud in India – A review. *International Journal of Mineral Processing* 118: 43–55.

Schlesinger, M.E., King, M.J., Sole, K.C. and Davenport, W.G. (2011). *Extractive Metallurgy of Copper*, 5th edn. Elsevier, Oxford, UK.

Shantgiri, S.S. (2015). The environmental law. *Indian Streams Research Journal* 5(9): 1–5.

Simate, G.S. and Ndlovu, S. (2014). Acid mine drainage: Challenges and opportunities. *Journal of Environmental Chemical Engineering* 2(3): 1785–1803.

Skousen, J. (1995). Acid mine drainage. *Green Lands* 25(2): 52–55.

Skousen, J., Rose, A., Geidel, G., Foreman, J., Evans, R. and Hellier, W. (1998). *Handbook of Technologies for Avoidance and Remediation of Acid Mine Drainage*. The National Mine Land Reclamation Centre, West Virginia University, Morgantown, WV.

Tangstad, M. (2013). Manganese ferroalloys technology. In: Gasik, M. (ed.), *Handbook of Ferroalloys: Theory and Technology*. Butterworth-Heinemann, Elsevier, Oxford, UK, pp. 222–265.

Tanskanen, P. and Makkonen, H. (2006). Design of slag mineralogy and petrology: Examples of useful methods for slag composition and property design. *Global Slag Magazine* 5: 16–20.

Taylor, J., Roney, N., Harper, C., Fransen, M.E. and Swarts, S. (2006). *Toxicological Profile for Cyanide*. Agency for Toxic Substances and Disease Registry, Atlanta, GA.

Twidwell, L.G. and Dahnke, D.R. (2001). Treatment of metal finishing sludge for detoxification and metal value. *The European Journal of Mineral Processing and Environmental Protection* 1(2): 76–88.

Udayabhanu, S.G. and Prasad, B. (2010). Studies on environmental impact of acid mine drainage generation and its treatment: An appraisal. *Indian Journal of Environmental Protection* 30(11): 953–967.

UNEP (2000). Mining and sustainable development II: Challenges and perspectives. *Industry and Environment* 23: 1–96.

UNEP (2013). Metal recycling: Opportunities, limits, infrastructure. Available at http://www.unep.org/resourcepanel/Portals/24102/PDFs/Metal_Recycling_Full_Report.pdf. Accessed 6 June 2016.

US Department of Energy (2009). Lamps standards final rule TSD Appendix 3C (Rare Earth Phosphor Market). Available at http://www1.eere.energy.gov/buildings/appliance_standards/residential/pdfs/app_3c_lamps_standards_final_tsd.pdf. Accessed 4 November 2015.

US Department of Energy (2011). Critical materials strategy. Available at http://energy.gov/sites/prod/files/DOE_CMS2011_FINAL_Full.pdf. Accessed 4 November 2015.

Weil, B. (2012). Uranium mining and extraction from ore. Available at http://large.stanford.edu/courses/2012/ph241/weil2/. Accessed 2 February 2016.

WHO (2004). Hydrogen cyanide and cyanides: Human health aspects. Available at http://www.who.int/ipcs/publications/cicad/en/cicad61.pdf. Accessed 17 March 2016.

WHO (2006). Elemental speciation in human health risk assessment: Environmental Health Criteria 234. Available at http://www.inchem.org/documents/ehc/ehc/ehc234.pdf. Accessed 17 March 2016.

WHO (2010). Persistent organic pollutants: Impact on child health. Available at http://apps.who.int/iris/bitstream/10665/44525/1/9789241501101_eng.pdf. Accessed 17 March 2016.

World Steel Association (2012). Sustainable steel: At the core of green economy. Available at: https://www.worldsteel.org/dms/internetDocumentList/bookshop/Sustainable-steel-at-the-core-of-a-green-economy/document/Sustainable-steel-at-the-core-of-a-green-economy.pdf. Accessed 20 November 2015.

World Steel Association (2015a). Sustainable steel: Policy and indicators. Available at http://www.worldsteel.org/dms/internetDocumentList/bookshop/2015/Sustainability-Steel-2015/document/Sustainable%20Steel%20-%20Policy%20and%20Indicators%202015.pdf. Accessed 18 June 2016.

World Steel Association (2015b). Life cycle thinking in the circular economy. Available at http://www.worldsteel.org/steel-by-topic/life-cycle-assessment/Life-cycle-thinking-in-the-circular-economy.html. Accessed 24 November 2015.

Zdun, T. (2001). Modelling the hydrodynamics of collie mining void 5B. MSc dissertation. University of Western Australia, Crawley, Western Australia, Australia.

2

Legislation and Policies Governing the Environmental and Health Impacts of Metallic Waste

2.1 Introduction

Mining and metallurgical industries play an important role in the economies of many countries (Aubertin and Bussière, 2001). In fact, these industries are considered to be among the major pillars of world industry (Zheng and Kozinski, 1996). However, mining and metallurgical industries create severe environmental problems by discharging waste in the form of solid, liquid and gaseous emissions (Wiertz, 1999; Chan et al., 2008). According to the Fraser Institute (2012) 'waste' is a general term for materials which have little or no economic value. For example, the soil and rock that are removed in order to gain access to buried ore and the material (liquid, solids and gases) left behind after the ore has been processed to remove the valuable commodities, are considered to be waste materials.

Generally, mining and metallurgical wastes are heterogeneous materials consisting of ore, gangue, industrial minerals, metals, coal or mineral fuels, rock, loose sediment, mill tailings, metallurgical slag, roasted ore, flue dust, ash, processing chemicals and fluids (Hudson-Edwards et al., 2011). Table 2.1 gives an overview of the types of waste from the mining and metallurgical industry (Rankin, 2011; Fraser Institute, 2012; Singo, 2013). Detailed discussions of the generation, categorization and recycling of the waste produced in the extraction and beneficiation, pyrometallurgical and hydrometallurgical processes are contained in Chapters 3 through 7, respectively. The use and recycling of manufacturing/finishing waste and post-consumer waste are covered in Chapters 8 and 9, respectively.

Though most of the wastes are just rocks and inert materials that do not generate any significant environmental impact, some of the waste can produce environmental problems (Wiertz, 1999). Therefore, environmental laws and social awareness programmes are required with respect to the potential threat of some of the mining and metallurgical wastes to the environment (Tariq et al., 2006). Basically, the environmental and social regulations governing the mining and metallurgical industry are of considerable importance and revolve around environmental protection, protection of local communities and promoting business ethics (Phadke et al., 2004). However, due to a lack of consensus among stakeholders, this chapter takes a simplified approach to the legislation and policies governing the environmental and health impacts of waste from the mining and metallurgical industries. Firstly, the chapter focuses on general legislation and policies governing wastes in the form of solid, liquid and gas. Secondly, the chapter discusses the general human health implications of a selected number of pollutants.

TABLE 2.1
Types of Waste

Type of Waste	Comment
Overburden	Overburden includes the soil and rock that is removed to gain access to the ore deposits at open pit mines. It is usually piled on the surface at mine sites where it will not impede further expansion of the mining operation. Overburden generally has a low potential for environmental contamination and is often used at mine sites for landscape contouring and revegetation during mine closure.
Waste rock	Waste rock is a material that contains minerals in concentrations considered too low to be extracted at a profit. Waste rock is often stored in heaps or dumps on the mine site, but may be stored underwater with tailings if it contains a lot of sulphide minerals and has a high potential for acid mine drainage formation. Waste rock dumps are generally covered with soil and revegetated following mine closure, although there are cases of waste rock being re-mined due to an increase in mineral market prices or improvements in extraction technology.
Tailings	Tailings consist of finely ground rock mixed with process water that remains after the minerals of economic interest have been removed from the ore. In other words, tailings are finely ground rock and mineral waste products of mineral processing operations. Tailings can also contain leftover processing chemicals, and are usually deposited in the form of a water-based slurry into tailings ponds (sedimentation lagoons enclosed by dams built to capture and store the tailings), although offshore tailings disposal has been successful in some cases.
Slags	Slag is the glassy material that remains after metals, such as copper, have been separated during the smelting process. Slags are largely environmentally benign and are being used increasingly as aggregate in concrete and road construction, soil conditioning materials, stone casting, sources for recovery valuable metals, etc.
Wastewater	Wastewater is produced in a number of ways, and can vary in its quality and potential for environmental contamination. For example, mining operations use water for mineral processing and metal recovery, controlling dust and meeting the needs of workers on site. Classes of wastewater include: • *Mine water*: Refers to any surface or groundwater present at the mine site • *Mining water*: Water which has come into contact with any mine workings • *Mill water*: Water used to crush and grind ore, may contain dissolved minerals and/or metals • *Process water*: Water used in the chemical extraction of metals, commonly contains process chemicals • *Leachate*: Water which has trickled through solid mine wastes and may contain dissolved minerals, process chemicals and/or metals • *Effluent*: Mining, mill or process water which is being discharged into surface water, often after being treated • *Mine drainage water*: Surface or groundwater which flows or has the potential to flow off the mine site
Water treatment sludge	Sludge is produced at active water treatment plants used at some mine sites, and consists of the solids that had been removed from the water as well as any chemicals that had been added to improve the efficiency of the process. Although ways of recycling the sludge are being explored, the majority of sludge has little economic value and is handled as waste. Disposal of water treatment residues in underground mine workings is the least expensive option where it is permitted and environmentally safe. In extreme cases where the sludge is rich in cadmium or arsenic, it may be classified as hazardous waste and require special handling and disposal.
Gaseous waste	Gaseous wastes include particulate matter (dust) and sulphur oxides (SOx). The majority of emissions to the atmosphere are produced during high-temperature chemical processing such as smelting, and vary in their composition and potential for environmental contamination. Environmental control technologies such as gravity collectors, cyclones and electrostatic precipitators are capable of removing up to 99.7% of dust and fumes, and wet scrubbers typically remove 80%–95% of sulphur oxide emissions.

Sources: Rankin, W.J., *Minerals, Metals and Sustainability: Meeting Future Material Needs*, CRC Press, Leiden, the Netherlands, 2011; Fraser Institute, How are waste materials managed at mine sites?, 2012, Available at http://www.miningfacts.org/Environment/How-are-waste-materials-managed-at-mine-sites/, Accessed February 2016; Singo, N.K., An assessment of heavy metal pollution near an old copper mine dump in Musina, South Africa, MSc (Eng) dissertation, University of South Africa, Pretoria, South Africa, 2013.

2.2 Legislation and Policies

As already stated, mining and metallurgical wastes result from the extraction, beneficiation and further processing of metal and industrial mineral ores (US Congress, 1992), and its categories are shown in Table 2.1. Just like all other industrial wastes, mining and metallurgical waste is subject to extensive environmental regulations related to air and water quality, and materials handling and disposal practices (US Congress, 1988). In other words, with an objective of protecting the environment due to the production of waste from mining and metallurgical activities, regulations and policies have been enacted so as to control its disposal and/or recycling. It must be noted, however, that various countries have developed different approaches for managing waste which differ both in the legislative scope and effectiveness (Kumar and Singh, 2013). In this section, a general overview of the environmental management system (ISO 14000) is given first. This is followed by an analysis of legislations and policies from a selected number of countries, while the last subsection looks at the guidelines and/or best practices for managing mining and metallurgical waste.

2.2.1 ISO 14000 Environmental Management System

Introduced in 1996, ISO 14000 is a global term for a set of internationally recognized standards for environmental management, similar to ISO 9000 standards for quality assurance systems to control the goods produced and services supplied by companies (Environmental Innovations, 2004; Morris, 2004). The ISO 14000 standards represent a consensus agreement by national standards bodies around the world about the procedures that need to be followed in establishing an effective environmental management system (EMS) (Morris, 2004).

The ISO 14000 series include standards for (1) environmental management systems, (2) environmental auditing, (3) environmental performance evaluation, (4) environmental labels and declarations and (5) life cycle assessment. Table 2.2 is a summary of these standards and many others that fall under the ISO 14000 series. Within the ISO 14000 series of standards, the fundamental standard that prescribes good practice in environmental management is ISO 14001. ISO 14001 specifies the various requirements that have to be satisfied in setting up an effective EMS, such that the risks of pollution incidents and other forms of environmental damage through the operations and activities of a company are minimized (Morris, 2004). The other standards in the series are merely guides that give assistance in interpreting and implementing the various clauses found in ISO 14001 as described in Table 2.2.

It must be noted that the EMS itself does not dictate a level of environmental performance that must be achieved, but each company's EMS is tailored to the company's business and goals. An ISO 14000-compliant EMS is a continuous improvement process that has the following standard elements (Environmental Innovations, 2004):

Environmental policy – The environmental policy is a road map for improving a business environmental performance. The policy should be appropriate for the size and complexity of the operation, but generally should not exceed 1–2 pages. The policy should include a senior management commitment to continual improvement and pollution prevention and a commitment to follow applicable regulations, standards, guidelines and codes of practice.

TABLE 2.2

Summary of ISO 14000 Series Standards

Standard	Relevant Content
ISO 14001 *Environmental management systems*: specification with guidance for use	This sets out the requirements for establishing and maintaining an environmental management system that assures a company of conformance with the environmental objectives that it sets for itself and also with any relevant legislation. If desired, this can be audited and certified by national accrediting bodies.
ISO 14004 *Environmental management systems*: general guidelines on principles, systems and supporting techniques	This gives general guidance and practical advice on implementing or improving an EMS, and is supported by a number of examples.
ISO 14020 *Environmental labels and declarations*: general principles	This gives guidelines on the use of labels to indicate the environmental impact of products. It requires that labels should be accurate and verifiable. Such labelling is provided for the benefit of environmentally conscious customers, who can decide whether or not to purchase a product, according to its environmental performance.
ISO 14031 *Environmental management*: environmental performance evaluation guidelines	This is provided as a management tool to help companies to assess whether its environmental performance is meeting the targets set. The document is particularly useful to companies that have not implemented a formal EMS, since it includes guidance on identifying the environmental impact of operations and setting targets for reduction in environmental damage.
ISO 14040 *Environmental management*: life cycle assessment: principles and framework	This standard defines a suitable framework for conducting life-cycle assessments for products in terms of their environmental impact. Recommendations are made about setting environmental targets and defining relevant indicators of environmental performance for all stages in a product's life cycle, including raw material acquisition, product manufacture, product use and product disposal at the end of its useful life. ISO 14040 defines the four phases in life-cycle assessment as defining goals, inventory analysis, impact assessment and interpretation. Further guidance on these four phases are provided in ISO 14041–ISO 14043.
ISO 14050 Environmental management vocabulary	This defines all concepts, words and phrases used in ISO 14000 standards, to ensure that a common international understanding is achieved.
ISO 19011 Guidelines for quality and/or environmental management systems auditing	This mainly copies the guidance on system auditing contained in ISO 14010–14012, but does so without specific reference to environmental systems, so that it is applicable either to quality assurance systems (ISO 9000) or to environmental management systems. This single document supersedes ISO 14010, ISO 14011 and ISO 14012 (and also corresponding documents in the ISO 9000 series).

Source: Morris, A.S., *ISO 14000 Environmental Management Standards: Engineering and Financial Aspects*, John Wiley & Sons Ltd., Chichester, UK, 2004.

Planning – The organization should identify legal requirements and also consider how its activities interact with the environment. The organization should then develop a plan to reduce any adverse effects its operation may have on the environment. Ideally, the planning stage should involve employees from all levels within the business.

Implementation – The policy and plans to improve environmental performance must be documented and communicated to employees; responsibilities must be assigned and training may be required to ensure that staff are aware of the plan and able to perform any required duties related to it. Operation procedures (e.g. work procedures) may be required to identify how specific tasks are to be carried out.

Checking and correction – The organization should develop a process (or processes) to monitor and record how well it is doing with respect to the environmental plan and regulatory obligations. An internal or third party audit may be performed at this stage.

Management review – Senior management should review the EMS at regular intervals, for example, annually.

The EMS can provide significant benefits across a number of areas, and two broad categories of benefits that could be achieved from an effective EMS such as ISO 14001 are listed in Table 2.3.

In general, running an effective EMS can help a company in the following aspects (Matuszak-Flejszman, 2009): (1) better regulatory compliance – running an EMS will help ensure the legal environmental responsibilities are met and more easily managed on a day-to-day basis, (2) more effective use of resources – there will be policies and

TABLE 2.3

Economically and Non-Economically Quantifiable Benefits of an Effective Environmental Management System

Category of Benefits	Benefits
Economically quantifiable	• Raw materials saving, energy saving • Improvement in production system availability • Reduction of rejects • Reduction of waste treatment costs • Exploitation of rejects • Reduction of idle times • Public incentives • Health care • Insurance cost reductions • Increased in capacity • Decreased some logistics costs • Increase in resource usage efficiency
Non-economically quantifiable	• Company image • Liability and risk reduction • Market opportunities • Reduction of contamination risk • Better knowledge of job and production system • Product image toward customers • Relationship with authorities • Reduction of human risk • Flexibility of management system • Pro-active attitude to the environmental question • Compatibility with ISO 9001 and ISO 14001

Source: Matuszak-Flejszman, A., *Pol. J. Environ. Stud.*, 18(3), 411, 2009.

procedures in place that would help to manage waste and resources more effectively and reduce costs, (3) marketing – running an EMS can help to prove the business' credentials as an environmentally aware operation that has made a commitment to continual environmental improvement, (4) finance – it may be easier to raise investment from banks and other financial institutions, which are increasingly keen to see businesses controlling their environmental impact, (5) increased sales opportunities – large businesses and government departments may only deal with businesses that have an EMS and (6) lighter regulation – even if an EMS is not a regulatory requirement, by showing a commitment to environmental management, a business may benefit through less frequent site visits or reduced fees from environmental regulators.

2.2.2 Analysis of Legislation and Policies

2.2.2.1 United States of America

An extensive regulatory system has been developed to govern current mining operations in the United States, as well as to guide the cleanup of historical ones (Hudson et al., 1999). In fact, there are several dozens of federal environmental laws and regulations that cover all aspects of mining in the United States (NMA, 2015). The framework for these regulations is primarily based on federal laws dating back to the late 1960s. Table 2.4 lists some of the major federal laws and regulations governing the mineral industry. In addition, each state has laws and regulations that mining companies must follow. Ideally, regulatory standards established at state levels are commonly as strict as or more stringent than federal standards (Hudson et al., 1999).

2.2.2.2 European Union

In addition to general industrial regulations, the European Union (EU) has adopted some legislations specific to mining (Szczepanski, 2012). In other words, the EU has developed a set of environmental directives that have had a significant effect on the mining industries of member nations (McKinley, 2004). In fact, each country's environmental laws are derived from these directives. These regulations are concerned with the environmental impact of mining (especially waste and groundwater), as well as occupational health and safety (Szczepanski, 2012). Among the key directives are the Environmental Impact Assessment Directive (similar to the EIS requirements of the United States); the Water Framework Directive (addresses concerns similar to those of the United States' Clean Water Act) and the Waste Framework, Hazardous Waste, and Landfill Directives (all address concerns similar to those of the United States' Resource Conservation and Recovery Act [RCRA]).

2.2.2.3 Australia

In Australia, mining is regulated under a state-based regime with an overlay of federal regulation (HSF, 2012). In other words, mining activities in each of the states are governed by their respective Mining Acts and Mining Regulations. Examples include (1) the Mining Act 1978 (WA) and the Mining Regulations 1981 (WA) in Western Australia, (2) the Mineral Resources Act 1989 (QLD) and the Mineral Resources Regulation 2003 (QLD) in Queensland and (3) the Mining Act 1992 (NSW) and the Mining Regulation 2010 (NSW) in New South Wales.

TABLE 2.4

Major Federal Laws and Regulations Affecting the Mineral Industry in the United States

Law/Regulation	Relevant Content
National Environmental Policy Act (NEPA)	This Act was enacted in 1970. It requires federal agencies to prepare environmental impact statements (EIS) for major federal actions that may significantly affect the environment. These procedures exist to ensure that environmental information is available to public officials and citizens before actions are taken. NEPA applies to mining operations requiring federal approval.
Comprehensive Environmental Response, Compensation and Liability Act (CERCLA)	This law commonly known as Superfund was enacted in 1980. This law requires operations to report inventory of chemicals handled and releases of hazardous substances to the environment. Hazardous substances are broadly defined under CERCLA and have included mining, milling and smelter wastes that are currently excluded from regulation under RCRA. It requires clean-up of sites where hazardous substances are found. The Superfund program was established to locate, investigate and clean up the worst abandoned hazardous waste sites nationwide and is currently being used by the U.S. Environmental Protection Agency (EPA) to clean up mineral-related contamination at numerous locations.
Resource Conservation and Recovery Act (RCRA)	RCRA was passed into law in 1976. The goals of the law are to conserve energy and natural resources, reduce the amount of waste generated and ensure that wastes are managed to protect human health and the environment. RCRA gives EPA power to make and enforce regulations for managing many kinds of wastes. RCRA regulations apply to three kinds of waste management: municipal, solid waste landfills; hazardous waste generators and transporters, and treatment, storage and disposal facilities and underground tanks that store hazardous materials.
Federal Water Pollution Control Act (Clean Water Act) (CWA)	This Act commonly referred to as the Clean Water Act, came into effect in 1977. The act requires mining operations to meet standards for surface water quality and for controlling discharges to surface water. CWA-based regulations cover such mining-related situations as the disposal of mining-related waters, the pumping or draining of mine water to the surface, storm water runoff in mining operation areas and control of seeps from mine tailings impoundments.
Resource Conservation and Recovery Act (RCRA)	This Act enacted in 1976, regulates the generation, storage and disposal of solid waste and hazardous waste, using a 'cradle-to-grave' system, meaning that these wastes are governed from the point of generation to disposal. Ideally, the act focuses on preventing the release of hazardous wastes into the environment by providing for their management from generation to disposal.
Federal Land Policy and Management Act	Prevents undue and unnecessary degradation of federal lands.

(*Continued*)

TABLE 2.4 (*Continued*)

Major Federal Laws and Regulations Affecting the Mineral Industry in the United States

Law/Regulation	Relevant Content
Clean Air Act (CAA)	This act passed in 1970, authorizes regulations to address airborne pollution that may be potentially hazardous to human health or natural resources. Examples of mining-related situations that are covered by CAA-based regulations include dust emissions that accompany operations or tailings disposal in impoundments, exhaust emissions from heavy equipment and emissions from processing facilities, such as smelters.
Safe Drinking Water Act	Directs standards for quality of drinking water supplied to the public (States are primary authorities) and regulating underground injection operations.
Solid Waste Disposal Act	Regulates generation, storage and disposal of hazardous waste and manages solid, non-hazardous waste.
Toxic Substance Control Act (TSCA)	The TSCA, passed in 1977, focuses on controlling the development and application of new and existing chemical substances that present risk to health or environment. Chemicals and hazardous materials used in the processing of ore or ore concentrates, such as sodium cyanide solutions used in the leaching of gold ores, are regulated under TSCA.
Endangered Species Act	Lists threatened plants and animals; protection plans mandated.
Migratory Bird Treaty Act	Protects nearly all bird species.
Surface Mining Control and Reclamation Act	Regulates coal mining operations and reclamation.

Source: Hudson, T.L. et al., Metal mining and the environment, 1999, Available at http://www.americangeosciences.org/critical-issues/faq/what-are-regulations-mining-activities, Accessed May 2016.

While the said legislations form the bulk of regulation in the industry, approvals under various other legislative regimes are also considered, including the following key requirements:

1. Environmental approvals under the Environmental Protection Act 1986 (WA), Environmental Protection Act 1994 (QLD) or the Environmental Planning and Assessment Act 1979 (NSW) (the Environment Acts) and (if applicable) the Environment Protection and Biodiversity Conservation Act 1999 (Cth).
2. Compliance with the 'future act' provisions of the Native Title Act 1993 (Cth) (NTA) where the grant of a mining tenement would affect native title.

Furthermore, different activities may require different types of licences. For example, a number of licences may be required to construct and operate ore processing and tailings storage facilities, including, for example, in Western Australia, (1) a miscellaneous licence under the Mining Act 1978 (WA) and Mining Regulations 1981 (WA) for the purposes of constructing and using a tailings storage or transportation facility; (2) a general purpose lease under the Mining Act 1978 (WA) for the purpose of depositing or treating minerals or tailings obtained from land; (3) a works approval and environmental licence for waste, noise, odour or electromagnetic radiation under the Environmental Protection Act 1986 (WA) to operate a waste storage, treatment, handling or transport facility and to discharge waste or emit odour and (4) a licence to take water from surface and underground sources and watercourses under the Rights in Water and Irrigation Act 1914 (WA) to construct, enlarge, deepen or alter any artesian or non-artesian well, to take commercial quantities of water from specified sources or do anything that obstructs a watercourse or wetland or its bed and banks.

2.2.2.4 South Africa

More than 70% of the waste generated in South Africa comes from the mining industry (Lottermoser, 2010). The country's mining waste (e.g. mining residue deposits and stockpiles) is regulated in terms of the Mineral and Petroleum Resources Development Act (MPRDA) of 2002. However, the major framework laws in South Africa that deal with waste materials and management activities are varied, fragmented and rely on implementing regulations (Puder and Veil, 2004). Table 2.5 presents statutes that are relevant for waste management operations in South Africa.

2.2.2.5 China

China's mineral resources are extensive and diverse, and the country is now the world's top producer of a number of important metals and minerals (Greenovation Hub, 2014). Its mining activities are regulated under the Mineral Resources Law and various other Administrative Measures (HSF, 2012). These laws are administered by the Ministry of Land and Resources (MOLAR) and the various provincial, regional and municipal geology and mineral resources departments. Basically, two types of mining licences are available in China: (1) an exploration licence and (2) a mining licence.

It must be noted that while China's regulatory framework has developed significantly in recent years, gaps still remain (Greenovation Hub, 2014). Therefore, it has become clear that stronger and more detailed regulations (or mechanisms) need to be adopted in order to protect China's environment and sustainably manage the utilization of the

TABLE 2.5
South African Framework Laws

Law/Regulation	Relevant Content
Bill of Rights, Chapter 2 of the South African Constitution (Act No. 108 of 1996)	States in section 24 that everyone has the right to environmental protection through reasonable and other measures that prevent pollution, promote conservation and secure ecologically sustainable development and use of natural resources, while promoting justifiable economic and social development.
National Environmental Management Act (No. 107 of 1998)	This Act provides the framework and principles for sustainable development and sets national norms and standards for Integrated Environmental Management (Section 24) where all spheres of government and all organs of state must co-operate, consult and support one another. Section 28 of the Act also imposes a Duty of Care and remediation of environmental damage on any person who causes, has caused or may cause significant pollution or degradation of the environment. Furthermore, sections 32 and 33 of the Act provides for legal standing to enforce environmental laws and private prosecution, respectively.
Environment Conservation Act (No. 74 of 1983)	Concerned with littering, waste management and regulations, with an emphasis on land disposal sites.
National Water Act (No. 36 of 1998)	Contains wide provisions particularly related to responsibility for the integrity of water resources. The basis of water management at mines is therefore the mine water management hierarchy. This hierarchy is based on a precautionary approach and sets the following order of priority for mine water management actions: (1) pollution prevention, (2) water re-use or reclamation, (3) water treatment and (4) discharge.
Health Act (No. 63 of 1977)	Used by the Department of Water Affairs to determine buffer zones of proposed waste disposal sites.
Minerals Act (Act No. 50 of 1991)	The Act provides statutory requirements for enforcing environmental protection, the management of the environmental impacts and the rehabilitation of the affected prospecting and mining environment in South Africa.
Air Quality Act (Act No. 39 of 2004)	The purpose of the regulations is to prescribe general measures for the control of dust in all areas.

Source: Puder, M.G. and Veil, J.A., Review of waste management regulations and requirements from selected jurisdictions, 2004, Available at http://www.netl.doe.gov/kmd/cds/disk23/D-Water%20Management%20Projects/Produced%20Water%5CFEW%2049177%5CFEW%2049177%20Final%20Rpt.pdf, Accessed May 2016.

country's resources. A major problem within China is that the costs associated with breaking the law are sometimes lower than the costs of following the law. For example, paying fines may be cheaper than conducting high-quality and timely environmental impact assessments, paying for expensive waste treatment facilities or installing and using emissions reduction technology.

2.2.2.6 *India*

India is ranked fourth among the mineral-producing countries, behind China, the United States and Russia, on the basis of volume of production (FICCI, 2013). The GDP contribution of the mining industry varies from 2.2% to 2.5%, but going by the GDP of the total industrial sector, it contributes around 10%–11% (Metalworld, 2013).

The basic laws governing the mining sector in India consist of the Mines and Minerals (Development and Regulation) Act 1957 (MMDR Act) and the Mines Act 1952, together with the rules and regulations framed under them (HSF, 2012). A mine operation is also mandated to observe the provisions of the Atomic Energy Act 1962 insofar as they relate to atomic minerals.

The relevant rules in force under the MMDR Act are the Mineral Concession Rules 1960 (MCR), and the Mineral Conservation and Development Rules 1988 (MCDR). The provisions of the MCR are not applicable to atomic minerals and the provisions of the MCDR are not applicable to coal or any minerals declared as prescribed substances under the Atomic Energy Act 1962 or minor minerals. The MCR outline the procedures and conditions for obtaining mineral concessions in India. Under these rules, mining operations are to be undertaken in accordance with the duly approved mining plan. A modification of the approved mining plan during the operation of a mining lease also requires prior approval. The MCDR lays down guidelines for ensuring mining on a scientific basis, while at the same time conserving the environment.

The mining of minor minerals is separately notified and is governed by the state governments in accordance with the relevant Minor Mineral Concession Rules. The health and safety of the workers is governed by the Mines Rules 1955 created under the Mines Act 1952.

There are also separate laws along with associated rules and notifications that deal with the environmental requirements associated with the opening of the mine, the operation of the mine and the closure of the mine, including (1) the Water (Prevention and Control of Pollution) Act 1974 (amended in 1988), (2) the Air (Prevention and Control of Pollution) Act 1981 (amended in 1988), (3) the Environment (Protection) Act 1986, (4) the Indian Forest Act 1927, (5) the Forest (Conservation) Act 1980 (amended in 2010) and (6) the Wildlife (Protection) Act 1972 (amended in 1991).

2.2.3 Guidelines for Managing Mining and Metallurgical Waste

Both legislation and public awareness are becoming increasingly more important for the mining and minerals industry. As a result, companies have become more environmentally responsible, and want to be seen by the public as producers of 'clean' products via environmentally friendly processes (Schreuder, 2006). It must also be noted that the mining and metallurgical industry also plays a leading role in waste management and is one of few industries that recycle their own waste (Lottermoser, 2010). Detailed discussions of the generation, categorization and recycling of the waste produced in the beneficiation, pyrometallurgical and hydrometallurgical processes are contained in Chapters 3 through 7, respectively. The reuse and recycling of manufacturing/finishing waste and post-consumer waste are covered in Chapters 8 and 9, respectively.

This particular section discusses only the guidelines and/or best practices for managing mining and metallurgical waste. In the context of this section, these guidelines and/or best practices can be defined as the most efficient (requires the least amount of effort) and effective (best results) ways of accomplishing a task, based on repeatable procedures that have proven themselves over time for large numbers of people (Bernard, 2012). These guidelines on waste management have been developed at international, national and regional levels, and provide an advisory framework for best practices in mine waste management (van Zyl et al., 2002).

2.2.3.1 Waste Rock

As outlined in Table 2.1, during mining, large quantities of overburden or waste rock often need to be removed to expose the mineral to be mined. On the other hand, though the mineral processes are designed to extract as much marketable product(s) as possible, they leave behind a significant amount of tailings (discussed in more detail in the next section). The management of both wastes – tailings and waste rocks – is important in order to protect human health, safety and the environment. In principle, tailings and waste rock may be managed in a variety of ways, depending on their physical and chemical nature (e.g. projected volume, grain size distribution, density and water content, among other issues), the site topography, downstream receptors, climatic conditions, national regulation and the socio-economic context in which the mine operations and processing plant are located (IFC, 2007; Vogt, 2013). The following are some of the many options for managing tailings and waste rock (European Commission, 2009):

1. Dry-stacking of thickened tailings on land
2. Dumping of more or less dry tailings or waste rock onto heaps or hillsides
3. Backfilling of tailings or waste rock into underground mines or open pits or for the construction of tailings dams
4. Discarding of tailings into surface water (e.g. sea, lake and river) or groundwater
5. Use as a product for land use, for example, as aggregates, or for restoration
6. Discarding of slurried tailings into ponds

More specifically, Table 2.6 is a summary of techniques used in the management of waste rock for a selected number of minerals in Europe. In addition, the International Finance Corporation (IFC) has also given the following recommendations for the management of waste rock dumps (IFC, 2007):

1. Dumps should be planned with appropriate terrace and lift height specifications based on the nature of the material and local geotechnical considerations to minimize erosion and reduce safety risks.
2. Management of potentially acid generating (PAG) wastes should be undertaken.
3. Potential changes of geotechnical properties in dumps due to chemical or biologically catalysed weathering should be considered. This can reduce the dumped spoils significantly in grain size and mineralogy, resulting in high ratios of clay fraction and a significantly decreased stability towards geotechnical failure. These changes in geotechnical properties (notably cohesion and internal angle of friction) apply especially to facilities which are not decommissioned with a proper cover system, which would prevent precipitation from percolating into the dump's body. Design of new facilities has to provide for such potential deterioration of geotechnical properties with higher factors.

2.2.3.2 Tailings

Tailings are basically what is left after physical and/or chemical separation of the target minerals (e.g. gold, silver or copper) from the ore (Vogt, 2013; Barclays, 2015). Tailings include the fine grained, silt-sized particles (1–600 μm) (Suvio et al., 2016) from ore milling and processing operations (Barclays, 2015) and the residues of chemical reagents – all as

TABLE 2.6

Waste Rock Management Techniques for a Selected Number of Minerals

Mineral	Waste Rock Characteristics	Waste Rock Management	Closure and After-Care
Base metals	Sometimes has ARD potential	In one case selective management of ARD and non-ARD waste-rock, sometimes used for dam construction, in one case backfill, collection of surface run-off	Vegetative cover, engineered cover to reduce ARD generation
Chromium			Backfilling of all waste-rock underground
Iron	No net ARD potential, possibly ammonium nitrate leachate	On heaps, at one site together with coarse tailings	Vegetative cover using soil and seeds, long-term seepage monitoring
Precious metals		On heaps, for dam construction or backfill into open pit	At one site covering with topsoil
Barytes		Sometimes sold as aggregate or backfill	
Fluorspar		Backfill	
Kaolin		Collection of surface run-off	
Limestone		Backfilling into old quarry	
Phosphate		Some uses as aggregate	Landscaping plans have been developed with local authorities and communities
Talc			Water drainage and vegetative cover
Coal		With coarse tailings on heaps, temporary heaps and later backfilling	Final heap design agreed with authorities and communities with the goal of creating landscape integrated structures

Source: European Commission, Reference document on best available techniques for management of tailings and waste-rock in mining activities, 2009, Available at http://eippcb.jrc.ec.europa.eu/reference/BREF/mmr_adopted_0109.pdf, Accessed May 2016.

part of a slurry (Vogt, 2013). It must be noted that as the composition of ores get more complicated, the amount of tailings per ton of metal produced rises (Suvio et al., 2016), leading to a real concern over the growing amount of tailings waste from mining and metallurgical operations. Table 2.7 gives estimates of the tonnages of a range of different ores and the percentage of the ore that remains as waste (Gardner and Sampat, 1998). As can be seen from Table 2.7, a very high proportion of tailings are generated, thus tailings must be managed properly to avoid adverse impacts (Hardygóra et al., 2004).

Tailings management strategies vary according to site constraints and the nature or type of the tailings (IFC, 2007; Vogt, 2013). Besides what has already been discussed (see Section 2.2.3.1), tailings management strategies should also consider how tailings are handled and disposed of during operation, in addition to permanent storage after decommissioning. Basically, tailings are physically and/or chemically treated before being transported by pipeline to a tailings management (or storage) facility (TMF) (Barclays, 2015). The TMFs are containment facilities and are most commonly either a dammed and backfilled natural valley or 'dry stack' stockpiles. In some locations where the landscape is flat, tailings are stored within constructed earth bunds.

TABLE 2.7

World Ore and Waste Production for Selected Metals in 1995

Metal	Ore Mined (Million Tonnes)	Proportion of Ore That Becomes Waste (%)[a]
Iron	25,503	60
Copper	11,026	99
Gold[b]	7,235	99.99
Zinc	1,267	99.95
Lead	1,077	97.5
Aluminium	856	70
Manganese	745	70
Nickel	387	97
Tin	195	99
Tungsten	125	99.75

Source: Gardner, G. and Sampat, P., Mind over matter: Recasting the role of materials in our lives, 1998, Available at http://www.worldwatch.org/system/files/WP144.pdf, Accessed May 2016.

[a] Excluding overburden.

[b] 1997 data.

It must be noted that despite their widespread usage, constructed earth bunds and/or tailings dams may impact the environment, if not properly designed, through the leaching of acids and/or heavy metals into surface and/or groundwater resources (IFC, 2007; Barclays, 2015). Other potential environmental impacts include the sedimentation of drainage networks, dust generation and the creation of potential geotechnical hazards associated with the selected management option (IFC, 2007; Barclays, 2015). Table 2.8 is a summary of techniques used in the management of tailings for a selected number of minerals in Europe.

2.2.3.3 Slag

Slag is the waste matter separated from metals during the smelting or refining of ore. As discussed in Chapters 4 through 6, slag produced during the smelting or refining processes might be recycled directly to a previous stage or processed in slag cleaning or fuming furnaces to remove the valuable metals. Depending on the leachate test, the resulting clean slag with low metal content is either reused or disposed of (ECCC, 2012).

Ideally, slags that do not pass leachate tests or for which commercial markets are not readily available cannot be reused and need to be disposed of in secure landfills. Often, in order to obtain environmental approvals, any leachate from slag disposal areas must be collected and treated.

2.2.3.4 Sludge

Sludge may come from wastes/residues of air pollution abatement systems or from liquid effluent treatment or from hydrometallurgical processes. The waste/residues from such processes may be managed as outlined in succeeding sections.

TABLE 2.8

Tailings Management Techniques for a Selected Number of Minerals

Mineral	Mineralogy	Mineral Processing	Tailings Characteristics	Tailings Management	Closure and After-Care
Aluminium	Al_2O_3 $SiO_2Fe_2O_3$ CaO TiO_2	Bayer process	Elevated pH, red mud: $d_{80} < 10$ µm, process sand: $d_{80} < 1000$ µm	Slurried or thickened	Dewater and dry cover, discharge treatment
Base metals	Mostly sulphides	Flotation, cyanide leaching for Au	d_{80}: 50–100 µm, often ARD potential	Slurried, usually large ponds: 35–1450 ha, some backfill (coarse fraction)	Dewater and dry cover or wet cover
Chromium	26% Cr_2O_3	Dense medium and magnetic separation	Containing Cr and Ni	Slurried	No plans
Iron	Phosphorous magnetite, iron carbonates	Magnetic separation, dense medium separation	No ARD potential, Kiruna: mostly SiO_2 and Fe_xO_y	Fines: slurried, coarse: heaps	Dewater and dry cover
Manganese	MnO_2	Only crushing	No tailings		
Precious metals	Complex sulphides, native gold, gossan, etc.	CN leaching, spirals, shaking table	Some have ARD potential, in case of CN leach: containing cyanide, complexed metals, cyanate, thiocyanate	Slurried, some backfill (coarse fraction), CN destruction	Dewater and dry cover, wet cover, raised groundwater table
Coal	Carbon, ash, sulphur	Coarse fractions in jigs or dense medium, flotation for fines	Clay, shale, sandstone, sulphides, some reagents, can be radioactive	Backfilling often too costly, coarse tailings on heaps or in old pits, fines in ponds, sold or filtered and to heaps	Landscape integrated heap design agreed with authorities and communities
Tungsten	(Fe, Mn)WO_4, $CaWO_4$	Flotation, dense medium separation, shaking tables	$d_{80} = 100$ µm, no ARD potential	Slurried, some backfill (coarse fraction)	Dewater and dry cover
Strontium		Dense medium and flotation		Coarse tailings are backfilled, flotation slurry in ponds	
Feldspar	Orthoclase, albite, anorthite	Sometimes none, otherwise optical separation, flotation, electrostatic or magnetic separation	Solids contain fine sands and micas, 10% iron oxides, some flocculants, process water: pH 4.5, some fluoride	Coarse tailings on heaps, slurries are backfilled or ponded	

(Continued)

TABLE 2.8 (*Continued*)

Tailings Management Techniques for a Selected Number of Minerals

Mineral	Mineralogy	Mineral Processing	Tailings Characteristics	Tailings Management	Closure and After-Care
Fluorspar	CaF_2 (in one case also PbS)	Dense medium separation, flotation	Mostly silica (90%), Fe and Al oxides	Backfilling and process water re-use, slurries mostly to ponds, in one case fine tailings into sea	After-care period of 10 years expected to monitor heavy metals, fund for closure/after-care costs
Limestone/calcium carbonate	97%–98% $CaCO_3$, <1% $MgCO_3$, <1% SiO_2	Limestone: washing; calcium carbonate: flotation, magnetic separation	Limestone: <0.25 mm	Slurries in ponds, in one case the pond is a former quarry, sometimes the slurry is dried and the tailings discarded onto heaps	Dewater and dry cover
Kaolin	Kaolinite, quartz, micas, feldspar residues	No comminution, magnetic separation, flotation	Fine sands and micas, <1% iron oxides, some flocculants, process water: pH 4.5, some phosphates, sulphates, foam inhibitor	Coarse tailings on heaps, slurries in ponds lined with clay, in one case dewatered fines are transferred to heaps	Dewater and dry cover
Phosphate	Apatite (10%), phlogopite mica (65%), carbonates (20%) and silicates (5%)	Flotation		Slurries in ponds	
Talc	Talc, carbonates, chlorites and sulphides	Often only comminution, sometimes flotation	Flotation tailings in ponds, which through dewatering become heaps		
Potash	Sylvinite carnallitite hard salt kainitite and other salts	Hot leaching, flotation, electrostatic separation, dense medium separation	Liquid and solid tailings, containing sodium chloride with other salts, clay and anhydrite	Solid tailings on heaps, liquid tailings into deep wells or surface waters, in one case marine discharge of liquids and solids, some solid tailings are backfilled	Heaps remain unchanged and dissolve over time
Barytes	$BaSO_4$	All techniques, e.g. jigging, dense medium, flotation		Often no tailings, fines as slurry, sometimes backfilled, coarse tailings on heaps or sold as aggregates	
Borates	B_2O_3	Dissolution, crystallisation, drying/cooling		Coarse tailings first on heaps and then backfilled, slurry in ponds	

Source: European Commission, Reference document on best available techniques for management of tailings and waste-rock in mining activities, 2009, Available at http://eippcb.jrc.ec.europa.eu/reference/BREF/mmr_adopted_0109.pdf, Accessed May 2016.

2.2.3.4.1 *Air Pollution Abatement Systems*

Most off-gas dusts collected in gas-cleaning systems are returned directly to the smelting furnaces. However, dust removed in a wet system (e.g. scrubber) forms a sludge which must typically be dewatered (e.g. dried and filtered) prior to recycling. Some residues from air pollution abatement systems are processed on site and sold as by-products. For example, SO_2-rich off-gases from roasters, smelters and/or converters can be used to produce sulphuric acid (H_2SO_4) in an acid plant or mercury removed from the gas stream prior to the acid plant can be used as a feed material for mercury production. Sludge with a lower valuable metals content that does not justify the additional costs to recycle or recover the metals, for example, some sludge from wet-based air pollution abatement systems (e.g. wet scrubbers), is disposed of into a landfill.

2.2.3.4.2 *Liquid Effluent Treatment*

Water is typically used in pyrometallurgical operations for the direct or indirect cooling of furnaces, lances and casting machines, thus wastewater is generated from such operations. If wet scrubbers are used for off-gas cleaning, wastewater is also generated. The wastewater generated in pyrometallurgical operations and wet scrubbing is treated to settle out solids and if the metal content in the sludge is high enough, it is recycled to be processed and/or reused. Wastes/residues from liquid effluent treatment that cannot be recycled or reused is disposed of.

2.2.3.4.3 *Hydrometallurgical Processes*

The leaching, purification and the electrolysis processes generate metals-rich solids. These can be recycled to the production process or sent for metals recovery in base metals facilities (e.g. for the production of precious metals, lead, copper and cadmium). However, some sludge from leaching processes is often classified as hazardous waste due to the leaching of elements such as cadmium, arsenic and lead, thus such sludge is normally disposed of in a hazardous waste facility, or in on-site residue areas or tailings ponds.

2.2.3.5 Air Emissions

Management of ambient air quality during mining and metallurgical operations is very important (IFC, 2007). Some of the major airborne pollutants generally attributed to mining and metallurgical operations and their associated abatement technologies are discussed here.

Dust: The principal sources of dust include blasting, exposed surfaces such as tailings facilities, stockpiles, waste dumps, haul roads and infrastructure (IFC, 2007). Some of the management strategies recommended by IFC (2007) include (1) dust suppression techniques (e.g. wetting down, use of all-weather surfaces, use of agglomeration additives) for roads and work areas, optimization of traffic patterns and reduction of travel speeds; (2) revegetation or covering of exposed soils and other erodible materials promptly; (3) opening up and clearing of new areas only when absolutely necessary; (4) revegetation of surfaces or otherwise rendering to be non-dust forming when inactive; (5) enclosing of storage for dusty materials or operation of such with efficient dust suppressing measures; (6) loading, transfer and discharge of materials with a minimum height of fall, shielding against the wind and consideration of the use of dust suppression spray systems and

(7) conveyor systems for dusty materials should be covered and equipped with measures for cleaning return belts.

A number of abatement technologies have also been developed to remove dust particulates, namely, electrostatic precipitators, bag or fabric filter plants and wet scrubbing (Schreuder, 2006). There are also many other technologies for removal of particulates, but are seldom efficiently applied for dust burdens below 100 mg/N m^3.

Acidic gases: The main acidic gases found in conventional flue-gas streams include sulphur dioxide (SO_2), sulphur trioxide (SO_3), hydrochloric acid (HCl) and hydrofluoric acid (HF). Abatement of these pollutants can be broadly classified into four classes (Schreuder, 2006): (1) dry scrubbing, (2) semi-dry scrubbing, (3) wet scrubbing and (4) adsorption technologies. For most of the acidic gases' abatement techniques stated previously, a reagent is required to effect the required abatement. Typically these are lime, hydrated lime and limestone. For smaller scrubbers, caustic and other reagents may be used. For adsorption technologies, a variety of activated carbons and catalysts may be used.

NOx: NOx abatement technologies include selective catalytic reduction (SCR) whereby NOx is destroyed over a catalyst with a small amount of ammonia (Schreuder, 2006). This can be done at moderate temperatures. Selective non-catalytic reduction (SNCR) also destroys NOx by injection of ammonia into the high temperature regions of the boiler or furnace.

Acid mist and aerosols: Acid mists and aerosols are typically removed with wet electrostatic precipitators (Schreuder, 2006). Wet electrostatic precipitators are highly efficient, have low power consumption and are extensively used in the non-ferrous metal industry and, iron and steel industry as follows:

Non-ferrous metal industry: Among other things to clean zinc-, lead- and cadmium oxide-laden exhaust air.

Iron and steel industry: Among other things to clean exhaust air coming from convertors, blast furnaces, cast houses, scarfing machines and blow-torch cutting-off machines, casting pits, desulphurization stands, tube lines and cupola melting furnaces.

2.3 Environmental and Health Impacts

The mining process begins with the extraction of the ore from the ground. The ore then goes through beneficiation, pyrometallurgical and/or hydrometallurgical processes, and finally ends up with metallic products that are ready for sale to the world market (Dudgeon, 2009). However, as already discussed, these mining and metallurgical processes produce many harmful wastes along the way. Some of the harmful wastes can be categorized as follows:

1. Leachates from dumping activities
2. Particulate matter
3. Gaseous emissions, for example SOx and NOx
4. Wastewater and effluents from mining and metallurgical processes

2.3.1 Leachates

Controlled and/or uncontrolled dumped wastes from the mining and metallurgical operations include, but are not limited to overburden, waste rock, slag and tailings. Over time, and when water is allowed to percolate through the dumped material, it produces a leachate, which is defined as the toxic aqueous effluent generated as a consequence of rainwater percolation through wastes, inherent moisture and chemical reactions occurring within the disposed waste (Kurniawan et al., 2006; Renou et al., 2008; Oulego et al., 2016). Leachate can vary in its composition, depending on the waste composition, degree of compaction, climate, moisture content and solubility of the waste constituents (Raghab et al., 2013). However, most of the dumped wastes from the mining and metallurgical operations contain heavy metals (European Commission, 2002). Heavy metals are elements with atomic density greater than 6 g/cm^3 (Gardea-Torresdey et al., 2005; Akpor and Muchie, 2010) or conventionally defined as elements with metallic properties and an atomic number greater than 20 (Tangahu et al., 2011). An alternative classification of metals based on their coordination chemistry, categorizes heavy metals as class B metals that come under non-essential trace elements, which are highly toxic elements (Nieboer and Richardson, 1980; Sharma and Agrawal, 2005). Most heavy metals are toxic because they tend to bioaccumulate in the human body (Kampa and Castanas, 2008; Lawley et al., 2008; Bini and Bech, 2014). Bioaccumulation means an increase in the concentration of a chemical in a biological organism over time, compared to the chemical's concentration in the environment (Kampa and Castanas, 2008; Bini and Bech, 2014).

The mobility of heavy metals (in the leachate) into the surrounding environment including water and soil depends on diverse environmental parameters such as pH, temperature, adsorption–desorption processes, complexation, uptake by biota, degradation processes and the intrinsic chemical characteristics of the waste (Sepúlveda et al., 2010). It is noted that though some heavy metals are biologically important, their release into water and soil is a serious concern for human health and the associated environment (Singo, 2013).

The following sections discuss the environmental and health effects of a selected number of heavy metals, namely, lead (Pb), mercury (Hg), cadmium (Cd), chromium (Cr), zinc (Zn) and arsenic (As). These are all naturally occurring metals which are often present in the environment at low levels (Martin and Griswold, 2009). Generally, each heavy metal has its own independent properties, but common to lead, mercury and cadmium is that neither has any known useful function in biological organisms (European Commission, 2002; Singh et al., 2011), and their accumulation over time in the bodies of animals can cause serious illness (Singh et al., 2011).

2.3.1.1 Lead

Lead is a highly toxic metal whose widespread use has caused extensive environmental contamination and health problems in many parts of the world (Jaishanka et al., 2014). In humans, depending upon the level and duration of exposure, lead can result in a wide range of biological effects ranging from inhibition of enzymes to the production of marked morphological changes to death (European Commission, 2002). Essentially, it can affect every organ and system in the body (Martin and Griswold, 2009), and its impact on children is more pronounced than in adults (Martin and Griswold, 2009). Lead influences the nervous system and slows down the nervous response, which influences children's learning abilities and behaviour (European Commission, 2002).

The long-term exposure of adults to lead can result in decreased performance in some tests that measure functions of the nervous system; weakness in fingers, wrists or ankles; small increases in blood pressure and anaemia (Martin and Griswold, 2009). In pregnant women, high levels of exposure to lead may cause miscarriage. Lead exposure has also been linked to hormonal changes, reduced fertility in men and women, menstrual irregularities, delays in puberty onset in girls, memory loss and mood swings; nerve, joint and muscle disorders and cardiovascular, skeletal, kidney and renal problems (Singo, 2013).

In the environment, lead is known to be toxic to plants, animals and microorganisms (European Commission, 2002). It disturbs various physiological processes in plants (Jaishanka et al., 2014). For example, a plant with a high lead concentration accelerates the production of reactive oxygen species, causing lipid membrane damage that ultimately leads to damage of chlorophyll and photosynthetic processes and suppresses the overall growth of the plant (Jaishanka et al., 2014; Najeeb et al., 2014). In general, inorganic lead compounds are of lower toxicity to microorganisms than are trialkyl- and tetraalkyl lead compounds (European Commission, 2002).

In all species of tested animals, including non-human primates, lead has been shown to cause adverse effects in several organs and organ systems, including the blood system, central nervous system, kidney and the reproductive and immune systems (European Commission, 2002). In communities of aquatic invertebrates, some populations are more sensitive than others and community structure may be adversely affected by lead contamination (European Commission, 2002). However, populations of invertebrates from polluted areas can show more tolerance to lead than those from non-polluted areas.

2.3.1.2 Mercury

Mercury exists mainly in three forms: metallic elements, inorganic salts and organic compounds, each of which possesses a different level of toxicity and bioavailability (European Commission, 2002; Jaishanka et al., 2014). Generally, organic forms are much more toxic than the inorganic forms, and organic mercury is known to bioaccumulate due to the body's inability to process and eliminate it (Holum, 1983; Singo, 2013).

Methylmercury represents the most toxic form of mercury to humans, and the general population is primarily exposed to methylmercury through consumption of fish and fish products (European Commission, 2002; Jaishanka et al., 2014). Generally, aquatic species including fish take up mercury (in all the three forms) and it gets transformed into methylmercury within the organisms (Jaishanka et al., 2014). Methylmercury has been found to have adverse effects on several organ systems in the human body as well as in animals, namely, the central nervous system (mental retardation, deafness, blindness, impairment of speech, etc.) and the cardiovascular system (blood pressure, heart-rate variety and heart diseases) (European Commission, 2002).

Exposure to high levels of mercury can permanently damage the brain, kidneys and developing foetuses (Martin and Griswold, 2009). For example, effects on brain functions may result in irritability, shyness, tremors, changes in vision or hearing and memory problems. Short-term exposure to high levels of metallic mercury vapours may cause lung damage, nausea, vomiting, diarrhoea, increases in blood pressure or heart rate, skin rashes and eye irritation (Martin and Griswold, 2009).

Research has shown that the central nervous system and the kidneys of animals are the organs most vulnerable to damage from methylmercury and inorganic mercury exposure (AMAP, 1998; European Commission, 2002). Effects include neurological impairment, reproductive effects, liver damage and significant decreases in intestinal absorption.

Birds fed with inorganic mercury show a reduction in food intake, thus resulting in poor growth (WHO, 1991; European Commission, 2002). Plants are generally insensitive to the toxic effects of mercury compounds though it is accumulated in taller plants, especially in perennials (Boening, 2000; European Commission, 2002).

2.3.1.3 Cadmium

Cadmium occurs naturally in ores together with zinc, lead and copper, and its compounds are used as stabilizers in PVC products, colour pigment, several alloys and, now most commonly, in rechargeable nickel–cadmium batteries (Järup, 2003). However, cadmium is the seventh most toxic heavy metal (Jaishanka et al., 2014) and is a known human carcinogen (Martin and Griswold, 2009). Once cadmium metal is absorbed by humans, it accumulates inside the body (particularly in the kidney and the liver) throughout life, and cadmium distributed in the environment will remain in soils and sediments for several decades (Jaishanka et al., 2014). Subsequently, plants gradually take up the metal from the soil which accumulate in them and concentrate along the food chain and ultimately reach the human body.

Damage to the kidney is probably the most serious health effect of cadmium to humans (European Commission, 2002). For example, the accumulation of cadmium in the kidney (in the renal cortex) leads to dysfunction of the kidney with impaired reabsorption of, for instance, proteins, glucose and amino acids. Ingesting very high levels of cadmium severely irritates the stomach, leading to vomiting and diarrhoea (Martin and Griswold, 2009). Other effects of cadmium exposure are disturbances in calcium metabolism (leading to fragile bones), hypercalciuria, formation of renal stones and lung damage.

Chronic cadmium exposure produces a wide variety of acute and chronic effects in mammals similar to those seen in humans (European Commission, 2002). For example, kidney damage and lung emphysema are the primary effects of high cadmium in the body. Kidney damage has also been reported in wild colonies of pelagic sea birds having high cadmium levels. Cadmium is also recognized for its adverse influence on the enzymatic systems of cells, oxidative stress and for inducing nutritional deficiency in plants (Irfan et al., 2013; Jaishanka et al., 2014).

2.3.1.4 Chromium

Chromium is the seventh most abundant element on earth (Katz and Salem, 1994; Mohanty and Patra, 2011, 2013; Jaishanka et al., 2014), and also ranks 21st in abundance in the earth's crust (McGrath, 1995; Barnhart, 1997). It occurs in a number of oxidation states in the environment ranging from Cr^{2+} to Cr^{6+}, but the most commonly occurring forms of chromium and of biological relevance are trivalent (Cr^{3+}) and hexavalent (Cr^{6+}) (European Commission, 2002; Rodríguez et al., 2007; Mohanty and Patra, 2013; Jaishanka et al., 2014). Though both states are toxic to animals, humans and plants (European Commission, 2002; Mohanty and Patra, 2013), there is a great difference between Cr^{3+} and Cr^{6+} with respect to toxicological and environmental effects, and they must always be considered separately (European Commission, 2002). In general, Cr^{3+} is considerably less toxic while Cr^{6+} has been demonstrated to have a number of adverse effects ranging from causing irritation to cancer. In other words, Cr^{6+} compounds are toxins and known human carcinogens, whereas Cr^{3+} is an essential nutrient in moderate quantities (Martin and Griswold, 2009).

Hexavalent chromium is toxic to microorganisms – a property utilized in chromium-based biocides (European Commission, 2002). Chromium can make fish more susceptible to infection; high concentrations can damage and/or accumulate in various fish tissues and in invertebrates such as snails and worms.

Although chromium is present in all plants, it has not been proved to be an essential element for plants (European Commission, 2002). The presence of excess chromium in the soil beyond the permissible limit is destructive because it greatly affects the biological processes of various plants such as maize, wheat, barley, cauliflower, watermelon and vegetables and enters the food chain on consumption of these plant materials (Jaishanka et al., 2014). Common features due to chromium phytotoxicity are reduction in root growth, leaf chlorosis (which is similar to that of iron deficiency), inhibition of seed germination, depressed biomass and necrosis (Ghani, 2011; Jaishanka et al., 2014).

2.3.1.5 Zinc

Zinc is an essential mineral of exceptional biological and public health importance (Hambidge and Krebs, 2007). For example, it plays a role in immune function; protein and DNA synthesis and cell division; wound healing; normal growth and development during the fetal stage, childhood and adolescence; is required for normal senses of taste and smell and is an antioxidant (Egwurugwu et al., 2013). However, taking too much zinc into the body through food, water or dietary supplements can also affect health (Roney et al., 2005; Plum et al., 2010).

It must be noted, however, that the levels of zinc that produce adverse health effects are much higher than the recommended dietary allowances (RDAs) for zinc of 11 mg/day for men and 8 mg/day for women (Roney et al., 2005). If large doses of zinc (10–15 times higher than the RDA) are taken by mouth even for a short time, stomach cramps, nausea and vomiting may occur (Roney et al., 2005; Plum et al., 2010). Ingesting high levels of zinc for several months may cause anaemia, lethargy, dizziness and decreased levels of high-density lipoprotein cholesterol and damage the pancreas (Porea et al., 2000; Roney et al., 2005; Plum et al., 2010). Taking up large doses of supplemental zinc over extended periods of time is also associated with copper deficiency whose symptoms include hypocupremia, impaired iron mobilization, anaemia, leukopenia, neutropenia, decreased superoxide dismutase, ceruloplasmin and cytochrome-c oxidase (Prohaska, 1990; Fiske et al., 1994; Plum et al., 2010).

Generally, zinc is also an essential nutrient for the growth of plants, but its intake in higher concentrations, just like in humans, is toxic (Davisearter and Shuman, 1993; Rout and Das, 2003). The general symptoms of zinc toxicity in plants include stunting of shoot, curling and rolling of young leaves, death of leaf tips and chlorosis (Rout and Das, 2003). It is also noted that sensitive terrestrial plants die when soil zinc levels exceed 100 mg/kg or when plant zinc content exceeds 178 mg/kg dry weight (Eisler, 1993). Furthermore, significant adverse effects of zinc on growth, survival and reproduction occur in representative sensitive species of aquatic plants, protozoans, sponges, molluscs, crustaceans, echinoderms, fish and amphibians at nominal water concentrations between 10 and 25 μg Zn/L (Eisler, 1993).

Livestock and small laboratory animals are comparatively resistant to zinc, however, excessive zinc intake through inhalation or oral exposure can have drastic effects on survival, metabolism and well-being (Eisler, 1993).

2.3.1.6 Arsenic

Arsenic is classified chemically as a metalloid, having both properties of a metal and a non-metal; however, it is frequently referred to as a metal (Chou et al., 2007). Arsenic is the 20th most abundant element in the earth's crust (Eisler, 1988; Mandal and Suzuki, 2001; Jaishanka et al., 2014), 14th in seawater, 12th in the human body (Eisler, 1988) and is a component in more than 245 minerals (Mandal and Suzuki, 2001). These are mostly ores containing sulphide, along with copper, nickel, lead, cobalt or other metals (Mandal and Suzuki, 2001). It is released into the environment through metal production; use of pesticides; burning of fossil fuels, particularly coal and waste disposal (Singo, 2013). Humans are mostly exposed to arsenic through the intake of food and drinking water, but other sources of exposure include soil and ambient air (Järup, 2003; Singo, 2013). Over the years, arsenic has been a great concern because of its chronic and epidemic effects on human, plant and animal health (Hughes et al., 2011; Hasanuzzaman et al., 2015). However, its effects depend on the chemical form of the arsenic, the nature of the surrounding environment and their own particular biological sensitivity (Gomez-Caminero et al., 2001), and the inorganic forms of arsenic such as arsenite and arsenate compounds are the most lethal to the environment and living creatures (Jaishanka et al., 2014).

Ingested arsenic compounds are readily absorbed in the gastrointestinal tract and distributed throughout the body, including foetuses (Singo, 2013), and very high levels of ingested arsenic increases the risk of cancer in internal organs like the bladder, liver and lungs (WHO, 1988; Aposhian et al., 2003; Chou et al., 2007). Long-term exposure via ingestion can also lead to skin thickening or discolouration and cancer of the skin (Aposhian et al., 2003; Singo, 2013). Deliberate consumption of arsenic in case of suicidal attempts or accidental consumption by children may also result in cases of acute poisoning (Saha et al., 1999; Mazumder, 2008; Jaishanka et al., 2014). Arsenic is, essentially, a protoplastic poison since it affects primarily the sulphydryl group of cells causing malfunctioning of cell respiration, cell enzymes and mitosis (Gordon and Quastel, 1948; Jaishanka et al., 2014). Other effects of arsenic exposure include a decrease in blood cell production, blood vessel damage, numbness in feet and hands, nausea and diarrhoea (Singo, 2013).

In the environment, the two forms of inorganic arsenic, arsenate and arsenite, are easily taken up by the cells of the plant root (Finnegan and Chen, 2012), thus inhibiting root extension and proliferation (Hasanuzzaman et al., 2015). Actually, in general, arsenic can severely inhibit plant growth by slowing or arresting cell expansion or biomass accumulation. Arsenic can also actively react with enzymes and proteins, disrupting biochemical functions of cells and thus, seriously hamper processes such as photosynthesis, respiration, transpiration and plant metabolism (Meharg and Hartley-Whitaker, 2002; Hasanuzzaman et al., 2015). Arsenic is also responsible for the generation of reactive oxygen species within the cell that may cause lipid peroxidation and protein oxidation, thus severely affecting cellular and subcellular organelles, and can even damage the DNA (Hasanuzzaman et al., 2015).

Mammals are exposed to arsenic primarily by the ingestion of naturally contaminated vegetation and water or through human activities (Eisler, 1988). General signs of arsenic toxicosis in warm-blooded organisms by inorganic and organic arsenic include intense abdominal pain, staggering gait, extreme weakness, trembling, salivation, vomiting, diarrhoea, fast and feeble pulse, prostration, collapse and death (Selby et al., 1977; Eisler, 1988). Signs of inorganic arsenite poisoning in birds include muscular incoordination, debility, slowness, jerkiness, falling hyperactivity, fluffed feathers, drooped eyelid, huddled position, unkempt appearance, loss of righting reflex, immobility, seizures and death (Hudson et al., 1984; Eisler, 1988).

Aquatic microorganisms show a wide range of sensitivities to arsenic species, with arsenite generally being more toxic than arsenate (Gomez-Caminero et al., 2001). Arsenic compounds give rise to acute and chronic effects in individuals, populations and communities at concentrations ranging from a few micrograms to milligrams per litre, depending on the species, time of exposure and end-points measured. These effects include lethality; inhibition of growth, photosynthesis and reproduction and behavioural effects (Gomez-Caminero et al., 2001).

2.3.2 Particulate Matter

Particulate matter is a complex widespread air pollutant, consisting of a mixture of extremely small solid particles and liquid droplets containing acids, organic chemicals, metals and soil or dust particles (WHO, 2003; Anderson et al., 2012). It is categorized by size, which is based on the aerodynamic equivalent diameter (AED) (Anderson et al., 2012). The AED is the diameter of a sphere with the density of water (1000 kg/m^3) that would have the same settling velocity as the particle (Lindeburg, 2013). The AED fractions have traditionally been based on how the particles are generated and where they deposit in human airways (i.e. how deep they can penetrate the respiratory system): <10, <2.5 and <0.1 μm (PM_{10}, $PM_{2.5}$ and $PM_{0.1}$, respectively) (Anderson et al., 2012). In such categorization, the distribution of particulate matter sizes is classified as follows: (1) the inhalable fraction (<10 μm AED) can be breathed into the nose and mouth, (2) the thoracic fraction (<2.5 μm AED) can enter the larger lung airways and (3) the respirable fraction (<0.1 μm AED) can penetrate beyond terminal bronchioles into the gas exchange regions (Lindeburg, 2013). Alternatively, particles with a diameter between 2.5 and 10 μm (PM2.5–10) are categorized as 'coarse', less than 2.5 μm as 'fine' and less than 0.1 μm as 'ultrafine' particles. These particles are derived from various sources and by various mechanisms as shown in Table 2.9.

Air pollution, in general, and particulate matter, in particular, has been associated with significant adverse health effects leading to increased morbidity and mortality (Araujo and Nel, 2009). Generally, the constituents that are likely to affect human health include (1) emissions from combustion of fossil and biomass fuels, (2) particles generated

TABLE 2.9

Classification of Particles Based on Size

Particle	Aerodynamic Diameter (μm)	Sources	Mode of Generation	Atmospheric Half-Life
Coarse particles ($PM_{2.5-10}$)	2.5–10	Suspension from disturbed soil (farming, mining, unpaved roads), construction, plant and animal fragments	Suspension from disturbed soil (farming, mining, unpaved roads), construction, plant and animal fragments	Minutes to hours
Fine particles ($PM_{2.5}$)	<2.5	Power plants, oil refineries, wildfires, residential fuel combustion, tailpipe and brake emissions	Gas-to-particle conversion by condensation, coagulation (accumulation mode)	Days to weeks
Ultrafine particles ($PM_{0.1}$)	<0.1	Fuel combustion (diesel, gasoline) and tailpipe emissions from mobile sources (motor vehicles, aircrafts, ships)	Fresh emissions, secondary photochemical reactions (nucleation mode)	Minutes to hours

Source: Araujo, J.A. and Nel, A.E., *Part. Fibre Toxicol.*, 6, 24, 2009.

by high temperature industrial processes such as smelting, (3) products of chemical reactions in the atmosphere such as SO_4^{2-} and NO_3^- and (4) fine particles from soil and other sources (Pope, 2000; Davidson et al., 2005). However, this section will only deal with particulates emanating from mining and metallurgical operations. Therefore, firstly, the section only gives a general overview of the effects of particulate matter on human health. Secondly, only the effects of silica, asbestos, coal particulates and smelter dust on human health (i.e. pneumoconiosis) are analysed in detail. In fact, silicosis (caused by silica), coal worker pneumoconiosis (caused by coal dust) and asbestosis (caused by asbestos fibres) are the three most common types of pneumoconiosis (Chong et al., 2006). However, it should also be noted that airborne particulate matter is also responsible for a number of effects apart from human health, such as alterations in visibility and climate (Davidson et al., 2005).

As stated already, only a snapshot of the general effects of particulate matter on human health is given here. Though there is no single parameter that could explain most of the effects of particulate matter (Kampa and Castanas, 2008), generally, the health effects of particulate matter are thought to be strongly associated with particle size and surface, composition and concentration (Davidson et al., 2005; Kampa and Castanas, 2008). Inhalable particles that are small enough to penetrate the thoracic region of the respiratory system may have the following short-term (hours to days) and long-term (months to years) human health effects: (1) respiratory and cardiovascular morbidity, such as aggravation of asthma and respiratory symptoms, and (2) mortality from respiratory and cardiovascular diseases and from lung cancer (WHO, 2003; Davidson et al., 2005). Apart from the cardiovascular and the respiratory system, other organs may also be affected (Kampa and Castanas, 2008). Furthermore, in addition to the classical risk factors such as serum lipids, smoking, hypertension, ageing, gender, family history, physical inactivity and diet, some data have shown that air pollution as an important additional risk factor for atherosclerosis (Araujo and Nel, 2009). It must be noted, however, that inhaled particulate matter can remain in the lungs for years without producing symptoms, abnormal physical signs or interference with lung function (Kirkham's and Joshi, 2013).

The following parts of this section discuss different types of inhalable particulates, namely, silica, asbestos, coal dust and smelter dust, and the other types of pneumoconiosis diseases are given in Table 2.10.

At this point it is important to give some details (though briefly) about pneumoconiosis. Pneumoconiosis is an occupational lung disease that is caused by the accumulation of inhaled particulates and involves a reaction of the tissues in the lung (Chong et al., 2006). In other words, pneumoconiosis refers to a family of respiratory diseases caused by chronic inhalation of organic or inorganic particulates over a prolonged period (Tamparo, 2016). The original term was 'pneumonokoniosis' which was coined by Zenker in 1866, but was modified by Proust in 1874 to 'pneumoconiosis' which means 'dusty lung' (Kirkham's and Joshi, 2013). Pneumoconiosis may be clinicopathologically classified as either fibrotic or non-fibrotic, depending on the presence or absence of fibrosis. Silicosis, coal worker pneumoconiosis, asbestosis, berylliosis and talcosis are examples of fibrotic pneumoconiosis. Siderosis, stannosis and baritosis are non-fibrotic forms of pneumoconiosis that result from the inhalation of iron oxide, tin oxide and barium sulphate particles, respectively.

It must be noted, however, that the exposure of the respiratory tract to different inhalable particulates may produce the same pathology because there are only limited numbers of patterns of reaction of lung tissue to inhaled particulates (Honma, 1992). Accordingly, the general pathogenesis of pneumoconiosis as discussed by Kirkham's and Joshi (2013) is as follows: After the inhalation of particulate matter, the alveolar macrophages converge upon extra-cellular particles and engulf them. If the number of particles is large, the

TABLE 2.10

Other Types of Pneumoconiosis

Type of Pneumoconiosis	Comment
Berylliosis	Berylliosis, or chronic beryllium disease, is a chronic allergic-type lung response and chronic lung disease caused by exposure to beryllium and its compounds. Acute beryllium disease causes nonspecific inflammatory reaction. Histologically, chronic beryllium disease may mimic sarcoidosis. The chest radiograph shows hilar lymphadenopathy and increased interstitial markings.
Talcosis	This is caused by exposure to talc dust, usually during talc mining or milling. Talc pneumoconiosis also can lead to lung fibrosis.
Siderosis	This pneumoconiosis, also known as welder's lung or silver polisher's lung, is caused by inhaling iron particles. Although siderosis often looks abnormal on a chest x-ray, it does not usually cause any symptoms.
Stannosis	Stannosis is a condition in which tin-oxide is deposited in the lung tissue after inhalation. Cut-surface of the lungs reveals numerous tiny (1–3 mm), grey-black dust macules, soft to touch and not raised above the cut-surface of the lung. Macrophages containing tin-oxide dust particles are present in alveolar walls and spaces, perivascular lymphatics and interlobular hilar nodes. Radiologically, it presents with numerous small, very dense opacities scattered evenly throughout the lung fields and may even be somewhat larger at 2–4 mm diameter and more fluffy or irregular in outline than those of siderosis. Kerley's lines are often clearly defined and dense linear opacities are seen in the upper lung zones.
Baritosis	Baritosis, a benign type of pneumoconiosis, is caused by long-term exposure to barium dust. Inhaled particulate matter remains in the lungs for years without producing symptoms, abnormal physical signs or interference with lung function. Owing to the high radio-opacity of barium, the discrete shadows in the chest radiograph are extremely dense. The discrete opacities in baritosis clear slowly over the years.
Aluminosis	Lung diseases induced by aluminium dust are very rare. Inhalation of dusts containing metallic and oxidised aluminium is associated with the development of pulmonary fibrosis. Aluminium lung is characterized as diffuse interstitial fibrosis that is mainly located in the upper and middle lobes of the lung. In advanced stages, it is characterised by subpleural bullous emphysema with an increased risk of spontaneous pneumothorax.
Hard metal disease	Hard metal is an alloy of tungsten carbide, cobalt and occasionally other metals, such as titanium, tantalum, chromium and nickel. This composite material has hardness almost like that of diamond; and is used to make machine parts that require high heat resistance, or to make tools used for drilling, cutting, machining or grinding. Of the hard metal compositions, cobalt whether free or in alloy form, is allergenic and cytotoxic and capable of provoking release of a fibrogenic agent from macrophages. The disease has been known by various names, such as hard metal pneumoconiosis, tungsten carbide pneumoconiosis, hard metal lung, giant cell interstitial pneumonitis and cobalt lung. Three types of disorder are attributed to cobalt exposure: acute (in the form of asthma), sub-acute (fibrosing alveolitis) and chronic in the form of diffuse interstitial fibrosis of the giant cell variety. One important aspect of hard-metal lung disease is that the disease may occur after a short duration of exposure, thus suggesting that individual susceptibility, rather than cumulative exposure, plays a major role.
Mixed-dust fibrosis	Mixed-dust pneumoconiosis is defined as a pneumoconiosis caused by concomitant exposure to silica and less fibrogenic dusts, such as iron, silicates and carbon. The silica is usually at a lower concentration than what causes silicotic nodules to occur. Microscopically, a stellate shape characterises the mixed-dust fibrotic nodule and has a central hyalinised collagenous zone surrounded by linearly and radially arranged collagen fibres admixed with dust-containing macrophages. The radiological findings of a mixed-dust pneumoconiosis include a mixture of small, rounded and irregular opacities. Honeycombing is also seen.

(Continued)

TABLE 2.10 (*Continued*)

Other Types of Pneumoconiosis

Type of Pneumoconiosis	Comment
Flock worker's lung	Flock worker's lung is an industrial lung disease. It is exclusive to employees of the rotary cut synthetic materials industry. Rotary-cut nylon, polyester, rayon and other synthetic fibres produce a powder of short fibres that are then adhesive coated to fabrics and other objects to produce a velvety surface. Flock production is associated with an increased risk in workers of developing a chronic interstitial lung disease characterised by a lymphocytic bronchiolitis, bronchiolocentric nodular and diffuse lymphocytic interstitial infiltrates and variable interstitial fibrosis.
Kaolin (china clay)	This pneumoconiosis is caused by inhaling kaolin, an ingredient used in the manufacture of ceramics, paper, medicines, cosmetics and toothpaste. Workers who mine, mill or bag kaolin are at risk.

Source: Kirkham's, V.S. and Joshi, J.M., *Ind. J. Chest Dis. Allied Sci.*, 55, 25, 2013.

elimination mechanism fails and dust containing macrophages collect in the interstitium especially in perivascular and peribronchiolar regions. If these aggregates remain in situ, *type 1* pneumocytes grow over them so that they become enclosed and are then entirely interstitial in position. Depending on the amount of particles and cell accumulation, the alveolar walls either protrude into the alveolar spaces or obliterate them. At the same time, a delicate supporting framework of fine reticulin fibres develops between the cells and in the case of particles with fibrogenic potential, the proliferation of collagen fibres follows. Inert particles such as carbon, iron, tin and titanium remain within the macrophages in these lesions until these cells die at the end of their normal life span. The particles are then released and reingested by other macrophages. Some particle-laden macrophages continually migrate to lymphatics or to bronchioles where these are eliminated, and migration is increased by infection or oedema of the lungs.

2.3.2.1 Silica

Silica, or silicon dioxide (SiO_2), is a group IV metal oxide, which occurs naturally in both crystalline and amorphous forms (IARC, 2012). Actually, it is the name given to a group of minerals that contain chemically combined silicon and oxygen (IAPA, 2008). It may be free, in which case only SiO_2 is present, or combined, in which the SiO_2 is combined chemically with some other atom(s) or molecule(s) (IAPA). This difference is important to recognize, since the silica problem exists only with free silica. Figure 2.1 illustrates the interrelationship between the various forms of silica (IAPA, 2008). It must also be noted, however, that the most abundant and thermodynamically stable form of silica is α-quartz, and the term quartz is often used in place of the general term crystalline silica (IARC, 2012).

Silica has widespread industrial applications including use as a food additive (i.e. anti-caking agent), as a means to clarify beverages, control viscosity, as an anti-foaming agent, as a dough modifier and as an excipient in drugs and vitamins (Martin, 2007). Despite its well-documented vast industrial applications, silica has serious effects on human health (Merget et al., 2002). However, this section will concern itself only with the effects of crystalline silica on human health since it is the most studied. For example, occupational exposure to crystalline silica dust is associated with a number of respiratory diseases such as silicosis, tuberculosis, progressive pulmonary fibrosis, chronic obstructive pulmonary

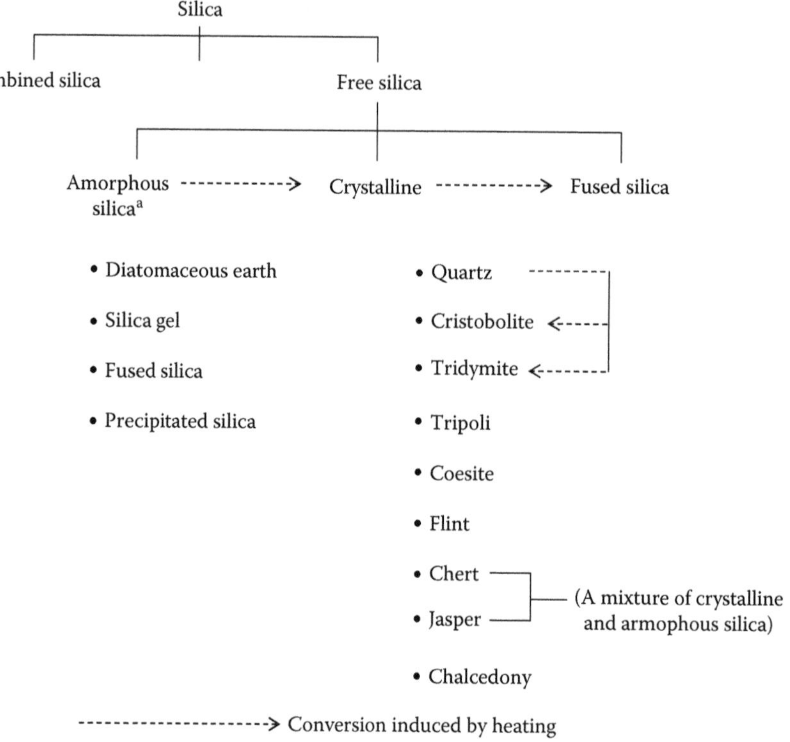

FIGURE 2.1
Interrelationships between the forms of silica. *Note:* [a]Commercially available products may contain appreciable quantities of crystalline-free silica. (From IAPA, Silica in the work place, 2008, Available at http://www.iapa.ca/pdf/Silica-in-the-workplace-FEB03.pdf, Accessed March 2016.)

disease (e.g. chronic bronchitis and emphysema) and lung cancer. Although the most common health effect of silica exposure is respiratory diseases, other diseases connected to respirable crystalline silica include systemic autoimmune diseases, such as rheumatoid arthritis, scleroderma, systemic lupus erythematosus and some of the small vessel vasculitides and renal diseases (Mulloy and Silver, 2016).

Among the stated health effects, only silicosis will be discussed further because of its prevalence. The pathogenesis of silicosis begins with the inhalation of one of the forms of crystalline silica particles (Mossman and Churg, 1998; Mulloy and Silver, 2016), and it is the result of the body's response to the presence of silica dust in the lung (IAPA, 2008). Ideally, the interaction between the silica particle and the alveolar macrophage (the main phagocytic cell in the alveolar space) starts the process of silicosis (Mulloy and Silver, 2016). When the respirable fraction of the silica dust penetrates the innermost part of the respiratory tract (i.e. alveoli or air sacs where the exchange of oxygen and carbon dioxide occur), they are removed by white blood cells known as macrophage. However, particles of free crystalline silica cause the macrophages to break open. The result is the formation of scar like patches on the surface of the alveolus. The formation of a large number of scars following prolonged exposure to silica particles causes the alveolar surface to become less elastic, and thus reduces the transfer of gases. The ongoing inflammation and scarring in the form of nodular lesions in the upper lobes of the lungs is the histologic hallmark of silicosis (Rhoades and Pflanzer, 2003). In the early stages, macroscopically, one can

TABLE 2.11

Three Major Types of Silicosis

Type of Silicosis	Comment
Acute	Acute silicosis develops from inhaling large amounts of silica dust over a few days or months. Signs of the disease include shortness of breath, fever, cough and weight loss. Generally, people with acute silicosis have stable health, however for some it may lead quickly to death.
Chronic	Chronic silicosis is the most common type and occurs after many years of contact with low levels of silica dust in the air. There are two forms of chronic silicosis: simple or complicated. With *simple silicosis*, small solid or unclear nodules can be detected on a chest x-ray, however, individuals are asymptomatic. Long-term exposure to silica dust may lead to complicated silicosis. With *complicated silicosis*, also called progressive massive fibrosis (PMF), larger nodules can be detected on a chest x-ray. Some individuals may still be asymptomatic or initial symptoms may include shortness of breath with exercise, wheezing or sputum that causes coughing. Other lung diseases can aggravate the condition and severe complicated silicosis can result in heart disease with lung disease, called cor culmonale.
Accelerated	Accelerated silicosis is similar to the chronic type, however it forms more quickly. The lung scars can be detected sooner and nodules appear on a chest x-ray five years after the first exposure to silica dust. This type of silicosis occurs from exposure to large amounts of silica dust over a short time period and can progress quickly.

Source: IAPA, Silica in the work place, 2008, Available at http://www.iapa.ca/pdf/Silica-in-the-workplace-FEB03.pdf, Accessed March 2016.

identify small dark (in association with anthracosis) or pale nodules located predominantly in the upper lobes of lungs (Danciu et al., 2014; Cheepsattayakorn and Cheepsattayakorn, 2016). In the later stages of the disease, the fibrous nodules enlarge and merge, resulting in large areas of pulmonary fibrosis. These nodules, referred to as 'silica nodules' can cause central ischemic necrosis. Microscopically, silica nodules consist of concentric laminated collagen fibres and tend to become confluent, compressing adjacent alveoli. The silicosis disease is characterized by shortness of breath, cough, fever and cyanosis (bluish skin) (Rhoades and Pflanzer, 2003). Although there are several different clinical and pathologic varieties of silicosis (Mossman and Churg, 1998), only the three major types are outlined in Table 2.11.

2.3.2.2 Asbestos

Asbestos is a generic term for a group of six naturally occurring fibrous silicate minerals of the serpentine (curved) and amphibole (thin and straight) series that have been applied widely in the past in construction, automotive, textile and plastic industries (Hanley et al., 2001; IARC, 2012). Basically, when these minerals are crushed, they break into fibres rather than dust (Murphy et al., 1986). The different fibrous minerals include the serpentine mineral chrysotile (also known as 'white asbestos') and the five amphibole minerals – actinolite, amosite (also known as 'brown asbestos'), anthophyllite, crocidolite (also known as 'blue asbestos') and tremolite (IARC, 2012). All forms of asbestos are hazardous, and all can cause cancer, but the amphibole forms of asbestos are considered to be more hazardous to health than chrysotile (Hanley et al., 2001). This is perhaps due to the greater ability of the amphibole forms to survive in lung tissue, while chrysotile tends to disappear over time (Murphy et al., 1986). This section will only consider the clinical and pathologic features of asbestosis. However, other types of asbestos-related diseases are listed in Table 2.12.

TABLE 2.12

Types of Asbestosis

Types of Asbestosis	Comment
Pleural plaques	The pleura is the outer protective membrane of the lung. By breathing in asbestos this membrane can become scarred with the accumulation of asbestos dust. It can occur in one or both lungs. Usually there is no breathlessness or disability. In themselves plaques are not a serious injury but do worry most people. The reason for this is that there is a risk of more serious illnesses developing. Unfortunately it is not currently possible to claim compensation for this injury.
Pleural thickening	Pleural thickening is more serious than plaques because it can cause pain and breathlessness. Pleural thickening occurs when the asbestos fibres accumulate in or around any part of the pleura.
Pleural effusion	This is a build-up of fluid in the chest cavity. It is usually treated by way of a pleural aspiration i.e. drainage of the fluid from the chest cavity. A complete recovery would be expected although it can be the case that further illnesses do develop.
Asbestosis	This is the formation of fibrous or scar tissue in the lung itself. Symptoms associated with this include pain and discomfort when breathing, a dry cough and breathlessness. More often than not the development of symptoms will over the years increase and people often mistakenly associate the symptoms with old age rather than the disease itself. Asbestosis often gets worse with time. There is no cure for it.
Mesothelioma	This is a type of malignant tumour usually caused by inhalation of asbestos dust. It is a type of cancer of the pleura or peritoneum. Unfortunately despite advances in medical science there is no cure for this type of cancer. The symptoms associated with Mesothelioma include gradually increasing chest pain caused by a build-up of fluid around the lung, shortness of breath, tiredness, weight loss and lethargy. These diseases have no connection with smoking and some can be caused as a result of only short periods of exposure to asbestos dust.
Other asbestos related diseases	Modern medical research has now proved that there is a link between asbestos exposure and some cancers including lung, gastro intestinal and laryngeal cancer.

Source: Wyatt, J. and Ball, S., Asbestos brochure, Hamers Solicitors, Hull, UK, 2007, Available at http://www.hamers.com/wp-content/uploads/2014/10/Asbestos-Brochure.pdf, Accessed October 2016.

Asbestosis, also known as diffuse pulmonary fibrosis, is defined as bilateral diffuse interstitial fibrosis of the lungs caused by the inhalation of asbestos fibres (Murphy et al., 1986; Mossman and Churg, 1998). Though the precise molecular mechanism regulating asbestos-induced lung damage is not fully understood, it is believed that, once inhaled, the asbestos fibre triggers a persistent inflammatory response after being engulfed by alveolar macrophages (Lang, 2009). The response involves the generation of reactive oxygen and nitrogen species and the expression of cytokines, growth factors and chemokines, which together lead to activation of lung fibroblasts and overproduction of extracellular matrix (Lang, 2009). The resulting asbestos or ferruginous body which is usually recognized by its beaded necklace–like or drumstick-like appearance is a most characteristic index of asbestos exposure (Murphy et al., 1986).

Asbestosis is characterized by severe shortness of breath, which is generally the first symptom to appear (Davis et al., 1999; Merck, 2011; Charlevois, 2015). At first, this is only evident after exertion, but may eventually happen even during rest. Other symptoms include tightness in the chest, dry cough, loss of appetite, fatigue and – when the disease reaches an advanced stage – clubbing of the fingers. It must be noted that as the development of asbestosis is dose-dependent, symptoms appear only after a latent period of 20 years or longer (Kirkham's and Joshi, 2013). However, the latent period may be shorter after intense exposure.

2.3.2.3 Coal Dust

Coal, just like other fossil fuels such as oil and natural gas, is mainly used as a non-renewable source of energy (Simate et al., 2016). However, in the recent past, various types of coal have been increasingly used not only as fuels, but also as valuable materials in several remediation and pollution control processes (Ishaq et al., 2007; Anwar et al., 2009). Coal can also be used as a chemical feedstock instead of oil for the production of the petrochemical-based commodities that are so important to our modern way of life (Viljoen, 1979). Despite the various industrial applications, the mining, processing and transport of raw coal result in a large proportion of it breaking into dust. Basically, from mining to coal cleaning, from transportation to usage to disposal, coal releases countless volumes of dust into the air, land and water. Unfortunately, inhalation of respirable coal dust results in a respiratory disease called coal workers' pneumoconiosis (CWP), commonly called black lung disease (Castranova and Vallyathan, 2000; Finkelman et al., 2002). The name 'black lung' comes from the fact that those with the disease have lungs that look black instead of a healthy pink (Hair, 2016). The risk for CWP depends on the total dust burden in the lungs and is also related to the coal rank, which is based on its carbon content (anthracite has a higher rank than bituminous, followed by sub-bituminous and lignite) (Ross and Murray, 2004). In the higher ranking coal, there may be a greater relative surface area of the coal dust particles, higher surface-free radicals and higher silica content, thus CWP is more common in mines with a high grade of coal (Ross and Murray, 2004).

The pathogenesis of CWP is similar to other pneumoconioses such as silicosis and asbestosis (Albert et al., 2008), whereas, pneumoconiosis, as already discussed, refers to a family of respiratory diseases caused by chronic inhalation of organic or inorganic particulates (Tamparo, 2016). The following chain of events has been proposed for the initiation and progression of CWP: (1) inhaled coal dust concentrates at the bifurcations of the respiratory bronchioles; (2) local inflammation results in the accumulation of phagocytic cells (alveolar macrophages) that scavenge coal dust particles, forming lung lesions known as coal macules; (3) with further exposure, coal macules enlarge to form coal nodules; (4) as the lesions condense, surrounding tissue is torn forming scar emphysema and (5) connective tissue becomes associated with these lesions leading to progressive massive fibrosis (PMF) (Finkelman et al., 2002).

Depending on the extent (or severity) of the disease, there are two forms of CWP – simple CWP and complicated CWP (Castranova and Vallyathan, 2000). Simple CWP is characterized by the formation of black coal dust macules in the upper lobes of the lung (Castranova and Vallyathan, 2000; Chong et al., 2006). Microscopically, macules contain coal dust-laden macrophages with a fine network of reticulin and some collagen fibres (Castranova and Vallyathan, 2000). An increased exposure to coal dust results in the development of nodular lesions that are firm on palpation in contrast to non-palpable macules. Ideally, nodules are coal dust–laden fibrotic lesions with round or irregular borders and also have an irregular distribution of collagen fibres (Vallyathan et al., 2011). Nodules are classified on the basis of size as micronodules (<7 mm diameter) and macronodules, which range in diameter from 8 mm to 2 cm (Kleinerman et al., 1978; Castranova and Vallyathan, 2000). With chronic exposure to coal dust, nodules may converge and coalesce to produce lesions measuring larger than 2 cm with a fibrous nature. At this stage, the disease is called complicated CWP or PMF (Castranova and Vallyathan, 2000).

Respiratory symptoms associated with CWP include dyspnoea, cough, sputum (or phlegm) production, wheezing, pneumonia, bronchitis, asthma and emphysema (Marine et al., 1998; Beeckman et al., 2001; Wade et al., 2011).

In the environment, coal dust can find its way into the water and thus affect the water quality. For example, coal dust can harm the flora and fauna because it reduces the amount of oxygen available for clams, barnacles, mussels and crab larvae, hence impacting on fertility and growth of sea creatures. Coal dust can also cover the leaves of vegetation and reduce its photosynthesis capabilities (Ahearn, 2013). Furthermore, coal dust may also have toxic effects on wildlife that might eat the vegetation. Unfortunately, there has been very little research into the effects of coal and coal dust on vegetation, waterways and the ecosystems they support, mainly owing to the lack of adequate investigation.

2.3.2.4 Smelter Dust

Pyrometallurgical processes, which are discussed in detail in Chapters 4 through 6, use heat to separate desired metals from other materials. During pyrometallurgical processing, beneficiated ore (concentrated by crushing, grinding, floating and drying) and/or secondary materials are smelted and refined to produce intermediate products (Roto, 2011). Smelting specifically involves heating the charge with a reducing agent such as coke, charcoal or other reagents in specially designed furnace reactors. However, smelting processes are known to emit large quantities of air pollutants such as hydrogen fluoride, sulphur dioxide, oxides of nitrogen, offensive and noxious smoke fumes, vapours, gases and other toxins (Blacksmith Institute, 2016). A variety of heavy metals are also released by the facilities as fine particles and/or volatile compounds, either via off-gas systems or as 'fugitive' emissions from the operations. The generation of smelter pollutants is fully discussed in Chapters 4 through 6, and thus will not be discussed further in this chapter. This chapter only deals with environmental and health effects of 'smelter dust' – a term used for a collection of fine particulate emissions from smelters.

Smelter dust production is a worldwide problem in pyrometallurgical processes (Montenegro et al., 2008). Smelter dust contains condensate matter and fine particles of semi-melted concentrate which are transported with the off-gas. However, the composition of smelter dust depends on the type of the smelting furnace and raw materials charged, while the form of the elements depends on the conditions and the operational parameters in the smelting process (EPA, 1995). Of particular importance in this chapter are the health effects of metals contained in the smelter dust. Epidemiological and experimental studies have shown that the cardiovascular and respiratory systems are the most affected, but the effects on several other organs can also be significant (Kampa and Castanas, 2008; Genc et al., 2012; Fortoul et al., 2015).

Effect on respiratory system. Heavy metals can significantly affect the respiratory system because of their toxic potency. Irritation of the respiratory system followed by bronchoconstriction, asthma and dyspnoea are some of the symptoms of inhaling particulate matter containing heavy metals (Kampa and Castanas, 2008). Heavy metals such as cadmium, chromium, mercury and lead, which have already been discussed, may also cause fibrosis and damage to the lungs (Fortoul et al., 2015). The exact mechanisms involved in particulate matter exposures and subsequent lung damage have already been discussed under pneumoconioses diseases (silicosis, asbestosis and CWP).

Effect on cardiovascular system. There is epidemiological and clinical evidence that metal contaminants may play a role in the development of cardiovascular diseases (Solenkova et al., 2014). Typically, the inhalation of particulate matter containing metals into the lung leads to the constriction of pulmonary and systemic vessels (Huang and Ghio, 2006). Research results have shown that the inhalation of particulate matter leads to pulmonary inflammation with secondary systemic effects or, after translocation from the lung into the circulation, to direct toxic cardiovascular effects (Mills et al., 2008). In fact, the study by Riediker et al. (2004) has shown that apart from lung inflammation, particulate matter can induce systemic inflammatory changes that affect blood coagulation (or clotting). Subsequently, changes in blood clotting can obstruct (cardiac) blood vessels, leading to angina or even to myocardial infraction (Vermylen et al., 2005; Kampa and Castanas, 2008). Symptoms such as tachycardia, increased blood pressure and anaemia due to an inhibitory effect on haematopoiesis have also been observed as a consequence of pollution by heavy metal (Huang and Ghio, 2006; Kampa and Castanas, 2008).

Huang and Ghio (2006) explained how inhalation of particulate matter may result in direct or indirect cardiovascular effects. One possible mechanism is via the activation of lung cells producing inflammatory mediators. These mediators then trigger a wide array of secondary vasoactive signals. The particulate matter may also disrupt autonomic nervous system balance and promote vasoconstriction without causing lung inflammation. A third possibility is the direct effects on vascular reactivity, much like free-floating cigarette smoke causing acute vasoconstriction, increasing plasma endothelin levels and triggering endothelial dysfunction. Currently, there is little evidence to prove the permeation of intact particles of any size (i.e. coarse, fine and ultrafine fractions) across the alveolar–capillary membrane of the lung (Huang and Ghio, 2006). However, there are multiple routes by which metals can be transported into the systemic circulation. For example, metals can be internalized by resident alveolar macrophages and/or there may be interaction of the metal with the epithelial lining (alveolar or airway). Once inside the epithelial cells, further processing may occur with subsequent release of metals in the basal direction, resulting in systemic redistribution.

Effect on nervous system. The central nervous system is also a target organ for the detrimental effects of airborne heavy metal pollutants (Kampa and Castanas, 2008). The inhalation of heavy metals such as arsenic, lead, cadmium, mercury, manganese and vanadium have been found to induce brain damage and thus, result in mental and behavioural disorders (Fortoul et al., 2015). Neurotoxicity leading to neuropathies, with symptoms such as memory disturbances, sleep disorders, anger, fatigue, hand tremors, blurred vision and slurred speech, have been observed after arsenic, lead and mercury exposure (Ewan and Pamphlett, 1996; Ratnaike, 2003; Kampa and Castanas, 2008).

Other effects. Heavy metals can affect the urinary system by damaging the kidney, lead to renal cancer and may increase the risk of stone formation or nephrocalcinosis (Vamvakas et al., 1993; Järup, 2003; Kampa and Castanas, 2008). Some heavy metals such as vanadium may increase pancreatic enzymes and hypertrophic acinar cells, which results in an acute and chronic pancreatitis or lead to pancreatic cancer (Cano-Gutierrez et al., 2009; Hall et al., 2014; Fortoul et al., 2015). Exposure to some metals has also been associated with increased risk of diabetes (Fortoul et al., 2015). The retina of the eye may also be affected by the inhalation of metal particulates. For example, a study on mice showed morphological alterations in the photoreceptor layer and in the inner and outer nuclear layers, as well as a gradual rhodopsin pigment reduction in the photoreceptor layer and an increase of the oxidative stress biomarker – 4-hydroxinonenal – in the photoreceptor layer and in the inner and outer plexiform layers (Quezada, 2013; Fortoul et al., 2015). Maternal exposure to heavy metals such as lead, cadmium, mercury, manganese, chromium and nickel increases the risks of infertility, miscarriages, low birth weight and reduced foetus growth (Kampa and Castanas, 2008). In other words, these metals have female reprotoxic effects (Gerhard et al., 1998).

2.3.3 Gaseous Emissions

It has already been pointed out that the mining, mineral processing and metal production sectors generate and emit various types of toxic gases into the environment. The six standard air pollutants that have been extensively studied in urban populations are sulphur dioxide (SO_2), ozone (O_3), nitrogen dioxide (NO_2), carbon monoxide (CO), lead and particulates (McCarthy et al., 2001). A list of these air pollutants (excluding lead and particulates) and some of the other common gases from the mining–metallurgical processes and the nature of their effects on human health, the ecology and climate change are given in Table 2.13.

2.3.4 Wastewater and Effluent

Water is essential to most mining and metallurgical operations, as it is both a transport and a process medium in the processing of raw minerals (Slatter et al., 2009). However, process wastewaters discharged from mining and metallurgical operations are widely contaminated by various pollutants (Rubio et al., 2002). The various pollutants discharged in wastewater streams include metal ions, anions, oils, acid, residual chemicals and/or organic reagents and complexes or chelates (Stander et al., 1970; Rubio et al., 2002).

Table 2.14 shows the hazardous environmental and health effects of a selected number of pollutants in wastewater streams from mining and metallurgical operations. It must be noted that the environmental and health effects of heavy metals have been extensively covered under Section 2.3.1. However, a summary of the general effects of heavy metals is also included in Table 2.14.

2.3.5 Other Mining and Metallurgical Effects

The effects of the mining and metallurgical activities go beyond what has already been discussed in various sections of this chapter to include land degradation and noise pollution. For example, mining can cause physical disturbances to the landscape, creating eyesores such as waste rock piles and open pits (Barreto, 2007; Higginson and Wheeler, 2010; Chattopadhyay and Chattopadhyay, 2013). Such disturbances may contribute to the decline of wildlife and plant species in an area. In addition, it is possible that many of the pre-mining surface features cannot be replaced after the mining ceases. Mine subsidence (ground movements of the earth's surface due to the collapse of overlying strata into voids created by underground mining) can cause damage to buildings and roads. When the soil is removed during mining, vegetation is also removed thus exposing the soil to erosion. As a result, sediments, typically from increased soil erosion, cause siltation and the smothering of streambeds. This siltation affects fisheries, swimming, domestic water supply, irrigation and other uses of streams.

TABLE 2.13

Gaseous Emissions and Their Impacts

Type of Gas	Impact
Sulphur dioxide (SO_2)	*Health*: Sulphur dioxide causes upper respiratory irritation and can aggravate existing respiratory diseases, especially asthma. *Ecological*: Sulphur dioxide contributes to acid deposition, which causes acidification of dams and rivers, and damages trees and crops as well as buildings and statues. Leachates and percolates can contaminate subterranean aquifers.
Nitrogen oxides (NO_x, NO, NO_2)	*Health*: Exposure to nitrogen dioxide (NO_2) increases the risk of respiratory infections. *Ecological*: Nitrogen oxides play an important role in the atmospheric reactions that create ozone and contribute to acid deposition. Ozone can cause acidification of dams and rivers, damage trees and crops as well as buildings and statues, and also reduce visibility.
Ground-level ozone (O_3)	*Health*: Ozone at ground level is a major health concern. This gas damages lung tissue and reduces lung function. It also reduces resistance to colds and other infections. *Ecological*: Ozone is beneficial in the stratosphere because it shields the Earth from the Sun's harmful ultraviolet radiation. At ground level, it is a major component of smog, reducing visibility and impacting adversely on plant function and productivity. Ozone is also implicated in the deterioration of rubber, paints, plastics and textiles. *Climate change*: Ground-level ozone is a greenhouse gas, and modelling by NASA has shown O_3 to be responsible for between one-third and half of the observed warming trend in the Arctic during winter and spring.
Carbon monoxide (CO)	*Health*: When CO enters the blood stream, it reduces the delivery of oxygen to the body's tissues and cells, because the haemoglobin in the red blood cells has a higher affinity for CO than for oxygen.
Carbon dioxide (CO_2)	*Health*: Carbon dioxide constitutes a health risk only at concentrations high enough to displace oxygen and cause asphyxiation. *Climate change*: Carbon dioxide is a greenhouse gas.
Hydrogen sulphide (H_2S) and total reduced sulphur (TRS)	*Health*: At low concentrations, hydrogen sulphide is associated with malodour and mild respiratory ailments. As concentration increases, it can cause eye, nose and throat irritation, headache, nausea and vomiting. At high concentrations, pulmonary oedema may develop. Exposure to concentrations greater than 500 ppm may cause death. *Ecological*: Hydrogen sulphide can be oxidized in air to form sulphuric acid and elemental sulphur. In water, it can damage plants such as rice.
Methane (CH_4)	*Health*: Methane constitutes a health risk only at concentrations high enough to displace oxygen and cause asphyxiation. It is a serious hazard at explosive or combustible concentrations. *Climate change*: Methane is a greenhouse gas, with a global warming potential (GWP) of 23.
Volatile organic compounds (VOCs) including hydrocarbons	*Health*: Some VOCs are respiratory irritants, others cause malodour (e.g. limonene, amines, butyric acid), and some are carcinogens (such as benzene and methylene chloride). *Ecological*: VOCs participate in the complex chemical reactions whereby O_3 is formed at ground level.
Ammonia (NH_3)	*Health*: Ammonia is produced within the human body and contributes to the acid/base balance. Exposure to high concentrations can cause irritation to the eyes, nose and throat. *Ecological*: Ammonia in soil is a source of nitrogen for plants.

Source: Department of Environmental Affairs (DEA), State of air report 2005: A report on the state of air in South Africa, 2009, Available at https://www.environment.gov.za/sites/default/files/docs/stateofair_executive_summary.pdf, Accessed October 2016.

TABLE 2.14

Pollutants in Wastewater Streams and Their Impacts

Type of Pollutant	Impact	References
Heavy metals	The toxicity or poisoning of heavy metals results from the disruption of metabolic functions. Heavy metals disrupt the metabolic functions in two ways: (1) they accumulate in vital organs and glands such as the heart, brain, kidneys, bone and liver where they disrupt their important functions, and (2) they inhibit the absorption, interfere with or displace the vital nutritional minerals from their original place, thereby, hindering their biological functions. Plants experience oxidative stress upon exposure to heavy metals that leads to cellular damage and disturbance of cellular ionic homeostasis; thus disrupting the physiology and morphology of plants.	Simate and Ndlovu (2014)
Acidity	There is an indirect, but devastating effect of high acidity. For example, acidification increases bioconversion of mercury to methylmercury, which accumulates in fish, increasing the risk to toxicity in people who eat fish. When the soil pH is low; nitrogen, phosphorus and potassium are tied up in the soil and not available to plants. Calcium and magnesium, which are essential plant nutrients, may also be absent or deficient in low pH soils. At low pH, toxic elements such as aluminium, iron and manganese are also released from soil particles, thus increasing their toxicity. Furthermore, if soil pH is low, the activity of soil organisms that break down organic matter is reduced.	Goyer et al. (1985); Simate and Ndlovu (2014)
Residual organic chemicals	Adverse health effects of exposure to high levels of organic chemicals, particularly persistent organic pollutants, include cancer, damage to the nervous system, reproductive disorders, disruption of the immune system, allergies and hypersensitivity, sensory system, endocrine system and death. During the decomposition process of organic compounds, the dissolved oxygen in the receiving water may be used up at a greater rate than it can be replenished, causing oxygen depletion and having severe consequences for the biota. Organic effluents also frequently contain large quantities of suspended solids which reduce the light available to photosynthetic organisms and, on settling out, alter the characteristics of the river bed, rendering it an unsuitable habitat for many invertebrates.	WHO (2010); Lenntech (2016)
Organometallic compounds	Organometallic compounds are metal complexes containing at least one direct, covalent metal-carbon bond. Organification (e.g. methylation) of metals generally increases their toxicity by rendering them more lipid soluble and facilitating their crossing of lipid barriers such as the cell membrane or blood-tissue (e.g. blood-brain) barriers. For example, tin in its inorganic form is considered non-toxic but is toxic in organotin compounds. The toxicity of organometallic compounds is related to their accumulation degree in cells of living organisms and their actions include blocking metabolic processes, disorders of barrier functions of intestine epithelial cells and induction of the apoptosis of nerve cells.	WHO (2006); Engel et al. (2009)
Sulphates	There is no definite sulphate dose that would induce adverse health effects in adults, and thus there is no set of health-based standard for sulphate in drinking water. However, dehydration and diarrhoea are common side-effect following the ingestion of large amounts of sulphate containing compounds.	Fawell et al. (2004a)

(Continued)

TABLE 2.14 (*Continued*)
Pollutants in Wastewater Streams and Their Impacts

Type of Pollutant	Impact	References
Hydrogen sulphide	There is scantily available data on the oral toxicity of hydrogen sulphide. However, alkali sulphides irritate mucous membranes and can cause nausea, vomiting and epigastric pain following ingestion. The oral dose of sodium sulphide fatal to humans has been estimated at 10–15 g.	Fawell et al. (2003)
Nitrates	The toxicity of nitrate to humans is mainly attributable to its reduction to nitrite. The major biological effect of nitrite in humans is its involvement in the oxidation of normal haemoglobin (Hb) to methaemoglobinaemia (metHb), which is unable to transport oxygen to the tissues. Methaemoglobinaemia, causes cyanosis and, at higher concentrations, asphyxia. Other suggested effects include congenital malformations, goitre and cancer.	Cotruvo et al. (2011)
Cyanide	The health effects of large amounts of cyanide are similar, whether through breathing air, eating food or drinking water that contains it. The effects of acute cyanide exposure are dominated by central nervous system and cardiovascular disturbances. Typical signs of acute cyanide poisoning include tachypnoea, headache, vertigo, lack of motor coordination, weak pulse, cardiac arrhythmias, vomiting, stupor, convulsions and coma. Other symptoms include gastrointestinal tract disturbances, trauma and enlarged thyroid. Pathological findings may include tracheal congestion with haemorrhage, cerebral and pulmonary oedema, gastric erosions and petechiae of the brain meninges and pericardium.	WHO (2004); Taylor et al. (2006)
Fluoride	The most obvious health effect of excess fluoride exposure is dental fluorosis, which when mild includes white streaks, and when severe can include brown stains, pits and broken enamel. Elevated fluoride intakes can also have more serious effects on skeletal tissues.	Fawell et al. (2004b)

The environmental impacts of mining and metallurgical activities are also exacerbated by the water pollution problems caused by acid mine drainage (AMD). Basically, AMD is characterized by low pH water with elevated concentrations of sulphates, iron and non-ferrous metals whose profile depend on the originating mineral deposit types (Kontopouls, 1998; Sheoran and Sheoran, 2006; Obreque-Contreras et al., 2015). The occurrence of AMD and the recovery of a number of useful products from it are discussed in Chapter 3.

2.4 Concluding Remarks

The importance of mining and metallurgical industries to the economies of many countries cannot be overemphasized. These industries are pivotal since they create large-scale employment opportunities and generate revenue to governments through different forms of taxation. However, both developed and developing nations are increasingly concerned about the adverse impact of mining and metallurgical operations to the environment and human health. Basically, from mining and milling through hydrometallurgical and pyrometallurgical processing to refining, metal production can have significant adverse impacts on air quality, surface and ground water quality and land. In order to sustain and build on the contributions to development, the mining and metallurgical industries are seriously addressing environmental questions, particularly the relationship of mining and metallurgical processes to sustainable development. In fact, Chapters 3 through 9 of this book deal with issues of sustainable development – the 3 Rs – reduce, reuse, recycle. More importantly, national and/or international legislation and policies governing the environmental and health impacts of mining and metallurgical operations are in place in many countries. As a result, a good deal of attention is now being paid to the health and safety of people at the workplace and the local community. Furthermore, nowadays, the regulations that protect land, water and air resources from pollution are well developed for active mining and metallurgical sites, including substantial management for closure and/or post-closure activities.

References

Ahearn, A. (2013). Coal dust's environmental impacts. Available at http://www.opb.org/news/article/coal-dusts-environmental-impacts/. Accessed 22 March 2016.

Akpor, O.B. and Muchie, M. (2010). Remediation of heavy metals in drinking water and wastewater treatment systems: Processes and applications. *International Journal of the Physical Sciences* 5(12): 1807–1817.

Albert, R.K., Spiro, S.G. and Jett, J.R. (2008). *Clinical Respiratory Medicine*. Elsevier Health Sciences, London, UK.

Anderson, J.O., Thundiyil, J.G. and Stolbach, A. (2012). Clearing the air: A review of the effects of particulate matter air pollution on human health. *Journal of Medical Toxicology* 8(2): 166–175.

Anwar, J., Shafique, U., Salman, M., Zaman, W., Anwar, S. and Anzano, J.M. (2009). Removal of chromium (III) by using coal as adsorbent. *Journal of Hazardous Materials* 171: 797–801.

Aposhian, H.V., Zakharyan, R.A., Mihaela, D.A., Koplin, M.J. and Wollenberg, M.L. (2003). Oxidation and detoxification of trivalent arsenic species. *Toxicology and Applied Pharmacology* 193: 1–8.

Araujo, J.A. and Nel, A.E. (2009). Particulate matter and atherosclerosis: Role of particle size, composition and oxidative stress. *Particle and Fibre Toxicology* 6: 24.

Arctic Monitoring and Assessment Programme (AMAP) (1998). *AMAP Assessment Report: Arctic Pollution Issues*. AMAP, Oslo, Norway.

Aubertin, M. and Bussière, B. (2001). Meeting environmental challenges for mine waste management. *Geotechnical News*, September 21–26.

Barclays (2015). Environmental and social risk briefing: Mining and metals. Available at https://www.home.barclays/content/dam/barclayspublic/docs/Citizenship/mining-metals-guidance-note.pdf. Accessed 15 May 2016.

Barnhart, J. (1997). Occurrences, uses, and properties of chromium. *Regulatory Toxicology and Pharmacology* 26: S3–S7.

Barreto, A.M. (2007). War cry in Loliem over mega mining project. Available at https://www.mail-archive.com/goajourno@puggy.symonds.net/msg01245.html. Accessed 11 March 2016.

Beeckman, L.A.F., Wang, M.L., Petsonk, E.L. and Wagner, G.R. (2001). Rapid declines in FEV1 and subsequent respiratory symptoms, illnesses, and mortality in coal miners in the United States. *American Journal of Respiratory and Critical Care Medicine* 163: 633–639.

Bernard, P. (2012). *The IT Service Part 1 – The Essentials*. Van Haren Publishing, Zaltbommel, the Netherlands.

Bini, C. and Bech, J. (2014). *PHEs, Environment and Human Health: Potentially Harmful Elements in the Environment and the Impact on Human Health*. Springer, London, UK.

Blacksmith Institute (2016). Metals smelters and processing. Available at http://www.worstpolluted.org/projects_reports/display/61. Accessed 11 March 2016.

Boening, D.W. (2000). Ecological effects, transport, and fate of mercury: A general review. *Chemosphere* 40: 1335–1351.

Cano-Gutierrez, G.F.R., Montaño, L.C., Rodriguez-Lara, F.V. and Fortoul, T.I. (2009). Pancreatic changes and vanadium inhalation. *Current Topics in Toxicology* 6: 39–44.

Castranova, V. and Vallyathan, V. (2000). Silicosis and coal workers' pneumoconiosis. *Environmental Health Perspectives* 108: 675–684.

Chan, B.K.C., Bouzalakos, S. and Dedeney, A.W.L. (2008). Integrated waste and water management in mining and metallurgical industries. *Transactions of Nonferrous Metals Society of China* 18: 1497–1505.

Charlevois, A. (2015). Asbestos disease. Available at http://www.mesothelioma.com/asbestos-cancer/disease/. Accessed 22 March 2016.

Chattopadhyay, S. and Chattopadhyay, D. (2013). Mining industries and their sustainable management. In: Malhotra, R. (ed.), *Fossil Energy*. Springer, New York.

Cheepsattayakorn, A. and Cheepsattayakorn, R. (2016). Pulmonary crystalline silica exposure: The adverse effects. Available at http://medcraveonline.com/JLPRR/JLPRR-03-00067.pdf. Accessed 21 February 2017.

Chong, S., Lee, K.S., Chung, M.J., Han, J., Kwon, O.J. and Kim, T.S. (2006). Pneumoconiosis: Comparison of imaging and pathologic findings. *Radio Graphics* 26: 59–77.

Chou, S., Harper, C., Ingerman, L., Llados, F., Colman, J., Chappell, L., Osier, M., Odin, M. and Sage, G. (2007). *Toxicological Profile for Arsenic*. Agency for Toxic Substances and Disease Registry, Atlanta, GA.

Cotruvo, J., Fawell, J.K., Giddings, M., Jackson, P., Magara, Y., Ngowi, A.V.F. and Ohanian, E. (2011). Nitrate and nitrite in drinking-water. Available at http://www.who.int/water_sanitation_health/dwq/chemicals/nitratenitrite2ndadd.pdf. Accessed 15 March 2016.

Danciu, M., Mihailovici, M.S., Dima, A. and Cucu, C. (2014). *Atlas of Pathology*, 3rd edn. Available at http://www.pathologyatlas.ro/pneumoconiosis-silicosis.php. Accessed 19 March 2016.

Davidson, C.I., Phalen, R.F. and Solomon, P.A. (2005). Airborne particulate matter and human health: A review. *Aerosol Science and Technology* 39: 737–749.

Davis, G.S., Marcy, T.W. and Seward, E.A. (1999). *Medical Management of Pulmonary Diseases*. CRC Press, New York.

Davisearter, J.G. and Shuman, L.M. (1993). Influence of texture and pH of baolinitic soil on zinc fractions and zinc uptake by peanuts. *Soil Science* 155: 376–384.

Department of Environmental Affairs (DEA) (2009). State of air report 2005: A report on the state of air in South Africa. Available at https://www.environment.gov.za/sites/default/files/docs/stateofair_executive_summary.pdf. Accessed 7 October 2016.

Dudgeon, S. (2009). Copper mining: From the ground up. Available at http://faculty.virginia.edu/metals/cases/dudgeon3.html. Accessed 10 March 2016.

Egwurugwu, J.N., Ifedi, C.U., Uchefuna, R.C., Ezeokafor, E.N. and Alagwu, E.A. (2013). Effects of zinc on male sex hormones and semen quality in rats. *Nigerian Journal of Physiological Sciences* 28: 17–22.

Eisler, R. (1988). Arsenic hazards to fish, wildlife, and invertebrates: Synoptic review. Available at https://www.pwrc.usgs.gov/eisler/CHR_12_Arsenic.pdf. Accessed 22 March 2016.

Eisler, R. (1993). Zinc hazards to fish, wildlife, and invertebrates: A synoptic review. Available at https://www.pwrc.usgs.gov/eisler/CHR_26_Zinc.pdf. Accessed 22 March 2016.

Engel, G., Podolak, M. and Man, D. (2009). The effect of selected organometallic compounds on membrane voltage. *Polish Journal of Environmental Studies* 18(2): 207–211.

Environment and Climate Change Canada (ECCC) (2012). Guidance document for management of wastes from the base metals smelting sector. Available at https://www.ec.gc.ca/planp2-p2plan/default.asp?lang=En&n=D4D7479A-1&offset=6&toc=hide#section6_6. Accessed 3 May 2016.

Environmental Innovations (2004). Environmental management systems and ISO 14000: An overview. Available at https://www.novascotia.ca/nse/pollutionprevention/docs/EMS_fact-sheet.pdf. Accessed 5 June 2016.

EPA (1995). Compilation of air pollutant emission factors. Vol. I: Stationary point and area sources. Available at https://www3.epa.gov/ttn/chief/ap42/ch12/final/c12s03.pdf. Accessed 22 March 2016.

European Commission (2002). Heavy metals in waste. Available at http://ec.europa.eu/environment/waste/studies/pdf/heavy_metalsreport.pdf. Accessed 22 March 2016.

European Commission (2009). Reference document on best available techniques for management of tailings and waste-rock in mining activities. Available at http://eippcb.jrc.ec.europa.eu/reference/BREF/mmr_adopted_0109.pdf. Accessed 22 May 2016.

Ewan, K.B. and Pamphlett, R. (1996). Increased inorganic mercury in spinal motor neurons following chelating agents. *Neurotoxicology* 17(2): 343–349.

Fawell, J.K., Lund, U. and Mintz, B. (2003). Hydrogen sulfide in drinking-water. Available at http://www.who.int/water_sanitation_health/dwq/chemicals/en/hydrogensulfide.pdf. Accessed 17 March 2016.

Fawell, J.K., Ohanian, E., Giddings, M., Toft, P., Magara, Y. and Jackson, P. (2004a). Sulfate in drinking-water. Available at http://www.who.int/water_sanitation_health/dwq/chemicals/sulfate.pdf. Accessed 30 March 2016.

Fawell, J.K., Ohanian, E., Giddings, M., Toft, P., Magara, Y. and Jackson, P. (2004b). Fluoride in drinking-water. Available at http://www.who.int/water_sanitation_health/dwq/chemicals/fluoride.pdf. Accessed 6 March 2016.

FICCI (2013). Development of Indian mining industry – The way forward: Non-fuel minerals. Available at http://ficci.in/spdocument/20317/mining-industry.pdf. Accessed 15 May 2016.

Finkelman, R.B., Orem, W., Castranova, V., Tatu, C.A., Belkin, H.E., Zheng, B., Lerch, H.E., Maharaj, S.V. and Bates, A.L. (2002). Health impacts of coal and coal use: Possible solutions. *International Journal of Coal Geology* 50: 425–443.

Finnegan, P.M. and Chen, W. (2012). Arsenic toxicity: The effects on plant metabolism. *Frontiers in Physiology* 3(182): 1–18.

Fiske, D.N., McCoy, H.E. and Kitchens, C.S. (1994). Zinc-induced sideroblastic anemia: Report of a case, review of the literature, and description of the hematologic syndrome. *American Journal of Hematology* 46(2): 147–150.

Fortoul, T.I. et al. (2015). Health effects of metals in particulate matter. In: Nejadkoorki, F. (ed.), *Current Air Quality Issues*. In-Tech, Rijeca, Croatia.

Fraser Institute (2012). How are waste materials managed at mine sites? Available at http://www.miningfacts.org/Environment/How-are-waste-materials-managed-at-mine-sites/. Accessed 4 February 2016.

Gardea-Torresdey, J.L., Peralta-Videa, J.R., Rosa, G.D. and Parsons, J.G. (2005). Phytoremediation of heavy metals and study of the metal coordination by x-ray absorption spectroscopy. *Coordination Chemistry Reviews* 249: 1797–1810.

Gardner, G. and Sampat, P. (1998). Mind over matter: Recasting the role of materials in our lives. Available at http://www.worldwatch.org/system/files/WP144.pdf. Accessed 8 May 2016.

Genc, S., Zadeoglulari, Z., Fuss, S.H. and Genc, K. (2012). The adverse effects of air pollution on the nervous system. *Journal of Toxicology* 12: 1–23.

Gerhard, I., Monga, B., Waldbrenner, A. and Runnebaum, B. (1998). Heavy metals and fertility. *Journal of Toxicology and Environmental Health, Part A: Current Issues* 54(8): 593–611.

Ghani, A. (2011). Effect of chromium toxicity on growth, chlorophyll and some mineral nutrients of *Brassica juncea* L. *Egyptian Academic Journal of Biological Sciences* 2(1): 9–15.

Gomez-Caminero, A., Howe, P., Hughes, M., Kenyon, E., Lewis, D.R., Moore, M., Ng, J., Aitio, A. and Becking, G. (2001). Environmental health criteria for arsenic and arsenic compound. Available at http://apps.who.int/iris/bitstream/10665/42366/1/WHO_EHC_224.pdf. Accessed 19 March 2016.

Gordon, J.J. and Quastel, G.H. (1948). Effect of organic arsenicals on enzyme system. *Biochemical Journal* 42: 337–350.

Goyer, R.A. et al. (1985). Potential human health effects of acid rain: Report of a workshop. *Environmental Health Perspectives* 60: 355–368.

Greenovation Hub (2014). China's mining industry at home and overseas: Development, impacts and regulation. Available at http://www.ghub.org/cfc_en/wp-content/uploads/sites/2/2014/11/China-Mining-at-Home-and-Overseas_Main-report2_EN.pdf. Accessed 6 May 2016.

Hair, J. (2016). Coal miner black lung disease resurgence in Queensland to face Senate inquiry. Available at http://www.abc.net.au/news/2016-02-12/coal-miner-disease-resurgence-in-queensland-senate-inquiry/7161694. Accessed 1 March 2016.

Hall, T.C., Garcea, G., Webb, M.A., Al-Leswas, D., Metcalfe, M.S. and Dennison, A.R. (2014). The socioeconomic impact of chronic pancreatitis: A systematic review. *Journal of Evaluation in Clinical Practice* 20(3): 203–207.

Hambidge, K.M. and Krebs, N.F. (2007). Zinc Deficiency: A special challenge. *Journal of Nutrition* 137(4): 1101–1105.

Hanley, G.D., Kess, S., Stevens, Y.W., Wilbur, S. and Williams, M. (2001). *Toxicological Profile for Asbestos*. Agency for Toxic Substances and Disease Registry, Atlanta, GA.

Hardygóra, M., Paszkowska, G. and Sikora, M. (2004). Mine planning and equipment selection 2004. *Proceedings of the 13th International Symposium on Mine Planning and Equipment Selection*, Wroclaw, Poland, 1–3 September 2004. CRC Press, Boca Raton, FL.

Hasanuzzaman, M., Nahar, K., Hakeem, K.R., Öztür, M. and Fujita, M. (2015). Arsenic toxicity in plants and possible remediation. In: Hakeem, K.R., Sabir, M., Öztürk, M. and Mermut, A.R. (eds.), *Soil Remediation and Plants*. Elsevier, Amsterdam, the Netherlands.

Herbert Smith Freehills (HSF) (2012). Mining law in Asia. Available at http://www.herbertsmithfreehills.com/-/media/HS/Insights/Guides/PDFs/Asia%20guides/Mining%20Law%20in%20Asia.pdf. Accessed 1 May 2016.

Higginson, S. and Wheeler, A. (2010). Global markets international. Available at http://www.willis.com/documents/publications/industries/Mining_and_Metals/Willis_Mining_Market_Review_2010.pdf. Accessed 3 March 2016.

Holum, J.R. (1983). *Elements of General and Biological Chemistry*. John Wiley & Sons, New York.

Honma, K. (1992). Pathology of nonasbestos pneumoconiosis (silicosis and mixed dust pneumoconiosis). *Revista Española de Patología* 32(3): 392–394.

Huang, Y.C. and Ghio, A.J. (2006). Vascular effects of ambient pollutant particles and metals. *Current Vascular Pharmacology* 4(3): 199–203.

Hudson, R.H., Tucker, R.K. and Haegele, M.A. (1984). *Handbook of Toxicity of Pesticides to Wildlife*. U.S. Fish and Wildlife Service, Washington, DC.

Hudson, T.L., Fox, F.D. and Plumlee, G.S. (1999). Metal mining and the environment. Available at http://www.americangeosciences.org/critical-issues/faq/what-are-regulations-mining-activities. Accessed 2 May 2016.

Hudson-Edwards, K.A., Jamieson, H.E. and Lottermoser, B.G. (2011). Mine wastes: Past, present, future. *Elements* 7: 375–380.

Hughes, M.F., Beck, B.D., Chen, Y., Lewis, A.S. and Thomas, D.J. (2011). Arsenic exposure and toxicology: A historical perspective. *Toxicological Sciences* 123: 305–332.

IAPA (2008). Silica in the work place. Available at http://www.iapa.ca/pdf/Silica-in-the-workplace-FEB03.pdf. Accessed 9 March 2016.

IARC (2012). Arsenic, metals, fibres, and dusts: A review of human carcinogens. Available at http://www.ncbi.nlm.nih.gov/books/NBK304375/pdf/Bookshelf_NBK304375.pdf. Accessed 4 March 2016.

IFC (2007). Environmental, health and safety guidelines for mining. Available at http://www.ifc.org/wps/wcm/connect/1f4dc28048855af4879cd76a6515bb18/Final++Mining.pdf?MOD=AJPERES. Accessed 12 May 2016.

Irfan, M., Hayat, S., Ahmad, A. and Alyemeni, M.N. (2013). Soil cadmium enrichment: Allocation and plant physiological manifestations. *Saudi Journal of Biological Sciences* 20(1): 1–10.

Ishaq, M., Ahmad, I., Shakirullah, M., Rehman, H., Khan, M.A., Ahmad, I. and Rehman, I. (2007). Adsorption study of phenol on Lakhra coal. *Toxicological and Environmental Chemistry* 89(1): 1–6.

Jaishanka, M., Tseten, T., Anbalagan, N., Mathew, B.B. and Beeregowda, K.N. (2014). Toxicity, mechanism and health effects of some heavy metals. *Interdisciplinary Toxicology* 7(2): 60–72.

Järup, L. (2003). Hazards of heavy metal contamination. *British Medical Bulletin* 68: 167–182.

Kampa, M. and Castanas, E. (2008). Human health effects of air pollution. *Environmental Pollution* 151: 362–367.

Katz, S.A. and Salem, H. (1994). *The Biological and Environmental Chemistry of Chromium*. Wiley-VCH Publishers, New York.

Kirkham's, V.S. and Joshi, J.M. (2013). Pneumoconioses. *The Indian Journal of Chest Disease and Allied Science* 55: 25–34.

Kleinerman, J., Green, F.H.Y., Harley, R., Lapp, N.L., Laqueur, W., Naeye, R.L., Pratt, P., Taylor, G. and Wyatt, J. (1978). Pathology standards for coal workers' pneumoconiosis: Report of the pneumoconiosis committee of the College of American Pathologists to the National Institute for Occupational Safety and Health. *Archives of Pathology and Laboratory Medicine* 103: 375–431.

Kontopouls, A. (1998). Acid mine drainage control. In: Castro, S.H., Vergara, F. and Sánches, M.A. (eds.), *Effluent Treatment in the Mining Industry*. University of Concepcion, Concepcion, Chile, pp. 57–118.

Kumar, U. and Singh, D.N. (2013). E-waste management through regulations. *International Journal of Engineering Inventions* 3(2): 6–14.

Kurniawan, T.A., Lo, W.H. and Chan, G.Y.S. (2006). Physico-chemical treatments for removal of recalcitrant contaminants from landfill leachate. *Journal of Hazardous Materials* 129(1–3): 80–100.

Lang, F. (2009). *Encyclopedia of Molecular Mechanisms of Diseases*. Springer-Verlag GmbH, Berlin, Germany.

Lawley, R., Curtis, L. and Davis, J. (2008). *The Food Safety Hazard Guidebook*. Royal Society of Chemistry, London, UK.

Lenntech (2016). Effects of organic pollution on freshwater ecosystems. Available at http://www.lenntech.com/aquatic/organic-pollution.htm. Accessed 26 March 2016.

Lindeburg, M.R. (2013). *Mechanical Engineering Reference Manual for the PE Exams*. Professional Publications Inc., Belmont, CA.

Lottermoser, B. (2010). *Mine Wastes: Characterization, Treatment and Environmental Impacts*. Springer Science and Business Media, New York.

Mandal, B.K. and Suzuki, B.T. (2001). Arsenic round the world: A review. *Talanta* 58: 201–235.

Marine, W.M., Gurr, D. and Jacobsen, M. (1998). Clinically important respiratory effects of dust exposure and smoking in British coal miners. *American Review of Respiratory Diseases* 137: 106–112.

Martin, K.R. (2007). The chemistry of silica and its potential health benefits. *The Journal of Nutrition, Health & Aging* 11(2): 94–97.

Martin, S. and Griswold, W. (2009). Human health effects of heavy metals. *Environmental Science and Technology Briefs for Citizens* 15: 1–6.

Matuszak-Flejszman, A. (2009). Benefits of environmental management system in Polish companies compliant with ISO 14001. *Polish Journal of Environmental Studies* 18(3): 411–419.

Mazumder, G. (2008). Chronic arsenic toxicity and human health. *Indian Journal of Medical Research* 128(4): 436–447.

McCarthy, J.J., Canziani, O.F., Leary, N.A., Dokken, J. and White, K.S. (2001). *Climate Change 2001: Impacts, Adaptation, and Vulnerability*. Cambridge University Press, Cambridge, UK.

McGrath, S.P. (1995). Chromium and nickel. In: Alloway, B.J. (ed.), *Heavy Metals in Soils*, 2nd edn. Blackie Academic and Professional, Glasgow, UK.

McKinley, M.J. (2004). Mining: Pollution A to Z. Available at http://www.encyclopedia.com/topic/mining.aspx. Accessed 14 May 2016.

Meharg, A.A. and Hartley-Whitaker, J. (2002). Arsenic uptake and metabolism in arsenic resistant and non-resistant plant species. *Tansley Reviews – New Phytologist* 154: 29–43.

Merck (2011). *The Merck Manual Home Health Handbook*. John Wiley & Sons, London, UK.

Merget, R., Bauer, T., Küpper, H.U., Philippou, S., Bauer, H.D., Breitstadt, R. and Bruening, T. (2002). Health hazards due to the inhalation of amorphous silica. *Archives of Toxicology* 75(11–12): 625–634.

Metalworld (2013). Mining sector in doldrums. Available at http://metalworld.co.in/newsletter/2013/jan/analysis4-0113.pdf. Accessed 10 May 2016.

Mills, N.L., Donaldson, K., Hadoke, P.W., Boon, N.A., MacNee, W., Cassee, F.R., Sandström, T., Blomberg, A. and Newby, D.E. (2008). Adverse cardiovascular effects of air pollution. *Nature Clinical Practice Cardiovascular Medicine* 6(1): 36–44.

Mohanty, M. and Patra, H.K. (2011). Attenuation of chromium toxicity by bioremediation technology. *Reviews of Environmental Contamination and Toxicology* 210: 1–34.

Mohanty, M. and Patra, H.K. (2013). Effect of ionic and chelate assisted hexavalent chromium on mung bean seedlings (*Vigna radiata* L. *wilczek. var k*-851) during seedling growth. *Journal of Stress Physiology and Biochemistry* 9(2): 232–241.

Montenegro, V., Sano, H. and Fujisawa, T. (2008). Recirculation of Chilean copper smelting dust with high arsenic content to the smelting process. *Materials Transactions* 49(9): 2112–2118.

Morris, A.S. (2004). *ISO 14000 Environmental Management Standards: Engineering and Financial Aspects*. John Wiley & Sons Ltd., Chichester, UK.

Mossman, B.T. and Churg, A. (1998). Mechanisms in the pathogenesis of asbestosis and silicosis. *American Journal of Respiratory and Critical Care Medicine* 157: 1666–1680.

Mulloy, K.B. and Silver, K. (2016). Silica: The deadly dust. Available at http://www.elcosh.org/document/1426/668/d000803/5.html. Accessed 4 March 2016.

Murphy, R.L. et al. (1986). Diagnosis of nonmalignant diseases related to asbestos. *American Review of Respiratory Diseases* 134(2): 363–368.

Najeeb, U., Ahmad, W., Zia, M.H., Malik, Z. and Zhou, W. (2014). Enhancing the lead phytostabilization in wetland plant *Juncus effusus* L. through somaclonal manipulation and EDTA enrichment. Available at http://www.sciencedirect.com/science/article/pii/S1878535214000124. Accessed 21 February 2017.

National Mining Association (NMA) (2015). Federal environmental laws that govern U.S. mining. Available at http://www.nma.org/index.php/federal-environmental-laws-that-govern-u-s-mining. Accessed 19 May 2016.

Nieboer, E. and Richardson, D.H.S. (1980). The replacement of the nondescript term 'heavy metals' by biologically and chemically significant classification of metal ions. *Environmental Pollution* 1: 2–26.

Obreque-Contreras, J., Pérez-Flores, D., Gutiérrez, P. and Chávez-Crooker, P. (2015). Acid mine drainage in Chile: An opportunity to apply bioremediation technology. *Hydrology Current Research* 6(3): 1–8.

Oulego, P., Collado, S., Laca, A. and Díaz, M. (2016). Impact of leachate composition on the advanced oxidation treatment. *Water Research* 88: 389–402.

Phadke, H., Collins, A., Cotran, H., Glazer, S., Gorodniuk, A., Lavigne-Delville, J., Moin, N., Rodriguez, A., Rogers, J. and Vozza, G. (2004). Metals and mining: Research brief. Available at http://www.sasb.org/wp-content/uploads/2014/06/NR0302_MetalsMining_2014_06_24_Industry_Brief.pdf. Accessed 20 February 2016.

Plum, L.M., Rink, L. and Haase, H. (2010). The essential toxin: Impact of zinc on human health. *International Journal of Environmental Research and Public Health* 7: 1342–1365.

Pope, C.A. (2000). What do epidemiologic findings tell us about health effects of environmental aerosols? *Journal of Aerosol Medicine* 13(4): 335–354.

Porea, T.J., Belmont, J.W. and Mahoney, D.H. (2000). Zinc-induced anemia and neutropenia in an adolescent. *Journal of Pediatrics* 136: 688–690.

Prohaska, J.R. (1990). Biochemical changes in copper deficiency. *Journal of Nutritional Biochemistry* 1(9): 452–461.

Puder, M.G. and Veil, J.A. (2004). Review of waste management regulations and requirements from selected jurisdictions. Available at http://www.netl.doe.gov/kmd/cds/disk23/D-Water%20Management%20Projects/Produced%20Water%5CFEW%2049177%5CFEW%2049177%20Final%20Rpt.pdf. Accessed 18 May 2016.

Quezada, M.E.M. (2013). *Cambios en la histología de la retina de ratones expuestos a la inhalación de pentóxido de vanadio y la participación del estrés oxidante*. Universidad Nacional Autónoma de México, México City, México.

Raghab, S.M., Meguid, A.M.A.E. and Hegaziet, H.A. (2013). Treatment of leachate from municipal solid waste landfill. *Housing and Building National Research Center* 9: 187–192.

Rankin, W.J. (2011). *Minerals, Metals and Sustainability: Meeting Future Material Needs*. CRC Press, Leiden, the Netherlands.

Ratnaike, R.N. (2003). Acute and chronic arsenic toxicity. *Postgraduate Medical Journal* 79(933): 391–396.

Renou, S., Givaudan, J.G., Poulain, S., Dirassouyan, F. and Moulin, P. (2008). Landfill leachate treatment: Review and opportunity. *Journal of Hazardous Materials* 150: 468–493.

Rhoades, R.A. and Pflanzer, R.G. (2003). *Human Physiology*, 4th edn. Thomson, London, UK.

Riediker, M., Cascio, W.E., Griggs, T.R., Herbst, M.C., Bromberg, P.A., Neas, L., Williams, R.W. and Devlin, R.B. (2004). Particulate matter exposure in cars is associated with cardiovascular effects in healthy young men. *American Journal of Respiratory and Critical Care Medicine* 169: 934–940.

Rodríguez, M.C., Barsanti, L., Passarelli, V., Evangelista, V., Conforti, V. and Gualtieri, P. (2007). Effects of chromium on photosynthetic and photoreceptive apparatus of the alga *Chlamydomonas reinhardtii*. *Environmental Research* 105(2): 234–239.

Roney, N., Smith, C.V., Williams, M., Osier, M. and Paikoff, S.J. (2005). *Toxicological Profile for Zinc*. Agency for Toxic Substances and Disease Registry, Atlanta, GA.

Ross, M.H. and Murray, J. (2004). Occupational respiratory disease in mining. *Occupational Medicine* 54: 304–310.

Roto, P. (2011). Smelting and refining. Available at http://www.iloencyclopaedia.org/component/k2/135-smelting-and-refining-operations/smelting-and-refining. Accessed 12 March 2016

Rout, G.R. and Das, P. (2003). Effects of metal toxicity on plant growth and metabolism: I. Zinc. *Agronomie* 23: 3–11.

Rubio, J., Souza, M.L. and Smith, R.W. (2002). Overview of flotation as a wastewater treatment technique. *Minerals Engineering* 15: 139–155.

Saha, J.C., Dikshit, A.K., Bandyopadhyay, M. and Saha, K.C. (1999). A review of arsenic poisoning and its effects on human health. *Critical Reviews in Environmental Science and Technology* 29(3): 281–313.

Schreuder, D. (2006). Technologies for emissions reduction in the metallurgical and chemical process industries. Available at http://www.saimm.co.za/Conferences/Pt2006/181-190_Schreuder.pdf. Accessed 25 May 2016.

Selby, L.A., Case, A.A., Osweiler, G.D. and Hages, H.M. (1977). Epidemiology and toxicology of arsenic poisoning in domestic animals. *Environmental Health Perspectives* 19: 183–189.

Sepúlveda, A., Schluep, M., Renaud, F.G., Streicher, M., Kuehr, R., Hagelüken, C. and Gerecke, A.C. (2010). A review of the environmental fate and effects of hazardous substances released from electrical and electronic equipments during recycling: Examples from China and India. *Environmental Impact Assessment Review* 30: 28–41.

Sharma, R.J. and Agrawal, M. (2005). Biological effects of heavy metals: An overview. *Journal of Environmental Biology* 26(2): 301–313.

Sheoran, A.S. and Sheoran, V. (2006). Heavy metal removal mechanism of acid mine drainage in wetlands: A critical review. *Minerals Engineering* 19: 105–116.

Simate, G.S., Maledi, N., Ochieng, A., Ndlovu, S., Zhang, J. and Walubita, L.F. (2016). Coal-based adsorbents for water and wastewater treatment. *Journal of Environmental Chemical Engineering* 4(2): 2291–2312.

Simate, G.S. and Ndlovu, S. (2014). Acid mine drainage: Challenges and opportunities. *Journal of Environmental Chemical Engineering* 2(3): 1785–1803.

Singh, R., Gautam, N., Mishra, A. and Gupta, R. (2011). Heavy metals and living systems: An overview. *Indian Journal of Pharmacolology* 43(3): 246–253.

Singo, N.K. (2013). An assessment of heavy metal pollution near an old copper mine dump in Musina, South Africa. MSc (Eng) Dissertation, University of South Africa, Pretoria, South Africa.

Slatter, K.A., Plint, N.D., Cole, M., Dilsook, V., de Vaux, D., Palm, N. and Oostendorp, B. (2009). Water management in Anglo platinum process operations: Effect of water quality on process operations. *International Mine Water Conference*, Pretoria, South Africa, 19–23 October.

Solenkova, N.V., Newman, J.D., Berger, J.S., Thurston, G., Hochman, J.S. and Lamas, G.A. (2014). Metal pollutants and cardiovascular disease: Mechanisms and consequences of exposure. *American Heart Journal* 168: 812–822.

Stander, G.J., Henzen, M.R. and Funke, J.W. (1970). The disposal of polluted effluents from mining, metallurgical and metal fnishing industries, their effects on receiving water and remedial measures. *Journal of South African Institute of Mining and Metallurgy* 71(5): 95–103 (December).

Suvio, P., Palmer, J., del Olmo, A.G. and Kauppi, J. (2016). Holistic tailings management solutions. Available at file:///C:/Users/a0009328/Downloads/OTE_White_Paper_Tailings_management%20(3).pdf. Accessed 21 February 2017.

Szczepanski, M. (2012). Mining in the EU: Regulation and the way forward. Available at http://www.europarl.europa.eu/RegData/bibliotheque/briefing/2012/120376/LDM_BRI(2012)120376_REV1_EN.pdf. Accessed 11 May 2016.

Tamparo, C.D. (2016). *Diseases of the Human Body*. F.A. Davis Company, Philadelphia, PA.

Tangahu, B.V., Abdullah, S.R.S., Basri, H., Idris, M., Anuar, N. and Mukhlisin, M. (2011). A review on heavy metals (As, Pb, and Hg) uptake by plants through phytoremediation. *International Journal of Chemical Engineering* 2011: 1–31.

Tariq, M., Ali, M. and Shah, Z. (2006). Characteristics of industrial effluents and their possible impacts on quality of underground water. *Soil and Environment* 25(1): 64–69.

Taylor, J., Roney, N., Harper, C., Fransen, M.E. and Swarts, S. (2006). *Toxicological Profile for Cyanide*. Agency for Toxic Substances and Disease Registry, Atlanta, GA.

USA Congress (1988). *Copper Technology and Competitiveness*. U.S. Government Printing Office, Washington, DC.

US Congress (1992). *Managing Industrial Solid Wastes from Manufacturing, Mining, Oil and Gas Production, and Utility Coal Combustion*. U.S. Government Printing Office, Washington, DC.

Vallyathan, V., Landsittel, D.P., Petsonk, E.L., Kahn, J., Parker, J.E., Osiowy, K.T. and Green, F.H. (2011). The influence of dust standards on the prevalence and severity of coal worker's pneumoconiosis at autopsy in the United States of America. *Archives of Pathology and Laboratory Medicine* 135(12): 1550–1556.

Vamvakas, S., Bittner, D. and Koster, U. (1993). Enhanced expression of the protooncogenes c-myc and c-fos in normal and malignant renal growth. *Toxicology Letters* 67(1–3): 161–172.

van Zyl, D., Sassoon, M., Digby, C., Fleury, A.M. and Kyeyune, S. (2002). Mining for the future. Available at http://pubs.iied.org/pdfs/G00560.pdf. Accessed 13 May 2016.

Vermylen, J., Nemmar, A., Nemery, B. and Hoylaerts, M.F. (2005). Ambient air pollution and acute myocardial infarction. *Journal of Thrombosis and Haemostasis* 3: 1955–1961.

Viljoen, D.A. (1979). The importance of coal. Available at http://www.saimm.co.za/Journal/v079n16p493.pdf. Accessed 21 March 2016.

Vogt, C. (2013). International assessment of marine and riverine disposal of mine tailings. Available at http://www.craigvogt.com/links/Mine_Tailings_Marine_and_Riverine_Disposal.pdf. Accessed 27 May 2016.

Wade, W.A., Petsonk, E.L., Young, B. and Mogri, I. (2011). Severe occupational pneumoconiosis among West Virginian coal miners: One hundred thirty-eight cases of progressive massive fibrosis compensated between 2000 and 2009. *Chest* 39(6): 1458–1462.

WHO (1988). Environmental Health Criteria 61. World Health Organisation, International Programme on Chemical Safety (IPCS), Geneva, Switzerland.

WHO (1991). Inorganic mercury. Environmental Health Criteria 118. World Health Organisation, International Programme on Chemical Safety (IPCS), Geneva, Switzerland.

WHO (2003). Health effects of particulate matter: Policy implications for countries in eastern Europe, Caucasus and central Asia. Available at http://www.euro.who.int/__data/assets/pdf_file/0006/189051/Health-effects-of-particulate-matter-final-Eng.pdf. Accessed 22 March 2016.

WHO (2004). Hydrogen cyanide and cyanides: Human health aspects. Available at http://www.who.int/ipcs/publications/cicad/en/cicad61.pdf. Accessed 22 March 2016.

WHO (2006). Elemental speciation in human health risk assessment. Environmental Health Criteria 234. Available at http://www.inchem.org/documents/ehc/ehc/ehc234.pdf. Accessed 22 March 2016.

WHO (2010). Persistent organic pollutants: Impact on child health. Available at http://apps.who.int/iris/bitstream/10665/44525/1/9789241501101_eng.pdf. Accessed 22 March 2016.

Wiertz, J.V. (1999). Mining and metallurgical waste management in Chilean copper industry. *IWA Proceedings*, Sevilla, Spain, pp. 403–408.

Wyatt, J. and Ball, S. (2007). Asbestos brochure. Hamers Solicitors, Hull, UK. Available at http://www.hamers.com/wp-content/uploads/2014/10/Asbestos-Brochure.pdf. Accessed 6 October 2016.

Zheng, G.H. and Kozinski, J.A. (1996). Solid waste remediation in the metallurgical industry: Application and environmental impact. *Environmental Progress* 15(4): 283–292.

3

Mining and Beneficiation Waste Production and Utilization

3.1 Introduction

Mining operations consist of excavation (extraction in pits and underground mine workings) to remove ore; beneficiation units, such as mills and processing facilities for upgrading or concentrating the ore; refining facilities for further purification of the metal and manufacturing of finished products. However, mining operations generate extremely large quantities of wastes some of which is shown in Table 2.1 of Chapter 2. This chapter deals with mining waste resulting from extraction and beneficiation processes. The chapter also covers acid mine drainage, a phenomenon which occurs from the dissolution and mobilization of metals from waste rock and tailings into surface water and groundwater. Other types of waste such as that emanating from the refining and manufacturing processes are discussed in Chapters 4 through 8.

3.2 Mine Overburden and Waste Rock

Mining is largely divided into open-pit (surface) and underground mining. Open-pit is the most common technique in modern metal ore extraction. In open-pit metal extraction, the soil and vegetation are first removed from the mine site. Bulk ore is then blasted so as to loosen the bedrock. In underground mining, shafts are sunk into the ore deposits, passages are opened off the shaft and the ore is then broken up and brought to the surface.

Overburden and waste rock are the non-liquid wastes generated in the largest volumes by extraction activities (EPA, 1994). Overburden includes the soil and rock that is removed to gain access to the ore deposits in open pit mines (Fraser Institute, 2012). On the other hand, waste rock consists of rock that is mined, but contains minerals in concentrations considered too low to be extracted at a profit and is, therefore, removed ahead of processing (IIED, 2002).

Overburden and waste rocks have limited potential for reuse because of their high impurity content or remote location and are thus disposed of. However, some of it may be favourably located and of interest because of contained metals and minerals or inherent physical and chemical characteristics. Therefore, the utilization of overburden and waste rocks is an attractive alternative to disposal because disposal costs and potential pollution problems are reduced or even eliminated and resources are conserved. Waste rock and/or overburden could be used as landfill; in roadbed and dam construction; as railroad ballast; as an aggregate in concrete and asphalt mixes, and, for the finer sizes of waste, as construction sand; as an additive to concrete and asphalt mixes; in brick and block manufacture; in the manufacture of thermal insulation; for the recovery of valuable metals or minerals and as mineral fillers.

3.3 Mineral Beneficiation Waste

Mineral beneficiation is the first step after mining in the minerals and metal value chain. Mineral beneficiation consists of the manipulation of the physical properties of the ore to (1) produce desired particle sizes for a final product (2) liberate and concentrate the value mineral from the gangue material. The process consists of a number of steps, the first being comminution which involves size reduction of minerals by crushing and grinding. This process is normally followed by the concentration or enrichment step which is achieved by the separation of valuable minerals from the gangue using various techniques that are based on the physical and chemical properties of the minerals. Some of the techniques applied in the minerals concentration/enrichment step includes sorting which makes use of optical or other properties of the mineral or gangue. This could be done by hand or machines. Separation using density properties is also common; the methods include gravity concentrators and dense media separators. Separation can also be achieved using surface properties of the minerals, for example froth flotation. Magnetic and electrical conductivity properties can also be exploited to effect separation.

The initial comminution process is usually dry during crushing and is mostly wet during grinding. However, for most of the enrichment processes, the ore is normally handled in slurry form making dewatering one of the essential steps. The water removal step is undertaken using specific unit operations like thickening and filtration which prepare the ore material for downstream processing. Once the ore is concentrated, it can be processed further using chemical methods, which can include hydrometallurgy or pyrometallurgy to extract the valuable metals.

Figure 3.1 gives a general overview of the process flowsheet options from run-of-mine (ROM) ore to concentrate.

Some of the processes highlighted in Figure 3.1 are described briefly in the following section.

3.3.1 Mineral Beneficiation Methods and Techniques

This section gives a brief description of some of the typical mineral beneficiation methods and techniques. As has just been discussed, mineral beneficiation techniques involve both size reduction and concentration. Size reduction usually occurs through comminution processes that involve crushing and grinding. There are, however, a number of techniques and methods used in the enrichment or concentration process. The choice of the method applied is largely dependent on the chemical and physical characteristics of the material being treated.

3.3.1.1 Comminution

Comminution aims to reduce the size of solid particles so as to liberate the valuable minerals from the gangue and to increase the surface area for chemical reaction in subsequent processes (Wills and Napier-Munn, 2006). The process involves the use of crushers and

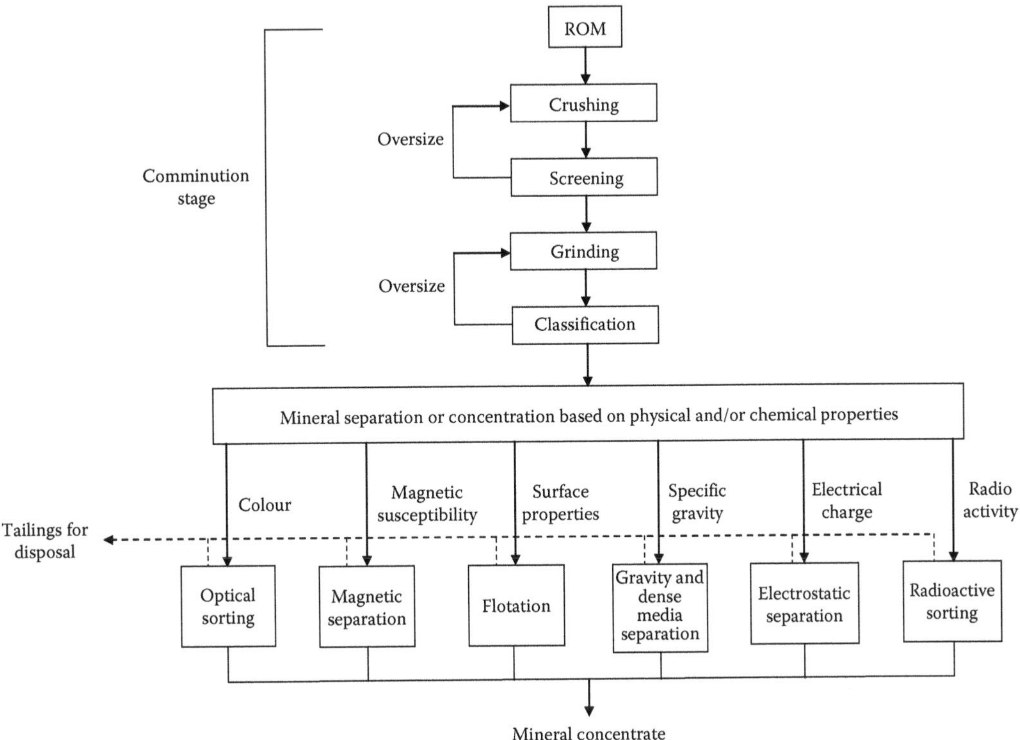

FIGURE 3.1
Generic flowsheet options from ROM to concentrate.

grinding mills. Crushing is normally the first mechanical size reduction stage following ore extraction and can process feed particle sizes as large as 1.5 m. Crushing can occur in several stages, that is, primary, secondary and tertiary crushing. Primary crushing is normally done underground to reduce the particle size of the ROM to a size that is suitable for transportation. The typical examples of primary crushers employed in mineral beneficiation include gyratory and jaw crushers. The product of the primary crusher is further processed in secondary or tertiary crushers, for example cone, roll and impact crushers (Wills and Napier-Munn, 2006).

Crushing equipment generally has a specific reduction ratio and this ultimately means each crushing stage has a size limitation in terms of the final products. This suggests that the number of crushing stages can be reduced or increased depending on the feed size accepted by the primary grinding stage (Wills and Napier-Munn, 2006). Smaller and finer-sized particles, mostly below 5–20 mm, are achieved in a subsequent step known as grinding (Wills and Napier-Munn, 2006).

According to Metso (2016), the grinding process serves two main objectives:

1. To liberate individual minerals of interest trapped in the matrix of gangue within the ore thereby opening them up for subsequent separation or enrichment processes.
2. To produce fines from mineral fractions so as to increase the specific surface area. The specific surface area is a very important component in the subsequent metal extraction processes, such as flotation if relevant and leaching as practiced in the hydrometallurgical plant.

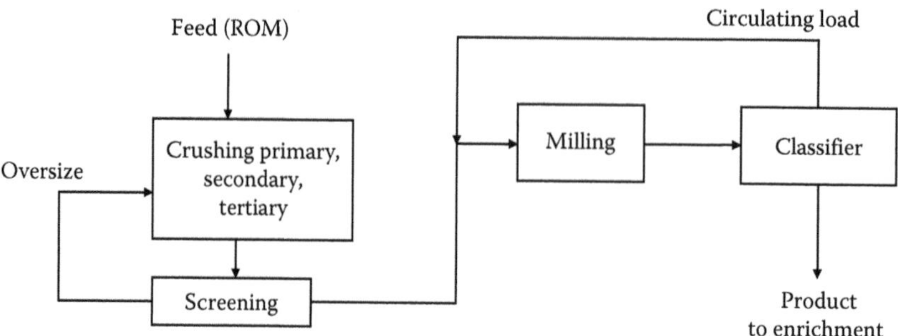

FIGURE 3.2
Typical comminution circuit.

Grinding takes place in equipment such as ball mills, rod mills and autogenous (AG) or semi-autogenous (SAG) mills; the name given to each milling machine is related to the grinding media used. For instance, in ball milling, grinding media is composed of steel balls, while in rod milling the grinding media is made up of steel rods. In AG milling the size reduction is accomplished through attrition effects taking place between the particles of the raw feed. In SAG milling the grinding media used is the feed plus about 4%–18% ball charges (ball diameter 100–125 mm). In general, the control of particle sizes in the comminution circuit is done through classifiers such as screens and cyclones. Figure 3.2 shows a typical comminution circuit.

After the liberation of all individual minerals in the ore feed by crushing and grinding, separation is undertaken using some of the typical classical methods as discussed in the subsequent sections.

3.3.1.1.1 Gravity and Dense Media Separation

Gravity concentration uses the differences in the size and shapes of particles and the specific gravities of minerals to enable separation by force of gravity or by centrifugal forces. Gravity concentrators are characterised by the applied medium (air or water) and are widely used in the processing and recycling industry. Dense medium separation on the other hand not only exploits the difference between the specific gravity of the particles, but also utilizes the variation in the effective specific gravity of the material of a fluid medium.

For separation to occur there has to be differences in the densities between the gangue material and the value mineral. However, the effectiveness of separation is not only dependent on the density of the minerals, but also on the size and shape of the particles. The larger the particle sizes, the higher the gravity separation efficiency. If a feed material has similar densities but different particle sizes, then classification is a more relevant type of separation, and the separation would be based on cut-off particle sizes. If a feed material has similar particle sizes, but different densities, then concentration is a more relevant basis for separation, and the separation would be based on cut-off density. If a feed material has different particle sizes and densities, then a combination of classification and concentration is more relevant, where a screen can be used to narrow the size ranges and then a concentration step can follow (Grewal, 2016). There are a number of gravity concentrators that are used in industry. These include separation in water media through the use of equipment such as jigs (coal, gold, chromite, galena), shaking tables (tin, copper, gold, lead, zinc, tungsten) and spirals (coal, beach sands, chromite, iron), and separation in a heavy medium such as in dense medium separation (DMS).

From practice, the particle sizes of the feed to the gravity concentrator must be closely controlled so that the size effect is eliminated and the separation is only dependent on the specific gravities (Wills and Napier-Munn, 2006).

Centrifugal gravity concentrators: Centrifugal gravity concentrators have dominated in the gold industry for a number of years although the technology has recently expanded to encompass the processing of other heavy minerals. The Knelson and Falcon concentrators are the two most predominant commercially applied units today for the recovery of fine particles of free gold, which do not require the application of the cyanidation process for recovery. The centrifugal concentrator consists of a riffled cone that spins at high speed to create forces that are in excess of gravity. The feed material in slurry form typically from a ball mill or cyclone discharge is introduced into the centre of the cone from above. During operation, the centrifugal force produced by rotation drives the solids towards the walls of the cone. The slurry migrates up along the wall where heavier particles are captured within the riffles (Grewal, 2016).

Jigs: The jig operates by dilating a bed containing the mixture of minerals under treatment allowing heavier particles to fall through. The process consists of two actions; one is the effect of hindered settling meaning that a heavier particle will settle faster than a lighter particle. The other action relies heavily on the density properties in which an upward flow of water separates the particles based on their densities. These two actions are combined in a jig by slurry pulses generated mechanically or by air. In the processing of most heavy minerals, the denser material would be the desired mineral and the rest would be discarded as float or tailings. Jigging is an efficient process even within close specific gravities, whereas with larger specific gravity differences, separation improves significantly.

Spirals: The spiral concentrator is one of the most common and cost-effective gravity concentrators and has found wide spread application in the processing of minerals such as coal and chromite ores. It uses a flowing film principle based on the size and specific gravity difference present in a suite of minerals. The device consists of one or more helical profiled troughs supported on a central column. As slurry travels down the spiral, high and low density particles are stratified and separated at the bottom end of the spiral with a set of adjustable bars, channels or splitters. The lighter minerals are recovered on the outer walls of the spirals and become more diluted as it contains the bulk of the liquid. The heavy minerals that form the concentrate part are selectively directed into the inside of the spiral surface through the use of adjustable product splitters.

Shaking tables: These consist of a cross stream of water which transports material over the table to riffles running perpendicular to the direction of feed. Particles build up behind each riffle and stratification occurs with heavier particles sinking to the bottom. The light particles are carried over each riffle to the tailings zone. The shaking action of the tables carries the heavy particles along the back of each riffle to the concentrate discharge. Shaking tables have been commonly used in the concentration of gold and the recovery of tin and tungsten minerals. The technique can also be used in conjunction with other gravity concentration equipment such as spirals and jigs, especially downstream to help generate a much cleaner product.

Dense Media Separation (DMS) is a beneficiation technology that takes place in fluid media possessing a density between that of the light and heavy fractions to be separated. The process results in two products, the floats and the sinks. Particles much more dense than the specific gravity of the separating medium sink while the lighter ones float. The process is more efficient when there is a significant distinct density difference between the valuable minerals and waste materials, thus allowing for a clearer cut-off between the two minerals being separated. Furthermore, the separation process works very well when the valuable material is liberated from the gangue minerals. Thus, for this reason, comminution processes tend to precede the DMS processes. The commonly used media for the separation process are powdered ferrosilicon, commonly known in industry as FeSi, and, magnetite (Fe_3O_4). This mixture acts as a high density medium and its density can be varied by changing the ratio of water and powder in the mixture. The popularity of ferrosilicon and magnetite in industrial application lie in their magnetic properties which make them easy to recover and re-circulate within the plant, thus saving on operational costs.

3.3.1.1.2 Magnetic Separation

The separation of minerals based on their magnetic susceptibility is a widely used technique that can be very efficient separation process. Magnetic separation is extensively used in the mining industry in the processing of, for example, iron ore, ilmenite, chromite, mineral sands, silica, fluorspar, kaolin and talc. The magnetic separation processes are generally classified into two (Balakrishnan et al., 2013; Grewal, 2016):

1. Strongly magnetic particles commonly classified as ferromagnetic, such as iron and magnetite that can be easily separated from other minerals by the application of a low intensity magnetic field.
2. Weakly magnetic particles, commonly classified as paramagnetic and diamagnetic. These are not magnetic but differ in how they interact with magnetic fields. Paramagnetic minerals are weakly attracted whereas diamagnetic minerals are weakly repelled along the lines of magnetic forces. These minerals require a high intensity magnetic field for separation. Typical examples are rutile, ilmenite and chromite.

The magnetic separation process is generally a low-cost method of recovery unless high intensity separators are required. The process can be accomplished under wet or dry conditions.

3.3.1.1.3 Flotation

Flotation is the process which utilizes differences in surface properties of wanted and unwanted minerals (Wills and Napier-Munn, 2006). In other words, the separation of the required minerals from the ore is achieved by the difference in surface properties (Gupta, 2003). According to Wills and Napier-Munn (2006), the process of recovering valuable minerals from the ore by flotation consists of mechanisms such as

- Attachment of valuable minerals to air bubbles
- Chemistry of carrying valuable particles in the pulp
- Physical attraction between particles in the froth that stick to air bubbles

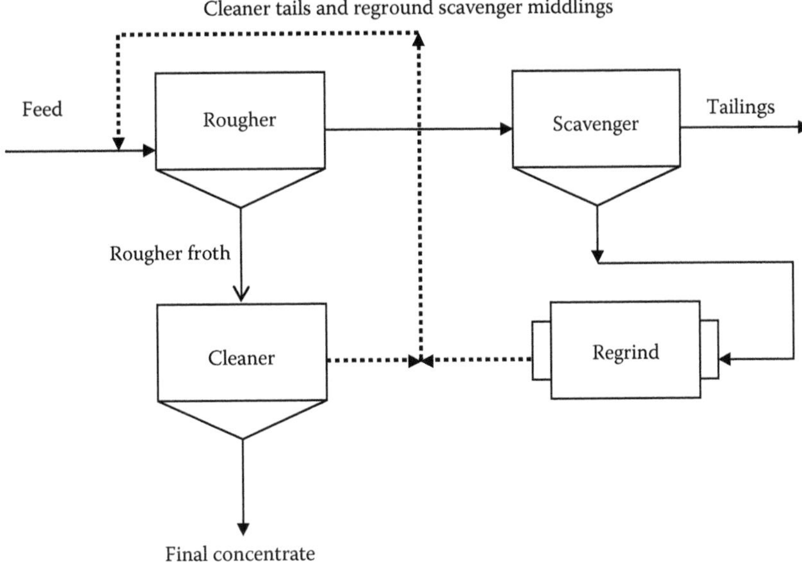

FIGURE 3.3
The basic flotation circuit of roughing, scavenging and cleaning.

Valuable mineral attachment to air bubbles is considered as the main mechanism during flotation (Wills and Napier-Munn, 2006). It follows after the separation of valuable minerals from gangue minerals. The separation of minerals depends on the degree of physical entrapment and entrainment (Gupta, 2003). The entrainment of gangue minerals affects the quality of valuable concentrate. In order to increase the recovery of the valuable concentrate, more than one flotation stage is required. An increase in number of flotation stages results in an increase in the recovery of valuable material which simultaneously results in a decreased production of tails.

Figure 3.3 shows a typical flotation circuit for the beneficiation and upgrading of metal concentrates.

The basic flotation circuit is divided into:

Roughing: Designed to remove the easily recoverable liberated valuable minerals.

Scavenging: The emphasis here is recovery and scavenging is designed to extract the entire remaining valuable minerals that are economically justifiable to recover. Scavenger concentrates are known as the middlings, that is, unliberated particles, and thus tend to go through the process of regrinding before being fed back into the flotation stream.

Cleaners: The emphasis here is on grade and cleaners are designed to clean the stream of any misplaced liberated waste particles. Since the grade is achieved at the expense of recovery, tailings are, therefore, recycled.

A number of flotation reagents are used in the process. These serve two purposes: to prepare the mineral surfaces for attachment to the air bubbles and to impart stability to the froth once particle attachment has occurred. Typical classes of reagents used in the plant are

- *Collectors*: These adsorb on the surfaces of the mineral rendering the minerals hydrophobic and thus, promoting the adhesion of the mineral surfaces to the air bubbles.

- *Modifiers*: These are divided into (1) depressants that inhibit the adhesion of mineral to air bubbles to promote selective flotation and (2) activators that work in conjunction with collectors and precondition the surface to facilitate the subsequent adsorption of collectors.
- *Frothers*: These help maintain a steady froth phase and prevent air bubble particle breakage once attachment has occurred.

The flotation circuit is clearly an enrichment process, upgrading the concentration of the valuable metals by separating them from the gangue material. This process results in the production of two major exit streams, the float concentrate that proceeds to the next processing stage (pyro or hydrometallurgical processing) and the waste material taken for disposal, for example, into the tailings dam.

3.3.2 Characterization of Tailings

Most beneficiation technologies as described in the preceding sections, especially those immediately upstream from the initial processing operation in a production sequence, generate a high volume of solid waste streams. Despite the fact that valuable constituents have been removed, the remaining material (tailings) is often earthen in character, physically and chemically similar to the raw ore fed into the processing circuit, except for the changes in particle sizes arising from the comminution processes. Tailings may also contain trace or substantial quantities of compounds added during the beneficiation process, such as flotation chemicals and media used in the DMS process.

The quality and composition of tailings can vary widely depending on the source material. Hence, when considering tailings material for utilization in alternative applications, attention should be given to the chemical and physical properties of the tailings as the properties play a significant role on the potential impacts on the environment of application. According to Bian et al. (2012), the mineralogical and chemical characteristics of mining wastes are, for example, useful in forecasting the leachability of potentially harmful compounds found within the waste material. The successful utilization of waste may be prevented if it is likely to create a potential or actual environmental hazard (Collins and Miller, 1979). For example, waste material largely sulphidic in nature cannot be considered for application in aggregates because of potential oxidation and subsequent acid production causing a reaction with cement and leading to concrete decay. Characterization of the waste and assessing any potential technical difficulties that could arise in its proposed environment of application in addition to establishing a market for the product are thus, an essential prerequisite. According to Collins and Miller (1979), the waste material should be evaluated in the same way and detail as a primary resource.

Table 3.1 provides chemical composition data for selected samples of copper, gold, iron and platinum group metals (PGMs).

In general most tailings materials contain a significant amount of silica, iron, alumina and other oxide minerals. Although the main chemical constituents are largely the same from one tailing stream to another of the same metallic component, the percentage quantity can vary slightly depending on the primary deposit and the efficiency of the processing methods. For example, in the work done by Malatse and Ndlovu (2015) in the production of bricks using gold tailings from the South African Witwatersrand Basin, the silica content was about 77.7%. This is quite high compared to the 51%–56% indicated by the gold tailings from India shown in Table 3.1. The composition of the tailings has a great

TABLE 3.1

Chemical Composition of Selected Tailings

	Tailings			
Composition %	**Gold**	**Copper**	**Iron**	**PGMs**
SiO_2	51.8–56	75.0	47.39	47.16
Fe_2O_3	10.2–18.9	3.60	24.82	8.23
CaO	7.6–8.4	0.16	8.85	7.81
Al_2O_3	8.2–11.9	12.16	7.42	16.90
MgO	6.3–8.6	0.49	0.097	13.31
Na_2O	–	4.29	0.32	1.33
K_2O	–	1.85	0.7	1.43
CuO		0.32	–	
TiO_2		0.013	–	0.24
Cr_2O_3				3.69
LOI	2.0–3.9	2.10	10.40	0.48
Tailings source	Kolar Gold Fields, Karnataka, India	Hindustan Copper Limited, Khetri, Rajasthan, India	Anshan of Liaoning Province, China	Waterval, South Africa
Reference	Roy et al. (2007)	Thomas et al. (2013)	Li et al. (2010b)	Amponsah-Dacosta and Reid (2014)

role to play in the choice of potential alternative application of the tailings. This will be highlighted in the subsequent section on the utilization of tailings.

3.3.3 Potential Utilization of Tailings

The large volumes of the tailings produced at mining operations are expensive to manage and are frequently cited as an obstacle in the environmental sustainability of mining. For example, tailings that contain a significant amount of iron sulphides play a role in the generation of one of the most environmentally damaging products, acid mine drainage. Acid mine drainage is discussed in detail in Section 3.4. Therefore, the drive for most mining operations at the moment is to reduce the costs associated with the development of processes and procedures that have to be put in place to make sure that the tailings discharged meet the required environmental standards. As a result, the mining industry is moving towards making use of its own waste materials. In fact, the concept of the utilization of tailings embraces the basic principles of increased reuse, recycling, reprocessing and converting these materials to value-added products as an alternative management strategy (Dean et al., 1986; Edraki et al., 2014; Liu et al., 2014). Furthermore, Dean et al. (1986) and Yellishetty (2008) observed that the utilization of mine tailings in the downstream economic activities and industries can alleviate the socio-economic issues of mine closure associated with loss of employment.

Some typical examples of tailings utilization include the following:

- Reprocessing to extract minerals and metals
- Sand-rich tailings mixed with cement used as backfill in underground mines
- Use of Mn-rich tailings in agroforestry, building and construction materials and coatings

- Cast resin products, glass, ceramics and glazes
- Cu-rich tailings as extenders for paints
- Fe-rich tailings mixed with fly ash and sewage sludge as lightweight ceramics
- Clay-rich tailings as an amendment to sandy soils and for the manufacturing of bricks, cement, floor tiles, sanitary ware and porcelains

The following sections look at some of the potential applications of copper, gold, PGMs and iron-rich tailings. Since Table 3.1 indicates a commonality in terms of the tailings chemical composition, it is also expected that there will be some similarities in the application of the tailings generated from the processing of different minerals.

3.3.3.1 Reprocessing of Tailings for Metal Recovery

The growing interest towards metal recovery from wastes is due to the depletion of mineral resources in the most industrialized countries along with the rapid development of technologies enabling the beneficiation of ores even from the poorest deposits. For example, the processing of refractory sulphide and complex material has always presented a challenge in the minerals and metal extraction industry, resulting in most of the valuable metals being left behind in the tailings. The major problems that have existed in the past with regards to the reprocessing of such tailings have been the availability of more efficient, effective and economical processes for obtaining high metal recoveries at high production rates from the once-processed material. However, advances in process technology has seen some of the challenges being overcome, leading to renewed interest in the processing of tailings in order to recover valuable metals. Reprocessing of such wastes brings potential financial and waste reduction benefits. In addition, higher metal prices and the stringent environmental regulations and associated penalties further justify the reprocessing of large volumes of tailings across the globe. For example, tailings that contain economic quantities of copper or gold, if reprocessed, could potentially have higher production rates than primary mining operations. The reprocessing of gold-bearing tailings in particular has increased in the recent past due to it being one of the few commodities that has maintained some semblance of price stability within the current prevailing harsh mining environment.

The key drivers for the extraction and reprocessing of metal-bearing tailings can be summarized as follows:

- Tailings are generally considered to require fewer resources when compared to conventional mining. The mining of raw ores requires a lot of capital investing and other related funding for mining of hard rock and crushing and grinding to achieve high metal recoveries, while waste dumps generally require just crushing and grinding as the treatment methods prior to metal extraction. In most cases, some of the existing facilities for the treatment of raw ores can be utilized for the processing of waste materials. This reduces costs related to capital equipment, labour, fuel and maintenance. For example, despite differences in grade, the reprocessing of old, finely ground mill tailings is probably much more economic than treating most newly mined ores. The mining of raw ores requires a lot of capital investing and other related funding, mining of hard rock and crushing and grinding to achieve high metal recoveries, while waste dumps generally require just crushing and grinding as the treatment methods prior to metal extraction. In addition, in most cases, some of the existing facilities for the treatment of raw ores can be utilized for the processing of these secondary resources, thus lowering the capital investment costs significantly.

- Advances in processing technology result in changes in patterns of metal extraction from both primary sources and waste or secondary material. As processing techniques become more cost-effective and efficient, it becomes more economically beneficial to recover valuable metals from discarded materials which were previously regarded as untreatable.
- The processing of tailings allows for additional production without increasing a mine's footprint and without requiring additional land permits.
- According to Binnemans et al. (2013), the reprocessing of residues from waste can also positively influence the availability of metals and consequently stabilize their prices in the world market.

Some practical examples in which tailings can be reprocessed so as to recover metals are highlighted in the subsequent sections.

3.3.3.2 Reprocessing of Copper Tailings

According to Gordon (2002), mills process ores that contain 0.5%–2% copper into concentrate containing 25%–35% copper leaving a residue (tailings) consisting of the gangue and the unseparated copper-bearing minerals. One of the primary challenges is to efficiently recover copper from the tailings without significant added capital and operating costs (Edraki et al., 2014). For example, Lutandula and Maloba (2013) used flotation to recover copper and cobalt by reprocessing tailings from oxidized ores in the Democratic Republic of the Congo (DRC). However, the excessive consumption of reagents in relation to tailings was found to be of major concern from a technical and economic standpoint.

Based on the challenges of using conventional tailings reprocessing techniques such as flotation, there has been a focus on applying bioleaching as an alternative reprocessing technique. Bioleaching has been the process of choice for the treatment of most copper sulphide tailings. The importance of this technique in the treatment of waste material is due to its associated low cost and its environmental friendliness. For example, research has been conducted on the potential for the application of an integrated complete bioprocessing route for copper extraction from flotation tailings generated at the Neves Covo mining site in Portugal (Duarte et al., 2006). These tailings constituted mainly metal sulphides, pyrite and chalcopyrite with a very high copper content. The researchers proposed bioleaching using thermophilic bacteria to recover copper and other valuable metals. This was followed by the application of sulphate-reducing bacteria to remediate the final discharge solutions. Figure 3.4 shows the typical flowsheet that would be expected.

The integrated process not only looks at value recovery through the treatment of the copper tailings generated from the beneficiation process, but also considers the bioremediation of the final effluent discharge solution so as to lessen its impact on the environment. As already stated, biohydrometallurgical processes are generally considered to be environmental friendly and cost-effective. Thus, an integrated process would work very well for the treatment of tailings materials which are considered as low-value feed.

Bioleaching has also been used at the Copper Mine Bor in Sebia. The process makes use of the abundant and highly acidic water of Lake Robule as a lixiviant to leach the copper

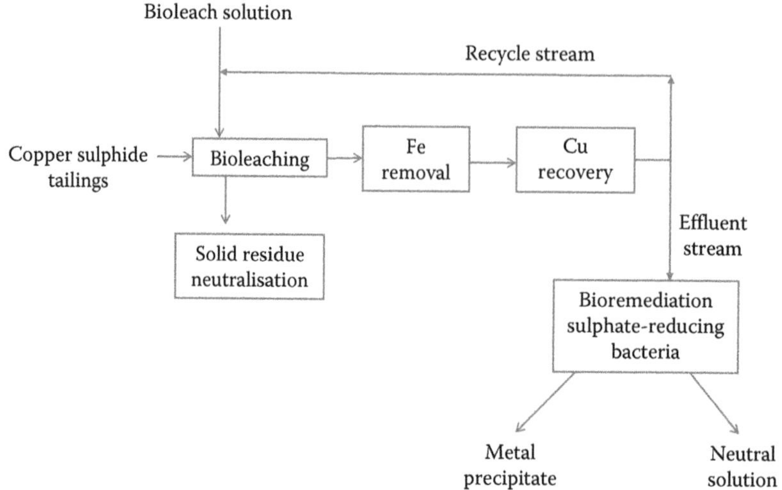

FIGURE 3.4
An integrated bioprocessing approach for the treatment of copper tailings.

flotation tailings (Stanković et al., 2015). The widespread open-pit acidic lakes in the mining region also present a potential source of acidophilic microorganisms that can be used for the bioleaching of copper. Studies by Stanković et al. (2015) showed the presence of acidophilic heterotrophic bacteria, *Acidiphilium cryptum*, and the autotrophic iron oxidizers *Leptospirillum ferrooxidans*, in the lake water. There was also a relatively small number of autotrophic iron- and sulphur-oxidizers, *Acidithiobacillus ferrooxidans*. The obtained results showed that bioleaching could be more efficient than inorganic acid leaching for copper extraction from flotation tailings with higher sulphide contents.

Fraser Alexander, one of the companies making inroads into the operation and management of tailings dams for precious metals, base metals and coal industries internationally, undertakes re-mining projects mostly through hydro-sluicing (Wilkins, 2013). The company operates a copper re-mining solution for Vedanta's Konkola copper mine near Chingola on the Zambian Copperbelt. For a significant amount of metal recovery to be realized from the processing of these tailings, the volumes of material being processed should, however, be extremely high (Wilkins, 2013).

The significant impact of mining can also be understood through the large quantity of copper mining tailings that have accumulated at the Morenci mine complex in Arizona (Tilton and Landsberg, 1997). When the depression in the price of copper in 1892 threatened many copper mining companies, the Morenci Mine implemented the leach–cementation process in 1893 to recover copper lost in mill tailings (Gordon, 2002). The leach–cementation process was seen to be a more cost-effective and simple to implement process at that particular time. This operation remained in place till late 1977. The process treated an average of 60 kt of the mill tailings containing about 0.2% copper per day, and recovered about 40% copper (McKinney and Graves, 1977). Copper was recovered from the resulting cement copper by feeding it into the smelter together with the sulphide concentrates. However, each kilogram of copper made by cementation consumed 3 of iron (Gordon, 2002). When the price of the iron metal used in the recycling process became high, the leach–cementation process became uneconomic. An alternative cost-effective and much more efficient technology could have prolonged the treatment of the tailings at the Morenci mine. In general, the leach–cementation process did not prove popular for companies that did not have smelter facilities as building a smelter is capital intensive.

According to Hancock and Pon (1999), the copper mining activities in Zambia from early 1913 have resulted in the generation of waste ponds that contain about 1.2×10^9 tonnes of copper tailings. It is, therefore, no wonder that one of the most well-known commercial copper tailings reprocessing plant in Southern Africa is the Nchanga tailings leach plant located in Zambia. The plant produces copper cathodes from concentrator tailings and reclaimed tailings. The process was initially based on the acid leach–cementation process. However, the advent of the solvent extraction technology led to the plant being upgraded in 1973/1974 which greatly increased the utility of hydrometallurgy in the extraction of copper (Chisakuta et al., 2005). Through the introduction of solvent extraction, the Nchanga mill in Zambia went from a copper loss to its tailings of about 54 kt a year, to recovering 100 kt of copper per year through the treatment of 10 million metric tonnes (Mt) of tailings per year (White, 1979).

Another company making significant progress in the processing of tailings is the Canadian company, Amerigo Resources, which focusses on old and new copper/molybdenum tailings on a very large scale from one of the world's great copper mines – Codelco's El Teniente, the world's largest underground copper mine in Chile (Henderson, 2013). Amerigo's wholly owned subsidiary, Minera Valle Central (MVC) is re-treating all the tailings from El Teniente's present production and has the right to treat high grade old tailings from a 200 Mt *in-situ* tailings impoundment, Colihues, which is located next to MVC's plant (Williams, 2010). There are two additional major tailings impoundments at El Teniente: the Cauquenes tailings dam, which contains an estimated 500 Mt of additional high grade tailings, and Barahona, which is similar in size to Colihues and located closer to El Teniente mine (Henderson, 2013). Amerigo hopes to obtain the rights to process tailings from these additional sources, although Colihues alone has sufficient material to keep the plant running at its existing production rate for around 20 years (Williams, 2010).

The Roan Tailings retreatment project in the DRC involves the reprocessing of oxide tailings produced by the mineral concentrator in Kolwezi. This concentrator processed high grade copper cobalt ores from nearby mines from 1952 and, thus, significant volumes of tailings were generated. The tailings were previously deposited in the Kingamyambo tailings dam and in the Musonoi river valley. The tailings dams have the potential to be the source of the world's largest cobalt producers at much lower costs. These tailings dams contain copper and cobalt at an average grade of 1.49% and 0.32% respectively, making these metals economically viable to recover.

One future prospect in copper tailings reprocessing in South Africa lies with Xtract Resources. The company has invested in the development of the Carolusberg and O'Kiep project, which looks at the potential processing of a sulphide copper tailings dam located in the Northern Cape close to the town of Springbok. The project is estimated to contain 33.8 Mt of sulphide tailings material that was mined between 1980 and 2010 by O'Kiep Copper Company. The Carolusberg tailings dam represents 28 Mt of material grading at 0.19% Cu and the O'Kiep tailings dam represents 5.8 Mt of material grading at 0.23% copper.

3.3.3.3 Reprocessing of Gold Tailings for Metal Recovery

The gold industry is a typical example where the reprocessing of mine tailings has been extensively reported. According to Wilkins (2013), the high cost of deep level mining combined with low recoveries, fractious labour in certain countries with South Africa being a typical example, has resulted in some mining companies exploring alternative avenues to maintain margins and to unlock profits hidden in substantial volumes of tailings. This is also largely driven by technological advances coupled with the rising price of gold over the past decade that has made it both possible and profitable for mining companies to extract microscopic gold from already worked rock. There has also been a trend towards the consolidation of smaller tailings dams into single tailings facilities in order to better manage reprocessing and rehabilitation. Reports by Nummi (2015) and Wilkins (2013) cite a number of examples showing the extensive reprocessing of tailings being undertaken by gold mining companies. Typical examples of projects in operation and/or being piloted include the following:

- DRD Gold is one of the first South African companies who, with the advent of new technologies, have abandoned traditional mining to focus on extracting gold from tailings. Recoveries of up to 40% of the gold from tailings have been reported.
- Mintails processes about 350,000 tonnes of tailings from its extensive gold tailings dams. It expects to recover 58 kg of gold per month and has declared that it has enough gold tailings to last until 2025.
- Goldfields has developed a new technology for the processing of about 12,000 tonnes of new tailings and 88,000 tonnes of old tailings a month.
- Gold One Group purchased an 83 million tonnes tailings dump from Rand Uranium at the beginning of 2012 with the intention of recovering uranium, gold and sulphur from the tailings dumps. The company estimates that 0.8 million ounces of gold and 34 million pounds of uranium will be recovered from the dump over a 17-year period.
- The Australian company, Carbine Resources, investigating the extraction of an estimated million ounces of gold left in the tailings from the Mt. Morgan mine, while also fixing a huge environmental problem at the former mine site (Calderwood, 2014). The mine is an open-pit and has, in the past few years, filled with water that has also turned acidic. Despite a water treatment plant being run at the mine, the pit at one time overflowed and spilled acidic water into a nearby river.
- Las Lagunas Gold Tailings is another retreatment project involving the application of the Albion oxidation technology for the reprocessing of previously unrecovered gold and silver refractory tailings from the Pueblo Viejo mine in the Dominican Republic (Barradas, 2012).

3.3.3.4 Reprocessing of Iron Tailings

The increasing demand for heavy construction material like steel and iron has led to the establishment of many iron ore mining activities, subsequently leading to the generation of a large amount of iron ore tailings. The iron ore tailings in most cases are fine materials, containing mostly silica together with some fines of iron oxides, alumina and other minor minerals (da Silva et al., 2014). This composition places these tailings as a potential raw feed for aggregated materials of mortar and concrete in the civil construction industry. Iron ore tailings have also served as secondary resource of iron ore. The reprocessing of iron tailings streams has been accomplished mostly through gravity, magnetic, flotation separation and direct reduction techniques.

Praes et al. (2013) studied the concentration by flotation of a fraction of iron ore from the magnetic separation tailings having an Fe content of about 35% and SiO_2 generally above 40%. The study was carried out in a flotation column at a pilot plant unit scale and produced a concentrate of Fe and SiO_2 content of about 66% and 1.0%, respectively, which met the standards required for commercialization.

Ajaka (2009) investigated the reprocessing of iron ore tailings produced in the Itakpe iron ore processing plant in Nigeria. The tailings contained up to 22% iron minerals, mostly natural fines in the ore and fines produced inevitably during comminution. Ajaka (2009) analysed the existing circuit and undertook specific recovery tests on the tailing material using simple hindered settling and flotation process for the recovery of iron minerals in the tailings. The results showed that concentrates of grades ranging from 41% to 62% could be attained with the selected processes.

Sakthivel et al. (2010) used iron ore tailings containing 15.98% Fe_2O_3, 83.36% SiO_2 and 0.44% Al_2O_3 to produce magnetite powder. The process involved the production of Fe(III) solution by the digestion of the tailings with HCl followed by the separation of the acid-insoluble residue. In the presence of $NaHB_4$ used as the reducing agent, Fe(III) solution was then used to synthesize magnetite via the formation of metallic iron and with $NaHB_4$ used as the reducing agent.

Although laboratory research and development on the recovery of iron from tailings is quite substantial, the industrial application remains limited due to the associated high production cost (Li et al., 2010a). The content of iron in the tailings is generally very low, making the reprocessing and reuse very uneconomic. Since the amount of iron recovered from the tailings is low, a large volume of the residue is attained after the iron removal process creating another problem of storage and secondary pollution. With these challenges in mind, Li et al. (2010a) investigated the comprehensive utilization of iron tailings by first recovering the iron from the tailings as magnetite. This was accomplished by using magnetizing roasting to reduce hematite in the iron ore tailings to magnetite followed by separation using a low intensity magnetic separation method. Using this process, a magnetic concentrate of 61.3% Fe at a recovery rate of 88.2% was achieved. The residue obtained after the iron removal process was then used to prepare a cementitious material (Li et al., 2010b). The results from the test work showed that up to 30% of raw material could be replaced by this residue, with 34% blast furnace slag, 30% clinker and 6% gypsum making the other balance to produce a cementitious material with mechanical properties comparable to those of ordinary Portland cement according to Chinese GB175-2007 standard. This approach allows for a more economic, environmentally friendly and comprehensive utilization of iron tailings (Argane et al., 2016).

On the commercial front, in 2009 Xstrata Copper announced the development of the Ernest Henry Mine (EHM) magnetite extraction plant as part of the life of mine extension program for Ernest Henry Mine and also as part of a strategic objective aimed at maximizing the value of the existing resources (Siliezar et al., 2011). The EHM magnetite plant extracts the magnetite from the copper concentrator's tailings stream. It produces approximately 1.2 Mt of magnetite concentrate per annum at full capacity for export to Asia, making EHM Queensland's first iron ore concentrate exporter.

Outotec has also developed the SLon® high gradient magnetic separator for the reprocessing of iron ore tailings in order to recover iron metal (Jain, 2015). The SLon® separator utilizes a combination of magnetic force, pulsating fluid and gravity to continuously separate the magnetic from the non-magnetic minerals (Jain, 2015). Outotec highlights some of the advantages of the equipment as the high beneficiation ratio, high recovery, adaptability to varying particle sizes and minimized matrix blocking. The feasibility for its application has been tested in iron ore processing companies in Brazil, the United States, Africa and Russia. The technology can be applied not only for the concentration of iron ore, but also for ilmenite, chromite ore and other paramagnetic materials.

Magnetation LLC is another company that has been making inroads into the processing of tailings from open-pit mines in the Minnesota area, USA (Magnetation, 2016).

The company uses a patented Rev3™ Separator technology to separate weakly magnetic particles from waste minerals with high efficiency and throughput. According to Magnetation (2016) the Rev3 process is designed specifically for high volume and high availability iron ore applications to produce high quality hematite concentrates. The product is then used as a primary raw material supply for North American blast furnaces in the steel making process.

3.3.3.5 Reprocessing of Chrome Tailings for PGMs and Chrome Recovery

South Africa accounts for 96% of the known global reserves of PGMs with the Bushveld Igneous Complex in South Africa being host to some of the world's major PGM ore deposits including the Merensky Reef, UG2 Reef and Platreef (Mohale et al., 2015; Sameera et al., 2016). South Africa's reputation as one of the leading producers of PGMs worldwide consequently makes it one of the largest producers of mine waste compared to other mining industries. The country's mining operations associated with the primary production of PGMs from deposits generates over 77 Mt of fresh tailings per annum (Vogeli et al., 2011). Though the incorporation of PGM tailings into building materials, road aggregates, agricultural applications, landfills, manufactured fillers and many other schemes have been proposed, these high-tonnage waste materials also contain elevated concentrations of valuable minerals and could, therefore, be regarded as an economic resource rather than a waste, provided suitable economically viable processes can be developed.

The reprocessing of chrome tailings to recover PGMs is one of the recent developments in the mining industry being applied to try to meet the ever-growing demand in the fuel cell technology, jewellery market in the Far East and the automobile catalytic converters. With the South African platinum industry being the world's largest producer of platinum and the world's second largest producer of palladium, it follows therefore, that a number of developments in tailings treatment processes have been initiated in the country. According to Van Niekirk and Viljoen (2005), the large volumes of tailings produced in the country also require active planning and management to prevent major environmental or social impacts such as tailings dam failures, for example the 1974 Bafokeng tailings disaster; acid mine drainage or other problems, such as dust and environmental health issues.

In December 2003, Anglo Platinum commissioned the first purpose built tailings retreatment facility for PGMs recovery in South Africa. The Western Limb Tailings Retreatment (WLTR) project near Rustenburg takes advantage of modern technology, such as fine grinding, to economically recover PGMs from material historically considered as waste (Buys et al., 2004). The flowsheet includes the recovery of tailings by high pressure water monitoring, ball milling, rougher flotation, rougher concentrate regrinding and cleaner/recleaner flotation. The rougher concentrate is reground in a mill allowing smeltable concentrate grades to be produced from the oxidized, slow floating tailings.

In 2013, the Two Rivers Platinum Mine, also located in South Africa, commissioned a recovery plant to reprocess its UG2 tailings to recover chromite (Implats, 2015; Sameera et al., 2016). The tailings dams at Two Rivers currently contain about 24 Mt of milled and processed ore with an average Cr_2O_3 grade of about 17%.

In 2014, Jubilee Platinum executed a Tailings Access Agreement with ASA Metals Proprietary Limited (ASA) and its subsidiary Limpopo-based Dilokong Chrome Mine (DCM) for the recovery of platinum group metals and chrome from tailings dams (Odendaal, 2014). The agreement offered Jubilee the option to construct, in partnership with ASA, a dedicated chrome and PGM processing plant for the treatment of the tailings on ASA's DCM property. The agreement gave Jubilee exclusive access to an estimated 800,000 t of PGM-bearing chromite tailings (Odendaal, 2014).

Sylvania Platinum Limited in South Africa operates the Sylvania Dump Operations (SDO), which focuses on the re-treatment of PGM-rich chrome tailings material from mines in the region in order to recover chrome and PGM concentrate (Sylvania Platinum Report, 2016). The SDO currently includes seven fully operational chrome tailings processing complexes on the western limb of the Bushveld Igneous Complex (BIC) and a few others on the eastern limb.

The PGMs from tailings constitute an important secondary resource of PGMs for the global market and it is clear that South Africa is playing a significant role in this development. Although this section has highlighted a myriad of the commercial developments in South Africa, it is important to note that the recovery of PGMs from tailings is also being practiced in other platinum-producing countries such as Russia and Zimbabwe (Glaister and Mudd, 2010).

3.3.3.6 Tailings Utilization in the Construction Industry

The large amount of mine tailings generated in the mining industry each year provides a large new source of raw material for utilization in the construction and building industry. The utilization of mine tailings as construction material can reduce the volume of disposable mine tailings and thus, saving the land required for the disposal impoundments and the related high monetary costs. It can also lead to a reduction in the environmental and ecological costs and lower the demand for virgin materials that would need quarrying, thus, prolonging the life of natural resources. The following section looks at the potential use of tailings in different sectors of building and construction.

3.3.3.6.1 Utilization as Aggregates for Concrete Mix

Aggregates make up about 70%–80% of a concrete mix (Shetty et al., 2014). As the natural granite quarries for aggregates are gradually decreasing, there is a need for alternative materials to be used as aggregates in concrete. If mine tailings are considered as a partial or complete replacement of natural aggregates in concrete, a majority of these tailings could be recycled and used sustainably, thereby turning the tailings into useful resource and providing cheaper alternatives in concrete production (Ugama et al., 2014). However, many mineral processing waste materials have limited potential for use as aggregates because of their fineness, variable metal compositions, leachability of some of the trace metals and the propensity for acid generation, especially for sulphide-containing materials. Further to this, these potential raw material resources tend to be found in remote locations, that is, away from aggregate markets (Collins and Miller, 1979). However, when the location and material property characteristics are favourable, coarse tailings may be suitable for use in the production of cement concrete aggregate, asphalt aggregate, flowable fill aggregate. In addition, if these tailings are segregated properly, both fine and coarse aggregates for concrete can be obtained.

3.3.3.6.1.1 Iron Ore Tailings The processing activities associated with the iron ore beneficiation is such that it results in the tailings with particle sizes ranging from fine to coarse (Kuranchie et al., 2014). Therefore, by proper segregation of the tailings, the iron ore mining companies could incorporate comprehensive utilization of the tailings in their operation, leading to cleaner production and sustainable development (Haibin and Zhenling, 2010). Research has shown that the most comprehensive use of iron tailings is through the production of building and construction materials such as concrete aggregate, as a substitute for Portland cement and as bricks.

Iron ore tailings have been used as a replacement material in the making of Portland cement (Li et al., 2010b). The researchers used the residue obtained after iron removal (Li et al., 2010a) from iron ore tailings to produce cement of high quality standard. Luo et al. (2016) further successfully utilized iron ore tailings to replace clay as alumina silicate raw material for the production of Portland cement clinker. The key to the success of this application was found to be the rich iron (III) oxide content in the form (Fe_2O_3) which can be effectively used as raw material in cement clinker production (Oluwasola et al., 2014).

Kuranchie et al. (2015) studied the feasibility of using Australian iron ore tailings as both fine and coarse aggregates in making concrete. The authors evaluated the technical and environmental characteristics of the concrete with the main aim being the recycling and addition of economic value to the iron ore tailings for use as cheaper alternative materials for concrete aggregates. The compressive strength of the concrete with iron ore tailings aggregates at 28 days showed an improvement of 11.56% over the concrete with conventional aggregates. However, the split tensile strength exhibited by the concrete with tailings aggregates at 28 days was slightly lower than for the concrete with conventional aggregates by 16%. This was due to higher quantity of fines in the iron ore tailings as compared with the natural sand in the control mix. Nevertheless, the tensile strength increased favourably with ageing. In addition, the concrete with tailings aggregates had a lower potential for corrosion and a low vulnerability to acid attack due to high pH values of the resulting mix.

Ugama et al. (2014) investigated the strength characteristics of control concrete and concrete made with iron ore tailings and determined the optimum replacement level for iron ore tailings content in concrete beam production to be about 20%. It was observed that concrete workability decreased with an increase in the percentage of iron ore tailings in the mix. This was attributed to the fineness and therefore, large surface area of the iron ore tailings which resulted in the need for a large amount of water to wet all the particles in the mix. The iron ore tailings were classified as a heavy aggregate and would therefore, produce concrete pieces of heavier weight. The high density of the waste was related to the high content of iron, which could not be removed by a magnetic separation process. It was further noted that these iron tailings would be suitable as an aggregate for the production of concrete for paving directly on the ground due to higher self-weight and it being less subject to deformation due to external loads.

Kumar et al. (2014) also used the iron ore tailings as a replacement material for fine aggregates in cement concrete pavements. Their results showed that 40% iron ore tailings could be used as replacement to give an optimum compressive strength. It was suggested that the concrete so-generated could be used for pavements and village roads.

The information given in this section suggests that the utilization of iron ore tailings as a raw material in the construction industry is a more attractive and effective solution for management of these tailings. This approach not only utilises a large amount of the iron ore tailings but also provides an added advantage of protecting the natural resources.

3.3.3.6.1.2 Copper Tailings Copper ore tailings are classified as a good quality class-N natural pozzolana and can, thus, be used as a pozzolanic material in the construction industry (Prahallada and Shanthappa, 2014). According to Prahallada and Shanthappa, (2014), copper ore tailings utilization in concrete production has a number of environmental benefits such as increasing the life of pavements and structures by improving the stability of the soil, reduction in the adverse air emissions when used in copper tailing bricks and stabilized blocks, etc. In the studies by Prahallada and Shanthappa, (2014), the suitability of copper ore tailings as an additive mixture in the preparation of concrete was tested by replacing the ordinary Portland cement in different percentages. The results showed that the replacements of ordinary Portland cement with copper ore tailings was safe up to 20% considering average minimum field strength. If characteristic strength was considered, replacement copper ore tailings up to 30% could be considered as safe.

Thomas et al. (2013) looked at the use of copper tailings as a partial replacement of natural river sand in the preparation of a concrete mix. The results indicated that copper tailings could be utilized for the partial replacement of up to 60% of natural fine aggregates, with the water–cement ratios of 0.4, 0.45 and 0.50. The so-produced copper tailings concrete exhibited good strength and durability characteristics and was further found to be suitable for applications in a number of construction activities.

Onuaguluchi and Eren (2012a) investigated the potential application of copper tailing as an additive in concrete mix. The results indicated that copper tailings have a slight negative impact on the slump, porosity and setting time of concrete mixtures. However, when compared to the control specimen, there was a noticeable improvement in mechanical strength and abrasion resistance and reduced chloride permeability. The conclusion was that a minimum of 5% copper tailings could be introduced into concrete as an environment-friendly and zero cost material. Onuaguluchi and Eren (2016) further investigated the time to initiation of corrosion and deterioration in copper tailings blended concrete mixtures. The results showed that the use of copper tailings as an additive in concrete was more effective in delaying the commencement of corrosion compared to when the copper tailings were used to replace some of the cement raw material. The researchers further undertook an economic analysis by comparing the cost efficiency of using copper tailings either as a cement replacement material or as a cement additive in concrete. Based on the corrosion performance and cost efficiency of mixtures, utilization as a cement additive was found to be the best reuse option for copper tailings in concrete.

The work by Huang et al. (2012) looked at the utilization of Chinese skarn-type copper tailings to prepare autoclaved aerated concrete (AAC). The AAC samples were prepared by completely substituting lime by skarn-type copper tailings and blast furnace slag. The results showed that AAC samples with a dry density of 610.2 kg/m^3 and compressive strength of 4.0 MPa could be prepared at a laboratory scale. The patented work by Koumal (1994) also focussed on the use of copper tailings from the Arizona copper mine to produce AAC which could be further utilized as building material for the production of building blocks and prefabricated walls. The mine tailings material was used as a substitute for the processed silica sand during the formation of the AAC building material.

3.3.3.7 Tailings Utilization in Brick Making

The conventional production of bricks usually utilizes clay and shale as the source materials and requires high temperature (900°C–1000°C) kiln firing (Muller et al., 2008). Clay and shale are sourced through quarrying operations which are energy intensive, adversely affect the landscape, and can release a large quantity of waste materials (Bennet et al., 2013; Zhang, 2013). The bricks are produced by mixing ground clay with water, forming the clay into the desired shape followed by drying and firing. The major disadvantage of the conventional high temperature kiln firing process is that it not only consumes significant amounts of energy but also releases a substantial quantity of greenhouse gases (Zhang, 2013). Furthermore, in many areas of the world, there is already a shortage of natural resource material for the conventional brick making processes which has led to restrictions in terms of clay brick production. For instance, countries such as China

TABLE 3.2

Typical Composition of the Clay Material Used in Brick Making

Component	Weight %	Notes
SiO_2	50–60	Prevents cracking, shrinking and warping and thus imparts uniform shape to the bricks
Al_2O_3	20–30	Imparts plasticity to the clay so that it can be moulded
Fe_2O_3	5–6	Helps to fuse sand and also imparts a red colour to the bricks
CaO	2–5	Helps to fuse sand and also prevents the shrinkage of raw bricks
MgO	<1	Imparts a yellow tint to the bricks and decreases shrinkage

Sources: Chen, Y. et al., *Constr. Build. Mater.*, 25, 2107, 2011; Bennet, J.M. et al., *Eur. Int. J. Sci. Technol.*, 2(5), 133, 2013; Mueller, H. et al., *Greenbrick Making Manual*, Hillside Press, Kathmandu, Nepal, 2008, Available at http://www.ecobrick.in/resource_data/KBAS100046.pdf, Accessed May 2016.

have started to limit the use of bricks made from clay and have instead been actively advocating for the development of eco-friendly building materials and processes (Zhang, 2013) in order to protect the environment and sustain development.

The use of mine tailings to produce bricks offers an alternative raw material for the building industry. The popularity in the application of these tailings for the production of bricks is largely due to the fact that the composition of these waste materials is closely similar to that of the typical clay material used in conventional brick making. Table 3.2 shows the typical composition of the clay material that is used in brick making.

It can be noted that the tailings material in Table 3.1 primarily have a high composition of silica, alumina, calcium oxide and hematite which are all essential constituents in the brick making process as shown in Table 3.2. In addition, the source material must have some plasticity that can allow it to be shaped or moulded when it is mixed with water.

The use of mine tailings in the production of bricks and blocks was investigated as early as 1979 by Collins and Miller as part of an intensive research into waste utilization in the United States. Copper, lead, zinc and iron ore tailings from various U.S. sources were evaluated in the production of cemented blocks. Base metal mine tailings were on the other hand, also evaluated for the production of concrete blocks during tailings recycling research in Melbourne, Australia (Struthers, 1999). These blocks were shown to be superior in compressive strength compared to conventional blocks. However, Struthers (1999) further observed that tailings are often too fine to work successfully in a commercial high-speed production system. They would instead, be suited to slower, intermediate technology methods and are ideal for the production of complex concrete formwork.

The production of bricks using mine tailings can be divided into three general categories based on the production methods – firing, cementing and geopolymerization.

3.3.3.7.1 *Production of Bricks through Firing*

In the firing process, the waste material is substitute partially or entirely for clay and follows the traditional method of kiln firing (Malaste and Ndlovu, 2015). Chen et al. (2011) studied the feasibility of utilizing hematite tailings from the Western Hubei Province of China as the main raw material in the production of fired bricks. Clay and Class F fly ash were used as additives to improve the brick quality. It was found that the percentage of tailings used could be up to 84% of the total weight. The firing temperature applied ranged from 980°C to 1030°C over a period of 2 h. The mechanical strength, water absorption and other physical properties and durability of the reddish fired specimens were found to conform to the Chinese Fired Common Bricks Standard (GB/T5101-2003).

Gold tailings from Kolar Gold Fields, in Karnataka, India were also evaluated for the production of bricks through the firing method (Roy et al., 2007). In this work, different amounts of gold mill tailings ranging from 0% to 75% were mixed with black cotton soils or red soils. After drying, the bricks were fired in an electric furnace at three different temperatures, 750°C, 850°C and 950°C. The results indicated that 65%, 75%, 50% and 45% of tailings could be used to produce bricks that pass the quality assessment in terms of compressive strength, water absorption and linear shrinkage.

3.3.3.7.2 Production of Bricks through Cementing

The production of bricks through the cementing method relies on the cementing properties or cementing reactions within the waste material itself or other added cementing material(s) such as ordinary Portland cement (OPC) and lime. In this process, there is no kiln firing involved. The process depends on the presence of C–S–H and C–A–S–H phases for matrix formation and strength (Zhao et al., 2012; Zhang, 2013). If the process is based on the waste material providing the cementing properties, then the waste material has to contain a large amount of calcium or any calcium-containing material (Zhao et al., 2012).

The work by Onuaguluchi and Eren (2012b) showed that copper tailings have the potential to enhance the durability properties of cement-based materials. Comparative tests showed that the addition of copper tailings to concrete led to higher compressive strengths of the produced concrete blocks compared to that of copper tailings–free concrete blocks. The production of bricks by Morchhale et al. (2006) showed that a high compressive strength and lower water absorption capacity of the bricks could be obtained with an increase in the OPC content. The work by Prahallada and Shanthappa, (2014) further showed that the replacement of OPC with copper tailings was safe up to only 30% as thereafter, a further increase in percentage tailings led to a decrease in the strength of the concrete blocks.

Fang et al. (2011) studied the utilization of copper tailings with a low silica content to partially replace sand in the production of autoclaved sand–lime bricks. The copper mine tailings were mixed with river sand and sand powder at different proportions. The mixture was then pressed in a mould and autoclaved under a pressure of 20 MPa, forming moulded bricks. The results showed that in order to produce autoclaved sand–lime bricks meeting the China National Standard for Mu 15 sand–lime brick, the proportion of the copper tailings in the brick batch should be limited to below 50% (% by mass). However, appropriate proportions of river sand and sand powder had to be added to compensate for the low silica content.

In addition to making fired bricks, Roy et al. (2007) tested the viability of a mixture of gold mill tailings, OPC, black cotton soils and red soils in different proportions to make cement bricks. The bricks were cured by immersing them in water for different periods of time followed by the determination of the compressive strengths. The most significant parameter levels for bricks of acceptable standard were found to be 20% cement and 14 days of curing. Gold mine tailings have also been used to produce autoclaved calcium silicate bricks (Jain et al., 1983). The bricks are cured under saturated steam, and in the process, lime reacts with silica grains to form a cementing material known as calcium silicate hydrate. Malatse and Ndlovu (2015) also tested the feasibility of using gold tailings to produce bricks through the cementing method. Different ratios of gold tailings, cement and water were used. The bricks were cured under different environments, sun dried, oven dried and water cured for different periods of time. The tailings bricks were found to require more cement in order to give compressive strengths comparable to that of commercial bricks. This was explained as being due to the possible lack of plasticity in the gold tailings used. Although a cost analysis done by both Malatse and Ndlovu (2015) and Roy et al. (2007) showed that bricks made from tailings cost more than conventional bricks because of the higher quantity of cement used, the tailings brick manufacturing process was found to be more conservative in water consumption. Overall, the results indicated that gold mine tailings have a high potential to substitute for the natural materials currently used in brick making.

Zhao et al. (2012) investigated the possibility of using hematite tailings as the main raw material for the production of high strength autoclaved bricks. The results indicated that a mixture of 70% hematite tailings, 15% lime and 15% sand and an autoclave pressure and time of 1.2 MPa and 6 h, respectively, produced bricks with mechanical strength and durability conforming to the China Autoclaved Lime–Sand Brick Standard (GB11945-1999) for MU20 autoclaved bricks.

From the analysis given in this section on the production of bricks using waste material through the cementing process, it is apparent that the optimum strength and durability is mostly obtained using a higher content of OPC or lime. This is a disadvantage since cement and lime production is expensive; a lot of energy is consumed and a large amount of greenhouse gases are also generated during the production process (Zhao et al., 2012). There is thus, a need to investigate cheaper alternative additives that have a high plasticity or binding properties that can be used in the place of cement (Malatse and Ndlovu, 2015).

3.3.3.7.3 Production of Bricks through Geopolymerization

Geopolymerization is a technology that relies on the chemical reaction of amorphous silica and alumina-rich solid with a high alkaline solution at ambient or slightly elevated temperatures to form an amorphous to semi-crystalline aluminosilicate inorganic polymer or geopolymer (Zhao et al., 2012). During the synthesizing process, silicon–aluminium bonds are formed that are chemically and structurally comparable to those binding the natural rocks (Bennet et al., 2013), giving geopolymer binder–based bricks advantages such as rapid strength gain and good durability, especially in acidic environments. Previous research has shown that geopolymers are cheap to produce, especially since they can be made by mixing naturally occurring materials (i.e. material in possession of a high Si–O–Al framework) with NaOH and water (Davidovits, 1988; Smith and Comrie, 1988). This process consumes less energy, releases less greenhouse gases and is much more environmentally friendly than conventional brick making (Ahmari and Zhang, 2012).

The geopolymerization technology has been applied in the utilization of copper tailings from Mission Mine operations of ASARCO LLC in Tucson, Arizona in the United States for the production of eco-friendly bricks (Ahmari and Zhang, 2012). These copper mine tailings are rich in silica and alumina and can thus, be used as a feed material for the production of geopolymer bricks. The procedure for producing the bricks involves mixing the tailings with an alkaline solution (NaOH), forming the brick by compressing the mixture within a mould under specific pressure and curing the brick at a slightly elevated temperature. Research by Ahmari and Zhang (2012) has indicated that by carefully selecting the right level of parameters such as different initial water content, NaOH concentration, forming pressure and curing temperature, copper mine tailings can be used to produce eco-friendly geopolymer bricks that meet ASTM requirements.

Research into geopolymer bricks has also incorporated the use of copper mine tailings and cement kiln dust (CKD) (Ahmari and Zhang, 2013; Bennet et al., 2013). The work by Ahmari and Zhang (2013) showed that a significant improvement of unconfined compressive strength and durability could be attained when CKD was mixed with the copper tailings. The enhancement of unconfined compressive strength and durability was attributed to the improving effect of CKD on the dissolution of aluminosilicate species, formation of $CaCO_3$ and integration of Ca into the geopolymer gel (Ahmari and Zhang, 2013). Water absorption was however, found to, slightly increase due to the hydration of Ca in the added CKD.

Kiventerä et al. (2016) looked at the use of sulphidic gold mine tailings from a site in Northern Finland together with granulated blast furnace slag (GBFS) for the generation of a geopolymer product which could be used in the construction industry. The tailings were activated using NaOH solution and the GBFS was used as a co-binder. The results showed that the alkaline activation of the pure mine tailings in the presence of the GBFS co-binder allowed the production of specimens with sufficient compressive strength for potential use as a backfill in mining sites or raw material in the construction industry.

Although much research has been conducted on the application of mine tailings in the production of bricks, the commercial production of bricks from waste materials is still very limited. The possible reasons are related to the methods for producing bricks from waste materials, the potential in-use contamination from the applied waste materials, the absence of relevant standards, the economic potential and the slow acceptance of waste materials–based bricks by industry and public (Zhang, 2013). According to Zhang (2013), there is a significant need for further research and development, not only on the techno-economic and environmental aspects but also on standardization, government policy and public education related to waste recycling and sustainable development before a wide production and application of bricks from waste can be implemented.

3.3.3.8 Other Potential Applications

3.3.3.8.1 Copper Tailings

Copper mine tailings have also been tested in the manufacture of tiles and glass-ceramic products. Marghussian and Maghsoodipoor (1999) investigated the production of unglazed tiles by mixing copper tailings with other raw materials. They observed that tiles containing about 40% copper tailings fired at 1025°C for 60 min showed good acid resistance and mechanical properties. In a related work by Çoruh et al. (2006), copper flotation waste was vitrified at 850°C for 2 h. The results showed that glass and glass-ceramic products of acceptable quality and durability could be produced by the vitrification process. Microstructure materials with improved physical and mechanical properties and with a wide range of potential applications especially in the construction industry were obtained after heat treatment at temperatures such as 650°C and 750°C.

Copper tailings have also been investigated for potential utilization and value addition as an extender for the production of paints (Saxena and Dhimole, 2016). Saxena and Dhimole (2016) indicated that copper tailings waste had a good potential to be used as an extender in paints with respect to oil absorption, pH and specific gravity. The copper tailings were found to be superior in enhancing the physicochemical properties such as film-like hardness, impact resistance, abrasion resistance and adhesion compared to the conventional extender.

3.3.3.8.2 Iron Tailings

Ceramic floor and wall tiles are some of the value-added products where industrial wastes such as iron ore tailings can be used as raw materials and as an inexpensive source of alkaline earth oxides for fast fired tile bodies. As already noted, the iron ore tailings are fine material composed mostly of silica, together with some fines of iron oxides, alumina and other minor minerals. This composition is similar to that of the clays used in the ceramic industry (da Silva et al., 2014). Further, since the tailings are in powder form, they require less grinding time, providing a significant saving on energy costs. In addition, If tailings have a high silicon, potassium and sodium content they could be used as raw materials for glass production. These are all alternative approaches to a cost-effective solution in managing tailings of iron ore and controlling environmental pollution.

Das et al. (2000, 2012) and da Silva et al. (2014) have reported on the possibility of using iron ore tailings as one of the raw materials for ceramic floor and wall tile bodies. Iron ore tailings used in this study contained high percentages of silica and this was noted to contribute advantageously to favourable properties in the development of ceramic tile compositions. The iron ore tailings could be used to a maximum of 40 wt.% and the high silica content was found to favour the formulation of tile body compositions. These tiles were found to have high strength and hardness compared to conventional tiles and also noted to conform to most of the EN standards.

3.3.3.8.3 PGM Tailings

Beside the reprocessing of the PGM tailings for precious metal recovery; recent research has shown the feasibility of the utilization of the tailings generated during the processing of PGM ores in South Africa to potentially sequester significant amounts of CO_2 through the mineral carbonation of the waste material. A scoping study by Vogeli et al. (2011) on the potential application of the Bushveld PGM tailings in South Africa showed that these tailings alone have the potential capacity to store 14 Mt of CO_2 per annum. The characteristics of the Bushveld PGM tailings are potentially suited for mineral carbonation due to the vast quantities of fine-grained Ca–Mg–Fe-bearing silicate waste material (Meyer et al., 2014). The main advantage of mineral carbonation compared to the other proposed CO_2 storage technologies is that it currently represents the only known form of permanent storage with environmentally benign carbonation products (Lackner et al., 1995; Hietkamp et al., 2008).

3.3.4 Summary

The review and information given in this section indicates that there are many opportunities for the utilization of tailings in metal producing and alternative manufacturing industries. These envisaged modes of use of the tailings could significantly reduce the volumes of tailings and could contribute to the sustainability of the mining sector and reduce penalties imposed on mines for waste accumulation. It has also been noted that the largest volume of waste utilization is being realized in the building and construction industry. This is largely because not much material pretreatment is required before its application. However, there has also been a significant increase in the reprocessing of waste materials for metal recovery arising possibly due to the increase in the monetary value of the metal and also the development of new technologies that can treat much lower grade material.

3.4 Acid Mine Drainage

3.4.1 Formation of Acid Mine Drainage

Acid mine drainage (AMD) is a severe environmental pollutant that can contaminate surrounding soil, groundwater and surface water (Higginson and Wheeler, 2010; Pozo-Antonio et al., 2014), thus affecting plant and animal life (Chattopadhyay and Chattopadhyay, 2013). Basically, AMD is characterized as low pH water with elevated concentrations of sulphates, iron and non-ferrous metals whose profile depends on the originating mineral deposit types (Kontopouls, 1988; Sheoran and Sheoran, 2006; Obreque-Contreras et al., 2015). The AMD may be generated from mine waste rocks or tailings or mine structures such as active or closed or abandoned pits and underground workings. A generalized conceptual model of sources, pathways and receiving environments is shown in Figure 3.5. These sources, pathways and receiving environments vary by commodity, climate, mine facility and mine phase (INAP, 2012). It must also be noted that the causes of AMD are not only limited to the mining industry, but can also occur where earthmoving processes have taken place such as in highway and tunnel construction and other deep excavations (Skousen, 1995; Skousen et al., 1998; Zdun, 2001; INAP, 2012). Basically, any activity that disturbs mineralized materials can lead to AMD (Simate and Ndlovu, 2014). In fact, the phenomenon is termed acid rock drainage (ARD) when it occurs naturally without human intervention (Chattopadhyay and Chattopadhyay, 2013).

The formation of AMD is a function of many factors including geology, hydrology, temperature and location (Chattopadhyay and Chattopadhyay, 2013). However, the chemistry of AMD generation is fairly straightforward (Costello, 2003; Manders et al., 2009) and results from a series of complex geochemical and microbial reactions when water comes in contact with sulphide-containing minerals such as pyrite (FeS_2) in the presence of oxygen (Sheoran et al., 2011). Ideally, AMD is produced through a series of reactions when minerals containing reduced forms of sulphur (S) are exposed to oxygen and water (Obreque-Contreras et al., 2015).

Generally, the process that produces AMD is well-illustrated and explained by using the reaction pathways of the oxidation of FeS_2, which is one of the most common sulphide minerals (Banks et al., 1997; Akcil and Koldas, 2006; Ruihua et al., 2011). According to Akcil and Koldas (2006), the initial and most important pyrite (or sulphide) oxidation

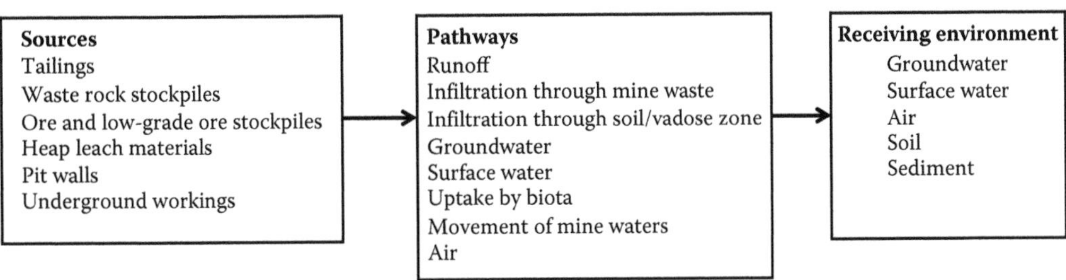

FIGURE 3.5
Generalized conceptual model of sources, pathways, receiving environment and impact at a mine or processing site. (From International Network for Acid Prevention (INAP), Global acid rock drainage guide (GARD Guide), 2012, Available at http://www.gardguide.com, Accessed March 2016; Simate, G.S. and Ndlovu, S., *J. Environ. Chem. Eng.*, 2, 1785, 2014.)

reaction that occurs in the presence of atmospheric oxygen produces dissolved iron, sulphate and hydrogen (Equation 3.1).

$$2FeS_2 + 7O_2 + 2H_2O \rightarrow 2Fe^{2+} + 4SO_4^{2-} + 4H^+ \tag{3.1}$$

Oxygen dissolved in water can also result in pyrite oxidation, but due to its limited solubility in water, this process is much less prominent (INAP, 2012). Reaction 3.1 can also occur 'biologically' (i.e. mediated through microorganisms), and such reactions have been discussed by many researchers (Eligwe, 1988; Boon et al., 1995; Fowler et al., 2001). Nevertheless, such mechanisms have been replaced by two 'indirect' mechanisms (termed the thiosulphate mechanism and polysulphide mechanism) with no evidence for a 'direct' biologically mediated process (Schippers et al., 1996; Schippers and Sand, 1999; Hanford and Vargas, 2001; Sand et al., 2001). Figure 3.6 is an illustration of the thiosulphate and polysulphide mechanisms. The thiosulphate mechanism involves solely the chemical reaction of ferric iron with acid-insoluble metal sulphides (FeS_2, MoS_2 and WS_2) producing thiosulphate, while the polysulphide reaction mechanism involves the attack of acid soluble sulphides (ZnS, NiS, CoS, $CuFeS_2$ and PbS) by ferric iron and protons (Schippers and Sand, 1999). Readers are referred to Schippers et al. (1996), Schippers and Sand (1999), Hanford and Vargas (2001) and Sand et al. (2001) for detailed discussions of the two mechanisms.

In a sufficiently oxidizing environment (dependent on O_2 concentration, pH greater than 3.5, bacterial activity), ferrous iron (Fe^{2+}) released in Equation 3.1 may be oxidized to ferric iron (Fe^{3+}) according to the following reaction (Blowes et al., 2003; Akcil and Koldas, 2006):

$$4Fe^{2} + O_2 + 4H^+ \leftrightarrow 4Fe^{3+} + 2H_2O \tag{3.2}$$

If oxygen is low, reaction 3.2 will not occur until the pH reaches 8.5 (Fripp et al., 2000). In general under many conditions, reaction 3.2 is the rate-limiting step in pyrite oxidation because the conversion of ferrous to ferric is slow at pH values below 5 under abiotic

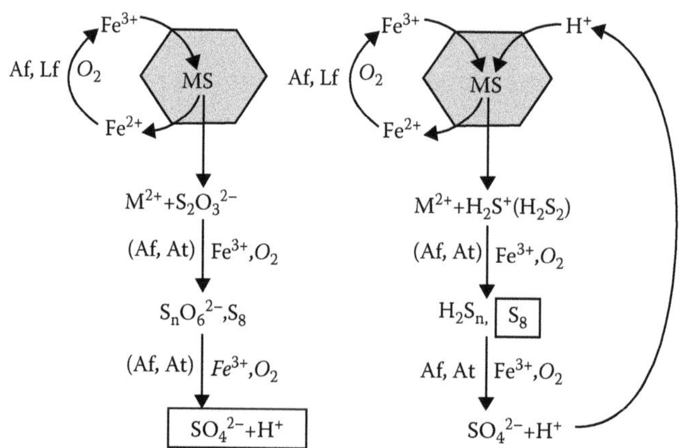

FIGURE 3.6
Hypothetical scheme for the oxidation of pyrite or metal sulphide by iron- and sulphur-oxidizing acidophilic bacteria producing sulphuric acid. (From Schippers, A. et al., *Appl. Environ. Microbiol.*, 62(9), 3424, 1996; Schippers, A. and Sand, W., *Appl. Environ. Microbiol.*, 65, 319, 1999; Simate, G.S. and Ndlovu, S., *J. Environ. Chem. Eng.*, 2, 1785, 2014.)

conditions (Skousen et al., 1998). The oxidation of Fe^{2+} to Fe^{3+} can also occur biologically and such reactions have been extensively discussed in literature (Schippers et al., 1996; Schippers and Sand, 1999; Simate and Ndlovu, 2014).

Aqueous Fe^{3+} ions produced in reaction 3.2 can also oxidize pyrite according to the following reaction:

$$FeS_2 + 14Fe^{3+} + 8H_2O \rightarrow 15Fe^{2+} + 2SO_4^{2-} + 16H^+ \quad (3.3)$$

Reaction 3.3 is considerably faster (2–3 orders of magnitude) than the reaction of pyrite with oxygen and generates substantially more acid per mole of pyrite oxidized (Ritchie, 1994; Dold, 2010; INAP, 2012). In fact, Fe^{3+} ions resulting from the oxidation of Fe^{2+} ions are recognized as a more potent oxidant than oxygen even at circumneutral pH (Zdun, 2001). Luther (1987) attributed this effect to the more efficient electron transfer of Fe^{3+} compared to oxygen. This molecular orbital study by Luther (1987) is consistent with pyrite oxidation data obtained by McKibben and Barnes (1986) and Moses et al. (1987). However, it must be noted that reaction 3.3 is limited to conditions in which significant amounts of dissolved Fe^{3+} ions occur (i.e. acidic conditions) (INAP, 2012). Studies have shown that at pH values between 2.3 and 3.5, Fe^{3+} ions formed in reaction 3.2 may precipitate as $Fe(OH)_3$ (and to a lesser degree as, jarosite, $H3OFe_3(SO_4)_2(OH)_6$), leaving little Fe^{3+} in solution (for pyrite oxidation) while simultaneously lowering pH (Blowes et al., 2003; Akcil and Koldas, 2006) as shown in reaction 3.4.

$$Fe^{3+} + 3H_2O \leftrightarrow 4Fe(OH)_3(s) + 3H^+ \quad (3.4)$$

If pH is less than 2, Fe^{3+} ion hydrolysis products like $Fe(OH)_3$ are not stable and Fe^{3+} remains in solution (Dold, 2010), thus it is available for oxidizing additional pyrite.

It must be noted that other sulphide minerals are also susceptible to oxidation thus, releasing elements such as aluminium, arsenic, cadmium, cobalt, copper, mercury, nickel, lead and zinc into the water flowing through the mine waste (Blowes et al., 2003). However, not all sulphide minerals generate acidity when being oxidized (INAP, 2012). As a general rule, iron sulphides (pyrite, marcasite, pyrrhotite), sulphides with molar metal/sulphur ratios < 1 and sulphosalts (e.g. enargite) generate acid when they react with oxygen and water. Sulphides with molar metal/sulphur ratios = 1 (e.g. sphalerite, galena, chalcopyrite) tend not to produce acidity when oxygen is the oxidant (INAP, 2012). However, when aqueous Fe^{3+} ions are the oxidant, all sulphides are capable of generating acidity. It must also be noted that the oxidation rates vary among sulphide minerals, and research studies have shown that reactivity decreases in the order marcasite → pyrrhotite → sphalerite-galena → pyrite-arsenopyrite → chalcopyrite → magnetite (Kwong and Ferguson, 1990; Jambor, 1994; Lapakko, 2002; Dold, 2010). The oxidation of these minerals will not be discussed any further in this chapter as they have been extensively covered elsewhere (e.g. see Blowes et al., 2003; Dold, 2010; Simate and Ndlovu, 2014).

3.4.2 Prevention and Mitigation of Acid Mine Drainage

Environmental damage or pollution associated with AMD is mainly characterized by a decrease in pH and/or elevated concentrations of heavy metals in nearby waters and soils (Costello, 2003). The hazardous effects of such low pH and heavy metals on the health of humans, wildlife, plants and aquatic species have already been discussed in Chapter 2. As AMD and its hazardous effects threaten the environment, a number of measures aimed at preventing (or minimizing) and mitigating its impact on the environment have been developed (Johnson and Hallberg, 2005). Two categories of AMD control technologies are (1) source control (or prevention) techniques and (2) migration control and/or treatment (or mitigation) techniques.

3.4.2.1 Source Control Techniques

Source control techniques are preventive methods that are directed towards controlling the formation of AMD at the source (Egiebor and Oni, 2007; Luptakova et al., 2010). They are based on preventing oxygen and/or water into the system (Skousen et al., 1998; Kuyucak, 1999, 2002; Johnson and Hallberg, 2005) or inhibiting the activity of sulphide-oxidizing microorganisms (Kleinmann, 2006). In other words, preventive techniques act on any of the three elements that are essential in the formation of acidic waters: oxygen, water and oxidizing microorganisms (Pozo-Antonio et al., 2014). As shown by reactions 3.6 through 3.9, oxygen and water are two of the three principal reactants (Skousen et al., 1998). Therefore, removing water (for example) by pumping it out before it contacts pyritic material may minimize the formation of acidic products. In addition, the control and prevention methods which aim to eliminate sulphide oxidation can be achieved by separating sulphide minerals from the waste (Kuyucak, 2002). Table 3.3 shows some of the various source control techniques that have been evaluated to prevent or minimize the generation of AMD (Kuyucak, 2002; Johnson and Hallberg, 2005). Ideally, these methods can be categorized as (1) barrier methods, methods that are intended to isolate sulphide minerals from weathering reactants or from the hydrologic transport system; (2) chemical methods, methods that alter the composition of AMD solutions, limit reactant availability or serve to passivate the sulphide mineral surfaces so as to limit oxidation and (3) bacterial inhibition methods, methods that represent a class of chemical treatments that disrupt the biologically catalysed cyclic

TABLE 3.3
Source Control Measures

Control Measure	Description
Control of water migration	Water is intercepted and diverted away to prevent it from passing through the waste materials with the potential to form AMD.
Flooding/sealing of underground mines	Sealing the mine to prevent water and air infiltration.
Underwater storage of mine tailings	Because oxygen has a very low solubility and diffusion rate through water, almost four orders of magnitude less than in air, the oxidation of reactive wastes can be minimized by deposing and storing sulphide-bearing waste underwater.
Land-based storage in sealed waste heaps	Dry covers, caps and seals (incorporating an organic layer) used to isolate or encapsulate sulphide-bearing waste, thus limiting the access of either oxygen or water, or both.
Blending of mineral wastes	Blending acid-generating and acid-consuming materials.
Application of anionic surfactants	Use of anionic surfactants as bactericide inhibit bacterial activity that catalyse the conversion of ferrous iron and, subsequently, prevent acid generation.
Coating of certain mine wastes	Coating involves the leaching of waste with a phosphate solution containing hydrogen peroxide. Hydrogen peroxide oxidizes the surface portion of the pyrite and releases iron oxides so that phosphate precipitation forms a passive surface coating.

Sources: Kuyucak, N., *Can. Inst. Min. Metallur. Bull.*, 95(1060), 96, 2002; Johnson, D.B. and Hallberg, K.B., *Sci. Total Environ.*, 338, 3, 2005.

oxidation processes (Kleinmann, 2006). Nevertheless, despite several years of research, attempts to prevent AMD generation have proven to be practically and extremely difficult (Dinardo et al., 1991; Johnson and Hallberg, 2005).

3.4.2.2 Migration Control and/Treatment Methods

Migration control and/or treatment methods or processes generally require pH adjustment, oxidizing or reducing (redox) conditions and/or stabilization of wastes. The migration control and/or treatment techniques can be classified as (1) traditional methods and (2) innovative methods. These two categories are also broadly termed as mitigation techniques because they are directed towards the treatment of the resulting drainage (Egiebor and Oni, 2007; Luptakova et al., 2010; Simate and Ndlovu, 2014). In other words, these are corrective techniques that are used to purify the acidic waters produced (Pozo-Antonio et al., 2014). They are conceptually opposed to the preventive techniques, whose main objective is to avoid the formation of the acidic waters. Furthermore, corrective techniques are applied without distinction to underground or surface mines (Pozo-Antonio et al., 2014). The migration control techniques have been extensively studied and reported elsewhere (Gazea et al., 1996; Skousen et al., 1998; Fripp et al., 2000; Costello, 2003; Johnson and Hallberg, 2005; Kalin et al., 2006; Egiebor and Oni, 2007; Caraballo et al., 2011) and thus are only discussed briefly in this chapter.

3.4.2.2.1 *Traditional Techniques*

Traditional or conventional techniques are methods that follow the pattern of an ordinary wastewater treatment plant, and are often referred to as 'active' treatment methods (Costello, 2003). These methods rely on conventional, well-recognized technology to raise pH or create redox conditions. Basically, these treatment systems involve treating AMD with alkaline chemicals to raise pH, neutralize acidity and precipitate metals (Skousen et al., 1998).

A variety of methods fall in this category of 'active' techniques, but the predominant one is the ODAS (O = oxidation, DA = dosing with alkali and S = sedimentation), which is similar to that of traditional wastewater treatment plants (Younger et al., 2002; Costello, 2003; Trumm, 2008, 2010). Although the most common order of treatment in industrial wastewater treatment systems is ODAS, for the treatment of AMD, the most common order followed is DAOS (Younger et al., 2002; Trumm, 2008). Dosing with alkali is typically the first step in AMD treatment followed by oxidation and sedimentation (Trumm, 2008). Ideally, the DAOS steps are followed in AMD treatment because oxidation rates for dissolved metals in reduced form such as Fe^{2+} are strongly influenced by pH (Stumm and Morgan, 1996), therefore, it is beneficial to raise the pH prior to the oxidation step in the treatment of AMD (Trumm, 2008). In other words, the main goal of the dosing with alkali step is to add enough neutralizing agent so as to raise the pH and to lower the concentrations of dissolved metals by forming metal hydroxides and oxyhydroxides (Trumm, 2008, 2010). On the other hand, the goal of the oxidation step is to ensure reduced metals such as Fe^{2+} and Mn^{2+} are oxidized to Fe^{3+} and Mn^{4+}, respectively, so that they can form hydroxide, oxide and carbonate precipitates and be removed from AMD (Skousen et al., 2000; Younger et al., 2002). However, the oxidation step may not be necessary if the metals are already highly oxidized through the dosing with alkali step (Trumm, 2008, 2010). Sometimes a pretreatment step such as sedimentation precedes DAOS so as to reduce the concentration of total suspended solids (TSS) which can affect the treatment system performance (Trumm, 2008).

Figure 3.7 is a basic flowsheet for selecting a site-specific 'active' treatment system for AMD (Trumm, 2008, 2010). As can be seen from Figure 3.7, the selection of such a system is influenced by a number of factors, including TSS content, Mn concentration (mg/L), flow rate (L/s), Fe concentration (mg/L) and the available land area (Trumm, 2008). Once an 'active' treatment system has been selected, a computer program such as AMDTreat can be used to design specific components of the system and to determine potential costs (Rajaram et al., 2001; Means et al., 2003; Trumm, 2008, 2010).

Other traditional or 'active' methods that are occasionally used for AMD treatment include, (1) sulphidization, (2) biosedimentation, (3) sorption and ion exchange and (4) membrane processes like filtration and reverse osmosis (Younger et al., 2002; Costello, 2003; Trumm, 2010).

3.4.2.2.2 Innovative Techniques

Innovative is defined in the Cambridge English Dictionary as 'featuring new methods or advanced and original', and according to Costello (2003), a variety of 'passive' treatment methods have become the most dominant innovative treatment techniques. Passive treatment systems do not require continuous chemical inputs, but take advantage of naturally occurring chemical and biological processes to treat AMD (Skousen et al., 1998). In other words, 'passive' treatments techniques are methods that treat AMD (or wastewater in general) using enhanced natural processes, in-situ, and require minimal upkeep (Hedin et al., 1994; Younger et al., 2002; Costello, 2003).

Based on the chemistry of AMD generation (i.e. an oxidation process, which results in the dominant contaminant, iron, being present in two states, Fe^{2+} and Fe^{3+}), remediation of AMD using 'passive' technologies can be classified into two broad categories: oxidizing and reducing strategies (Trumm, 2010). Ideally, the choice between the two strategies is typically based on the water chemistry (mainly dissolved oxygen [DO] content and Fe^{2+}/Fe^{3+} ratio). For AMD which is highly oxidized (DO level at saturation and all iron as Fe^{3+}), the oxidizing strategy is most appropriate; for AMD with low DO and all iron existing as Fe^{2+}, the reducing strategy is usually recommended. However, site limitations, such as available land area, climate and topography, may limit the use of certain systems (Trumm, 2010).

Oxidizing systems remove iron from the AMD by continuing the oxidation process so that all Fe^{2+} is oxidized to Fe^{3+}, and once the pH has been raised sufficiently, the iron precipitates out of the AMD as ferric hydroxide ($Fe(OH)_3$). Typical remediation systems that employ the oxidizing strategy are open limestone channels (OLCs), open limestone drains (OLDs), limestone leaching beds (LLBs), slag leaching beds (SLBs) and diversion wells (DWs) (Trumm, 2010). The OLCs and DWs typically require a steep topography in order to generate the necessary aeration and to prevent armouring of limestone by metal hydroxides, which can inhibit the dissolution of limestone (Ziemkiewicz et al., 1997; Trumm, 2010).

As for the reducing systems, the AMD oxidation process is reversed, such that iron and sulphate are reduced, forming compounds like FeS_2, FeS and H_2S (Trumm, 2010). In this way, dissolved iron and sulphate are all removed from the AMD at the same time. Typical remediation systems that employ the reducing strategy are anaerobic wetlands; anoxic limestone drains (ALDs); bacteria-based sulphate-reducing bioreactors (SRBRs), also known as biogeochemical reactors (BGCRs) and successive alkalinity producing systems (SAPS), also known as vertical flow wetlands (VFWs) or reducing and alkalinity producing systems (RAPS) (Trumm, 2010).

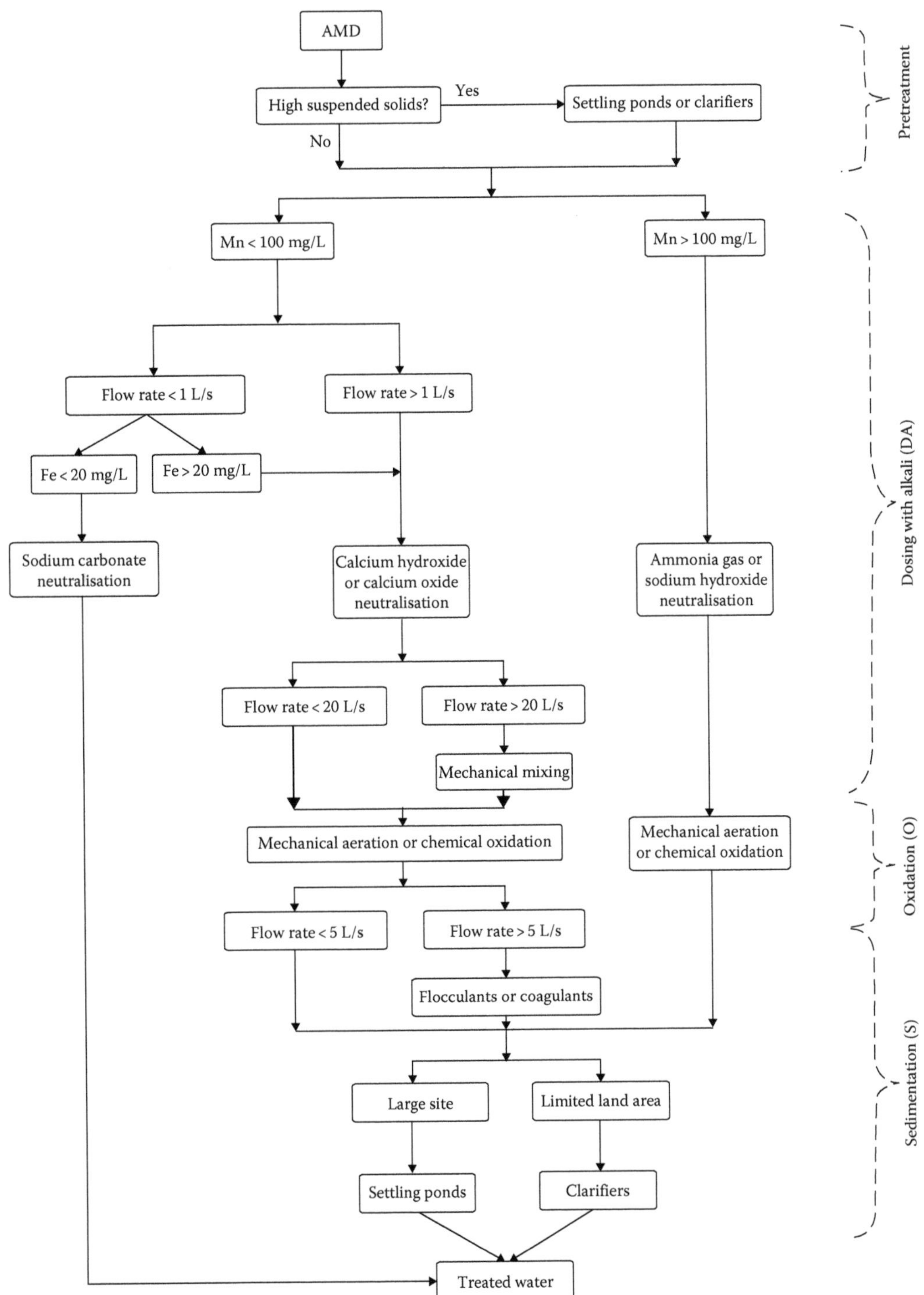

FIGURE 3.7
Flowsheet for designing a site-specific active treatment system for AMD. (From Trumm, D.A., Selection of active treatment systems for acid mine drainage, 2008, Available at http://www.crl.co.nz/downloads/geology/2008/Trumm%202008%20AusIMM%20Active%20Systems.pdf, Accessed March 2016; Trumm, D.A., *NZ J. Geol. Geophys.*, 53(23), 195, 2010.)

3.4.3 Opportunities in Acid Mine Drainage

While a wide range of technologies are available for treating AMD before discharge, most of these technologies consider AMD as a nuisance that needs to be quickly disposed of after minimum required treatment. However, in the recent past, there has been an emerging worldwide paradigm towards environmental responsibility and sustainable development. In fact, the economic sustainability of any AMD remediation system is a factor that is becoming increasingly critical in decision-making (Johnson and Hallberg, 2005). Indeed, significant attention should now be directed towards the recovery of industrially and economically useful materials (Simate and Ndlovu, 2014). In fact, when the value of treated water and by-products exceeds the cost of treatment, it is feasible to create enterprises that will provide economic benefits while dealing with the environmental problem (Simate and Ndlovu, 2014). Therefore, this section of Chapter 3 focuses on the recovery of valuable and saleable products such as metals, water, acid and pigments from AMD.

3.4.3.1 Recovery of Metals

The removal of metals as hydroxide precipitates from AMD using alkaline reagents (e.g. NaOH, etc.) has been the most widely used treatment method (Johnson and Hallberg, 2005; Balintova and Petrilakova, 2011). However, this technique is slowly losing popularity due to large volumes of hazardous concentrated sludge generated that require further treatment and controlled final disposal (Macingova and Luptakova, 2012). The other drawback of hydroxide precipitation is that selective extraction of metals is very difficult (Simate and Ndlovu, 2014).

A recent and suitable alternative method recovers metals from AMD in the form of sulphides. The use of sulphides not only allows the production of effluents with metal concentrations in the order of magnitude of ppm and ppb, but also gives the possibility of precipitation at low pH and selective precipitation for metal reuse (Sampaio et al., 2009). Sulphide precipitation can be effected using either solid (FeS and CaS), aqueous (Na_2S, NaHS, NH_4S) or gaseous sulphide sources (H_2S) (Lewis, 2010). There is also the possibility of using the degeneration reaction of sodium thiosulphate ($Na_2S_2O_3$) as a source of sulphide for metal precipitation (Lewis, 2010). In a study by Sampaio et al. (2009), Cu was continuously and selectively precipitated from Zn using Na_2S. Selective precipitation was based on the control of pS ($= -\log [S^{2-}]$) and pH. Here, having the solubility product defined as $K_{SP} = (Me^{2+})(S^{2-})$, it means that different sulphide concentrations (S^{2-} potentials) are required to precipitate different metals. Therefore, the addition of sulphide to selectively precipitate heavy metals can be controlled using an ion selective electrode for sulphide (S^{2-}), a so-called pS electrode (Sampaio et al., 2009). In this study, the selective precipitation of Cu from Zn was achieved at pS and pH of 25 and 3, respectively.

Another promising approach to the sulphide precipitation method uses sulphate-reducing bacteria (SRB) at anaerobic conditions to generate sulphide. The generation of sulphide by SRB is a favourable method that eliminates safety concerns related to the handling, transportation and storage of precipitating agents. The other advantage of the process is that it uses sulphate which is also present in the AMD. Using discrete chemical and biological stages, Macingova and Luptakova (2012) selectively removed Fe, Cu, Al, Zn and Mn from AMD in a selective sequential precipitation (SSP) process. In this process, Fe was completely removed in two steps by oxidizing ferrous iron using hydrogen followed by precipitation using sodium hydroxide (Macingova and Luptakova, 2012). In the initial stage, a partial precipitation of Fe was achieved due to a decrease in pH during iron oxidation. In the biological stage, sulphate-reducing bacteria were used to produce hydrogen

TABLE 3.4

Conditions and Results of Selective Sequential Precipitation Process

Step	1	2	3	4	5	6
pH	2.8	3.7	3.7	5.0	5.0	9.5
Reagent	H_2O_2	NaOH	H_2S	NaOH	H_2S	NaOH
Removed metals	Fe	Fe	Cu	Al, Zn	Zn	Mn
Proportion (%)	99.99	99.99	99.99	98.94:1.05	99.99	99.99

Sources: Macingova, E. and Luptakova, A., *Chem. Eng. Trans.*, 28, 109, 2012; Simate, G.S. and Ndlovu, S., *J. Environ. Chem. Eng.*, 2, 1785, 2014.

sulphide that was later transported to the contactor filled with AMD (in the chemical stage) where the precipitation of metal sulphide occurred. After the filtration of precipitate, the filtrate pH was adjusted to a higher value using sodium hydroxide. The pH adjustment simultaneously precipitated metal hydroxides. After the filtration of metal hydroxide precipitates, the filtrate was returned to the contactor and the whole process was repeated at the higher pH value. Table 3.4 shows the results obtained (Macingova and Luptakova, 2012; Simate and Ndlovu, 2014). As can be seen from the table, the selective recovery of various metal precipitates was achieved.

Adsorption is another method used for the recovery of heavy metals from aqueous solutions and has evolved as the preferred method in many applications (Rao et al., 2007; Nagpal et al., 2013). Most of the adsorbents used in these studies are highly porous materials, providing adequate surface area for adsorption (Hu et al., 2006). However, with the advent of nanotechnology, various types of nanomaterials with large surface areas and small diffusion resistance have been developed and are now receiving considerable attention in water treatment (Zhang, 2003; Hu et al., 2006; Simate, 2012; Simate et al., 2012). For example, several studies have been carried out to assess the technical feasibility of various kinds of raw and surface oxidized carbon nanotubes (CNTs) for the sorption of various metals from aqueous solutions (Rao et al., 2007). The sorption/desorption studies showed that CNTs could be regenerated and reused consecutively several times without significant loss in adsorbent capacity signifying its appropriateness for commercial application.

Some of the metals such as uranium, thorium, rare earth elements (REEs), gold, silver, PGMs, chromium, copper, zinc, nickel, cobalt and tungsten have been recovered and purified on an industrial scale by means of ion exchange (Hubicki and Kołodyńska, 2012). Ion exchange is defined as the exchange of ions between the substrate and surrounding medium (Hubicki and Kołodyńska, 2012). Depending on the type of functional groups, ion exchangers may be classified as strongly acidic, for example, sulphonate ($-SO_3H$); weakly acidic, for example, carboxylate (–COOH); strongly basic, for example, quaternary ammonium ($-N^+R_3$) and weakly basic, for example, tertiary and secondary amine ($-N^+R_2H$ and $-N^+RH_2$) (Hubicki and Kołodyńska, 2012). There are also amphoteric exchangers, which, depending on the pH of the solution, may exchange either cations or anions. These ion exchangers are also called bipolar electrolyte exchange resins (BEE) or zwitterionic ion exchangers (Nesterenko and Haddad, 2000; Hubicki and Kołodyńska, 2012).

In recent past, a method termed cyclic electrowinning/precipitation (CEP) which combines precipitation and electrowinning was developed (Brown University, 2011). In this method, the concentration of metal cations is increased in the first stage followed by electrowinning to recover and separate the metals from wastewater in the second stage. Basically, the operations are as follows: In the first unit the metal-laden water is fed into a tank in which an acid (e.g. H_2SO_4) or a base (e.g. NaOH) is added to change the water pH, thus effectively separating the water from the metal precipitate which settles at the bottom. The clear water is siphoned off and more contaminated water is brought in. The pH swing is applied again, first re-dissolving the precipitate and then re-precipitating all the metals, thus increasing the metal concentration each time (Brown University, 2011). This process is repeated until the concentration of the metal cations has reached a point where electrowinning can be efficiently employed. When that point is reached it is taken to a second device called a spouted particulate electrode, where metal cations are converted to stable metal solids, so that they can be removed (Brown University, 2011). The tested heavy metals were cadmium, copper and nickel. The results showed that cadmium, copper and nickel were reduced to 1.50, 0.23 and 0.37 ppm, respectively, way below the maximum contamination levels allowed by the EPA (Brown University, 2011). The main advantage of this process is that sludge is continuously formed and re-dissolved within the system so that none is left as an environmental contaminant.

Studies by Cheng et al. (2007, 2011) showed the development of an AMD fuel cell based on fuel cell technology, which was capable of abiotic electricity generation from synthetic AMD solutions. Test works conducted using synthetic solutions of similar composition and constituent as typical AMD solutions showed that the device (AMD fuel cell) could efficiently remove dissolved iron from the solution while also generating electricity at power levels similar to conventional microbial fuel cells (MFCs). Though the power and current yields were relatively low, improvements in fuel cell technology will possibly lead to more efficient power generation in the future.

3.4.3.2 Recovery of Water

Water is one of the most important substances on earth (Simate, 2016). All plants and animals need water to survive. If there was no water there would be no life on earth. Therefore, it is imperative that the millions of litres of water in AMD are recovered. To respond to this need, a large number of research studies have been devoted to the recovery of water from AMD. For example, the removal of heavy metals from AMD using the technologies discussed earlier simultaneously recovers water (Simate and Ndlovu, 2014). However, the processes have several disadvantages which include pH dependence, which means that the removal of the mixture of heavy metals cannot be achieved at a single pH level. Therefore, other technologies which are not pH-dependent, such as electrodialysis, microfiltration, ultrafiltration, nanofiltration, reverse osmosis and membrane distillation, have been used to recover water from AMD (Simate and Ndlovu, 2014).

Among the different technologies, electrodialysis has been tested and proven to be an effective technology for water recovery from acidic solutions (Wisniewski and Wisniewska, 1999; Cifuentes et al., 2006, 2009; Agrawal and Sahu, 2009). Electrodialysis is a membrane separation process based on the selective migration of aqueous ions through an ion-exchange membrane as a result of an electrical driving force (Rodrigues et al., 2008). It allows the separation of anions and metallic cations, thus electrodialysis possesses the advantages of removing contaminant metals and simultaneously recovering water (Buzzi et al., 2013). A study by Buzzi et al. (2013) has shown that electrodialysis is suitable for recovering water from AMD with contaminant removal efficiencies that are greater than 97%. The recovery of water has also been studied in other acidic systems containing various metals (Wisniewski and Wisniewska, 1999; Agrawal and Sahu, 2009; Cifuentes et al., 2009; Benvenut et al., 2013). In all these studies electrodialysis has been found to be an effective method for water recovery.

Membrane distillation is another technique that has been used to recover water, and concentrate acid and metal values from mining wastewater and process solutions (Kesieme et al., 2012). This process is a combination of the conventional distillation and membrane separation processes (Shirazi et al., 2014a). Unlike other membrane separation processes which are driven by absolute pressure difference (e.g. reverse osmosis, nanofiltration and microfiltration) or an electrical potential gradient (e.g. electrodialysis) or concentration gradient (e.g. dialysis), membrane distillation is a thermally driven separation process that utilizes hydrophobic, microporous membranes as a contactor (Shirazi et al., 2014b). The driving force in the membrane distillation is the vapour pressure difference induced by the temperature difference across the hydrophobic membrane (Alkhudhiri et al., 2012). In other words, the temperature difference existing across the membrane results in a vapour pressure difference, thus vapour molecules are transported from the high vapour pressure side to the low vapour pressure side through the pores of the membrane (Tomaszewska, 2000).

Individual and/or combinations of microfiltration, ultrafiltration, nanofiltration and reverse osmosis membrane processes have also been studied for the recovery of both water and metals (Ahn et al., 1999; Garba et al., 1999; Zhong et al., 2007). For example, Vaclav and Eva (2005) used the reverse osmosis technique to recover good quality water from three different sources of AMD. Günther and Mey (2008) evaluated different water treatment technologies, including the membrane technology with the objective of treating AMD from a coal mine. The results showed that only the biological sulphate removal process and reverse osmosis membranes could be used to produce a high water recovery cost-effectively. Bhagwan (2012) studied the use of a high recovery precipitating reverse osmosis (HiPRO) process for the recovery of low salinity water from mine waters. The main advantage of the developed process is that it makes use of reverse osmosis to concentrate the water and produce supersaturated brine from which the salts can be released in a simple precipitation process.

3.4.3.3 Recovery of Sulphuric Acid

AMD is predominantly acidic due to high concentrations of sulphuric acid (Simate and Ndlovu, 2014). In fact, over the past five decades, assiduous efforts have been directed at remediating AMD through acid removal so as to reduce the impact of the acidic water on the environment and produce water suitable for reuse (Johnson and Hallberg, 2005; Simate and Ndlovu, 2014; Nleya et al., 2016). Indeed, apart from the production of reusable water and saleable metals, the recovery of sulphuric acid would also be used to offset the treatment costs (Simate and Ndlovu, 2014). Table 3.5 is a summary of the methods used to recover sulphuric acid from various wastewater solutions, including AMD, showing the recoveries, advantages and disadvantages of the processes (Nleya et al., 2016). Furthermore, a study by Nleya et al. (2016) critically evaluated the technical and economic feasibilities of the processes in Table 3.5 for application to AMD, and the results of the study are given in Table 3.6. Based on the technical and economic feasibility results in Table 3.6, the freeze crystallization and acid retardation processes are expected to be the most suitable technologies for acid recovery from AMD. Several studies on these processes have been conducted in the past, and their fundamentals are well documented (Etter and Langill, 2006; Kim, 2006; Özdemir et al., 2006; Tjus et al., 2006; Agrawal and Sahu, 2009). Furthermore, their use in commercial applications is widely recognized.

Another process, not included in Tables 3.5 and 3.6, which may be used to recover or concentrate sulphuric acid, is the membrane distillation process. Moreover, Tomaszewska (2000) used membrane distillation to recover hydrochloric acid from spent pickling liquors containing residual acid.

TABLE 3.5

Summary of Methods Used for Sulphuric Acid Recovery

Method	Solution Content	H_2SO_4 Recovery	Advantages	Disadvantages
Rectification	H_2SO_4, nitro compounds	R = 98.3%	Recovery of high purity acid	High energy consumption High operating cost
Diffusion dialysis	H_2SO_4, AlH_2SO_4, Fe, V H_2SO_4, Ni H_2SO_4, rare earth sulphates	R = 82%–90% R = 84% R = 80% R =70%–80%	High acid recovery Low pay back period Strong salt rejection	Not efficient at low acid concentration
Electrodialysis	H_2SO_4, NiH_2SO_4, Fe H_2SO_4, Cu, Sb, As H_2SO_4, Fe, Na	R = 80%–90% R = 90% R = Up to 99% –	Clean acid product Reduced solid waste for disposal	High operating cost Membrane fouling
Acid retardation	H_2SO_4, FeH_2SO_4, Fe H_2SO_4, Ni	R = 74%–96% R = 96% R = 70%–95%	Low operating cost High acid recovery Small equipment size and space	Increases product volume High consumption of fresh water Dilute acid product
Crystallization	H_2SO_4, Fe		Low cost Reduced waste for disposal	Risk of scale formation in crystallizer. Increased energy consumption
Solvent extraction	H_2SO_4, CuH_2SO_4, Fe, Mn H_2SO_4, Zn	E = 75%–79% E = 90% E = 90%	Can manage great volumes of solutions with high contents of toxic solutes Clean acid product Only physical separation High throughput with compact equipment	Chemicals used are hazardous Pretreatment is required to remove impurities Difficulties in stripping from Cyanex 923 Co-extraction of Fe and Zn

Source: Nleya, Y. et al., *J. Clean. Prod.*, 113, 17, 2016.
Note: R, recovery; E, percentage extraction.

TABLE 3.6

Summary of Capital and Operating Expenditures of Proposed Process Routes

Cost (US$)	Rectification	Diffusion Dialysis	Electrodialysis	Solvent Extraction	Crystallization	Acid Retardation
Fixed capital cost	1,193,165	381,780	1,523,478	442,956	407,268	307,973
Total estimated CAPEX	1,372,140	439,047	1,752,000	509,399	468,368	354,169
Total estimated annual OPEX	427,789	247,752	266,809	125,124	154,798	124,810
Estimated annual revenue	≈9,800	≈9,800	≈9,800	≈9,800	≈10,320	≈9,800

Source: Nleya, Y. et al., *J. Clean. Prod.*, 113, 17, 2016.

3.4.3.4 Uses of AMD Treatment Sludge

Despite widespread usage, one of the main drawbacks of the active treatment methods is the use of large amounts of chemicals needed to reduce metals to an acceptable level before discharge. Other disadvantages are its excessive sludge production that requires further treatment, increased costs, slow metal precipitation, poor settling, the aggregation of metal precipitates and the long-term environmental impacts of sludge disposal (Kurniawan et al., 2006). Therefore, in recent years, due to several disadvantages of AMD treatment as stated, concerted effort has been made to investigate the potential use of the sludge which would otherwise be disposed of in a landfill at significant costs. The succeeding sections focus on and cite some of the applications of the AMD treatment sludge, especially with reference to the publication by Simate and Ndlovu (2014).

3.4.3.4.1 Production of Iron Pigments

Studies have shown that the sludge obtained from AMD can be considered for the production of inorganic pigments (Hedin, 1998, 2003; Marcello et al., 2008; Michalkova et al., 2013), and magnetic particles like ferrites (Wei et al., 2008). To produce commercially usable iron oxides as raw material for the production of pigments, additives to ceramics, etc., the treatment of AMD using a two-step selective precipitation process was developed (Hedin, 1998, 2003). The two-step process that uses magnesium oxide and sodium hydroxide results in the ferrous and ferric oxyhydroxide sludge that can be thermally transformed to basic ferric pigment. A study by Hedin (2003), however, indicated that while the end product was of high quality, the costs associated with the process made the materials more costly to produce than mined oxides, although this may be offset when considering the high cost of hydrous ferric oxide disposal. Marcello et al. (2008) investigated the use of hydrous ferric oxides from active coal mine drainage treatment as pigment within ceramic tile glaze. Favourable results occurred when the ferrous hydrous oxides was blended with an industrial standard pigment.

Research by Cheng et al. (2007, 2011) has also shown that fuel cell technologies was not only used for simultaneous AMD treatment and power generation, but also generated iron oxide particles of sizes appropriate for use as pigments and other applications. As already discussed in Section 3.4, a fuel cell called an AMD fuel cell based on an MFC was developed and used during the process. During treatment, ferrous iron was oxidized in the anode chamber under anoxic conditions, while oxygen was reduced to water at the cathode. Ferrous iron was completely removed through oxidation to insoluble ferric iron and precipitated at the bottom of the anode chamber. The particle diameter of the iron oxides could be controlled by varying the conditions in the fuel cell, especially current density, pH and initial ferrous iron concentration. Upon drying, the iron oxide particles were then transformed to goethite (α-FeOOH).

3.4.3.4.2 Building and Construction-Related Materials

Many constituents in sludge are the same as those used in cement manufacturing. For example, calcite, gypsum, silica, Al, Fe and Mn are common raw materials for cement. Therefore, the components that make up AMD treatment sludge such as gypsum, calcite and ferrihydrite can be utilized as raw materials in the manufacturing of construction materials and other products. Levlin (1998) indicated that in order to avoid waste disposal, inorganic components in sludge can be used for the production of building materials. Research by Lubarski et al. (1996), found that the high aluminium content of sludge produced from the treatment of acidic drainage at some coal and gold mines could be used for the production of aluminous cement. Other studies have suggested that sludge can replace up to as much as 30% Portland cement in blended cement (Tay and Show, 1994). In these options, the utilization of sludge reduces the mining of raw materials for the production of building material.

3.4.3.4.3 Adsorbents in Industrial Wastewater Treatment

The standard treatment technology for the removal of impurities such as phosphorus from municipal wastewater involves dosing the wastewater with coagulants such as alum, ferric chloride or lime (Metcalf and Eddy, 1991; Parsons and Smith, 2008), but the costs of these reagents often make their use impractical or uneconomic for the low concentrations and high volumes often encountered in most wastewater. Therefore, it is necessary that alternative coagulants are sought. Given the wide range of studies that successfully removed phosphorous from wastewaters using iron and/or aluminium hydroxide sludges, Wei et al. (2008) hypothesized that the AMD sludge containing a mixture of iron and aluminium hydroxide precipitates would be a suitable medium for the adsorption of dissolved orthophosphate from solution. In other words, sludges produced by the neutralization of AMD could be used for phosphorus sequestration as it is composed primarily of aluminium and iron hydrous oxides, the same chemical forms produced when alum or ferric chloride is added to wastewater at near-neutral pH. In addition, research by Sibrell et al. (2009) and Sibrell and Tucker (2012) has shown that dried AMD sludge, or residuals can be used as a low-cost adsorbent to efficiently remove phosphorus from agricultural and municipal wastewaters. The phosphorus that has been adsorbed by the AMD sludge can later be stripped from the sludge and recycled into fertilizer, and the mine drainage sludge can be regenerated and reused for a number of additional applications.

Studies by Edwards and Benjamin (1989) have also shown that the iron ferrihydrite component of AMD sludge from lime treatment plants can be used as a highly effective adsorbent for the removal of metals from water streams. Similarly metal hydroxide sludge has also been used to remove carcinogenic dyes from wastewater (Netpradit et al., 2003).

3.5 Concluding Remarks

Mining and beneficiation operations produce several tonnages of waste annually. This waste is composed of broken rocks from open pits and underground mines; coarse mill rejects from screening and heavy media separation processes and fine mill tailings from screening, magnetic separation, flotation processes, etc. The volumes of waste produced in each category are dependent on the geologic characteristics of an ore body, the type of mining (underground versus open pit) and the metal being mined as well as the size of the mining operation. For example, less mine waste is produced from underground mining operations than from open-pit mining operations. Furthermore, less tailings are produced from underground mining as ore grades are higher and tonnages lower. Some of this waste is inert and consequently unlikely to be a significant environmental hazard apart from smothering river beds and the risk of collapse if stored in large quantities. Other fractions, in particular those generated by the non-ferrous metal mining industry, may contain large quantities of dangerous substances, such as heavy metals. However, the sheer volume and chemical composition of mining and beneficiation waste mean that where it is put and how it is managed is very critical.

This chapter discussed various examples of utilizing mining and beneficiation waste. Basically, the utilization of waste is an attractive alternative to disposal because disposal costs and potential pollution problems are reduced or even eliminated and resources are conserved. Some of the industrial uses of overburden and/or waste rock, for example, include use as an aggregate for asphalt roads, road fill, commercial sand-blasting and anti-skid surfacing of bridges. Other waste such as tailings can be used in the recovery of metals, manufacture of building and construction materials, glass and ceramics.

The chapter also showed that acid generation from the oxidation of sulphide minerals, such as pyrite and pyrrhotite, in mining and beneficiation wastes is one of the most significant issues facing the mining industry. The chapter highlighted some of the basic approaches that are normally employed in the controlling or mitigation of AMD, (1) source control where one of the major components of acid drainage formation is eliminated making the possibility of acid generation very small; (2) migration control where acid drainage can still occur, but measures are implemented that controls the migration of the generated acid and (3) treatment where the acid drainage is intercepted and treated. Measures 2 and 3 are interrelated, and thus are normally used interchangeably.

References

Agrawal, A. and Sahu, K.K. (2009). An overview of the recovery of acid from spent acidic solutions from steel and electroplating industries. *Journal of Hazardous Materials* 171(1–3): 61–75.

Ahmari, S. and Zhang, L. (2012). Production of eco-friendly bricks from copper mine tailings through geopolymerization. *Construction and Building Materials* 29: 323–331.

Ahmari, S. and Zhang, L. (2013). Utilisation of Cement Kiln Dust CKD to enhance mine tailings-based geopolymer bricks. *Construction and Building Materials* 40: 1002–1011.

Ahn, K.H., Song, K.G., Cha, H.Y. and Yeom, I.T. (1999). Removal of ions in nickel electroplating rinse water using low-pressure nanofiltration. *Desalination* 122(1): 77–84.

Ajaka, E.O. (2009). Recovering fine iron minerals from Itakpe Iron ore process tailing ARPN. *Journal of Engineering and Applied Sciences* 4(9): 17–28.

Akcil, A. and Koldas, S. (2006). Acid mine drainage (AMD): Causes, treatment and case studies. *Journal of Cleaner Production* 14: 1139–1145.

Alkhudhiri, A., Darwish, N. and Hilal, N. (2012). Membrane distillation: A comprehensive review. *Desalination* 287: 2–18.

Amponsah-Dacosta, M. and Reid, D.L. (2014). Mineralogical characterization of selected South African mine tailings for the purpose of mineral carbonation. In: Sui, W., Sun, Y. and Wang, C. (eds.),

Annual International Mine Water Association Conference – An Interdisciplinary Response to Mine Water Challenges, China University of Mining and Technology Press, Xuzhou, China, pp. 686–689.

Argane, R., Benzaazoua, M., Hakkou, R. and Bouamrane, A. (2016). A comparative study on the practical use of low sulphide base-metal tailings as aggregates for rendering masonry mortars. *Journal of Cleaner Production* 112: 914–925.

Balakrishnan, M., Batra, V. and Hargreaves, J. (2013). Waste from metal processing industries. In: Hargreaves, J.S.J., Pulford, I.D., Balakrishnan, M. and Batra, V.S. (eds.), *Conversion of Large Scale Waste into Value-Added Products*. CRC Press, Boca Raton, FL, pp. 23–68.

Balintova, M. and Petrilakova, A. (2011). Study of pH influence of selective precipitation of heavy metals from acid mine drainage. *Chemical Engineering Transactions* 25: 1–6.

Banks, D., Younger, P.L., Arnesen, R.T., Iversen, E.R. and Banks, S.B. (1997). Mine-water chemistry: The good, the bad and the ugly. *Environmental Geology* 32: 157–174.

Barradas, S. (2012). Las Lagunas tailings retreatment facility. Available at: http://www.miningweekly.com/article/las-lagunas-tailings-retreatment-facility-dominican-republic-2012-09-07. Accessed 5 February 2016.

Bennet, J.M., Sudhakar, M. and Natarajan, C. (2013). Development of coal ash – GGBS based geopolymer bricks. *European International Journal of Science and Technology* 2(5): 133–139.

Benvenut, T., Rodrigues, M.A., Krapf, R.S., Bernardes, A.M. and Zoppas-Ferreira, J. (2013). Electrodialysis as an alternative for treatment of nickel electroplating effluent: Water and salts recovery. *Fourth International Workshop, Advances in Cleaner Production*, São Paulo, Brazil, 22–24 May.

Bhagwan, J. (2012). Turning acid mine drainage water into drinking water: The eMalahleni water recycling project. Available at http://www.reclaimedwater.net/data/files/247.pdf. Accessed 17 March 2016.

Bian, Z., Miao, X., Lei, S., Chen, S., Wang, W. and Struthers, S. (2012). The challenges of reusing mining and mineral processing wastes. *Science* 337(6095): 702–703.

Blowes, D.W., Ptacek, C.J., Jambor, J.L. and Weisener, C.J. (2003). The geochemistry of acid mine drainage. *Treatise on Geochemistry* 9: 149–204.

Boon, M., Hansford, G.S. and Heijne, J.J. (1995). The role of bacterial ferrous iron oxidation in the bio-oxidation of pyrite. In: Vargas, T., Jerez, C.A., Wertz, J.V. and Toledo, H. (eds.), *Biohydrometallurgical Processing*, Vol. 1. University of Chile, Santiago, Chile, pp. 153–181.

Brown University (2011). Novel device removes heavy metals from water. Available at http://news.brown.edu/pressreleases/2011/12/cep. Accessed 23 March 2016.

Buys, S., Rule, C. and Curry, D. (2004). *The Application of Large Scale Stirred Milling to the Retreatment of Merensky Platinum Tailings*. CIM, Montreal, Canada, pp. 1–15.

Buzzi, D.C., Viegas, L.S., Rodrigues, M.A.S., Bernardes, A.M. and Tenório, J.A.S. (2013). Water recovery from acid mine drainage by electrodialysis. *Minerals Engineering* 4: 82–89.

Caraballo, M.A., Macías, F., Rötting, T.S., Nieto, J.M. and Ayora, C. (2011). Long term remediation of highly polluted acid mine drainage: A sustainable approach to restore the environmental quality of the Odiel river basin. *Environmental Pollution* 159: 3613–3619.

Chattopadhyay, S. and Chattopadhyay, D. (2013). Mining industries and their sustainable management. In: Malhotra, R. (ed.), *Fossil Energy*. Springer, New York.

Chen, Y., Zhang, Y., Chen, T., Zhao, Y. and Bao, S. (2011). Preparation of eco-friendly construction bricks from hematite tailings. *Construction and Building Materials* 25: 2107–2111.

Cheng, S., Dempsey, B.A. and Logan, B.E. (2007). Electricity generation from synthetic acid-mine drainage (AMD) water using fuel cell technologies. *Environmental Science and Technology* 41(23): 8149–8153.

Cheng, S., Jang, J.H., Dempsey, B.A. and Logan, B.E. (2011). Efficient recovery of nano-sized iron oxide particles from synthetic acid-mine drainage (AMD) water using fuel cell technologies. *Water Research* 45(1): 303–307.

Chisakuta, G.C., Banda, M. and Chabinga, M. (2005). Process developments at tailings leach plant. *The Third Southern African Conference on Base Metals*, Kitwe, Zambia, The South African Institute of Mining and Metallurgy, South Africa, 2005, pp. 379–390.

Cifuentes, L., García, I., Arriagada, P. and Casas, J.M. (2009). The use of electrodialysis for metal separation and water recovery from $CuSO_4$–H_2SO_4–Fe solutions. *Separation and Purification Technology* 68(1): 105–108.

Cifuentes, L., García, I., Ortiz, R. and Casas, J.M. (2006). The use of electrohydrolysis for the recovery of sulphuric acid from copper-containing electrolytes. *Separation and Purification Technology* 50: 167–174.

Collins, R.J. and Miller, R.H. (1979). Utilization of mining and mineral processing wastes in The United States. *Minerals and Environment* 1(1): 8–19.

Çoruh, S., Osman, N.E. and Cheng, T. (2006). Treatment of copper industry waste and production of sintered glass-ceramic. *Waste Management and Research* 24: 234–241.

Costello, C. (2003). Acid mine drainage: Innovative treatment technologies. Available at https://brownfieldstsc.org/pdfs/amdinnovativetrttech_03.pdf. Accessed 2 March 2016.

Das, S.K., Ghosh, J., Mandal, A.K., Sing, N. and Gupta, S. (2012). Iron ore tailing: A waste material used in ceramic tile compositions as alternative source of raw materials. *Transactions of the Indian Ceramic Society* 71(1): 21–24.

Das, S.K., Kumar, S. and Ramachandrarao, P. (2000). Exploitation of iron ore tailing for the development of ceramic tiles. *Waste Management* 20(8): 725–729.

da Silva, F.L., Araújo, F.G.S., Teixeiraa, M.P., Gomes, R.C. and von Krüger, F.L. (2014). Study of the recovery and recycling of tailings from the concentration of iron ore for the production of ceramic. *Ceramics International* 40: 16085–16089.

Davidovits, J. (1988). Geopolymers of the first generation: SILIFACE Process. In: Davidovits, J. and Orlinski, J. (eds.), *Proceedings of the First International Conference on Geopolymer '88*, Vol. 1, Compiegne, France, 1–3 June, pp. 49–67.

Dean, K.C., Froisland, L.J. and Shirts, M.B. (1986). Utilization and stabilization of mineral waste. Bulletin 688, Department of Interior Bureau of Mines, United States, Washington DC, Available at http://digital.library.unt.edu/ark:/67531/metadc12830/. Accessed 4 April 2016.

Dinardo, O., Kondos, P.D., MacKinnon, D.J., McCready, R.G.L., Riveros, P.A. and Skaff, M. (1991). Study of metals recovery/recycling from acid mine drainage. MEND Treatment Committee, Mineral Sciences Laboratories, Ottawa, Ontario, Canada. Available at http://mend-nedem.org/wp-content/uploads/2013/01/3.21.1a.pdf. Accessed 7 March 2016.

Dold, B. (2010). Basic concepts in environmental geochemistry of sulphide mine waste management. In: Kumar, E.S. (ed.), *Waste Management*. In-Tech, Rijeka, Croatia, pp. 173–198.

Duarte, J.C., Estrada, P., Beaumont, H., Sitima, M. and Pereira, P. (2006). Biotreatment of tailings for metal recovery. *International Journal of Mine Water* 9(1–4): 193–206.

Edraki, M., Baumgartl, T., Manlapig, E., Bradshaw, D., Franks, D.M. and Moran, C.J. (2014). Designing mine tailings for better environmental, social and economic outcomes: A review of alternative approaches. *Journal of Cleaner Production* 84: 411–420.

Edwards, M. and Benjamin, M. (1989). Regeneration and reuse of iron hydroxide adsorbents in the treatment of metal-bearing wastes. *Journal of the Water Pollution Control Federation* 61(4): 481–490.

Egiebor, N.O. and Oni, B. (2007). Acid rock drainage formation and treatment: A review. *Asia-Pacific Journal of Chemical Engineering* 2: 47–62.

Eligwe, C.A. (1988). Microbial desulphurization of coal. *Fuel* 67: 451–458.

EPA (1994). Background for NEPA reviewers: Non-coal mining operations. Available at https://www.epa.gov/sites/production/files/2014-08/documents/non-coal-mining-background-pg.pdf. Accessed 19 June 2016.

Etter, K. and Langill, P.D. (2006). Acid recycling-process and design notes on hot-dip galvanizing. Available at http://www.galvanizeit.org/images/uploads/memberGalvanizingNotes/Acid Recycling,KurtEtter(GalvanizingNotes,2006Feb).pdf. Accessed 14 March 2016.

Fang, Y., Gu, Y., Kang, Q., Wen, Q. and Dai, P. (2011). Utilization of copper tailing for autoclaved sand–lime brick. *Construction and Building Materials* 25: 867–872.

Fowler, T.A., Holmes, P.R. and Crundwell, F.K. (2001). On the kinetics and mechanism of the dissolution of pyrite in the presence of *Thiobacillus ferrooxidans*. *Hydrometallurgy* 59: 257–270.

Fraser Institute (2012). How are waste materials managed at mine sites? Available at http://www.miningfacts.org/Environment/How-are-waste-materials-managed-at-mine-sites/. Accessed 25 February 2017.

Fripp, J., Ziemkiewicz, P.F. and Charkavorki, H. 2000. Acid mine drainage treatment. EMRRP-SR-14. Available at http://el.erdc.usace.army.mil/elpubs/pdf/sr14.pdf. Accessed 29 March 2016.

Garba, Y., Taha, S., Gondrexon, N. and Dorange, G. (1999). Ion transport modelling through nanofiltration membranes. *Journal of Membrane Science* 160(2): 187–200.

Gazea, B., Adam, K. and Kontopoulos, A. (1996). A review of passive systems for the treatment of acid mine drainage. *Minerals Engineering* 9(1): 23–42.

Glaister, B.J. and Mudd, G.M. (2010). The environmental costs of platinum PGM mining and sustainability: Is the glass half-full or half-empty? *Minerals Engineering* 23(5): 438–450.

Gordon, R.B. (2002). Production residues in copper technological cycles. *Resources, Conservation and Recycling* 36: 87–106.

Grewal, I. (2016). Introduction to mineral processing. Available at http://met-solvelabs.com/library/articles/mineral-processing-introduction. Accessed 11 April 2016.

Günther, P. and Mey, W. (2008). Selection of mine water treatment technologies for the eMalahleni (Witbank) water reclamation project. Available at http://www.ewisa.co.za/literature/files/122%20Gunther.pdf. Accessed 12 March 2016.

Gupta, C.K. (2003). *Chemical Metallurgy: Principles and Practice.* Wiley-VCH Verlag GmbH & Co. KGaA, Weinheim, Germany. 831pp.

Haibin, L. and Zhenling, L. (2010). Recycling utilization patterns of coal mining waste in China. *Resources, Conservation and Recycling* 54: 1331–1340.

Hancock, B.A. and Pon, M.R.L. (1999). Mineral processing/environment, health and safety. *Copper 99, Proceedings of the Fourth International Mineral Processing Conference*, Vol. II, TMS, Warrendale, PA.

Hanford, G.S. and Vargas, T. (2001). Chemical and electrochemical basis of bioleaching processes. *Hydrometallurgy* 59: 135–145.

Hedin, R. (1998). Potential recovery of iron oxides from coal mine drainage. *Nineteenth Annual West Virginia Surface Mine Drainage Task Force Symposium*, West Virginia University, Morgantown, WV.

Hedin, R.S. (2003). Recovery of marketable iron oxide from mine drainage in the USA. *Land Contamination and Reclamation* 11(2): 93–97.

Hedin, R.S., Narin, R.W. and Kleinmann, R.L.P. (1994). Passive treatment of coal mine drainage. Available at http://www.arcc.osmre.gov/resources/impoundments/BoM-IC-9389-PassiveTreatmentofCoalMineDrainage-Hedinetal1993.pdf. Accessed 18 March 2016.

Henderson, R.D. (2013). Minera valle central operation Rancagua, Region VI, Chile 43-101 technical report. Available at http://amerigoresources.com/_resources/reports/tech_report_april_8_2014.pdf. Accessed 3 April 2016.

Hietkamp, S., Engelbrecht, A., Scholes, B. and Golding, A. (2008). The potential for sequestration of carbon dioxide in South Africa. Carbon capture and storage in South Africa. Available at http://playpen.meraka.csir.co.za/~acdc/education/CSIR%20conference%202008/Proceedings/CPO-0027.pdf. Accessed 2 April 2016.

Higginson, S. and Wheeler, A. (2010). Global markets international. Available at http://www.willis.com/documents/publications/industries/Mining_and_Metals/Willis_Mining_Market_Review_2010.pdf. Accessed 18 March 2016.

Hu, J., Chen, G. and Lo, I.M.C. (2006). Selective removal of heavy metals from industrial wastewater using maghemite nanoparticle: Performance and mechanisms. *Journal of Environmental Engineering* 132(7): 709–715.

Huang, X., Ni, W., Cui, W., Wang, Z. and Zhu, L. (2012). Preparation of autoclaved aerated concrete using copper tailings and blast furnace slag. *Construction and Building Materials* 27: 1–5.

Hubicki, Z. and Kołodyńska, D. (2012). Selective removal of heavy metal ions from waters and waste waters using ion exchange methods. In: Kilislioğlu, A. (ed.), *Ion Exchange Technologies*. In-Tech, Rijeka, Croatia.

Implats. (2015). Mineral resources and mineral reserves statement 2015. Available at http://financialresults.co.za/2015/implats-minerals-report-2015/chromium.php. Accessed 9 March 2016.

International Institute for Environment and Development (IIED) (2002). Mining for the future – Appendix A: Large volume waste working paper. Available at http://pubs.iied.org/pdfs/G00883.pdf. Accessed 11 March 2016.

International Network for Acid Prevention (INAP) (2012). Global acid rock drainage guide (GARD Guide). Available at http://www.gardguide.com. Accessed 19 March 2016.

Jain, R. (2015). *Iron Ore Tailings Reprocessing to Maximize Ore Production Iron Ore Conference 2015*, Perth, Western Australia, Australia, 13–15 July 2015. Available at http://www.ironore2015.ausimm.com.au/Media/ironore2015/presentations/S7B%20-%20Jain.pdf. Accessed 22 April 2016.

Jain, S.K., Garg, S.P. and Rai, M. (1983). Autoclaved calcium silicate bricks from gold mine tailings. *Research and Industry* 28: 170–172.

Jambor, J.L. (1994). Mineralogy of sulfide-rich tailings and their oxidation products. In: Jambor, J.L. and Blowes, D.W. (eds.), *Short Course Handbook on Environmental Geochemistry of Sulfide Mine-Waste*. Mineralogical Association of Canada, Nepean, Ontario, Canada, pp. 201–244.

Johnson, D.B. and Hallberg, K.B. (2005). Acid mine drainage remediation options: A review. *Science of the Total Environment* 338: 3–14.

Kalin, M., Fyson, A. and Wheeler, W.N. (2006). The chemistry of conventional and alternative treatment systems for the neutralization of acid mine drainage. *Science of the Total Environment* 366: 395–408.

Kesieme, U.K., Milne, N., Aral, H., Cheng, C.Y. and Duke, M. (2012). Novel application of membrane distillation for acid and water recovery from mining waste waters. In: McCullough, C.D., Lund, M.A. and Wyse, L. (eds.), *International Mine Water Association Symposium 2012: Mine Water and the Environment*, Red Hook, New York.

Kim, K.J. (2006). Purification of phosphoric acid from waste acid etchant using layer melt crystallization. *Chemical Engineering and Technology* 29(2): 271–276.

Kiventerä, J., Golek, L., Yliniemi, J., Ferreira, V., Deja, J. and Illikainen, M. (2016). Utilization of sulphidic tailings from gold mine as a raw material in geopolymerization. *International Journal of Mineral Processing* 149: 104–110.

Kleinmann, R.L.P. (2006). At source control of acid mine drainage. Available at https://www.imwa.info/bibliographie/09_14_085-096.pdf. Accessed 8 March 2016.

Kontopouls, A. (1988). Acid mne drainage control. In: Castro, S.H., Vergara, F. and Sánches, M.A. (eds.), *Effluent Treatment in the Mining Industry*. University of Concepcion, Concepcion, Chile, pp. 57–118.

Koumal, G. (1994). Method of environmental clean-up and producing building material using Cu mine tailing waste material. US patent number US5286427A.

Kumar, B.N.S., Suhas, R., Shet, S.U. and Srishaila, J.M. (2014). Utilization of iron ore tailings as replacement to fine aggregates in cement concrete pavements. *International Journal of Research in Engineering and Technology* 3(7): 369–376.

Kuranchie, F.A., Shukla, S.K. and Habibi, D. (2014). Utilisation of iron ore mine tailings for the production of geopolymer bricks. *International Journal of Mining, Reclamation and Environment* 30(2): 92–114.

Kuranchie, F.A., Shukla, S.K., Habibi, D. and Mohyeddin, A. (2015). Utilisation of iron ore tailings as aggregates in concrete. *Cogent Engineering* 2: 1–11.

Kurniawan, T.A., Chan, G.Y.S., Lo, W.H. and Babel, S. (2006). Physico-chemical treatment techniques for wastewater laden with heavy metals. *Journal of Chemical Engineering* 118: 83–98.

Kuyucak, N. (1999). Acid mine drainage prevention and control options. Available at http://www.imwa.info/docs/imwa_1999/IMWA1999_Kuyucak_599.pdf. Accessed 6 March 2016.

Kuyucak, N. (2002). Acid mine drainage prevention and control options. *Canadian Institute of Mining and Metallurgy Bulletin* 95(1060): 96–102.

Kwong, Y.T.J. and Ferguson, K.D. (1990). Water chemistry and mineralogy at Mt. Washington: Implications to acid generation and metal leaching. In: Gadsby, J.W., Malick, J.A. and Day, S.J. (eds.), *Acid Mine Drainage: Designing for Closure*. Bitech Publishers, Vancouver, British Columbia, Canada, pp. 217–230.

Lackner, K.S., Wendt, C.H., Butt, D.P., Joyce, E.L. Jr and Sharp, D.H. (1995). Carbon dioxide disposal in carbonate minerals. *Energy* 20: 1153–1170.

Lapakko, K. (2002). Metal mine rock and waste characterisation tools: An overview. Available at http://pubs.iied.org/pdfs/G00559.pdf. Accessed 17 March 2016.

Levlin, E. (1998). Sustainable sludge handling – Metal removal and phosphorus recovery. *Proceedings of a Polish-Swedish Seminar*, Nowy Targ, Poland, 1–2 October, pp. 73–82.

Lewis, A.E. (2010). Review of metal sulphide precipitation. *Hydrometallurgy* 104: 222–234.

Li, C., Sun, H., Bai, J. and Li, L. (2010a). Innovative methodology for comprehensive utilization of iron ore tailings. Part 1. The recovery of iron from iron ore tailings using magnetic separation after magnetizing roasting. *Journal of Hazardous Materials* 174: 71–77.

Li, C., Sun, H., Bai, J. and Li, L. (2010b). Innovative methodology for comprehensive utilization of iron ore tailings. Part 2: The residues after iron recovery from iron ore tailings to prepare cementitious material. *Journal of Hazardous Materials* 174: 78–83.

Liu, W., Chen, X., Li, W., Yu, Y. and Yan, K. (2014). Environmental assessment, management and utilization of red mud in China. *Journal of Cleaner Production* 84: 606–610.

Lubarski, V., Levlin, E. and Koroleva, E. (1996). Endurance test of aluminous cement produced from water treatment sludge. *Vatten* 52(1): 39–42.

Luo, L., Zhang, Y., Bao, S. and Chen, T. (2016). Utilization of iron ore tailings as raw material for portland cement clinker production. *Advances in Materials Science and Engineering* 2016: 1–6. http://dx.doi.org/10.1155/2016/1596047

Luptakova, A., Balintova, M., Jencarova, J., Macingova, E. and Prascakova, M. (2010). Metals recovery from acid mine drainage. *Nova Biotechnologica* 10(1): 23–32.

Lutandula, M.S. and Maloba, B. (2013). Recovery of cobalt and copper through reprocessing of tailings from flotation of oxidised ores. *Journal of Environmental and Chemical Engineering* 1(4): 1085–1090.

Luther, I.G.W. (1987). Pyrite oxidation and reduction: Molecular orbital theory considerations. *Geochimica et Cosmochimica Acta* 51: 3193–3199.

Macingova, E. and Luptakova, A. (2012). Recovery of metals from acid mine drainage. *Chemical Engineering Transactions* 28: 109–114.

Magnetation LLC (2016). Iron ore concentrate. Available at https://www.magnetation.com/. Accessed 24 April 2016.

Malatse, M. and Ndlovu, S. (2015). The viability of using the Witwatersrand gold mine tailings for brickmaking. *Journal of the Southern African Institute of Mining and Metallurgy* 115(4): 321–327.

Manders, P., Godfrey, L. and Hobbs, P. (2009). Acid mine drainage in South Africa: Briefing note 2009/02. Available at http://www.csir.co.za/nre/docs/BriefingNote 2009_2_AMD_draft.pdf. Accessed 12 March 2016.

Marcello, R.R., Galato, S., Peterson, M., Riella, H.G. and Bernardin, A.M. (2008). Inorganic pigments made from the recycling of coal mine drainage treatment sludge. *Environmental Management* 88: 1280–1284.

Marghussian, V.K. and Maghsoodipoor, A. (1999). Fabrication of unglazed floor tiles containing Iranian copper slag. *Ceramics International* 25: 617–622.

McKibben, M.A. and Barnes, H.L. (1986). Oxidation of pyrite in low temperature acidic solutions: Rate laws and surface textures. *Geochimica et Cosmochimica Acta* 50: 1509–1520.

McKinney W.S. and Graves R.D. (1977). New copper hydrometallurgy practices and processes. In: Kislali A.S. (ed.), *Cento Symposium on Mining Methods, Beneficiation and Smelting of Copper Ores*, Ankara, Turkey.

Means, B., McKenzie, B. and Hilton, T. (2003). A computer-based model for estimating mine drainage treatment costs. Available at https://wvmdtaskforce.files.wordpress.com/2016/01/03-means03.pdf. Accessed 6 March 2016.

Metcalf & Eddy Inc (1991). *Wastewater Engineering – Treatment, Disposal, and Reuse.* McGraw-Hill, Singapore, Singapore.

Metso (2016). Basics in minerals processing. Available at http://www.metso.com/miningandconstruction/MaTobox7.nsf/DocsByID/EAE6CA3B8E216295C2257E4B003FBBA6/$File/Basics-in-minerals-processing.pdf. Accessed 30 April 2016.

Meyer, N.A., Vögeli, U.J., Becker, M., Broadhurst, J.L. and Franzidis, J.-P. (2014). Carbonation of PGM mine tailings for CO_2 storage in South Africa: A case study. *Minerals Engineering* 59: 45–51.

Michalkova, E., Schwarz, M., Pulisova, P., Masa, B. and Sudovsky, P. (2013). Metals recovery from acid mine drainage and possibilities for their utilization. *Polish Journal of Environmental Studies* 22(4): 1111–1118.

Mohale, S., Masetlana, T.R., Bonga, M., Ikaneng, M., Dlambulo, N., Malebo, L. and Mwape, P. (eds.) (2015). *South Africa's Mining Industry 2013–2014*. Department of Mineral Resources, Pretoria, South Africa, p. 276.

Moses, C.O., Nordstrom, D.K., Herman, J.S. and Mills, A.L. (1987). Aqueous pyrite oxidation by dissolved oxygen and ferric iron. *Geochimica et Cosmochimica Acta* 1: 1561–1571.

Mueller, H., Maithy, S., Prajapati, S., Bhatta, A.D. and Shrestha, B.L. (2008). *Greenbrick Making Manual*. Hillside Press, Kathmandu, Nepal. Available at http://www.ecobrick.in/resource_data/KBAS100046.pdf. Accessed 27 May 2016

Nagpal, G., Bhattacharya, A. and Singh, N.B. (2013). Removal of copper (II) from aqueous solution by Khangar. *International Journal of Engineering Research and Technology* 2(12): 3624–3631.

Nesterenko, P.N. and Haddad, P.R. (2000). Zwitterionic ion-exchangers in liquid chromatography. *Analytical Sciences* 16(6): 565–574.

Netpradit, S., Thiravetyan, P. and Towprayoon, S. (2003). Application of 'waste' metal hydroxides for adsorption of azo reactive dyes. *Water Research* 37: 763–772.

Nummi, E. (2015). From tailings to treasure? A new mother lode: Miners make money reprocessing tailings. Available at https://www.thermofisher.com/blog/mining/from-tailings-to-treasure-a-new-mother-lode/. Accessed 20 April 2016.

Nleya, Y., Simate, G.S. and Sehliselo, N. (2016). Sustainability assessment of the recovery and utilisation of acid from acid mine drainage. *Journal of Cleaner Production* 113: 17–27.

Obreque-Contreras, J., Pérez-Flores, D., Gutiérrez, P. and Chávez-Crooker, P. (2015). Acid mine drainage in Chile: An opportunity to apply bioremediation technology. *Hydrology Current Research* 6(3): 1–8.

Odendaal, N. (2014). Jubilee to build independent chrome, PGM processing plant. Available at http://www.miningweekly.com/article/jubilee-to-build-independent-chrome-pgm-processing-plant-2014-06-09. Accessed 15 March 2016.

Oluwasola, E.A., Hainin, M.R., Aziz, M.M.A., Yaacob, H. and Warid, M.N.M. (2014). Potential of steel slag and copper mine tailings construction materials. *Materials Research Renovation* 18: 250–254.

Onuaguluchi, O. and Eren, O. (2012a). Copper tailings as a potential additive in concrete: Consistency, strength and toxic metal immobilization properties. *Indian Journal of Engineering and Materials Sciences* 19(2): 79–86.

Onuaguluchi, O. and Eren, O. (2012b). Cement mixtures containing copper tailings as an additive: Durability properties. *Materials Research* 15(6): 1029–1036.

Onuaguluchi, O. and Eren, O. (2016). Reusing copper tailings in concrete: Corrosion performance and socioeconomic implications for the Lefke-Xeros area of Cyprus. *Journal of Cleaner Production* 112: 420–429.

Özdemir, T., Öztin, C. and Kincal, N.S. (2006). Treatment of waste pickling liquors: Process synthesis and economic analysis. *Chemical Engineering Communications* 193: 548–563.

Parsons, S.A. and Smith, J.A. (2008). Phosphorus removal and recovery from municipal wastewaters. *Elements* 4: 109–112.

Pozo-Antonio, S., Puente-Luna, I., Lagüela-López, S. and Veiga-Ríos, M. (2014). Techniques to correct and prevent acid mine drainage: A review. *Dyna* 81(184): 73–80.

Praes, P.E., de Albuquerque, R.O. and Luz, A.F.O. (2013). Recovery of iron ore tailings by column flotation. *Journal of Minerals and Materials Characterization and Engineering* 1: 212–216.

Prahallada, M.C. and Shanthappa, B.C. (2014). Use of copper ore tailings – As an excellent pozzolana in the preparation of concrete. *International Journal of Advanced Research in Engineering and Applied Sciences* 3(3): 1–10.

Rajaram, V., Glazer, A. and Coghlan, G. (2001). Methodology for estimating the costs of treatment of mine drainage. Available at http://www.maden.org.tr/resimler/ekler/ee1bc7fa5da061b_ek.pdf. Accessed 3 March 2016.

Rao, G.P., Lu, C. and Su, F. (2007). Sorption of divalent metal ions from aqueous solution by carbon nanotubes: A review. *Separation and Purification Technology* 58: 224–231.

Ritchie, A.I.M. (1994). Sulfide oxidation mechanisms: Controls and rates of oxygen transport. In: Jambor, J.L. and Blowes, D.W. (eds.), *Short Course Handbook on Environmental Geochemistry of Sulfide Mine-Waste*. Mineralogical Association of Canada, Nepean, Ontario, Canada, pp. 201–244.

Rodrigues, M.A.S., Amado, F.D.R., Xavier, J.L.N., Streit, K.F., Bernardes, A.M. and Zoppas-Ferreira, J. (2008). Application of photoelectrochemical-electrodialysis treatment for the recovery and reuse of water from tannery effluents. *Journal of Cleaner Production* 16: 605–611.

Roy, S., Adhikari, G.R. and Gupta, R.N. (2007). Use of gold mill tailings in making bricks: A feasibility study. *Waste Management Research* 25: 475–482.

Ruihua, L., Lin, Z., Tao, T. and Bo, L. (2011). Phosphorus removal performance of acid mine drainage from wastewater. *Journal of Hazardous Materials* 190: 669–676.

Sakthivel, R., Vasumanthi, N., Sahu, D. and Mishra, B.K. (2010). Synthesis of magnetite powder from iron ore tailings. *Powder Technology* 201: 187–190.

Sameera, M., vander Merwe, E.M., Alterman, W. and Doucet, F.J. (2016). Process development for elemental recovery from PGM tailings by thermochemical treatment: Preliminary major extraction studies using ammonium sulphate as extracting agent. *Waste Management* 50: 334–345.

Sampaio, R.M.M., Timmers, R.A., Xu, Y., Keesman, K.J. and Lens, P.N.L. (2009). Selective precipitation of Cu from Zn in a pS controlled continuously stirred tank reactor. *Journal of Hazardous Materials* 165: 256–265.

Sand, W., Gehrke, T., Jozsa, P.-G. and Schippers, A. (2001). Bio(chemistry) of bacterial leaching – Direct versus indirect bioleaching. *Hydrometallurgy* 51: 115–175.

Saxena, M. and Dhimole, L.K. (2016). Utilisation and value addition of copper tailing as an extender for development of paints. *Journal of Hazardous Materials B* 129: 50–57.

Schippers, A., Jozsa, P.G. and Sand, W. (1996). Sulphur chemistry in bacterial leaching of pyrite. *Applied and Environmental Microbiology* 62(9): 3424–3431.

Schippers, A. and Sand, W. (1999). Bacterial leaching of metal sulphides proceeds by two indirect mechanism via thiosulphate or via polysulphides and sulphur. *Applied and Environmental Microbiology* 65: 319–321.

Sheoran, A.S. and Sheoran, V. (2006). Heavy metal removal mechanism of acid mine drainage in wetlands: A critical review. *Minerals Engineering* 19: 105–116.

Sheoran, V., Sheoran, A.S. and Tholia, N.K. (2011). Acid mine drainage: An overview of Indian mining industry. *International Journal of Earth Sciences and Engineering* 4(6): 1075–1086.

Shetty, K.K., Nayak, G. and Vijayan, V. (2014). Use of red mud and iron tailings in self compacting concrete. *International Journal of Research in Engineering and Technology* 3: 111–114.

Shirazi, M.M.A., Kargari, A. and Tabatabaei, M. (2014a). Evaluation of commercial PTFE membranes in desalination by direct contact membrane distillation. *Chemical Engineering and Processing* 76: 16–25.

Shirazi, M.M.A., Kargari, A., Tabatabaei, M., Ismail, A.F. and Matsuura, T. (2014b). Assessment of atomic force microscopy for characterization of PTFE membranes for membrane distillation (MD) process. *Desalination and Water Treatment* 2014: 1–10.

Sibrell, P.L., Montgomery, G.A., Ritenour, K.L. and Tucker, T.W. (2009). Removal of phosphorus from agricultural wastewaters using adsorption media prepared from acid mine drainage sludge. *Water Research* 43(8): 2240–2250.

Sibrell, P.L. and Tucker, T.W. (2012). Fixed bed sorption of phosphorus from wastewater using iron oxide-based media derived from acid mine drainage. *Water, Air and Soil Pollution* 223: 5105–5117.

Siliezar, J., Stoll, D. and Twomey, J. (2011). Unlocking the value in waste and reducing tailings: Magnetite production at Ernest Henry Mining. *Iron Ore Conference*, Perth, Western Australia, Australia, 2011. Available at http://www.isamill.com/EN/Downloads/Downloaded%20Technical%20Papers/Ernest%20Henry%20Iron%20Ore%202011%20Paper.pdf. Accessed 13 March 2016.

Simate, G.S. (2012). The treatment of brewery wastewater using carbon nanotubes synthesized from carbon dioxide carbon source. PhD thesis. University of the Witwatersrand, Johannesburg, South Africa.

Simate, G.S. (2016). The use of carbon nanotubes in the treatment of water and wastewater. In: Thakur, V.J. and Thakur, M.K. (eds.), *Chemical Functionalization of Carbon Nanomaterials: Chemistry and Applications*. CRC Press, Boca Raton, FL.

Simate, G.S., Iyuke, S.E., Ndlovu, S. and Heydenrych, M. (2012). The heterogeneous coagulation and flocculation of brewery wastewater using carbon nanotubes. *Water Research* 46(4): 1185–1197.

Simate, G.S. and Ndlovu, S. (2014). Acid mine drainage: Challenges and opportunities. *Journal of Environmental Chemical Engineering* 2: 1785–1803.

Skousen, J. (1995). Acid mine drainage. *Green Lands* 25(2): 52–55.

Skousen, J., Rose, A., Geidel, G., Foreman, J., Evans, R. and Hellier, W. (1998). *Handbook of Technologies for Avoidance and Remediation of Acid Mine Drainage*. The National Mine Land Reclamation Centre, West Virginia University, Morgantown, WV.

Skousen, J.G., Sextone, A. and Ziemkiewicz, P.F. (2000). Acid mine drainage control and treatment. In: Barnhisel, R.I., Darmody, R.G. and Daniels, L. (eds.), *Reclamation of Drastically Disturbed lands*, Agronomy Monograph Number 41, Madison, WI, pp. 131–168.

Smith, J.W. and Comrie, D.C. (1988). Geopolymeric building materials in third world countries. In: Davidovits, J. and Orlinski, J. (eds.), *Proceedings of the First International Conference Geopolymer '88*, Vol. 1, Compiegne, France, 1–3 June 1988, pp. 89–92.

Stanković, S., Morić, I., Pavić, A., Vojnović, S., Vasiljević, B. and Cvetković, V. (2015). Bioleaching of copper from samples of old flotation tailings (Copper Mine Bor, Serbia). *Journal of the Serbian Chemical Society* 80(3): 391–405.

Struthers, S. (1999). Constructive recycling of mine wastes. *Proceedings of the Fourth International Conference on Tailings and Mines Waste*, Fort Collins, CO, 1999, pp. 17–25.

Stumm, W. and Morgan, J.J. (1996). *Aquatic Chemistry: Chemical Equilibria and Rates in Natural Waters*. Wiley-Interscience, New York.

Sylvania Platinum Report. (2016). Available at http://www.sylvaniaplatinum.com/b/b_i.php. Accessed 21 April 2016.

Tay, J.H. and Show, K.-Y. (1994). Municipal wastewater sludge as cementitious and blended cement materials. *Cement and Concrete Composites Journal* 16(1): 39–48.

Thomas, B.S., Damare, A. and Gupta, R.C. (2013). Strength and durability characteristics of copper tailing concrete. *Construction and Building Materials* 48: 894–900.

Tilton J.E. and Landsberg H.H. (1997). Innovation, productivity growth, and the survival of the US copper industry. Discussion Paper 97–41. Washington DC: Resources for the Future. 32pp.

Tjus, K., Bergström, R., Fortkamp, U., Forsberg, K. and Rasmuson, A. (2006). Development of a recovery system for metals and acids from pickling baths using nanofiltration and crystallization. Available at http://www3.ivl.se/rapporter/pdf/B1692.pdf. Accessed 28 March 2016.

Tomaszewska, M. (2000). Membrane distillation – Examples of applications in technology and environmental protection. *Polish Journal of Environmental Studies* 9(1): 27–36.

Trumm, D.A. (2008). Selection of active treatment systems for acid mine drainage. Available at http://www.crl.co.nz/downloads/geology/2008/Trumm%202008%20AusIMM%20Active%20Systems.pdf. Accessed 10 March 2016.

Trumm, D.A. (2010). Selection of active and passive treatment systems for AMD flow charts for New Zealand conditions. *New Zealand Journal of Geology and Geophysics* 53(23): 195–210.

Ugama, T.I., Ejeh, S.P. and Amartey, D.Y. (2014). Effect of iron ore tailing on the properties of concrete. *Civil and Environmental Research* 6: 7–13.

Vaclav, P. and Eva, G. (2005). Desalting of acid mine drainage by reverse osmosis method – Field tests. Available at https://www.imwa.info/docs/imwa_2005/IMWA2005_052_Pisa.pdf. Accessed 23 March 2016.

Van Niekirk, H.J. and Viljoen, M.J. (2005). Causes and consequences of the Merriespruit and other tailings – Dam failures. *Land Degradation and Development* 16(2): 201–212.

Vogeli, J., Reid, D.L., Becker, M., Broadhurst, J. and Franzidis, J.P. (2011). Investigation of the potential for mineral carbonation of PGM tailings in South Africa. *Minerals Engineering* 24: 1348–1356.

Wei, X., Viadero, R.C. and Bhojappa, S. (2008). Phosphate removal by acid mine drainage sludge from secondary effluents of municipal wastewater treatment plants. *Water Research* 42(13): 3275–3284.

White, L. (1979). Zambia. *Engineering Minerals Journal* 180: 146–159.

Wilkins, B. (2013). The changing faces of tailings re-treatment. *Modern Mining* August 2013: 58–63.

Williams, L. (2010). Teniente tailings provide a good revenue stream. *International Mining* January 2010: 58–60.

Wills, B. and Napier-Munn, T.J. (2006). *Mineral Processing Technology: An Introduction to the Practical Aspects of Ore Treatment and Mineral Recovery*, 7th edn. Butterworth-Heinemann (Elsevier), Burlington, MA, 456pp.

Wisniewski, J. and Wisniewska, G. (1999). Water and acid recovery from the rinse after metal etching operations. *Hydrometallurgy* 53(2): 105–119.

Yellishetty, M. (2008). Reuse of iron ore mineral wastes in civil engineering constructions: A case study. *Resources, Conservation and Recycling* 52(11): 1283–1289.

Younger, P.L., Banwart, S.A. and Hedin, R.S. (2002). *Mine Water: Hydrology, Pollution, Remediation*. Kluwer Academic Press, Dordrecht, the Netherlands.

Zdun, T. (2001). Modelling the hydrodynamics of collie mining void 5B. MSc dissertation. University of Western Australia, Perth, Western Australia, Australia.

Zhang, L. (2013). Production of bricks from waste materials – A review. *Construction and Building Materials* 47: 643–655.

Zhang, W. (2003). Nanoscale iron particles for environmental remediation: An overview. *Journal of Nanoparticle Research* 5: 323–332.

Zhao, Y., Zhang, Y., Chen, T., Chen, Y. and Bao, S. (2012). Preparation of high strength autoclaved bricks from hematite tailings. *Construction and Building Materials* 28: 450–455.

Zhong, C.M., Xu, Z.L., Fang, X.H. and Cheng, L. (2007). Treatment of acid mine drainage (AMD) by ultra-low-pressure reverse osmosis and nanofiltration. *Environmental Engineering Science* 24(9): 1297–1306.

Ziemkiewicz, P.F., Skousen, J.G., Brant, D.L., Sterner, P.L. and Lovett, R.J. (1997). Acid mine drainage treatment with armoured limestone in open limestone channels. *Journal of Environmental Quality* 26: 1017–1024.

4

Ferrous Metals Waste Production and Recycling

4.1 Introduction

Steel is one of the most ubiquitous commodities used in modern day society. It is the cornerstone and key driver of the global economy with irreplaceable applications in construction, mining, transport and energy sectors. According to the World Steel Association (2012), about 215 kg of steel was used per capita in the world in 2011, and over two million people were employed in the steel industry directly. Despite the current commodity slump being experienced in the global market, the global crude steel production reached 1.662 billion tonnes in 2014, signifying a 1.2% increase from 2013 (World Steel Association, 2012, 2015a). Generally, the growth in the global iron and steel industry is optimistic, mostly due to increased demand for infrastructure as a result of increased urbanization. Figure 4.1 summarizes the global crude steel production in the period from 1970 to 2014 (World Steel Association, 2012, 2015a).

The crude steel produced from the conventional iron and steel making processes is further refined to produce stainless steels and and other special grades of steels for applications in various industry sectors. The global stainless steel production exceeded 40 million metric tonnes (Mt) in 2014, at an estimated compounded annual growth of 6% per annum (ISSF, 2015). Actually, stainless steel exhibited a more robust growth of 5.44% compounded annual growth rate (CAGR) in the period 1980–2014 compared to other major engineering metals such as carbon steel (2.46%), aluminium (3.71%), zinc (2.35%), copper (2.80%) and lead (2.16%) (ISSF, 2015). In essence, the growth and demand in stainless steel products is expected to remain bullish in the long term, buoyed by global mega-trends in specialized end uses, such as infrastructure, transportation, automobiles, chemicals and those in the petrochemical and energy sectors (Outokumpu, 2013; ISSF, 2015). The long-term growth of, and demand in, stainless steel products is expected to remain bullish, buoyed by global mega-trends in specialized end uses, such as infrastructure, transportation, automobiles, chemicals and those in the petrochemical and energy sectors (Outokumpu, 2013). It must be noted that though the sustained global growth in specialized stainless steel products provides growth opportunities for global producers, at the same time, a considerable amount of pressure is exerted in terms of high grade raw materials supply, process optimization, waste mitigation and holistic environmental management practices. Therefore, mitigating the supply chain risks has become a top priority for global steel producers, and one of the most feasible action plans is to find alternative sources of high grade resources by tapping into metallurgical wastes as secondary sources of raw materials.

The focus of this chapter is to (1) provide an overview of the unit processes in the smelting and refining of iron and steel, (2) highlight the types and categories of metallurgical wastes produced in the respective unit processes, (3) provide an overview of the recycling

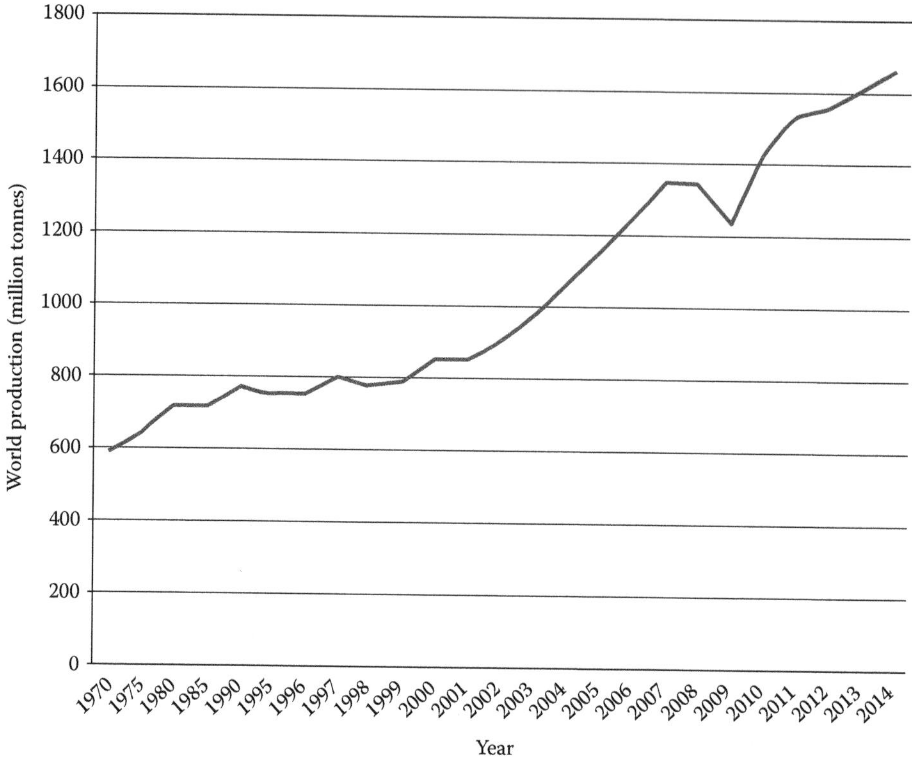

FIGURE 4.1
Global crude steel production. (From World Steel Association, Press Release, 2015a, Available at https://www.worldsteel.org/media-centre/press-releases/2015/World-crude-steel-output-increases-by-1.2–in-2014.html, Accessed 20 November 2015, p. 4.)

and/or reuse opportunities of the various metallurgical wastes and (4) discuss the current practices and trends in terms of waste reduction, recycling and reuse in the iron and steel sector. Finally, the chapter discusses the typical challenges and binding constraints associated with such waste valorization measures.

4.2 Production of Iron and Steel

Most of the steel consumed globally is produced via two main routes, *viz.* the integrated smelting and refining process involving blast furnace and basic oxygen furnace processes (BOFs) and the electric arc-basic oxygen furnace (EAF-BOF) smelting and refining route (World Steel Association, 2012). In the first process route, the blast furnace is the source of a significant amount of the hot metal charged into the BOF with a small contribution coming from the direct charging of steel scrap and other forms of metallic charge in the BOF. On the other hand, the electric arc-basic oxygen process utilizes mostly solid steel scrap and other forms of metallic charges, which are melted in electric furnaces.

On average, the integrated blast furnace process requires about 1400 kg of iron ore, 800 kg of metallurgical grade coal, 300 kg of limestone and 120 kg of steel scrap to produce

1000 kg of hot metal (World Steel Association, 2012). Typically, the hot metal produced from the blast furnace process contains about 4–4.5 wt.% carbon, 0.15–2.5 wt.% manganese, 0.5–4 wt.% silicon, less than 0.2 wt.% phosphorus and 0.025–0.25 wt.% sulphur (World Steel Association, 2012). On the other hand, the electric arc-basic oxygen process route requires, on average, 880 kg of steel scrap, 300 kg of iron metallics, 16 kg of coal and 64 kg of limestone per ton of hot metal produced to produce steel suitable for further processing in downstream refining processes (World Steel Association, 2012).

In general, the chemistry of hot metal produced from the blast furnace and the electric arc processes varies greatly depending on the quality of raw materials, composition of steel scrap and other metallic additives. In addition, the choice of the process option between the blast furnace integrated process and electric arc furnace-basic oxygen process routes also depends on factors such as raw material availability, raw material quality, cost and availability of energy, existing shop logistics, capital and market economics and environmental regulations.

Section 4.2 covers the production of iron and steel and is divided into six sections, namely, (1) the blast furnace ironmaking process (Section 4.2.1), (2) alternative ironmaking processes (Section 4.2.2), (3) the electric arc steelmaking process (Section 4.2.3), (4) the basic oxygen steelmaking process (Section 4.2.4), (5) the stainless steel refining process (Section 4.2.5) and, finally, (6) the solidification and casting of steel (Section 4.2.6).

4.2.1 Blast Furnace Ironmaking Process

The blast furnace is still the principal route for the production of liquid iron, contributing a significant amount of the global iron production (World Steel Association, 2012). The preference for the blast furnace process is due to its superior heat and mass transfer efficiencies and the high productivity levels achieved under continuous steady state operation, and the century-long operation and continual optimization of the process has resulted in highly efficient large-scale operating facilities (Biswas, 1981; Sano et al., 1997). In the integrated iron and steelmaking process, the blast furnace unit process is integrated with the basic oxygen steelmaking, stainless steel refining, solidification and casting processes and metal finishing processes as shown in Figure 4.2.

In essence, the blast furnace is a countercurrent gas–solid-liquid reactor in which iron oxides are reduced into iron, the iron and gangue are melted and the liquid metal is separated from gangue (Biswas, 1981; Sano et al., 1997). The main raw materials in the blast furnace process include lumpy iron ore containing about 63%–65% T-Fe; sinter; pellets; coke; fluxes (limestone and dolomite); and to some extent pulverized coal, oxide wastes and metallic reverts (Biswas, 1981; Sano et al., 1997). The solid charge materials are charged in alternate layers at the top of the furnace through a carefully designed charging sequence.

During descent in the blast furnace shaft through the different physicochemical zones of the furnace, the iron oxides are reduced and subsequently melted, and the liquid iron separates from the slag in the hearth (Biswas, 1981). The liquid slag and hot metal, superheated above their respective melting points, are subsequently tapped from the furnace at temperatures around 1480°C–1550°C (Biswas, 1981; Sano et al., 1997). The thermal energy and reducing potential requirement for the blast furnace process is supplied from the combustion of coke and other fuel additives in the tuyères zone. The combustion of coke in the temperature range of 2000°C–2200°C in the tuyères zone produces a gas mixture comprising of CO, H_2 and N_2, which then take part in the chemical reduction of ferrous charge materials, as well as in the gas-solids heat transfer (Sano et al., 1997). Table 4.1 shows the physicochemical reactions and conditional raw materials requirements in the blast furnace.

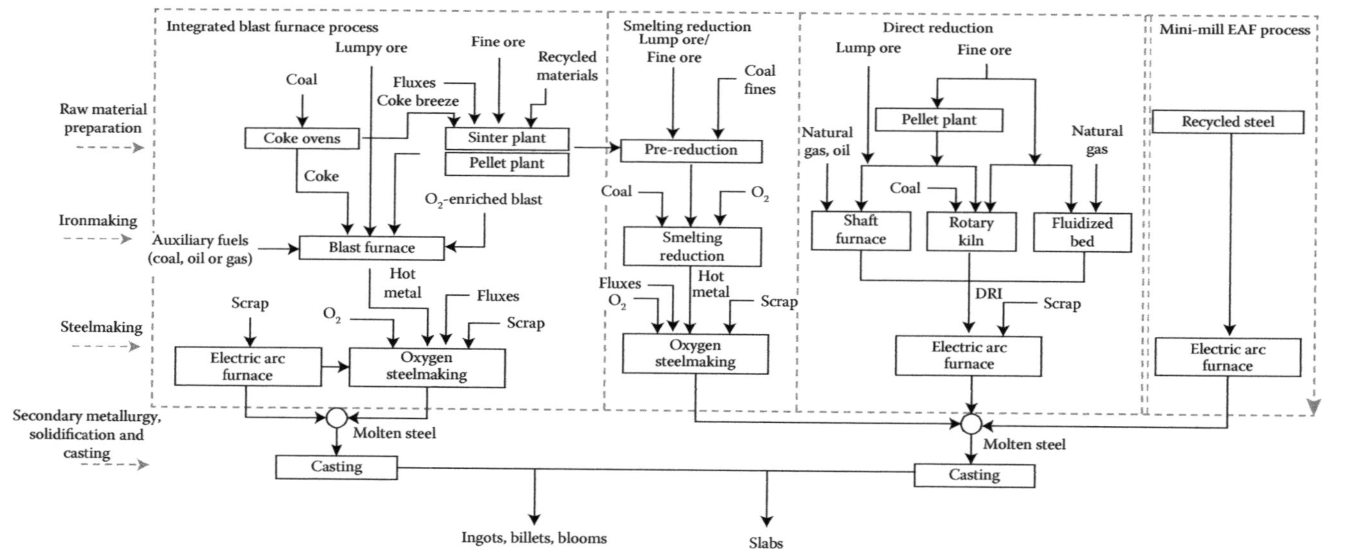

FIGURE 4.2
Schematic representation of the integrated iron and steel process. (From Remus, R. et al., Best available techniques (BAT) reference document for iron and steel production, Industrial Emissions Directive 2010/75/EU, European Commission JRC Reference Report, 2013, p. 9.)

TABLE 4.1
Physicochemical Reactions and Conditional Raw Materials Requirements in the Blast Furnace

Zone	Reactions	Raw Materials and Quality Requirements		
		Sinter and Pellets	**Coke**	**Conditional Requirements**
Granular zone	• Charging, drying, preheating	• Size uniformity • Tumble index • Compressive strength	• Size consistency • Mechanical stability	• Low fines ratio • High compressive strength • Low reduction degradation index
	$3Fe_2O_3 + CO = 2Fe_3O_4 + CO_2$	• Low temp. disintegration		
	$Fe_3O_4 + CO = 3FeO + CO_2$ $FeO + CO = Fe + CO_2$ $C + CO_2 = 2CO$	• Reducibility • Swelling • Compression strength after reduction	• Coke reactivity • Coke Strength after Reaction (CSR) • High temperature strength	• Low swelling index • High softening temperature • High reducibility
Cohesive zone	$FeO + CO = Fe + CO_2$ $C + CO_2 = 2CO$ • Gas-metal reactions	• Softening and Melting • Direct reduction		• High permeability • High reducibility • High temperature of pressure loss elevation and meltdown temperature
Active coke zone	a. Gas-metal reactions			
Stagnant coke zone	b. Gas-metal and slag-metal reactions			• Small slag volume
Raceway	$2C + O_2 = 2CO$ $H_2O + C = H_2 + CO$		• Combustibility	• High slag fluidity

4.2.1.1 Physicochemical Functions of the Blast Furnace

The blast furnace process is a very complex process involving multiple phases which undergo various chemical and physical phenomena. The basic metallurgical functions of the blast furnace include (Biswas, 1981; Sano et al., 1997):

1. Chemical reduction of iron oxides, combustion and gasification of carbon, melting and separation of slag and metal and the refining of hot metal through slag–metal reactions.
2. Heat transfer through exchange of heat between gas, solids and liquids as the charge material descends down the shaft towards the hearth. The heat requirement, especially in the cohesive zone and the bosh sections of the furnace, affects the overall coke consumption of the process. In essence, the cohesive zone extends from the bosh to the lower shaft and is affected by parameters such as the coke quality (mechanical strength and abrasion resistance), burden mechanical properties, the extent of indirect reduction (FeO/Fe ratio in the charge) and other parameters, such as charge composition and melting characteristics.
3. Aerodynamic function of maintaining the permeability of the charge materials. The heat and mass transfer occur mainly in a granular bed, and therefore, the efficient operation of the blast furnace is strongly connected to the fluid–particulates flow (Biswas, 1981; Wright et al., 2011).

The mechanical flow of gases in the blast furnace depends on factors such as the resistance encountered by the ascending gases, while the chemical and thermal utilization potential depends on the uniformity, intimacy and time of contact with the charge (Biswas, 1981). In other words, the optimum gas distribution and gas utilization through the blast furnace charge, itself a function of the charge permeability enhanced by properly controlled quality of raw materials, is considered an important parameter in the efficient blast furnace operation. For example, a non-uniform distribution or high gas flow near the walls may result in the overheating of the blast furnace walls, leading to increased rate of refractory erosion and heat losses (Biswas, 1981). On the other hand, a high gas flow through the central region resulting from channelled gas flow through the solid charge may reduce the chemical and thermal utilization efficiencies of the ascending countercurrent blast furnace gases (Biswas, 1981).

The countercurrent conditions account for the outstanding thermal and chemical efficiency of the blast furnace process. The reducing gases rich in CO produced from the combustion of coke and to some extent auxiliary fuels, such as pulverized coal at the tuyéres zone, ascend through the descending blast furnace charge, exchange thermal and chemical enthalpies with the burden and impart the necessary reducing power (Biswas, 1981; Sano et al., 1997; Wright et al., 2011). The gas distribution and gas velocity profile depends on the charge distribution and shape of the cohesive zone, and these factors affect the overall thermal efficiency and utilization of the reduction potential of the blast furnace gas (Biswas, 1981; Sano et al., 1997).

Several models governing the efficient blast furnace operation have widely been accepted within the scientific community, and in particular, the graphical mass balance models proposed by Rist and Meysson (Rist and Meysson, 1966; Biswas, 1981; Sano et al., 1997). These models were developed to elucidate the complex physicochemical phenomena experienced in the blast furnace and clearly emphasize the blast furnace process's inflexibility in terms of raw material specifications and the need for a strict control of the operating regime within the furnace.

4.2.1.2 Raw Materials for the Blast Furnace

Iron ore, sinter, pellets and metallurgical coke: The main raw materials include lumpy iron ore (63%–65% Fe), sinter (58% Fe, 5% SiO_2, 9% CaO, 1%–2% MgO and 1% Al_2O_3), pellets, prereduced charge, coke (88% C, 0.3% H_2O and 10% mineral matter), limestone and dolomite, oxygen-enriched blast and to some extent pulverized coal (80% C, 5% H_2, 6% O_2, 1% N_2, 1% S and 5% mineral matter) (Biswas, 1981; Sano et al., 1997). Other raw material additives in the blast furnace include metallic scrap, direct reduced iron, recycled oxide wastes, ferro-cokes and carbon composite agglomerates.

The raw material quality plays a critical role in the efficient operation of the blast furnace. The blast furnace process exerts tight raw material constraints in terms of chemical composition for the correct control of pig iron composition and slag and the requisite physical strength to maintain the burden permeability during reduction and melting (Biswas, 1981; Sano et al., 1997). Reduction disintegration, a phenomenon commonly observed in blast furnace burden feed, such as lumpy ores, sinters and pellets, is normally used as a key parameter for evaluating the suitability of ferrous charge materials in the blast furnace (Biswas, 1981; Sano et al., 1997).

Naturally, fines are inherently generated during the mining and handling of iron ores. Such fines require to be agglomerated before charging into the blast furnace. The raw material agglomerates charged in the form of sinter, pellets and briquettes are essential in the efficient operation of the blast furnace. Key operational parameters, such as blast furnace permeability and raw material reducibility are achieved by carefully controlling the composition and physical characteristics before charging in the blast furnace (Ball et al., 1973; Biswas, 1981; Sano et al., 1997). As discussed in Section 4.2.1, the blast furnace is a countercurrent gas–solid reactor, and the best possible contact between the solids and the reducing gas is obtained with a permeable burden to allow for high gas flow rate and uniform gas flow with minimum channelling of the gas (Poveromo, 2006). According to Poveromo (2006), the primary purpose of agglomeration is to improve burden permeability and the gas–solid contact, thereby resulting in increased rate of reduction of oxides, as well as reducing the blast furnace coke rake. In addition to possessing high reducibility and low reduction disintegration properties, the typical blast furnace agglomerates must also contain a high iron content; a minimum level of impurity elements such as sulphur and phosphorus and high strength to withstand mechanical degradation during reduction and melting (Ball et al., 1973; Biswas, 1981; Sano et al., 1997).

4.2.1.3 Iron Ores

Iron ore represents the main source of iron in the blast furnace process. Different categories of iron ores exist, namely, magnetite, hematite, goethite, limonite, ilmenite and siderite among others (Biswas, 1981; Poveromo, 1998). Furthermore, iron ores for the blast furnace process can be categorized as high grade (more than 65 wt.% iron), medium grade (62–65 wt.% iron) and low grade (less than 62 wt.% iron) (Yang et al., 2014).

4.2.1.4 Blast Furnace Sinter

According to Lu and Ishiyama (2016), sintering is the most economic and widely used agglomeration process to prepare iron ore fines for blast furnace use. The sinter plant is an essential step in the integrated iron and steelmaking process and is typically used in the agglomeration of iron ores and fine-grained recycled iron-bearing materials such as dusts, sludge and mill scales (Lanzerstorfer, 2015). In essence, sintering is an agglomeration process achieved through the combustion of coke fines mixed with moist iron ore fines around 1300°C–1480°C in a sinter grate, resulting in the incipient fusion of ore particles containing interspaced hematite and magnetite phases (Ball et al., 1973; Biswas, 1981; Remus et al., 2013; Lanzerstorfer, 2015; Lu and Ishiyama, 2016). The typical raw materials for the sinter process consist of a mixture if fine ores, coke breeze, fluxes (e.g. lime and olivine) and recycled iron-bearing materials from downstream operations such as coarse dust and sludge from off-gas cleaning systems (Ball et al., 1973; Biswas, 1981; Remus et al., 2013; Lu and Ishiyama, 2016). Coke breeze is used as fuel to supply the thermal energy required for the incipient fusion of feed materials (Lanzerstorfer et al., 2015) and is usually consumed at a rate of about 44–64 kg/ton of sinter (Remus et al., 2013).As the raw materials mixture is ignited in the sinter grate, a series of solid state and heterogeneous physicochemical reactions take place between the melt, solids and gaseous phases in the sintering zone to produce a product possessing the requisite size and strength characteristics suitable for feeding into the blast furnace (Ball et al., 1973; Biswas, 1981).

Basically, there are three types of sinter products, namely, acid or non-fluxed sinter, self-fluxing sinter and super-fluxed sinter (Ball et al., 1973; Biswas, 1981). In acid or non-fluxed sinters, no basic fluxes (limestone and dolomite) are added, while self-fluxing sinters contain a sufficient amount fluxes required to react with acid gangue components of ore, such as SiO_2 and Al_2O_3. On the other hand, super-fluxed sinters contain more than the stoichiometric flux requirements to provide the desired final slag characteristics (Ball et al., 1973; Biswas, 1981). In most cases, the self-fluxing and super-fluxed sinters are most commonly preferred by modern blast furnace operators because (1) the additional fluxes, incorporated into the sinter charge rather than charging separately, reduce the melting temperatures of the burden to produce process slag possessing lower viscosity and liquidus temperatures, and hence permitting a smoother furnace operation through better uniform charge composition (2) the improved high temperature properties of self-fluxed sinter facilitate the smooth operation and improved productivity of the blast furnace (Ball et al., 1973; Biswas, 1981).

Based on the stringent raw material demands, sinter quality is of utmost importance to the operation of the blast furnace operation. Some of the most commonly used qualitative measures of sinter quality are summarized in Table 4.2, and include, inter alia,

TABLE 4.2

Desirable Properties of Blast Furnace Sinter

Chemistry		Physical and Metallurgical Properties	
FeO	5.0%–6.0%	ISO strength (ISO 3271)	>6.3 mm: 70%–80%
Mn	≈0.2 wt.% (as low as possible)	Grain structure	
P	≈0.4 wt.% (as low as possible)	<5 mm or <6 mm	Max. 5%
SiO_2	5.0–5.5 wt.%	<10 mm	Max. 30%
Al_2O_3	1.0–1.3 wt.%	>50 mm	Max. 10%
TiO_2	– (as slow as possible)	Disintegration in the static test RDI	
$Na_2O + K_2O$	<0.08 wt.% (as slow as possible)	RDI: <2.8 mm	Max. 20%–30%
CaO	8–10 wt.%	RDI: <3.15 mm	Max. 35%
MgO	1.5–2.0 wt.%	ISO 4696 : <3.15 mm	Max. 30%–35%
CaO/SiO_2	>1.8	Reducibility (ISO 4695)	1.4–1.6%/min

Source: Adapted from Poveromo, J.J., Agglomeration processes: Pelletizing and sintering, in Kogel, J.E., Trivedi, N.C., Barker, J.M., and Krukowski, S.T. (eds.), *Industrial Minerals and Rocks: Commodities, Markets and Uses*, 7th edn., Society for Mining, Metallurgy and Exploration, http://www.smenet.org/, 2006, p. 1394.

size distribution, cold strength and resistance to abrasion disintegration, reduction–disintegration and softening and melting properties (Ball et al., 1973; Biswas, 1981; Kortmann et al., 1992; Poveromo, 2006).

4.2.1.5 Blast Furnace Pellets

Blast furnace pellets are produced by the balling of iron ore fines that are too fine to be charged into a sintering process. Such fines usually considered too fine for the sinter plant; due to the resultant poor permeability the sintering process are usually consumed in the production of blast furnace pellets (Ball et al., 1973; Biswas, 1981), and the process is highly desirable for the agglomeration of finely sized concentrates because they are normally of a size that can be formed into a green ball with little difficulty (Poveromo, 2006). According to de Paula Vitoretti and de Castro (2013), the pelletization process is one of the important steps to produce agglomerates of high quality with additional benefits of recycling the ultra-fines within the steelmaking industry.

Essentially, the pelletization process involves two steps, namely, the balling of moist fine ore (containing a minimum of total iron of 64 wt.%), fuel and organic and/or inorganic binder additives to produce green pellets and the firing and/or induration in travelling grate at around 1200°C–1350°C to harden the pellets to achieve the requisite mechanical resistance and appropriate metallurgical characteristics required in the blast furnace ironmaking process (Ball et al., 1973; Biswas, 1981; de Paula Vitoretti and de Castro, 2013). Other raw material additives, such as limestone and or dolomite, are also added as fluxes to control the pellets' basicity and provide bonding phases. Generally, coal or coke breeze fines are added as solid fuel supplements during the pellets induration stage (Ball et al., 1973; Biswas, 1981; Mohamed et al., 2010; Remus et al., 2013; Lanzerstorfer, 2015). The process flow for the conventional pelletizing process used in the production of blast furnace pellets is highlighted in Figure 4.3 (Remus et al., 2013).

Compared to conventional lumpy ore charge, blast furnace pellets usually possess superior properties in terms of nominal iron content, uniform size distribution and porosity and high mechanical and abrasive strengths (Ball et al., 1973; Biswas, 1981; Remus et al., 2013). The proper control of the desirable physical and metallurgical properties of blast furnace pellets, summarized in Table 4.3, is critical to the productivity and energy efficiency of the blast furnace ironmaking process (Ball et al., 1973; Biswas, 1981; Kortmann et al., 1992; Poveromo, 2006; Remus et al., 2013).

4.2.1.6 Metallurgical Coke

The coke making process involves the high temperature carbonization of metallurgical grade coal in the absence of oxygen in coke oven batteries (Diez et al. 2002; Remus et al., 2013). The quality of coke produced depends on key parameters such as coking time and the quality of coal used (Diez et al., 2002; Remus et al., 2013). In general, the coke yield varies between 700 and 800 kg dry coke/ton of dry coal and about 140–200 kg of coke oven gas per ton dry coal (Remus et al., 2013).

Coke constitutes a major cost driver per ton of hot metal produced, and is one of the most indispensable raw materials in the blast furnace ironmaking process. The coke consumption rate is often used as a parameter to predict the economics of charging particular types of ores in the blast furnace process and inevitably determines the operational regimes and productivity of the process (Ball et al., 1973; Biswas, 1981). Coke plays three major functions in the efficient operation of the blast furnace process, vis-à-vis, thermal, chemical and mechanical functions.

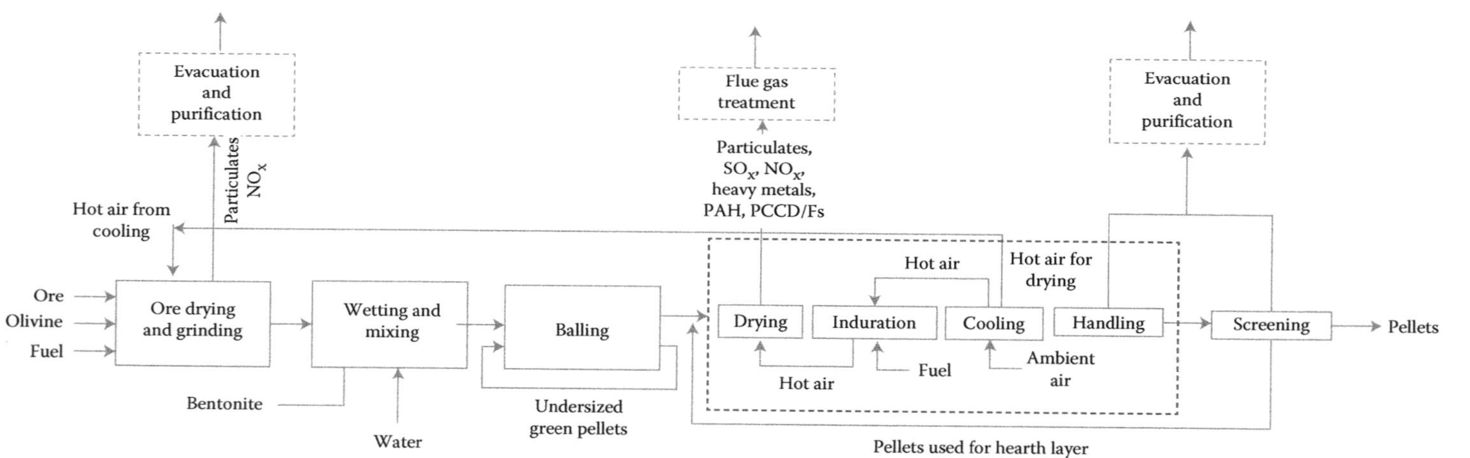

FIGURE 4.3
Process flowsheet for pelletization of iron ore fines. (From Remus, R. et al., Best available techniques (BAT) reference document for iron and steel production, Industrial Emissions Directive 2010/75/EU, European Commission JRC Reference Report, 2013, p. 182.)

TABLE 4.3

Desirable Properties of Blast Furnace Pellets

Physical Properties		Metallurgical Properties	
Grain structure		Low-temperature disintegration, dynamic test (SEP 1771/82)	
>16 mm	Max. 5%	>6.3 mm	Min. 95%
8–16 mm	Min. 85%	<0.5 mm	Max. 5%
<6.3 mm	Max. 5%	Reduction properties under load (ISO 7992)	
Cold compression strength: Pellets 10–12.5 mm (ISO 4700)		80% degree of reduction	Max. 15 mm WG
Average	Min. 2,500 N	Reducibility R40 value (ISO 4695)	Min. 0.8%/min
<2000 N	Max. 10%	Swelling (ISO DP 4698)	Max. 20%
<1500 N	Max. 5%		

Source: Adapted from Poveromo, J.J., Agglomeration processes: Pelletizing and sintering, in Kogel, J.E., Trivedi, N.C., Barker, J.M. and Krukowski, S.T. (eds.), *Industrial Minerals and Rocks: Commodities, Markets and Uses*, 7th edn., Society for Mining, Metallurgy and Exploration, http://www.smenet.org/, 2006, p. 1394.

Coke acts as the fuel to provide the thermal enthalpy required for driving the endothermic reduction reactions and the thermal energy for melting the reduced iron and slag (Ball et al., 1973; Biswas, 1981; Sano et al., 1997). The calorific value of coke is a function of the amount of fixed carbon contained, that is, the higher the amount of fixed carbon, the higher the calorific value of the coke product. Coke also provides the chemical potential required for the reduction of iron oxides, acts as the source of carbon required for the direct reduction of residual iron oxides and metalloids, as well as providing the carbon required for carburization of the liquid metal (Biswas, 1981; Gudenau et al., 1990; Sano et al., 1997; Murakami and Nagata, 2003; Ohno et al., 2004).

The aerodynamic functions of the blast furnace process highlighted in Section 4.2.1 are predicated on the mechanical properties of the coke bed in the blast furnace. In general, such mechanical properties are controlled by the quality of the coking coal used to produce the coke product (Diez et al., 2002). In addition, coke acts as a spacer in the cohesive zone to increase the charge permeability, gas distribution and liquid flow and to facilitate the smooth descent of the charge material during the reduction, softening and melting of iron and slag (Biswas, 1981; Diez et al., 2002). Important chemical properties of coke include moisture, the amount of fixed carbon and ash and the amount of tramp elements such as sulphur, phosphorus and alkali compounds. In general, the chemical characteristics of coke have a significant effect on the energy requirements, blast furnace operation, hot metal quality and refractory lining (Biswas, 1981).

The quality, cost and availability of metallurgical coal present a major raw material constraint in the blast furnace process. Although the use of auxiliary fuels has gained traction as alternatives in reducing the coke rate per 1 ton hot metal, there is no satisfactory reductant material that can fully or partially replace the coke in the blast furnace (Biswas, 1981; Diez et al., 2002). Metallurgical coking coals must produce, when coked, the specified mechano–chemical stability properties, such as those measured by the coke strength after reaction in CO_2 (CSR), to support the blast furnace charge and enhance aerodynamic characteristics. The CSR parameter measures the potential of coke to break down into smaller sizes in the high temperature and high CO/CO_2 gas ratio environment that exists in the lower parts of the blast furnace (Diez et al., 2002; ASTM D5341, n.d.). Coking coals which, when coked, achieve a high CSR value are considered premium and are highly regarded in the blast furnace process. As a result, good quality coking coals are required to produce high grade metallurgical coke (Diez et al., 2002). Nevertheless, only certain types of coals, ranked as coking or bituminous coals, possess the right thermos-plasticity and physico-chemical properties suitable for coke making, with the majority of coals being applicable only as coal blends to improve blast furnace productivity (Ball et al., 1973; Biswas, 1981; Diez et al., 2002). Diez et al. (2002) proposed that proper coal selection and blend composition based on coal rank, rheological properties, petrology and ash chemistry represent one of the major factors used to control the mechanical and chemical coke properties. Because metallurgical coke is such an indispensable raw material and major cost driver in the blast furnace iron making, securing long-term supplies of premium grade coking coals has always been a top priority for most blast furnace producers globally.

4.2.1.7 Carbon Composite Agglomerates and Ferro-Cokes

Until recently, sinters, pellets and cokes constituted the conventional raw material agglomerates charged in the blast furnace. However, extensive research has been carried out over the years to incorporate pre-reduced charge and fuel containing agglomerates in the blast furnace. Of note are the early attempts to incorporate iron ore fines, flue dust and coal blends to produce agglomerates which are commonly referred to as carbon-composite agglomerates (CCA) or ferro-cokes (Russell et al., 1955; Granger et al., 1959; Ball et al., 1973). These documented pioneer studies have demonstrated the possibility of producing carbon composite agglomerates, suitable for use as blast furnace charge material by mixing flue dust and/or iron ore fines with highly volatile non-coking coals (Russell et al., 1955; Granger et al., 1959; Ball et al., 1973). More recent studies have focused on optimizing the blast furnace operation based on pre-reduced carbon composite agglomerates as auxiliary charge (Kunimoto et al., 2006), and on engineering the reactivity of these carbon composite agglomerates in the blast furnace (Anyashiki et al., 2009; Sato et al., 2013). These comprehensive studies have proved, through extensive modelling and pilot testing, the satisfactory application of utilizing carbon composite agglomerates, ferro-cokes and normal coke blends in the blast furnace process (Kunimoto et al., 2006; Anyashiki et al., 2009; Sato et al., 2013). As a result, the adoption of carbon composite agglomerates as blast furnace burden blends has made it possible to increase the use of non-coking coals, iron ore fines and recycled iron bearing fines as indirect feed materials in the blast furnace process (Russell et al., 1955; Granger et al., 1959; Ball et al., 1973; Kunimoto et al., 2006; Anyashiki et al., 2009; Sato et al., 2013).

4.2.1.8 Resource Constraints in the Iron and Steelmaking Process

In the past decade, the increase in the demand of steel has resulted in rapid deterioration in the quality of iron ore shipped to consumers. This is mainly due to the depletion of good quality ore bodies, and the deliberate decision by iron ore and coking coal producers to supply the market via volumes driven production strategies (Zervas et al., 1996a,b; Anameric and Kawatra, 2009; Gordon and Kumar, 2013; Meijer et al., 2013; Lungen et al., 2015). The decline in the quality of iron ore and coking coal has an obvious impact on the blast furnace operation, for example, the lower iron content and higher levels of impurities would result in higher slag rate, which in turn would directly impact the coke rate and the productivity of the process (Meijer et al., 2013; Lungen et al., 2015). Naturally, the quality of starting raw materials also affects the quality of the final refined steel product, the overall productivity of the ironmaking process and the profitability of downstream steel refining processes.

The iron and steel process is a raw material intensive process, and material utilization efficiency is one of the most effective methods for achieving a more sustainable operation in this sector (Suopajärvi et al., 2013; World Steel Association, 2015b). Because raw materials consumption and cost of energy are some of the major cost drivers, increasing the resource productivity has become a strategic business objective adopted by global iron and steel producers in order to gain global competitiveness. Given the recent global dynamics in iron and steel sector, the emphasis nowadays has now shifted towards (1) process intensification to maximize the utilization efficiency of raw materials and/or other resources, (2) developing alternative and more flexible unit processes to augment the amount of iron and steel products produced using the blast process (discussed in Section 4.2.2) and (3) finding alternative uses of iron and steelmaking by-products, either internally within the steel plant itself or eternal to the steel plant as a separate product or process stream.

4.2.2 Alternative Ironmaking Processes

The physicochemical phenomena and the blast furnace charge characteristics discussed in Section 4.2.1 highlighted the sensitivity of the blast furnace process to raw material quality. The blast furnace is a capital and energy intensive process which requires strict control of agglomerated charge materials such as metallurgical cokes and sinters. Due to concerns over supply constraints of high grade resources, the high cost of coking coals and the environmental impact of coking and sinter plants, the global iron and steel industry is gradually shifting from heavy reliance on the traditional blast furnace process towards more nascent and environmentally friendly ironmaking technologies (Zervas et al., 1996a,b; IAE Clean Coal Centre, 2004; Anameric and Kawatra, 2009; Gordon and Kumar, 2013; Meijer et al., 2013; Lungen et al., 2015).

Given the flexibility to utilize low-grade raw materials such as iron ore fines and non-metallurgical coal, these alternative ironmaking technologies are gradually taking centre stage, especially in meeting the needs of small-sized local and regional markets (Zervas et al., 1996a,b; IAE Clean Coal Centre, 2004; Anameric and Kawatra, 2009; Gordon and Kumar, 2013; Meijer et al., 2013). Furthermore, the dominance of electric arc furnace (EAF) melting is expected to increase due to lower capital costs and higher process flexibility compared to the blast furnace process, thereby creating a demand for new iron feedstocks to augment the solid steel scrap (Zervas et al., 1996a,b; Gordon and Kumar, 2013; Meijer et al., 2013; World Steel Association, 2014). As a result, the demand for high-quality virgin metallic charges, such as those produced from some of the alternative ironmaking processes, to augment steel scrap melting in the EAF is also expected to grow (Zervas et al., 1996a,b; IAE Clean Coal Centre, 2004; Kopfle and Hunter, 2008; Anameric and Kawatra, 2009; Kikuchi et al., 2010; Meijer et al., 2013; World Steel Association, 2014).

Generally, alternative ironmaking processes can be categorized into two broad categories, namely, (1) direct reduction ironmaking processes, where the iron ore or other iron oxide material is reduced to metallic iron at temperatures below the melting point of iron to produce a solid product commonly known as direct reduced iron (DRI), and (2) smelting reduction processes in which a liquid metal product with chemistry similar to that of blast furnace hot metal is produced (Zervas et al., 1996a,b; Feinman, 1999; IAE Clean Coal Centre, 2004; Kopfle and Hunter, 2008; Anameric and Kawatra, 2009; Gordon and Kumar, 2013; Meijer et al., 2013). Another variation of the smelting reduction processes involves the direct steelmaking process, where the objective is to produce liquid steel directly from oxide ores.

While the blast furnace is expected to remain the world's principal source of iron units for the steelmaking process as long as adequate supplies of suitable iron ores and coking coals remain available at competitive cost, these alternative ironmaking processes are

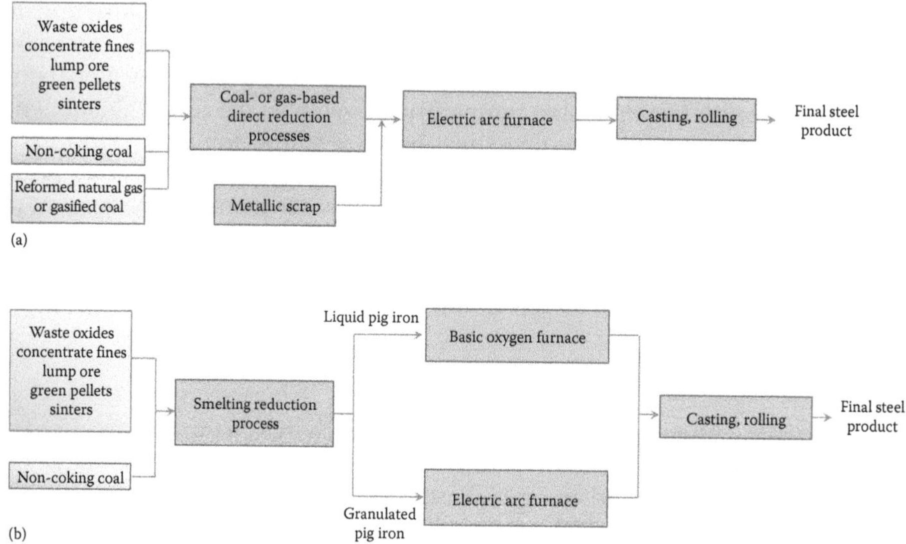

FIGURE 4.4
Schematic representation of primary and alternative ironmaking processes. (a) Direct reduction processes; (b) smelting reduction processes. (Adapted from Remus, R. et al., Best available techniques (BAT) reference document for iron and steel production, Industrial Emissions Directive 2010/75/EU, European Commission JRC Reference Report, 2013, p. 521.)

slowly gaining traction as serious contenders to the blast furnace process, particularly for small-sized local and regional markets (Fruehan, 1993; Zervas et al., 1996a,b; Feinman, 1999; IAE Clean Coal Centre, 2004; Anameric and Kawatra, 2009; Gordon and Kumar, 2013; Remus et al., 2013). Figure 4.4 shows a schematic representation of the primary and alternative ironmaking processes (Remus et al., 2013).

4.2.2.1 Direct Reduction Ironmaking Processes

The direct reduction processes can be categorized according to the type of reactor employed, namely, shaft furnaces (e.g. Midrex® gas-based process), rotary kilns (e.g. SN/RN process), rotary hearth furnaces (RHFs) (e.g. FASTMET®/FASTMELT®, Inmetco™/RedIron, and ITMK3®), and fluidized bed reactors (e.g. Circofer®) (Feinman, 1999; Barozzi, 1997; McCelland, 2002; Kopfle and Hunter, 2008; Tateishi et al., 2008; Kikuchi et al., 2010; Tanaka, 2015). Furthermore, the direct reduction process can also be classified according to the nature of the reductant, namely, (1) gas-based direct reduction ironmaking processes utilizing natural gas or gasified coal as reductants, and (2) coal-based direct reduction ironmaking processes which utilize coal fines as a reducing agent (Feinman, 1999; Kopfle and Hunter, 2008; Kikuchi et al., 2010; Tanaka, 2015). In most cases, the coal-based processes are based on the carbon-composite agglomerates discussed in Section 4.2.1. On the other hand, natural gas-based processes mostly utilize CH_4, CO and H_2 gases produced by reforming natural gas in steam-filled catalyst tubes according to the reaction (Feinman, 1999):

$$C_nH_{(2n+2)} + nH_2O \rightarrow nCO + (2n+1)H_2 \tag{4.1}$$

The reduction of iron ore in gas-based reduction processes is accomplished by the same chemical reactions that occur in the blast furnace (Feinman, 1999; Tanaka, 2015):

$$Fe_2O_3 + 3CO \rightarrow 2Fe + 3CO_2 \tag{4.2}$$

$$Fe_2O_3 + 3H_2 \rightarrow 2Fe + 3H_2O \tag{4.3}$$

$$3Fe + CH_4 \rightarrow Fe_3C + 2H_2 \tag{4.4}$$

The major part of the direct reduced iron produced globally is consumed as a substitute for metallic scrap in the electric arc steelmaking, mainly because the direct reduced iron derived from virgin iron units contains lower levels of impurities than metallic scrap (Feinman, 1999; McCelland, 2002; IAE Clean Coal Centre, 2004; Kopfle and Hunter, 2008; Tateishi et al., 2008; Kikuchi et al., 2010; Tanaka, 2015). In essence, the direct reduced iron substitution of steel scrap functions as diluents to contaminants commonly found in steel scrap, thereby improving the overall quality of steel. However, direct reduced iron may contain a high amount of gangue, which increases the energy consumption and volume of slags in the electric steelmaking process. The additional energy consumption may counteract the diluent benefits achieved in steelmaking and is particularly so in areas with high electric power costs (IAE Clean Coal Centre, 2004). Ideally, direct reduction processes are favourable in areas which have a limited supply of steel scrap, abundant reserves of inexpensive natural gas (applies to gas-based processes), non-coking coals (for coal-based processes), relatively cheap electricity for EAF melting and access to suitable iron ore fines and/or iron bearing waste materials (Feinman, 1999).

4.2.2.2 Direct Smelting Reduction Processes

Direct smelting reduction processes have been developed as alternatives to hot metal production in the blast furnace (Zervas et al., 1996a,b; Feinman, 1999; IAE Clean Coal Centre, 2004; Kopfle and Hunter, 2008; Anameric and Kawatra, 2009; Gordon and Kumar, 2013; Meijer et al., 2013). Generally, the development and commercialization of smelting reduction processes highlighted in Figure 4.4 and Table 4.4 has been precipitated by the need for flexibility and ability to produce smaller quantities of hot metal directly from iron oxide feed stocks, preferably low grade ore and/or without agglomeration and the flexibility to utilize non-coking coals as reducing-carburizing agent (Zervas et al., 1996a; Anameric and Kawatra, 2009; Meijer et al., 2013).

Several smelting reduction processes have been adopted commercially to overcome the challenges inherent to the blast furnace process. Some of the processes commercially investigated include, among others, processes such as Corex®, DIOS, AISI, ROMELT®, HIsmelt®, Redsmelt®, ITMK3 and Plasmamelt (Feinman, 1999; IAE Clean Coal Centre, 2004; Anameric and Kawatra, 2009; Tanaka, 2015). Some of these processes have demonstrated a high degree of integration to provide hot liquid metal directly to the EAF; a practice that has proved to increase productivity, reduce electric power consumption and lower electrode and refractory consumption during the EAF operation (Feinman, 1999; IAE Clean Coal Centre, 2004; Anameric and Kawatra, 2009). In addition, the popularity of some of the smelting reduction processes has also been buoyed by their dual capability to utilize both virgin ore and recycled oxide wastes as process feed raw materials.

TABLE 4.4
Generic Characteristics of the Blast Furnace and Smelting Reduction Processes

Properties	Processes		References
	Blast Furnace	**Smelting Reduction**	
Products	Liquid pig iron for basic oxygen steelmaking. Granulated pig iron for coolant and diluent feed in EAF and basic oxygen steelmaking. Granulated feed in ferrous foundry operations.	Liquid hot metal for basic oxygen steelmaking. Granulated pig iron for coolant and diluent feed in EAF and basic oxygen steelmaking. Granulated feed in ferrous foundry operations.	Zervas et al. (1996a,b); Anameric and Kawatra (2009); Gordon and Kumar (2013); Meijer et al. (2013)
Raw materials	Self-reducing pellets, lump ore, sinters, briquettes.	Self-reducing pellets, lump ore, fine ore and oxide wastes.	Zervas et al. (1996a,b); Anameric and Kawatra (2009); Gordon and Kumar (2013); Meijer et al. (2013)
Reductants	Metallurgical grade coke, pulverized coal injection.	Non-coking coals, coke, coke breeze.	Zervas et al. (1996a,b); IAE Clean Coal Centre (2004); Anameric and Kawatra (2009); Gordon and Kumar (2013); Meijer et al. (2013)
Operating temperatures	Around 1535°C.	Around 1300°C.	Zervas et al. (1996a,b); Anameric and Kawatra (2009)
Advantages	High productivity and control of hot metal chemistry. Sulphur removal in the blast furnace reduces the need for external desulphurization. Low costs per nut due to large economies of scale. Well-researched (reductant rate, pulverized coal injection, greenhouse gas emissions) and stable process in terms of established operational philosophies.	Flexibility in capacity and easy to integrate and used in conjunction with mini-mills to produce pig iron at a lower cost and lower installed capacity. Can be integrated to augment productivity from the blast furnace process. High flexibility in terms of raw materials and reductants.	Zervas et al. (1996a,b); IAE Clean Coal Centre (2004); Anameric and Kawatra (2009); Gordon and Kumar (2013); Meijer et al. (2013)
Challenges	High initial capital requirements. Environmental emissions from ancillary processes such as coke ovens and sinter plant. Low raw material flexibility due to high dependence on high quality raw materials such as iron ores, sinter, pellets and coke. Viability dependent on large economies of scale. Lack operational flexibility.	High refractory wear rates and costs in commercial smelting reduction most processes. High iron losses due to reactive FeO-rich slag. Additional hot metal desulphurization stages required. Need to control char in slag for optimum reduction and slag foaming rates. High consumption rate of coal and oxygen. Effects of fluctuations in raw material chemistry on hot metal chemistry.	Zervas et al. (1996a,b); IAE Clean Coal Centre (2004); Anameric and Kawatra (2009); Gordon and Kumar (2013); Meijer et al. (2013)

4.2.3 Electric Arc Steelmaking Processes

EAFs are operated by remelting metallic scrap, fluxes and other sources of metallic iron such as direct reduced iron (Rentz et al., 1997; World Steel Association, 2012; Industrial Efficiency Technology Database, n.d. Remus et al., 2013). Currently, EAF melting accounts for about 30% of global liquid steel production (World Steel Association, 2012). In fact, as new investments in the global installations of integrated blast furnace processes slow down, the EAF processes are gaining traction to increase their share in the production of carbon steels and alloy steels (World Steel Association, 2012).

In the EAF process, metallic charges in the form of steel scrap, fluxes and other additives are charged into the EAF and melted at around 1600°C to produce varying grades of carbon and stainless steel (Rentz et al., 1997; Jones et al., 1998; World Steel Association, 2012; Remus et al., 2013; Industrial Efficiency Technology Database, n.d.). The electrical energy for melting is supplied via graphite electrodes, and the heat generated by the arc is used to melt the metallic charges (Jones et al., 1998; Industrial Efficiency Technology Database, n.d). Additional chemical energy can also be supplied through specially designed oxy-fuel burners and oxygen lances (Rentz et al., 1997; Jones et al., 1998; Remus et al., 2013). Principally, the steelmaking process via the EAF comprises of the following steps (Rentz et al., 1997): (1) handling of inputs and preparation of the furnace, (2) batch charging, (3) melting and (4) oxidation refining (e.g. decarburization and dephosphorization treatment), even though the advent of secondary refining processes has now limited the role of EAFs to high efficiency melting to produce the liquid steel (Jones et al., 1998). Figure 4.5 shows the section view of the direct current EAF (Industrial Efficiency Technology Database, n.d.).

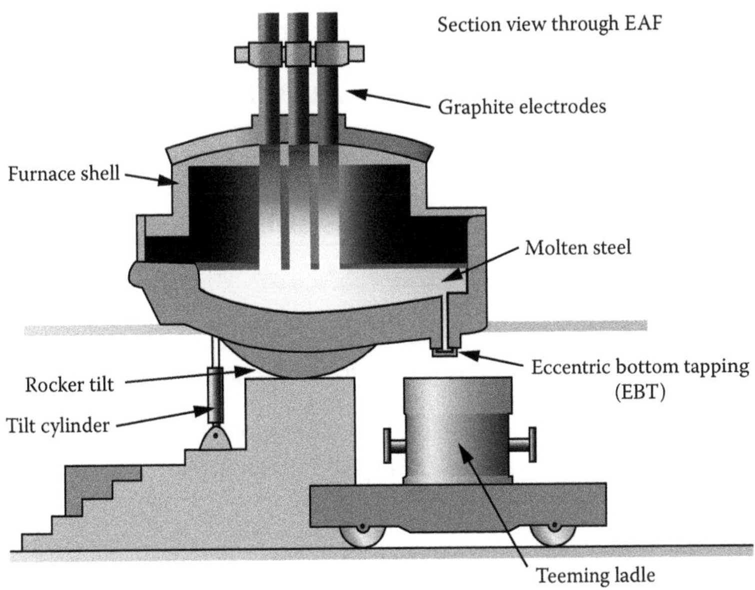

FIGURE 4.5
Section view of the direct current EAF. (From Industrial Efficiency Technology Databasem, Electric arc furnace, Available at http://ietd.iipnetwork.org/content/electric-arc-furnace, Accessed 30 December 2015, n.d.)

4.2.4 Basic Oxygen Steelmaking Process

The oxygen steelmaking process, also commonly referred to as the basic oxygen steel making process, is used to rapidly refine a charge of molten pig iron and ambient scrap into steel of desired carbon content and temperature by using industrial grade oxygen (Miller et al., 1998; Worrel et al., 2010). In general, the main raw materials for the oxygen steelmaking process include (1) hot metal from the blast furnace; (2) liquid metal from the electric arc melting furnaces; (3) fluxes in the form of burnt lime, silica and dolomite and (4) steel scrap and other metallic charge additions such as direct reduced iron and pre-reduced charge (Miller et al., 1998; World Steel Association, 2012). In the basic oxygen process, oxygen is blown at supersonic velocities to oxidize the impurities present in the liquid iron. According to Miller et al. (1998), the commercial success of the oxygen steelmaking process is due to two main factors: (1) the process is autogenous as the oxidation reactions during the blow provide all the necessary enthalpies to melt the fluxes and steel scrap and achieve the desired temperature of the steel product, and (2) the process is capable of refining steel at high production rates due to fast reaction achieved by the large surface area available for reactions.

In general, the hot metal from the blast furnace process typically contains about 4–4.5 wt.% carbon, 0.15–2.5 wt.% manganese, 0.5–4 wt.% silicon, less than 0.2 wt.% phosphorus, and 0.025%–0.25% sulphur (World Steel Association, 2012; Remus et al., 2013). Therefore, the role of the oxygen steelmaking process is to reduce the carbon from about 4 wt.% to less than 1 wt.%, to reduce or control the sulphur and phosphorus, and finally, to raise the temperature of the liquid steel to1600°C–1700°C (Miller et al., 1998). The basic oxygen process is an autothermic process, that is, the energy required for melting the fluxes and raising the temperatures of steel scrap and hot metal to steelmaking temperatures is provided by the exothermic oxidation reactions of the various reactive elements in the liquid metallic charge materials. The main chemical reactions taking place in the basic oxygen process are summarized in Table 4.5. In most cases, temperatures as a result of these exothermic reactions can become too high such that that steel scrap, iron ore, mill scale and other materials are usually added as coolants to control the process temperatures in the desired range of 1600°C–1700°C (Miller et al., 1998; Remus et al., 2013). In general, the required quantities

TABLE 4.5

Main Chemical Reactions Taking Place in the Basic Oxygen Process

Oxidation Process	Chemical Reactions	By-Product
Decarburization reactions	$[C] + [O] \rightarrow CO_{(gas)}$	Off gas
	$2[C] + O_{2(gas)} \rightarrow 2CO_{(gas)}$	
	$[CO] + [O] \rightarrow CO_{2(gas)}$	
Oxidation of impurities from steel	$[Si] + 2[O] + 2[CaO] \rightarrow (2CaO \cdot SiO_2)$	Slag
	$[Mn] + [O] \rightarrow (MnO)$	
	$2[P] + 5[O] + 3(CaO) \rightarrow (3CaO \cdot P_2O_5)$	
Oxidation of alloying elements in steel and slag-metal-gas reactions	$[Fe] + [O] \rightarrow (FeO)$	Slag and off-gas
	$2[Cr] + 3[O] \rightarrow (Cr_2O_3)$	
	$(FeO) + [C] \rightarrow [Fe] + CO_{(g)}$	
	$(FeO) + CO_{(g)} \rightarrow [Fe] + CO_{2(g)}$	

Source: Adapter from Remus, R. et al., Best available techniques (BAT) reference document for iron and steel production, Industrial Emissions Directive 2010/75/EU, European Commission JRC Reference Report, 2013, p. 367.

of hot metal, steel scrap, oxygen and fluxes vary according to their compositions and temperatures and also on the desired chemistry and temperature of the steel to be tapped (Miller et al., 1998).

Generally, fluxes are added early on in the oxygen blow to control the sulphur and phosphorus distribution between metal and slag and also to protect the furnace refractory lining (Miller et al., 1998). The oxides formed from the oxidation reactions of silicon, manganese, iron and phosphorus react with these fluxes to form liquid slag that then floats on top of the liquid steel and is removed during intermittent tapping of the furnace (Miller et al., 1998; Remus et al., 2013). During the blowing stage, foamy slag emulsions are formed which enhances quick reactions and rapid mass transfer of elements from metal and gas phases to the slag (Miller et al., 1998). Depending on the quality of hot metal charged and the steelmaking practice involved, about 100–150 kg of slags are produced in a BOF per ton of liquid steel produced (Das et al., 2007; Remus et al., 2013).

Typically, the composition of slag in the BOF process consists of about 35–45 wt.% CaO, 11–17 wt.% SiO_2, 1–6 wt.% Al_2O_3, 2–9 wt.% MgO, 16–26 wt.% FeO and 2–6 wt.% MnO, and the typical mineralogical phases consist mainly of dicalciumsilicate (Ca_2SiO_4), dicalcium ferrite ($Ca_2Fe_2O_5$) and wustite ($(Fe_{1-x-y}Mg_xMn_y)O_2$) (EUROSLAG, 2015). The BOF slag constitutes one of the major waste products generated during the steelmaking process. The physicochemical characteristics and the recycling and reuse opportunities of these types of slags are discussed in detail in Sections 4.6 and 4.7.

Basically, there are three variations of oxygen steelmaking processes depending on the way the oxygen gas is introduced into the liquid bath, and these include (Miller et al., 1998; JFE 21st Century Foundation, 2003): (1) top blowing, (2) bottom blowing and (3) combined blowing processes. In the top blown basic process, oxygen is introduced from the top of the furnace via a water cooled lance, and the agitation created by the supersonic oxygen jet results in the formation of the necessary slag-metal emulsion and vigorous bath flows to sustain the rapid gas–metal–slag reactions (Miller et al., 1998). In the bottom-blown converters, oxygen is introduced via several natural gas or propane cooled tuyéres installed at the bottom of the vessel. The main advantage of bottom blowing over top blowing is that all the oxygen is introduced through the bottom and passes through the bath and slag, thereby creating vigorous bath stirring and formation of slag emulsion (Miller et al., 1998).

A third variation involves combined blowing, which basically incorporates both the merits of top and bottom blowing. Combined blowing processes involve a combination of oxygen blowing from the top of the furnace with bottom stirring using argon or nitrogen gas through specially designed tuyéres located through the bottom lining of the reactor (Miller et al., 1998). In essence, combined blowing processes have widely been adopted in the commercial converting of steel, mainly due to their associated higher yield, increased vessel life and reduced oxygen and flux consumption (Sano et al., 1997; Miller et al., 1998; Remus et al., 2013). Figure 4.6 shows a schematic of a combined blowing basic oxygen converter (Remus et al., 2013).

4.2.5 Stainless Steel Refining Processes

The refining processes of steel, also commonly referred to as secondary metallurgy or ladle processes, are usually the final process steps in producing different grades of carbon steel, stainless steels and high purity steels (Patil et al., 1998; Remus et al., 2013). Depending on the grade and specification of the steels, a series of refining and alloy addition steps are carried out under carefully controlled melt atmospheres in refining ladles. Generic liquid steel treatment steps include the final desulphurization, addition of deoxidizing agents and alloying elements

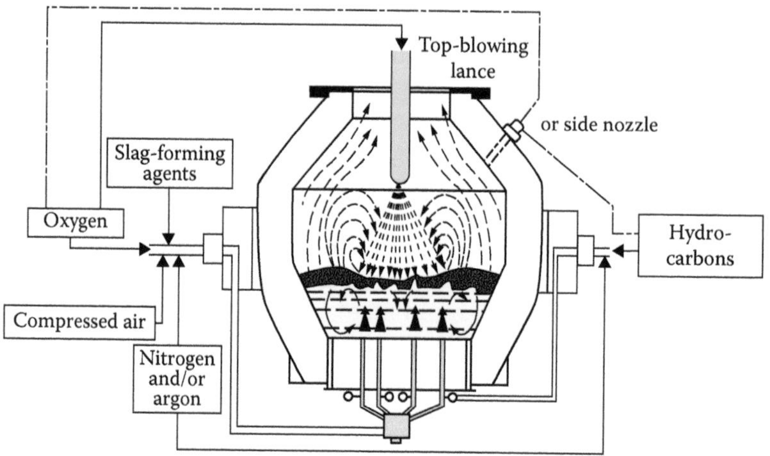

FIGURE 4.6
Schematic of combined blowing basic oxygen converter. (From Remus, R. et al., Best available techniques (BAT) reference document for iron and steel production, Industrial Emissions Directive 2010/75/EU, European Commission JRC Reference Report, 2013, p. 358.)

in to adjust the chemical composition of the finished steel, degassing, adjusting the liquid steel temperatures to casting temperatures, the removal of non-metallic inclusions, and finally the near net shape casting and solidification (Oeters, 1975; Weeks, 1975; Zhang and Thomas, 2003; Wang et al., 2009; Remus et al., 2013; Harada et al., 2014; Reis et al., 2014). Generally, the main objectives of the secondary metallurgy processes include chemical and temperature homogenization of the liquid steel to close chemical analysis and casting temperature tolerances, deoxidation and degassing control of steel and controlling the purity of steel by removing non-metallic inclusions (Oeters, 1975; Weeks, 1975; Zhang and Thomas, 2003; Zhang and Thomas, 2006; Wang et al., 2009; Remus et al., 2013; Harada et al., 2014; Reis et al., 2014).

Stainless steels typically contain between 10 and 30 wt.% chromium (Patil et al., 1998). Varying amounts of alloying elements, such as nickel, molybdenum, niobium and titanium among others, are also added to achieve the desired mechanical and chemical characteristics of the final steel product (Patil et al., 1998). In general, stainless steels can primarily be classified as austenitic, ferritic, martensitic, duplex or precipitation hardening grades (Patil et al., 1998). Generally, several process options are available for the refining of the different stainless steel grades with the individual choice being influenced by factors such as by raw material availability, desired final product characteristics, existing shop logistics and capital economics (Patil et al., 1998; Total Materia, 2008a,b).

Technically, stainless steel refining processes can either be duplex or triplex types. The duplex process involves the melting mixture of scrap, ferroalloys and other raw materials in an EAF to produce the liquid steel of desired chemical composition (Patil et al., 1998; Total Materia, 2008a; Remus et al., 2013). The liquid steel which contains controlled amounts of chromium, nickel and other alloying elements is then charged to refining converters, such as the argon-oxygen decarburization (AOD) converter to produce low carbon stainless steel which is then tapped into teeming ladles (Patil et al., 1998; Total Materia, 2008a,b). On the other hand, the triplex refining process involves coupling the EAF melting process, a pre-blowing converter and a vacuum refining unit (Patil et al., 1998; Total Materia, 2008a,b). In the triplex process, the molten steel undergoes treatment for final decarburization, homogenization and flotation of inclusions before the ladle is taken to the teeming operation (Patil et al., 1998; Total Materia, 2008a,b).

In the decarburization treatment of stainless steel, care must be taken to minimize the oxidation of chromium (Sano et al., 1997; Patil et al., 1998; Turkdogan and Fruehan, 1998). When oxygen is injected into a stainless steel melt, chromium and iron are oxidized to form oxides as depicted in Equations 4.5 and 4.6 (Turkdogan and Fruehan, 1998):

$$2[\mathrm{Cr}]_{\mathrm{melt}} + \frac{3}{2}\mathrm{O}_{2(g)} \rightarrow (\mathrm{Cr_2O_3})_{\mathrm{slag}} \tag{4.5}$$

$$[\mathrm{Fe}]_{\mathrm{melt}} + \frac{1}{2}\mathrm{O}_{2(g)} \rightarrow (\mathrm{FeO})_{\mathrm{slag}} \tag{4.6}$$

The dissolved carbon then reduces the chromium and iron oxides according to Equations 4.7 and 4.8 (Turkdogan and Fruehan, 1998):

$$(\mathrm{Cr_2O_3})_{slag} + 3[\mathrm{C}]_{\mathrm{dissolved}} \rightarrow 2[\mathrm{Cr}]_{\mathrm{melt}} + 3\mathrm{CO}_{(g)} \tag{4.7}$$

$$(\mathrm{FeO})_{\mathrm{slag}} + [\mathrm{C}]_{\mathrm{dissolved}} \rightarrow [\mathrm{Fe}]_{\mathrm{melt}} + \mathrm{CO}_{(g)} \tag{4.8}$$

Technically, control techniques to minimize chromium oxidation may involve the following (Turkdogan and Fruehan, 1998): (1) controlling the temperature to manipulate the equilibrium carbon content; (2) dilution, particularly in the argon-oxygen decarburization process, where inert gases, such as nitrogen or argon, are injected to lower the partial pressure of CO in the bath and hence allow higher chromium content to be in equilibrium with lower carbon content; (3) vacuum refining to remove the CO and force the equilibrium of Equation 4.6 to the left and (4) careful manipulation of slag.

The different process steps for refining stainless steel based on the EAF are shown in Figure 4.7. The complexity of these processes, largely dependent on the grade and composition of steel to be produced, increases widely from the production of various grades of carbon steels and stainless steels to high alloyed specialty alloys, where the control of the composition and physicochemical properties of dissolved impurities and gases such as

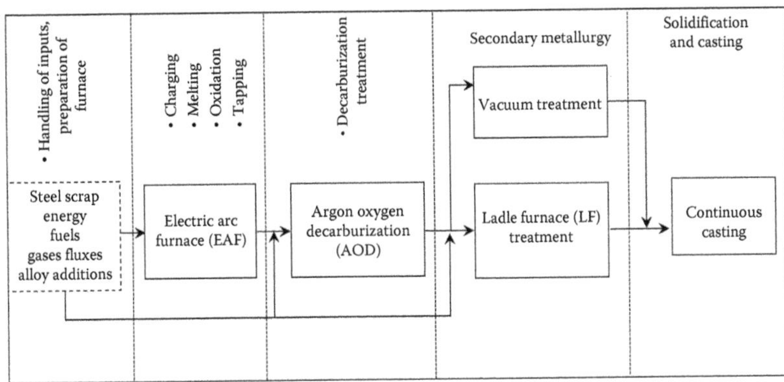

FIGURE 4.7
Overview of stainless steel melting and refining processes. (Adapted from Remus, R. et al., Best available techniques (BAT) reference document for iron and steel production, Industrial Emissions Directive 2010/75/EU, European Commission JRC Reference Report, 2013, p. 420.)

carbon, nitrogen and hydrogen, is crucial without the excessive oxidation loss of alloying elements (Rentz et al., 1997; Steins et al., 2008; Total Materia, 2008b; Remus et al., 2013).

4.2.6 Solidification and Casting of Steel

The main objective of the smelting and refining processes described so far is to produce liquid steel of the right chemical composition and temperature conditions. According to Louhenkilp (2014), secondary steelmaking and casting are the central process steps which have a strong influence on the final quality of the steel products. The processing of liquid steel in ladles and tundish, and the final solidification in the mold consist of a series of complex chemical, physical and thermal phenomena (Remus et al., 2013; Louhenkilp, 2014). As such, the control of such phenomena is predicated on knowledge of the transient and dynamic changes in the steel chemistry, temperature, flow conditions and interactions of steel with slag, refractory materials and/or mould wall (Louhenkilp, 2014). In general, once the final quality of the liquid steel has been achieved, it is then cast and solidified into semi-finished products such as billets, blooms or slabs, for subsequent processing in the metal manufacturing and finishing mills (Kozak and Dzierzawski, n.d.).

Basically, two process options exist in the solidification and casting of the liquid steel into semi-finished products, and these are classified as ingot casting and continuous casting (Vijayaram, 2012; Remus et al., 2013; Kozak and Dzierzawski, n.d.). As opposed to ingot casting where the molten steel is poured and solidified batch-wise into one or more molds, the continuous casting process has been widely adopted in large-scale commercial processes, mainly due to improved productivity, product quality and cost efficiency (Okumura, 1994; Thomas, 2001; Kozak and Dzierzawski, n.d.).

In general, continuous casting transforms the molten metal into a solid on a continuous basis, and is considered to be one of most efficient ways of solidifying large volumes of liquid metal into simple shapes for subsequent processing (Thomas, 2001; Vijayaram, 2012; Remus et al., 2013). In the continuous casting process, the liquid steel from the EAF and/or BOF is tapped into preheated ladles for secondary metallurgy treatment before being transferred into the tundish of the continuous casting machine (Vijayaram, 2012; Remus et al., 2013). The liquid steel flows out of the ladle into the tundish and finally into water cooled mold via submerged entry nozzles, and the solidification process begins in the mold and continues through the different zones of cooling while the strand is continuously being withdrawn at controlled casting speeds (Vijayaram, 2012; Vertnik and Šarler, 2014; Kozak and Dzierzawski, n.d.). Finally, the solidified strand is straightened, cut and discharged for intermediate storage, or is hot charged into subsequent downstream manufacturing processes such as rolling and forging (Vijayaram, 2012; Kozak and Dzierzawski, n.d.). Some of the applicable metal manufacturing and finishing processes are discussed further in Chapter 8. Figure 4.8 shows the schematic diagram of a typical continuous casting plant (Vertnik and Šarler, 2014).

Naturally, liquid steel contains solid non-metallic inclusions, such as alumina and silica inclusions, formed as deoxidation products in Al-killed and Si-killed steels (Zhang and Thomas, 2006). One of the important aspects of continuous casting is the ability to remove these non-metallic inclusions from the liquid steel through flotation and/or dissolution in the mould slag (Thomas, 2001; Zhang and Thomas, 2006; Brandaleze et al., 2012; Louhenkilp, 2014; Reis et al., 2014). Mould fluxes are added to the top free surface of the liquid steel in the mold in order to (1) provide the thermal and chemical insulation for the molten steel by forming a floating liquid slag layer that prevents the liquid steel from freezing and oxidation, (2) provide strand lubrication and control of mould heat transfer in horizontal direction, (3) provide a solid slag film for the optimum heat extraction and

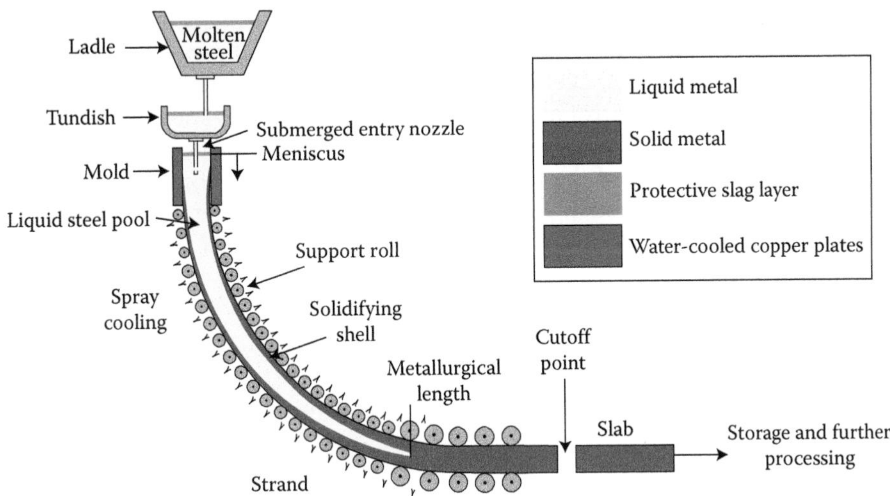

FIGURE 4.8
Schematic diagram of a continuous casting plant. (Reprinted from *Eng. Anal. Bound. Elem.*, 45, Vertnik, R. and Šarler, B., Solution of a continuous casting of steel benchmark test by a meshless method, 45–61. Copyright 2014, with permission from Elsevier.)

(4) remove the non-metallic inclusions by dissolution or incorporation into the slag (Thomas, 2001; Zhang and Thomas, 2006; Brandaleze et al., 2012; Louhenkilp, 2014; Reis et al., 2014; Mills, 2016).

Brandaleze et al. (2012) defined mould fluxes as synthetic slags consisting of a complex mix of oxides, minerals and carbonaceous materials. The typical composition of such fluxes is highlighted in Table 4.6. Furthermore, Mills (2016) categorized the mould

TABLE 4.6
Typical Composition of Mould Fluxes

Category	Compounds	Composition (wt.%)
Glass formers	SiO_2	17–56
	Al_2O_3	0–13
	B_2O_3	0–19
	Fe_2O_3	0–6
Basic oxides or modifiers	CaO	22–45
	MgO	0–10
	BaO	0–10
	SrO	0–5
Alkalis	Na_2O	0–25
	Li_2O	0–5
	K_2O	0–2
Fluidizing	F	2–15
	MnO	0–5
Melting control	C	2–20

Source: Brandaleze, E. et al., Mould fluxes in the steel continuous casting process, in Srinivasan, M. (ed.), INTECH, Available at http://www.intechopen.com/books/science-and-technology-of-casting-processes/mould-fluxes-in-the-steel-continuous-casting-process, Accessed 21 February 2016, 2012, p. 206.

powders as synthetic slags consisting of a blend of slag network formers (SiO_2 and Al_2O_3), slag network breakers (CaO and MgO), fluxes (Na_2O, K_2O, CaF_2 and B_2O_3), carbon particles and sometimes exothermic agents, such as Ca/Si or Fe/Si to reduce the vertical heat flux in the mould. Careful control of the mould flux parameters, such as the melting rate and the ability of the molten fluxes to flow and absorb inclusions from the steel, is critical to achieving the right quality of the cast products (Thomas, 2001; Brandaleze et al., 2012). According to Thomas (2001), such parameters depend on the chemical composition and are governed by time-dependent thermodynamics of the process (Thomas, 2001).

4.3 Refractory Materials in the Iron and Steelmaking Process

4.3.1 General Considerations

In general, refractory materials are composite structural materials used in large volumes to line commercial furnaces such as steelmaking vessels, glass tanks and cement kilns, where the thermomechanical and thermochemical properties of the refractory lining materials are critical (Lee and Zhang, 1999; Lee et al., 2004). Hubble (1999) further defined refractory materials in the steelmaking process as primary materials used by the steel industry in the internal linings of furnaces for making iron and steel, in ladles for holding and transporting liquid metal and slag, in furnaces for heating steel before further processing and in the flues or stacks through which hot gases are conducted. By nature, refractory materials have high melting points and/or insulating properties in their pure or oxide states (melting temperature range 1815°C–3315°C), are able to withstand high temperatures in service (260°C–1850°C) and are able to maintain their structural integrity when exposed to different degrees of mechanical, thermal and chemical stress and strain from solids, liquids and gases (Hubble, 1999; Lee and Zhang, 1999; Harbison-Walker, 2005). As a result, the selection and application of the refractory materials in steelmaking unit processes is based on the thermomechanical, thermochemical and chemical characteristics of refractory materials under severe service conditions (Hubble, 1999; Lee and Zhang, 1999; Harbison-Walker, 2005).

Several naturally occurring raw materials are used in production of high quality refractory materials (Hubble, 1999; Lee and Zhang, 1999; Harbison-Walker, 2005; Guèguen et al., 2014). Guèguen et al. (2014) categorized the chemical and temperature specifications, as well as the main geographical locations of some of the most important raw materials. Basically, some important refractory materials are limited to certain geographical regions, which in essence, tend to affect the cost and market dynamics. For example, the major geographical regions possessing significant economic deposits of some of the specialist refractory materials include South Africa (a major supplier of zirconia and chromite), Australia (a major supplier of zircon) and China (a major supplier of alumina silicate, magnesia and graphite).

4.3.2 Classification of Refractory Materials

According to Harbison and Walker (2005), refractory materials are classified as basic, high alumina, silica, fireclay and insulating refractories. There are also special classes of refractory materials consumed in steelmaking, and these include silicon carbide (SiC), graphite, zircon, zirconia and fused cast refractories (Harbison-Walker, 2005). Steelmaking refractory materials range from dense (up to 90% density) to low density (10% density) fibrous thermal insulation materials (Lee et al., 2004), and may be sold as prefabricated bricks or sold as unshaped monolithic hydraulic setting castables, plastic refractories, ramming mixes and gunning mixes (Hubble, 1999; Harbison-Walker, 2005). Depending on the type of application, shaped refractory bricks are usually manufactured from powder mixtures by combining many different material types such as ceramic/oxide powders, graphite flakes and polymer resins, that are usually fired to achieve the requisite ceramic bond (Lee et al., 2004). On the other hand, unshaped monolithic refractory materials can be installed and shaped in situ. For example, castable monolithic materials are often sold as dry mixes which are then mixed with water and shaped by casting or vibrating to form a hydraulic cement chemical bond at ambient temperatures.

From a chemical standpoint, refractory materials can be classified as basic (mostly containing dolomite and burnt lime), acidic (mostly containing silicates and aluminosilicates) and neutral (mostly containing alumina) (Hubble, 1999; Harbison-Walker, 2005; Garbers-Craig, 2008). According to Harbison-Walker (2005), the common types of refractory materials used in the iron and steel industry can be categorized according to chemical composition into acid-based alumina-silicate types consisting of (1) fireclay (super-duty, high duty and low duty types), high alumina and conventional and super-duty silica; (2) basic refractory types (magnesite, magnesite chrome and chrome-magnesite) and (3) special refractory materials (zircon and silicon carbide).

Naturally, the chemical nature of the liquid metal, liquid slag and off-gases in the iron and steelmaking unit processes plays a critical role on the selection and application of the refractory materials in the respective unit processes (Hubble, 1999; Lee et al., 2004; Harbison-Walker, 2005). For example, basic refractory materials which exhibit resistance to corrosive reactions with chemically basic slags, dusts and fumes at elevated temperatures are best used in basic chemical environments such as in the basic oxygen steelmaking process (Hubble, 1999).

Complex refractory systems are used in modern steelmaking unit processes to provide long and safe operational campaigns necessary to ensure furnace availability and permit continuous operation (Dzermejko et al., 1999; Hubble, 1999). Technically, the iron and steel industry consumes over 60% of the refractory materials produced globally (Liu and Zhou, 2003; Lee et al., 2004). In the iron and steel manufacture, the refractory materials are used in applications ranging from the ironmaking processes, such as the blast furnace, hot stove and coke ovens, to the converter steelmaking and secondary refining furnaces, solidification continuous casting processes and reheating furnaces in the hot rolling process (Kayama et al., 2008). Table 4.7 summarizes the major types and areas of high usage of refractory materials in the steel plant (Dzermejko et al., 1999; Hubble, 1999). According to Hubble (1999), the selection of refractory materials best suited to each application is of utmost importance, as the refractory materials are not only expensive, but any failure in these materials can be catastrophic, with significant economic losses being incurred in the form of loss of production, equipment and the product itself. Furthermore, Hubble (1998) asserts that the type of refractory materials selected for any application has a significant effect on the overall energy consumption of the process as well as the quality of the final product produced.

TABLE 4.7

Major Areas of High Usage of Refractory Materials in the Steel Plant

Process Area	Type of Refractory Product Used	Desirable Refractory Features
a. Ironmaking Refractories		
Coke oven batteries	Silica bricks ($SiO_2 \geq 94.5$ wt.%)	Volume stability and chemical resistance
Blast furnace stack and belly furnace lining	90 wt.% Al_2O_3, Al_2O_3–Cr_2O_3 brick, 60–70 wt.% Al_2O_3, often tar-impregnated	Resistance to chemical attack (alkali), abrasion and thermal shock. Shift towards non-oxide refractories such as SiC and graphite based bricks
Blast furnace bosh	Conventionally baked , hot pressed or synthetic carbon, semi-graphite and graphite	Conductive, resistance to alkali and slag chemical attack and thermal shock
Blast furnace stove-hot/air system	Creep-resistant alumina brick (60–80 wt.% Al_2O_3)	Creep resistance, volume stability
Blast furnace cast house (troughs, runners, spouts)	Al_2O_3-SiC-C or high alumina castable, plastics, ram mixes	Volume stability, metal and slag erosion resistance. Longer life/high quality Al_2O_3 types
Hot metal transfer/treatment (ladles, mixers, stirrers)	Burned and impregnated high-Al_2O_3 products (70–80 wt.% Al_2O_3), Al_2O_3–SiC–C brick for metal treatment	Thermal cycle resistance. Increased adoption Al_2O_3–SiC–C types for metal treatment
b. Steelmaking Refractories		
EAF hearth	Monolithic magnesite (90–97 wt.% MgO)	High thermal, chemical and impact resistance. Use of conductive bricks/ monolithic types for DC arc furnaces
EAF slagline	Resin bonded magnesia-carbon brick (10–20 wt.% C)	Resistance to thermochemical attack by slag, oxidation resistance, thermal resistance. Operational strategy to include build-up of protective slag layer on slagline
EAF upper sidewall	Magnesia-graphite brick (10–20 wt.% C)	Slag corrosion resistance, thermal resistance (flame impingement from oxygen lances and oxy-fuel burners), impact resistance (impingement by scrap during charging) Higher quality bricks based on fused magnesia grain over sintered magnesite grain in standard bricks

(Continued)

TABLE 4.7 (*Continued*)

Major Areas of High Usage of Refractory Materials in the Steel Plant

Process Area	Type of Refractory Product Used	Desirable Refractory Features
EAF tap-hole	Magnesia-carbon bricks (for top-hole sleeves); MgO gunning mix	Thermochemical resistance to liquid slag and metal flow during tapping. Metallic additions to MgO-C tap-hole bricks to provided erosion and oxidation resistance
EAF runners	Al_2O_3-SiC-C monolithics; MgO-based castable	Metal and slag resistance. Improved thermal shock resistance
EAF roof and exhaust ductwork	High alumina (70–90 wt.% Al_2O_3) precast bricks; low moisture, high Al_2O_3 (40–60 wt. %) gunning mix for delta sections	Resistance to thermal shock (during charging, melting and discharging); resistance to slag-carryover and slag abrasion from particulate-laden gases. Combination of better thermal shock and abrasion resistance
BOF cone	Standard-quality MgO–C containing anti-oxidants; Pitch-bonded MgO brick	Oxidation and thermo-mechanical resistance
BOF charge pad	Pitch impregnated burned MgO brick, High strength MgO-Graphite refractories containing anti-oxidants	Mechanical impact and abrasion resistance from scrap and hot metal
BOF slagline	Premium grade MgO–C refractories containing fused MgO and anti-oxidants	Thermo-mechanical and slag corrosion resistance
BOF bottom and stadium (bottom stirred)	High strength standard quality MgO-C bricks containing anti-oxidants, Pitch-impregnated burned MgO	Resistance to erosion my metal, slag and gases, thermos-mechanical strength (thermal gradients between gas-cooled tuyeres and surrounding lining)

Source: Adapted from Hubble, D.H., Steel plant refractories, in Wakelin, D.H. (ed.), *The Making, Shaping and Treatment of Steels: Steelmaking and Refining*, vol. 2, 11th edn., The AISE Steel Foundation, Pittsburgh, PA, 1999, p. 227.

4.4 Overview of Waste Generation in the Iron and Steelmaking Processes

In the quest to produce steel using the unit processes described in the preceding sections, a lot of waste is inherently produced. While the detailed descriptions of the unit processes given previously show that balancing productivity and efficiency of operation is a top priority for producers, special consideration to the environmental effects is now taking centre stage. Nowadays, environmental stewardship is a top priority for metal producers globally, with a lot of research being focused on process integration to minimize the generation of metallurgical wastes, and to recycle and reuse the metallurgical wastes either within the metal production processes or in processes outside the steel plant.

Generic wastes produced in unit processes in the production of iron and steel include process slag, which constitute about 90% of total waste produced, and metallurgical dusts and sludge (World Steel Association, 2014). Figure 4.8 shows the generic categories of solid by-products by steelmaking process (World Steel Association, 2014). Furthermore, the various types of wastes produced from the steel plants can further be grouped into specific categories: blast furnace ironmaking slag dusts and sludge, basic oxygen steelmaking converter slag, electric arc steelmaking slag and mill scale and pickling sludge as shown in Figure 4.9 (Pazdej and Vogler, 1995; Das et al., 2007; Remus et al., 2013).

The amount of metallurgical wastes generated in unit processes depends on factors such as quality of raw materials used, technology status of the process, operating philosophy,

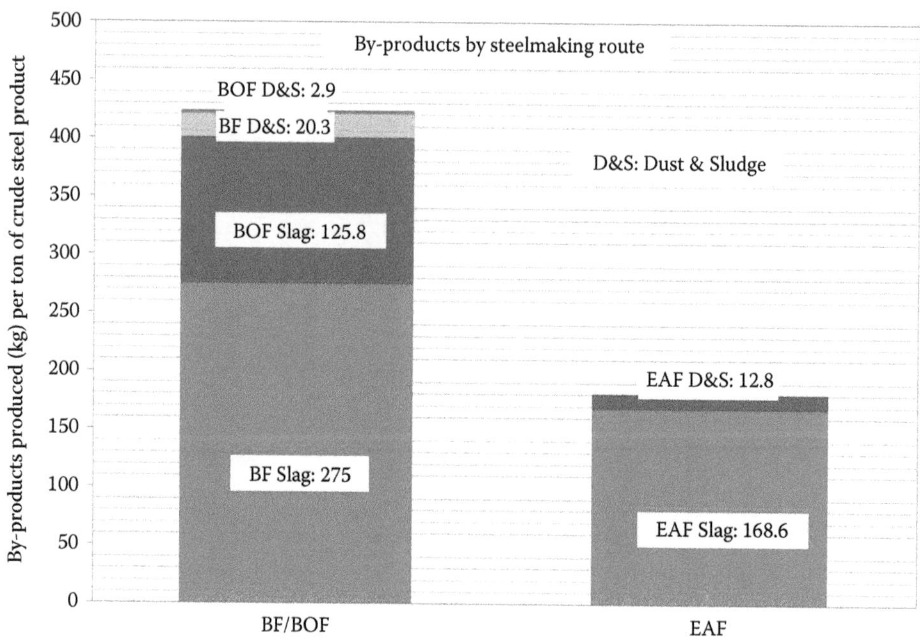

FIGURE 4.9
Generic solid waste products by steelmaking process. (From World Steel Association, Steel industry by-products, Available at https://www.worldsteel.org/publications/fact-sheets/content/01/text files/file/document/Fact_By-products_2014.pdf, Accessed 20 November 2015, 2014.)

final product mix and the extent of process integration involved to minimize the generation of wastes (Pazdej and Vogler, 1995; Das et al., 2007; Remus et al., 2013; Lobato et al., 2015). If untreated, these metallurgical wastes can pose serious hazards to the environment and health as a result of the toxic metals contained in the wastes. On the other hand, these metallurgical wastes can also be potential sources of secondary metals that can be converted into useful by-products if they are treated and recycled properly. Table 4.8 shows the typical categories and amounts of wastes per unit product generated from iron and steel smelting and refining unit processes.

TABLE 4.8

Typical Categories and Amount of Wastes Generated from Unit Processes

Unit Process	Typical Waste	Waste Quantities, kg/ton of Product	References
Raw aterial agglomeration	Coke oven gas (COG)	280–450 Nm^3/ton coke	Remus et al. (2013)
	Coke dust/ breeze	15.7–298 g/ton coke	Remus et al. (2013)
	Sinter blending dusts	0.5–37.7 g/ton sinter	Remus et al. (2013)
	Sinter secondary dusts	14.5–40 g/ton sinter	Remus et al. (2013)
	Sinter sludge	Not available	—
	Sinter cooling	14–212 g/ton sinter	Remus et al. (2013)
	Pelletization dust	14–150 g/ton pellets	Remus et al. (2013)
Blast furnace process	Blast furnace dust	7–45 kg/ton HM	Pazdej and Vogler (1995); Das et al. (2007)
	Blast furnace sludge	6 kg/ton HM	Pazdej and Vogler (1995); Das et al. (2007); Remus et al. (2013)
	Blast furnace slag	200–400 kg/ton HM	Pazdej and Vogler (1995); Das et al. (2007); Remus et al. (2013); Lobato et al. (2015)
	Refractory bricks	5.9 kg/ton HM	Remus et al. (2013)
Basic oxygen process	BOF slag	85–200 kg/ton liquid steel	Das et al. (2007); Remus et al. (2013)
	Desulphurization slag	3–40 kg/ton liquid steel	Remus et al. (2013)
	BOF Sludge	15–16 kg/ton liquid steel	Das et al. (2007); Remus et al. (2013); Lobato et al. (2015)
	BOF dusts	14–143 kg/ton liquid steel	Remus et al. (2013)
	BOF Refractories	Not available	—
Electric arc smelting	EAF dust	15–20 kg/ton liquid steel	Das et al. (2007); Remus et al. (2013); Suetens et al. (2014); Lobato et al. (2015)
	EAF slag	100–180 kg/ton liquid steel	Das et al. (2007); Remus et al. (2013); Lobato et al. (2015)
	Refractory bricks	2–25 kg/ton liquid steel	Remus et al. (2013)
Stainless steel refining	EAF dust	10–30 kg/ton liquid steel	Remus et al. (2013)
	Secondary refining slag	9–15 kg/ton liquid steel	Remus et al. (2013)
	AOD slag	~160 kg/ton liquid steel	Remus et al. (2013)
	Casting process slag	4–5.7 kg/ton liquid steel	Remus et al. (2013)
Casting and rolling mills	Mill scale	35–40 kg/ton of steel	Das et al. (2007); Lobato et al. (2015)
	Mill sludge	12 kg/ton steel	

4.5 Categories of Metallurgical Wastes Generated in Iron and Steelmaking Unit Process

4.5.1 Metallurgical Dusts and Sludge

Metallurgical dusts are defined as particulate waste produced in the dry cleaning systems of process gases produced from the various unit processes in the production of iron and steel (Das et al., 2007; Remus et al., 2013; Lobato et al., 2015). On the other hand, metallurgical sludge is obtained in the wet gas cleaning systems used to treat the furnace off-gases (Lobato et al., 2015). The categories and typical amounts of metallurgical dusts and sludge generated from iron and steelmaking processes are shown in Table 4.8. Sections 4.5.1.1 to 4.5.1.3 discusses in detail the generation of metallurgical dust and sludges in the different unit processes used in the production of iron and steel products.

4.5.1.1 Generation of Particulate Dust from the Coke Production Process

In general, the coking process involves the high temperature carbonization of metallurgical grade coal in the absence of oxygen to yield about 700–800 kg dry coke per ton of dry coal (Diez et al., 2002; Remus et al., 2013). Typically, the gas produced from the coke making process varies from 140 to 200 kg of coke oven gas per ton of dry coal (Remus et al., 2013). During the coke making process, several waste streams and particulate dust emissions are also produced. Typical waste emissions from the coke production process include the conventional pollutants in the form of particulate matter (PM); gaseous emissions containing sulphur dioxide (SO_2) and nitrogen oxides (NO_x) and numerous organic compounds containing polycyclic organic matter (POM), volatile organic compounds (VOCs) and polycyclic aromatic hydrocarbons (PAH) (EPA, 2008). The major sources of dust and gaseous emissions in the coke making process include the coal handling and preparation, coke oven battery operation and the handling and conveying of the coke product. Figure 4.10 shows the process flowsheet and material flow in a by-product recovery coke making process (Atkinson and Kolarik, 2001; Remus et al., 2013).

Coal handling and preparation: This stage involves normal handling operations such as coal pulverization, transportation and conveying, storage and stockpiling and charging operations. The generation of coal fines during handling and preparation is inevitable, resulting in the production of harmful airborne coal dust emissions (Remus et al., 2013). As discussed in Chapter 2, prolonged exposure to airborne respirable dust results in pneumoconiosis, a chronic lung disease that is associated with inhalation and deposition of dust in human lungs.

Coke oven battery operation: Coke oven battery operation involves coal charging, heating and firing, coking, coke pushing and coke quenching. This stage is characterized by particulate dust and gaseous emissions, particularly from the coke quenching, leakages and fugitive emissions (Atkinson and Kolarik, 2001; Remus et al., 2013). Typically, the categories of waste from the coke oven battery operation include PM, VOCs, PAHs, methane, ammonia, hydrogen sulphides, carbon monoxide, sulphur oxides and hydrogen cyanide (Atkinson and Kolarik, 2001; Remus et al., 2013).

Coke oven gas (COG) treatment: Although COG is technically not categorized as a waste material, it still constitutes a major by-product in the coke making process. If not recovered properly, coke oven gas can result in material and energy losses from the coking process (Atkinson and Kolarik, 2001). Because the gas contains high calorific value due to the presence of hydrogen and hydrocarbons, the standard procedure is to recycle it internally as fuel

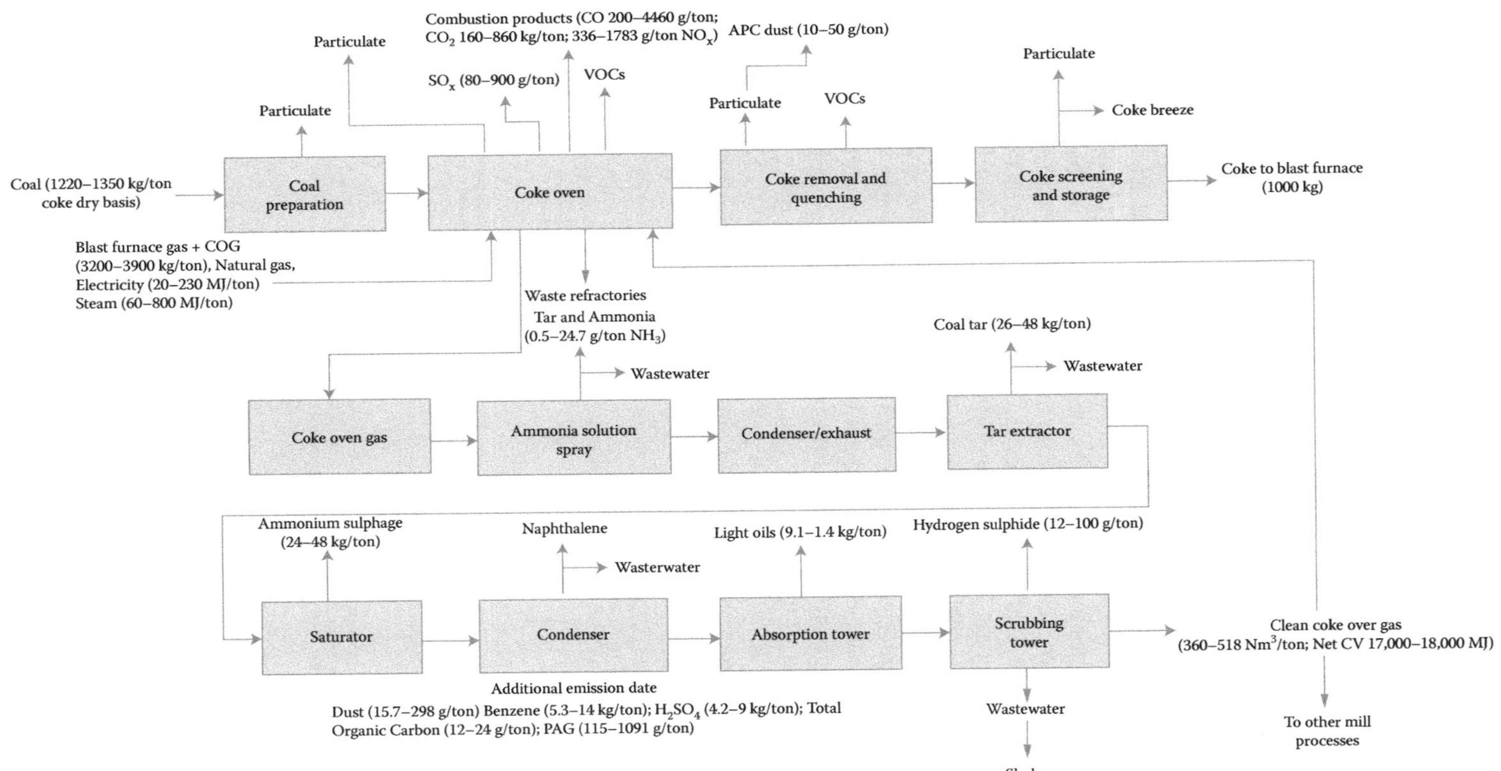

FIGURE 4.10
Process flowsheet and material flow in the coke making process. (From Atkinson, M. and Kolarik, R., *Steel Industry Technology Roadmap*, The American Iron and Steel Institute, Washington, 2001; Remus, R. et al., Best available techniques (BAT) reference document for iron and steel production, Industrial Emissions Directive 2010/75/EU, European commission JRC Reference Report, 2013, p. 233.)

in coke oven batteries, blast furnace, sinter plant or reuse for power generation. According to Remus et al. (2013), the average yield of coke oven gas from European coke producers is in the range of 280–450 m^3/ton dry coal, and the gas typically contains about 39–65 vol.% H_2, 20–42 vol.% CH_4 and calorific value in the range 17.4–20 MJ/Nm^3 (Remus et al., 2013). In non-recovery processes where the by-products from the coke oven batteries are not recovered, the raw coke oven gas is usually combusted directly to generate electricity and provide heat for the coking process (Atkinson and Kolarik, 2001; Remus et al., 2013).

4.5.1.2 Generation of Dust from the Raw Material Agglomeration Processes

The importance of raw material agglomerates in the efficient operation of the blast furnace was highlighted in Section 4.2. The raw material agglomerates in the form of sinter and pellets play an integral part if the integrated iron and steelmaking process (Lanzerstorfer, 2015). Notwithstanding the criticality of these agglomerates in the blast furnace operation, the agglomeration process is considered one of the major sources of particulate dusts and gaseous emissions within the integrated iron and steelmaking process (Atkinson and Kolarik, 2001; Buekens et al., 2001; Kasai et al., 2001; Das et al., 2007; Ooi and Lu, 2011; Remus et al., 2013; Lanzerstorfer, 2015). This section discusses the generation and categories of particulate waste materials produced in the raw materials agglomeration process, with particular emphasis being given on the sinter plant.

Figure 4.11 shows the material flow and typical categories of waste streams generated from the sinter plant. In most cases, the particulate dusts from the sinter plant are usually recovered by electrostatic precipitators and/or mechanical de-dusting processes, and are recycled back as feed material for the sinter plant along with the other fine-grained residues (Atkinson and Kolarik, 2001; Ball et al., 1973; Das et al., 2007; Remus et al., 2013; Lanzerstorfer, 2015). The internal recycling of sinter dust is practiced by most integrated operators, mainly because the main fraction of the dust consists of iron oxides (Ball et al., 1973;

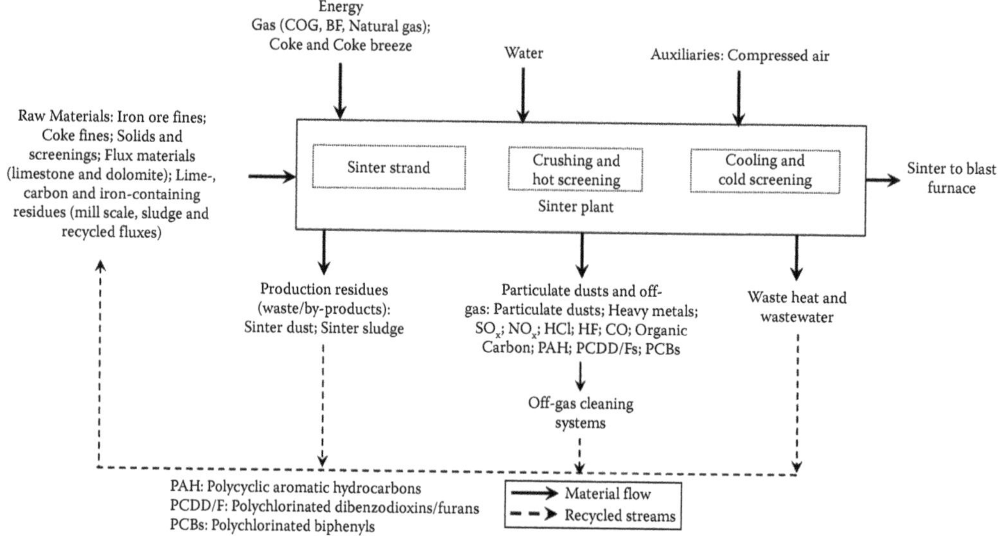

FIGURE 4.11
Material flow and dust generation within the sinter plant. (Adapted from Remus, R. et al., Best available techniques (BAT) reference document for iron and steel production, Industrial Emissions Directive 2010/75/EU, European Commission JRC Reference Report, 2013, p. 233.)

Biswas, 1981; Atkinson and Kolarik, 2001; Yadav et al., 2002; Mohamed et al., 2010; Gavrilovski et al., 2011; Remus et al., 2013; Niesler et al., 2014; Lanzerstorfer et al., 2015). However, the direct recycling of sinter dust has its own challenges, such as the build-up of alkali and volatile metal compounds, which tend to compromise the quality of the sinter product and adversely affect the operation of the blast furnace (Atkinson and Kolarik, 2001; Remus et al., 2013; Trinkel et al., 2015).

Table 4.8 categorized the different types and amounts of wastes generated from typical raw material agglomeration processes (Das et al., 2007; Remus et al., 2013). Major sources of waste particulate dust emissions include undersize material from screening processes, and particulate dusts from raw material blending, secondary de-dusting and gas cleaning units. According to Atkinson and Kolarik (2001), the wind-box exhaust is considered as the primary source of particulate dust emissions in the sinter plant. These particulate dusts consist of mainly iron oxides, sulphur oxides, carbonaceous compounds and chlorides, while those emitted at the discharge end mainly consist of iron and calcium oxides. Typical components of sinter dust include Fe, C, S, SiO_2, Al_2O_3, CaO and MgO (Steiner, 1976; Atkinson and Kolarik, 2001). In addition, Remus et al. (2013) characterized the typical composition of sinter particulate dust from several European sinter plants as containing 43.7–49.9 wt.% total iron, 2.9–25.8 wt.% chlorine, 0.58–31.6 wt.% sodium oxides, 2.9–6.12 wt.% carbon, 7.55–7.83 wt.% calcium, 3–9.07 wt.% potassium and 0.22–4.07 wt.% sulphur.

Sinter plant gaseous emissions: Despite the criticality of sinter in the blast furnace iron-making process, the sinter plant is synonymous with the production of toxic gaseous emissions in the form of polychlorinated-p-dibenzodioxins and dibenzofurans (PCDD/Fs) (Buekens et al., 2001; Kasai et al., 2001; Fisher et al., 2004; Ooi and Lu, 2011). Technically, the production of these PCDD/Fs and other organometallic compounds occur as a result of high temperature combustion reactions and volatilization of organic compounds in sinter raw materials and recycled oily mill scale additives (Buekens et al., 2001; Kasai et al., 2001; Das et al., 2007; Ooi and Lu, 2011; Remus et al., 2013). Basically, the formation of PCDD/Fs is proposed to occur via two mechanistic pathways (Vogg and Stieglitz, 1986; Buekens et al., 2001; Kasai et al., 2001; Ooi and Lu, 2011): (1) the de novo synthesis, where PCDD/Fs are presumed to form as a result of the combustion of non-chlorinated organic aromatic matter such as polystyrene, cellulose, lignin and coal in the presence of chlorine donors and a suitable metal catalyst, and (2) precursor pathway, where the PCDD/Fs are formed heterogeneously from structurally similar precursors, such as chlorobenzene and chlorophenols, either in the gas phase or on the surface of particles. From several empirical tests using pilot and laboratory scale sintering processes, it is generally accepted that the formation of PCDD/Fs in the sinter plant is due to the de novo reaction (Buekens et al., 2001; Kasai et al., 2001; Ooi and Lu, 2011).

Generally, the sinter plant functions not only to agglomerate iron ore fines for the blast furnace process, but also to recycle the different particulate waste streams produced in the iron and steelmaking process (Atkinson and Kolarik, 2001; Das et al., 2007; Ooi and Lu, 2011; Remus et al., 2013). Naturally, the toxic gaseous emissions from the sinter plant present challenges in controlling environmental pollution as well as workplace hygiene (Ooi and Lu, 2011). The permissible emission limits of PCDD/Fs in metallurgical processes require elaborate off-gas cleaning systems guided by the Stockholm Convection for Persistent Organic Pollutants (United Nations Environmental Program, 2011). As such, a number of off-gas treatment methods have been proposed to reduce the emission of PCDD/Fs in sintering processes, and these include a combination of wet and dry off-gas scrubbing systems, exhaust gas recirculation, selective catalytic reduction and activated carbon adsorption (Ooi and Lu, 2011; Yu et al., 2012; Remus et al., 2013).

4.5.1.3 Particulate Dust Generation from Iron and Steel Unit Processes

The proper management of particulate dusts and sludges collected from the gas cleaning systems of steelmaking processes is a growing global concern in terms of environmental and health considerations. The blast furnace, EAF and BOF gas cleaning systems constitute a major source of particulate dust emissions within the iron and steelmaking process (Zeydabadi et al., 1997; Yadav et al., 2002; Das et al., 2007; Remus et al., 2013; Lobato et al., 2015). The composition of these wastes varies depending on the process design and configuration, operational procedures and the type of product produced (Das et al., 2007). Basically, these particulate dusts and sludges consist of oxides of iron, calcium, silicon, manganese, aluminium, varying concentrations of volatile toxic metal compounds such as zinc, lead, cadmium, as well as alkaline metal compounds (Zeydabadi et al., 1997; Yadav et al., 2002; Remus et al., 2013; Lobato et al., 2015). This section discusses in detail the different categories of particulate dusts and sludges produced in the different unit processes.

Blast furnace dust and sludges: On average, the blast furnace process produces about 7–45 kg of dust and 6 kg of sludge per ton of hot metal, and the total dust and sludge emissions constitute about 1%–4% of the hot metal production (Pazdej and Vogler, 1995; Atkinson and Kolarik, 2001; Das et al., 2007; Remus et al., 2013). These dusts and sludges are generated from the scrubbing and cooling of flue gases during the ironmaking process. Once the off-gases have been cleaned and cooled, they are usually combusted in the blast furnace stoves to preheat the incoming cold air injected into the furnace and/or used as fuel in other parts of the plant (Atkinson and Kolarik, 2001). Generally, the cleaning of flue gases involves the removal of large particulates via dry dust collection systems to yield blast furnace dust, followed by a wet gas cleaning process for fine particulate removal to produce sludges (Street, 1997; Atkinson and Kolarik, 2001). As shown in Table 4.9, the

TABLE 4.9

Physicochemical Characteristics of a Typical Blast Furnace Flue Dust

Constituent	Sample I (%)	Sample II (%)
Carbon	29.90	33.62
Fe_2O_3	51.10	49.50
SiO_2	6.31	8.30
Al_2O_3	5.12	2.54
CaO	4.90	1.96
MgO	0.88	1.33
Pb	0.024	0.019
Zn	0.042	0.028
MnO	0.58	0.02
K_2O	1.22	0.154
Na_2O	0.47	0.07
T-Fe	35.7	34.62
Bulk density (g cm^{-3})	1.42	1.32
Specific gravity	2.59	2.56
Porosity	45.17	48.53

Source: Reprinted from *Resour. Conserv. Recycl.*, 50, Das, B., Prakash, S., Reddy, P.S.R. and Misra, V.N., An overview of utilization of slag and sludge from steel industries, 40–57. Copyright 2007, with permission from Elsevier.

physicochemical characteristics of blast furnace dusts and sludges fundamentally consist of oxides of iron; calcium; silicon; manganese; aluminium and varying concentrations of toxic elements such as zinc, lead, cadmium and alkaline metals compounds, as well as carbon in the form of coke breeze (Street, 1997; Zeydabadi et al., 1997; Atkinson and Kolarik, 2001; Yadav et al., 2002; Das et al., 2007; Remus et al., 2013; Lobato et al., 2015).

It is generally agreed that the physicochemical characteristics of a blast furnace consist of complex physical mixtures of fine irregularly shaped particles consisting entirely of heterogeneous mixtures of fine feed material entrained in the ascending gases in the blast furnace (Itoh and Fieser, 1982; Street, 1997; Zeydabadi et al., 1997; Atkinson and Kolarik, 2001; Yadav et al., 2002; Gupta et al., 2005; Das et al., 2007; Remus et al., 2013; Lobato et al., 2015; Trinkel et al., 2015). Generally, the particle size, composition, heterogeneity and mineralogy of the blast furnace particulate dusts are dependent, not only on the specific process conditions, but also on the method of collection such as the use of dry or wet dust collection systems (Itoh and Fieser, 1982; Street, 1997; Trinkel et al., 2015).

As shown in Table 4.8, zinc constitutes one of the major constituent components in the blast furnace particulate dust and sludges. Basically, the amount of zinc contained in blast furnace dusts can be as high as 100–150 g/ton of hot metal produced (Pazdej and Vogler, 1995; Remus et al., 2013). The zinc and alkali metal compounds which are introduced into the blast furnace as parts of raw materials are volatilized from the high temperature zones of the furnace and accrete on to the refractory material on the cooler upper parts of the blast furnace, and the dust particles in the off-gas act as nuclei for the vaporized zinc to condense on (Biswas, 1981; Itoh and Fieser, 1982; Street, 1997; Trinkel et al., 2015). An over-limit zinc loading in the blast furnace is detrimental to blast furnace operation, particularly due to potential damage to refractory materials thereby affecting the blast furnace campaign life and productivity (Biswas, 1981; Ma, 2016). According to Biswas (1981), the formation of scaffolds from the volatile accreted materials, especially in the upper and middle sections of the blast furnace, results in the slagging and softening attack of refractory materials, and thereby resulting in the accelerated wear of the refractory materials in the affected regions. As a result, the presence of toxic and undesirable elements, such as zinc, lead and alkali metal oxides, presents serious challenges to the direct recycling of blast furnace flue dust back into the steel plant (Pazdej and Vogler, 1995; Zeydabadi et al., 1997; Remus et al., 2013).

EAF dusts: In the electric arc process, steel charge and slag-forming fluxes are charged and heated to melting temperatures to produce liquid steel and slag of varying compositions (Remus et al., 2013). In the process, the electric arc process also generates a lot of particulate dust, typically in the range 15–20 kg/ton of liquid steel produced (Das et al., 2007; Remus et al., 2013; Suetens et al., 2014; Lobato et al., 2015). According to Suetens et al. 2014, over 7.1 Mt of EAF dusts were generated globally in 2010, and the rate of generation is expected to increase due to the increased recycling volumes of steel scrap. In general, the particulate dust tends to contain both valuable and toxic metal elements such as iron, zinc, magnesium, manganese, calcium, zinc and lead (Nyirenda, 1991; Street, 1997; Yadav et al., 2002; Dutra et al., 2006; Das et al., 2007; Remus et al. 2013; Lobato et al., 2015). During the melting process, volatile elements, such as zinc, lead and cadmium, present in the feed materials are volatilized from the steel bath and other high temperature sections of the furnace and concentrated as complex compounds in the EAF off-gas (Nyirenda, 1991; Atkinson and Kolarik, 2001; Yadav et al., 2002; Dutra et al., 2006; Das et al., 2007; Remus et al., 2013). In most operations, the major source of such volatile elements in the EAF process is through the charging of galvanized steel scrap and/or the charging of unsorted scrap materials. (Nyirenda, 1991; Atkinson and Kolarik, 2001; Remus et al., 2013). As a result, the concentration of zinc and other toxic volatile metal elements in EAF flue dusts is a major

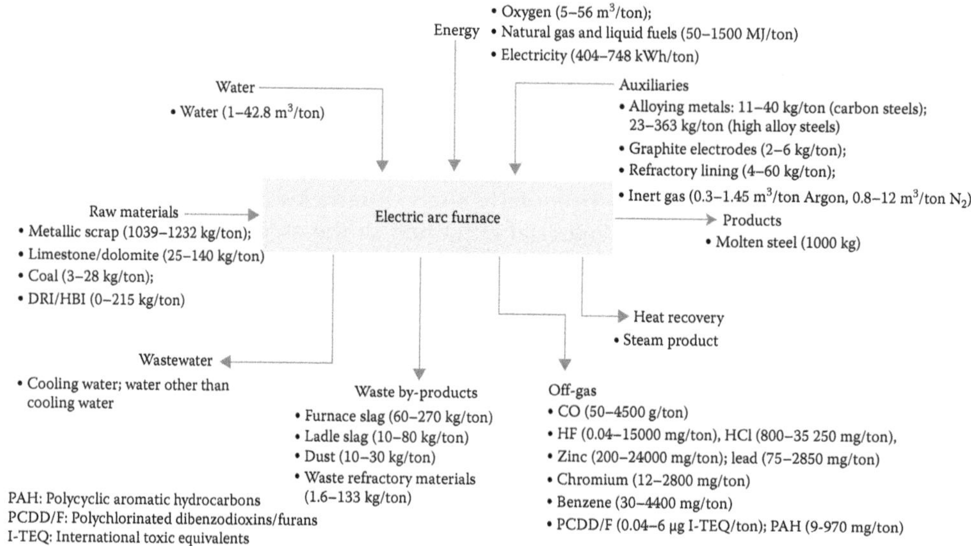

FIGURE 4.12
Material flow in the EAF. (From Atkinson, M. and Kolaric, R., *Steel Industry Technology Roadmap*, The American Iron and Steel Institute, Washington, 2001; Remus, R. et al., Best available techniques (BAT) reference document for iron and steel production, Industrial Emissions Directive 2010/75/EU, European Commission JRC Reference Report, 2013, p. 428.)

concern for global operators. Figure 4.12 shows a schematic highlight of the material flows per ton of liquid steel based on the EAF process practice within some selected EU member states (Remus et al., 2013).

Mechanism of formation of EAF dust: The majority of the EAF particulate dust emissions are generated during the melting, oxygen blowing cycles and the general operation of the furnace (Nyirenda, 1991; Li and Tsai, 1993; Guézennec et al., 2005; Remus et al., 2013). During the melting and oxygen blowing cycles, the volatile elements are volatilized from the liquid metal bath into the off-gas where they condense and are reoxidized to form oxides and complex compounds (Nyirenda, 1991; Li and Tsai, 1993; Guézennec et al., 2005). The detailed mechanism of formation of dusts during EAF melting was discussed in detail by Li and Tsai (1993) and Guézennec et al. (2005). These researchers proposed that the particulate dust formation process can take place in two steps, namely, the emission of dust precursors such as metal vapours, metal droplets and solid particles inside the furnace, followed by the conversion of the precursors into dust by agglomeration and physicochemical transformations (Li and Tsai, 1993; Guézennec et al., 2005). Furthermore, Guézennec et al. (2005) proposed that the emission of the dust precursors involved any one of the following mechanistic steps: (1) the localized volatilization at the hot spots in the arc zone, the oxygen jet zone, as well as in the CO bubbles; (2) the projection of droplets at the impact points of the arc or of the oxygen jet on the steel bath resulting in particles whose chemical composition resembles that of slag; (3) the projection of fine droplets by bursting of CO bubbles as a result of the decarburization of the steel bath; (4) the bursting of droplets in contact with an oxidizing atmosphere within the furnace via a phenomenon commonly referred to as bubble-burst mechanism and (5) direct fly-off of solid particles during the introduction of charge materials such as scrap, slag formers and recycled dust into the furnace. Based on the proposed mechanisms of formation, the extensive characterization of EAF dust samples by Guézennec et al. (2005) identified the following morphologies: (1) homogenous spheres whose composition corresponded to that of either zinc-enriched slag or steel bath and (2) heterogeneous spheres made up of a slag phase and a steel phase enriched with zinc, exhibiting an iron-rich dendritic structure encapsulated within a vitreous phase.

Physicochemical characteristics of EAF dusts: The chemical composition and mineralogical distribution of elements in ultrafine EAF dust was also highlighted in several studies (Nyirenda, 1991; Li and Tsai, 1993; Sofilic et al., 2004; Havlík et al., 2006; Machado et al., 2006; Rao, 2011; Remus et al., 2013). The extensive particle size characterization of particulate dusts revealed that some of the samples exhibited bimodal distribution in the particle size range around 40 μm for ultra-fine particles, and 100–125 μm for fine-to-coarse particles (Li and Tsai, 1993; Delalio et al., 1998; Sofilic et al., 2004). In general, the composition of the particulate dust varies widely depending on the type of scrap used, the type of steel product produced and the process conditions used (Nyirenda, 1991; Li and Tsai, 1993; Sofilic et al., 2004). For example, the dusts from the primary melting of steel scrap to produce carbon steels would be richer in zinc and lead, whereas those produced from the secondary remelting of steel would contain a relatively low amount of volatile elements but higher in alloying elements such as chromium, nickel, manganese and molybdenum (Nyirenda, 1991; Remus et al., 2013).

Understanding the physicochemical and morphological behaviour of EAF dusts is important in developing effective dust recycling and stabilization processes. Extensive characterization of EAF dusts by x-ray diffraction methods revealed the predominance of magnetite (Fe_3O_4); complex spinel ferrite types ($(Mn_xZn_yFe_{1-x-y})Fe_2O_4$), such as $ZnO–Fe_2O_3$ and $(Mn, Zn)Fe_2O_4–Fe_2O_4–Fe_3O_4$); basic oxides such ZnO and PbO; and other complex phases such as $Ca[Zn(OH)_3]_2$, $ZnCl_2 \cdot 4Zn(OH)_2$) (Nyirenda, 1991; Li and Tsai, 1993; Sofilic et al., 2004; Havlík et al., 2006; Machado et al., 2006; Rao, 2011; Remus et al., 2013). As shown from these studies, the physicochemical characteristics of the EAF particulate dusts are very complex, and would differ not only from operation to operation but also within the different heats. Since the EAF dusts collected in the off-gas treatment installations must be treated to detoxify the toxic metal elements contained therein and to recover any valuable metals present, the success of such interventions are controlled by the quality of the raw materials charged, furnace operation procedures and intimate knowledge of the process from which the dusts are generated (Pelino et al., 2002; Dutra et al., 2006; Hong-Xu et al., 2010; Lobato et al., 2015).

The contribution of EAF melting to the global steel production is gradually increasing, and this would inevitably result in an increase in the amount of EAF dusts generated annually (Remus et al., 2013; Suetens et al., 2014; World Steel Association, 2014). Due to the presence of leachable toxic metals such as Zn, Cd, Cu, Cr (VI) and Pb, among others, electric arc process dusts and sludges are classified as hazardous wastes in most countries. As a result, most jurisdictions globally have banned the disposal of untreated particulate dust materials in landfills. The tight environmental regulations, coupled with the challenges associated with the direct recycling of the EAF dusts and sludges back into the steel plant, have prompted serious research to develop processes to either recover these metals present and/ or to pretreat and stabilize the toxic metal elements before the disposal of the wastes in landfills (Lopez and Lopez-Delgado, 2002; Pelino et al., 2002; Rao, 2011).

Basic oxygen process dust and sludges: Generally, BOFs are used to refine liquid iron into steel. As discussed in Section 4.2.4, the BOF process utilizes hot liquid metal from the blast furnace process and/or liquid steel from electric arc melting furnaces (Miller et al., 1998; Worrell et al., 2010; Remus et al., 2013). Steel scrap and other metallic charges are also added as auxiliary metal additions and as coolants in the BOF. Despite the ubiquity of the

BOF in steel refining, the process also generates a tremendous amount of metallurgical wastes, mainly in the form of process slag, and the particulate dusts and sludges captured from the off-gas dry and wet cleaning systems, respectively.

Basically, the BOFs generate a significant amount of metal-containing dust especially during the charging of steel scrap and hot metal, the oxygen blowing stages and the tapping of slag and liquid steel (Miller et al., 1998; Remus et al., 2013). The rate of generation of the aforementioned waste streams varies greatly depending on the operating conditions such as the quality of raw materials charged, furnace operation procedure and grade of the steel product (Street, 1997; Das et al., 2007; Ma, 2016). On average, the basic oxygen process generates about 15–16 kg of sludge and 14–143 kg of dust per ton of liquid steel, respectively (Das et al., 2007; Remus et al., 2013; Lobato et al., 2015). The particulate dusts and sludges are typically rich in Fe; CaO; MgO; and varying amounts of zinc, manganese, nickel, chromium and other metal compounds (Nyirenda, 1991; Wu, 1999; Remus et al., 2013; Ma, 2016). The presence of toxic heavy metal and volatile elements in the basic oxygen process particulate dusts and sludges entails that these waste materials are designated as hazardous materials in whose disposal in an untreated form is banned in most jurisdictions globally (Wu, 1999; Atkinson and Kolarik, 2001; Kelebek et al., 2004; Das et al., 2007; Trung et al., 2011; Veres et al., 2015; Ma, 2016). In general, the chemistry and composition of the BOF off-gas waste can be directly correlated to the chemistry of the metallic charge used in the process, the amount and composition of alloying elements added and type of steel scrap used. Figure 4.13 shows a schematic representation of the material flows per ton

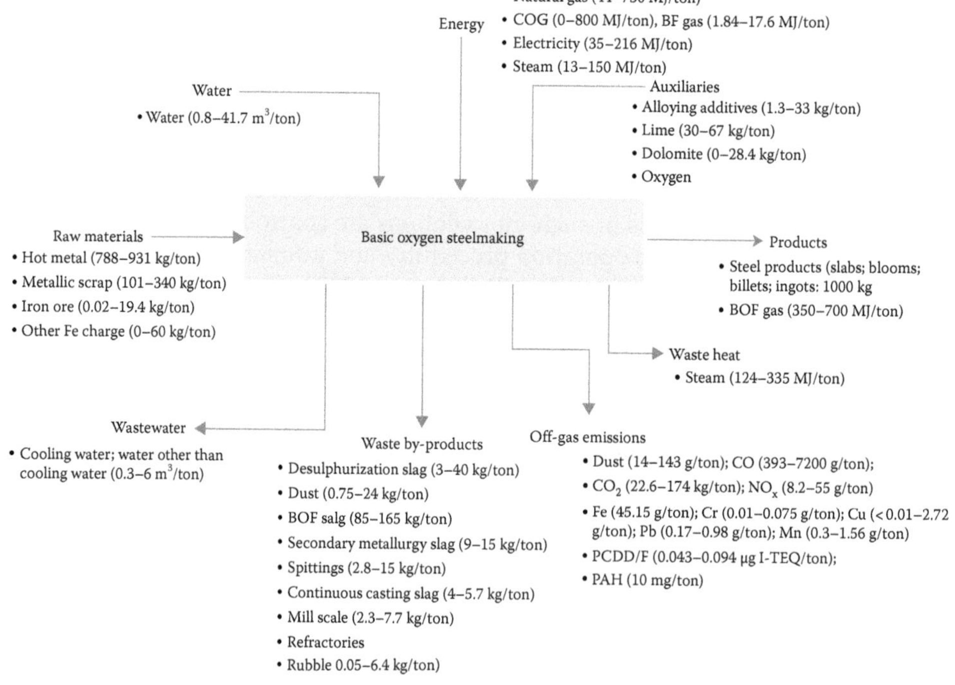

FIGURE 4.13
Material flow for basic oxygen steelmaking process. (From Remus, R. et al., Best available techniques (BAT) reference document for iron and steel production, Industrial Emissions Directive 2010/75/EU, European Commission JRC Reference Report, 2013, p. 367.)

of liquid steel based on standard basic oxygen steelmaking practice within some selected EU member states (Remus et al., 2013).

Formation of basic oxygen furnace dust and sludges: Since the basic oxygen process is operated at process temperatures in excess of 1600°C, volatile elements such as zinc, lead and cadmium present in the charge are volatilized into the furnace off-gas gas, and collect as fine dust particles in the dry and wet off-gas cleaning systems (Wu, 1999; Kelebek et al., 2004; Das et al., 2007; Ma, 2016). The speciation of volatile metals in particulate dusts and sludges in the basic oxygen process is a major concern, and has been extensively investigated by several researchers (Wu, 1999; Kelebek et al., 2004; Das et al., 2007; Tateishi et al., 2008; Trung et al., 2011; Remus et al., 2013; Veres et al., 2015). In particular, the main source of zinc in the BOF is galvanized steel scrap additions, which entails that the zinc content within the basic oxygen process dusts and sludges would vary depending on the extent of steel scrap addition (Kelebek et al., 2004).

Physicochemical characteristics of basic oxygen furnace dust and sludges: According to Wu (1999) and Veres et al. (2015), zinc is usually present in BOF flue dust mainly as franklinite (zinc ferrite spinel $ZnFe_2O_4$), calcium ferrite ($CaFe_2O_4$), zincite (ZnO), magnetite (Fe_3O_4), hematite (Fe_2O_3) and some elemental iron. Based on these studies, the morphological and mineralogical characteristics of BOF flue dust and sludges are very complex due to the high temperature and turbulent hydrodynamic conditions in the basic oxygen process. Furthermore, the dusts and sludges have been observed to consist of ultra-fine to fine-grained spherical particles formed from the volatilization and the subsequent condensation of the metals (Wu, 1999; Kelebek et al., 2004). Kelebek et al. (2004) applied both x-ray diffraction and scanning electron microscopy methods to characterize the morphologies of typical BOF sludge samples. The results showed that the cross-sectional composition of 40 μm diameter spherical particle sample exhibited the characteristic of a metallic iron core inside the particle which was coated by a slightly agglomerated ferrous oxide layer and agglomerated fine particles of zinc ferrite.

The complexity of the physicochemical properties of the BOF dusts and sludges presents technical, economic and sustainability challenges to the recycling and reuse of these particulate wastes (Nyirenda, 1991; Street, 1997; Ma, 2016). For example, the direct recycling of the dust and sludges containing iron and alloying elements into the steel plant is usually constrained by the potential recycle of the deleterious metal elements which subsequently result in the build-up of impurities in liquid steel (Nyirenda, 1991; Wu, 1999; Yadav et al., 2002; Kelebek et al., 2004; Das et al., 2007; Remus et al., 2013; Ma, 2016).

Stainless steel refining dusts and sludges: Stainless steels contain a relatively high amount of alloying elements such as chromium, nickel, molybdenum, niobium and titanium among others, which are added to engineer the desired mechanical and chemical characteristics of the final stainless steel product (Patil et al., 1998; Total Materia, 2008a). The stainless steel refining processes can either be duplex, that is the melting of steel scrap in an EAF followed by AOD converter refining, or triplex, where the EAF melting and decarburization refining are coupled to additional processes such as ladle and/or vacuum converter treatment (Patil et al., 1998; Total Materia, 2008a).

In particular, stainless steel refining dusts and sludges constitute important waste products collected in the furnace off-gas cleaning system (Ma and Garbers-Craig, 2009). Approximately, about 30–50 kg of particulate dust is generated per ton of liquid stainless steel produced in the combined electric arc melting, decarburization converting, grinding shop and rolling mill processes (Ma and Garbers-Craig, 2009; Huaiwei and Xin, 2011; Singhal and Rai, 2015).

In the quest to produce high-value products, the stainless steel refining process inadvertently produces a wide range of hazardous wastes in the form of stainless steel slag, particulate dusts, mill scale and stainless steel pickling sludges (Das et al., 2007; Ma and Garbers-Craig, 2009; Remus et al., 2013; Singhal and Rai, 2015). In particular, stainless steel refining process dusts and sludges constitute important waste products collected in the furnace off-gas cleaning system (Ma and Garbers-Craig, 2009). Approximately, about 30–50 kg of particulate dust is generated per ton of liquid stainless steel produced in the combined electric arc melting, decarburization converting, grinding shop and rolling mill processes (Ma and Garbers-Craig, 2009; Huaiwei and Xin, 2011; Singhal and Rai, 2015).

The chemical composition and physicochemical characteristics of the stainless steel refining process dusts vary considerably based on the operational parameters, raw materials used, grade and desired chemistry of the steel produced and the type of furnace equipment used (Ma and Garbers-Craig, 2006a,b,c, 2009; Laforest and Duchesne, 2006a,b; Dominguez et al., 2010; Singhal and Rai, 2015). Table 4.10 shows the non-normalized typical compositions of different stainless steel dusts. The different categories and amounts of waste dusts and sludges generated from the stainless steel refining unit processes were summarized in Table 4.8. Figure 4.14 further categorizes the input and output material flow for the stainless steelmaking process (Remus et al., 2013).

Globally, special attention is being given to stainless steel refining process wastes, with particular emphasis on particulate dusts, sludges and slags, due to the presence of toxic metal species such as Cr (VI), nickel and molybdenum (Ma and Garbers-Craig, 2006a,b,c, 2009; Huaiwei and Xin, 2011; Singhal and Rai, 2015). However, the stainless steel refining process wastes inevitably also contain a significant amount of entrained valuable metals, such as Fe,

TABLE 4.10

Typical Compositions of Different Stainless Steel Dusts

Component (wt.%)	Ma and Garbers-Craig (2009)	Laforest and Duchesne (2006a)	Dominguez et al. (2010)	Singhal and Rai (2015)
Fe_2O_3	43.4	49.56	37.1	23–30.0
CaO	12.9	6.59	16.4	27–36.0
Cr_2O_3	14.6	15.90[a]	15.4	4–9.0
SiO_2	4.81	5.76	3.1	5–7.0
NiO	2.79	5.21[a]	2.8	0–0.5
ZnO	4.49	6.50[a]	10.4	<0.5
TiO_2	0.08	0.16	na	na
Al_2O_3	0.40	0.43	0.1	na
MnO	5.08	5.88	6.8	2–5.0
MgO	5.44	4.25	3.1	9–11.0
PbO	0.39	1.51[a]	1.6	na
Na_2O	0.60	1.01	na	na
MoO_3	1.35	na	0.8	na
K_2O	0.97	0.48	0.7	na
P_2O_5	0.04	0.04	na	na
V_2O_5	0.09	na	na	na
CuO	na	na	0.5	na
Loss on ignition	−0.21	3.67	na	na

Note: na, not available.

[a] Oxides recalculated from given elemental analysis.

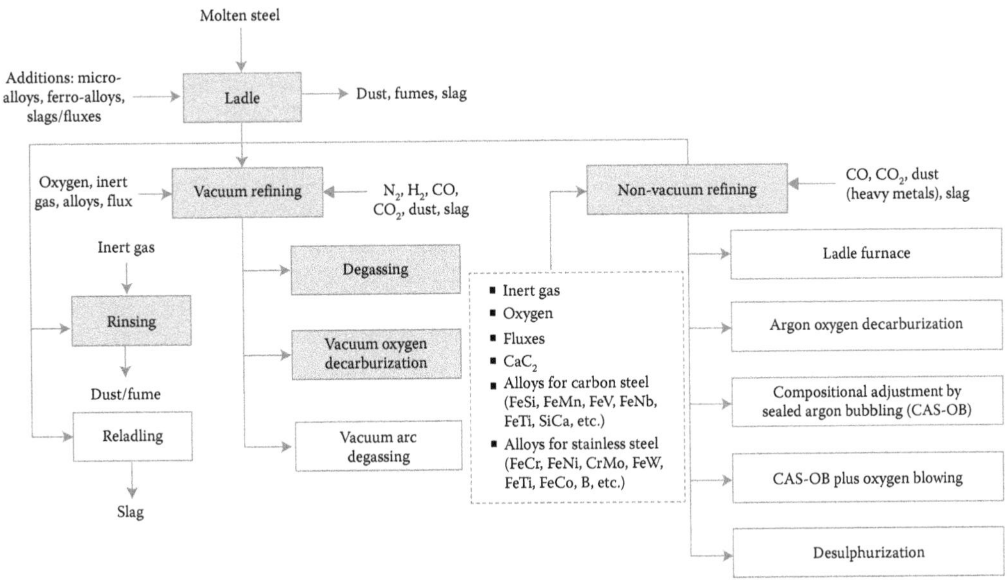

FIGURE 4.14
Input and output material flow for the stainless steel making process. (From Remus, R. et al., Best available techniques (BAT) reference document for iron and steel production, Industrial Emissions Directive 2010/75/EU, European Commission JRC Reference Report, 2013, p. 426.)

Ni, Cr and Mn, as well as toxic metal species such as Cr (VI), Zn and Pb (Laforest and Duchesne, 2006a,b; Ma and Garbers-Craig, 2009; Huaiwei and Xin, 2011; Singhal and Rai, 2015).

Mechanism of formation of stainless steel refining process dusts: The mechanism of the formation of stainless steel dust is similar to that of EAF dusts described by Li and Tsai (1993) and Guézennec et al. (2005). However, Ma and Garbers-Craig (2006a,b) proposed a different mechanism of formation of the stainless steel refining process dust to involve any of the following mechanistic steps: (1) direct entrainment of charge materials in the off-gas, particularly for charge materials such as CaF_2, SiO_2, Ni, Cr and CaO, (2) evaporation and/or volatilization of elements in the melting bath or other high temperature zones of the furnace, followed by the oxidation and/or solidification of these elements in the off gas dust and (3) the ejection of slag and metal by spitting or bursting of gas bubbles such as carbon monoxide and or argon gas from inert gas purging.

Based on the stated mechanisms, zinc, in particular, was observed to coat on the surface of the dust particles, and was presumed to have vaporized first before condensing on the dust particles (Kelebek et al., 2004). Furthermore, it is proposed that spherical metal particles and slag particles containing both primary and quench spinel crystals, enter the off gas duct by bursting of gas bubbles (Ma and Garbers-Craig, 2006a,b). The third mechanistic step proposed by Ma and Garbers-Craig (2006a,b) is most likely to become more pronounced during the gas blowing stage.

Physicochemical characteristics of stainless steel refining dust: The different mechanisms of formation of the stainless steel refining process dusts affect the morphological and crystalline behaviour of stainless steel refining dusts. The extensive characterization of crystalline and morphological phases present in the stainless steel dusts was extensively discussed by several authors (Nolasco-Sobrinho et al., 2003; Ma and Garbers-Craig, 2006a,b; Huaiwei and Xin, 2011). In most cases, the morphological and mineralogical compositions of the stainless steel dusts are very complex. The typical crystalline phases in stainless steel dusts, as reported by Ma and Garbers-Craig (2006a,b), range from simple basic oxides such as chromium oxide (CrO) and iron oxides (Fe_2O_3) to complex spinel-containing phases such as $FeCr_2O_4$, (Mg, Fe, Mn, Cr)$_3O_4$, Fe_3O_4, $MnFe_2O_4$ and $ZnFe_2O_4$. Other compounds and oxides such as CaF_2, CaO, SiO_2 and $CaCO_3$, formed from the direct entrainment of charge materials into furnace off-gas, also constitute the major crystalline phases found in stainless steel dusts (Nolasco-Sobrinho et al., 2003; Ma and Garbers-Craig, 2006a,b). Minor phases also observed include pure metallic particles rich in iron, zinc, chromium and nickel; discrete oxide phases such as NiO, MgO, PbO, SiO_2 and ZnO; varying amounts of metal chlorides such as $PbCl_2$, $ZnCl_2$, KCl and aCl and complex mixture of sulphates and hydrates such as $ZnCl_2 \cdot 4Zn(OH)_2$ and $Ca(OH)_2$ (Nolasco-Sobrinho et al., 2003; Ma and Garbers-Craig, 2006a,b).

Due to the presence of toxic metal elements such as Cr (VI), Zn, Ni, Mo and Pb, among others, stainless steel dusts and sludges are typically classified as hazardous waste material in most countries (Ma and Garbers-Craig, 2006a,b,c, 2009; Singhal and Rai, 2015). Cr (VI) is of particular environmental and health concern, not only because of its carcinogenicity, but also its solubility in water and high mobility in soils (Ma and Garbers-Craig, 2006c; Huaiwei and Xin, 2011). The toxicity and hazardous nature of stainless steel particulate dusts means that these waste materials cannot simply be stockpiled and/or disposed of in landfills without prior treatment and/or stabilization. Based on the aforementioned, detailed discussions highlighting the recycling, stabilization and solidification opportunities for the stainless steel particulate dusts are covered in detail in Section 4.6.

4.5.2 Iron and Steelmaking Slags

Metallurgical slags are produced as by-products from the various iron and steel smelting and refining unit processes. According to the World Steel Association (2012), more than 400 million tonnes of iron and steel slags are produced each year. The main categories of slags from these unit processes can be classified into two broad categories, namely, blast furnace ironmaking slags and steelmaking slags. On the other hand, steelmaking slags can also be further classified according to the specific unit process from which they are generated, namely, BOF, EAF and ladle furnace slags (Zheng and Kozinski, 1996; Shen and Forssberg, 2003; Reuter et al., 2004; Das et al., 2007; Yildirim and Prezzi, 2011; Yi et al., 2012; Lobato et al., 2015). The rate of slag generation in the iron and steelmaking process varies depending on the process, with about 200–400 kg of slag being generated per ton of hot metal in the blast furnace (Das et al., 2007; Horii et al., 2013; Lobato et al., 2015), 130–180 kg of slag per ton of liquid steel in the EAF process (Lobato et al., 2015), while the basic oxygen process generates about 150–200 kg/ton of molten steel (Lobato et al., 2015). As a result of the differences in process chemistry and operating conditions, the physicochemical characteristics of slags generated from the respective unit processes differ greatly. Figure 4.15 shows the ternary phase system for the different types of slags produced in various pyrometallurgical processes (Huaiwei and Xin, 2011).

4.5.2.1 Functions and Properties of Slags

Slags play a crucial role in the efficient smelting and refining processes in iron and steel unit processes. Depending on the physicochemical properties of slags and the specific

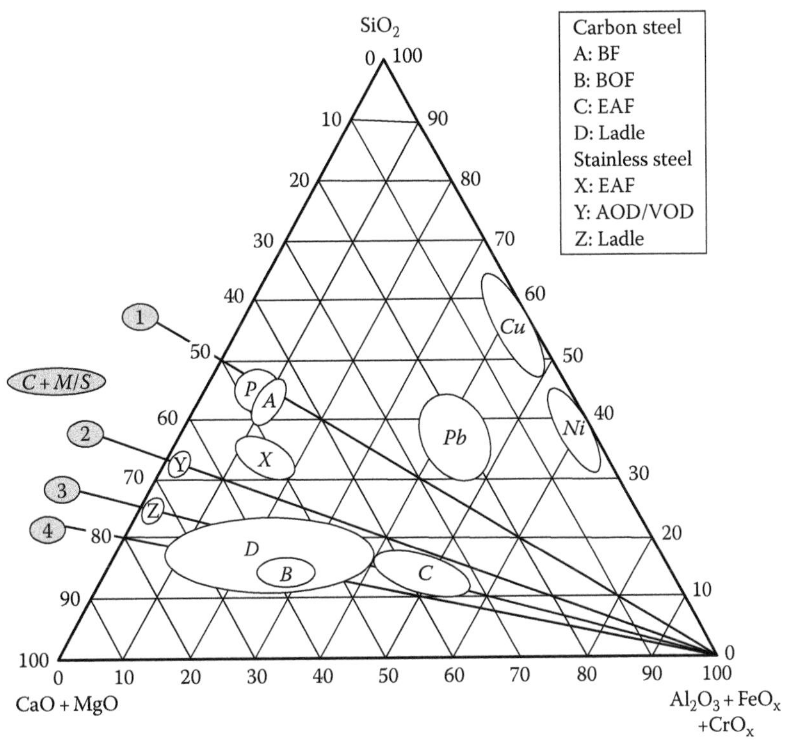

FIGURE 4.15
Slag system used in different iron and steelmaking unit processes. (Reprinted from *Resour. Conserv. Recycl.*, 55, Huaiwei, Z. and Xin, H., An overview for the utilization of wastes from stainless steel industries, 745–754. Copyright 2011, with permission from Elsevier.)

unit processes involved, process, slags play the following unique roles: (1) control of liquid metal bath chemistry through controlled slag–metal–gas reactions (Sano et al., 1997; Dippenaar, 2004; Fruehan, 2004), (2) insulation of the liquid metal bath from atmospheric reactions such as re-oxidation and nitrogen pick-up (Sano et al., 1997; Dippenaar, 2004; Fruehan, 2004), (3) refining role in the removal of impurities such as sulphur, phosphorus and non-metallic inclusions from the liquid metal bath (Sano et al., 1997; Fruehan, 2004) (4) thermal insulation to prevent heat losses from the liquid metal bath (Sano et al., 1997; Fruehan, 2004) and (5) in a slag bath, protection of the furnace refractory lining from coming into direct contact with the liquid metal bath, thereby reducing the contamination of the liquid bath through refractory–metal interactions (Oeters, 1975; Weeks, 1975; Lee et al., 2001; Zhang and Thomas, 2003; Wang et al., 2009; Harada et al., 2014).

According to Sano et al. (1997), iron and steelmaking slags must possess certain minimum properties for them to carry out the aforementioned unique functions, namely, the slags must possess liquidus temperatures lower than process temperatures to ensure the complete formation of liquid slag at iron and steelmaking temperatures. A typical example is blast furnace ironmaking slags which possess a narrow liquidus temperature in the range of 1350°C–1400°C versus the blast furnace operational temperatures of around 1500°C–1550°C (Sano et al., 1997). In addition, good slag fluidity and low slag viscosity properties ensure complete metal–slag immiscibility and reduce metal entrainment and also to ensure maximum flow properties during tapping (Sano et al., 1997). Lastly, the refining role of steelmaking slags is determined by their capacity to retain impurities through slag–metal–gas reactions (Sano et al., 1997; Fruehan, 2004), and in this case, the slags must play a refining role through removing impurities such as sulphur, phosphorus and non-metallic inclusions from the liquid metal bath.

Careful control of the physicochemical properties of slag is important for the efficient operation of smelting and refining processes. The control of physicochemical properties of slags in the iron and steelmaking process is usually done by controlled addition of slag formers or fluxes, and these fluxes in turn affect the amount and chemical composition of the of slag produced (Sano et al., 1997; Fruehan, 2004). In other words, the chemical composition and the amount of slag the generated per unit process in iron and steelmaking process tend to vary depending on the process in which they are generated, and also on quality of raw materials used and the type of slag formers and other fluxes used to constitute the slag.

4.5.2.2 Blast Furnace Slags

Blast furnace slags are formed as a by-product during the production of hot metal by thermochemical reduction in the blast furnace (EUROSLAG, 2015), with about 200–400 kg of blast furnace slags being produced per ton of hot metal using the blast furnace process (Lobato et al., 2015). According to Biswas (1981) and Sano et al. (1997), blast furnace slags belong to the $CaO–SiO_2–Al_2O_3–MgO$ system and may contain minor components such as MnO, FeO, Na_2O and K_2O, among others. Typically, blast furnace ironmaking slags have liquidus temperatures in the range of 1350°C–1400°C and a basicity ratio (CaO/SiO_2) in the range of 1.2–1.3 (Biswas, 1981; Sano et al., 1997). The process slags tapped from the blast furnace may also contain some entrained metallic globules due to insufficient slag–metal separation during tapping (Biswas, 1981; Sano et al., 1997). Because the primary function of the blast furnace is to absorb gangue and act as a recipient of non-metallic impurities, the amount and composition of slag is largely governed by the initial composition of the raw materials, the ash content in the coke and pulverized coal injections, process chemistry requirements and the furnace operation procedure (Dippenaar, 2004; EUROSLAG, 2015). Several authors have investigated the typical chemical composition of iron and steelmaking slags, and their findings are summarized in Table 4.11.

The method used to cool the tapped molten slag determines the mineralogy and different morphologies of slag products produced (Reuter et al., 2004; EUROSLAG, 2006; Remus et al., 2013; EUROSLAG, 2015). Basically, the tapped blast furnace slags are usually cooled and solidified in four different ways (Dippenaar, 2004; Reuter et al., 2004; EUROSLAG, 2015; Nippon Slag, n.d): (1) solidification in slag pots or pouring pads to form air-cooled slags, (2) quenching in water to form granulated slags, (3) cooling and solidifying with water in a pelletizing drum to form pelletized slag and (4) cooling by the controlled addition of water, air or steam to form expanded and/or slag wool with high porosity and low bulk density for use as thermal insulation materials. In most cases, the slow-cooled and granulated slags represent the majority of slag processing options for most steel plants (EUROSLAG, 2015; Nippon Slag, n.d.).

According to EUROSLAG (2015) and Reuter et al. (2004), the mineral composition of slow-cooled blast furnace slags consist of crystallized Ca–Al–Mg silicates, especially melilite ($Ca_2MgSi_2O_7–Ca_2Al_2SiO_7$) and merwinite ($Ca_3MgSi_2O_8$). In contrast, little or no crystallization takes place in granulated slags, resulting in the formation of glassy

TABLE 4.11

Typical Composition of Blast Furnace Slags

Composition							
CaO	SiO_2	Al_2O_3	MgO	MnO	FeO	S	Reference
36–45	33–42	10–16	3–16	≤2.0	na	na	Zheng and Kozinski (1996)
34.7	35.5	9.9	14.6	na	0.6	1.6[a]	Shi (2002)
35–42	35–40	8–15	8–9	0.3–1.0	0.5–0.8	0.7–1.5	Reuter et al. (2004)
35–42	33–38	10–15	7–12	≤1.0	≤1.0	1–1.5	EUROSLAG (2015)
41.7	33.8	13.4	7.4	0.3	0.5	0.8	Nippon Slag (n.d.)
33.7	35.1	10.6	14.4	0.5	0.4	1.3	Johnson (1999)

Note: na, not available.

[a] Calculated from amount of SO_3 given.

amorphous phases as a result of the vitreous solidification process (Reuter et al., 2004; Horii et al., 2013). According to Horii et al. (2013), granulated slag tend to be vitreous in nature, has a chemical composition and hydraulic properties similar to those of Ordinary Portland Cement, and hardens when reacted with water. On the other hand, air-cooled slag is similar in appearance and properties to crushed basalt stone (Horii et al., 2013).

Despite the two different types of solidified slags being produced from the same process, the distinct properties and characteristics arising from the two different processing methods may provide distinct reuse opportunities. For example, granulated slag is mostly reused as a raw material in the production of cement, while air-cooled slag has found different applications in civil construction as aggregate additives in the production of concrete and also as paving materials where they are used as substitutes for natural crushed stone (Shi, 2002, 2004; Reuter et al., 2004; Das et al., 2007; Horii et al., 2013; EUROSLAG, 2015; Lobato et al., 2015; Nippon Slag, n.d.).

4.5.2.3 Steelmaking Slags

In contrast to blast furnace ironmaking slags, steelmaking slags are by-products from the melting of steel scrap to produce liquid steel in EAFs, and from the converting and refining of steel in oxygen steelmaking and ladle furnaces (Shi, 2004; Yildirim and Prezzi, 2011). Specifically, steelmaking slags can further be categorized according to the type of steel produced, for example, carbon steel and stainless steel slags, or according to the type of the steelmaking process generating them, such as pretreatment slag, BOF slag, EAF slag, ladle refining slag and casting residue (Yi et al., 2012).

In the basic oxygen process, hot metal from the blast furnace, steel scrap and fluxes in the form of lime and dolomite are charged into the furnace (Patil et al., 1998; Shi, 2004; Yildirim and Prezzi, 2011; Remus et al., 2013). The impurities dissolved in liquid metal such as silicon, phosphorus, manganese, sulphur and phosphorus, are oxidized and combine with the lime and dolomite fluxes to form BOF slag (Shi, 2004; Remus et al., 2013). At the end of the oxidation stage, liquid steel and slag form immiscible layers and are separately tapped into ladles and slag pots, respectively. The composition of the hot metal and scrap charged, and the chemical reactions occurring during the oxidation of impurities, determine the final chemical composition, quantity of slag produced per ton of liquid steel produced and the eventual mineralogy of the solidified BOF slags (Shi et al., 2004; Yildirim and Prezzi, 2011; Remus et al., 2013). Due to the nature of impurities in the crude liquid steel and the refining requirements in the BOF process, the slags tend to have a high basicity ratio (2.4 < CaO/SiO_2 < 4.0), a higher FeO content as a result of the oxidizing conditions and blow regimes and a more variable chemical composition when compared to blast furnace slags (Sano et al., 1997; Atkinson and Kolarik, 2001; Reuter et al., 2004; Remus et al., 2013). Depending on the degree of process integration and final product specifications, the steel liquid steel produced from the basic oxygen process can either undergo further refining in secondary refining or ladle furnaces, or can be sent directly for solidification and casting into semi-finished products (Shi, 2004; Yildirim and Prezzi, 2011; Remus et al., 2013).

As opposed to the basic oxygen process, the EAF can be operated independent of the blast furnace. In other words, the electric arc process relies on solid steel scrap, fluxes and slag formers and the ferroalloy charges that are melted by passing an electric current through the charge mixture via graphite electrodes (Patil et al., 1998). In instances where the EAF assumes both the smelting and refining roles, oxygen and other oxy-gas mixtures are injected into the molten steel through specially designed lances to oxidize the impurities dissolved in the crude liquid steel. In this case, the oxidation products react with fluxes and slag formers to form the final EAF slag (Shi, 2004; Yildirim and Prezzi, 2011; Remus et al., 2013). When the desired composition and temperature of the steel is achieved, the molten steel and liquid slag are eventually tapped into separate ladles and slag pots, respectively. In this case, and the liquid steel can either be sent for further refining in secondary refining or ladle furnace units, or can be sent for casting and downstream manufacturing processing (Patil et al., 1998; Shi, 2004; Yildirim and Prezzi, 2011; Remus et al., 2013). In most cases, the EAF process slag is usually sent to slag processing and/or metal recovery units before final disposal in the slag yard (Patil et al., 1998; Shi, 2004; Yildirim and Prezzi, 2011; Remus et al., 2013).

As highlighted in Section 4.2.5, secondary ladle refining is the final stage in the refining of different grades of stainless steels. The most notable function of secondary metallurgy or ladle refining processes is to precisely control the final composition and temperature of the refined liquid steel before casting (Patil et al., 1998). The ladle refining process includes unit processes such as final decarburization, desulphurization and degassing to remove dissolved oxygen, nitrogen and hydrogen (Patil et al., 1998; Shi, 2004; Yildirim and Prezzi, 2011). During the reefing process, desulphurization agents (such as Mg, CaSi and CaC_2) and deoxidation agents (such as Si and Al) are injected into the steel bath, and the reaction products such as CaS, $3CaO \cdot P_2O_5$, SiO_2 and Al_2O_3 either dissolves into, or are absorbed by, carefully engineered slag and are eventually skimmed off the surface of liquid steel (Patil et al., 1998; Shi, 2004; Yildirim and Prezzi, 2011). In this regard, the careful control and manipulation of the composition of ladle refining slag is an important parameter in controlling the final product quality (Patil et al., 1998).

Physicochemical characterization of steelmaking slags: The production rates of different kinds of steelmaking slags are highlighted in Table 4.8. In addition, the chemical composition of steelmaking slags varies considerably depending on process conditions and desired quality of steel, as highlighted in Table 4.12.

Although there are some similarities between basic oxygen and EAF slags due to somewhat similar refining process conditions, ladle refining slags, especially those produced in the production of stainless steels, tend to be uniquely different and vary greatly depending on the targeted grades of stainless steels produced (Patil et al., 1998; Dippenaar, 2004; Shi, 2004; Yildirim and Prezzi, 2011). As shown in Table 4.12, ladle refining slags tend to

TABLE 4.12

Chemical Composition of Basic Oxygen, Electric Arc and Ladle Furnace Slags

	Typical composition (wt.%)											
Slag type	CaO	SiO_2	Al_2O_3	MgO	FeO	Fe_2O_3	T-Fe	MnO	P_2O_5	TiO_2	Cr_2O_3	Reference
BOF	45–60	10–15	1–5	3–13	7–20	3–9	na	2–6	1–4	na	na	Yi et al. (2012)
BOF	39.40	11.97	2.16	9.69	30.23	na	na	2.74	1.00	0.40	0.20	Yildirim and Prezzi (2011)
BOF	47.71	13.25	3.04	6.37	na	24.36	na	2.64	1.47	0.67	0.19	Waligora et al. (2010)
BOF	47.9	12.2	1.2	0.8	26.3	na	na	0.3	3.3	na	na	Das et al. (2007)
BOF	45	11.1	1.9	9.6	10.7	10.9	–	3.1	–	–	–	Tossavainen et al. (2007)
BOF	30–55	8–20	1–6	5–15	10–35	na	na	2–8	0.2–2	0.4–2		Shi (2004)
BOF	42–55	12–18	≤3	≤8	na	na	na	≤5	≤2	na	≤10	Zheng and Kozinski (1996)
EAF	30– 50	11–20	10–18	8–13	8–22	5–6	na	5–10	2–5	na	na	Yi et al. (2012)
EAF	45.5	32.2	3.7	5.2	3.3	1.0	na	2.0	na	0.13	0.05	Tossavainen et al. (2007)
EAF	46.9	33.5	2.30	6.22	1.07	0.36	1.56	2.60	0.02	0.16	2.92	Shen et al. (2004)
EAF	39–45	24–32	3–7.5	8–15	1–6	na	na	0.4–2	≤0.16	na	29.2	Shi (2004)
EAF	25–40	10–17	4–7	4–15	na	na	na	≤6	≤1.5	na	≤3	Zheng and Kozinski (1996)
Ladle	55	15	12.5	7.5	na	2.1	na	0.36	na	0.33	0.01	Setien et al. (2009)
Ladle	42.5	14.2	22.9	12.6	0.5	1.1	na	0.2	na	840 ppm	2700 ppm	Tossavainen et al. (2007)
AOD	54.10	26.50	4.91	6.30	1.18	0.63	2.03	1.02	0.02	0.98	1.83	Shen et al. (2004)
AOD	47.6	31.24	1.66	3.65	na	na	1.75	1.6	na	0.22	5.07	Shen and Forssburg (2003)
AOD	45.5	26.3	9.7	7.3	na	na	3.5	2.1	na	na	6.3	Shen and Forssburg (2003)
AOD	46–48	26–31	1.7–10	3.7–7.3	na	na	1.8–2.	1.4–2.1	na	0.2–1.1	3.6–6.3	Lopez et al. (1997)

Note: na, not available.

contain higher amounts of alloying elements such as chromium and nickel, which are intentionally added to control the mechanical and thermochemical properties of the different grades of the solidified steel products (Patil et al., 1998; Reuter et al., 2004; Shi, 2004; Tossavainen et al., 2007; Yildirim and Prezzi, 2011; Remus et al., 2013).

The main phases in steelmaking slags are in the form of complex oxides and silicates (Shen et al., 2004; Shi, 2004; Tossavainen et al., 2007; Waligora et al., 2010; Huaiwei and Xin, 2011; Yildirim and Prezzi, 2011). In general, the slags produced in the EAF and argon-oxygen decarburization processes are characterized by the presence of Cr–Fe–O, Mg–Mn–Cr–Al–Fe–O and Cr–Fe–Ni spinel solid solutions (Shen et al., 2004; Tossavainen et al., 2007; Huaiwei and Xin, 2011). In particular, Waligora et al. (2010) carried out extensive characterization of basic oxygen process slags using optical microscopy, scanning electron microscopy with x-ray microanalysis (SEM-EDS) and x-ray diffraction analyses methods, and their findings, based on slag samples with composition highlighted in Table 4.13, are shown in Figure 4.16. Basically, Figure 4.16 confirms the presence of β-Ca_2SiO_4, a heterogeneous calcium silicate phase containing micro-inclusions; dicalcium ferrite phases resembling $Ca_2(Al,Fe)_2O_5$ and $Ca_2Fe_2O_5$; wustite (FeO) and entrained metallic iron particles (Waligora et al., 2010).

As shown in Table 4.13 and Figure 4.16, the mineralogical and morphological phases observed are often very complex, and such characteristics are often controlled by factors such as the chemical composition of slag cooling rate of slag and process from which the slag is generated (Tossavainen et al., 2007; Yildirim and Prezzi, 2011). According the Yildirim and Prezzi (2011), knowledge of the chemical, morphological and mineralogical properties of steelmaking slag is essential, mainly because their recycling and reutilization potential is closely linked to these properties. In this regard, Shen et al. (2004) studied the physicochemical and mineralogical of stainless steel slag with the objective towards metal recovery. The main focus of this study was to evaluate the chemical and mineral phase

TABLE 4.13

Mineralogical Phases in Steelmaking Slags

Unit	Method	Mineralogical Phases	Reference
BOF	XRD	Ca_3SiO_5; α-Ca_2SiO_4; $Ca_2Fe_2O_5$; β-Ca_2SiO_4; FeO–MnO–MgO solid solution; MgO; wustite solid solution	Tossavainen et al. (2007)
BOF	XRD	$3CaO \cdot SiO_2$; γ-Ca_2SiO_4; β-Ca_2SiO_4; $2CaO \cdot Fe_2O_3$; CaO; MgO; $4CaO \cdot Al_2O_3 \cdot Fe_2O_3$; CaO-FeO-MnO-MgO solid solution	Shi (2004)
BOF	SEM	β-Ca_2SiO_4; $Ca_2(Al,Fe)_2O_5$; $Ca_2Fe_2O_5$; FeO; CaO; Metallic-Fe	Waligora et al. (2010)
EAF	XRD	Ca_2SiO_4; Ca_3SiO_5; CaO–FeO–MnO–MgO solid solution; $Ca_3MgSi_2O_8$; tetra-calcium aluminoferrite; free CaO	Yi et al. (2012)
EAF	XRD	$FeCr_2O_4$; $FeFe_2O_4$; Ni–Cr–Fe solid solution; Ca_2SiO_4; SiO_2; $Ca_{14}Mg_2(SiO_4)_8$; $Ca_3Mg(SiO_4)_2$; $Ca_2MgSi_2O_7$; $Ca_2Al_2SiO_7$; Fe–Cr; Fe–Ni	Shen et al. (2004)
EAF	XRD, SEM	$Ca_3Mg(SiO_4)_2$; β -Ca_2SiO_4; $(Mg,Mn)(Cr,Al,Fe)_2O_4$ spinel solid solution; $CaAl_2SiO_6$; (Fe,Mg,Mn)O wustite type solid solution; $Ca_2(Al,Fe)_2O_5$	Tossavainen et al. (2007)
EAF	XRD	$FeCr_2O_4$; $FeFe_2O_4$; Ni–Cr–Fe; Ca_2SiO_4; $Ca_{14}Mg_2(SiO_4)_8$; $Ca_3Mg(SiO_4)_2$; $Ca_2MgSi_2O_7$; SiO_2	Huaiwei and Xin (2011)
AOD	XRD, SEM	β-Ca_2SiO_4;MgO; γ-Ca_2SiO_4; $Ca_{12}Al_{14}O_{33}$; $Ca_2Al_2SiO_7$	Tossavainen et al. (2007)
AOD	XRD	$FeCr_2O_4$; $FeFe_2O_4$; Ni–Cr–Fe solid solution; Ca_2SiO_4; CaF_2; $Ca_{14}Mg_2(SiO_4)_8$; Ca_2SiO_4; $Ca_4Si_2O_7F_2$; MgO; Fe–Cr; Fe–Ni	Shen et al. (2004)

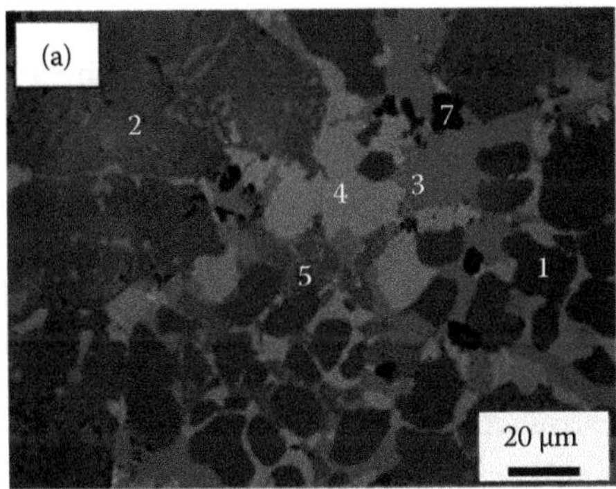

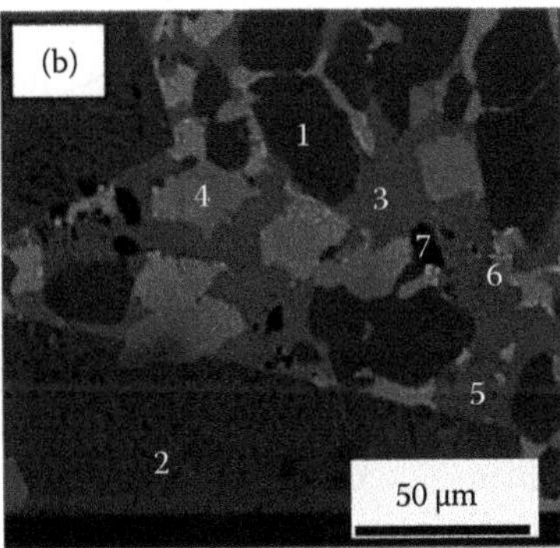

FIGURE 4.16
Optical microscopy (a) and backscattered scanning electron micrographs (b) of the mineral phases contained in the LD steel slag showing (1) homogeneous dicalcium silicate, (2) heterogeneous calcium silicate phase containing micro-inclusions, (3) dicalcium ferrite, (4) iron oxide, (5) lime, (6) entrained metallic iron and (7) porosity. (Reprinted from *Mater. Char.*, 61, Waligora, J., Bulteel, D., Degrugilliers, P., Damidot, D., Potdevin, J.L., and Measson, M., Chemical and mineralogical characterization of LD converter steel slags: A multi-analytical techniques approach, 39–48. Copyright 2010, with permission from Elsevier.)

composition analyses, particle size distribution and Ni–Cr distribution, among other important physicochemical properties of stainless steel slags. Their findings reiterated the possibility of recovering most of the Cr and Ni from slag by regrinding to further liberate the entrained metals, followed by gravity and/or magnetic separation processing (Shen et al., 2004).

As opposed to the blast furnace slags which have found widespread applications in the civil engineering sector (EUROSLAG, 2015; Lobato et al., 2015), the presence of relatively high amounts of toxic metal elements such as chromium and nickel often limits the direct utilization of steelmaking slags as a construction materials (Shi, 2004; Ma and Garbers-Craig, 2006). As such, protracted efforts to recover these valuable metals, to detoxify the toxic metal elements in steelmaking slags before disposal and/or reuse in alternative processes, have been extensively highlighted in Section 4.7.

4.6 Recycling and Reuse of Iron and Steelmaking Wastes

Due to the increased environmental awareness, global, steel producers are being forced to proactively adopt environmentally sustainable business models in their operations. According to the World Steel Association (2015b), the steel industry is an integral part of a circular economy model which promotes zero waste through the reuse and recycling of by-product materials. In this regard, the steel industry has made significant progress in finding new markets and applications for steel by-products and wastes such as process slags and process gases such as coke oven gas, blast furnace gas and BOF gas, tar and benzene (Atkinson and Kolarik, 2001; World Steel Association, 2015b). Based on the life cycle assessment methodology developed by the World Steel Association (2015b), the recycling and/or reuse of waste materials and by-products produced in the iron and steel sector accounts for a significant amount of energy and raw material savings being realized in the iron and steel sector. For example, the World Steel Association (2015b) estimated that over 1400 kg of iron ore, 740 kg of coal and 120 kg of limestone can be saved for every ton of steel scrap recycled into new steel.

Section 4.5 highlighted in detail the different types of metallurgical wastes generated in the production of iron and steel. Despite metallurgical wastes sometimes being a challenge for global steel producers, increasing the recycling and reuse potential of these wastes can also be the panacea to environmental problems currently being caused by the of wastes disposal in landfills. In addition, metallurgical wastes can also be potential sources of secondary metals, and can easily be converted into useful by-products if they are treated, reused and recycled properly (Atkinson and Kolarik, 2001; World Steel Association, 2015b). Despite the several benefits associated with the recycling of metallurgical wastes, the low intrinsic monetary value of these chemically and morphologically complex waste materials also limits their full scale recycling and reuse (Atkinson and Kolarik, 2001; World Steel Association, 2015b). Based on the aforementioned, following sections highlight the potential and current recycling and or reuse options for different categories of metallurgical wastes generated in the production of iron and steel products.

As highlighted in Table 4.8, metallurgical wastes produced from the iron and steelmaking processes can be broadly categorized into raw material agglomeration wastes (i.e. coke making, sintering and pelletization process dusts and sludges), blast furnace ironmaking wastes (i.e. blast furnace flue dust and sludge, slag and spent refractory materials), BOF wastes (i.e. converter dusts and sludges, slag and spent refractory bricks), EAF waste (i.e. dusts and sludges, slag and spent refractory bricks), stainless steel refining wastes (i.e. dusts, sludges and slag) and casting and rolling wastes (such as mill scale, sludge and cast house dust and slag) (Nyirenda, 1991; Pazdej and Vogler, 1995; Das et al., 2007; Gavrilovski et al., 2011; Remus et al., 2013; World Steel Association, 2014; Lobato et al., 2015). In addition to the previously mentioned categories of wastes, the emissions of polychlorinated dibenzo-p-dioxins (PCDDs), polychlorinated dibenzofurans (PCDFs) and other chlorinated aromatic compounds from iron and steelmaking processes, particularly in the coke making and sintering processes, are also of serious environmental and health concern (Buekens et al., 2001, Kasai et al., 2001; Remus et al., 2013). Since the generation of PCDDs and PCDFs in iron and steelmaking was not focus of this work, interested readers are advised to refer to specialized texts and journal articles for more details.

According to Gavrilovski et al. (2011), the iron and steelmaking wastes contain a significant amount of iron, fluxes and combustible components, and therefore, can be considered as secondary raw materials whose increased utilization and recycling can result in the reduction of production costs. Although these wastes contain valuable components that can be directly recycled back into the steel plant, they are also contaminated with deleterious and harmful elements, such as zinc, sulphur and phosphorus, which inhibit their direct recycling and or reuse without prior treatment (Atkinson and Kolarik, 2001; Remus et al., 2013). Despite these challenges, the recycle and reuse of metallurgical wastes from the iron and steelmaking process has gained traction over the years, with the steelmakers striving to reduce the amount of waste generated, develop economic recycling processes and/or finding alternative markets for the by-products produced in the steelmaking process (World Steel Association, 2015b). Sections 4.6.1–4.6.10 highlights some of specific interventions currently being implemented by the iron and steel making industry in an effort to reduce, recycle and reuse the different metallurgical wastes.

4.6.1 Recycling and Reuse of Raw Materials Agglomeration Dusts and Sludges

In the integrated blast furnace ironmaking process, sinter, coke and pellets constitute a significant amount of ferrous and reductant agglomerate materials that are charged into the process. During the production of such agglomerates, particulate wastes are inherently generated, and these include, inter alia, dusts and sludges generated from the coke making, sintering and pelletizing processes. In most cases, the dusts and sludges that are generated from the raw materials agglomeration processes are characteristically rich in iron and carbon, and are typically recycled back in the steel plant (Atkinson and Kolarik, 2001; Yadav et al., 2002; Remus et al., 2013). While on-going process improvement initiatives are aimed at reducing the generation of these types of waste materials, the recycling and reuse of raw materials agglomeration wastes is of special interest to steel plant operators, mostly as a form of raw materials replacement and reduction of disposal costs (Atkinson and Kolarik, 2001; Yadav et al., 2002; Gavrilovski et al., 2011; Remus et al., 2013).

Specifically, the following approaches are used in the commercial recycling and reuse of raw material agglomeration wastes such as coke breeze and coke fines, coke oven gas and sinter and pellet plant dusts and sludges.

1. Coke breeze and coke fines are the main fuel additives in the sintering process (Ball et al., 1973; Biswas, 1981; Mohamed et al., 2010; Lanzerstorfer et al., 2015). In general, the $-3 + 1$ mm size fraction is considered the optimum particle size of the coke breeze suitable for the sintering process (Mohamed et al., 2010). However, too fine-grained particles compromise the permeability of the sinter mix, and as a result, the coke breeze particles less than 1 mm require granulation prior to charging in the sintering process to achieve higher process efficiency, shorter sintering duration and fuel saving. (Ball et al., 1973; Biswas, 1981; Mohamed et al., 2010; Niesler et al., 2014).
2. Coke oven gas is usually rich in hydrogen, methane and carbon monoxide and often contains a high calorific value in the range 17–20 MJ/Nm^3 (Remus et al., 2013). In most cases, the coke oven gas is usually consumed within the steel plant as fuel in the coke ovens, is used as a reductant and/or is combusted to generate electricity to power the steel plant (Atkinson and Kolarik, 2001; Remus et al., 2013; Suopajärvi et al., 2013). The coke oven gas generated from coke batteries equipped with by-product recovery systems is used to produce chemicals by recovering by-products such as light oil, tar, naphthalene, phenols and ammonia (Atkinson and Kolarik, 2001; Remus et al., 2013). Other innovative applications of coke oven residues have also been explored as well. For example, Suopajärvi et al. (2013) investigated the utilization of coking plant residues as a reducing agent in the blast furnace with the objective of increasing their utilization potential in iron making process. Their findings, based on injecting coke plant liquid by-product, tar and solid residues, coke sand and coke dust in the blast furnace, provided an innovative opportunity to reduce the coke consumption by recycling a mixture of oil, coal tar and coke fines to the blast furnace at constant oil injection ratios (Suopajärvi et al., 2013).
3. In most cases, particulate dusts and sludges generated in the sinter and pellets plants are either recycled back into the respective processes or are agglomerated into cold-bonded pellets or briquettes to produce materials suitable for charging in the blast furnace, BOF or EAF processes (Su et al., 2004; Singh et al., 2014; Pal et al., 2015). In most cases, these types of waste are usually mixed with other steel plant wastes, such as blast furnace flue dust, sludge from direct reduction plants, basic oxygen converter sludge and mill scale, and then agglomerated to constitute auxiliary raw material charge in the steel plant (Su et al., 2004; Remus et al., 2013; Singh et al., 2014; Pal et al., 2015). However, the recycling of sinter and pellet plant dusts and sludges has its own challenges, particularly on the build-up of deleterious and volatile elements and compounds, which may eventually affect the sinter chemistry and mechanical properties (Mohamed et al., 2010; Lanzerstorfer et al., 2015).

In general, sinter plants have traditionally been used to recycle by-products in the integrated steel plant, and is still the most popular process option for the recycling of most particulate dusts produced in the raw materials agglomeration processes (Ball et al., 1973; Biswas, 1981; Atkinson and Kolarik, 2001; Yadav et al., 2002; Mohamed et al., 2010; Gavrilovski et al., 2011; Remus et al., 2013; Niesler et al., 2014; Lanzerstorfer et al., 2015). As discussed in Section 4.2.1.4, the sinter feed consists of iron ore fines, coke breezes and fine-grained recycled iron-containing materials such as dusts, sludges and mill scales (Mohamed et al., 2010; Lanzerstorfer, 2015). Although the recycling of raw materials agglomeration wastes in the sinter plant could have significant cost advantages, several challenges are also associated with the charging of too fine-grained particles into the sinter plant, and these include compromised plant productivity and quality of product (Yadav et al., 2002; Lanzerstorfer et al., 2015).

4.6.2 Recycling and Reuse of Blast Furnace Ironmaking Dusts and Sludges

Depending on the type and amount of charge materials, the blast furnace process generates dusts and sludges in the range of 7–45 kg/ton of hot metal produced (Pazdej and Vogler, 1995; Atkinson and Kolarik, 2001; Das et al., 2007; Remus et al., 2013). These types of wastes are primarily composed of carbon; oxides of iron; calcium; silicon; manganese; aluminium, and varying concentrations of volatile elements such zinc, lead and alkaline metal compounds (Zeydabadi et al., 1997; Atkinson and Kolarik, 2001; Yadav et al., 2002; Das et al., 2007; Remus et al., 2013). In the integrated steel plants, blast furnace dusts and sludges are usually recycled back either into the sinter plant as additive sources of iron and carbon for the sintering operations or in the blast furnace as briquetted charge (Ball et al., 1973; Biswas, 1981; Atkinson and Kolarik, 2001; Remus et al., 2013; Lanzerstorfer et al., 2015). However, for non-integrated plants operating without sinter plants, the dusts and sludges are sometimes mixed with other by-product residues and briquetted, and recycled back into the blast furnace or as additives in the electric arc or BOFs (Atkinson and Kolarik, 2001). The in-process recycling of blast furnace dusts and sludges into the sinter plant can also be problematic since the residues can also affect the sinter chemistry, sinter strength and sinter plant productivity (Atkinson and Kolarik, 2001; Remus et al., 2013; Lanzerstorfer et al., 2015). In addition, the strict adherence to maximum allowable concentrations of zinc, lead and alkali metals in materials charged into the blast furnace also limits the direct recyclability of these particulate wastes into the blast furnace (Atkinson and Kolarik, 2001; Remus et al., 2013; Lanzerstorfer et al., 2015). In essence, the uncontrolled recycling of the dust and sludge residues in the blast furnace may result in the build-up of these elements in the blast furnace stack, which in turn may result in operational problems such as scaffolding, as well as damage to refractories as a result of the chemical attack by these reactive volatile metals (Biswas, 1981; Atkinson and Kolarik, 2001).

In the recent years, several processes have been developed to overcome some of the challenges associated with the in-process recycling of blast furnace dusts and sludges. The recent developments in the cold briquetting of dusts with coke breeze for direct return to the blast furnace (Atkinson and Kolarik, 2001; Pal et al., 2015), and the development of rotary hearth furnace (RHF)-based reduction processes (Atkinson and Kolarik, 2001; Oda et al., 2002; Tateishi et al., 2008; Hendrickson and Iwasaki, 2013) have demonstrated that the iron units in the blast furnace dusts and sludges can be economically recovered for reintroduction into the mainstream iron and steelmaking process.

4.6.3 Recycling and Reuse of Blast Furnace Ironmaking Slags

Blast furnace slags belong to the $CaO–SiO_2–Al_2O_3–MgO$ system, and basically consist of oxides of silica and alumina from iron-bearing charge and the lime and magnesia charged as fluxes (Biswas, 1981; Sano et al., 1997; Zheng and Kozinski, 1996; Shi, 2002; Dippenaar, 2004; Reuter et al., 2004; Das et al., 2007; Horii et al., 2013; Remus et al., 2013; EUROSLAG, 2015; Lobato et al., 2015; Nippon Slag, n.d.). In general, about 200–400 kg of slag are produced per ton of hot metal produced using the blast furnace process (Horii et al., 2013; Lobato et al., 2015). The mineralogical and physicochemical characteristics of both air-cooled and granulated blast furnace slags were discussed in Section 4.5. Notwithstanding their utility and potential as recyclable and reusable materials, the approaches to managing blast furnace slags often vary from operation to operation, and are mostly affected by the in-house recycling capabilities and the availability of alternative uptake markets (Atkinson and Kolarik, 2001). Generally, blast furnace slags can be recycled back into the steel plant as premelted fluxes, and to some extent, as supplementary sources of iron (Reuter et al., 2004). Nevertheless, blast furnace slags contain relatively low amounts of iron even though the recovery of metallic iron has not always been prioritized (Shen and Forssberg, 2003). To date, a significant amount of blast furnace slags have found widespread commercial applications outside the steel plant, such as in the manufacture of cement, aggregates, concrete and other applications in the construction industry (Shi, 2002; Reuter et al., 2004; Das et al., 2007; Horii et al., 2013; Remus et al., 2013; EUROSLAG, 2015; Lobato et al., 2015; Nippon Slag, n.d.).

4.6.3.1 Granulated Blast Furnace Slags as Construction Materials

As discussed in Section 4.5, granulated blast furnace slags posses cementitious properties, and have chemical composition and hydraulic properties similar to Portland cement. As a result, these types of slags are commonly used as high value alternatives or additives to conventional Portland cement (Reuter et al., 2004; Horii et al., 2013; EUROSLAG, 2015). When ground to the right particle sizes, the chemical composition and non-crystalline nature of blast furnace slags enable them to form cement-like hydration products when reacted with water (Reuter et al., 2004; Kumar et al., 2008). According to Kumar et al. (2008), the use of mechanically activated granulated blast furnace slag as a partial replacement of clinker is a well-established process, and has resulted in lower raw material and energy consumption in cement production, as well as improvement in the properties of the final cement product. In addition, the granulated blast furnace slags have found other applications such as (1) in the production of ceramic materials due to their relatively high hardness and mechanical strength (Mostafa et al., 2010), (2) in the preparation of glass ceramic materials from controlled crystallization of the slag materials (Liu et al., 2009) and (3) as soil corrective additives in agriculture (Das et al., 2007).

In addition to the already commercially proven applications of granulated blast furnace slags in the manufacture of Portland cement and other construction materials, the potential application of granulated slags as active filler raw materials for the production of high-value geopolymers for fire resistant applications has also been explored by Cheng and Chiu (2003). In these studies, Cheng and Chiu (2003) proposed an innovative approach to fabricate fire-resistant geopolymers based on the geopolymerization behaviour, physical and mechanical properties and fire resistance characteristics of granulated blast furnace slag–based geopolymers (Cheng and Chiu, 2003).

4.6.3.2 Granulated Blast Furnace Slags as Adsorbent Materials

The use of blast furnace slags as adsorbents has gained attention in recent years. For example, the possibility of applying granulated blast furnace slags as adsorbents in acid mine drainage and wastewater remediation has been explored by several researchers, mainly due to their high specific surface area and high reactivity (Johnson, 1999; Liu and Shih, 2004; Kostura et al., 2005; Sasaki et al., 2008; Wajima, 2015). In particular, Kostura et al. (2005) investigated the use of both crystalline and amorphous blast furnace slags as phosphorus adsorbents in wastewater solutions. Their findings, based on adsorption kinetics measurements, showed that the sorption of phosphorus on both types of slags was based on a pseudo-second-order reaction model and Langmuir adsorption isotherm. Since acid mine drainage is problematic in most countries such as South Africa where mining activities are prevalent, the use of granulated blast furnace slags in acid mine drainage remediation is a promising area that requires further research and pilot scale validation tests (Johnson, 1999; Kostura et al., 2005; Sasaki et al., 2008; Simate and Ndlovu, 2014; Wajima, 2015).

4.6.3.3 Granulated Blast Furnace Slags as Geopolymeric Materials

The use of blast furnace slags as geopolymeric materials applied in the stabilization and/or solidification of toxic and hazardous waste has been investigated by several researchers (Kindness et al., 1994; Rodríguez-Piñero et al., 1998; Laforest and Duchesne, 2005). In particular, Kindness et al. (1994) investigated the ability of Portland cement and blended cements containing blast furnace slag in physically and chemically immobilizing the chromium ions from wastewater solutions. The studies were based on investigating the different mixtures of Ordinary Portland Cement, high alumina cement and ground granulated blast furnace slag on the immobilization of chromium from solutions containing Cr (VI) and Cr (III) ions. Kindness et al. (1994) reported that slag-containing cements were the most effective at immobilizing Cr (VI) ions, probably by reducing the Cr (VI) to Cr (III) and the eventual immobilization of Cr (III) ions as solid solutions in aluminate phases. In other studies, Duchesne and Laforest (2004) and Laforest and Duchesne (2005) investigated the interaction mechanism of chromium (VI) ions with granulated blast furnace slag and Ordinary Portland Cement based on the absorption isotherms. These studies concluded that the granulated slags were more efficient than cement at Cr (VI) fixation at lower initial solution concentration, while the opposite was true at higher initial concentrations (Duchesne and Laforest, 2004; Laforest and Duchesne, 2005). Based on post-leaching characterization tests using scanning electron microscopy and x-ray diffraction methods, Laforest and Duchesne (2005) identified the $CaCrO_4$ and $CaCrO_4 \cdot 2H_2O$, calcium silicate hydrates (C-S-H) and the calcium aluminate phases, as the chromium bearing phases. The solidification and stabilization of hazardous wastes using blast furnace slags, particularly alkali-activated granulated blast furnace slags, is further discussed in Section 4.7.

4.6.3.4 Slow-Cooled Blast Furnace Slags

Unlike granulated slags, slow-cooled blast furnace slags generally possess little or no cementitious properties (Reuter et al., 2004; Shi, 2004; Das et al., 2007; Lobato et al., 2015). In general, slow-cooled blast furnace slags mostly consist of crystalized Ca–Al–Mg silicates, especially melilite ($Ca_2MgSi_2O_7$–$Ca_2Al_2SiO_7$) and merwinite ($Ca_3MgSi_2O_8$), whose properties and appearance resemble that of crushed natural basalt stone (Shi and Qian, 2000; Horii et al., 2013; EUROSLAG, 2015). As a result, slow-cooled slags have found widespread applications as substitute materials for crushed natural basalt stone, mostly for use as road base coarse materials, concrete aggregates, ballasts and light weight building materials (Shi and Qian, 2000; Horii et al., 2013; EUROSLAG, 2015). In addition, these slow-cooled blast furnace slags are also widely used as raw materials in the production of glass composites, mineral wool, lime fertilizer and soil stabilization conditioners (Reuter et al., 2004; EUROSLAG, 2006; Remus et al., 2013).

4.6.4 Recycling and Reuse of Steelmaking Dusts and Sludges

As discussed in Section 4.5, steelmaking dusts and sludges typically consist of oxides of iron, calcium, silicon, manganese and aluminium; alloying elements such as chromium, nickel and molybdenum and varying concentrations of volatile metal compounds such as zinc, lead and cadmium (Zeydabadi et al., 1997; Yadav et al., 2002; Ma and Garbers-Craig, 2006; Laforest and Duchesne, 2006a,b; Ma and Garbers-Craig, 2009; Dominguez et al., 2010; Remus et al., 2013; de Araujo and Schalch, 2014; Lobato et al., 2015; Singhal and Rai, 2015; Ma, 2016). In particular, the stainless steel refining process produces metallurgical dust and sludges richer in oxides of alloy elements than any other types of dust and sludges produced in the iron and steel manufacturing process (Denton et al., 2005; Laforest and Duchesne, 2006a,b; Ma and Garbers-Craig, 2006a,b,c, 2009; Dominguez et al., 2010; Singhal and Rai, 2015). Despite these types of metallurgical wastes being rich in iron and alloying elements, the direct recycling of steelmaking dusts and sludges into the steel plant is generally constrained by the potential build-up of deleterious and volatile metal elements such as zinc and lead in the process (Nyirenda, 1991; Wu, 1999; Yadav et al., 2002; Kelebek et al., 2004; Das et al., 2007; Remus et al., 2013; Ma, 2016). For example, excessive zinc loading is detrimental to the operation of iron and steelmaking reactors, due to the potential chemical attack on the reactor refractory materials, which eventually reduces the reactor campaign life and productivity (Biswas, 1981; Ma, 2016). According to Ma (2016), the zinc concentration in steelmaking dust and sludges is often too high to directly recycle the waste materials back into the iron and steelmaking process, but is also too low for direct utilization of the waste materials as raw materials for zinc recovery.

Due to the presence of toxic metal species such as chromium (mostly as Cr (VI) species) and zinc, steelmaking dusts and sludges are classified as hazardous waste materials which require special treatment before disposal (Ma and Garbers-Craig, 2006a,b,c; Laforest and Duchesne, 2006a,b; Ma and Garbers-Craig, 2009; Dominguez et al., 2010; Singhal and Rai, 2015; Ma, 2016). As a result, steelmaking companies are faced with ever increasing challenges to manage these types of wastes in an economically viable and environmentally sustainable manner. Currently, some of the most common approaches to managing the environmental challenges associated with the steelmaking dusts and sludges include recovering the metallic elements contained and/or thermally vitrifying the waste residue before disposal (Atkinson and Kolarik, 2001; Nyirenda, 1991; Yadav et al., 2002; Das et al., 2007; Huaiwei and Xin, 2011; Remus et al., 2013).

To date, several processes have been investigated to recover the various metallic elements contained in steelmaking dusts and to make the particulate dusts amenable for safe disposal. In order to recover the valuable meta elements in the particulate dusts, several researchers have investigated the feasibility of solid state direct reduction and/or dezincification of steelmaking dusts to produce direct reduced iron based on static bed, rotary kiln and/or rotary hearth technologies (Nyirenda, 1991; Barozzi, 1997; Oda et al., 2002; Sugitatsu and Miyahara, 2004; Tateishi et al., 2008; Gao et al., 2012; Haga et al., 2012; de Araujo and Schalch, 2014). Other processes have also been developed based on the high temperature smelting reduction processing of metal oxides in dust in order to recover the valuable metal elements such as Fe, Cr, V and Ni by concentrating them in the liquid metal bath (Barcza and Nelson, 1990; Schoukens et al., 1996; Hasegawa et al., 1998; Hara et al., 2000; von Scheele, 2004; Denton et al., 2005; Takano et al., 2005). Such smelting reduction processes eventually produces a barren slag byproduct for safe disposal in landfills. Based on the aforementioned approaches, several processes have been commercialized globally, and these processes are summarized in Table 4.14 and are discussed further in Sections 4.6.4.1 and 4.6.4.2.

4.6.4.1 Direct Reduction Processing of Agglomerated Dust

The direct recycling of steelmaking dusts is hampered by the presence of volatile metal elements, particularly the different zinc compounds. As discussed in Section 4.5, the zinc compounds in steelmaking dusts exist mainly as zincite (ZnO) and spinel franklinite ($ZnFe_2O_4$), and hence would require special treatment and/or dezincification processing to remove the zinc and recover it as a separate product (Nyirenda, 1991; Li and Tsai, 1993; Sofilic et al., 2004; Havlík et al., 2006; Machado et al., 2006; Pickles, 2009; Remus et al., 2013; Suetens et al., 2014). Potential pre-treatment processes include the carbothermic reduction processing of agglomerated dusts to break down the complex zinc compounds present in steelmaking dusts. In fact, the solid state reduction and dezincification processes are cited as the promising technologies to recover valuable metals and the zinc contained in steelmaking dusts (Nyirenda, 1991; Gao et al., 2012; Haga et al., 2012). Such processes involve the solid state reduction of oxides by solid carbonaceous material, carbon monoxide gas and/or hydrogen gas, and the resultant volatilization removal of the volatile metals in the fine or agglomerated dust.

In the presence of carbon, metal oxides and compounds of zinc in the steelmaking dust are reduced according to Equations 4.9 through 4.13 (Peng et al., 2009; Pickles, 2009; Haga et al., 2012):

$$3ZnFe_2O_{4(s)} + C_{(s)} \rightarrow 3ZnO_{(s)} + 2Fe_3O_{4(s)} + CO_{(g)} \tag{4.9}$$

$$Fe_3O_{4(s)} + C_{(s)} \rightarrow 3FeO_{(s)} + CO_{(g)} \tag{4.10}$$

$$ZnO_{(s)} + C_{(s)} \rightarrow Zn_{(g)} + CO_{(g)} \tag{4.11}$$

$$ZnO_{(s)} + CO_{(g)} \rightarrow Zn_{(g)} + CO_{2(g)} \tag{4.12}$$

$$FeO_{(s)} + C_{(s)} \rightarrow Fe_{(s)} + CO_{(g)} \tag{4.13}$$

The reduced metal elements are recovered as direct-reduced iron, while the volatilized elements are recovered from the furnace fumes (Nyirenda, 1991; Gao et al., 2012). According to Nyirenda (1991), the main objectives of direct reduction processes is to volatize the elements such as zinc, lead and sulphur present in the feed materials, and obtain a direct reduced product containing a high percentage of metallized iron. In particular, Gao et al. (2012) investigated a laboratory scale gas-based reduction of EAF dust using both H_2 and CO gases at temperatures below 1000°C. The process proposed by Gao et al. (2012) involves

TABLE 4.14

Processes for Metal Recovery and Recycling of Electric Arc Furnace Dusts

Process	Process Description	Temperatures	Inputs	Products	References
RHF (Low Zn EAF dust)	Rotary hearth furnace for solid state carbothermic reduction and de-zincification of pelletized steelmaking dusts	1250–1300	Coke, coal, binder, natural gas, pelletized fine dust	Crude zinc oxide rich off-gas, direct reduced iron	Ichikawa and Morishige (2002); Oda et al. (2002); Oda et al. (2006); Tateishi et al. (2008); Peng et al. (2009); Haga et al. (2012)
RHF (High Zn EAF dust)		1250–1300			
FASTMET		1200–1400			Tateishi et al., (2008); Sugitatsu and Miyahara (2004)
INMETCO	Pelletized steelmaking dust is reduced with carbon is rotary hearth furnace. Secondary dust is processed to recover zinc	1250–1300			Barcza and Nelson (1990); Barozzi 1997
Primus®	Multi-hearth furnace and EAF for recycling steelmaking dust using coke or coal fines	1000–1100	Iron ore fines, coal	Crude zinc oxide off-gas (Zn > 55 wt.%), HBI charged directly into electric arc furnace	Roth et al. (2001); Suetens et al. (2014); Degel et al. (2015); Paul Wurth (n.d.).
Coke-Packed Bed Shaft Furnace	Coke-packed bed shaft furnace smelting reduction process for steelmaking dusts	1500–1600	Coke, coal fines, natural gas, fine dust, fluxes, O_2-enriched blast	Crude zinc oxide rich off-gas, hot metal, slag	Hasegawa et al. (1998); Hara et al. (2000); Suetens et al. (2014)
Oxyfines™	Oxy-fuel based process for smelting fines dust	1600–1700	Coke, natural gas, fine dust, fluxes, oxyfuel	Crude zinc oxide rich off-gas, hot metal, slag	von Schéele (2004)
Enviroplas™	DC–arc plasma furnace smelting reduction of steelmaking dusts		Coke, natural gas, fine dust, fluxes, air, electricity	Crude zinc oxide rich off-gas, molten metal and slag	Denton et al. (2005); Schoukens et al. (1996)
OxyCup® Process	Reduction and melting of agglomerated steelmaking dusts in shaft furnace	1500–1600	Coke, scrap, self-reducing briquetted oxide wastes	Crude zinc oxide rich off-gas, molten metal and slag	Kesseler and Erdmann (2007); Suetens et al. (2014); Holtzer et al. 2015; Küttner GMBH (n.d.)

the solid state reduction of the metal oxides contained in the dust, followed by wet magnetic separation of the metallized Fe from gangue, the recovery and enrichment of zinc as zinc oxide in off-gas fumes and the production of a metal-depleted non-magnetic gangue component which can safely be landfilled or used as building material. Some of the commercial applications of the solid state direct reduction processes of agglomerated dusts are further discussed under rotary hearth technologies.

Rotary hearth processes for recycling agglomerated dust: Rotary hearth furnace (RHF) processes have proved to be the most promising technology in the solid state recycling process of steelmaking dusts. Global companies such as Nippon Steel Corporation Ltd, Midrex Technologies, Kobe Steel Ltd and INMETCO have successfully commercialized steelmaking dust recycling technologies based on the RHF technologies (Barozzi, 1997; Oda et al., 2002; Sugitatsu and Miyahara, 2004; Tateishi et al., 2008; Gao et al., 2012; Haga et al., 2012). These processes have been applied in both the solid state carbothermic reduction and the de-zincification of briquetted and/or pelletized steelmaking dusts to produce a metallized direct-reduced iron product suitable for the direct recycling into the steel plant (Barozzi, 1997; Oda et al., 2002; Sugitatsu and Miyahara, 2004; Tateishi et al., 2008; Gao et al., 2012; Haga et al., 2012).

In general, the RHF consists of flat, refractory hearth rotating inside a high temperature doughnut shaped tunnel kiln (Barozzi, 1997; McCelland, 2002; Tateishi et al., 2008; Haga et al., 2012), wherein the composite feed agglomerates consisting of mixtures of pelletized steelmaking dust and carbonaceous reductant materials are evenly placed in layers on the hearth (Tateishi et al., 2008; Haga et al., 2012). Burners fired with natural gas, fuel oil or pulverized coal are located above the hearth to provide the heat for the process, and the pellets are heated mainly by the heat radiated from the furnace walls and the counterflow off-gases (Barozzi, 1997; McCelland, 2002; Tateishi et al., 2008; Haga et al., 2012). As the furnace rotates at a constant speed, the continuously charged agglomerates are heated and moves first into the heating zone, and then the reducing zone, where oxides of iron and other elements are reduced and finally recovered as direct reduced iron at the discharge end (Barozzi, 1997; McCelland, 2002; Tateishi et al., 2008; Haga et al., 2012). During the reduction process, the oxides of volatile metals, such as zinc, lead and cadmium, are reduced and volatilized into the furnace atmosphere where they are reoxidized and collected as oxides in the gas cooling and cleaning sections (Barozzi, 1997; McCelland, 2002; Tateishi et al., 2008; Haga et al., 2012).

In recent years, the rotary hearth technologies have been successfully applied in the production of high quality direct reduced products suitable for direct recycling into the steel plant (Barozzi, 1997; Oda et al., 2002; Sugitatsu and Miyahara, 2004; Tateishi et al., 2008; Gao et al., 2012; Haga et al., 2012). However, the commercialization of RHF technologies was impaired by operational challenges such as the adherence of oxide wastes to the hearth refractory surfaces, the regeneration of fines during reduction and the alkali metal intrusion into refractory materials (McCelland, 2002; Tateishi et al., 2008). To date, most commercially available RHF technologies have managed to achieve a high degree of metallization and de-zincification as a result of the high operating temperatures (1200°C–1400°C), as well as the degree of uniformity of the agglomerated steelmaking dust and reductants before charging into the process (Barozzi, 1997; Oda et al., 2002; Sugitatsu and Miyahara, 2004; Tateishi et al., 2008; Haga et al., 2012). Based on pilot tests conducted under the FASTMET® process conditions, Tateishi et al. (2008) reported that, on average, 73%–88% metallization and 91%–98% dezincification were achieved from EAF dust–coal agglomerates initially containing about 31%–33 wt.% Fe and 17–19 wt.% Zn.

Multiple-hearth-electric smelter technology for recycling agglomerated dust: This process combines the solid state reduction of steelmaking dust in a multi-hearth furnace followed by the hot charging of the resulting briquetted iron in a specially designed EAF to recover the valuable metal products (Roth et al., 2001; Suetens et al., 2014; Degel et al., 2015; Paul Wurth, n.d). In this case, the multiple-hearth furnace is charged with steelmaking dust fines and coal fines, and the coal fines basically function as the reductant and source of thermal energy. The energy generated during the post-combustion of CO, volatilized oils and the volatile components of coal is sufficient to maintain the required process temperatures in the range 1000°C–1100°C (Roth et al., 2001; Suetens et al., 2014; Degel et al., 2015; Paul Wurth, n.d). The volatile metal oxides are reduced from the feed material, volatilized into the oxidizing furnace off-gas where they are further reoxidized to form dust particles, and are finally recovered in bag filters. The volatilization of the volatile elements in the charge, coupled with the carbothermic reduction of metal oxides, results in the formation of a highly metallized iron concentrate which is then recycled by melting in the EAF.

According to Degel et al. (2015), the multiple- hearth-electric arc furnace process is suitable for treating dusts and sludges from oxygen steelmaking and EAF, and is capable of producing hot metal of quality similar to that produced from the blast furnace process. Currently, several commercial plants based on the Primus® technology are in operation globally at various capacities, with one installed at Dragon Steel (Taiwan) being proposed to possess a nominal recycling capacity of 120,000 tonnes/year on a dry basis (Degel et al., 2015).

4.6.4.2 Smelting Reduction Processing of Steelmaking Dust

In general, smelting reduction processes can easily be applied in the processing of metal oxides in steelmaking dust by recovering the valuable metal elements such as Fe, Cr, V and Ni in the liquid metal bath while producing an innocuous barren slag by-product (Barcza and Nelson, 1990; Sarmar et al., 1996; Schoukens et al., 1996; Hasegawa et al., 1998; Hara et al., 2000; von Scheele; 2004; Denton et al., 2005; Takano et al., 2005). In essence, smelting reduction processes produce hot metal product saturated with carbon, which can either be cast into semi-finished products, or can be used as feed in the EAF or BOFs. Other product streams include (1) a barren slag product that can be used in civil and construction applications, (2) off-gas containing considerable fuel value and (3) bag house sludge containing high enough levels of zinc that can be used as a viable raw material for zinc production. To date, several smelting reduction processes have been developed to recycle steel plant dusts, and these include the coke-packed shaft furnace smelting reduction process developed by Kawasaki Steel (Hasegawa et al., 1998; Hara et al., 2000), the OXYFINES™ process developed by Linde AG (von Schéele, 2004), the OxyCup® process developed by Küttner GMBH & Co and ThyssenKrupp AG (Kesseler and Erdmann, 2007; Li et al., 2014; Holtzer et al., 2015; Küttner GMBH, n.d.) and the Enviroplas™ DC-arc plasma process developed by Mintek (Barcza and Nelson, 1990; Jones et al., 1993; Schoukens et al., 1996; Denton et al., 2005).

Coke-packed bed smelting reduction process: The potential to recycle steelmaking dusts based on the smelting reduction process in a shaft furnace with a coke-packed bed was highlighted by Hasegawa et al. (1998) and Hara et al. (2000). In this process, the coke-packed bed shaft furnace process is based on the smelting reduction of non-agglomerated dust fines injected through specially designed feeding chutes, and involves the instantaneous fusion of the injected materials in the high temperature (1500°C–1600°C) region of the furnace (Hasegawa et al., 1998; Hara et al., 2000). Basically, the advantage of the coke-packed bed shaft furnace smelting reduction process lies in its ability to utilize the fine dust materials without the need for agglomeration and in its capability to treat refractory metal oxides by utilizing the high temperatures attained in the furnace.

Oxyfines™ Process: Other researchers have proposed a cost-effective process for the recycling of steel plant dusts based on the Oxyfines™ process, where an oxy-fuel burner is used to augment the melting of fine dusts in an EAF (von Schéele, 2004). In the Oxyfines™ technology, the waste dust materials are agglomerated into solid aggregates of suitable sizes before being charged in the EAF. In general, this process has been applied in the recycling of a number of both dry and wet dust materials produced in the production of various ferroalloys, iron powder dust and rejects, dust from carbon and stainless steel production processes and dust materials from blast furnaces and steelmaking converters (von Schéele, 2004).

OxyCup® Process: This process is based on the traditional shaft/cupola furnace and has been proved to be effective in the processing of a wide range of ferrous wastes (Kesseler and Erdmann, 2007; Li et al., 2014; Holtzer et al., 2015; Küttner GMBH, n.d.). In essence, the OxyCup® technology can be used to fully recycle different types of metallurgical wastes such as slag crusts, metallurgical dusts, coke breeze, carbon dust sludges and mill scale generated from the different stages in the iron and steelmaking process, to produce hot metal of quality similar to that produced in the blast furnace (Kesseler and Erdmann, 2007; Holtzer et al., 2015). In this process, self-reducing briquettes obtained from agglomeration of particulate dust, sludges and other iron-bearing wastes are melted in a hot-blast countercurrent reactor operated enriched with up to 5%–15% oxygen (Kesseler and Erdmann, 2007; Holtzer et al., 2015; Küttner GMBH, n.d.). In addition to briquetted metallurgical wastes and fluxes, solid metallic scrap can also be charged and melted in the OxyCup® furnace. The formation of sponge iron from the reduction of iron oxides is usually achieved at around 1450°C, and the resulting reduced iron, slag components and other metallic charges undergo the final melting in the hearth. The hot metals containing about 4 wt.% carbon, and the liquid slag, are tapped from the furnace at around 1500°C via an iron–slag siphon system.

Because the OxyCup® process is based on a cupola furnace design that is similar to a miniaturized blast furnace, it typically combines both the reduction and melting functions in a single unit (Küttner GMBH & Co, n.d.). Furthermore, the process is capable of recycling the zinc and alkali-containing metallurgical wastes of unknown analysis, which in essence, tend to be problematic materials in the iron and steel manufacturing process. In this case, zinc oxide particles are formed from the carbothermic reduction of zinc-containing compounds such as $ZnFe_2O_4$ at high temperatures, and some of the zinc oxides are further reduced to zinc metal. In the high temperature zones of the furnace, the reduced zinc gets volatilized from the metal bath and is entrained into the flue gases where it is eventually re-oxidized and carried out of the furnace as fine dust in the flue gases (Holtzer et al., 2015; Küttner GMBH & Co, n.d). In general, the zinc content of the furnace dust can be as high as 30 wt.%, which is sufficiently high for zinc recovery (Holtzer et al., 2015).

4.6.5 Plasma-Based Processing of Steelmaking Dusts

4.6.5.1 Fundamentals of Thermal Plasma Processing

Plasma is considered to be the fourth state of matter in the sequence solid–liquid–gas–plasma (Boulos, 1991; Jones et al., 1993; Taylor and Pirzada, 1994; Gomez et al., 2009). In general, plasma consists of a mixture of electrons, ions and neutral particles, even though it is overall considered to be electrically neutral (Boulos, 1991; Jones et al., 1993; Taylor and Pirzada, 1994; Gomez et al., 2009). Plasma technology involves the creation of a sustained electrical arc by the passage of electric current through a gas, via a phenomenon commonly referred to as electrical breakdown, and significant amount of heat that is generated as a result of the electrical resistivity across the system strips away electrons from the gas molecules resulting in an ionized gas stream (Gomez et al., 2009).

Thermal plasmas constitute one of the most important applications of plasma technology in the processing of materials (Gomez et al., 2009). In thermal plasmas, the degree of ionization is measured by the proportion of atoms that have either gained or lost electrons and is typically controlled by temperature conditions (Jones et al., 1993; Gomez et al., 2009). In the recent years, the processing of materials using thermal plasmas has received a lot of attention in wide range of applications such as (Gomez et al., 2009) (1) coating techniques such as plasma spraying, wire arc spraying and thermal plasma chemical vapor deposition, (2) synthesis of fine powders, (3) metals processing applications such as clean melting and re-melting in industrial furnaces and (4) treatment of hazardous waste materials. In essence, the application of thermal plasma in the processing of engineering materials has gained traction due to their significant advantages such as the ability to attain high temperatures, high intensity and non-ionizing radiation and the resulting high energy intensity (MacRae, 1992; Taylor and Pirzada, 1994; Gomez et al., 2009). As a result, the potential of plasma technology in the various industrial applications lies in its ability to generate high temperatures and high heat fluxes while maintaining control of the chemical potential of the gaseous atmosphere (MacRae, 1992).

In general, thermal plasma-generating devices convert electrical energy to thermal energy (MacRae, 1992; Taylor and Pirzada, 1994). A high temperature plasma arc is generated between at least two electrodes, that is, a cathode from which electrons are emitted and an anode at which the electrons are absorbed (Jones et al., 1993). According to Gomez et al. (2009), the majority of plasma arc generators used in materials processing use direct current (DC) rather than alternating current (AC) mainly due to the fact that the DC devices (1) can afford a more stable operation and better control, (2) require a minimum of only two electrodes, (3) have a lower electrode consumption rate, (4) result in slightly lower refractory wear, (5) have lower power consumption and (6) generate less flicker and noise.

Furthermore, DC plasma devices can be categorized based on the plasma production methods, that is, either as transferred or as non-transferred arc plasma torches (Boulos, 1991; Jones et al., 1993; Taylor and Pirzada, 1994; Gomez et al., 2009). In the case of transferred arc plasma devices, one of the electrodes is located outside the torch, and in most cases, is usually the material to be heated. In other words, the transferred arc plasmas tend to be characterized by a relatively large physical separation between the cathode and the anode (Jones et al., 1993; Taylor and Pirzada, 1994; Gomez et al., 2009). In the case of non-transferred plasma devices, a high-temperature plasma arc produced interacts with the flowing gas to produce a hot jet into which the material to be processed can be injected for in-flight processing such as melting and vaporization (Gomez et al., 2009).

To date, plasma furnaces have been successfully applied not only in the primary production of metals and other engineering materials, but also in the melting and recycling of a variety of waste materials (Jones et al., 1993; Neuschütz, 1996; Gomez et al., 2009). The extensive applications of plasma furnaces in the recycling of metal scrap, metallurgical residues and fly ashes from municipal solid waste and sewage sludge incineration processes were highlighted in several studies (MacRae, 1992; Jones et al., 1993; Neuschütz, 1996; Gomez et al., 2009). For example, Jones et al. (1993) reported that plasma furnaces are well suited to the treatment of wastes due to high temperature chemical processing capabilities, and the ability to directly utilize fine feed materials. In addition, electrically based thermal plasma processes allow for a sealed reactor design, high–energy density operation at low oxygen potentials, reduced gas volumes and suitably high off-gas temperatures (Jones et al., 1993). Because the power input is not limited by the electrical conductivity of the materials being processed, the feed rate and power can be controlled independently

(Jones et al., 1993). Other process advantages of utilizing plasma furnaces in the processes of metallurgical wastes, particularly particulate dusts, include: (1) the ability to process ultra-fine dusts without prior agglomeration, especially when the central feeding through a hollow graphite electrode is used, and (2) the ability to condense volatile metals such as zinc and lead directly from the furnace off-gas (Jones et al., 1993).

Enviroplas™ Process for treating steelmaking dust: The Enviroplas™ process is a DC arc furnace developed by Mintek (RSA) to treat solid metallurgical wastes (Barcza and Nelson, 1990; Jones et al., 1993; Schoukens et al., 1996; Abdel-latif, 2002; Denton et al., 2005). According to Jones et al. (1993) and Denton et al. (2005), the process is capable of treating EAF dusts, alloy-steel dusts and/or mixtures of both. Basically, the Enviroplas™ process is based on the high temperature carbothermic reduction of metal oxides in the waste dusts to produce an innocuous liquid slag and molten crude metal (Jones et al., 1993; Denton et al., 2005). In essence, some of the technical advantages of the Enviroplas™ process hinge on the following (Denton et al., 2005): (1) an open hearth operation which negates the necessity for the costly agglomeration processing of fines; (2) fast kinetics and recoveries achieved from the process, with the actual results being closer to thermodynamic predictions as a result of the even temperature distribution and improved slag/reductant contact and (3) optimized chemistry and greater flexibility due to tighter control of the reagent mix and power.

In addition, the Enviroplas process is capable of processing fine steelmaking dusts and sludges, particularly those that contain a high amount of zinc oxides (Jones et al., 1993; Denton et al., 2005). Based on the extensive pilot scale tests conducted by the researchers at Mintek, the following process benefits have been demonstrated: (1) the potential to process ultra-fine dusts directly without prior agglomeration by feeding the fine materials through a hollow graphite electrode; (2) the potential to condense volatile metal elements, specifically zinc, directly from the furnace off-gases, (3) the potential to strip the slag phase of oxides of chromium and other alloying elements present and (4) the concomitant generation of a valuable metal alloy (Barcza and Nelson, 1990; Jones et al., 1993; Schoukens et al., 1996; Abdel-latif, 2002; Denton et al., 2005).

4.6.6 Hydrometallurgical Processing of Steelmaking Dusts

Generally, two process options exist for the recycling and recovery of valuable metals from secondary raw materials such as steelmaking wastes, and these are classified as pyrometallurgical and hydrometallurgical processes. The pyrometallurgical recycling processing of steelmaking wastes such as dusts and sludges was extensively covered in the preceding sections. The major advantages of the pyrometallurgical processes lie in their ability to economically recycle steelmaking dusts containing high amounts of complex zinc compounds existing as ZnO and $ZnFe_2O_4$. Despite the several advantages offered by most of the high temperature processes discussed so far, most of these processes have inherent disadvantages such as the following elements (Barret et al., 1992; Trung et al., 2011): (1) high sensitivity to economies of scale, (2) high thermal energy requirements and elaborate dust collection systems and (3) the requirement for additional processing steps to recover the valuable metal elements.

In the recent years, research on the hydrometallurgical processing of steelmaking wastes to recover valuable metals in solution and produce an environmentally benign residue has received widespread attention (Nyirenda, 1991; Barret et al., 1992; Jha et al., 2001; Leclerc et al., 2003; Kelebek et al., 2004; Havlic et al., 2005; Havlík et al., 2006; Rao, 2011; Trung et al., 2011; Cantarino et al., 2012; Havlík et al., 2012; Montenegro et al., 2013). This notable upward trajectory towards the development of novel hydrometallurgical processes to recover these valuable metals from steelmaking wastes has been driven by the need to process complex raw materials and the ability to develop cost-effective processes that meet the ever-increasing legislative demands towards environmental protection (Trung et al., 2011). According to Trung et al. (2011), hydrometallurgical processes are gaining more prominence mainly because (1) they have higher flexibility in the operation and required scale of plants, (2) the processes are more economical because of lower capital and operating costs and (3) they pose little or no environmental challenges associated with off-gases, dusts and noise. Despite these process advantages, there are also process challenges associated with these hydrometallurgical processes, that is, the leaching methods are dependent not only on the type of the waste materials being processed but also on the physical, physicochemical, chemical and mineralogical properties as well (Trung et al., 2011; Havlík et al., 2012).

The high content of zinc as well as the complex chemical and morphological composition of the steelmaking dusts and sludges has been cited as one of the key challenges affecting the recycling potential of such materials in the steel plant. As a result, the chemical and morphological form in which the zinc exists in steelmaking dusts is an essential indicator of the efficiency of the dust processing method (Havlík et al., 2012). As discussed in Section 4.5.1.3, it is difficult to correctly predict the form in which the zinc compounds are present due to the complexity in the chemical and mineralogical characteristics of steelmaking dusts. According to Leclerc et al. (2003), most of the zinc speciates as franklinite ($ZnFe_2O_4$) in steelmaking dusts, and these compounds tend to be stable and insoluble in most acidic, alkaline and chelating media under mild conditions, thereby creating problems for zinc recovery based on conventional hydrometallurgical processes.

In general, hydrometallurgical processes can be divided into acidic and alkaline leaching processes, either at ambient conditions or elevated at temperatures and pressure (Havlík et al., 2012). After the leaching process, the dissolved metals are recovered from the solution using various techniques such as selective precipitation, cementation, solvent extraction, ion exchange and electrolysis. Detailed discussions on the fundamentals of hydrometallurgical processes are also covered in Chapter 7.

Acid leaching processes: The leaching of steelmaking dusts using different kinds of acidic solutions was studied by several researchers, with most of these researchers citing sulphuric acid (H_2SO_4) and hydrochloric acid (HCl) as the most common acidic lixiviants used (Nyirenda, 1991; Jha et al., 2001; Havlik et al., 2005; Trung et al., 2011; Havlík et al., 2012; Montenegro et al., 2013). According to these researchers, the reactions taking place in the sulphuric acid leaching of steelmaking dusts are depicted in Equations 4.14 through 4.19:

$$ZnO_{(s)} + H_2SO_{4(aq)} \rightarrow ZnSO_{4(aq)} + H_2O_{(l)} \tag{4.14}$$

$$ZnFe_2O_{4(s)} + 4H_2SO_{4(aq)} \rightarrow ZnSO_{4(aq)} + Fe_2\left(SO_4\right)_{3(aq)} + 4H_2O_{(l)} \tag{4.15}$$

$$ZnFe_2O_{4(s)} + H_2SO_{4(aq)} \rightarrow ZnSO_{4(aq)} + Fe_2O_{3(s)} + H_2O_{(l)} \tag{4.16}$$

$$ZnFe_2O_{4(s)} + H_2SO_{4(aq)} + H_2O_{(l)} \rightarrow ZnSO_{4(aq)} + 2Fe\left(OH\right)_{3(s)} \tag{4.17}$$

$$3ZnFe_2O_{4(s)} + 2H_2SO_{4(aq)} \rightarrow ZnO \cdot 2ZnSO_{4(aq)} + Fe_2O_{3(s)} + 2H_2O_{(l)} \tag{4.18}$$

$$Ca\left[Zn\left(OH\right)_3\right]_2 \cdot 2H_2O_{(s)} + 3H_2SO_{4(aq)} \rightarrow CaSO_4\left(aq\right) + 2ZnSO_{4(aq)} + 8H_2O_{(l)} \tag{4.19}$$

Due to the heterogeneous and chemically complex nature of the steelmaking dusts, the co-dissolution of other compounds present in the dust can also occur according to Equations 4.20 through 4.22 (Trung et al., 2011; Havlík et al., 2012):

$$CaCO_{3(s)} + H_2SO_{4(aq)} \rightarrow CaSO_{4(aq)} + H_2O_{(l)} + CO_{2(g)} \quad (4.20)$$

$$CaO_{(s)} + H_2SO_{4(aq)} \rightarrow CaSO_{4(aq)} + H_2O_{(l)} \quad (4.21)$$

$$CaFe_2O_{4(s)} + 4H_2SO_{4(aq)} \rightarrow CaSO_{4(aq)} + Fe_2\left(SO_4\right)_{3(aq)} + 4H_2O_{(l)} \quad (4.22)$$

Basically, different Eh-pH models have been proposed as effective tools in controlling these co-dissolution reactions (Trung et al., 2011). The thermodynamic models based on the Eh-pH stability regions of the Zn–S–H_2O and Fe–S–H_2O systems proposed by Trung et al. (2011) clearly highlights the stability area of Zn^{2+} ions over the limited stability area of Fe^{2+} or Fe^{3+} ions in the solution, and hence provides practical opportunities for the selective leaching of zinc over iron and the separation of the zinc and iron directly in the solution by controlling the pH of the lixiviant.

In addition to the sulphuric acid leaching, hydrochloric acid has also been found to be effective in the leaching of steelmaking dusts containing zinc compounds (Nyirenda, 1991; Baik and Fray, 2000; Jha et al., 2001; Langová et al., 2009). The dissolution process using hydrochloric acid as a lixiviant can occur according to Equations 4.23 through 4.26 (Jha et al., 2001; Langová et al., 2009):

$$ZnFe_2O_{4(s)} + 2HCl_{(aq)} \rightarrow ZnCl_{2(aq)} + Fe_2O_{3(s)} + H_2O_{(l)} \quad (4.23)$$

$$ZnFe_2O_{4(s)} + 8HCl_{(aq)} \rightarrow ZnCl_{2(aq)} + FeCl_{3(aq)} + 4H_2O_{(l)} \quad (4.24)$$

$$ZnO_{(s)} + 2HCl_{(aq)} \rightarrow ZnCl_{2(aq)} + H_2O_{(l)} \quad (4.25)$$

$$Zn_{(s)} + 2HCl_{(aq)} \rightarrow ZnCl_{2(aq)} + H_{2(g)} \quad (4.26)$$

Due to the co-dissolution of iron-bearing compounds, hematite residue can also be formed from the precipitation from $FeCl_3$ generated in solution, according to Equation 4.27.

$$2FeCl_{3(aq)} + 3H_2O_{(l)} \rightarrow Fe_2O_{3(s)} + 6HCl_{(aq)} \quad (4.27)$$

Baik and Fray (2000) proposed that hydrochloric acid is an effective lixiviant for the leaching of EAF dusts, giving a high zinc recovery with minimum jarosite formation. However, Havlík et al. (2012) contends that despite the problem of jarosite formation and limited dissolution kinetics during the atmospheric leaching, sulphuric acid is still the most widely used and preferred reagent in the leaching of steelmaking dusts, mainly due to its low price and relatively well known process dynamics for the electro-recovery of the zinc from sulphate solutions.

Alkaline leaching process: The alkaline leaching of steelmaking dust using concentrated sodium hydroxide solution has proved to be an effective reagent in the selective leaching of zinc, lead and other toxic heavy metal compounds, without significantly co-dissolving the iron (Jha et al., 2001; Dutra et al., 2006; Rao, 2011; Cantarino et al., 2012). Equations 4.28 and 4.29 show the typical reactions taking place in the dissolution of zinc and lead compounds using sodium hydroxide (Jha et al., 2001; Rao, 2011):

$$ZnO_{(s)} + 2NaOH_{(aq)} \rightarrow Na_2ZnO_{2(aq)} + H_2O_{(l)} \tag{4.28}$$

$$PbO_{(s)} + 2NaOH_{(aq)} \rightarrow Na_2PbO_{2(aq)} + H_2O_{(l)} \tag{4.29}$$

The co-dissolution of silica and the subsequent precipitation of the silicates from the solution are also significant in the sodium hydroxide leaching of steelmaking dusts, and the co-dissolution reactions in shown in Equations 4.30 through 4.31 given by

$$SiO_{2(s)} + 2NaOH_{(aq)} \rightarrow Na_2SiO_{3(aq)} + H_2O_{(l)} \tag{4.30}$$

$$Na_2SiO_{3(aq)} + Ca(OH)_{2(s)} \rightarrow CaSiO_{3(s)} + 2NaOH_{(aq)} \tag{4.31}$$

As shown in Equations 4.30 through 4.31, the zinc is solubilized zinc zincate, and is recovered from solution by electrowinning methods. During the electrowinning stage, the zinc is deposited while the NaOH lixiviant is regenerated:

$$Na_2ZnO_{2(aq)} + 2H_2O_{(l)} \rightarrow Zn_{(s)} + 4NaOH_{(aq)} + O_{2(g)} \tag{4.32}$$

Based on studies conducted by Cantarino et al. (2012), the sodium hydroxide leaching of BOF sludges in the temperature range 300°C–450°C achieved over 90% zinc removal with practically no iron extraction taking place. According to Dutra et al. (2006), the low iron content of the resulting solution makes the process attractive, as the purification processes required in the downstream electrowinning recovery of zinc become simplified.

4.6.7 Solidification and Stabilization (S/S) of Steelmaking Dusts

The recycling and reuse of steelmaking wastes using pyrometallurgical processes and the recovery of metals from steelmaking dusts using hydrometallurgical processes have been discussed in detail so far. The main objective behind the development and commercialization of these metal recovery processes is to recover the valuable metals in steelmaking wastes and the resultant production of an environmentally benign residue that can safely be disposed of in landfills. However, metal recovery is not always profitable due to poor economies of scale and/or complex chemical and mineralogical compositions of the steelmaking wastes (Machado et al., 2011). As a result, other alternative approaches to managing the hazardous steelmaking wastes involving the stabilization and/or solidification (S/S) of the hazardous components using cementitious, pozzolanic and geopolymeric additive materials prior to disposal in landfills have also been developed and adopted as a practical alternative to managing these toxic and complex waste materials (Trussell and Spence, 1994; Fernández et al., 2003; Laforest and Duchesne, 2006b; Fernández-Pereira et al., 2007; Salihoglu et al., 2007; Giergiczny and Król, 2008; Tang et al., 2008; Bulut et al., 2009; Ma and Garbers-Craig, 2009; Luna-Galiano et al., 2010; Machado et al., 2011;

Stathopoulos et al., 2013; Bayraktar et al., 2015). Although the stabilization and solidification of steelmaking wastes using cementitious, pozzolanic and geopolymeric materials does not fall within any one of the mainstream recycling and/or reuse categories focused on in this book, a detailed analysis of managing steelmaking wastes using these methods is critical from a holistic waste management point of view.

4.6.7.1 Fundamentals of Waste Stabilization and Solidification

According to Conner and Hoeffner (1998), the processes and techniques for hazardous waste S/S have matured into an accepted and important part of environmental technology. In essence, the S/S of hazardous waste materials and contaminated soils is an attractive technology to reduce their toxicity and facilitate handling prior to disposal in landfills (Wiles, 1987; Trussell and Spence, 1994; Conner and Hoeffner, 1998; Batchelor, 2006; Malviya and Chaudhary, 2006; Zhou et al., 2006; Chen et al., 2009). Malviya and Chaudhary (2006) proposed that the ideal material for the solidification and/or stabilization process should not only solidify the hazardous waste by chemical means, but should also insolubilize, immobilize, encapsulate, destroy, sorb or otherwise interact with the selected waste components. Furthermore, Chen et al. (2009) proposed that the broad objectives of S/S processes are to achieve and maintain the desired physical properties of the disposed waste materials, and to chemically stabilize or permanently bind the contaminants.

4.6.7.2 Solidification

Solidification refers to processes and techniques that encapsulate the waste into a monolithic solid of high structural integrity (Conner and Hoeffner, 1998). Conner and Hoeffner (1998) defined the solidification process to involve physically or mechanically binding the toxic waste to in order restrict the surface area exposed to leaching and/or by isolating the toxic elements within an impervious capsule. Basically, the solidification process can be divided into two categories, namely: (1) microencapsulation, involving the encapsulation of fine waste particles, and (2) macro encapsulation, involving the encapsulation of agglomerated waste particles (Wiles, 1987; Conner and Hoeffner, 1998). In addition, Wiles (1987) reported that solidification may or may not involve a chemical bond between the toxic contaminant and the additive.

4.6.7.3 Stabilization

In contrary, stabilization refers to those techniques that reduce the hazard potential of waste by converting the contaminants into less soluble, less mobile or less toxic forms without necessarily changing the physical and handling characteristics of the waste materials (Conner and Hoeffner, 1998). In this context, stabilization includes chemical processes that use chemical reactions to transform the toxic component into a new non-toxic compound or substance (Wiles, 1987; Conner and Hoeffner, 1998; Batchelor, 2006). According to Batchelor (2006), the typical chemical stabilization methods may include (1) precipitation reactions, where pH control can result in metal contaminants forming insoluble hydroxides or mixed hydroxide solids and (2) oxidation–reduction reactions, especially for those contaminants that exist in multiple redox states. Chromium is a typical example of a hazardous element that is stabilized via an oxidation–reduction reaction is chromium, mainly because it tends to be more mobile and toxic in the hexavalent (Cr^{+6}) oxidation state than in the trivalent (Cr^{+3}) state (Conner and Hoeffner, 1998; Rodríguez-Piñero et al., 1998; Batchelor, 2006; Bulut et al., 2009; Ma and Garbers-Craig, 2009).

4.6.7.4 Stabilization and Solidification of Steelmaking Wastes Using Cementitious Systems

Portland cement is used extensively as a primary binder in the solidification and/or stabilization of a wide variety of steelmaking dusts (Johnson, 2004; Batchelor, 2006; Laforest and Duchesne, 2006b; Malviya and Chaudhary, 2006; Zhou et al., 2006; Chen et al., 2007; Salihoglu et al., 2007; Giergiczny and Król, 2008; Bulut et al., 2009; Chen et al., 2009; Ma and Garbers-Craig, 2009; Luna-Galiano et al., 2010). In cementing systems, heavy metals can be precipitated as hydroxides, carbonates, sulphates and silicates (Glasser, 1997; Conner and Hoeffner, 1998; Johnson, 2004; Shi and Fernández-Jiménez, 2006; Chen et al., 2007). Shi and Fernández-Jiménez (2006) reported that the solidification and stabilization of contaminants by cement materials includes the following aspects: the (1) chemical fixation of contaminants by chemical interactions between the hydration products of the cement and the contaminants, (2) physical absorption of the contaminants on the surface of hydration products of the cements and (3) physical encapsulation of the contaminated waste as a result of the low permeability of the hardened paste. In addition, the immobilization of heavy metal ions in cementitious systems was proposed to follow any of the following mechanistic steps (Glasser, 1997; Johnson, 2004): heavy metal ions (1) precipitated in the alkaline cement matrix as oxides, mixed oxides or as other solid phases, (2) sorbed or precipitated onto the surfaces of cement minerals or (3) incorporated into hydrated cement materials.

According to Chen et al. (2007), the overall process of cement hydration and immobilization of heavy metals in cement systems include a combination of complex solution processes, interfacial phenomena, and solid state reactions. In particular, the immobilization process using cementitious materials has been mostly attributed to the hydration reactions of Ca_3SiO_5 (C_3S) in the presence of water to form calcium silicate hydrates (C–S–H), a major component of hardened cement paste (Glasser, 1997; Johnson, 2004; Zhou et al., 2006; Chen et al., 2007; Giergiczny and Król, 2008; Chen et al., 2009). Based on any one of the different mechanistic steps proposed by Glasser (1997) and Johnson (2004), the calcium silicate hydrates formed from the hydration reactions can be directly attributed for the immobilization of the heavy metal ions such as zinc, lead and chromium present in steelmaking dusts. In this case, the immobilization process was proposed to occur by forming single hydroxides such as $Zn(OH)_2$, $Pb(OH)_2$ and/or double hydroxides such as $CaZn_2(OH)_6 \cdot 2H_2O$, $Ca_2Cr(OH)_7 \cdot 3H_2O$ (Johnson, 2004; Laforest and Duchesne, 2005; Chen et al., 2007; Chen et al., 2009).

Zhou et al. (2006) reported on the attributes of cementitious systems that make them suitable for waste encapsulation and immobilization, and these include (1) its ability to readily acting as a barrier to the migration of toxic heavy metal ions, (2) the incorporation of many toxic heavy metal ions into solid solution within the hydrated calcium silicates, (3) low liquid and gas permeabilities in the hardened state, (4) relatively easy processing which is suitable for remote locations and (5) low price and ready availability. In most cases, the selection of cementitious systems for use in immobilizing the toxic heavy metal ions is benchmarked against internationally standardized test methods such as the toxicity characteristics leaching procedure (TCLP), synthetic precipitation leaching procedure (SPLP) and ASTM D3987-85 methods (Conner and Hoeffner, 1998; Mills, n.d.; ASTM D3987-85, 2004; Batchelor, 2006; Townsend et al., 2006; Bulut et al., 2009; Ma and Garbers-Craig, 2009).

4.6.7.5 Stabilization of Steelmaking Waste Using Geopolymer Systems

The stabilization and solidification of hazardous wastes using geopolymerization techniques has gained widespread attention in recent years (van Jaarsveld et al., 1996;

van Jaarsveld et al., 1998; Pereira et al., 2001; Fernández-Pereira et al., 2007; Luna et al., 2007; Zhang et al., 2008a,b; Luna-Galiano et al., 2010; Luna et al., 2014). According to van Jaarsveld et al. (1998), geopolymers represent a group of materials that are formed by reacting silica-rich and alumina-rich solids with solutions of alkali or alkali salts, resulting in a mixture of gels and crystalline compounds that eventually harden into a new strong matrix. Generally, geopolymers are also referred to as alkali-activated aluminosilicate binders or alkali-activated cements, and are considered as the synthetic analogues of natural zeolite materials (van Jaarsveld et al., 1996; van Jaarsveld et al., 1998; Pereira et al., 2001; Fernández-Pereira et al., 2007; Luna et al., 2007; Zhang et al., 2008; Luna-Galiano et al., 2010; Luna et al., 2014). The stabilization and solidification of hazardous waste by encapsulation using geopolymeric systems is similar to that of cementitious binders (van Jaarsveld et al., 1996; van Jaarsveld et al., 1998; Pereira et al., 2001; Fernández-Pereira et al., 2007; Luna et al., 2014). In general geopolymeric materials are considered to possess several advantages over ordinary Portland cement in the immobilization of hazardous wastes, and these include their improved chemical and physical properties such as structural integrity, low permeability, high compressive strength and durability. As a result, these characteristics have increased the preference for geopolymeric materials in the immobilization of hazardous and toxic wastes.

To date, several researchers have studied the immobilization of toxic and hazardous metal ions using a variety of geopolymer-based materials, namely, (1) geopolymeric materials based on alkali-activated coal fly ash (Vempati et al., 1995; van Jaarsveld et al., 1998; Pereira et al., 2001; Goretta et al., 2004; Duxson and Provis, 2008; Zhang et al., 2008a,b; Fernández-Pereira et al., 2001; Izquierdo et al., 2009; Kumar et al., 2010; Luna-Galiano et al., 2010; Luna et al., 2014), (2) geopolymeric materials based on naturally occurring alkali-activated aluminosilicates in the form of metakaolin and calcined clays (Lecomte et al., 2003; Lecomte et al., 2006; Duxson and Provis, 2008; Li et al., 2010) and (3) geopolymeric materials based on alkali activated granulated blast furnace slags (Lecomte et al., 2003; Goretta et al., 2004; Zhang et al., 2007; Duxson and Provis, 2008; Zhang et al., 2008b; Kumar et al., 2010; Li et al., 2010; Huang et al., 2015; Robayo et al., 2016). These studies have demonstrated the novel opportunities for the utilization of waste materials such as the fly ash and granulated blast furnace slags in the stabilization and solidification of toxic and hazardous waste materials.

Furthermore, the possibility of solidifying and stabilizing EAF dusts using coal fly ash and granulated blast-furnace slags based geopolymeric materials has been investigated extensively (Pereira et al., 2001; Duchesne and Laforest, 2004; Laforest and Duchesne, 2005; Fernández-Pereira et al., 2007; Duxson and Provis, 2008; Fernández-Pereira et al., 2009; Luna-Galiano et al., 2010). These studies focused on the different types of alkali-activated coal fly ash materials (Classes C & F) as the main sources of silica and alumina, and the findings highlighted that the calcium silicate hydrates (C–S–H) and calcium aluminate hydrates (in the form of ettringite ($6CaO \cdot Al_2O_3 \cdot 3SO_3 \cdot 32H_2O$)), solids formed in the long-term leaching experiments, were associated with a significant reduction in the leachate concentration of toxic elements (Pereira et al., 2001; Johnson, 2004; Fernández-Pereira et al., 2007; Zhang et al., 2008a; Fernández-Pereira et al., 2009; Izquierdo et al., 2009).

As discussed in Section 4.6, blast furnace slags constitute a typical waste material produced from the production of iron and steel using the blast furnace process. Technically, granulated blast furnace slags belong to the $CaO–SiO_2–MgO–Al_2O_3$ system and are broadly described as a mixture of phases with composition resembling gehlenite ($2CaO \cdot Al_2O_3 \cdot SiO_2$) and akermanite ($2CaO \cdot MgO \cdot 2SiO_2$), as well as a depolymerized calcium aluminosilicate glass (Sano et al., 1997; Duxson and Provis, 2008; Li et al., 2010). According to Duxson and

Provis (2008) and Shi et al. (2006), the reactivity of different slags in alkali-activated materials is a function of composition and glass structure, among other factors. Furthermore, Duxson and Provis (2008) proposed that the degree of depolymerization largely controls the reactivity of the granulated blast furnace slags.

Fundamentally, alkali-activated granulated blast furnace slags are considered to possess good geopolymeric properties needed for the stabilization of hazardous and toxic waste materials. As a result, the efficacy of alkali-activated granulated blast furnace slags as geopolymeric materials for the immobilization of toxic metal species contained in the EAF dusts has also been extensively investigated (Kindness et al., 1994; Rodríguez-Piñero et al., 1998; Lecomte et al., 2003; Duchesne and Laforest, 2004; Goretta et al., 2004; Laforest and Duchesne, 2005; Zhang et al., 2007; Duxson and Provis, 2008; Zhang et al., 2008b; Kumar et al., 2010; Li et al., 2010; Huang et al., 2015; Robayo et al., 2016). For example, Kindness et al. (1994) investigated the immobilization of toxic metal species in the EAF dusts based on the different mixtures of ordinary Portland cement, high alumina cement and ground granulated blast furnace slag. The granulated blast furnace slags were observed to significantly increase the immobilization of chromium species from solutions containing Cr (VI) and Cr (III) ions (Kindness et al., 1994). Furthermore, Laforest and Duchesne (2005) compared the interaction of chromium (VI) ions with granulated blast furnace slag and ordinary Portland cement based on absorption isotherms, and concluded that the granulated slags were more efficient than ordinary Portland cement at Cr (VI) fixation at lower initial solution concentrations, while the opposite was true at higher initial concentrations. The effect of initial concentration of Cr (VI) species on the different adsorption behavior between ordinary Portland cement and granulated blast furnace slag was also confirmed by scanning electron microscopy analysis tests as shown in Figure 4.17 (Laforest and Duchesne, 2005).

4.6.7.6 Thermal Vitrification of Steelmaking Dusts

Thermal vitrification of hazardous waste is a well-established high temperature fusion technology that involves the conversion of hazardous waste into a stable and homogenous silicate glass phase, with or without the addition of glass-forming additives (Mikhail et al., 1996; Pelino et al., 2002; Colombo et al., 2003; Kavouras et al., 2007; Tang et al., 2008; Ma and Garbers-Craig, 2009; Machado et al., 2011; Stathopoulos et al., 2013). According to Colombo et al. (2003), the main advantages of the vitrification process include (1) the ability of inorganic glass phases to incorporate large amounts of heavy metal ions by chemically bonding them inside the inorganic amorphous network; (2) the ability to obtain glassy phases that are inert towards most chemical and biological agents, thereby enabling the bound toxic wastes to be disposed of in landfills or used in other applications such as in the construction of roads, pavements and embankments; (3) the ability of the vitrification processes to accept wastes of different composition and form, such as mixtures of sludges and solids; (4) vitrification is a mature technology based on well-documented glass forming systems and (5) vitrification results in a large reduction in the volume of waste.

To date, extensive research has been conducted on the thermal vitrification of toxic steelmaking dusts (Mikhail et al., 1996; Barbieri et al., 2002; Pelino et al., 2002; Kavouras et al., 2007; Tang et al., 2008; Ma and Garbers-Craig, 2009; Machado et al., 2011; Stathopoulos et al., 2013). In particular, the studies investigated the potential of thermal vitrification and the utilization of steelmaking dust, clay mixtures and/or glass cutlets in the production of structural ceramics via controlled crystallization processes (Mikhail et al., 1996; Barbieri et al., 2002; Pelino et al., 2002; Kavouras et al., 2007; Tang et al., 2008; Ma and Garbers-Craig, 2009; Machado et al., 2011; Stathopoulos et al., 2013). For example,

FIGURE 4.17
SEM micrographs of solid granulated blast furnace slag (GBFS) and Ordinary Portland Cement (OPC) samples in contact with a 3,000, 15,000 and 50,000 mg/L of Cr (IV) solution for 28 days. (Reprinted from *Cement Con. Res.*, 35, Laforest, G. and Duchesne, J., Immobilization of chromium (IV) evaluated by binding isotherms for ground granulated blast furnace slag and ordinary Portland cement, 2322–2332. Copyright 2005, with permission from Elsevier.)

Ma and Garbers-Craig (2009) proposed a process of stabilizing Cr (VI) in stainless steel plant dust by using silica-rich clay which was sintered on a 50 mass % dust: 50 mass % clay at 1100°C. These findings have demonstrated the efficacy of stabilizing the toxic metal elements contained in steelmaking dusts by structural incorporation into a stable solid state lattice (Mikhail et al., 1996; Barbieri et al., 2002; Pelino et al., 2002; Kavouras et al., 2007; Tang et al., 2008; Ma and Garbers-Craig, 2009; Machado et al., 2011; Stathopoulos et al., 2013). Furthermore, the vitrification products were characterized based on mechanical and standard toxicity characteristic measurements (TCLP), and the toxicity characteristic measurements of the glass-ceramic products were found to fall within leachate regulatory limits for most jurisdictions (Barbieri et al., 2002; Pelino et al., 2002; Kavouras et al., 2007; Machado et al., 2011; Stathopoulos et al., 2013).

4.6.8 Recycling and Reuse of Steelmaking Slags

As discussed in Section 4.5, steelmaking slags are produced as a by-product from the converting and refining process of liquid iron into steel in the BOF, the melting of scrap to produce liquid steel in EAFs, and the refining of liquid steel in ladle furnaces (Zheng and Kozinski, 1996; Patil et al., 1998; Atkinson and Kolarik, 2001; Shen and Forssberg, 2003; Dippenaar, 2004; Reuter et al., 2004; Shi, 2004; Das et al., 2007; Yildirim and Prezzi, 2011; Yi et al., 2012; Remus et al., 2013; Lobato et al., 2015). According to Horii et al. (2013), the liquid steelmaking slags are tapped from these furnaces in the temperature range 1300°C–1700°C and are then cooled and solidified under various conditions of slow-cooling or granulation processes.

Naturally, the chemical composition of steelmaking slags differ from that of blast furnace ironmaking slags. For example, BOF slags contain higher amounts of iron oxides, phosphorus and sulphur, while the iron content of the blast furnace slags is generally lower due to the reducing nature of the blast furnace ironmaking process (EUROSLAG, 2015). On the other hand, steelmaking slags produced in the electric arc and ladle furnace processes usually contain elevated amounts of alloy additives such as chromium, manganese, molybdenum and nickel (Patil et al., 1998; Reuter et al., 2004; Shen et al., 2004; Shi, 2004; Tossavainen et al., 2007; Huaiwei and Xin, 2011; Yildirim and Prezzi, 2011; Remus et al., 2013; Ma and Houser, 2014). The chemical composition and mineralogical phases of steelmaking slags were highlighted in Tables 4.14 and 4.15, respectively. As discussed in Section 4.5.2.3, there is a great deal of heterogeneity in terms of chemical and mineralogical characteristics of steelmaking slags, and such complexity presents serious challenges to the utilization, recycle and reuse of these types of slags. In addition, the low intrinsic monetary value coupled with the strict environmental control requirements of steelmaking slags also presents challenges to their utilization and disposal (Atkinson and Kolarik, 2001). As such, this section particularly focuses on the recycling and reuse opportunities and challenges of steelmaking slags typically produced from the basic oxygen steelmaking, electric arc and ladle furnace processes.

4.6.8.1 Electric Arc, Basic Oxygen and Ladle Furnace Slags

Several researchers proposed that steelmaking slags, in particular BOF slags, can be used as secondary sources of raw materials in the steel plants where they are generated (Atkinson and Kolarik, 2001; Reuter et al., 2004; Shen et al., 2004; Das et al., 2007; Yi et al., 2012; Remus et al., 2013; World Steel Association, 2015b). For example, Atkinson and Kolarik (2001) and

Reuter et al. (2004) proposed that the BOF steelmaking slags can be recycled directly into the sinter plant and blast furnace processes in order to utilize the fluxing compounds CaO and MgO in the slag, to recover the contained chemical energy, as well as recovering the valuable metal elements such as iron and manganese. However, BOF slags typically contain a high amount of phosphorus and most of the phosphorus present in the recycled slag would revert to the hot metal in the blast furnace (Atkinson and Kolarik, 2001). As a result, the amount of phosphorus in slag practically limits the quantity of BOF slags that can be recycled back into the sinter plant and blast furnace. Unlike BOF slags which possess a slightly higher recycling potential within the steel plant, EAF slags have limited in-house recycling opportunities due to the highly variable chemical and physical properties (Atkinson and Kolarik, 2001; Reuter et al., 2004).

In general, steelmaking slags can be utilized as starter fluxing materials in the steelmaking processes due to their properties resembling pre-melted fluxes. Basically, the pre-melted fluxes produced from recycled steelmaking slags are easier to re-melt when compared to raw fluxes, can easily be engineered to form balanced lime/slag melting rates, as well controlling basicity fluidity of slags (Li et al., 1995; Atkinson and Kolarik, 2001; Reuter et al., 2004; Cavalloti et al., 2007; Dahlin et al., 2012). Based on the aforementioned, the recycling of ladle furnace slags as pre-melted fluxes in the steel has been investigated extensively by Dahlin et al. (2012) and Cavalloti et al. (2007). For example, Dahlin et al. (2012) investigated the effect of ladle furnace slag addition on the performance of 114t Linz Donawitz lance bubbling equilibrium (LD-LBE) converter based on process parameters such as slag weight and quality of steel. Based on standard blowing practice, the studies indicated the potential to reduce the amount of lime addition without affecting the phosphorus and sulphur refining capacity, and the enhanced early slag formation process during the initial stages of the blow (Dahlin et al., 2012). On the other hand, Cavalloti et al. (2007) investigated the metallurgical and economic advantages of recycling ladle furnace white slags in the EAF, and concluded that the recycling of ladle furnace-white slags resulted in: (1) improved the slag foaming properties of the resulting EAF slags, (2) reduced consumption of refractory materials, (3) better heat transfer to the steel bath and (4) reduced dumping costs of ladle furnace slags.

Fundamentally, the heterogeneous and complex chemical properties of steelmaking slags as well the relatively small volumes generated have not inspired much research into possible economic recycling options. Nevertheless, extensive research has been conducted on the recovery processing of metallic elements entrained in solidified slags. Basically, stainless steel refining slags, in particular, may contain an elevated amount of chromium and nickel, and as such, the recovery of these metals from stainless steel slag is not only significant economically but is also beneficial to the environment (Shen and Forssberg, 2003; Ma and Garbers-Craig, 2006). Because these types of slags can contain as much as 10–40 wt.% entrained metallic iron and varying amounts of entrained alloy elements, it is economically advantageous to recover these metal elements from the slag with the objective of reusing the recovered elements as a substitute for metallic scrap and virgin alloying elements (Atkinson and Kolarik, 2001; Shen and Forssberg, 2003; Ma and Garbers-Craig, 2006; Horii et al., 2013). As such, the recovery of valuable metals entrained in steelmaking slags is discussed in detail in Section 4.6.8.2.

4.6.8.2 Recovery of Entrained Metallic Elements from Steelmaking Slags

The recovery of entrained metallic alloying elements from steelmaking slags has been a subject of extensive research. Of particular interest are studies that were carried out to investigate the possibility of using magnetic separation to recover entrained metallic elements from various steelmaking slags (Topkaya et al., 2004; Alanyali et al., 2006; Ma and Houser, 2014; Menad et al., 2014). For example, Ma and Houser (2014) applied a weak magnetic separation process coupled to a selective particle size screening of various steelmaking slag fines to produce iron-rich products with sufficiently high iron content and low level of impurities. Furthermore, Topkaya et al. (2004) proposed a slag processing unit for the reprocessing of steelmaking slags based on the crushing and classification of the slag into various sizes and magnetic separation processes. In addition, Menad et al. (2014) proposed a process flowsheet to recover the high grade iron from a Linz-Donawitz furnace slag for the subsequent recycling of the metallic product in other metallurgical processes and the non-magnetic portion of the slag being consumed as feed in cement manufacturing. In other studies, Alanyali et al. (2006) investigated the effect of pre-engineered parameters such as the slag grain size and drum-blade gap on the magnetic separation of steelmaking slags and concluded that the recovery of magnetic grains increased by reducing the grain size of slags and also by increasing the gap between the blades and drum of the magnetic separator. These studies demonstrated the potential to successfully recover and reutilize the metallic elements entrained in steelmaking slags, while opening up other possibilities for the utilization of the non-magnetic components of the steelmaking slag as a partial replacement for clinker in cement and aggregates manufacturing (Topkaya et al., 2004; Ma and Houser, 2014; Menad et al., 2014; Alanyali et al., 2006).

Metal recovery from steelmaking slags by smelting reduction: In addition to crushing and magnetic separation to recover entrained metallic components from steelmaking slags, other options have been investigated to recover the value metals using smelting reduction processes (Nakasuga et al., 2004; Das et al., 2007; Ye and Lindvall, 2012; Ye et al., 2013). In an effort to understand effective slag treatment processes, Nakasuga et al. (2004) investigated the recovery of chromium in the form of Cr_2O_3 from stainless steelmaking slags by iron melts containing carbon, aluminium and silicon in the temperature range 1400°C–1600°C. The authors concluded that the addition of fluxes such as Al_2O_3 and SiO_2 to the slag increased the recovery rate of chromium by promoting the formation of liquid slag at the early stages of reaction (Nakasuga et al., 2004). In addition, Ye and Lindvall (2012) and Ye et al. (2013) investigated the recovery of vanadium from steelmaking slags based on two pilot scale processes, namely, the In-Plant Byproduct Melting (IPBM) and the Vanadium recovery from LD slag (ViLD) processes. These pilot scale processes involved the use of a 3 MW DC arc furnace to melt the waste materials and concentrate the valuable metals in the metal phase while the reduced slag was modified according end-user requirements (Ye and Lindvall, 2012; Ye et al., 2013).

4.6.8.3 Utilization of Steelmaking Slags as Construction Materials

As discussed earlier, steelmaking slags are usually subjected to metal recovery prior to potential applications within or outside the steel plant. After the post-processing recovery of entrained metals, a small proportion of these steelmaking slags can be recycled within the steel plant, while significant amounts of the slag can either be sold to markets outside the steel plant, or are just landfilled (Atkinson and Kolarik, 2001; Remus et al., 2013). To date, steelmaking slags have been extensively characterized and have found widespread applications in the construction industry, mainly as road construction aggregates, hydraulic engineering construction and as cementing agent additives in the manufacture of Portland cement (Shi and Qian, 2000; Motz and Geiseler, 2001; Shen and Forssberg, 2003; Shi, 2004; Yildirim and Prezzi, 2011; Yi et al., 2012).

The application of slags is closely related to their technical properties, that is, processed aggregates for construction applications must possess the minimum technical properties such as bulk density, shape, resistance to impact and crushing, strength, water absorption, volume stability and resistance to abrasion and polishing (Motz and Geiseler, 2001; Shen and Forssberg, 2003). As a result, extensive research has been carried out over the years to benchmark the performance of steelmaking slags against that of natural aggregates (Shi and Qian, 2000; Motz and Geiseler, 2001; Shi, 2004; Manso et al., 2006; Reddy et al., 2006; Yildirim and Prezzi, 2011; Yi et al., 2012). As shown in Table 4.15, the bulk density of steelmaking slags (>3.2 g/cm³) qualifies their use as construction materials for hydraulic engineering applications (Motz and Geiseler, 2001), while other parameters such as compressive strength and resistance to impact are comparable to those of granite and natural aggregates, respectively (Shen and Forssberg, 2003). It is clear from these studies that steelmaking slags can further be processed to produce aggregates of high quality comparable to those of natural aggregates.

Despite the complexity and variability of the chemical and mineralogical characteristics of steelmaking slags cited in Tables 4.15 and 4.16, the presence of cementitious phases such as

TABLE 4.15

Technical Properties of Processed BOF and EAF Slags Benchmarked against Established Aggregates

Characteristics	BOF-Slag	EAF-Slag	Granite	Flint Gravel
Bulk density (g/cm³)	3.3	3.5	2.5	2.6
Shape-thin and elongated pieces (%)	<10	<10	<10	<10
Impact value (%/wt.)	22	18	12	21
Crushing value (%/wt.)	15	13	17	21
10% fines	320	350	260	250
Polished stone value	58	61	48	45
Water absorption (%/wt.)	1	0.7	<0.5	<0.5
Resistance to freeze-thaw (%/wt.)	<0.5	<0.5	<0.5	<1
Binder adhesion (%)	>90	>90	>90	>85

Source: Reprinted from *Waste Manage.*, 21, Motz, H. and Geiseler, J., Products of steel slag: An opportunity to save natural resources, 285–293. Copyright 2001, with permission from Elsevier.

TABLE 4.16

Reactivity, Basicity and Mineral Composition of Steel Slags

Hydraulic Reactivity	Type of Steel Slag	Basicity		Major Mineral Phase
		CaO/SiO_2	$CaO/(SiO_2+P_2O_5)$	
Low	Olivine	0.9–1.4	0.9–1.4	Olivine; CaO–FeO–MnO–MgO phase; Merwinite
	Merwinite	1.5–2.7	1.4–1.6	Merwinite; $2CaO \cdot SiO_2$; CaO–FeO–MnO–MgO phase
Medium	Dicalcium silicate		1.6–2.4	$2CaO \cdot SiO_2$; $2CaO \cdot SiO_2$; CaO–FeO–MnO–MgO phase
High	Tri-calcium silicate	>2.7	>2.4	$3CaO \cdot SiO2$; $2CaO \cdot SiO_2$; $2CaO \cdot Fe_2O_3$; $4CaO \cdot Al_2O_3Fe_2O_3$; CaO–FeO–MnO–MgO phase

Source: Shi, C., *J. Mater. Civil Eng.*, 16(3), 230, 2004.

tri-calcium silicate ($3CaO \cdot SiO_2$), dicalcium silicate ($2CaO \cdot SiO_2$), tetra-calcium aluminoferrite ($4CaO \cdot Al_2O_3Fe_2O_3$) and di-calcium ferrite ($2CaO \cdot Fe_2O_3$), has buoyed the research on the applications of steelmaking slags as sustainable alternatives for conventional construction materials (Shi and Qian, 2000; Motz and Geiseler, 2001; Reuter et al., 2004; Shi, 2004; Tossavainen et al., 2007; Setién et al., 2009; Waligora et al., 2010; Yildirim and Prezzi, 2011; Yi et al., 2012). It is generally agreed that the cementitious properties of steelmaking slags increase with increasing slag basicity and the rate of cooling on solidification (Shi and Qian, 2000; Shi, 2004; Yi et al., 2012). Shi (2004) summarized the relationship among basicity, main mineral phases and hydraulic reactivity of steelmaking slags, and the results are shown in Table 4.16.

Table 4.16 clearly shows that the hydraulic reactivity of steelmaking slags increases with increase in the slag basicity. However, the amount of free CaO and free MgO from unreacted raw materials and the precipitation reactions during slag solidification also increase with an increase in the basicity of steel slag (Motz and Geiseler, 2001; Shi, 2004). According to Motz and Geiseler (2001) and Horii et al. (2003), the amount of free CaO and free MgO is the most important component for the utilization of steel slags for civil engineering purposes, particularly with regard to their volume stability. When in contact with water, these mineral phases react to form hydroxides, and depending on the amount of free CaO and or free MgO, the hydration reactions can cause volume increases, disintegration and the loss of strength of the engineered slag product (Shi and Qian, 2000; Motz and Geiseler, 2001; Shen and Forssberg, 2003; Horii et al., 2013).

To overcome the problem of volumetric changes due to hydration reactions of free CaO and free MgO, Horii et al. (2013) proposed countermeasures based on ageing treatment processing at the steel plant to ensure the complete hydration of slag before it is shipped to consumers. The main representative methods of ageing treatment include (Horii et al., 2013): (1) normal ageing treatment, whereby the hydration reaction is allowed to take place under natural conditions of rainfall in the slag yard, and (2) accelerated ageing treatment, where the hydration reaction is induced to complete in a shorter period of time. Furthermore, the accelerated ageing treatment may involve processes such as (Horii et al., 2013): (1) steam ageing treatment using high temperature steam, (2) hot water ageing treatment by immersing the slag in hot water and (3) high pressure ageing treatment whereby the slag is reacted with steam in a pressurized vessel.

4.6.9 Recycling and Reuse of Continuous Casting Slags

Mills (2016) and Brandaleze et al. (2012) characterized the mold fluxes as synthetic slags consisting of oxides such as SiO_2, CaO, Na_2O, Al_2O_3 and MgO. As shown in Table 4.6, a similarity can be drawn between the chemical composition of continuous casting and steelmaking slags. As a result, it would suffice to say that the recycling and reuse applications of continuous casting process slags would be similar to those of steelmaking slags described in the preceding sections. Furthermore, the production and valorization opportunities of the different waste streams in the metal forming, metal finishing and foundry processes is covered in Chapter 8.

4.6.10 Recycling and Reuse of Iron and Steelmaking Refractory Materials

Section 4.5 discussed the different types of refractory materials consumed in the iron and steelmaking process. The refractory materials are exposed to severe in-service thermochemical and thermomechanical conditions, and this results in the gradual degradation and wear of these materials over time (Hubble, 1999; Dzermejko et al., 1999; Lee and Zhang, 1999;

Harbison-Walker, 2005; Garbers-Craig, 2008; Hanagiri et al., 2008). When the degree of damage to the refractory lining is such that stable furnace operation can no longer be ensured, the refractory lining is broken down and discarded before a new refractory lining is assembled (Hanagiri et al., 2008; Arianpour et al., 2010; Bradley and Hutton, 2014). According to Arianpour et al. (2010), the degraded refractory material that has reached the end of its life is replaced with new material manufactured from natural raw materials, and the spent refractory materials are usually disposed of in landfills. However, the recycling of spent steel plant refractory materials is gaining global attention, not only due to environmental considerations, but also from a resource efficiency point of view (Fang et al., 1999; Othman and Nour, 2005; Conejo et al., 2006; Hanagiri et al., 2008; Arianpour et al., 2009; Bradley and Hutton, 2014).

Landfilling has always been the most preferred way of managing spent refractory materials. To date, the disposal in landfills has been driven by factors such as high costs associated with the recycle and reuse of spent linings, perceived poor quality of reused materials due to the presence of contaminants, and the low costs traditionally associated with landfilling (Fang et al., 1999; Bennet and Kwong, 2000). However, the changing environmental awareness and regulations, the increasing cost of waste disposal in landfills, corporate citizenship associated with zero waste initiatives, and the economic incentives associated with raw materials savings achievable from the use of recycled refractory materials has caused an increased interest in the recycling and reuse of spent refractory materials (Fang et al., 1999; Othman and Nour, 2005; Hanagiri et al., 2008; Bradley and Hutton, 2014). In addition, the economic benefits associated with the recycling and reuse of these spent refractories is making the recycling process more preferred than landfilling (Bradley and Hutton, 2014).

4.6.10.1 Classification of Spent Refractory Materials

Spent refractory materials constitute one of the most important by-products generated in a steel plant. According to Bradley and Hutton (2014), recycled refractory materials can be classified into two primary categories, namely, magnesia-based and alumina-based refractories. Table 4.17 shows the recycled refractory material classification based on the abundance of generation (Bradley and Hutton, 2014).

In any given process, several types of refractory materials are used in the different parts of the furnaces (Hanagiri et al., 2008). In other words, the spent refractory materials that are dismantled from the furnace consist of a mixture of different types of refractory materials, and as a result, simply collecting and using them directly as recycled materials is undesirable from the refractory material quality point of view (Hanagiri et al., 2008). As a result, the recycling process of refractory materials may require sophisticated but consistent processes to sort and separate the used refractory material from the adhering metal, slag and/or altered parts (Hanagiri et al., 2008; Guèguen et al., 2014).

TABLE 4.17

Classification of Commonly Recycled Refractory Materials

Magnesia Based	Alumina Based
Magnesia brick	Firebrick (35% Al_2O_3)
Magnesia-carbon (>4% Al_2O_3)	Mid alumina (55% Al_2O_3)
Magnesia carbon (<4% Al_2O_3)	Bauxite brick (75% Al_2O_3)
Mag-chrome	Andalusite brick
Alumina-magnesite	Alumina-mag carbon

Source: Adopted from Bradley, M. and Hutton, T., *Refract. Eng.*, 22, 2014.

In general, the sorting and characterization of spent refractory materials allows for the segregation of similar refractory materials based on mineral type. For economical beneficiation, the dismantled refractory materials can be categorized into broad groups such as MgO-based, Al_2O_3-based, carbon-containing and carbon-free refractories (Hanagiri et al., 2008; Bradley and Hutton, 2014; Guèguen et al., 2014). Currently, characterization techniques such as visual examination, microscopy, chemical analysis and x-ray diffraction methods are the most commonly used techniques in identifying the physico-chemical changes in spent refractory materials (Guèguen et al., 2014). Other researchers have reported on the growing preference on the use of more sophisticated and economical sorting techniques such as sensor-based sorting (Guèguen et al., 2014). Once the sorting and characterization has been completed, the spent refractories are then crushed, screened and beneficiated to remove contaminants using techniques such as magnetic separation, leaching, flotation and differential density separation (Fang et al., 1999; Hanagiri et al., 2008; Bradley and Hutton, 2014; Guèguen et al., 2014).

4.6.10.2 Overview of Current Refractory Materials Recycling and Reuse Practices

Currently, most of the spent refractory materials are used for low-value applications such as construction material and landfill management (Fang et al., 1999; Othman and Nour, 2005; Hanagiri et al., 2008; Guèguen et al., 2014). Recently, the spent refractory materials have also found acceptable applications as slag conditioners in steelmaking (as desulphurizing and slag foaming additives), raw materials for cement manufacture and aggregates for refractory concretes (Atkinson and Kolarik, 2001; Kwong and Bennet, 2002; Conejo et al., 2006; Arianpour et al., 2010; Luz et al., 2013). Technically, the recycling of spent refractory materials has its own advantages, such as (Guèguen et al., 2014): (1) the environmental aspect by reducing pollution and protection of natural raw materials resources and (2) cost stabilization and/or reduction in manufacturing new refractories.

Spent refractory materials as substitute for virgin materials: In the recent years, spent refractory materials have found potential applications as a replacement for virgin raw materials consumed in the production of new refractory products. According to Bradley and Hutton (2014), typical applications for recycled refractory materials include medium range castables and precast shapes, tundish spray and furnace gunning repair products. Furthermore, Bradley and Hutton (2014) reported that there has been a protracted shift towards the recycling of spent refractory materials in the manufacture of virgin refractory bricks and monolithic refractories, where the recycled materials are incorporated as a percentage addition to control some critical parameters such as the stability during the high temperature firing process.

To date, extensive research has been focused on the application of spent refractory materials in the manufacture of monolithic refractory products for non-critical applications (Kendall, 1994; Noga, 1994; Hanagiri et al., 2008; Schutte, 2010; Ikram-Ul-Haq et al., 2012; Bradley and Hutton, 2014). Kendall (1994) and Noga (1994) reported that high-alumina brick can be employed to enrich the alumina content of firebrick, while lower grades of alumina can be included in monolithics such as gunning mixes and castables. Other researchers have investigated the recycling of chamotte, and alusite and magnesia spinel spent refractory materials in the manufacture of monolithic refractories and highlighted the possibility of manufacturing products containing up to 70%–80% recycled materials for use in non-critical and low temperature applications (Schutte, 2010). Furthermore, Bradley and Hutton (2014) reported that reclaimed alumina-based refractory materials can be recovered and added to medium-high alumina-based castables with minimum compromise to product quality. In the case of lower alumina castables (<40 wt.% Al_2O_3) for non-critical applications, it is also possible to manufacture products based on 100% recycled materials (Bradley and Hutton, 2014).

According to Bradley and Hutton (2014), magnesia-based spent refractory materials can be used to manufacture chemically bonded monolithics for tundish lining spray, gunning repair mixes for EAFs, basic oxygen converter and ladle slagline. Furthermore, Arianpour et al. (2010) studied the microstructural and corrosion behaviour of monolithic refractories produced from spent MgO-C refractory aggregates, and their results proved that using up to 30 wt.% of recycled aggregates had little or no negative effects on properties of magnesia-based refractories. However, when considering using the virgin and spent MgO–C blends in monolithic refractory applications, it is important assess the anti-oxidant content (e.g. Al, Si, Mg, Al/Mg alloy, SiC, B_4C etc.) of the spent refractory materials as the presence of oxy-carbides and/or oxy-nitrides can cause hydration cracking within the virgin product (Zhang et al., 2001; Bradley and Hutton, 2014). Other important parameters to consider in the recycling of spent MgO–C refractories include the carbon content, loss on ignition (LOI) as well as the spinel content of the reclaimed materials (Bradley and Hutton, 2014).

Application of spent basic refractory materials as slag conditioners: Spent MgO–C refractory materials have found wide applications as slag conditioners in the steel plant (Atkinson and Kolarik, 2001; Kwong and Bennet, 2002; Conejo et al., 2006; Arianpour et al., 2010; Luz et al., 2013; Bradley and Hutton, 2014). Due to their high MgO content, spent MgO–C refractory materials have been adopted as good alternatives to dolomite in controlling slag foaming characteristics in electric arc and BOFs (Kwong and Bennet, 2002; Conejo et al., 2006; Arianpour et al., 2010; Avelar et al., 2012; Luz et al., 2013). The fundamentals of slag foaming practices using recycled basic refractory materials in electric arc furnaces are discussed separately in Section 4.6.10.3.

4.6.10.3 Fundamentals of Slag Foaming Practice

Slag foaming is one of the most widely applied techniques in the field of slag engineering, particularly in the smelting and refining of steel in electric arc, basic oxygen and ladle furnaces (Ogawa et al., 1993; Pretorius and Nunnington, 2002; Luz et al., 2013). The practice of slag foaming in these processes has been widely adopted in order to protect the refractory materials from the high radiative energy intensity generated by the electrodes, decrease the electrical noise levels, improve productivity and energy efficiency and lower nitrogen pick-up in the steel (Kwong and Bennet, 2002; Avelar et al., 2012; Luz et al., 2013).

Foamy slags are obtained by the reaction of FeO with carbon to generate gas bubbles (Kwong and Bennet, 2002; Conejo et al., 2006). Technically, Ogawa et al. (1993) divided the slag foaming phenomenon into three principal stages, namely, the (1) formation of CO bubbles at the slag/metal interface and their detachment, (2) rise of the bubbles in the slag layer and their accumulation under the free surface and (3) coalescence of bubbles in the foam followed by the rupture of bubble films at the top of surface of the slag. Basically, the principal reactions for the gas bubble generations are given in Equations 4.33 through 4.36 (Pretorius and Nunnington, 2002; Avelar et al., 2012; Luz et al., 2013):

$$Fe_{(l)} + \frac{1}{2}O_{(s)} \rightarrow FeO_{(l)} \tag{4.33}$$

$$FeO_{(s/l)} + C_{(s)} \rightarrow Fe_{(l)} + CO_{(g)} \tag{4.34}$$

$$FeO_{(l)} + CO_{(g)} \rightarrow Fe_{(l)} + CO_{2(g)} \quad (4.35)$$

$$CO_{2(g)} + C_{(s)} \rightarrow 2CO_{(g)} \quad (4.36)$$

According to Luz et al. (2013), one of the major requirements for suitable slag foaming is the right balance between the refractory oxides (CaO and MgO) and fluxing compounds (SiO_2, Al_2O_3, FeO, MnO and CaF_2) in the liquid slag. In other words, the overall chemical composition of the slag directly defines the slag basicity, surface tension, viscosity and other important parameters at high temperatures (Luz et al., 2013). In addition, the presence of suspended second phase particles was observed to have a much greater impact on foaming than surface tension and viscosity (Kwong and Bennet, 2002; Pretorius and Nunnington, 2002; Luz et al., 2013). According to Pretorius and Nunnington (2002) and Luz et al. (2013), the ideal slags in CaO–MgO–SiO_2 system at 1600°C are not completely liquid or watery, but are saturated with respect to CaO and or MgO, that is, typically containing suspended second phase particles Ca_2SiO_4 and MgO·FeO, respectively. In this case, the second phases serve as gas nucleation sites, which lead to a large number of small gas bubbles in the foaming slag (Pretorius and Nunnington, 2002).

Benefits of utilizing spent basic refractory materials as slag conditioners: In general, the use of MgO and CaO-saturated liquids described earlier not only positively affect the slag foam generation but also result in a decrease in the refractory wear, as most of the lining materials applied in the smelting and refining reactors are MgO–C based bricks (Kwong and Bennet, 2002; Pretorius and Nunnington, 2002; Conejo et al., 2006; Luz et al., 2013). In practice, lime and dolomite fluxes are usually added to adjust the slag basicity and to achieve dual saturation in a CaO–MgO–SiO_2 slag system under steel production conditions (Pretorius and Nunnington, 2002; Conejo et al., 2006; Avelar et al., 2012; Luz et al., 2013). Due to the high concentrations of MgO and C in spent MgO–C refractory materials from steelmaking reactors such as electric arc, basic oxygen and ladle furnaces, it is technically and economically beneficial to consider the full scale recycling of the spent MgO–C refractory materials as slag conditioners in the steelmaking process (Pretorius and Nunnington, 2002; Conejo et al., 2006; Avelar et al., 2012; Luz et al., 2013). In addition, extensive plant trial tests conducted by Conejo et al. (2006) proved that the recycling of spent MgO–C refractory materials in the EAF melting of direct reduced iron not only allowed for faster slag saturation and suitable conditions to induce slag foaming but also reduced the consumption of virgin fluxes and electrical energy, and hence, serves to transform a potential waste material into a valuable resource.

4.7 Trends and Challenges to Waste Reduction in the Iron and Steelmaking Process

Section 4.6 highlighted the various processes applied in the recycling and reuse of the various metallurgical wastes produced from the integrated iron and steelmaking process. It is apparent that various interventions to minimize the disposal of untreated wastes are being adopted, albeit with varying degrees of success, towards the global agenda of zero waste by recycling and reusing these metallurgical wastes. However, the recycling and reuse interventions highlighted so far are also fraught with challenges such as the low intrinsic monetary value of the various wastes and the additional costs of investment required in acquiring the necessary competences and degree of process integration. In some cases, process constraints, such as the compromised product quality from recycled raw materials and the build-up of toxic and deleterious elements in the recycled streams, are one of the major binding constraints in in-plant recycling and reuse of metallurgical wastes such as dusts and sludges.

Lastly, the iron and steel industry is focused on reducing the amount of waste generated from their operations. These strategies include, among others, process integration to (1) facilitate the reduction of waste formation at sources, (2) recover valuable components of wastes and vitrify the barren waste component before disposal, (3) develop synergistic markets for the reuse of iron and steelmaking wastes within or outside the steel plant, for example, the application of metallurgical wastes in civil engineering, ceramics and other alternative industries. Based on the extensive literature evaluated so far, Table 4.18 highlights the some of the strategies currently being adopted and the challenges being faced in the reduction, utilization and recycling of metallurgical wastes in the iron and steelmaking processes.

4.8 Concluding Remarks

This chapter provided a technical overview of the unit processes in the smelting and refining of iron and steel and highlighted the different types and categories of metallurgical wastes produced in the respective unit processes. Generally, the typical wastes produced in the production processes of iron and steel products were categorized as (1) raw material agglomeration wastes from the production sinter, cokes and pellets; (2) blast furnace dusts and slags; (3) steelmaking dusts and slags and (4) spent refractory materials. Furthermore, the chapter provided an overview of the recycling and valorization opportunities and challenges of the various metallurgical wastes from the perspective of the unit processes producing them, and discussed the current practices and trends in terms of waste reduction, recycling and reuse in the iron and steel sector. Finally, the chapter discussed the typical challenges and binding constraints associated with such waste mitigation measures.

Basically, the production of steel products was broken down into two main categories, that is, the integrated blast furnace-basic oxygen smelting and refining process, and the EAF-BOF smelting and refining route. The key aspects pertinent to these two process categories were evaluated in terms of raw materials, process chemistry and technological considerations. In addition to the physicochemical, thermochemical and technical aspects of the blast furnace ironmaking process, Chapter 4 also discussed the recent developments in the production of crude molten steel, that is: (1) the protracted shifts towards process intensification to maximize the utilization efficiency of raw materials, (2) developing alternative and more flexible unit processes to augment the iron and steel products produced via the blast furnace process and (3) finding alternative uses, either endogenously or exogenously to the process, of the byproduct streams that are produced in the integrated iron and steelmaking process.

The key drivers in the production and recycling of the different waste streams from the ironmaking process were linked to their respective unit processes. Generally, the holistic management and recycling initiatives of the particulate dusts produced in the raw material agglomeration and blast furnace ironmaking process were documented. The valorization

TABLE 4.18

Potential Strategies for Waste Reduction, Utilization and Recycling in the Iron and Steelmaking Processes

Waste	Trends in Waste Reduction, Recycling and Reuse	Challenges and Binding Constraints
Coke plant emissions (Coke oven gas, coke dust/ breeze)	Process integration to recover by-products from COG, combustion for fuel the coking reactor or power generation By-product recovery systems (as shown in Figure 4.10) Process integration using secondary dedusting, electrostatic precipitation, bag filters to recover dust from waste streams Briquetting and recycling in sinter production and as a reductant in steel mills	Elaborate off-gas handling systems required Fugitive emissions may still be problematic Build-up of PCDD/Fs in gas recycle streams
Sinter dusts Pelletization dusts	Process integration using secondary dedusting, electrostatic precipitation, bag filters to recover dust from waste streams Partial or total recirculation of the induration waste gases Briquetting and recycling of fines in sinter plant and steel mills	Build-up of mercury, zinc, PCDD/Fs, alkali and lead compounds in recycle streams
Blast furnace off-gas	Scrubbing and internal recycling of off-gas for post process combustion fuel and power generation Process integration–elaborate dust catchers and scrubbers Limit amount of charged recycled materials such as galvanized and unsorted scrap	Build-up of mercury, zinc, PCDD/Fs, alkali and lead compounds in recycle streams may limit recycling potential.
Blast furnace particulate dust	Process integration -dust catchers and wet scrubbers Briquetting, sintering and recycling into sinter and carbon composite agglomerates Limit recycle of recycled materials and galvanized steel scrap	Build-up of alkali (Na, K) and volatiles metals (Zn, Cd, Pb) in furnace off gas Increased refractory attack due to accretion and chemical attack by circulating alkali metals
Blast furnace slag	Recycled internally as pre-melted flux material in sinter, blast furnace, basic oxygen furnace and other smelting processes Granulated and used as cement additive and civil engineering aggregates Raw material and process optimization to reduce slag volumes	The amounts of tramp elements (P&S) limit volumes of slag recycled into steel plant Slags contain entrained metals and may require additional recovery processes before reuse Low intrinsic monetary value of spent makes it uneconomic to reuse distant from source

(Continued)

TABLE 4.18 (*Continued*)

Potential Strategies for Waste Reduction, Utilization and Recycling in the Iron and Steelmaking Processes

Waste	Trends in Waste Reduction, Recycling and Reuse	Challenges and Binding Constraints
Refractory bricks	Recyclable as refractory materials in non-critical processes and low temperature applications Recycled as fluxes in smelting operations (slag foamers, particularly basic refractories) Reuse in the production of cement and construction aggregates	Low intrinsic monetary value of spent refractory bricks makes it uneconomic to reuse distant from source Concerns on quality may limit adoption as non-critical refractory materials
Steelmaking dusts	Process integration – dust catchers and wet scrubbers Briquetting, sintering and recycling into sinter and carbon composite agglomerates Reprocessing for metal recovery Limit recycle of galvanized scrap and materials high volatile metal compounds by incorporating steel scrap pre-treatment processes Stabilization/solidification using cementious systems Continuous operation in semi-closed or closed furnaces (see Chapter 5)	Build-up of volatile metal compoundsExpensive to set up additional metal recovery processes Additional investment in metal recovery processes Long-term leachability behaviour of toxic metal elements in stabilized/solidified dusts largely unknown
Steelmaking slags	Metal recovery and reuse barren slag product as cementious material Hot stage processing to immobilize toxic metal species in stable phases (see Chapter 5) Stabilize/solidification using cementious systems	Low intrinsic monetary value of spent makes it uneconomic to reuse distant from source Long-term leachability of toxic metal species (e.g. Cr and Zn) in stabilized/solidified slag largely unknown

potential of ironmaking slags was also discussed in terms of metal recovery, in-process recycling and the reuse opportunities as cementitious and pozzolanic materials in applications exogenous to the integrated steelmaking plant. Typically, blast furnace slag properties are granulated in order to engineer the solidified slag properties desirable for use as construction material additives.

Furthermore, the raw materials, process chemistry, technology and operation of the typical unit processes utilized in the refining of steel products were addressed. The different unit processes used in the refining of steel products, particularly the refining the different grades of stainless steel products, were categorized into oxygen steelmaking processes and refining unit processes, incorporating, but not limited to, argon-oxygen decarburization processes and secondary metallurgy or ladle refining processes. The different physico-chemical and thermochemical phenomena in steel refining, as well as the waste streams produced in the unit refining processes, were also discussed in detail.

The anthropogenic challenges presented by steelmaking dusts and slags, particularly due to the presence of toxic and hazardous heavy metal compounds, were presented. Initially, the different waste streams and their generation mechanisms were discussed in detail. The valorization potential of steelmaking wastes was also discussed in terms of metal recovery, in-process recycling challenges, as well as the reuse opportunities within and outside the steelmaking plant. In particular, an extensive review of the potential technologies and processes applicable to the economic recycling of, and metal recovery from, the different steelmaking wastes was highlighted. Since not all the waste may be recycled, the potential applications of solidification, stabilization and vitrification treatment of the hazardous waste were also highlighted.

References

Abdel-latif, M.A. (2002). Fundamentals of zinc recovery from metallurgical wastes in the Enviroplas process. *Minerals Engineering* 15: 945–952.

Alanyali, H., Cöl, M., Yilmaz, M. and Karagöz, S. (2006). Application of magnetic separation to steelmaking slags for reclamation. *Waste Management* 26: 1133–1139.

Anameric, B. and Kawatra, S.K. (2009). Direct iron smelting reduction processes. *Mineral Processing and Extractive Metallurgy Review* 30(1): 1–51.

Anyashiki, T., Fukada, K. and Fujimoto, H. (2009). Development of carbon iron composite process. JFE Technical Report 13.

Arianpour, F., Kazemi, F. and Fard, F.G. (2010). Characterization, microstructure and corrosion behaviour of magnesia refractories produced from recycled refractory aggregates. *Minerals Engineering* 23: 273–276.

ASTM D3987-85 (2004). Standard test method for shake extraction of solid waste with water. Available at http://www.astm.org/DATABASE.CART/HISTORICAL/D3987-85R04.htm. Accessed 20 February 2016.

ASTM D5341 (n.d.). Standard test method for measuring Coke Reactivity Index (CRI) and Coke Strength after Reaction (CSR). Available at http://www.astm.org/Standards/D5341.htm. Accessed 20 December 2015.

Atkinson, M. and Kolarik, R. (2001). *Steel Industry Technology Roadmap*. The American Iron and Steel Institute, Washington, DC.

Avelar, T.C., Veiga, A.F., Gueguen, E. and Oliveira, J.R. (2012). Recycling practices of crushed MgO-C bricks and dolomite sinter fines used a slag conditioning additive in the EAF. *WIT Transactions on Ecology and the Environment* 163: 259–270.

Baik, D.S. and Fray, D.J. (2000). Recovery of zinc from electric arc furnace dust by leaching with aqueous hydrochloric acid, plating of zinc and regeneration of electrolyte. *Mineral Processing and Extractive Metallurgy: Transactions of the Institution of Mining and Metallurgy* 109C: C121–C128.

Ball, D.F., Dartnell, J., Grieve, A. and Wild, R. (1973). *Agglomeration of Iron Ores*. American Elsevier Publishing Company, New York.

Barbieri, L., Corradi, A. and Lancellotti, I. (2002). Thermal and chemical behaviour of different glasses containing steel fly ash and their transformation into glass-ceramics. *Journal of European Ceramic Society* 22: 1759–1765.

Barcza, N.A. and Nelson, L.R. (1990). Technology for the treatment of steel plant dusts. Paper submitted for the launch of World Steel Review 1 (1), Mintek Project No. 19090/1A/. Paper Number 8104, Johannesburg, pp. 1–26.

Barozzi, S. (1997). Metal recovery from steel wastes by the INMETCO process. *Steel Times* 225: 191–192.

Barret, E.C., Nenniger, E.H. and Dziewinski, J. (1992). A hydrometallurgical process to treat carbon steel electric arc furnace dust. *Hydrometallurgy* 30: 59–68.

Batchelor, B. (2006). Overview of waste stabilization. *Waste Management* 26: 689–698.

Bayraktar, A.C., Avşar, E., Toröz, I., Alp, K. and Hanedar, A. (2015). Stabilization and solidification of electric arc furnace dust originating from steel industry by using low grade MgO. *Archives of Environmental Protection* 41(4): 62–66.

Bennet, J.P. and Kwong, K.S. (2000). Spent refractory recycling/reuse efforts in the steel and aluminium industries. *Fourth International Symposium on Recycling of Metals and Engineered Materials*, The Minerals, Metals and Materials Society, Pittsburgh, Pennsylvania, pp. 1367–2000.

Biswas, A.K. (1981). *Principles of Blast Furnace Ironmaking – Theory and Practice*. Cootha Publishing House, Brisbane, Queensland, Australia.

Boulos, M.I. (1991). Thermal plasma processing. *IEEE Transactions on Plasma Science* 19(6): 1078–1089.

Bradley, M. and Hutton, T. (2014). Recycling steel plant refractories. *The Refractories Engineer*, September Issue: 22–25.

Brandaleze, E., Di Gresia, G., Santini, L., Martín, A. and Benavidez, E. (2012). Mould fluxes in the steel continuous casting process. In: Srinivasan, M. (ed.). *Science and Technology of Casting Processes*, INTECH. Available at http://www.intechopen.com/books/science-and-technology-of-casting-processes/mould-fluxes-in-the-steel-continuous-casting-process. Accessed 21 February 2016.

Buekens, A., Stieglitz, L., Hell, K., Huang, H. and Segers, P. (2001). Dioxins from thermal and metallurgical processes: Recent studies for the iron and steel industry. *Chemosphere* 42: 729–735.

Bulut, U., Ozverdi, A. and Erdem, M. (2009). Leaching behaviour of pollutants in ferrochrome arc furnace dust and its stabilization/solidification using ferrous sulphate and Portland cement. *Journal of Hazardous Materials* 162: 893–898.

By Products Report (2014). World Steel Association. Available at http://www.worldsteel.org/publications/fact-sheets.html. Accessed 24 November 2015.

Cantarino, M.V., de Carvalho Filho, C. and Mansur, M.B. (2012). Selective removal of zinc from basic oxygen furnace sludges. *Hydrometallurgy* 111–112: 124–128.

Cavalloti, P.L., Mapelli, C., Memoli, F. and Pustorino, M. (2007). Recycling of LF-white slag. *La Metallurgica Italiana* 99: 41–48.

Chen, Q.Y., Hills, C.D., Tyrer, M., Slipper, I., Shen, H.G. and Brough, A. (2007). Characterization of products of tricalcium silicate hydration in the presence of heavy metals. *Journal of Hazardous Materials* 147: 817–825.

Chen, Q.Y., Tyrer, M., Hills, C.D., Yang, X.M. and Carey, P. (2009). Immobilization of heavy metal in cement-based solidification/stabilization: A review. *Waste Management* 29: 390–403.

Cheng, T.W. and Chiu, J.P. (2003). Fire resistant geopolymers produced by granulated blast furnace slag. *Minerals Engineering* 16: 205–210.

Colombo, P., Brisatin, G., Bernado, E. and Scarinci, G. (2003). Inertization and reuse of waste materials by vitrification and fabrication of glass-based products. *Current Opinion in Solid State and Materials Science* 7: 225–239.

Conejo, A.N., Lule, R.G., Lopez, F. and Rodriguez, R. (2006). Recycling MgO-C refractory in electric arc furnaces. *Resources, Conservation and Recycling* 49: 14–31.

Conner, J.R. and Hoeffner, S.L. (1998). A critical review of stabilization/solidification technology. *Critical Review in Environmental Science and Technology* 28(4): 397–462.

Dahlin, A., Tilliander, J., Eriksson, J. and Jönsson, P.G. (2012). Influence of ladle slag additions on BOF process performance. *Ironmaking and Steelmaking* 39(5): 378–385.

Das, B., Prakash, S., Reddy, P.S.R. and Misra, V.N. (2007). An overview of utilization of slag and sludge from steel industries. *Resources, Conservation and Recycling* 50: 40–57.

de Araújo, J.A. and Schalch, V. (2014). Recycling of electric arc furnace dust for use in steelmaking. *Journal of Materials Research and Technology* 3(3): 274–279.

Degel, R., Fröhling, C., Hansmann, T., Kappes, H. and Barozzi, S. (2015). Zero waste concept in steel production. *METEC & Second ESTAD*, Dusseldorf, Germany, pp. 1–9.

Delalio, A., Bajger, Z., Balaz, P., Castro, F., Magalhaes, J. and Curilla, J. (1998). Characterization and pre-treatment of steelmaking dusts to recover valuable products. *Acta Metallurgica Slovaca* 4(1): 55–59.

Denton, G.M., Barcza, N.A., Scott, P.D. and Fulton, T. (2005). EAF stainless steel dust processing. *John Floyd International Symposium on Sustainable Developments in Metals Processing*, Melbourne, Australia, pp. 273–284.

de Paula Vitoretti, F. and de Castro, J.A. (2013). Study of the induration phenomena in single pellet to travelling grate furnace. *Journal of Materials Research Technology* 2(4): 315–322.

Diez, M.A., Alvarez, R. and Barriocanal, C. (2002). Coal for metallurgical coke production: Predictions of coke quality and future requirements for coke making. *International Journal of Coal Geology* 50: 389–412.

Dippenaar, R. (2004). Industrial uses of slag (the use and reuse of iron and steelmaking slags). *VII International Conference on Molten Slags, Flues and Salts*, The Southern African Institute of Mining and Metallurgy, Johannesburg, South Africa, pp. 57–70.

Dominguez, M.I., Romero-Sarria, F., Centeno, M.A. and Odriozola, J.A. (2010). Physicochemical characterization and use of wastes from stainless steel mill. *Environmental Progress & Sustainable Energy* 29(4): 471–480.

Duchesne, J. and Laforest, G. (2004). Evaluation of the degree of Cr ions immobilization by different binders. *Cement and Concrete Research* 34: 1173–1177.

Dutra, A.J.B., Paiva, P.R.B. and Tavares, L.M. (2006). Alkaline leaching of zinc from electric arc furnace steel dust. *Minerals Engineering* 19(5): 478–485.

Duxson, P. and Provis, J.L. (2008). Designing precursors for geopolymer cements. *Journal of American Ceramic Society* 61(12): 3864–3869.

Dzermejko, A.J., Baret, D.F. and Hubble, D.H. (1999). Ironmaking refractory systems. In: Wakelin, D.H. (ed.), *The Making, Shaping and Treatment of Steels: Steelmaking and Refining*, Vol. 2, 11th edn. The AISE Steel Foundation, Pittsburgh, PA.

Environmental Protection Agency (2008). Coke production. Emission Factor Documentation for AP-42. Available at http://www3.epa.gov/ttn/chief/ap42/ch12/bgdocs/b12s02_may08.pdf. Accessed 15 December 2015.

EUROSLAG (2006). Legal status of slags: Position paper. Available at http://www.euroslag.org/fileadmin/_media/images/Status_of_slag/Position_paper_Jan_2006.pdf. Accessed 9 January 2016.

EUROSLAG (2015). Slag: A high grade product out of a high quality controlled industry. The European Association Representing Metallurgical Slag Producers and Processors. Available at http://www.euroslag.com/products/properties/. Accessed 15 December 2015.

Fang, H., Smith, J.D. and Peaslee, K.D. (1999). Study of spent refractory waste recycling from metal manufacturers in Missouri. *Resources, Conservation and Recycling* 25: 111–124.

Feinman, J. (1999). Direct reduction and smelting processes. In: Fruehan, R.J. INTECH. *The Making, Shaping and Treatment of Steels: Steelmaking and Refining*, Vol. 2. The AISE Steel Foundation, Pittsburgh, PA.

Fernández, A.I., Chimenos, J.M., Raventos, N., Miralles, L. and Espiell, F. (2003). Stabilization of electrical arc furnace dust with low-grade MgO prior to land filling (ed.). *Journal of Environmental Engineering* 129(3): 275–279.

Fernández-Pereira, C., Luna-Galiano, Y., Rodríguez-Piñero, M.A. and Vale Parapar, J. (2007). Long and short-term performance of a stabilized/solidified electric arc furnace dust. *Journal of Hazardous Materials* 148: 701–707.

Fisher, R., Anderson, D.R., Wilson, D.T., Aries, E., Hemfrey, D. and Fray, T.A.T. (2004). Effect of chloride on the formation of PCCD/Fs and WHO-12 PCBs in iron ore sintering. *Organohalogen Compounds* 66: 1116–1123.

Fruehan, R.J. (1993). Challenges and opportunities in the steel industry. *Iron and Steel Magazine* 3: 59–64.

Fruehan, R.J. (2004). Unique functions of slags in steelmaking. *VII International Conference on Molten Slags, Fluxes and Salts*, The Southern African Institute of Mining and Metallurgy, Johannesburg, South Africa, pp. 263–270.

Gao, J.T., Li, S.Q., Zhang, Y.L. and Chen, P.Y. (2012). Experimental study on the solid state recovery of metallic resources from EAF dust. *Ironmaking and Steelmaking* 39(6): 446–453.

Garbers-Craig, A.M. (2008). Presidential address: How cool are refractory materials? *The Journal of Southern African Institute of Mining and Metallurgy* 108: 491–505.

Gavrilovski, M., Kamberovic, Z., Filipovic, M. and Majinski, N. (2011). Optimization of integrated steel plant recycling: Fine grain remains and by-products synergy. *Strojarstvo* 53(5): 359–365.

Giergiczny, Z. and Król, A. (2008). Immobilization of heavy metals (Pb, Cu, Cr, Cd, Mn) in the mineral additions containing composites. *Journal of Hazardous Materials* 160: 247–255.

Glasser, F.P. (1997). Fundamental aspects of cement solidification and stabilization. *Journal of Hazardous Materials* 52(2–3): 151–170.

Gomez, E., Amutha Rani, D., Cheeseman, C.R., Deegan, D., Wise, M. and Boccaccini, A.R. (2009). Thermal plasma technology for the treatment of wastes: A critical review. *Journal of Hazardous Materials* 161: 614–626.

Gordon, Y. and Kumar, S. (2013). Selection of ironmaking technology: Principles and risks. *Transactions of the Indian Institute of Metals* 66(5–6): 501–513.

Goretta, K.C., Chen, N., Gutierrez-Mora, F., Routbort, J.L., Lukey, G.C. and van Deventer, J.S.J. (2004). Solid-particle erosion of a geopolymer containing fly ash and blast furnace. *Wear* 256: 714–719.

Granger, G., Lach, M.J. and Mitchell, J. (1959). *AIME Blast Furnace, Coke Oven, and Raw Materials Conference*, American Institute of Mining, Metallurgical and Petroleum Engineers, New York, pp. 152–160.

Gudenau, H.W., Mulanza, J.P. and Sharma, D.G.R. (1990). Carburization of hot metal by industrial and special cokes. *Steel Research* 61: 97–104.

Guèguen, E., Hartenstein, J. and Fricke-Begemann, C. (2014). Raw materials challenges in refractory application. *Proceedings of Berliner Konferenze Mineralischer Nebenproducke und Abflälle*, Berlin, Germany, pp. 489–501.

Guézennec, A.G., Huber, J.C., Patisson, F., Sessiecq, P., Birat, J.P. and Ablitzer, D. (2005). Dust formation in Electric Arc Furnace: Birth of the particles. *Powder Technology* 157: 2–11.

Gupta, S., Sahajwalla, V., Chaubal, P. and Youmans, T. (2005). Carbon structure of coke at high temperatures and its influence on coke fines in blast furnace. *Metallurgical and Materials Transactions* 36B: 385–394.

Haga, T., Katoh, K. and Ibaraki, T. (2012). Technical developments for saving natural resources and increasing material recycling. Nippon Steel Report 101.

Hanagiri, S., Mtsui, T., Shimpo, A., Aso, S., Inuzuka, T., Matsuda, T., Sasaki, S. and Nakagawa, H. (2008). Recent improvement of recycling technology for refractories. Nippon Steel Technical Report 98, pp. 93–98.

Hara, Y., Ishiwata, N., Itaya, H. and Matsumoto, T. (2000). Smelting reduction process with a coke packed bed for steelmaking dust recycling. *ISIJ International* 40(3): 231–237.

Harada, A., Maruoka, N., Shibata, H., Zeze, M., Asahara, N., Huang, F. and Kitamura, S. (2014). Kinetic analysis of compositional changes in inclusions during ladle refining. *ISIJ International* 54(11): 2569–2577.

Harbison-Walker (2005). *Handbook of Refractory Practice*. Harbison-Walker Refractory Company, Pittsburgh, PA.

Hasegawa, S., Kokubu, H. and Hara, Y. (1998). Development of a smelting reduction process for recycling steelmaking dust. Kawasaki Technical Report 38, Tokyo, Japan, pp. 32–37.

Havlík, T., Kukurugya, F., Orac, D. and Parilak, L. (2012). Acidic leaching of EAF steelmaking dust. *World of Metallurgy-ERZMETALL* 65(1): 48–56.

Havlík, T., Souza, B.V., Bernades, A.M., Scheider, I.A.H. and Miškufová, A. (2006). Hydrometallurgical processing of carbon steel EAF dust. *Journal of Hazardous Materials* B135: 311–318.

Havlik, T., Turzakova, M., Stopic, S. and Friedrich, B. (2005). Atmospheric leaching of EAF dust with diluted sulphuric acid. *Hydrometallurgy* 77: 41–50.

Hendrickson, D.W. and Iwasaki, I. (2013). Production of iron from metallurgical waste. US Patent 8535411 B2.

Holtzer, M., Kimita, A. and Roczniak, A. (2015). The recycling of materials containing iron and zinc in the OxyCup process. *Archives of Foundry Engineering* 15(1): 126–130.

Hong-Xu, L., Yang, W. and Da-Qiang, C. (2010). Zinc leaching from electric arc furnace dust in alkaline medium. *Journal of Central South University Technology* 17: 967–971.

Horii, K., Kitano, Y., Tsutsumi, N. and Kato, T. (2013). Processing and reusing technologies for steelmaking slags. Nippon Steel Report 104, pp. 123–129.

Huaiwei, Z. and Xin, H. (2011). An overview for the utilization of wastes from stainless steel industries. *Resources, Conservation and Recycling* 55: 745–754.

Huang, X., Yu, L., Li, D.W., Shiau, Y.C., Li, S. and Liu, K.X. (2015). Preparation and properties of geopolymer from blast furnace slag. *Materials Research Innovations* 19(10): 413–419.

Hubble, D.H. (1999). Steel plant refractories. In: Wakelin, D.H. (ed.), *The Making, Shaping and Treatment of Steels: Steelmaking and Refining*, Vol. 2, 11th edn. The AISE Steel Foundation, Pittsburgh, PA.

Ichikawa, H. and Morishige, H. (2002). Effective use of steelmaking dust and sludge by use of rotary hearth furnace. Nippon Steel Report 86, pp. 35–38.

IEA Clean Coal Centre (2004). Use of coal in direct ironmaking processes. PF 04-10. Available at http://www.iea-coal.org.uk/documents/81118/5528/Use-of-coal-in-direct-ironmaking-processes. Accessed 25 January 2016.

Ikram-Ul-Haq, M., Khanna, R., Kongkarat, S. and Sahajwalla, V. (2012). Chemical interactions innAl_2O_3-C/Fe system at 1823K: Implications for refractory recycling. *ISIJ International* 52(10): 1801–1808.

Industrial Efficiency Technology Database (n.d.). Electric arc furnace. Available at http://ietd.iipnetwork.org/content/electric-arc-furnace. Accessed 30 December 2015.

ISSF (2015). Stainless steel in figures 2015. International Stainless steel Forum. Available at http://www.worldstainless.org/statistics. Accessed 1 November 2015.

Itoh, Y. and Fieser, A. (1982). Zinc removal from blast furnace dust. *Iron and Steel Engineer* 59: 33–36.

Izquierdo, M., Querol, X., Davidovits, J., Antenucci, D., Nugteren, H. and Fernández-Pereira, C. (2009). Coal fly ash-slag based geopolymers: Microstructure and metal leaching. *Journal of Hazardous Materials* 166: 516–566.

JFE 21st Century Foundation (2003). Manufacturing process for iron and steel. Available at http://www.jfe-21st-cf.or.jp/chapter_2/2a_1.html. Accessed 5 December 2015.

Jha, M.K., Kumar, V. and Singh, R.J. (2001). Review of hydrometallurgical recovery of zinc from industrial wastes. *Resources, Conservation and Recycling* 33: 1–22.

Johnson, C.A. (2004). Cement stabilization of heavy metal-containing wastes. In: *Energy, Waste, and the Environment: A Geochemical Perspective*. Geological Society Special Publications (Editors: Giere, R. and Stille, P.) 236, pp. 595–606.

Johnson, L. (1999). Blast furnace slag as phosphorus sorbents: Column studies. *The Science of the Total Environment* 229: 89–97.

Jones, J.A.T., Bowman, B. and Lefrank, P.A. (1998). Electric furnace steelmaking. In: Fruehan, R.J. (ed.), *The Making, Shaping and Treatment of Steels: Steelmaking and Refining*, Vol. 2. The AISE Steel Foundation, Pittsburgh, PA.

Jones, R.T., Curr, T.R. and Barcza, N.A. (1993). Developments in plasma furnace technology. *High Intensity Pyrometallurgy, Institution of Mining and Metallurgy*, London: 1–18, Available at http://www.mintek.co.za/Pyromet/Files/PlasmaDev-OCR.pdf. Accessed 20 April 2016.

Kasai, E., Aono, T., Tomita, Y., Takasaki, M., Shiraishi, N. and Kitano, S. (2001). Macroscopic behaviour of dioxins in the iron ore sintering plants. *ISIJ International* 41(1): 86–92.

Kavouras, P., Kehagias, T., Tsilika, I., Kaimakamis, G., Chrissafis, K., Kokkou, S., Papadopoulos, D. and Karakostas, T. (2007). Glass ceramic materials from electric arc furnace dust. *Journal Hazardous Materials* A139: 424–429.

Kayama, T., Asano, K., Ebisama, H. and Ueno, K. (2008). Recent technology of refractory production. Nippon Steel Report 98, pp. 29–34.

Kelebek, S., Yoruk, S. and Davis, B. (2004). Characterization of basic oxygen furnace dust and zinc removal by acid leaching. *Minerals Engineering* 17: 285–291.

Kendall, T. (1994). Recycling in refractories: What prices waste? *Industrial Minerals* 323: 32–40.

Kesseler, K. and Erdmann, R. (2007). OxyCup® slag: A new product for demanding markets. Techforum 1. ThyssenKrupp AG, pp. 23–30.

Kikuchi, S., Ito, S., Kobayashi, I., Tsuge, O. and Tokuda, K. (2010). ITMK3® Process. *Kobelco Technology Review* 29: 7784.

Kindness, A., Macias, A. and Glasser, F.P. (1994). Immobilization of chromium in cement matrices. *Waste Management* 14(1): 3–11.

Kopfe, J. and Hunter, R. (2008). Direct reduction's role in the world steel industry. *Ironmaking and Steelmaking* 35(4): 254–259.

Kortmann, H., Lungen, H.B. and Ritz, V. (1992). Quality requirements for burden materials and testing methods used in German. *ISS Ironmaking Proceedings*, Warrendale, PA.

Kostura, B., Kulveitová, H. and Leško, J. (2005). Blast furnace slags as sorbents of phosphate from water solutions. *Water Research* 39: 1795–1802.

Kozak, B. and Dzierzawski, J. (n.d.). Continous casting of steel: Basic principles. Available at https://www.steel.org/making-steel/how-its-made/processes/processes-info/continuous-casting-of-steel---basic-principles.aspx?siteLocation=88e232e1-d52b-4048-9b8a-f687fbd5cdcb. Accessed 25 February 2016.

Kumar, S., Kumar, R., Bandopadhyay, A., Alex, T.C., Kumar, B.R., Das, S.K. and Mehrotra, S.P. (2008). Mechanical properties of granulated blast furnace slag and its effect on the properties and structure of Portland slag cement. *Concrete and Concrete Composites* 30: 679–685.

Kumar, S., Kumar, R. and Mehrotra, S.P. (2010). Influence of granulated blast furnace slag on the reaction, structure and properties of fly ash based geopolymer. *Journal of Materials Science* 45: 607–615.

Kunimoto, K., Fujiwara, Y., Takamoto, Y. and Onuma, T. (2006). Blast furnace ironmaking using pre-reduced iron ore. Nippon Steel Technical Report 94, Tokyo, Japan, pp. 133–138.

Küttner GMBH (n.d.). Shaft furnace technology for scrap and waste recycling. Available at http://www.kuettner.de/Files/Filer/Downloads/en/Kuettner_shaft-furnace.pdf. Accessed 18 January 2016.

Kwong, K.S. and Bennet, J.P. (2002). Recycling practices of spent MgO-C refractories. *Journal of Minerals and Materials Characterization and Engineering* 1(2): 69–78.

Laforest, G. and Duchesne, J. (2005). Immobilization of chromium (IV) evaluated by binding isotherms for ground granulated blast furnace slag and ordinary Portland cement. *Cement and Concrete Research* 35: 2322–2332.

Laforest, G. and Duchesne, J. (2006a). Characterization and leachability of electric arc furnace dust made from remelting of stainless steel. *Journal of Hazardous Materials* B135: 156–164.

Laforest, G. and Duchesne, J. (2006b). Stabilization of electric arc furnace dust by the use of cementious materials: Ionic competition and long-term leachability. *Cement and Concrete Research* 36: 1628–1634.

Langová, Š., Leško, J. and Matýsek, D. (2009). Selective leaching of zinc from zinc ferrite with hydrochloric acid. *Hydrometallurgy* 95: 179–182.

Lanzerstorfer, C. (2015). Application of air classification for improved recycling of sinter plant dust. *Resources, Conservation and Recycling* 94: 66–71.

Lanzerstorfer, C., Bamberger-Strassmayr, B. and Pilz, K. (2015). Recycling of blast furnace dust in the iron ore sintering process: Investigation of coke breeze substitution and the influence of off-gas emissions. *ISIJ International* 55(4): 758–764.

Leclerc, N., Meux, E. and Lecuire, J.M. (2003). Hydrometallurgical extraction of zinc from zinc ferrites. *Hydrometallurgy* 70: 175–183.

Lecomte, I., Henrist, C., Liégeois, M., Maseri, F., Rulmont, A. and Cloots, R. (2006). (Micro)-structural comparison between geopolymers, alkali activated slag cement and Portland cement. *Journal of European Ceramic Society* 26: 3789–3797.

Lecomte, I., Liégeois, M., Rulmont, A. and Cloots, R. (2003). Synthesis and characterization of new polymeric composites based on kaolin or white clay and on ground-granulated blast furnace slag. *Journal of Materials Research* 18(11): 2571–2579.

Lee, W.E. and Zhang, S. (1999). Melt corrosion of oxide and oxide-carbon refractories. *International Materials Review* 44(3): 77–104.

Lee, W.E., Zhang, S. and Karakus, M. (2004). Refractories: Controlled microstructure composites for extreme environments. *Journal of Materials Science* 39: 6675–6685.

Li, C., Sun, H. and Li, L. (2010). A review: The comparison between alkali-activated slag (Si+Ca) and metakaolin (Si+Al) cements. *Cement and Concrete Research* 40: 1341–1349.

Li, C.L. and Tsai, M.S. (1993). Mechanism of spinel ferrite dust formation in EAF steel making. *ISIJ International* 33(2): 284–290.

Li, H.J., Suito, H. and Tokuda, M. (1995). Thermodynamic analysis of slag recycling using a slag regenerator. *ISIJ International* 35(9): 1079–1088.

Liu, C.F. and Shih, S.M. (2004). Kinetics of the reaction of iron blast furnace/hydrated lime sorbents with SO_2 at low temperatures: Effects of sorbent preparation conditions. *Chemical Engineering Science* 59: 1001–1008.

Liu, H., Lu, H., Chen, D., Wang, H., Xu, H. and Zhang, R. (2009). Preparation and properties of glass-ceramics derived from blast furnace slag by a ceramic-sintering process. *Ceramics International* 35(8): 3181–3184.

Liu, J. and Zhou, N. (2003). The refractories – Worldwide and China. *The American Ceramic Society Bulletin* 82: 9601–9606.

Lobato, N.C.C., Villegas, E.A. and Mansur, M.B. (2015). Management of solid wastes from steelmaking and galvanizing processes – A review. *Resources, Conservation and Recycling* 102: 49–57.

Lopez, F.A. and Lopez-Delgado, A. (2002). Enhancement of electric arc furnace dust by recycling to electric arc furnace. *Journal of Environmental Engineering* 128(12): 1169–1174.

Lopez, F.A., Lopez-Delgado, A. and Balcazar, N. (1997). Physicochemical and mineralogical properties of EAF and AOD slags. *Second European Oxygen Steelmaking Congress*, Taranto, Italy, pp. 417–426.

Louhenkilp, S. (2014). Continuous casting of steel. In: Seetharaman, S., McLean, A., Guthrie, A. and Sridhar, S. (eds.), *Treatise on Process Metallurgy, Vol. 4: Industrial Processes*. Elsevier, Amsterdam, the Netherlands.

Lu, L. and Ishiyama, O. (2016). Recent advances in iron ore sintering. *Mineral Processing and Extractive Metallurgy: Transactions of the Institutions of Mining and Metallurgy* 125C(3): 132–139.

Luna, Y., Querol, X., Antenucci, D., Jdid, E.D., Fernández-Pereira, C. and Vale, J. (2007). Immobilization of a metallurgical waste using fly ash-based geopolymers. *Proceedings of the World Coal Fly Ash*, Covington, KY, pp. 1–17.

Luna, Y., Arenas, C.G., Cornejo, A., Leiva, C., Vilches, L.F. and Fernández-Pereira, C. (2014). Recycling of by-products from coal-fired power stations into different construction materials. *International Journal of Energy and Environmental Engineering* 5: 387–397.

Luna-Galiano, Y., Fernández Pereira, C. and Vale, J. (2010). Waste stabilization/solidification (S/S) of EAF dust using fly ash-based geopolymers. Influence on carbonation on the stabilized solids. *Coal Combustion and Gasification Products* 2: 1–8.

Lungen, H. B., Noldin, J.H. Jr. and Schmole, P. (2015). Trends in ironmaking given the new reality of iron ore and coal resources. *METEC & 2nd European Steel Technology and Application Days*, Dusseldorf, Germany, pp. 1–6.

Luz, A.P., Vivaldini, D.O., Lopez, F., Brant, P.O.R.C. and Pandolfelli, V.C. (2013). Recycling MgO-C refractories and dolomite fines as slag foaming conditioners: Experimental and thermodynamic evaluations. *Ceramics International* 39: 8079–8085.

Ma, G. and Garbers-Craig, A.M. (2006a). A review on the characteristics, formation mechanism, and treatment processes of Cr (IV)-containing pyrometallurgical wastes. *Journal of the Southern African Institute of Mining and Metallurgy* 106: 753–763.

Ma, G. and Garbers-Craig, A.M. (2006b). Cr (IV) containing electric arc furnace dusts and filter cake from stainless steel waste treatment plant: Part 1 – Characteristics and microstructure. *Iron and Steelmaking* 33(3): 229–237.

Ma, G. and Garbers-Craig, A.M. (2006c). Cr (IV) containing electric arc furnace dusts and filter cake from a stainless steel waste treatment plant: Part 2 – Formation mechanism and leachability. *Iron and Steelmaking* 33(3): 238–244.

Ma, G. and Garbers-Craig, A.M. (2009). Stabilization of Cr (IV) in stainless steel plant dust through sintering using silica-rich clay. *Journal of Hazardous Materials* 169: 210–216.

Ma, N. (2016). Recycling of basic oxygen furnace steelmaking dust by in-process separation of zinc from the dust. *Journal of Cleaner Production* 112: 4497–4504.

Ma, N. and Houser, J.B. (2014). Recycling of steelmaking slag fines by weak magnetic separation coupled with selective particle size screening. *Journal of Cleaner Production* 82: 221–231.

McCelland, J. (2002). Not all RHFs are created equal: A rotary hearth furnace primer. Midrex Technology Report, 2nd Quarter 2002, pp. 3–14: Available at http://www.midrex.com/assets/user/media/DFM2002Q2.pdf. Accessed 20 April 2016.

Machado, A.T., Valenzuela-Diaz, F.R., de Souza, C.A.C. and de Andrade Lima, L.R.P. (2011). Structural ceramics made with clay and steel dust pollutants. *Applied Clay Science* 51: 503–506.

Machado, J.G.M.S., Breham, F.A., Moraes, C.A.M., dos Santos, C.A., Vilela, A.C.F. and da Cunha, J.B.M. (2006). Chemical, physical, structural and morphological characterization of EAF dust. *Journal of Hazardous Materials* 136(3): 953–960.

MacRae, D.R. (1992). Plasma-arc technology for ferroalloys, Part II. *Proceedings of the Sixth International Ferroalloys Congress, INFACON 6*, Johannesburg, pp. 21–35.

Malviya, R. and Chaudhary, R. (2006). Factors affecting hazardous waste solidification/stabilization: A review. *Journal of Hazardous Materials* B137: 267–276.

Manso, J.M., Polanco, J.A., Losanez, M. and Gonzalez, J.J. (2006). Durability of concrete made from EAF slag as aggregate. *Cement and Concrete Aggregates* 28: 528–534.

Meijer, K., Zeilstra, C., Teerhuis, C., Ouwehand, M. and van der Stel, J. (2013). Developments in alternative ironmaking. *Transactions of the Indian Institute of Metals* 66(5): 475–481.

Menad, N., Kanari, N. and Save, M. (2014). Recovery of high grade iron compounds from LD slag by enhanced magnetic separation techniques. *International Journal of Mineral Processing* 126: 1–9.

Mikhail, S.A., Turcotte, A.M. and Aota, J. (1996). Thermo-analytical study of EAF dust and its vitrification product. *Thermochimica Act* 287: 71–79.

Miller, T.W., Jimenz, J., Sharan, A. and Goldstein, D.A. (1998). Refining of stainless steels. In: Fruehan, R.J. (ed.), *The Making, Shaping and Treatment of Steels: Steelmaking and Refining*, Vol. 2. The AISE Steel Foundation, Pittsburgh, PA.

Mills, C. (n.d.). Metal leaching test procedures. Available at http://technology.infomine.com/enviromine/ard/acid-base%20accounting/metal_leaching.htm. Accessed 20 February 2016.

Mills, K.C. (2016). Structure and properties of slags used in the continuous casting of steel: Part 2. Specialist mould powders. *ISIJ International* 56(1): 14–23.

Mohamed, F.M., El-Hussiny, N.A. and Shalabi, M.E.H. (2010). Granulation of coke breeze fines for using in the sintering process. *Science of Sintering* 42: 193–202.

Montenegro, V., Oustadakis, P., Tsakiridis, P.E. and Agatzini-Leonardou, S. (2013). Hydrometallurgical treatment of steelmaking electric arc furnace dusts (EAFD). *Metallurgical and Materials Transactions* 44B: 1058–1069.

Mostafa, N.Y., Shaltout, A.A., Abdel-Aal, M.S. and El-Maghraby, A. (2010). Sintering mechanism of blast furnace slag-kaolin ceramics. *Materials and Design* 31(8): 3677–3682.

Motz, H. and Geiseler, J. (2001). Products of steel slag: An opportunity to save natural resources. *Waste Management* 21: 285–293.

Murakami, T. and Nagata, K. (2003). New ironmaking process from the viewpoint of carburization and iron melting at low temperature. *Mineral Processing and Extractive Metallurgy Review* 24: 253–267.

Nakasuga, T., Nakashima, K. and Mori, K. (2004). Recovery rate of chromium from stainless slag by iron melts. *ISIJ International* 44(4): 665–672.

Neuschütz, D. (1996). Plasma processing of dusts and residues. *Pure & Applied Chemistry* 68(5): 1159–1165.

Niesler, M., Stecko, J., Blacha, L. and Oleksiak, B. (2014). Application of fine-grained coke breeze fractions in the process of iron ore sintering. *METABK* 53(1): 37–39.

Nippon Slag Association (n.d.). Chemical characteristics of iron and steel slag. Available at http://www.slg.jp/e/slag/character.html. Accessed 15 December 2015.

Noga, J. (1994). Refractory recycling developments. *Ceramic Engineering and Science Proceedings* 15(2): 74–77.

Nolasco-Sobrinho, P.J., Espinosa, D.C.R. and Tenorio, J.A.S. (2003). Characterization of dusts and sludges generated during stainless steel production in Brazilian industries. *Ironmaking and Steelmaking* 30(1): 11–17.

Nyirenda, R.L. (1991). The processing of steelmaking flue dust: A review. *Minerals Engineering* 4(7–11): 1003–1025.

Oda, H., Ibaraki, T. and Abe, Y. (2006). Dust recycling system by the rotary hearth furnace. Nippon Steel Technical Report 94, pp. 147–152.

Oda, H., Ibaraki, T. and Takahashi, M. (2002). Dust recycling technology by the rotary hearth furnace. Nippon Steel Technical Report 86, pp. 30–34.

Oeters, F. (1975). *Review of Mathematical Modelling for Steel Making and Solidification*. The Metals Society, Amsterdarm, the Netherlands, pp. 97–101.

Ogawa, Y., Huin, D., Gaye, H. and Tokumitsu, N. (1993). Physical model of slag foaming. *ISIJ International* 33(1): 224–232.

Ohno, K., Miki, T. and Hino, M. (2004). Kinetic analyses of iron carburization during smelting reduction. *ISIJ International* 48(10): 1368–1372.

Okumura, H. (1994). Recent trends and future prospects of continuous casting technology. Nippon Steel Technical Report 61, pp. 9–14.

Ooi, T.C. and Lu, L. (2011). Formation and mitigation of PCCD/Fs in iron ore sintering. *Chemosphere* 85: 291–298.

Othman, A.G.M. and Nour, W.M.N. (2005). Recycling of spent magnesite and ZAS bricks for the production of new basic refractories. *Ceramics International* 31: 1053–1059.

Outokumpu. (2013). *Handbook of Stainless Steel*. Outokumpu Stainless AB, Degerfors, Sweden, pp. 8–10.

Pal, J., Ghorai, S. and Das, A. (2015). Development of carbon composite iron ore micro-pellets by using micro fines of iron ore and carbon-bearing materials in iron making. *International Journal of Minerals, Metallurgy and Materials* 22(2): 132–140.

Patil, B.V., Chan, A.H. and Choulet, R.J. (1998). Refining of stainless steels. In: Fruehan, R.J. (ed.), *The Making, Shaping and Treatment of Steels: Steelmaking and Refining*, Vol. 2. The AISE Steel Foundation, Pittsburgh, PA.

Paul Wurth (n.d.). Recycling Technologies. Available at http://www.paulwurth.com/Our-Activities/Recycling-technologies/Primus. Accessed 18 January 2016.

Pazdej, R. and Vogler, R. (1995). Treatment of BF and BOF dust and sludges. *European Symposium on Environmental Control in Steel Industry*, CZ-Praha, Czech Republic, pp. 207–228.

Pelino, M., Karamanov, A., Pisciella, P., Crisucci, P. and Zonetti, D. (2002). Vitrification of EAF dusts. *Waste Management* 22: 945–949.

Peng, C., Zhang, F., Li, H. and Guo, Z. (2009). Removal behaviour of Zn, Pb, K, and Na from cold bonded briquettes of metallurgical dust in simulated RHF. *ISIJ International* 49(12): 1874–1881.

Pereira, C.F., Rodríguez-Piñero, M. and Vale, J. (2001). Solidification/stabilization of electric arc furnace dust using coal fly ash: Analysis of the stabilization process. *Journal of Hazardous Materials* B82: 183–195.

Pickles, C.A. (2009). Thermodynamic modelling of the multiphase pyrometallurgical processing of electric arc furnace dust. *Minerals Engineering* 22: 977–985.

Poveromo, J.J. (1998). Iron ores. In: Wakelin, D. (ed.), *The Making, Shaping and Treatment of Steels: Ironmaking Volume*. The AISE Steel Foundation, Pittsburgh, PA.

Poveromo, J.J. (2006). Agglomeration processes: Pelletizing and sintering. In: Kogel, J.E., Trivedi, N.C., Barker, J.M. and Krukowski, S.T. (eds.), *Industrial Minerals and Rocks: Commodities, Markets and Uses*, 7th edn. Society for Mining, Metallurgy and Exploration, Littleton, Colorado (http://www.smenet.org/).

Pretorius, E.B. and Nunnington, R.C. (2002). Stainless steel slag fundamentals: From furnace to tundish. *Ironmaking and Steelmaking* 29(2): 133–139.

Rao, S.R.R. (2011). *Resource Recovery and Recycling from Metallurgical Wastes*. Elsevier, Amsterdam, the Netherlands.

Reddy, A.S., Pradhan, R.K. and Chandra, S. (2006). Utilization of basic oxygen furnace slag in the production of hydraulic cement binder. *International Journal of Mineral Processing* 79: 98–105.

Reis, B.H., Bielefeldt, W.V. and Vilela, A.C.F. (2014). Absorption of non-metallic inclusions by steel-making slags – A review. *Journal of Materials Research and Technology* 3(2): 179–185.

Remus, R., Mononet, M.A.A., Roudier, S. and Sanch, L.D. (2013). Best available techniques (BAT) reference document for iron and steel production. Industrial Emissions Directive 2010/75/EU. European Commission JRC Reference Report.

Rentz, O., Hähre, S. Jochum, W.R. and Spengler, T. (1997). Report on Best Available Techniques (BAT) in the Electric Steelmaking Industry (Final Draft). French German Institute for Environmental Research (Research Project 109 05 006). Available at https://www.umweltbundesamt.de/sites/default/files/medien/publikation/long/2488.pdf. Accessed 10 June 2016.

Reuter, M., Xiao, Y. and Boin, U. (2004). Recycling and environmental issues of metallurgical slags and salt fluxes. *VII International Conference on Molten Slags, Fluxes and Salts*, The Southern African Institute of Mining and Metallurgy, Johannesburg, pp. 349–356.

Rist, A. and Meysson, N. (1966). A dual graphical representation of the blast furnace mass and heat balance. *AIME-ISS, Ironmaking Proceedings*, Philadelphia, pp. 88–98.

Robayo, R.A., Mejía de Gutiérrez, R. and Gordillo, M. (2016). Natural pozzolan-and granulated blast furnace slag-based geopolymers. *Materials De Construction* 66(321): 1–8.

Rodríguez-Piñero, M., Fernández-Pereira, C., Ruiz de Elvira Francoy, C. and vale Parapar, J.F. (1998). *Journal of the Air and Waste Management Association* 48: 1093–1099.

Roth, J.L., Frieden, R., Hansmann, T., Monai, J. and Solvi, M. (2001). PRIMUS: New process for recycling by-products and producing virgin iron. *Revue de Metallurgie* 98(11): 987–996.

Russell, C.C., Whitstone, P. and Ligget, R.P. (1955). *AIME Blast Furnace, Coke Oven and Raw Materials Conference*. American Institute of Mining Metallurgical and Petroleum Engineers, New York, pp. 93–121.

Salihoglu, G., Pinarli, V., Salihoglu, N.K. and Karaca, G. (2007). Properties of steel foundry electric arc furnace dust solidified/stabilized with Portland cement. *Journal of Environmental Management* 85: 190–197.

Sano, N., Lu, W.K. and Riboud, P.V. (1997). *Advanced Physical Chemistry for Process Metallurgy*, 1st edn. Academic Press, San Diego.

Sasaki, K., Nukina, S., Wilopo, W. and Hirajima, T. (2008). Removal of arsenate in acid mine drainage by a permeable reactive barrier bearing granulated blast furnace slag: Column study. *Materials Transactions* 49(4): 835–844.

Sato, T., Sumi, H., Fujimoto, H., Anyashiki, T. and Sato, H. (2013). Process for producing ferro-coke for metallurgy. European Patents (EP 2,543,716 A1).

Schoukens, A.F.S., Abdel-latif, M.A., Freeman, M.J. and Barcza, N.A. (1996). The Enviroplas™ process for the recovery of zinc, chromium, and nickel from steel-plant dust. *Electric Furnace Conference Proceedings*, pp. 341–351.

Schutte, M. (2010). Refractory recycling earning your environmental brownie points. *The South African Institute of Mining and Metallurgy, Refractories 2010 Conference*, Johannesburg, pp. 75–86.

Setien, J., Hernandez, D. and Gonzalez, J.J. (2009). Characterization of ladle furnace basic slag for use as a construction material. *Construction and Building Materials* 23: 1788–1794.

Shen, H. and Forssberg, E. (2003). An overview of recovery of metals from slags. *Waste Management* 23: 933–949.

Shen, H., Forssberg, E. and Nordstrom, U. (2004). Physicochemical and mineralogical properties of stainless steel slags oriented to metal recovery. *Resources, Conservation and Recycling* 40: 245–271.

Shi, C. (2002). Characteristics and cementious properties of ladle slag fines from steel production. *Cement and Concrete Research* 32: 459–462.

Shi, C. (2004). Steel slag – Its production, processing, characteristics, and cementious properties. *Journal of Materials in Civil Engineering* 16(3): 230–236.

Shi, C. and Fernández-Jiménez, A. (2006). Stabilization/solidification of hazardous and radioactive wastes with alkali-activated cements. *Journal of Hazardous Materials* B137: 1656–1663.

Shi, C., Krivenko, P.V. and Roy, D.M. (2006). *Alkali Activated Cements and Concretes*. Taylor & Francis, Abingdon, UK, p. 36.

Shi, C. and Qian, J. (2000). High performance cementing materials from industrial slags: A review. *Resources, Conservation and Recycling* 29: 195–207.

Simate, G.S. and Ndlovu, S. (2014). Acid mine drainage: Challenges and opportunities. *Journal of Environmental Chemical Engineering* 2: 1785–1803.

Singh, P.K., Katiyar, P.K., Kumar, A.L., Chaithnya, B. and Pramanik, S. (2014). Effect of sintering performance on the utilization of blast furnace solid wastes as pellets. *International Conference on Advances in Manufacturing and Materials Engineering. Procedia Materials Science* 5: 2468–2477.

Singhal, L.K. and Rai, N. (2015). Conversion of entire dusts and sludges generated during manufacture of stainless steels into value added products. *Transactions of the Indian Institute of Metals* 69: 1–7.

Sofilic, T., Rastovcan-Mioc, A., Cerjan-Stefanovic, S., Vovosel-Radovic, V. and Jenko, M. (2004). Characterization of steel mill electric arc furnace dust. *Journal of Hazardous Materials* B109: 59–70.

Stathopoulos, V.N., Papandreou, A., Kanellopoulou, D. and Stournaras, C.J. (2013). Structural ceramics containing electric arc furnace dust. *Journal of Hazardous Materials* 262: 91–99.

Steiner, B.A. (1976). Air pollution control in the iron and steel industry: Review. *International Metals Review* 209: 171–192.

Steins, J., Dimitrov, S., Speiss, J. and Hackl, A. (2008). Experiences with the usage of alloyed pig iron for the stainless steel production. Siemens Metals Technologies, Siemens VAI. Available at http://www.insg.org/presents/Mr_Spiess_Apr08.pdf. Accessed 28 September 2016.

Street, J.S. (1997). A metallurgical investigation of the bath smelting of composite organic and ferrous wastes. Doctoral thesis. Department of Materials Engineering. University of Wollongong. Wollongong, New South Wales, Australia. Available at http://ro.uow.edu.au/cgi/viewcontent.cgi?article=2479&context=theses. Accessed 10 June 2016.

Su, F., Lampinen, H.O. and Robinson, R. (2004). Recycling of sludge and dust to the BOF converter by cold bonded pelletizing. *ISIJ International* 44(4): 770–776.

Suetens, T., Van Acker, K., Blanpain, B., Mishra, B. and Apelian, D. (2014). Moving towards better recycling options for electric arc furnace dust. *Journal of Materials* 66(7): 1119–1121.

Sugitatsu, H. and Miyahara, I. (2004). Method of producing stainless steel by re-using waste material of a stainless steel producing process. European patent EP 1,375,686 A1.

Suopajärvi, H., Salo, A., Paananen, T., Mattila, R. and Fabritius, T. (2013). Recycling of coking plant residues in a Finnish Steelworks – Laboratory study and replacement ratio calculation. *Resources* 2: 58–72.

Takano, C., Cavllante, F.L., dos Santos, D.M. and Mourão, M.B. (2005). Recovery of Cr, Ni and Fe from dust generated in stainless steelmaking. *Mineral Processing and Extractive Metallurgy, Transactions of the Institute of Mining and Metallurgy* 114: C201–C2016.

Tanaka, H. (2015). Resources trend and use of direct reduced iron in steelmaking process. *Kobelco Technology Review* 33: 17.

Tang, M.T., Peng, J., Peng, B., Yu, D. and Tang, C.B. (2008). Thermal solidification of stainless steelmaking dust. *Transactions of Nonferrous Metallurgical Society of China* 18: 202–206.

Tateishi, M., Fujimoto, H., Harada, T. and Sugitatsu, H. (2008). Development of EAF dust recycling and melting technology using the coal-based FASTMELT Process. MIDREX RHF Technologies Report, pp. 9–15. Available at http://www.midrex.com/assets/user/media/Development_of_EAF_Dust_Recycling.pdf. Accessed 16 January 2016.

Taylor, P.R. and Pirzada, S.A. (1994). Thermal plasma processing of materials: A review. *Advanced Performance Materials* 1: 35–50.

Thomas, B.G. (2001). Continuous casting. In: Yu, K.O. (ed.), *Modelling for Casting and Solidification*, 1st edn. CRC Press, New York, pp. 499–540.

Topkaya, Y., Sevinc, N. and Günaydin, A. (2004). Slag treatment at Kardemir integrated iron and steel works. *International Journal of Mineral Processing* 74: 31–39.

Tossavainen, M., Engstrom, F., Yang, Q., Menad, N., Lidstrom Larsson, M. and Bjorkman, B. (2007). Characteristics of steel slag under different cooling conditions. *Waste Management* 27: 1335–1344.

Total Materia (2008a). Production of stainless steel: Part 2. Available at http://www.totalmateria.com/page.aspx?ID=CheckArticle&site=KTS&NM=220. Accessed 4 December 2015.

Total Materia (2008b). Production of stainless steel: Part 3. Available at http://www.totalmateria.com/page.aspx?ID=CheckArticle&site=KTS&NM=237. Accessed 4 December 2015.

Townsend, T., Dubey, B. and Tolaymat, T. (2006). Interpretation of synthetic precipitation leaching procedure (SPLP) results for assessing risk to groundwater from land applied granular waste. *Environmental Engineering Science* 23(1): 239–251.

Trinkel, V., Mallow, O., Thaler, C., Schenk, J., Rechberger, H. and Fellner, J. (2015). Behaviour of chromium, nickel, lead, zinc, cadmium, and mercury in the blast furnace: A critical review of literature data and plant investigations. *Industrial and Engineering Chemistry Research* 54: 11759–11771.

Trung, Z.H., Kukurugya, F., Takacova, Z., Orac, D., Laubertova, M., Miskufova, A. and Havlik, T. (2011). Acidic leaching of zinc and iron from basic oxygen furnace sludge. *Journal of Hazardous Materials* 192: 1100–1107.

Trussell, S. and Spence, R.D. (1994). A review of solidification/stabilization interferences. *Waste Management* 14(6): 507–519.

Turkdogan, E.T. and Fruehan, R.J. (1998). Fundamentals of iron and steel making. In: Fruehan, R.J. (ed.), *The Making, Shaping and Treatment of Steels: Steelmaking and Refining*, Vol. 2. The AISE Steel Foundation, Pittsburgh, PA.

United Nations Environmental Program (2011). *Conference of the Parties to the Stockholm Convection on Persistent Organic Pollutants Fifth Meeting*, Geneva, Switzerland. UNEP/POPS/COP.5?INF/4. Available at http://www.popsalgerie.com/en/uploads/webmaster/english/UNEP-POPS-COP.5-INF-4.English.pdf. Accessed 15 May 2016.

van Jaarsveld, J.G.S., van Deventer, J.S.J. and Lorenzen, L. (1996). The potential of geopolymeric materials to immobilize toxic metals: Part 1, Theory and applications. *Minerals Engineering* 10(7): 659–669.

van Jaarsveld, J.G.S., van Deventer, J.S.J. and Lorenzen, L. (1998). Factors affecting the immobilization of metals in geopolymerized flyash. *Metallurgical and Materials Transactions* 29B: 283–291.

Vempati, R.K., Mollah, Y.A.M., Chinthala, A.K. and Cocke, D.L. (1995). Solidification/stabilization of toxic metal wastes using coke and coal combustion products. *Waste Management* 15(5/6): 433–440.

Veres, J., Sepelak, V. and Hredzak, S. (2015). Chemical, mineralogical and morphological characterization of basic oxygen furnace dust. *Mineral Processing and Extractive Metallurgy (Transactions of the Institutions of Mining and Metallurgy, C)* 124(1): 18.

Vertnik, R. and Šarler, B. (2014). Solution of a continuous casting of steel benchmark test by a meshless method. *Engineering Analysis with Boundary Elements* 45: 45–61.

Vijayaram, T.R. (2012). Continuous casting technology for ferrous and non-ferrous foundries. *Metalworld* 11(3): 40–44.

Vogg, H. and Stieglitz, L. (1986). Thermal behaviour of PCDD/PCDF in fly ash from municipal incinerators. *Chemosphere* 15(12): 1373–1378.

von Schéele, J. (2004). Oxyfines™ technology for the re-melting of fines, dust and sludge. *Proceedings of the 10th International Ferroalloys Congress*, Cape Town, South Africa, pp. 678–686.

Wajima, T. (2015). Stabilization of mine waste using iron and steel making slag to prevent acid mine drainage. *International Conference on Advances in Environment Research* 87: 48–53.

Waligora, J., Bulteel, D., Degrugilliers, P., Damidot, D., Potdevin, J.L. and Measson, M. (2010). Chemical and mineralogical characterization of LD converter steel slags: A multi-analytical techniques approach. *Materials Characterization* 61: 39–48.

Wang, C., Nuhfer, N.T. and Sridhar, S. (2009). Transient behavior of inclusion chemistry, shape, and structure in Fe-Al-Ti-O melts: Effect of titanium source and laboratory deoxidation simulation. *Metallurgical and Materials Transaction B* 40B: 1005–1021.

Weeks, R. (1975). *Dynamic Model for the BOS*. The Metals Society, Armsterdam, the Netherlands, pp. 103–116.

Wiles, C.C. (1987). A review of solidification/stabilization technology. *Journal of Hazardous Materials* 14: 5–21.

World Steel Association (2012). Sustainable steel: At the core of green economy. Available at https://www.worldsteel.org/dms/internetDocumentList/bookshop/Sustainable-steel-at-the-core-of-a-green-economy/document/Sustainable-steel-at-the-core-of-a-green-economy.pdf. Accessed 20 November 2015.

World Steel Association (2014). Steel industry by-products. Available at https://www.worldsteel.org/publications/fact-sheets/content/01/text files/file/document/Fact_By-products_2014.pdf. Accessed 20 November 2015.

World Steel Association. (2015a). Press release. Available at https://www.worldsteel.org/media-centre/press-releases/2015/World-crude-steel-output-increases-by-1.2--in-2014.html. Accessed 20 November 2015.

World Steel Association (2015b). Life cycle thinking in the circular economy. Available at http://www.worldsteel.org/steel-by-topic/life-cycle-assessment/Life-cycle-thinking-in-the-circular-economy.html. Accessed 24 November 2015.

World Steel in Figures 2014. World Steel Association. Available at https://www.worldsteel.org. Accessed 25 November 2015.

Worrell, E., Blinde, P., Neelis, M., Blomen, E. and Masane, E. (2010). Energy effieincy improvement and cost saving opportunities for the US iron and steel industry: An ENERGY STAR® guide for the energy and plant managers. U.S. Environmental Protection Agency. Available at http://www.energystar.gov/ia/business/industry/Iron_Steel_Guide.pdf?25eb-abc5. Accessed 20 November 2015.

Wright, B., Zulli, P., Zhou, Z.Y. and Yu, A.B. (2011). Gas-solid flow in an ironmaking blast furnace: Physical modelling. *Powder Technology* 208: 86–97.

Wu, L.M. (1999). Characteristics of steelmaking flue dust. *Ironmaking and Steelmaking* 26(5): 372–377.

Yadav, U.S., Das, B.K., Kumar, A. and Sandhu, H.S. (2002). Solid waste recycling through sinter-status at Tata Steel. *Forum on Environment and Energy in Clean Technologies for Metallurgical Industries*, Indian Institute of Metals, Jamshedpur, India, pp. 81–94.

Yang, Y., Raipala, K. and Holappa, L. (2014). Ironmaking. In: Seetharaman, S., McLean, A., Guthrie, A. and Sridhar, S. (eds.), *Treatise on Process Metallurgy, Vol. 4: Industrial Processes*. Elsevier, London, UK.

Ye, G. and Lindvall, M. (2012). Pilot experiences of Swerea MEFOS on slag recycling. *Ninth International Conference on Molten Slags, Fluxes and Salts*, Beijing, China. Available at http://www.pyrometallurgy.co.za/MoltenSlags/.

Ye, G., Lindvall, M. and Magnusson, M. (2013). Swerea MEFOS perspectives on metal recovery from slags. *Proceedings of the Third International Slag Valorisation Symposium*, Leuven, Belgium, pp. 147–160.

Yi, H., Xu, G., Cheng, H., Wang, J., Wan, Y. and Chen, H. (2012). An overview of utilization of steel slag. *The Seventh International Conference on Waste Management and Technology. Procedia Environmental Sciences* 16: 791–801.

Yildirim, I.Z. and Prezzi, M. (2011). Chemical, mineralogical and morphological properties of steel slag. *Advances in Civil Engineering* 2011: 1–13.

Yu, Y., Zheng, M., Li, X. and He, X. (2012). Operating condition influences on PCDD/Fs emissions from sinter pot tests with hot flue gas recycling. *Journal of Environmental Sciences* 24(5): 875–881.

Zervas, T., McMullan, J.T. and Williams, B.C. (1996a). Developments in iron and steel making. *International Journal of Energy Research* 20: 69–91.

Zervas, T., McMullan, J.T. and Williams, B.C. (1996b). Direct smelting and alternative processes for the production of iron and steel. *International Journal of Energy Research* 20: 1103–1128.

Zeydabadi, B.A., Mowla, D., Shariat, M.H. and Kalajahi, J.F. (1997). Zinc recovery from blast furnace flue dust. *Hydrometallurgy* 47(1): 113–125.

Zhang, J., Provis, J.L., Feng, D. and van Deventer, J.S.J. (2008a). Geopolymers for immobilization of Cr^{6+}, Cd^{2+} and Pb^{2+}. *Journal of Hazardous Materials* 157: 587–598.

Zhang, L. and Thomas, B.G. (2003). State of the art in evaluation and control of steel cleanliness. *ISIJ International* 43(3): 271–291.

Zhang, L. and Thomas, B.G. (2006). State of the art in the control of inclusions during ingot casting. *Metallurgical and Materials Transaction* 37B: 733–761.

Zhang, S., Marriott, N.J. and Lee, W.E. (2001). Thermochemistry and microstructures of MgO-C refractories containing various antioxidants. *Journal of the European Ceramic Society* 21: 1037–1047.

Zhang, Y., Sun, W., Chen, Q. and Chen, L. (2007). Synthesis and heavy metal immobilization behaviours of slag based geopolymer. *Journal of Hazardous Materials* 143: 206–213.

Zhang, Y.J., Zhao, Y.L., Li, H.H. and Xu, D.L. (2008b). Structure characterization of hydration products generated by alkaline activation of granulated blast furnace slag. *Journal of Materials Research* 43: 7141–7147.

Zheng, H.H. and Kozinski, J.A. (1996). Solid waste remediation in the metallurgical industry: Application and environmental impact. *Environmental Progress* 15(4): 283–292.

Zhou, Q., Milestone, N.B. and Hayes, M. (2006). An alternative to Portland cement for waste management encapsulation – The calcium sulphonate cement system. *Journal of Hazardous Materials* 136: 120–129.

5

Ferroalloys Waste Production and Utilization

5.1 Introduction

Ferroalloys refer to the various master alloys of iron with a high proportion of one or more other elements such as chromium, manganese, nickel, molybdenum and silicon (Eric, 2014). In general, ferroalloys are mainly used in steelmaking as alloying agents in order to obtain an appropriate chemical composition of liquid steel (Karbowniczek et al., 2012; Pande et al., 2012). In particular, ferroalloys impact distinct mechanical and chemical properties in steels, and these include (1) resistance to corrosion, (2) hardness and tensile strength at high temperatures, (3) wear and abrasion resistance and (4) creep strength (Holappa, 2010; Pande et al., 2010; Pande et al., 2012; Gasik, 2013). In addition, ferroalloys also play important roles in the refining and solidification of various grades of steels, and these include, inter alia, in the refining, deoxidation and control of inclusions and precipitates (Cha et al., 2008; Pande et al., 2010; Pande et al., 2012; Gasik, 2013; Holappa and Louhenkilpi, 2013; Sen Gupta, 2015). In fact, the advent of ferroalloy metallurgy, in particular ferrochromium, ferronickel, ferromolybdenum and ferrovanadium, among others, has been credited for the development of various ubiquitous grades of stainless and high-specialty steels (Gasik, 2013).

According to Eric (2014), the principal ferroalloys include ferrochromium (FeCr), ferromanganese (FeMn), ferronickel (FeNi), ferrosilicon (FeSi), ferromolybdenum (FeMo), ferrotitanium (FeTi), ferrovanadium (FeV) and ferrotungsten (FeW). Furthermore, ferroalloys can be classified into two main categories, vis-à-vis, bulk or major ferroalloys, and special or noble ferroalloys. Bulk ferroalloys, such as ferrochromium, ferrosilicon, ferromanganese, silicomanganese and ferronickel, are produced in large quantities in electric furnaces and are exclusively consumed in the steelmaking and foundry processes (European Commission, 2014; Sen Gupta, 2015). On the other hand, special or noble ferroalloys, such as ferrovanadium, ferromolybdenum, ferrotungsten, ferrotitanium, ferroboron and ferroniobium, are produced in small quantities and are mostly used in special non-ferrous alloy processes and chemicals industries (European Commission, 2014; Gasik, 2014; Sen Gupta, 2015).

The growth of the ferroalloys industry has largely been driven by advances in the steel industry and the growing demand for high-performance materials in various applications (Holappa, 2010; Eric, 2014; Gasik, 2014; Bedinger et al., 2015; Leont'ev et al., 2016). Generally, the global production of ferroalloys is dominated by a few countries. For example, China, South Africa, India, Russia and Kazakhstan were ranked as the major global producers of ferroalloys in 2010, contributing over 80% of the world's ferroalloys production (Corathers et al., 2010; Bedinger et al., 2015). In essence, the growth in ferroalloys production is determined by the demand for special grades of steels and other engineered alloy products steel (Leont'ev et al., 2016). The global production of bulk ferroalloys increased from 18 Mt in the 1990s to a peak of about 43 Mt in 2010, a trend that is firmly connected to the peak in the growth of steel and stainless steel manufacturing (Holappa, 2010; Eric, 2014; Bedinger et al., 2015). As such, the ferroalloys industry has grown to become a key and indispensable supplier to the steel industry, and its growth is firmly connected to the time-dependent production rates of steel (Holappa, 2010; European Commission, 2014; Gasik, 2014; Leont'ev et al., 2016; Sen Gupta, 2015).

The criticality of the various grades of ferroalloys in various industrial applications has been highlighted so far. Despite the ubiquity of this special group of alloys, their production is associated with the generation of potentially toxic wastes. This chapter covers the technical aspects of production of the various ferroalloys, the categories and types of wastes generated in the various ferroalloys production processes, and the reduction, reuse and recycling of the different types of wastes. In order to holistically deal with ferroalloys production wastes, this chapter links the different types and amounts of wastes to the processes in which they are generated.

5.2 Ferroalloys Production Processes

5.2.1 Introduction

Generally, ferroalloys are produced by melting in electric furnaces, where ferroalloy-bearing raw materials, reductants, iron additions and fluxes are charged and smelted to separate the molten ferroalloys from slag (Gasik, 2014). Depending on the type of raw materials used, the production of ferroalloys can be classified into two broad categories, namely, primary and secondary ferroalloys (European Commission, 2014). Primary ferroalloys are principally produced by either the carbothermic or metallothermic reduction of oxide ores or concentrates. On the other hand, the secondary processes involve the melting of recycled ferroalloys and ferroalloys-containing iron scrap materials to produce new stock of ferroalloys (Eric, 2014; European Commission, 2014). To date, the carbothermic reduction process, in which carbon in the form of metallurgical coke, coal or charcoal, is used as a thermal and reducing agent, is the most dominant process (Holappa, 2010; European Commission, 2014). In contrast, the metallothermic reduction is mainly carried out with either silicon (silicothermic reduction to produce metal alloy plus silicon oxide) or aluminium (aluminothermic reduction to produce metal alloy plus aluminium oxide) as the reducing agents (European Commission, 2014). Some of the typical ferroalloys production processes are summarized in Table 5.1.

5.2.2 Production of Bulk Ferroalloys

Generally, a mixture of ferroalloy ores, fluxes and carbon reductants are charged and melted in electric furnaces equipped with carbon electrodes (Gasik, 2013; Eric, 2014). The electrical energy necessary for the endothermic reactions is supplied via the carbon electrodes, and the low-voltage, high-current arcs created at the electrode tips create zones of high-temperatures and low-oxygen potentials (Gasik, 2013). The radiation from the arc zone impinges directly on the feed material, thereby allowing for an efficient transfer of energy from the arc to the feed materials (Gasik, 2013).

Most of the electric furnaces utilized in the production of ferroalloys are of the submerged arc (SAF) type, in which the bulk of the energy is transferred to the feed material by resistance heating due to the flow of electric current through the furnace charge (Yang et al.,

TABLE 5.1

Typical Ferroalloys Production Processes

Process	Typical Products	Comments
1. *Carbothermic smelting reduction*		
Submerged arc furnace (SAF)	Silvery iron (15–22 wt.% Si), ferrosilicon (50 wt.% Si, 65–75 wt.% Si), silicon metal, silicon/manganese/zirconium (SMZ), high-carbon ferrochrome, high-carbon ferromanganese, silicon manganese, ferrochrome/silicon, ferrosilicon (90 wt.% Si)	Smelting in a refractory-lined cup-shaped steel shell by submerged graphite electrodes.
2. *Metallothermic smelting reduction*		
Silicon reduction	Low-carbon ferrochrome, low-carbon ferromanganese, medium-carbon ferromanganese	Molten charge material is reduced exothermically by the addition of silicon, aluminium or combination of the two.
Aluminium reduction	Chromium metal, ferrotitanium, ferrocolumbium, ferrovanadium	
Mixed aluminothermal/silicothermal	Ferromolybdenum, ferrotungsten	

Source: Reprinted from *Treatise on Process Metallurgy, Vol. 3: Industrial Process, Part A*, Eric, R.H., Production of ferroalloys, p. 527. Copyright 2014, with permission from Elsevier.

2004; Gasik, 2013; Eric, 2014). As a result, the temperature in the zones around the electrode tips is very high compared to other zones of the furnace. Depending on the electrical, physical and chemical reactions taking place, typical ferroalloys production furnaces can be dissected into five physicochemical phenomena zones, namely, (1) the gas region between the furnace roof and the charge top, which consists mostly of off-gas rich in carbon monoxide, (2) solids-packed bed region in the upper part of the furnace where the charge is heated and reduced in the solid state by the countercurrent gases, (3) smelting region around the electrode tips or arc zone above the slag layer packed with a solid coke bed, (4) molten slag layer formed by the reaction between the fluxes and the oxides in the ore, and the concomitant dissolution of the unreduced oxides into slag to be further reduced with the carbon dissolved in alloy phase or by solid carbon in the coke zone and (5) molten alloy layer at the bottom part of the furnace with relatively uniform temperatures (Yang et al., 2004).

According to Gasik (2013), the submerged arc furnaces used in the production of bulk ferroalloys are based on similar design and general principles of operation. However, the geometry, electrical design, and other parameters for a particular furnace and application tend to differ based on the type of commodity and specific requirements of the process. A detailed discussion on the submerged arc furnaces commonly used in the production of bulk ferroalloys processes was summarized by Gasik (2013) and Eric (2014), and hence will not be covered in detain in this chapter.

5.2.3 Production of Noble Ferroalloys

Typical examples of special or noble ferroalloys include ferrovanadium, ferromolybdenum, ferrotungsten, ferrotitanium, ferroboron and ferroniobium, among others (European Commission, 2014; Gasik, 2014; Sen Gupta, 2015). Unlike bulk ferroalloys, special or noble ferroalloys are usually produced and consumed in small quantities in the steelmaking, non-ferrous metals production and in chemicals industries (Sen Gupta, 2015). Special ferroalloys are typically produced using either the carbothermic or the metallothermic reduction processes shown in Table 5.1 (Eric, 2014). However, such processes tend to demand stricter control of the process parameters than in those required in the production of bulk ferroalloys. Extensive details on the specific processes applied in the production of special ferroalloys are discussed in several other scholarly texts (Eric, 2014; European Commission, 2014). As a result, the processes and types of wastes in the production of noble ferroalloys will not be discussed further in this chapter.

5.2.4 Overview of Waste Production in Ferroalloys Industry

The mining, beneficiation and smelting of ferroalloy ores tend to generate a significant amount of potentially toxic and hazardous wastes (Rai et al., 1989; Richard and Bourg, 1991; EPA, 1995; Barnhart, 1997a,b; Petersen et al., 2000; Papp and Lipin, 2001; Shanker et al., 2005; Ma and Garbers-Craig, 2006a,b,c; Oliveira, 2012). The typical wastes produced from these unit processes include process slags, particulate dusts and sludges, beneficiation residues and leach residues, among others (Petersen et al., 2000).

As discussed in Chapter 2, the generation of metallurgical wastes in the ferroalloys production processes is a contentious issue in most jurisdictions. Despite the absolute necessity of the ferroalloys, particularly in the steelmaking and chemicals industries, the production of ferroalloys result in the generation of a significant amount of solid, liquid and gaseous wastes. For example, particulate dusts and sludges are generated from the several activities along with the ferroalloys production chain, and raw material handling, smelting, slag and alloy tapping and product handling (Environmental Protection Agency (EPA), 1995; Eric, 2014). The smelting of ferroalloys in electric furnaces has been cited as the largest potential source of particulate dust and organic gaseous emissions (Environmental Protection Agency (EPA), 1995). In this case, the particulate dusts, carbon monoxide gases and other volatile organic compounds produced in the reaction between metal oxides in charge and reductants in the high-temperature zones of these furnaces are emitted into the environment in the form of stack gases (Environmental Protection Agency (EPA), 1995). In fact, it is proposed that the particulate emissions from the electric furnaces account for approximately more than 90% of the total particulate emissions in the ferroalloys industry (Environmental Protection Agency (EPA), 1995; Eric, 2014).

Traditionally, the waste materials generated in the production of ferroalloys have been disposed of in landfills (Petersen et al., 2000). However, environmental concerns associated with the disposal of ferroalloys production wastes in landfills, particularly the potential leachability of toxic heavy metal species such as chromium, nickel, manganese, lead and zinc compounds, among others, have resulted in a protracted shift towards developing holistic management practices incorporating the recycling and/or reuse of these waste materials as alternatives measures to disposal in landfills (Richard and Bourg, 1991; Barnhart, 1997a,b; Language et al., 2000; Shanker et al., 2005; Ma and Garbers-Craig, 2006a,b,c; Tanskanen and Makkonen, 2006; Olsen et al., 2007; Shen et al., 2007a,b; Durinck et al., 2008; Norval and Oberholster, 2011; Groot et al., 2013; Kumar et al., 2013; Tangstad, 2013a,b; Albertsson et al., 2014). In view of the aforementioned challenges, several waste mitigation and process performance-enhancing technologies have been widely adopted in the ferroalloys industries, vis-à-vis (Language et al., 2000), (1) production of ferroalloys in closed furnaces to allow for the in-process utilization of CO- and organic-carbon-rich off-gases, (2) metal recovery from slags and slag dumps by physical separation techniques and the utilization of the resultant barren slag by-product as a multipurpose aggregate for the construction industry, (3) agglomeration and/or prereduction of feed to increase the alloy recovery rate and decrease the overall energy consumption and (4) in-process recycling of the particulate dusts and sludges extracted from the cleaning of furnace off-gases.

The following sections cover in detail the properties, occurrence, uses and the processes used in the production of selected bulk ferroalloys from their oxide ore materials. Traditionally, the emphasis in the ferroalloys industry has been on process improvements based on the technical aspects of the different raw materials utilization, improving the energy and equipment efficiencies in the respective processes and process improvements focused on enhancing the quality of products. However, the core ferroalloys production technology based on the submerged arc furnaces has more or less matured (Gasik, 2013; Eric, 2014). As a result, the current focus of research is on reducing the carbon footprint, improving the energy efficiency, as well as on improving the ecological, health and environmental aspects in the production of ferroalloys. Thus, in the recent years, the ferroalloys industry has not only focused on the installation of new process technologies, but also focused on the process-integrated recycling of waste materials as well as improving the potential for reuse of waste materials in other industries external to the sector (Person, 1971; Petersen et al., 2000; Beukes et al., 2010; Basson and Daavittila, 2013; Tangstad, 2013a,b; Nelson, 2014). In other words, the ferroalloys producers have fully embraced the design for environment philosophy, which consciously integrates the environmental considerations into the design stages of products, processes and equipment.

Based on the various legislations governing the toxic and hazardous wastes covered in Chapter 2, the production and disposal of waste materials in the ferroalloys industry is a sensitive matter in most countries, and is particularly driven by the high toxicity and environmental hazards posed by the heavy metals containing compounds present in the wastes. According to Gasik (2013), the main environmental focus in the production of ferroalloys is to (1) reduce emissions of particulate dusts from stacks and tapping processes in order to reduce the contamination of the environment from the hazardous components contained therein, (2) reduce the amount of greenhouse gas emissions and (3) select more energy-efficient production routes or processes.

The various waste mitigation interventions by the ferroalloys industries highlighted so far clearly highlight the deliberate shift towards adopting the design for environment principles, particularly through the selection of technology-related options that reduce overall energy consumption and the carbon footprint (Gasik, 2013; Eric, 2014). For example, the available technology-related options include the use of large and closed furnaces, and the utilization of CO gas in the process off-gas for preheating of the furnace feed, ore sintering and co-generation of electricity (Petersen et al., 2000; Gasik, 2013; Eric, 2014; European Commission, 2014).

Based on the generic aspects in the processing of ferroalloys highlighted so far, the following sections discuss in detail the properties, occurrence, industrial uses, processing options and waste generation in the production of selected bulk ferroalloys. Furthermore, the detailed process requirements and equipment specifications of the different ferroalloy commodities are also discussed.

5.3 Production of Chromium and Its Alloys

5.3.1 Chemical and Physical Properties of Chromium

Chromium is a transition metal element with atomic weight 52.00, electronic structure [Argon]$3d^5 4s^1$ and a body-centred cubic (bcc) lattice structure which is stable throughout the range of solid chromium stability (Papp and Lipin, 2001; Gasik, 2013; Los Alamos, n.d.). In its pure state, chromium metal has melting and boiling points of 1907°C and 2671°C, respectively, and exhibits paramagnetic properties in its elemental form (Papp and Lipin, 2001; Los Alamos, n.d.). The ability of chromium to resist corrosion and accept a high surface polish has made it almost ubiquitous as a coating material (Papp and Lipin, 2001).

Chromium can exhibit a wide range of oxidation states, with the zero (Cr^0), divalent (Cr^{2+}), trivalent (Cr^{3+}) and hexavalent (Cr^{6+}) oxidation states being the most common. Among these oxidation states, the trivalent oxidation state is considered the most energetically stable (Barnhart, 1997a,b; Royal Society of Chemistry, 2012; Los Alamos, n.d.). Naturally occurring chromium tends to exist mostly in the trivalent state, usually as a combination with iron and/or other metal oxides (Barnhart, 1997a, b). According to Barnhart (1997a), the following statements can be generalized on the chemical characteristics of chromium that govern its fundamental use in many commercial applications as well as on its behaviour in the environment and effect on human health: (1) the dominant naturally occurring form of chromium is the trivalent state, (2) other forms tend to be converted to the trivalent oxidation state when in contact with the natural environment and (3) even when exposed to environments where it is not thermodynamically stable, the trivalent oxidation state of chromium is very slow to react.

Despite its health and environmental potency, chromium is also an essential nutrient for plant and animal if present in small quantities. However, the consequential environmental effects of chromium have also increased in recent years as a result of its wide anthropogenic uses (Shanker et al., 2005; Oliveira, 2012). When accumulated at high levels, chromium is associated with health problems such as nausea, ulcerations and cancer. In addition, the hexavalent state of chromium is considered the most toxic form due to its high solubility and mobility in soil and water (Rai et al., 1989; Richard and Bourg, 1991; Shanker et al., 2005; Oliveira, 2012). On the other hand, the trivalent state has low solubility and low reactivity, resulting in low mobility in the environment, and hence relatively low toxicity in living organisms (Rai et al., 1989; Richard and Bourg, 1991; Barnhart, 1997a,b; Shanker et al., 2005; Ma and Garbers-Craig, 2006a,b,c). The environmental and health effects of chromium and its compounds were covered in detail in Chapter 2.

5.3.2 Industrial Uses of Chromium

Chromium and its compounds have multifarious industrial and commercial uses in modern society (Shanker et al., 2005; Koleli and Demir, 2016). For example, different compounds of chromium have found a wide range of uses in the chemical, metallurgical and refractory industries (McGrath, 1995; Papp and Lipin, 2001; Shanker et al., 2005; Oliveira, 2012; Gasik, 2013; Los Alamos, n.d.; Koleli and Demir, 2016). Despite the ubiquity of the various industrial applications highlighted in Figure 5.1, chromium and its compounds are also associated with serious anthropogenic effects which, if not managed properly, could lead to serious environmental problems.

5.3.2.1 Metallurgical Uses of Chromium as an Alloying Element

Chromium is one of the most versatile and widely used alloying elements in the production of various grades of stainless steels, where it is used to impart corrosion and oxidation resistance, act as a mild hardenability agent, improve wear resistance and to promote the retention of useful strength at elevated temperatures (Bristow et al., 2000; Outokumpu, 2013; Gasik, 2013; Eric, 2014; Koleli and Demir, 2016). Stainless steels are alloys of steels typically containing a minimum of 10.5 wt.% chromium as an alloying element, with the corrosion resistance and other desirable properties increasing as the chromium content

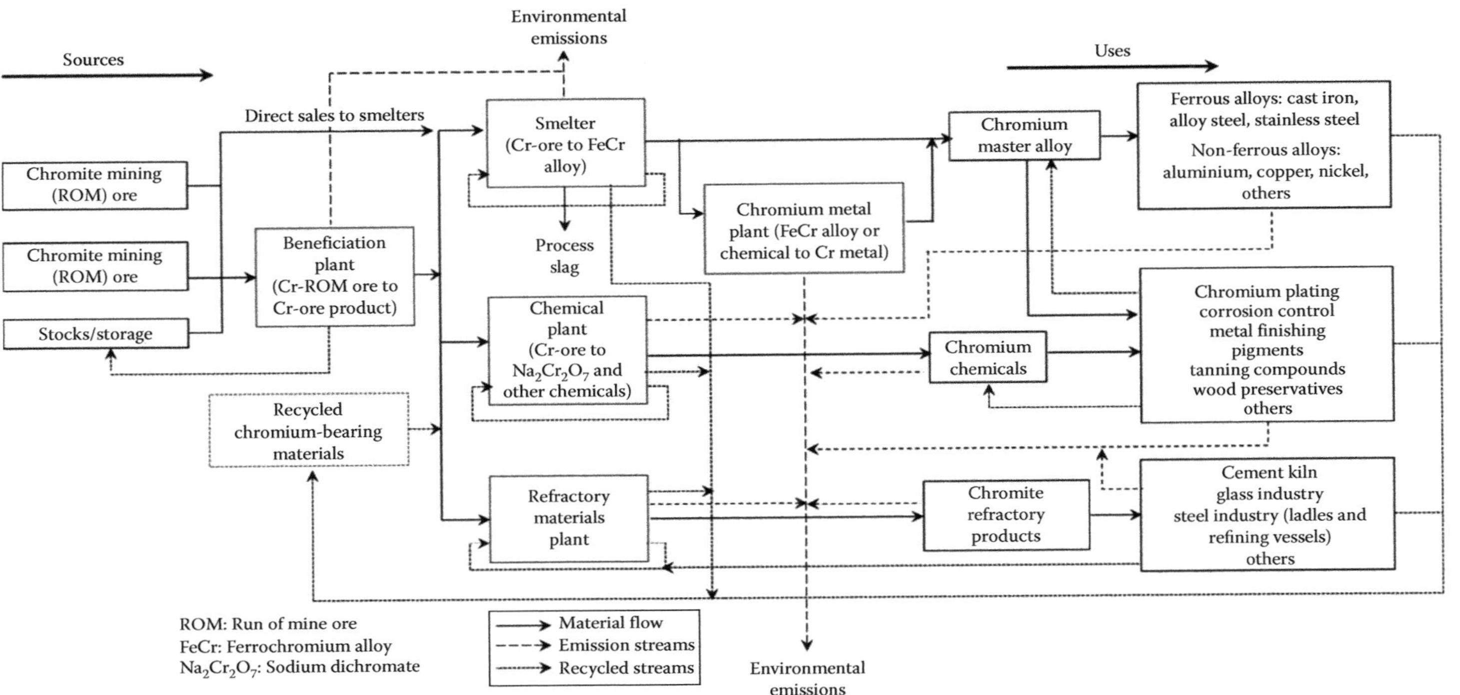

FIGURE 5.1
Chromium material flow of chromium from chromite ore beneficiation to end use of chromium products. (Reprinted from *Environmental Materials and Waste: Resource Recovery and Pollution Prevention*, Koleli, N. and Demir, A., Chromite, Chapter 11, 252. Copyright 2016, with permission from Elsevier.)

increases (Bristow et al., 2000; Papp and Lipin, 2001, 2006; McGuire, 2008; Gasik, 2013; Outokumpu, 2013). In fact, chromium is an irreplaceable constitute in these types of steels.

Typically, the different classes of stainless steels are categorized according to the typical content of chromium as well as the major phases in the final solidified product, inter alia, (1) austenitic stainless steels, which can further be subdivided into Cr–Mn grades (approx.18 wt.% Cr and 8 wt.% Ni), Cr–Mn grades (approx. 17 wt.% Cr, 4 wt.% Ni and 7 wt.% Mn), Cr–Ni–Mo grades (approx. 17 wt.% Cr, 10–13 wt.% Ni and 2–3 wt.% Mo), high-performance austenitic grades (17–25 wt.% Cr, 14–25 wt.% Ni and 3–7 wt.% Mo) and high-temperature austenitic grades (17–25 wt.% Cr and 8–20 wt.% Ni); (2) standard ferritic stainless steels (approx. 10.5 wt.% to 30 wt.% Cr); (3) martensitic and precipitation hardening stainless steels typically containing 12–14 wt.% chromium and (4) duplex stainless steels which consist of about 50% ferrite and 50% austenite phases and contain approx. 20–25.4 wt.% Cr, 1.4–7 wt.% Ni and 0.3–4 wt.% Mo (McGuire, 2008; Gasik, 2013; Outokumpu, 2013). It is proposed that over 70% of all the chromium used in steelmaking is consumed in the production of various grades of these stainless steels (Bristow et al., 2000).

5.3.2.2 Chromium Uses in High-Temperature Refractory Materials

Chromium, in the form of chromium oxide, plays an undoubtedly a niche role in the production of high temperature refractory materials (Papp and Lipin, 2001; Harbison-Walker, 2005; International Chromium Development Association, 2011a; Koleli and Demir, 2016). Chrome-based refractory materials are widely used in various high-temperature applications, such as in the smelting and refining of non-ferrous metals, cement production, glass industry and iron and steel production (Harbison-Walker, 2005; International Chromium Development Association, 2011a). Furthermore, granular refractory chromite is widely used in the foundry industry for ferrous and non-ferrous castings due to certain desired properties, viz, dimensional and thermal characteristics such as low linear thermal expansion, high density and high thermal conductivity properties (International Chromium Development Association, 2011a). According to the International Chromium Development Association (2011a), the dimensional and thermal characteristics of granular chromite materials are desirable in promoting the dimensional stability and rapid solidification of the castings.

In addition, chromium oxides are typically used in conjunction with other metal oxides to impart the thermal and chemical properties of refractory materials (Harbison-Walker, 2005; International Chromium Development Association, 2011a). Typical chrome-containing refractory materials include magnesite-chrome ($MgO \cdot Cr_2O_3$ spinel), chrome-fired and chrome-magnesite refractories (Harbison-Walker, 2005). In general, magnesite-chrome-based refractory materials, produced by sintering chrome and magnesia fines, constitute the most common type of chrome-based refractory materials (Harbison-Walker, 2005). In general, chrome-based refractory materials possess the following attributes: (1) good thermal stability, (2) high load softening point, (3) high temperature resistance above 1990°C and high resistance to alkali chemical attack and (4) low wettability by most molten metals and high melting temperatures (2150°C) (Harbison-Walker, 2005; International Chromium Development Association, 2011a). Further details on the fundamentals of refractory materials were covered briefly in Chapter 4.

5.3.2.3 Chromium Uses in Chemicals Industry

Chromium-based compounds have found widespread applications in the chemicals industry, particularly in the form of chromite (naturally occurring chromium-containing ore, $FeCr_2O_4$)

for the production of industrial chemicals such as chromic acid, sodium dichromate ($Na_2Cr_2O_7$) and other chromium chemicals and pigments. In addition, the various chromium-based chemicals, though consumed in small quantities, have found widespread industrial applications in niche areas such as in catalysts, corrosion inhibitors, metal plating and finishing, pigments and leather tanning compounds and in the production of high-purity chromium metal (Papp and Lipin, 2001, 2006; International Chromium Development Association, 2011a; Koleli and Demir, 2016). Chromic acid is one such typical example of an industrial chemical that is widely used in metal finishing and electroplating processes where metal surfaces are coated with a thin layer of chromium for decorative or for corrosion inhibition purposes.

5.3.3 Occurrence of Chromium

As stated in Chapter 2, chromium is the 7th most abundant element on earth but ranks 21st in abundance in the earth's crust. Chromium occurs naturally in ultramafic and serpentine rocks as chromite ($FeCr_2O_4$), which is a spinel crystal structure-compound containing trivalent chromium, iron and small amounts of other divalent and trivalent metallic oxides such as aluminium and magnesium (Barnhart, 1997a; International Chromium Development Association, 2011b; Oliveira, 2012; Gasik, 2013; Eric, 2014). A detailed overview of the different chromium minerals and the thermodynamics of chromium-containing systems are highlighted elsewhere (Gasik, 2013; Eric, 2014).

Chromite ore is a naturally occurring and commercially important source of chromium metal. Natural chromite has a spinel structure consisting of chromium and iron oxides ($(Fe^{2+}, Mg)O \cdot (Cr, Fe^{3+}, Al)_2O_3$), with varying amounts of magnesium and aluminium (Gu and Wills, 1988; International Chromium Development Association, 2011b; Murthy et al., 2011). In general, the different chromium-containing spinels exist as $FeO \cdot Fe_2O_3$, $FeO \cdot Cr_2O_3$, $MgO \cdot Cr_2O_3$ and $MgO \cdot Al_2O_3$ in solid solution with gangue components such as serpentine ($MgO \cdot SiO_2 \cdot nH_2O$) (Gu and Wills, 1988; Papp and Lipin, 2001; Eksteen et al., 2002; International Chromium Development Association, 2011b; Gasik, 2013; Eric, 2014). Depending on the degree of substitution within the spinel lattice structure, large variations are often observed in the total and relative amounts of Cr and Fe in the spinel lattice of the different chrome ore deposits (International Chromium Development Association, 2011b). Technically, these variations affect the chromite ore grade, not only in terms of the Cr_2O_3 content but also in the Cr:Fe ratio. The Cr: Fe ratio is often used as a commercial indicator of the chromium content in the ores and the final Cr content of the ferrochromium alloys produced (International Chromium Development Association, 2011b).

In general, a high Cr–Fe ratio is advantageous for the production of ferrochromium alloys with high chromium content (European Commission, 2014). In addition, the variation in the spinel substitution in the chromite also affects the refractory index (RI) of the ore, which is a measure of the reducibility of the ore (International Chromium Development Association, 2011b):

$$\text{R.I.} = \frac{\left(\text{wt.\%}Cr_2O_3 + MgO + Al_2O_3\right)}{\left(T - Fe\ \text{as}\ FeO\right) + SiO_2} \tag{5.1}$$

The global resources of chromite ore are estimated to be in the excess of 11 billion metric tonnes (U.S. Geological Survey, 2015). About 95% of the world's chromium resources are geographically concentrated in Kazakhstan and Southern Africa (U.S. Geological Survey, 2015). In fact South Africa is estimated to hold more than 70% of the world's known deposits followed by Zimbabwe's Great Dyke chrome deposits (Nafziger, 1982; Gu and Wills, 1988).

However, the Great Dyke ores have a high Cr_2O_3 content (40–50 wt.% Cr_2O_3) and Cr:Fe ratio > 3.2, which makes them directly suitable for the smelting of commercial high carbon ferrochromium alloys (Gasik, 2013). Other countries possessing commercial-grade chromite ores include Turkey, Finland, India, Brazil, China and Russia (International Chromium Development Association, 2011c).

Generally, the bulk of the world's commercial chromite deposits are found in two forms, namely, stratiform and podiform deposits (Gu and Wills, 1988; International Chromium Development Association, 2011c; Murthy et al., 2011). Stratiform deposits, such as those found in the Bushveld Igneous Complex of South Africa, are associated with large mafic or ultramafic layered intrusions (Gu and Wills, 1988; International Chromium Development Association, 2011c; Murthy et al., 2011). On the other hand, the podiform deposits are usually in the shape of pods, ranging from pea-sized nodules to large bodies hundreds of metres in extent, and typically have a highly irregular geometry and a high Cr:Fe ratio. Podiform chromite deposits are considered an important source for chromite, particularly for high-chromium, low-Al_2O_3 ores used in metallurgical applications, and for high-Al_2O_3, low-chromium ores used in refractory materials (Mosier et al., 2012). Kazakhstan, India and Finland predominantly exploit podiform deposits, while the chromite ore mineralization in Zimbabwe's Great Dyke has both the stratiform and podiform deposits (Gu and Wills, 1988; International Chromium Development Association, 2011c; Murthy et al., 2011).

Generally, commercial chromite ores mostly fall into two broad categories, namely, (1) high-silica hard lumps from ultramafic igneous rocks containing serpentine as the principal gangue, and overall Cr/Fe ratio of approximately 3, and (2) ferruginous friable ores from limonite deposits rich in secondary iron oxides (Eric, 2014). In addition, chromite ores can further be classified into commercial and non-commercial types, as shown in Table 5.2 (Gasik, 2013).

Traditionally, chromite ores were conventionally classified based on the chromite composition, type of deposit and principal use such as metallurgical grade (typically over 40 wt.% Cr_2O_3 and Cr:Fe ratio 3:1), chemical grade (typically over 40 wt. Cr_2O_3 and Cr:Fe ratio 1.5) and refractory grade (30–40 wt.% Cr_2O_3 and relatively high alumina (25–32 wt.% Al_2O_3)) (Gu and Wills, 1988; Papp and Lipin, 2001).

5.3.4 Beneficiation of Chromite Ores

The presence of gangue in naturally occurring chromite deposits dictates that the run-of-mine chromite ore be beneficiated to grades that are marketable and economic to smelt in electric furnaces. Typical beneficiation processes broadly include size reduction processes

TABLE 5.2

Classification of Chromite Ores

Category	Type	Wt.% Cr_2O_3	Wt.% SiO_2
Commercial	High	58–62	0.3–3
		50–58	3–8
	Medium	38–50	8–15
	Low	25–38	15–24
Non-commercial	Low	<25	>25

Source: Reprinted from *Handbook of Ferroalloys: Theory and Technology*, Gasik, M.M., Basics of Ferroalloys, Gasik, M.M. (ed.), Butterworth-Heinemann/Elsevier, Oxford, UK, p. 283. Copyright 2013, with permission from Elsevier.

(crushing and grinding), concentration processes (gravity and magnetic separation, and dense media separation) and agglomeration processes (pelletizing, sintering and briquetting) (Gu and Wills, 1988; Papp and Lipin, 2001). In addition, the concentration of chrome-bearing ores by flotation processing has also been widely investigated (Gu and Wills, 1988). Broadly, the main objective of the beneficiation processing is to render the ore concentrate physically, granulometrically and chemically suitable for subsequent processing (Papp and Lipin, 2001; Murthy et al., 2011). Typically, the concentration processes target to remove the gangue components, such as serpentine ($MgO \cdot SiO_2 \cdot nH_2O$), thereby increasing the desirable characteristics of the ore such as increasing the chrome oxide content, Cr:Fe ratio and/or alumina content (Papp and Lipin, 2001). The demand on the beneficiation process depends on factors such as ore source/mineralization and end-market requirements.

The amenability to beneficiation processing of chromite ores depends on the physicochemical properties of the ore. For example, the relative density of chromite depends on its composition, and varies from about 4.1 to 5.1 (Gu and Wills, 1988). Depending on the Cr_2O_3 content, the values tend to be higher than that of the matrix rock constituents such as magnesium-rich orthosilicates (S.G. 3.2), orthopyroxenes (S.G. 3.2–3.5) and serpentine (S.G. 2.5–2.6) (Gu and Wills, 1988). As a result, gravity separation methods tend to be the main methods of concentrating most of such chromite ores (Gu and Wills, 1988). In fact, the high-silica hard chromite ore lumps from igneous rocks containing $MgO \cdot SiO_2 \cdot nH_2O$ as the principal gangue material are generally amenable to beneficiation by gravity methods (Eric, 2014).

Furthermore, chromite ores can be concentrated based on their magnetic properties (Owada and Harada, 1985). Chromite ores exhibit paramagnetic properties at ambient temperatures, and their magnetic susceptibilities tend to have an apparent positive correlation with their Fe^{2+} content (Owada and Harada, 1985). According to Owada and Harada (1985), the distribution of magnetic ions tends to be non-uniform within the chromite crystal structure and, to some extent, ferromagnetic properties are generated in parts of high Fe^{2+} concentration. As a result, chromite can thus be concentrated as a magnetic product from the non-magnetic gangue minerals using high-intensity magnetic separation.

5.3.4.1 Agglomeration of Chromite Ore Fines

In general, chromite ore is used in the ferrochrome production processes in the form of lumpy ore, fines and concentrates (Gasik, 2013; Eric, 2014; European Commission, 2014). While the lumpy chromite ore is generally hard and compact, ferruginous chromite ores are typically friable and, as a result, tend to generate fines during mining, beneficiation and ore handling processes (Murthy et al., 2011; Agarwal et al., 2014). Naturally, chromite fines cannot be directly used for smelting in electric or submerged arc furnaces due to the difficulty in charging and the poor permeability of the solids packed bed during reduction (Yang et al., 2004; Gasik, 2014; Agarwal et al., 2014). According to Agarwal et al. (2014), the direct charging of fines into furnaces tends to result in these fines being fluidized into furnace off-gases, thereby resulting in compromised productivities and risks of explosions. To circumvent these challenges, agglomeration of the chromite fines by briquetting, sintering or pelletizing is usually carried out in order to granulometrically convert the chromite ore fines and concentrates into materials that can safely be charged in electric furnaces (Murthy et al., 2011; European Commission, 2014). Thus, the agglomeration of fine feed materials is critical from a furnace operation and productivity point of view. In fact, agglomerated charge gives a permeable charge that allows for the enhanced heat transfer, gas–solid reactions and the general flow of gases within the solids packed zone of the furnace (Yang et al., 2004). The fundamentals of the typical agglomeration processes were also covered in detail in Chapter 4.

In the pelletization process, chrome ore is usually ground to below 75 μm and balled into green pellets. After the balling process, the green pellets are hardened and sintered via a carbothermic reaction at around 1300°C–1400°C and then charged, together with lumpy ore and briquettes, into the submerged arc furnace for ferroalloys production (Singh et al., 2008; Nandy et al., 2009; European Commission, 2014). The resultant pellet properties play a vital role in the submerged arc furnace operation to improve charge permeability and to obtain superior conditions for the smelting reduction processes (Singh et al., 2008). In particular, the porosity achieved by the use of pelletized charge influences the flow of gases in the solids packed bed during the smelting reduction (Singh et al., 2008). However, it is proposed that the physical properties of the pellets tend to be inconsistent, and as such, ferrochromium alloy producers are slowly focusing on the sintering processing of chromite fines (Nandy et al., 2009; Wang et al., 2013). In this regard, the direct sintering of chromite ore fines is slowly receiving widespread attention as a cost-effective process of agglomeration (Deqing et al., 2008; Nandy et al., 2009; Wang et al., 2013; Agarwal et al., 2014).

As discussed in Chapter 4, the agglomeration of ore fines via the sintering process is an established science in most pyrometallurgical processes. According to Agarwal et al. (2014), the sintering process of friable chromite fines has several advantages in the operation of the submerged arc furnace, namely (Deqing et al., 2008; Nandy et al., 2009; Wang et al., 2013; Gasik, 2013; Agarwal et al., 2014): (1) uniform size distribution in the solids packed bed, (2) better metallurgical performance in terms of reducibility at high temperatures, (3) improved electrical resistivity of chromite sinter at high temperatures to facilitate heating by charge bed resistance (Joule effect) and (4) lower electricity consumption per unit compared to lumpy ore and/or cold-bonded briquettes as a result of the more open structure in sinter. Despite the proven benefits of charging sinter in the SAF, the sintering of chromite ore fines is still facing challenges, mainly because the melting point of the chromium spinel is too high to form the molten phases required for bonding process during the sintering (Deqing et al., 2008; Nandy et al., 2009; Agarwal et al., 2014).

In general, the formation of bonding phases is critical to the development of the desired properties such as tumble strength and abrasion resistance. Based on laboratory pot sintering tests, Deqing et al. (2008) proposed that the consolidation mechanism of chromite sinters was controlled by the formation of olivine liquid phases and the solid-state recrystallization of chromium oxides. According to Deqing et al. (2008), agglomeration of chromite sinter mainly consists of the 20% olivine liquid-phase consolidation, supplemented by the solid-state reaction and recrystallization of Cr_2O_3 at peak sintering temperatures of 1450°C–1500°C. Due to the refractoriness of the spinel chromite phases, the observed 20% olivine liquid phases were much lower than the typical ranges of 30–40% observed in conventional sintering of for ordinary iron ores (Deqing et al., 2008). Deqing et al. (2008) contend that it is the olivine liquid phases that act as a filler material among the mineral particles and contribute to the good strength and high yield of chromite sinter.

5.3.5 Smelting of Chromite Ores

In general, beneficiated chromite ores are smelted carbothermically in submerged arc furnaces, where the chromite ore, carbonaceous reducing agents in the form of metallurgical coke, coal or charcoal and fluxes (mostly quartz) are charged and melted at around 1650°C–1700°C to produce liquid ferrochromium alloy and slag (Downing, 1975; Yang et al., 2004; Gasik, 2013; Eric, 2014). The supernatant slag separates from the liquid ferrochromium alloy and tapped into slag pots for further processing, while the molten alloy is either granulated or is poured into moulds, cooled and crushed into sizes specified by customers (International Chromium Development Association, 2011d).

Chromium-containing raw materials are either charged directly as chrome ore lumps or are agglomerated to form briquettes, pellets or sinter. The smelting reduction reactions are highly endothermic, and as a result, ferrochromium smelting furnaces are associated with high electrical energy consumption (Yang et al., 2004; International Chromium Development Association, 2011d; Gasik, 2013; Eric, 2014). Basically, the common desirable characteristics of submerged arc furnaces in chromite smelting include (Yang et al., 2004; International Chromium Development Association, 2011d; Gasik, 2013; Eric, 2014) (1) relative ease of control provided the charge is well sorted to maintain a permeable solids packed bed, (2) self-regulating with power input determining the rate of consumption of charge and (3) possibility of preheating and prereduction of the burden by the hot ascending gases.

In addition, the design of submerged arc furnaces can be open, semi-closed or closed. Closed furnace designs possess better thermal efficiency and have the ability to utilize the sensible energy in the furnace off-gases (International Chromium Development Association, 2011d; Gasik, 2013; Eric, 2014). Furthermore, the chemical enthalpy of the furnace off-gas can also be used for heating the charge by combusting the gases in the furnace freeboard (Kapure et al., 2007). Further details on the electrical-thermal-chemical phenomena in submerged arc furnaces were discussed in detail by Gasik, 2013 and Eric, 2014.

5.3.5.1 Carbothermic Reduction of Chromite Ores

The high-temperature carbothermic reduction of chromium oxides with carbon was reported in several studies (Soykan et al., 1991; Uslu and Eric, 1991; Hino et al., 1998; Eksteen et al., 2002; Hayes, 2004; Yang et al., 2004; Eric, 2014; Eric and Demir, 2014). The reduction of chromite ores in the submerged arc furnaces occurs via both the gaseous- and solid-state reactions in the different zones of the submerged arc furnaces, as shown in Figure 5.2 (Hayes, 2004; Yang et al., 2004).

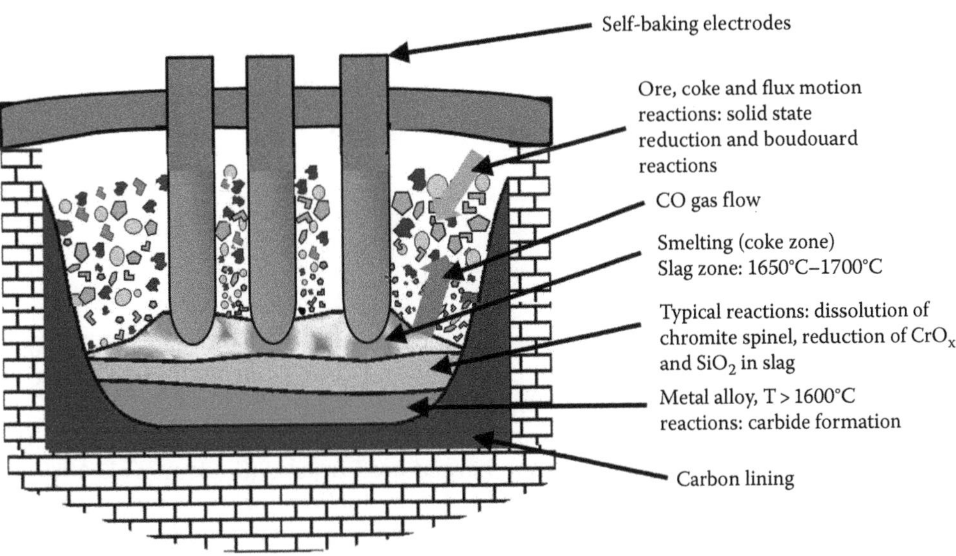

FIGURE 5.2
Schematic overview of zones in a typical submerged arc furnace in chromite ore smelting. (From Yang, Y. et al., Analysis of transport phenomena in submerged arc furnace for ferrochrome production, *Proceedings of the 10th International Ferroalloys Congress*, Cape Town, South Africa, 2004, http://www.pyrometallurgy.co.za/InfaconX/, Accessed 6 April 2016, p.13.)

Simplistically, the solid-state carbothermic reduction of chromite in the solids-packed zone and the Boudouard reaction are shown in Equations 5.2 through 5.4 (Yang et al., 2004):

$$FeCr_2O_4 + 4CO_{(g)} \rightarrow 2Cr + Fe + 4CO_{2(g)} \quad \text{(Solid-state reduction with CO gas)} \tag{5.2}$$

$$CO_{2(g)} + C \rightarrow 4CO_{(g)} \quad \text{(Boudouard reaction)} \tag{5.3}$$

$$FeCr_2O_4 + 4C \rightarrow 2Cr + Fe + 4CO_{(g)} \quad \text{(Overall reaction)} \tag{5.4}$$

The high temperature zone is also characterized by the concomitant dissolution of the unreduced oxides in slag. The reduction of dissolved chromite in molten slag takes place at the slag-metal interface by the by solid carbon in the coke zone and/or by the dissolved carbon in the alloy phase, and the resulting CO gas is released from the slag surface onto the solids packed zone to affect further reduction in the charge (Uslu and Eric, 1991; Eksteen et al., 2002; Yang et al., 2004; Eric, 2014; Eric and Demir, 2014). As discussed under Section 5.3.3, chromite ores consist of spinels of $FeCr_2O_4$, such as $FeO \cdot Cr_2O_3$, where the FeO can be replaced by MgO and the Cr_2O_3 by Al_2O_3 in the crystal structure. Typically, the $MgO \cdot Cr_2O_3$ spinel melts at 2350°C, and hence silica is usually added as flux to form lower melting point slag phases (Eksteen et al., 2002). The typical slag-forming reactions can be summarized as shown in Equations 5.5 and 5.6 (Eksteen et al., 2002):

$$(Fe,Mg)O(Cr,Al)O_{3(s,l)} + C_{(s)} \rightarrow (FeO)_{slag} + 2(CrO)_{slag} + (MgO \cdot Al_2O_3)_{slag} + CO_{(g)} \tag{5.5}$$

$$(FeO)_{slag} + (CrO)_{slag} + (Al_2O_3)_{slag} + (MgO)_{slag} + (SiO_2)_{flux} + (CaO)_{flux} \rightarrow \text{Slag} \tag{5.6}$$

Consequently, the iron oxides and chromium oxides dissolved in slag are reduced by carbon and carbon monoxide to form metallic iron, metallic chromium and chromium carbides. Equations 5.7 through 5.9 summarize the final reduction reactions taking place in the slag (Eksteen et al., 2002):

$$(FeO)_{slag} + C_{(s)} \rightarrow [Fe]_{metal} + CO_{(g)} \tag{5.7}$$

$$7(CrO)_{slag} + 10C_{(s)} \rightarrow [Cr_7C_3]_{metal} + 7CO_{(g)} \tag{5.8}$$

$$(SiO_2)_{slag} + 2C_{(s)} \rightarrow [Si]_{metal} + 2CO_{(g)} \tag{5.9}$$

Due to the differences in densities of the liquid alloy (~6.8 g/cm³) and slag (2.5–2.8 g/cm³) the reduced alloy in slag separates from the slag and collects in the hearth of the submerged arc furnace (Niemelä and Kauppi, 2007).

Naturally, the amounts of iron oxides and gangue in the chromite ore have a significant effect on the reduction reactions in the submerged arc furnaces (Soykan et al., 1989; Eric, 2014). From the extensive studies on the solid-state reduction of chromite ores conducted by Soykan et al. (1991) at 1416°C, the iron oxides tend to be more readily reduced by carbon, and as a result, a chromite ore rich in iron oxides will have high reducibility at relatively low temperatures. Conversely, the effect of gangue minerals has a significant effect on the reducibility of chromite ore (measured as the refractoriness index shown in Equation 5.1), and hence will have a significant impact on the operating temperature regimes of the smelting zone (Soykan et al., 1991; International Chromium Development

Association, 2011b; Eric, 2014). In most cases, silica, lime or dolomite fluxes are usually added to stoichiometrically react with the gangue components in the ore, thereby controlling the liquidus temperatures of the slag (Eric, 2014).

5.3.5.2 Classification of Ferrochromium Alloys

Fundamentally, the ferrochromium alloys are alloys of iron and chromium, with lesser amounts of dissolved carbon and silicon (International Chromium Development Association, 2011d). In addition, the amount of impurities such as sulphur and phosphorus should be as slow as possible. Broadly, ferrochromium alloys produced from the carbothermic reduction of chromite in SAF can be classified according to their chromium content into either high carbon ferrochromium alloys (typical contents of $60 \le Cr \le 70$ wt.% and $4 \le C \le 6$ wt.%, respectively) or charge chrome alloys ($50 \le Cr \le 55$ wt.% and $6 \le C \le 8$ wt.%) (Gasik, 2013; Eric, 2014). Furthermore, ferrochromium alloys can also be classified depending on the amount of dissolved carbon and silicon in the alloy, viz, high-carbon ferrochrome (HCFeCr), medium-carbon ferrochrome (MCFeCr), low-carbon ferrochrome (LCFeCr) and ferrosilicon chromium (FeSiCr) (European Commission, 2014).

5.3.5.3 High-Carbon Ferrochromium Alloys

High-carbon ferrochromium alloys are typically produced by the carbothermic smelting reduction of chromite ores in three-phase submerged arc furnaces (Downing, 1985; Bason and Daavittila, 2013; Eric, 2014; European Commission, 2014). Typically, the submerged arc furnaces are equipped with Söderberg electrodes and utilize the resistance heating to melt the charge (Yang et al., 2004; Bason and Daavittila, 2013; Eric 2014). Further technical details on the Söderberg electrode system are covered elsewhere and as a result will not be dealt with in this chapter (Beukes et al., 2013; Shoko et al., 2013).

The high-carbon ferrochrome smelting process is highly energy intensive. On average, the energy consumption in a high-carbon ferrochrome process is in the range of 4000–4500 kWh/ton FeCr (Eric, 2014). As a result, some operations have incorporated preheating and/or prereduction facilities to utilize the CO gas from the smelting process in a bid to reduce the overall energy consumption of the process (European Commission, 2014). According to Niemelä et al. (2004), the CO gas from submerged arc furnace typically contains about 75–90 vol.% CO, 2–15 vol.% H_2, 2–10 vol.% CO_2 and 2–7 vol.% N_2, and this off-gas can be combusted to produce energy in the range of 10–12 MJ/Nm^3 or 2–2.3 MWh/ton FeCr. Figure 5.3 shows the typical process schematic for producing high-carbon ferrochrome in a closed submerged arc furnace using a mixture of lumpy ore and sintered pellets (European Commission, 2014).

According to Gasik (2013), HCFeCr represents a group of ferroalloys with composition between 60 and 70 wt.% chromium, between 4 and 6 wt.% carbon and between 2 and 3 wt.% silicon. The melting temperature of high-carbon ferrochromium alloy is approximately 1550°C, and is tapped from the furnace at temperatures around 1580°C–1620°C (Karbowniczek et al., 2012). In general, the production of HCFeCr alloys requires chrome-bearing ores with a high Cr: Fe ratio, typically greater than 2, while lower Cr: Fe ratios in the range of 1.5–1.6 are typically consumed in the production of charge chrome alloys (Bason and Daavittila, 2013; Gasik, 2013).

As discussed in Section 5.3.2.1, chromium is a critical alloying element in the production of various grades of stainless steels. The development of argon-oxygen and vacuum-oxygen decarburization technologies in the refining process of stainless steels has enabled

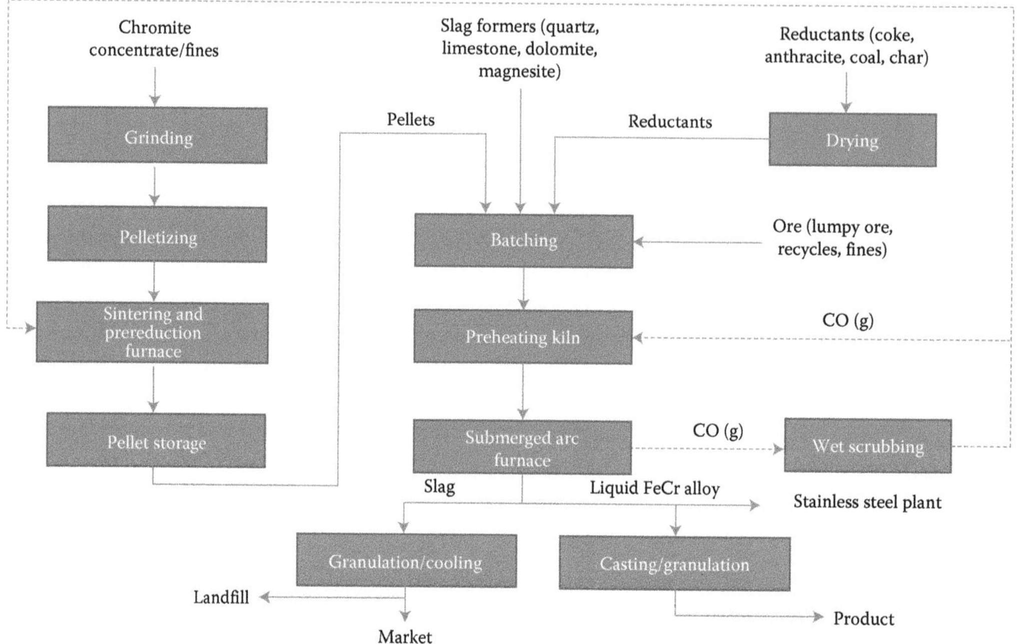

FIGURE 5.3
High-carbon ferrochrome production using closed submerged arc furnace. (From European Commission, Ferroalloys, Best Available Techniques (BAT) Reference Document for the Non-ferrous Metals Industries, Industrial Emissions Directive 2010/75/EU (Integrated Pollution Prevention and Control), 2014, Available at http://eippcb.jrc.ec.europa.eu/reference/BREF/NFM_Final_Draft_10_2014.pdf, Accessed 10 March 2016, p. 788.)

the use of high-carbon ferrochromium alloys, as a means of introducing chromium into steel with much improved process economics (Patil et al., 1998; Gasik, 2013). In fact, a significant amount of the high-carbon ferrochromium alloys produced globally are consumed in the production of stainless steel (Patil et al., 1998; Holappa, 2010; International Chromium Development Association, n.d.; Gasik, 2013; Outokumpu, 2013; Leont'ev et al., 2015).

Despite the criticality of HCFeCr alloys in the production of stainless less steels, the level of impurities in these alloying charges has an adverse effect on the final quality of steel (Holappa and Louhenkilpi, 2013; Holappa, 2004; Pande et al., 2010; Pande et al., 2012). In particular, parameters such as the type and quality of chromite ores, slag metallurgy and/or technology, and the quality of reductants used in the production of high-carbon ferrochromium alloys, have an effect on the composition of the final alloy (Bason and Daavittila, 2013). Naturally, deleterious impurities such as phosphorus and sulphur can be introduced into the smelting process via the type and quality of reductants used. Typically, the amount of dissolved sulphur and phosphorus in high carbon ferrochromium and charge chrome alloys should be below 0.015 and 0.02 wt.%, respectively (Gasik, 2013). In addition, metallic element impurities such as Al, Ca, Ti and V can also have special limitations in steelmaking due to their natural tendency to from non-metallic inclusions or precipitates such as nitrides and carbides in steels (Holappa and Louhenkilpi, 2013; Pande et al., 2010, 2012).

The amount of dissolved carbon in the high-carbon ferrochromium alloys can also present challenges in downstream refining process of steel, particularly in the production of high alloy stainless steels, superalloys and low-carbon to ultralow-carbon steels (Sano et al., 1997; Patil et al., 1998; Turkdogan and Fruehan, 1998; Bhonde et al., 2007; Total Materia, 2008;

Pande et al., 2010, 2012). As discussed in Chapter 4, the amount of dissolved carbon in the final liquid steel is controlled by the decarburization processing using either the argon-oxygen decarburization (AOD) or vacuum-oxygen decarburization (VOD) reactors. Due to the high reactivity of chromium at high oxygen potentials, the decarburization treatment of stainless steels requires a balance between the amount of residual carbon and chromium alloying element oxidation loss to slag (Turkdogan and Fruehan, 1998). As a result, the need to balance the amount of residual carbon without the excessive oxidation loss of alloying elements has led to the increased demand for using low-carbon-containing alloy charges in steelmaking.

5.3.5.4 Medium-Carbon Ferrochromium Alloys

Typically, medium-carbon ferrochromium alloys contain 56–70 wt.% chromium and 1–4 wt.% carbon (Eric, 2014). This group of ferrochromium alloys can be produced by the decarburization process of high-carbon ferrochrome alloys, either by oxygen blowing in a converter or by the addition of chromite to the melt (Bhonde et al., 2007; Rick, 2010; Karbowniczek et al., 2012; Bason and Daavittila, 2013; European Commission, 2014). Some of the decarburization treatment processes of liquid high-carbon ferrochromium alloys are briefly discussed in the following sections.

5.3.5.4.1 Decarburization by Converter Oxidation

In essence, the production of medium-carbon ferrochrome alloys involves oxidizing the dissolved carbon using gaseous oxygen at process temperatures around 1700°C (Bhonde et al., 2007; Rick, 2010; Karbowniczek et al., 2012; Gasik, 2013; Beskow et al., 2015). The blowing oxygen-containing gas mixture is done through combinations of submerged annular tuyeres, top tuyeres or a water-cooled lance system (Rick, 2010). As in the basic oxygen steelmaking process discussed in Chapter 4, the submerged blowing process has several advantages, inter alia, the high bath temperatures achieved in bottom-blown converters provide conditions favourable for the high decarburization and chromium recovery rates (Rick, 2010; European Commission, 2014; Beskow et al., 2015). Rick (2010) gave a detailed description of converter decarburization processes to produce various product ranges, and the boundary conditions of the refining process are given in Figure 5.4.

Based on the assumption that the carbon in high-carbon ferrochrome alloy exists in the carbide form, the typical converter decarburization reactions are shown in Equations 5.10 and 5.11 (Bhonde et al., 2007; Rick, 2010; Karbowniczek et al., 2012; Gasik, 2013):

$$\left[Cr_7C_3\right]_{alloy} + \frac{3}{2}O_{2(g)} \rightarrow 7\left[Cr\right]_{alloy} + 3CO_{(g)} \tag{5.10}$$

$$\left[C\right]_{alloy} + \frac{1}{2}O_{2(g)} \rightarrow 2CO_{(g)} \tag{5.11}$$

In the process, the chromium and iron elements in the alloy are also oxidized and report to the slag phase as shown in Equations 4.5 and 4.6 (Turkdogan and Fruehan, 1998). The chromium oxides formed in slag can also further take part in the decarburization reaction at the slag–metal interface, as shown by Equation 5.12 (Karbowniczek et al., 2012):

$$\left[Cr_7C_3\right]_{alloy} + \left(Cr_2O_3\right)_{(slag)} \rightarrow 9\left[Cr\right]_{alloy} + 3CO_{(g)} \tag{5.12}$$

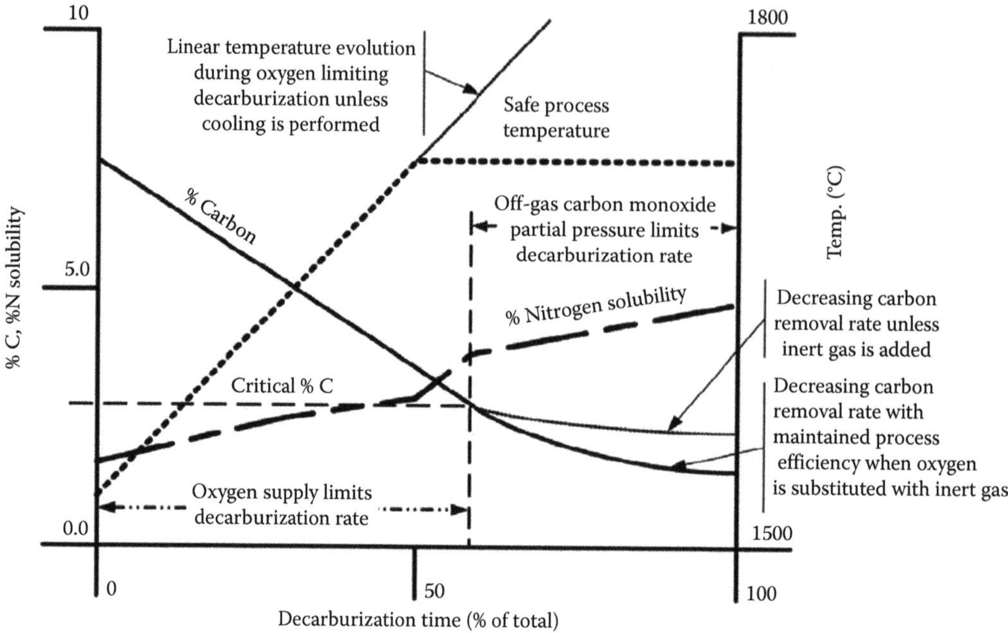

FIGURE 5.4
Process routes and products in the production of medium and low-carbon ferrochromium alloys. (From Rick, C.J., *J. South. Afr. Inst. Min. Metall.*, 110, 751, 2010.)

Controlling chromium oxidation loss to slag is the ultimate control strategy in converter decarburization processes. According to Rick (2010), chromium oxidation is the dominant reaction in the reactor at lower-carbon activities, unless measures are taken to lower the partial pressure of the formed carbon monoxide. Temperature as a parameter is also controlled such that it is neither too low for excessive chromium oxidation nor too high to prevent excessive wear of the converter refractory lining (Rick, 2010). Typically, converter decarburization processes utilize a Cr_2O_3–SiO_2–CaO–MgO slag system, and hence, these processes mostly utilize solid Cr_2O_3-rich slag during the decarburization stage, followed by the reduction of the Cr_2O_3 in slag using Si after the decarburization stage (Rick, 2010). As a result, the proper control of the process conditions during reduction stage is critical in reducing the oxidation losses of chromium in slag (Rick, 2010).

Notwithstanding the merits of the various converter decarburization processing, chromium is a reactive element, and any attempt to remove the carbon by simple oxidation also results in the oxidation of chromium (Bhonde et al., 2007; Wang et al., 2013). Based on whichever process that is chosen to produce medium-carbon ferrochrome alloys, it is desirable to minimize the loss of chromium to the slag phase in view of economic and environmental considerations (Wang et al., 2013).

The oxidation of chromium tends to result in the formation of a chromium oxide–rich refractory slag, which, in essence, will affect the liquidus temperatures of the process slags (Bhonde et al., 2007). Technically, the considerably high melting point and viscosity of high-chromium refining slags due to the accumulation of chromic oxides in the slag reduces the capacity of increasing the oxidizing capacity of decarburization slags (Bhonde et al., 2007). Furthermore, the decarburization process of high-carbon ferrochromium alloys by oxygen-containing gas mixtures also tend to be complicated by the nitrogen pickup in the alloy, particularly if nitrogen is used as the inter gas carrier in the process.

5.3.5.4.2 Decarburization by Chromite Additions

The typical decarburization reactions of high carbon ferrochroium alloy by chromite ore additions are shown in Equations 5.13 and 5.14 (Bhonde et al., 2007; Karbowniczek et al., 2012):

$$2[Cr_7C_3]_{alloy} + \frac{3}{2}(FeCr_2O_4)_{(s)} \rightarrow 17[Cr]_{alloy} + \frac{3}{2}[Fe]_{alloy} + 6CO_{(g)} \quad (5.13)$$

$$[Cr_7C_3]_{alloy} + 3(FeO)_{(s)} \rightarrow 7[Cr]_{alloy} + 3[Fe]_{alloy} + 3CO_{(g)} \quad (5.14)$$

Despite the potential benefits of using chromium-bearing materials as decarburization agents, this process is slowly being eclipsed by the converter processes in the production of medium-carbon ferrochromium alloys.

5.3.5.5 Low-Carbon Ferrochromium Alloys

Low-carbon ferrochromium alloys typically contain about 56–70 wt.% chromium and 0.015–1.0 wt.% carbon (Eric, 2014). According to Eric (2014), the low-carbon alloy grades are used mainly for the final adjustments in the composition of special steels and for the production of superalloys. In most cases, low-carbon ferrochromium alloys are produced by the silicothermic or aluminothermic reduction of oxides present in chromium ore or in high-chromium slag (Sano, 2004; Bhonde et al., 2007; Karbowniczek et al., 2012; Gasik, 2013).

The silicothermic reduction uses silicon, either in the form of SiCr or FeSi, as a reductant to produce a low-carbon ferrochromium alloy and an intermediate slag product (European Commission, 2014). In essence, the silicothermic reduction of a chromite-rich slag is shown in Equation 5.15 (Bhonde et al., 2007; Karbowniczek et al., 2012):

$$\frac{2}{3}(Cr_2O_3)_{(s)} + [Si]_{as FeSi or SiCr} + CaO \rightarrow \frac{4}{3}[Cr]_{alloy} + (2CaO \cdot SiO_2)_{slag} \quad (5.15)$$

Typically, the silicothermic reduction process is capable of producing an alloy with carbon content less than 0.03 wt.% (Bhonde et al., 2007). According to Sano (2004), different mechanisms of reduction are possible in the silicothermic reduction process, and these include, (1) the direct reaction of the ferrosilicon or other silicon-bearing reductant with slag, (2) the dissolution of silicon into the alloy followed by the dissolved silicon reacting with the slag at the slag–metal interface and (3) reaction mechanism involving a combination of (1) and (2).

The aluminothermic reduction can also be used in the production of pure chromium and its alloys (Sano, 2004; Bhonde et al., 2007; Karbowniczek et al., 2012; Gasik, 2013). The autothermic reactions, controlled by the addition rate of aluminium and/or other thermal admixtures to the charge, generate high enough process temperatures to keep the highly refractory Al_2O_3 + Cr_2O_3-rich slag sufficiently fluid. The reduction process of Cr_2O_3 takes place according to Equation 5.16 (Bhonde et al., 2007; Karbowniczek et al., 2012; Gasik, 2013; Eric, 2014):

$$2(Cr_2O_3)_{(s)} + 4Al + \rightarrow 4[Cr]_{alloy} + 2(Al_2O_3)_{slag} \quad (5.16)$$

In most cases, lime is added as a flux to reduce the liquidus temperatures and the viscosity of slag (Gasik, 2013). According to Gasik (2013), lime addition has a significant effect on the slag liquidus temperatures by forming calcium chromite-chromate phases,

$9CaO \cdot 4CrO_3 \cdot Cr_2O_3$, which is beneficial in the melting of lime-rich chromite slags. Furthermore, lime addition improves the reduction kinetics of the process by forming stable calcium aluminates in the slag, thereby reducing the activity of chromium oxide in slag (Gasik, 2013).

5.3.6 Waste Generation in Ferrochromium Production Processes

In general the mining, beneficiation and smelting of chromite ores generate a lot of toxic and hazardous wastes (Rai et al., 1989; Richard and Bourg, 1991; Barnhart, 1997a,b; Petersen et al., 2000; Papp and Lipin, 2001; Shanker et al., 2005; Ma and Garbers-Craig, 2006a,b,c; Bulut et al., 2009; Oliveira, 2012). The typical wastes produced from these unit processes include process slags, emission control dusts, beneficiation residues and leach residues, among others (Petersen et al., 2000). Of particular interest are process slags and particulate dusts which are generated in enormous quantities during the production of ferrochromium alloys, and contain significant amounts of entrained chromium compounds (Ma and Garbers-Craig, 2006a,b,c; Bulut et al., 2009; Beukes et al., 2010, 2012; Holappa, 2010). As discussed in Chapter 2, the production of waste in the ferrochromium industry is a sensitive matter in most countries due to the high toxicity levels and environmental hazards posed by chromium-containing compounds, particularly chromium in the hexavalent oxidation state.

The technical aspects of ferrochromium alloys production have been covered in the preceding sections. In order to holistically address the valorization requirements ferrochromium alloys wastes, it is important to categorize the different types and amounts of wastes generated in the production of these ferroalloy products. In this regard, the following sections discuss the generation and categories of typical solid waste materials, in particular, the process slag and particulate dusts produced from these processes. However, the typical waste materials produced in the generic mining and beneficiation processes are not discussed here as they were covered in detail in Chapter 3.

5.3.6.1 Particulate Dust Emissions in the Production of Ferrochromium Alloys

The agglomeration and smelting reduction processes of bearing ores result in the generation of both fine and coarse particulate dust emissions (Stegemann et al., 2000; Ma and Garbers-Craig, 2006a,b,c; Holappa, 2010). Typically, these particulate dusts contain significant amounts of potentially toxic chromium species and volatile metal-bearing compounds which present serious challenges in terms of metal recovery and separation, and the disposal of the barren wastes (Person, 1971; Cohen and Petrie, 1997; Petersen et al., 2000; Stegemann et al., 2000; Ma and Garbers-Craig, 2006a,b; Bulut et al., 2009; Sedumedi et al., 2009; Beukes et al., 2010). Technically, ferrochromium alloy industries emit fine particulate dusts containing chromium compounds through flue gas emissions, and such emissions end up settling down and polluting the environment (Sedumedi et al., 2009). The presence of hexavalent chromium compounds in such waste materials is a cause for environmental and health concerns, mainly because Cr (VI) is carcinogenic and highly soluble in water and has high mobility in soils (Rai et al., 1989; Richard and Bourg, 1991; Barnhart, 1997a,b; Shanker et al., 2005; Ma and Garbers-Craig, 2006a,b,c).

The environmental and health effects of chromium compounds were dealt with in Chapter 2. Furthermore, the biological uptake and biotransformation of hexavalent chromium in plants and humans were also highlighted in several studies (Richard and Bourg, 1991; Barnhart, 1997a,b; Stasinakis et al., 2003; Shanker et al., 2005; Sedumedi et al., 2009; Oliveira, 2012).

Technically, hexavalent chromium ions are capable of penetrating biological membranes of human cells, pass into the cytoplasm of cells where they get reduced to trivalent chromium and react with intracellular materials (Sedumedi et al., 2009). The toxicity effects of chromium in plants have been reported as well, and typically depend on the metal speciation, which in turn is responsible for its mobilization, subsequent uptake and resultant toxicity in plants (Shanker et al., 2005). According to Shanker et al. (2005), the effects of chromium toxicity in plants can be observed at multiple levels, i.e. from reduced yield, effects on leaf and root growth, to inhibition on enzymic activities and mutagenesis.

5.3.6.2 Particulate Dusts from the Agglomeration Processes of Chromite Ores

The typical agglomeration processes applicable to chromite ore fines were discussed in Section 5.3.4. Naturally, these processes require that the feed material undergo size reduction processes such as crushing and milling prior to the agglomeration processing (Riekkola-Vanhanen, 1999; Beukes and Guest, 2001; Nandy et al., 2009; Beukes et al., 2010; Agarwal et al., 2014). As a result, a significant amount of particulate dusts contaminated with the potentially toxic hexavalent forms of chromium are usually generated during such processes, particularly during the material handling, crushing and milling and agglomeration processes (Person, 1971; Riekkola-Vanhanen, 1999; Beukes and Guest, 2001; Beukes et al., 2010; European Commission, 2014). Furthermore, the crushing and screening processes employed in the sizing of sinter agglomerates are also a major source of these particulate wastes (Riekkola-Vanhanen, 1999).

Basically, wet milling is highly recommended over the dry milling in the processing of chrome-bearing materials (Person, 1971; Riekkola-Vanhanen, 1999; Beukes and Guest, 2001; Beukes et al., 2010; European Commission, 2014). Beukes and Guest (2001) reported that dry milling of chromium-bearing minerals, especially under atmospheric conditions, can lead to the formation of significant amounts of Cr (VI)-containing dusts. In the case that the dry milling processing cannot be avoided, the installation of dust prevention, extraction and suppression systems is mandatory in order to eliminate the dispersion of airborne dust particles (Person, 1971; Darcovich et al., 1997; Riekkola-Vanhanen, 1999; Beukes and Guest, 2001; Petavratzi et al., 2005; Beukes et al., 2010; European Commission, 2014). However, the dust mitigation systems may not always be sufficient, and there is always a danger of drying milling resulting in the dispersion of fugitive emissions into the workplace environment and surroundings (Person, 1971; Darcovich et al., 1997; Riekkola-Vanhanen, 1999; Petavratzi et al., 2005; Sedumedi et al., 2009; European Commission, 2014).

Since the different agglomeration processes discussed in Section 5.3.4 require typically fine feed materials the emission of particulate dusts is particularly problematic in the operation and product handling in these unit processes (Person, 1971; Riekkola-Vanhanen, 1999; Petavratzi et al., 2005; Beukes et al., 2010; Rao et al., 2010; European Commission, 2014). In particular, the handling and conveying of briquetted chromite fines also result in problematic generation of particulate dusts, mainly due to the poor compressive strengths of the briquettes as a result on the low curing temperatures inherent to the process (Riekkola-Vanhanen, 1999; Deqing et al., 2008; Agarwal et al., 2014; European Commission, 2014).

Typically, raw material agglomeration processes are fitted with dust extraction and abatement systems, either in the form of electrostatic precipitators, baghouses or scrubbers, to collect the coarse and fine particulate dusts from the exhaust gases (Person, 1971; Darcovich et al., 1997; Riekkola-Vanhanen, 1999; Petavratzi et al., 2005; European Commission, 2014). In most cases, the particulate dust materials are collected, blended with fresh feed and recycled back into the respective agglomeration processes (Riekkola-Vanhanen, 1999; European Commission, 2014). As discussed in Chapter 2, stringent environmental regulations restrict the disposal of untreated chromium-containing wastes in landfills, and as such, the current in-process recycling approaches not only increase the resource utilization efficiency in the raw material agglomeration processes, but also serve to reduce the environmental footprint of the respective agglomeration processes.

5.3.6.3 Particulate Dust Emission from Smelting of Chromite Ores

Typically, the dusts generated from ferrochrome plants consist of coarse dust collected by the cyclone separators and fine dust that is captured by the baghouse filters (Person, 1971; Riekkola-Vanhanen, 1999; Ma and Garbers-Craig, 2006a,c; Sedumedi et al., 2009; Beukes et al., 2010; Rao et al., 2010; Basson and Daavittila, 2013; Eric, 2014; European Commission, 2014). According to Ma and Garbers-Craig (2006c), about 18–25 kg of particulate dusts and/or sludges are generated by South African producers per ton of ferrochromium alloy produced. Overally, the South African ferrochromium industry generates about 100,000 tonnes of baghouse filter dust and sludges annually (Ma and Garbers-Craig, 2006c). Comparatively, the stainless steel industry generates about 18–33 kg of baghouse filter dust per ton of stainless steel produced, translating to about 24,000 tonnes of stainless steel dust produced annually (Ma and Garbers-Craig, 2006c).

5.3.6.3.1 Physicochemical Characteristics of Smelter Dusts

The physicochemical characteristics of particulate dusts emitted from the smelting of ferrochromium alloys vary from process to process, and depend mostly on the alloy grade produced, the raw materials used, the operational parameters and the type of furnaces used in the smelting process (Person, 1971; Cox, 1985; Ma and Garbers-Craig, 2006a,b,c; Nelson, 2014). According to Ma and Garbers-Craig (2006a,c), coarse particulate dusts are mostly oxide based and typically contain high levels of Cr, Si, Fe, Al, Mg and C. Furthermore, the mineral composition of the coarse particulate dusts is mostly in the form of fine chromite particles, carbon, silica and anorthite ($(Ca, Na)(Si, Al)_4O_8$) phases (Ma and Garbers-Craig, 2006a,c). On the other hand, the fine particulate dusts are mainly associated with silica, NaCl, ZnO and Mg_2SiO_4 phases and are particularly enriched in Si, Zn, Na, K, Mg, S and Cl species (Ma and Garbers-Craig, 2006a,c). Furthermore, Ma and Garbers-Craig (2006c) also observed the presence of small amounts of complex hydrates, in the form of $NaZn_4(SO_4)Cl(OH)_6 \cdot 6H_2O$ and $Zn_4SO_4(OH)_6 \cdot 5H_2O$, from the fine-dust samples.

The findings reported by Ma and Garbers-Craig (2006a,c) were also supported by Sedumedi et al. (2009) based on observations on smelter dust emitted from a typical South African producer. Based on extensive x-ray diffraction analyses, Sedumedi et al. (2009) reported the major components in the smelter dust to include the chromite ($FeO \cdot Cr_2O_3$), elemental Cr, forsterite (Mg_2SiO_4), halite (NaCl), periclase (MgO), quartz (SiO_2) and thenardite (Na_2SO_4) phases. The comparative findings by Ma and Garbers-Craig (2006a) on the mineralogical compositions of coarse and fine particulate dusts are summarized in Table 5.3.

5.3.6.3.2 Formation Hexavalent Chromium in Smelter Particulate Dusts

The general mechanism of formation of electric furnace dusts in the melting of carbon and stainless steels was discussed in Chapter 4. As discussed, the formation of particulate dusts in scrap-based electric arc furnace smelting is strongly related to the process and furnace

TABLE 5.3

Typical Mineralogical Compositions of Ferrochrome Coarse and Fine Dusts

Element	Coarse Dust	Fine Dust
Cr	Chromite (Fe,Mg)(Al,Fe,Cr)$_2$O$_4$	Chromite and FeCr metal
Cr (VI)	–	CrO_3, (K,Na)$_2$Cr$_2$O$_7$ or (K,Na)$_2$CrO$_4$
Si	Quartz (SiO_2) and anorthite (Ca,Na)(Si,Al)O$_8$	Quartz (SiO_2), forsterite (Mg_2SiO_4), (Mg,Fe,Mn,Cr)$_3$O$_4$ and $MgCr_2O_4$, and aluminium silicate ($Mg_3Al_2Si_3O_{12}$ and Al_2SiO_5)
Al	Chromite and anorthite (Ca,Na)(Si,Al)O$_8$	Chromite
Ca	Dolomite (Ca,Mg)(CO$_3$)$_2$ and anorthite	–
Zn	–	Zincite (ZnO), $NaZn_4(SO_4)Cl(OH)_6 \cdot 6H_2O$ and $Zn_4SO_4(OH)_6 \cdot 5H_2O$
Fe	Chromite and hematite	Chromite and FeCr metal
Mg	Dolomite and chromite	Chromite, forsterite, MgO and aluminium silicate ($Mg_3Al_2Si_3O_{12}$)
S	–	$NaZn_4(SO_4)Cl(OH)_6 \cdot 6H_2O$ and $Zn_4SO_4(OH)_6 \cdot 5H_2O$
Cl	–	Halite and $NaZn_4(SO_4)Cl(OH)_6 \cdot 6H_2O$
K	–	Halite and $NaZn_4(SO_4)Cl(OH)_6 \cdot 6H_2O$
C	Coal, coke and charcoal	Coal, coke and charcoal

Source: Ma, G. and Garbers-Craig, A.M., *J. South. Afr. Inst. Min. Metal.*, 106, 753, 2006a.

operation steps, and these include (Guézennec et al., 2005; Ma and Garbers-Craig, 2006a,b,c) (1) furnace charging, (2) melting, (3) refining, (4) slag foaming and (5) casting or tapping of alloy and slag. Despite some slight differences in the process conditions, the mechanism of formation of ferrochromium smelter dusts is somewhat similar to that in the EAF process.

The mechanism of formation of smelter dusts in semi-closed submerged arc furnaces was investigated by Ma and Garbers-Craig (2006a). Basically, the formation of smelter dusts was proposed to involve any one of the following mechanistic steps (Ma and Garbers-Craig, 2006a): (1) vaporization of elements or compounds from high temperature zones in the SAF, particularly halite (NaCl), cristobalite (SiO_2), periclase (MgO) and ZnO, as oxidation products of oxides that are fumed through reduction reactions with carbon or carbon monoxide, (2) ejection of slag and metal through the holes of electrodes, (3) direct fly-off of solid charge materials such as fine carbon-bearing particles and quartz and chromite particles that are captured in the off-gas and (4) phases and reaction products that form in the off-gas duct from species in the off-gas. Ma and Garbers-Craig (2006a) proposed the reaction products formed under the fourth mechanistic step to include anorthite ((Ca,Na)(Si,Al)O$_8$), aluminium silicates ($Mg_3Al_2Si_3O_{12}$ and Al_2SiO_5), forsterite (Mg_2SiO_4) and complex zinc-alkali hydrates such as $NaZn_4(SO_4)Cl(OH)_6 \cdot$6H2O and $Zn_4SO_4(OH)_6 \cdot 5H_2O$.

In particular Ma and Garbers-Craig (2006a,b,c) proposed that the fine smelter dusts can contain hexavalent chromium in the form of CrO_3, (K,Na)$_2$Cr$_2$O$_7$ or (K,Na)$_2$CrO$_4$ compounds. Since the conditions in the submerged arc furnace are highly reducing, with oxygen partial pressures typically in the range 10^{-8} atm, it is generally postulated that Cr (VI) is generated from the oxidation of Cr (III) when the dust is removed from the reducing furnace atmosphere to the cyclone separators and the baghouse filters where the oxygen potential is generally higher (Eksteen et al., 2002; Ma and Garbers-Craig, 2006a). In essence, it is proposed that the Cr (VI) in the dust can be formed by either the oxidation reaction as a result of increased partial pressure of oxygen or through reaction with alkali oxides in the dust according to the following reactions (Ma and Garbers-Craig, 2006a; Sedumedi et al., 2009):

$$2Cr_2O_{3(s)} + 3O_2 \rightarrow 4CrO_{3(s)} \tag{5.17}$$

$$2FeCr_2O_{4(s)} + 4Na_2CO_3 + 7O_{2(g)} \rightarrow 4Na_2CrO_{4(s)} + Fe_2O_{3(s)} + 4CO_2 \tag{5.18}$$

$$2FeCr_2O_{4(s)} + 4K_2CO_3 + 7O_{2(g)} \rightarrow 4K_2CrO_{4(s)} + Fe_2O_{3(s)} + 4CO_2 \tag{5.19}$$

$$4CaO_{(s)} + 3O_2 + 2Cr_2O_{3(s)} \rightarrow 4CaCrO_{4(s)} \tag{5.20}$$

Based on the reactions of Cr (VI) formation shown in Equations 5.17 through 5.20, it is important to note that controlling the oxygen potential and/or the presence of alkalis in feed in the submerged arc furnace operation can be an important operational strategy in controlling the formation of Cr (VI) (Ma and Garbers-Craig, 2006a; Sedumedi et al., 2009; Beukes et al., 2010; du Preez et al., 2015). Technically, furnace design also plays a key role in the generation of Cr (VI) in submerged arc furnace smelting (Person, 1971; Beukes et al., 2010; Nelson, 2014). For example, furnaces are more likely to generate less amounts of Cr (VI) than semi-open and open furnaces, mainly because the open and semi-closed furnaces have a partially oxidizing environment due to ambient air entrainment below the furnace roof (Person, 1971; Riekkola-Vanhanen, 1999; Ma and Garbers-Craig, 2006a; Beukes et al., 2010; Basson and Daavittila, 2013; Nelson, 2014).

The standard practice by most producers of ferrochromium alloys globally is to clean the furnace off-gases using venturi scrubbers before flaring the CO-rich off-gas (Niemelä et al., 2004; Ma and Garbers-Craig, 2006a; Sedumedi et al., 2009; Beukes et al., 2010; Basson and Daavittila, 2013; du Preez et al., 2015). However, the venturi scrubbers are capable of removing over 99% of particulate matter in the off-gas but very fine particles can remain entrained in the cleaned furnace off-gas, and as such, the entrainment of such particles can be problematic in controlling disperse particulate dust emissions from the operation of submerged arc furnaces (Niemelä et al., 2004; Ma and Garbers-Craig, 2006a; Sedumedi et al., 2009; Beukes et al., 2010; du Preez et al., 2015). Technically, dust particles smaller than 1 μm are theoretically difficult to remove from the off-gas with wet scrubbers (Niemelä et al., 2004; Beukes et al., 2010). This means that some Cr-containing particulate matter can also pass into the CO-rich off-gas stack, where they are oxidized during flaring to form Cr (VI) and are eventually dispersed into the surrounding environment (Sedumedi et al., 2009; Beukes et al., 2010; Du Preez et al., 2015). Du Preez et al. (2015) conducted a detailed parametric study on the generation of Cr (VI) during the flaring of CO-rich off-gas from closed ferrochromium submerged arc furnaces. In this study, the authors proposed a groundbreaking emission conversion factor to determine the amount of Cr (VI) levels to which communities surrounding smelters will be exposed, but further research is warranted to validate their findings. Nevertheless, the findings by Du Preez et al. (2015) provide the basis for holistic determination of latent hazards from fallout fine dusts in communities in which ferrochromium smelters are located.

Generally, a lot of figures on the amount and volumes of particulate dust emissions have been thrown around in open literature. However, such non-standardized figures can be misleading to generalize to most operations globally, due to the fact that the volume and composition of the off-gases emitted from discrete submerged arc furnace smelting operations depend not only on the furnace raw materials but also on the raw material pretreatment methods, the design of the furnace, the furnace controls and the metallurgical conditions of the process (Person, 1971; Beukes et al., 2010; Basson and Daavittila, 2013; Nelson, 2014). Basically, the nature and quantities of particulate dust emissions is driven by fundamental factors such as (Person, 1971; Beukes et al., 2010; Basson and Daavittila, 2013; Nelson, 2014): (1) type of furnace, that is, open, semi-closed or closed; (2) type of process, that is, whether continuous or batch processing, with batch processing significantly affecting emissions due to sudden temperature changes as the raw materials come into contact with hot furnace surfaces; (3) decrepitation and melting behaviour of raw materials, resulting in non-uniform descent of charge and hence gas channelling and by-passing; (4) operating techniques, such as operating at variable loads or with insufficient electrode submergence and (5) selection, design and maintenance practices of dust collection systems. According to Nelson (2014), the adoption of closed furnaces is most likely to dominate the ferroalloys industry due to a variety of reasons: (1) improved environmental capture due to less vapour and dust losses; (2) lower off-gas temperature and hence slightly improved specific energy consumption; (3) less infiltrate air requirements for combustion and cooling, hence less direct energy losses through smaller volume of off-gases and reduced energy consumption for operating downstream off-gas handling systems and (4) less dilution of the combustible chemical energy potential, in the form of CO and H_2, suitable for energy recovery and cogeneration.

Based on the aforementioned discussions, a holistic approach in the reduction of particulate emissions from submerged arc furnace operations should incorporate some of the following core strategies proposed by several researchers: (1) maximum consideration on mandating the use of closed furnaces in smelting (Person, 1971; Beukes et al., 2010; Basson and Daavittila, 2013; Gasik, 2013; Nelson, 2014); (2) minimizing as much as possible the direct charging of unagglomerated fines in smelting (Guèzennec et al., 2005; Ma and Garbers-Craig, 2006a); (3) reducing the amount of particulate matter emissions through the use of properly operating wet scrubbing systems (Niemelä et al., 2004; Beukes et al., 2010; Basson and Daavittila, 2013; Gasik, 2013; European Commission, 2014); (4) proper control of metallurgical process parameters, such as slag basicity and viscosity, temperature and the partial pressure of oxygen, which are critical in the potential generation of Cr (VI) both in the dusts and in process slags (Riekkola-Vanhanen, 1999; Beukes et al., 2010; Gasik, 2013); (5) complementing the current wet scrubber technologies currently in operation with more effective (though more expensive) technologies such as sintered plate filters (Niemelä et al., 2004; Beukes et al., 2010); (6) process-integrated recycling of particulate dusts (Niemelä et al., 2004; Ma and Garbers-Craig, 2006a; Beukes et al., 2010; European Commission, 2014) and (7) reducing the volumes of flared CO-rich off-gas by maximizing the reutilization of the off-gas within the process as a way of reducing the potential generation of Cr (VI) species during gas flaring (Niemelä et al., 2004; Beukes et al., 2010; du Preez et al., 2015). Actually, most of these proposed operational strategies have multiplier effects and latent benefits to the overall profitability of ferrochromium operations. As discussed in Section 5.2.4, consciously integrating the health and environmental considerations within the new project planning stages of ferrochromium smelters is fundamentally important in managing the plethora of challenges posed by the particulate dust emissions from these smelters.

5.3.7 Recycling and Utilization of Ferrochromium Smelter Particulate Dusts

As discussed in Section 5.3.6, the ferrochromium alloys industry produces particulate wastes that contains significant amounts of potentially hazardous chromium compounds. For example, investigations by Sedumedi et al. (2009) on dust samples obtained from a

South African–based smelter revealed the presence of significant amounts of Cr (VI), around 7800 μg/g, which are significantly higher than the maximum acceptable risk concentration of 20 μg/g allowed for waste disposal. The presence of both Cr^{3+} and Cr^{6+} species in ferrochromium smelter dusts was also confirmed in earlier studies by Cox et al. (1985). If untreated, the stockpiled or landfilled particulate dusts can result in serious anthropogenic problems, which are particularly exacerbated by the toxicity of the easily soluble and highly mobile hexavalent chromium species.

The following sections briefly discuss some of the potential approaches applicable in mitigating the anthropogenic effects of ferrochromium smelter dusts through recycling and reuse to reduce the amount of stockpiled material. Furthermore, the solidification and/or stabilization of the smelter dusts using cementitious and pozzolanic systems are also discussed.

5.3.7.1 Metal Recovery and Recycling of Ferrochromium Smelter Dusts

The potential for the partial or full recycling of ferrochromium smelter dusts was highlighted in several studies (Language et al., 2000; McClelland and Metius, 2003; von Scheele, 2004; Denton et al., 2005; Ma and Garbers-Craig, 2006a; Rao et al., 2010; Angadi et al., 2011; Basson and Daavittila, 2013). Furthermore, the potential of recovering the chromium values by physical beneficiation techniques such as gravity, flotation and magnetic separation has been proposed by several researchers (Ma and Garbers-Craig, 2006a; Rao et al., 2010; Angadi et al., 2011). However, the commercial applications of these physical beneficiation methods for recovering metal values from ferrochromium dusts are somewhat doubtful due to the fine particulate nature of the waste materials. As a result, extensive research has been conducted to increase the valorization potential by the direct recycling of these materials into the mainstream process (Riekkola-Vanhanen, 1999; Denton et al., 2005; Ma and Garbers-Craig, 2006a; Basson and Daavittila, 2013; European Commission, 2014).

In general, the particulate dusts can be recycled back into the process using agglomeration processes such as sintering, briquetting or pelletizing, or through the pneumatic direct injection of fine materials into the mainstream or separate melting processes (Riekkola-Vanhanen, 1999; Denton et al., 2005; Ma and Garbers-Craig, 2006a; Basson and Daavittila, 2013; European Commission, 2014). As discussed in Chapter 4, there are process limitations to the direct injection of dust materials into the melting units, mainly due to the high concentration and in-process build-up of volatile metal species. As a result, several processes utilizing the reduction of oxidic components, the removal of volatile metal species and the melting and separation of alloy to produce a barren slag product have been commercially tested to recover values from ferrochromium alloy dusts (Barozzi, 1997; McClelland and Metius, 2003; von Scheele, 2004; Denton et al., 2005; Kesseler and Erdmann, 2007; Tateishi et al., 2008; Haga et al., 2012).

Some of the high-temperature processes developed specifically to recover value metals from fine particulate dust materials were discussed in Chapter 4. Technically, some of these processes have commercially proven capabilities in processing fine particulate waste materials produced in the steelmaking processes, and have also been tested in the processing of ferrochromium smelter dusts. In particular, the successful application of some rotary hearth-based technologies (McClelland and Metius, 2003; Tateishi et al., 2008; Haga et al., 2012; Gao et al., 2012a), bath smelting reduction processes (von Scheele, 2004; Kesseler and Erdmann, 2007; Degel et al., 2015) and plasma-based smelting processes (Denton et al., 2005) in the recovery of chromium-rich alloy and producing a barren slag product during the recycling processing ferrochromium smelter dusts have been commercially demonstrated.

5.3.7.2 Stabilization and/or Solidification of Ferrochromium Smelter Dusts

In essence compounds containing Cr (VI) are believed to be responsible for most of the health problems associated with chromium compounds (Cox et al., 1985; Rai et al., 1989; Richard and Bourg, 1991; Barnhart, 1997a,b; Shanker et al., 2005; Ma and Garbers-Craig, 2006a,b; Oliveira, 2012). The focus of most of these studies has been on evaluating the toxicity effects of chromium compounds based on concentration, biogeochemical speciation and bioavailability. In particular, the speciation of chromium compounds, in the form of Cr^{3+} and Cr^{6+} species, in ferrochromium smelter dusts was confirmed in several studies (Person, 1971; Cox et al., 1985; Ma and Garbers-Craig, 2006a,b,c; Sedumedi et al., 2009; Beukes et al., 2010; du Preez et al., 2015). Although in theory, the chromium values can be recovered from ferrochromium smelter dusts and most of the dust can be recycled, the potential recovery processes and recycling processes may not be universally adopted for technical and economic reasons. In addition, some of the dust recycling processes may require additional investment in terms of new equipment and processes. As a result, other alternative approaches to dealing with ferrochromium smelter dusts, such as stabilization and solidification processes using cementitious and pozzolanic materials, have been extensively explored (Wiles, 1987; Trussell and Spence, 1994; Glasser, 1997; Conner and Hoeffner, 1998; Johnson, 2004; Batchelor, 2006; Malviya and Chaudhary, 2006; Zhou et al., 2006; Chen et al., 2009).

The fundamentals of stabilization and/or solidification processes for hazardous and toxic wastes were discussed in Chapter 4. Furthermore, the general aspects of solidification and stabilization of hazardous and toxic wastes using cementitious and pozzolanic systems were extensively highlighted by several researchers (Wiles, 1987; Trussell and Spence, 1994; Glasser, 1997; Conner and Hoeffner, 1998; Johnson, 2004; Batchelor, 2006; Malviya and Chaudhary, 2006; Zhou et al., 2006; Chen et al., 2009). In particular, extensive research focused on the stabilization and solidification of chromium-containing metallurgical wastes using cementitious and pozzolanic systems was conducted by several researchers (Kindness et al., 1994; Cohen et al., 1997; Cohen and Petrie, 1997, 2005; Rodríguez-Pinero et al., 1998; Laforest and Duchesne, 2006; Bulut et al., 2009; Ma and Garbers-Craig, 2009; Dhal et al., 2013). Generally, the performance of solidification and stabilization systems can be assessed predominantly based on (1) the characteristic leaching behaviour; (2) the permeability of the solidified monolith, which in essence will dictate the extent to which the fluid permeates through the waste form and access sites at which the contaminants are held and (3) the structural integrity of the monolithic mass.

Basically, cementious and pozzolanic materials are capable of effectively immobilizing the chromium species present in the smelter dusts (Kindness et al., 1994; Cohen et al., 1997; Cohen and Petrie, 1997, 2005; Rodríguez-Pinero et al., 1998; Laforest and Duchesne, 2006; Bulut et al., 2009; Ma and Garbers-Craig, 2009; Dhal et al., 2013). Because the ability of cement matrices in immobilizing the chromium-containing wastes depends on the oxidation state of the chromium species, the cementious materials generally provide a moderately oxidizing environment but the addition of reductants such as ferrous iron to has been reported to promote the reduction of the Cr (VI) contaminants for immobilization (Bulut et al., 2009). The addition of blast furnace slag cement blends has also been reported to provide the reducing conditions, particularly by releasing sulphides and other reduced sulphur compounds as well as significant quantities of the chemically reductive hydrated aluminates (Kindness et al., 1994; Rodríguez-Pinero et al., 1998; Bulut et al., 2009). Dhal et al. (2013) also proposed the adoption of bioremediation as an emerging tool to address the speciation of Cr (VI), potentially by utilizing the microbial reduction of Cr (VI) to Cr (III), in various chromium-containing waste materials.

Based on toxicity and leaching characteristic properties, the key findings from the extensive research on the solidification and stabilization of chromium-containing wastes can be summarized as follows (Kindness et al., 1994; Cohen and Petrie, 1997; Laforest and Duchesne, 2006; Bulut et al., 2009): (1) containment of metal species is dependent on the bonding structures within the cement matrix, and as such, there exists a relationship between the strength of solidification/stabilization composite and the leaching potential from the composite, (2) Cr (III) is stabilized either by chemical incorporation into the cement products or as solubilized or precipitated product in the pore solution, (3) the solubility of chromium and/or chromium-bearing phases is specially influenced by pH and (4) the Cr (VI) is soluble over the entire pH range.

5.3.8 Physicochemical Characteristics of Ferrochromium Alloy Slags

Process slags play a vital role in the submerged arc smelting of chromite ores (Eksteen et al., 2002; Jansson et al., 2002; Hayes, 2004; Holappa and Xiao, 2004; Gasik, 2013; Eric, 2014). Generally, ferrochromium slags act as a reactive media in the smelting reduction process, where the chromium oxides are reduced carbothermically in the liquid state to produce metallic chromium and ferrochromium alloys (Hino et al., 1998; Eksteen et al., 2002; Simbi and Tsomondo, 2002; Weber and Eric, 2006; Eric and Demir, 2014). According to Jansson et al. (2002), slag composition and properties have a significant effect on the reduction rate of chromite in slag.

To date, the liquid-state reduction of chromite ores has been studied extensively (Eksteen et al., 2002; Simbi and Tsomondo, 2002; Hayes, 2004; Sano, 2004; Weber and Eric, 2006; Liu et al., 2012; Eric and Demir, 2014). Simplistically, chromium oxides dissolve in the process slag until the point of saturation and where react with carbon in coke to form metallic droplets in slag (Simbi and Tsomondo, 2002). The metal droplets eventually coalesce and diffuse out of the slag to collect as liquid alloy melt in the furnace hearth (Maeda et al., 1981; Simbi and Tsomondo, 2002; Hayes, 2004; Liu et al., 2012). According to Hayes (2004), the key physicochemical phenomena involving the slag phase in the high-temperature zone of the submerged arc furnaces include (1) the dissolution of chromite spinel in slag, (2) the reduction of chromite in the slag phase to form alloy and (3) alloy/slag separation. Generally, the extent to which the chromite spinel is dissolved in the slag depends on the process conditions and slag characteristics, inter alia (Hayes, 2004): (1) process temperature, (2) slag composition, (3) slag volume, (4) oxygen partial pressure and (5) residence time in the high-temperature zone. Consequently, the slag properties have a significant effect on the smelting reduction process, and as such, slag control is of critical importance not only in maximizing the dissolution rate of chromite in the slag but also in enhancing the intensity of the liquid-state reduction process during the operation of submerged arc furnaces (Jansson et al., 2002; Simbi and Tsomondo, 2002; Hayes, 2004; Holappa and Xiao, 2004).

Depending on the type and quality of raw materials used, about 6 Mt of high-carbon ferrochromium slags are produced globally, corresponding to about 1.1–1.6 tonnes of slag per tonnes FeCr produced depending on the type of raw materials used (Niemelä and Kauppi, 2007). Generally, slags in the high-carbon ferrochromium alloy smelting processes belong to a multicomponent Al_2O_3–MgO–SiO_2 system, with minor components of CaO and CrO_x (Hayes, 2004; Holappa and Xiao, 2004; Reuter et al., 2004). Basically, the bulk compositions of high-carbon ferrochromium slags is in the range 26–30 wt.% SiO_2, 22–24 wt.% Al_2O_3, 24–26 wt.% MgO, 2–3 wt.% CaO, 2–4 wt.% FeO and 14–16 wt.% Cr_2O_3 (Harman and Rao, 2013; Niemelä and Kauppi, 2007). The composition of slag is an important control parameter that affects the partition of chromium between the alloy and slag phases (Rennie et al., 1972; Eric and Akyuzlu, 1990; Keene, 1995; Simbi and Tsomondo, 2002; Forsbacka and Holappa, 2004; Hayes, 2004; Holappa and Xiao, 2004; Niemelä and Kauppi, 2007).

Slag chemistry has a significant effect on the key functional properties of ferrochromium slags, and these include (Rennie et al., 1972; Muan, 1984; Eric and Akyuzlu, 1990; Keene, 1995; Riekkola-Vanhanen, 1999; Simbi and Tsomondo, 2002; Forsbacka and Holappa, 2004; Hayes, 2004; Holappa and Xiao, 2004; Niemelä and Kauppi, 2007; Mostafaee, 2011) (1) slag viscosity properties, which determine the physical separation and recovery of metal from slag; (2) slag electrical resistivity properties, which function to determine electrode penetration and power input into the furnace, with more resistive slags allowing for deeper penetration, higher-voltage operation and greater power inputs and (3) slag liquidus temperature, which is the key determinant of the operating temperatures and degree of superheat required to completely melt the slag. In general, the high-temperature characteristics of metallurgical slags have an effect on the process features such as foamability, chromium recovery and wear rate of refractory materials (Pretorius and Nunnington, 2002; Mostafaee, 2011). Furthermore, the metallurgical properties of the slags are strongly influenced by their high temperature microstructure, and the knowledge of pertaining to the chemical composition as well as the balance between the liquid and solid phases at the process temperatures is instrumental in developing a good slag practice (Mostafaee, 2011).

To date, the effect of key functional parameters of ferrochromium slags has been extensively documented (Kato and Minowa, 1969; Rennie et al., 1972; Eric and Akyuzlu, 1990; Keene, 1995; Forsbacka and Holappa, 2004; Hayes, 2004; Holappa and Xiao, 2004; Jahanshahi et al., 2004; Nakamoto et al., 2007). According to Holappa and Xiao (2004), slag liquidus temperature, viscosity and electrical conductivity represent some of the critical physicochemical properties of slags with the closest relation to the practical operation of a submerged arc furnaces for ferrochromium production. Based on the aforementioned, a good slag engineering practice involves the intentional addition of fluxes, in the form of quartz (SiO_2), lime (CaO) and dolomite ($MgO \cdot CaO$), as part of the submerged arc furnace feed in order to optimize the slag chemistry and properties during the smelting reduction of ferrochromium alloys (Rennie et al., 1972; Eric and Akyuzlu, 1990; Keene, 1995; Simbi and Tsomondo, 2002; Hayes, 2004; Holappa and Xiao, 2004; Gasik, 2013; Eric, 2014; Eric and Demir, 2014). Furthermore, the design of novel high temperature properties of the process slags is also vital in improving the post-process valorization potential and the environmental performance of such slags, particularly by affecting the speciation of entrained and dissolved chromium phases in the slag as well as the post-furnace solidification pathways and the resulting mineralogy of the solidified slags (Hayhurst, 1974; Tanskanen and Maknonen, 2006; Durinck et al., 2008; Albertsson, 2013; Panda et al., 2013; Albertsson et al., 2014). From an environmental point of view, understanding both the high temperature and the crystallization properties of ferrochromium slags is important in the development of crystal phases that affect the slag properties such the chromium distribution and stability within the slag, as well as the strength, refractoriness and bulk density of the barren slag product (Hayhurst, 1974; Onuma and Tohara, 1983; Zaldat et al., 1984; Barbieri et al., 1994; Albertsson, 2011, 2013; Samada et al., 2011; Jelkina et al., 2012; Li et al., 2013; Albertsson et al., 2014; Wu et al., 2014; Liao et al., 2016).

5.3.8.1 Mineralogical Phases of Ferrochromium Alloy Slags

Liquid slags are usually tapped from the submerged arc furnaces at around 1700°C and are either granulated using water jets or are allowed to cool slowly in slag yards (Niemelä and Kauppi, 2007). In general, solidified ferrochromium slags mostly consist of SiO_2, Al_2O_3 and MgO existing in different forms such as magnesia-spinel-forsterite-dicalcium silicate phases ($MgO–MgO \cdot Al_2O_3–2MgO \cdot SiO_2–2CaO \cdot SiO_2$) (Hayhurst, 1974; Zaldat et al., 1984; Tanskanen and Makkonen, 2006; Durinck et al., 2008; Albertsson, 2013). Basically, the mineralogical phases in ferrochromium slags are dependent on the chemical composition and the cooling characteristics and/or solidification pattern of the slag (Hayhurst, 1974; Niemelä and Kauppi, 2007; Tanskanen and Makkonen, 2006; Durinck et al., 2008; Albertsson, 2013; Li et al., 2013).

In general, solidified ferrochromium slags contain significant amounts of chromium and iron oxides in the form of partially altered chromites (PACs), and to some extent ferrochromium alloy and chromium carbides entrained in slag (Hayhurst, 1974; Hayes, 2004; Holappa and Xiao, 2004; Tanskanen and Makknonen, 2006; Durinck et al., 2008; Samada et al., 2011; Albertsson, 2013; Panda et al., 2013). In particular, some of the typical chromium-containing phases include (1) dissolved in slag phase as PACs (Hayes, 2004), (2) concomitant Cr_2O_3 and alloy phases dispersed in silicate phases (Holappa and Xiao, 2004; Panda et al., 2013) and (3) chromium bound in magnesiochromite ($MgO \cdot Cr_2O_3$) and/or chromite ($FeCr_2O_4$) spinel phases (Samada et al., 2011; Li et al., 2013). Earlier studies by Hayhurst (1974) also proposed the presence of spinel and forsterite phases as major phases, as well as chromium carbides (Cr_7C_3) and enstatite ($MgO \cdot SiO_2$) as minor phases in slow-cooled slags. In contrast, granulated ferrochromium slags have been reported to contain three major phases, namely (Niemelä and Kauppi, 2007; Panda et al., 2013), (1) amorphous glass phases, including those rich in magnesium and calcium silicates, (2) crystalline and zonal (Fe,Mg,Al,Cr) oxide spinel phases and (3) entrained metallic ferrochromium alloys. Furthermore, Tanskanen and Makkonen (2006) also confirmed the general phases of air-cooled industrial slags as typically containing partly crystalline and hypidiomorphic spinel $(Mg,Fe)(Fe,Al,Cr)_2O_4$, forsterite (Mg_2SiO_4) and pyroxene ($Mg_2(Cr,Al,Si)_2O_6$ crystals enclosed in a condensed and homogenous glass matrix.

5.3.9 Recycling and Reuse of Ferrochromium Alloy Slags

5.3.9.1 Factors Affecting Chromium Losses to Slag

As discussed in the preceding sections, chromium can be present in ferrochromium smelter slags as PACs dissolved in slag phase, concomitant Cr_2O_3 and alloy phases dispersed in silicate phases, and chromium bound in magnesiochromite ($MgO \cdot Cr_2O_3$) and/or chromite ($FeCr_2O_4$) spinel phases. Furthermore, the effect of slag chemistry variables such as liquidus temperature, slag viscosity and slag basicity on the partition of chromium between alloy and slag has also been highlighted. In addition to the effect of slag chemistry and liquidus temperature conditions, the chromium loss to slag can also be influenced by other factors such as the prevailing furnace atmosphere, mineralogical composition of the charge, and size and type of furnace (Holappa and Xiao, 2004). In essence, the effect of the specific thermochemical conditions on chromium recovery can be summarized as follows: (1) chromium loss to slag decreases with increasing slag basicity, and is particularly so at relatively low basicity after which the beneficial effects tend to level off when the basicity exceeds 1.4 (Muan, 1984; Eric and Akyuzlu, 1990; Akyuzlu and Eric, 1992; Holappa and Xiao, 2004), (2) the degree of superheat above the liquidus temperature of slag is important for the control of slag viscosity, and hence, the physical separation and recovery of the alloy from slag (Rekkola-Vanhanen, 1999), (3) the recovery of chromium is enhanced by the presence of Al_2O_3 (Holappa and Xiao, 2004), (4) the addition of MgO greatly enhances the recovery of chromium in melts with a basicity ratio greater than 0.80 (Holappa and Xiao, 2004) and (5) the detrimental effects of adding lime are twofold—(a) it increases the slag volume thereby causing extra losses, and (b) it raises the electrical conductivity of slags, which is detrimental to furnace operation (Holappa and Xiao, 2004).

As discussed in Section 5.3.8, the production process of ferrochromium alloys produces a discard slag which contains a significant amount of entrained chromium alloy, and the degree of entrainment is particularly exacerbated by poorly optimized superheat temperature and slag viscosity conditions during the smelting process (Sripriya and Murty, 2005). On average, the ferrochrome industry in South Africa generates between 1.1 and 1.9 tonnes of slags per ton FeCr, against the minimum regulatory requirement for disposal rate of 585 tonnes/ha/month, and this creates regulatory pressure to minimize the amount of waste produced during the various production processes (Department of Water Affairs and Forestry, 1998; Beukes et al., 2010; Biermann et al., 2012).

In general, ferrochromium slags are classified as hazardous waste materials in most jurisdictions due to the presence of potentially toxic chromium metal species such as Cr (VI) (Maeda et al., 1981; Cox et al., 1985; Rai et al., 1989; Richard and Bourg, 1991; Barnhart, 1997a,b; Rekkola-Vanhanen, 1999; Shen and Forssberg, 2003; Shanker et al., 2005). As a result, the benefits derived from recovering the entrained chromium from ferrochromium slags and utilizing the barren slag product in alternative industries are twofold, that is, increasing the environmental sustainability and resource utilization efficiency (Maeda et al., 1981; Visser and Barret, 1992; Mashanyare and Guest, 1997; Sripriya and Murty, 2005; Niemelä and Kauppi, 2007; Harman and Rao, 2013; European Commission, 2014; Bai et al., 2015).

The recovery of entrained chromium alloy from ferrochromium slags is discussed in the following sections. Furthermore, the stabilization of chromium in slag through engineered slag properties, and the potential application of ferrochromium slags in the construction industry are also discussed. However, the stabilization and solidification of chromium-containing wastes using cementitious and pozzolanic systems were discussed in Sections 4.6.7 and 5.3.7.2 and, as such, will not be discussed further.

5.3.9.2 Metal Recovery from Ferrochromium Smelter Slags

Some of the chromium that is entrained in slag and eventually finds its way into slag dumps. Over the years, the slag dumps have been regarded as a potential source of ferrochromium alloys, and protracted efforts have been channelled towards improving the economics of reclaiming the entrained ferrochromium alloys from the slag dumps (Maeda et al., 1981; Visser and Barret, 1992; Mashanyare and Guest, 1997; Rekkola-Vanhanen, 1999; Shen and Forssberg, 2003; Sripriya and Murty, 2005; Niemelä and Kauppi, 2007; Harman and Rao, 2013; European Commission, 2014; Bai et al., 2015). Technically, the focus on the recovery processes for the chromium entrained in slags can be categorized into two broad generic approaches, namely, (1) chromium recovery using physical beneficiation processes (Visser and Barret, 1992; Das et al., 1997; Mashanyare and Guest, 1997; Sripriya and Murty, 2005; Bai et al., 2015) and (2) chromium recovery using high-temperature smelting processes (Maeda et al., 1981; Shibata et al., 2002; Shen and Forssberg, 2003; Sano, 2004; Harman and Rao, 2013).

To date, several physical beneficiation processes have been successfully applied to produce directly saleable products from the reclamation processes. Basically, the physical beneficiation processes commonly applied include coupling the size reduction processing to liberate the alloy inclusions entrained in slag or a combination of any of the following processes (Visser and Barret, 1992; Das et al., 1997; Mashanyare and Guest, 1997; Rekkola-Vanhanen, 1999; Sripriya and Murty, 2005; Bai et al., 2015): (1) coarse and fine jigging processes, (2) magnetic separation of the ferromagnetic alloy and chromite inclusions, (3) dense medium separation processes and (4) separation using spirals (van Reenen et al., 2004). In essence, the prerequisite for the selection of these dump reclamation processes is to recover as much of the directly saleable metal as possible from the slag while producing a barren and environmentally benign slag that can be reused as a construction material additive (Visser and Barret, 1992).

In addition to entrained chromium metal alloy, it is reported that chromium can also exist in slag as undissolved chromite phases, particularly rich in non-stoichiometric CrO_x

and Cr_2O_3 (Xiao and Holappa, 1993). According to Xiao and Holappa (1993), the behaviour of dissolved chromium in slags is complicated by the existence of multivalence chromium and the high melting point of chromium oxide–containing slags, which, in essence, complicate the valorization potential of chromium-containing slags.

Technically, the recovery of chromium dissolved in slags by some of the physical beneficiation methods cited above is hampered by the complex behaviour of chromium-bearing phases in slags. As such, the recovery of dissolved alloy and chromite inclusions, and the resultant production of an environmentally benign barren slag product, using high-temperature processes have been considered as potential solutions to the valorization of chromium-bearing slags (Maeda et al., 1981; Shibata et al., 2002; Shen and Forssberg, 2003; Sano, 2004; Harman, 2013). In principle, the chromium in slag can be reduced carbothermically and recovered as chromium carbides or Fe–Cr–C alloy (Maeda et al., 1981; Shibata et al., 2002; Shen and Forssberg, 2003; Sano, 2004; Harman and Rao, 2013). Furthermore, the thermodynamics and reduction behaviour in the recovery of chromium from slags using different types of reductants have also been extensively investigated (Maeda et al., 1981, Shibata et al., 2002; Sano, 2004). For example, Sano (2004) extensively investigated the thermodynamic and reduction behaviour of the Cr_2O_3 in slags with carbon, aluminium and silicon reductants. Thermodynamically, aluminium would be the best reductant but its application tends to be limited by the high costs of metallic aluminium (Sano, 2004). As a result, process economics would greatly be improved by the use of carbon-saturated iron melts with small additions of silicon as ferrosilicon alloys (Sano, 2004).

5.3.9.3 Stabilization of Chromium Using Engineered Slag Properties

As discussed earlier, chromium in slags is present in slags in different oxidation states, particularly as entrained elemental chromium, and as dissolved, chromium species. If not stabilized, the chromium in slag can oxidize to form the environmentally potent hexavalent state upon disposal in landfills (Tanskanen and Makkonen, 2006; Durinck et al., 2008; Albertsson, 2013; Panda et al., 2013; Albertsson et al., 2014; Liao et al., 2016). Deliberate attempts to stabilize the leaching behaviour of chromium dissolved in slags by engineering the slag solidification behaviour have been conducted by several researchers. Of particular interest is the research focused on (1) deliberately stabilizing the chromium in the spinel crystal structure during the slag solidification stage (Hayhurst, 1974; Kilau and Shah, 1984; Barbieri et al., 1994; Tanskanen and Makkonen, 2006; Durinck et al., 2008; Samada et al., 2011; Albertsson, 2013; Li et al., 2013; Wu et al., 2014), (2) stabilization in glassy matrix and embedded crystalline phases (Liao et al., 2016) and, (3) stabilization using cementitious and pozzolanic systems (Panda et al., 2013).

As discussed in Section 5.3.8, understanding the mineralogical characteristics of slags is critical in developing the valorization potential of ferrochromium alloy slags. In essence, the leaching of chromium from stainless steel or ferrochromium slags can be controlled by the presence of spinel phases which bind the chromium in a stable structure (Kuhn and Mudersbach, 2004; Tanskanen and Makkonen, 2006; Durinck et al., 2007; Durinck et al., 2008; Albertsson et al., 2014; Wu et al., 2014). In recent years, the research on approaches to stabilize the chromium in these stable spinel phases has received widespread attention. To date, several studies have been conducted on the stabilization of chromium in various industrial slags systems such as $CaO–MgO–Al_2O_3–Fe_2O_3$ (Wu et al., 2014), $CaO–MgO–SiO_2–Cr_2O_3$ (Albertsson et al., 2014), $CaO–SiO_2–MgO–Al_2O_3–Cr_2O_3$ (Li et al., 2012), $CaO–MgO–Al_2O_3–SiO_2$ (Barbieri et al., 1994) and $CaO–SiO_2–Al_2O_3–MgO–Fe–Cr$ (Tanskanen and Makkonen, 2006; Durinck et al., 2008). These studies have proved the efficacy of stabilizing the different chromium compounds present in slag in stable spinel phases, particularly in the form of (Mg,Fe)(Al,Cr)$_2$O$_4$ spinels. In particular, Albertsson (2013) investigated the phase relationships and chromium partition in industrial and synthetic stainless steel refining slags with a view of stabilizing the chromium in stable spinel phases. Based on slag parameters such as slag basicity, slag heat treatment, oxygen partial pressure and alumina addition, the findings from this study can be summarized as follows: (1) slow cooling of slag and soaking at low temperatures, and oxygen partial pressure improved the spinel phase precipitation and reduced the chromium dissolved in the water-soluble matrix phases; (2) the heat treatment of slags with basicity ratio (CaO/SiO_2) greater than 1.4 at high partial pressure of oxygen resulted in potentially leachable chromium-containing solid solutions and (3) the addition of alumina to molten slag was effective in controlling the chromium partition between the spinel, slag matrix and metal phases at low oxygen partial pressure, mostly due to the amount of spinel phase solid solution $MgAl_2O_4$–$MgCr_2O_4$ increasing with increase in the alumina content.

Figure 5.5 shows the phase and mineralogical composition of air-cooled ferrochromium slags (Tanskanen and Makkonen, 2006). Based on the extensive studies on the

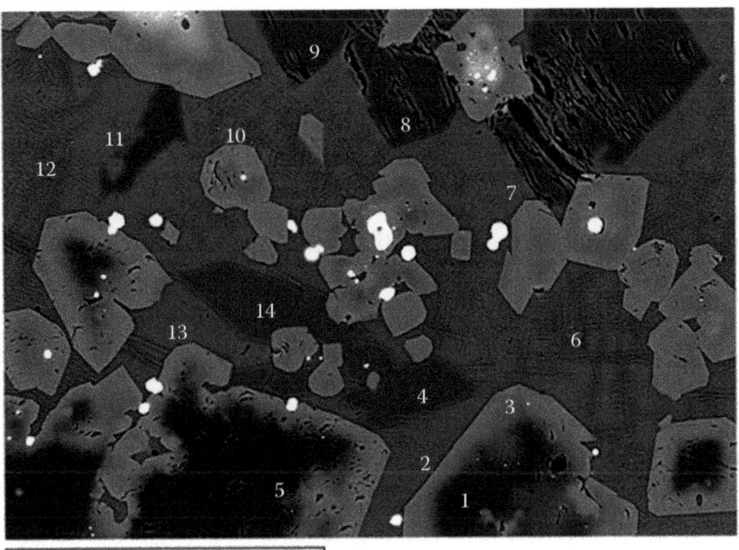

Point	Phase/zone	O	Mg	Al	Si	K	Ca	Cr	Fe
2	Glass phase	47.51	7.02	10.08	28.69	0.98	3.61	1.11	
3	Mg–Cr–Al–Spinel	40.08	15.59	27.59				15.82	0.94
4	Pyroxene	45.79	21.94	3.09	26.11			2.04	0.50
5	Mg–Al–Spinel	43.67	17.33	38.53					0.47
7	Glass phase	47.13	6.84	10.18	28.80	0.91	3.54	1.57	
8	Fosterite	43.10	34.01		20.30			1.60	0.99
13	Glass phase	47.46	5.70	11.11	28.42	1.20	3.73	1.28	
14	Pyroxene	45.50	21.62	3.46	26.05			2.24	0.53

FIGURE 5.5
Phase composition and mineralogy of air-cooled ferrochromium slags. (Adapted from Tanskanen, P. and Makkonen, H., *Glob. Slag Mag.*, 5, 16, 2006.)

mineralogical and physicochemical characterization of ferrochromium slags (containing approx. 8 wt.% Cr), Tanskanen and Makkonen (2006) proposed the existence of the chromium-deficient phases (in the form of glassy, forsterite and Mg–Al spinel phases) as well as chromium-rich Mg–Al–Cr spinel phases (Mg $(Al,Cr)_2O_4$), with the (Mg $(Al,Cr)_2O_4$) being the main binding and stabilization phases for the chromium in the slag (Tanskanen and Makkonen, 2006). Based on the findings from these studies, the additional protection against chromium leaching into the environment can further be augmented by manipulating the crystal structure of the slag phases by controlling the slag solidification pathways to further encapsulate the Cr-rich spinel grains into an amorphous glass phase (Kühn and Mudersbach, 2004; Tanskanen and Makkonen, 2006; Durinck et al., 2008).

The relationship between chromium leaching and the concentration of spinel forming oxides in the slag is given by the empirical factor sp, which relates the overall slag composition to the chromium leaching (Kühn and Mudersbach, 2004; Tanskanen and Makkonen, 2006; Durinck et al., 2007, 2008; Albertsson et al., 2014):

$$\text{Factor sp} = 0.2\text{MgO} + 1.0\text{Al}_2\text{O}_3 + \text{nFeO}_x - 0.5\text{Cr}_2\text{O}_3 \quad (5.21)$$

From Equation 5.21, n is a number between 1 and 4, and is dependent on the oxidation state of the slag. The leaching behaviour of chromium as a function of the factor sp is shown in Figure 5.6. As shown in Figure 5.6, when factor sp is below 5 wt.%, a high chromium leaching was observed, whereas the leaching chromium was low at factor sp above 5 (Kühn and Mudersbach, 2004; Durinck et al., 2008).

The extensive studies highlighted in this section have proved the synergist relationship of engineered slag properties in meeting both the metallurgical process and environmental considerations. In other words, the optimization of functional properties of solidified slag

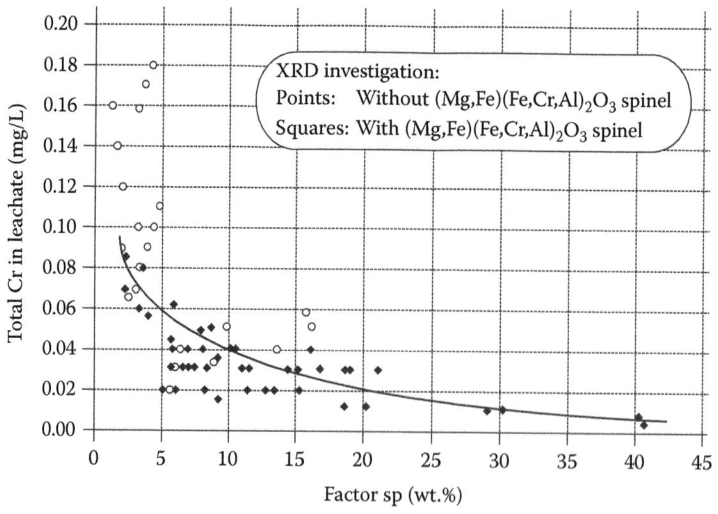

FIGURE 5.6
Chromium leaching as a function of factor sp. (From Kühn, M. and Mudersbach, D., Treatment of liquid EAF slag from stainless steelmaking to produce environmental friendly construction materials, *Proceedings of the Second International Conference in Iron and Steelmaking (SCANMET II)*, 2004, p. 369; Reprinted from *Resour., Conserv. Recycl.*, 52, Durinck, D., Engstrom, F., Arnout, S., Huelens, J., Jones, P.T., Bjorkman, B., Blanpain, B., and Wollants, P., Hot stage processing of metallurgical slags: Review, 1121–1131. Copyright 2008, with permission from Elsevier.)

for recycling and reuse purposes can essentially be engineered by manipulating the high-temperature properties as well as the solidification pathways (Tanskanen and Makkonen, 2006; Durinck et al., 2008; Albertsson et al., 2014; Wu et al., 2014).

5.3.9.4 Reuse of Ferrochromium Smelter Slags as Construction Materials

The valorization potential use of ferrochromium slags as construction materials has also been explored in several studies (Lind et al., 2001; Pillay et al., 2003; Zelić, 2005; Niemelä and Kauppi, 2007; Panda et al., 2013; Kumar et al., 2014). Panda et al. (2013) proposed that the slags from the production of high-carbon ferrochromium alloys possess the requisite mechanical and engineering properties for utilization as concrete aggregate materials. In essence, the slow-cooled ferrochromium slags crystallize to give a stable $CaO–MgO–Al_2O_3–SiO_2$ structured product with mechanical properties comparable to that of basalt (Zelić, 2005). In fact, the ferrochromium slag was observed to possess superior mechanical properties and was particularly found to be suitable for high-strength concrete brands (Zelić, 2005).

Unlike granulated blast furnace slags that have widely been accepted as additives in the construction industry, the adoption of ferrochromium slags as construction materials is still at its infancy, mainly due to the sensitivity surrounding the toxicity and leaching behaviour of the entrained and dissolved chromium from slags (Pillay et al., 2003; Shen et al., 2004; Tanskanen and Makkonen, 2006; Durinck et al., 2007, 2008; Panda et al., 2013). As discussed in the preceding sections, ferrochromium slags typically contain significant amounts of residual chromium which can cause serious challenges if solubilized into the environment. As a result, the valorization potential of ferrochromium slags is particularly dependent on the ability to immobilize the chromium in stable crystal phases (Lind et al., 2001; Pillay et al., 2003; Kühn and Mudersbach, 2004; Zelić, 2005; Tanskanen and Makkonen, 2006; Durinck et al., 2007, 2008; Niemelä and Kauppi, 2007; Panda et al., 2013).

5.3.9.5 Summary

This section discussed the raw materials, chemistry and process requirements in the production of various grades of ferrochromium alloys. The different industrial uses of chromium and its alloys in different industries were also highlighted. In particularly, this section highlighted how the parameters in the production and utilization of chromium and its compounds interact with each other in influencing the anthropogenic effects of chromium. Due to the presence of potentially toxic multivalent chromium species, the majority of the waste materials, particularly the particulate dusts and process slags, produced in the production of various grades of ferrochromium alloys are classified as hazardous waste materials.

The different strategies and the best practices currently being adopted in mitigating the anthropogenic effects of chromium-production processes were also highlighted in this section. As such, consciously integrating the health and environmental considerations in the design and conception of processes is fundamentally important in mitigating the plethora of challenges posed by the waste streams in the production of ferrochromium alloys. Furthermore, the best practices and best available technologies adopted in mitigating the environmental effects in the production of ferrochromium alloys were highlighted. In particular, this section highlighted the typical challenges associated with the limited valorization potential of the ferrochromium production wastes.

5.4 Production of Manganese and Its Alloys

5.4.1 Chemical and Physical Properties of Manganese

Manganese (Mn) is a transition metal element with atomic weight 54.938, electronic structure [Ar] $3d^5 4s^2$ and a body-centred cubic lattice (Greenwood and Earnshaw, 2012; Lenntech, n.d.; Los Alamos, n.d.). The melting and boiling points of pure manganese metal are 1244°C and 2061°C, respectively (Lenntech, n.d.; Los Alamos, n.d.). Manganese is an allotropic metal whose crystal structure changes with temperature. Generally, the typical allotropic phases include (1) the alpha phase (α-Mn) which is stable at room temperature up to 700°C and exhibits a complex cubic structure, (2) the gamma phase (β-Mn) with a complex cubic crystal structure in the temperature range of 700°C–1088°C, (3) the gamma phase (γ-Mn) with face-centred cubic structure and stable in the range of 1088°C–1139°C and (4) the delta phase (δ-Mn) which is stable above 1139°C up to the melting point of 1246°C of manganese (Cardarelli, 2008; Tangstad, 2013a; Downing, 2013). Typically, manganese is a chemically reactive metal that is very similar to iron, only that it is lighter and harder than iron (Cardarelli, 2008).

Manganese exhibits various oxidation states, from Mn (III) to Mn (VII), with the divalent state being the most stable (Corathers and Machamer, 2006). In addition, manganese occurs in three valence states in nature, that is, the divalent (Mn^{2+}), trivalent (Mn^{3+}) and quadrivalent (Mn^{4+}) states (Corathers and Machamer, 2006). At high temperatures, manganese tends to react vigorously with oxygen, sulphur and phosphorus, and in essence, such properties have enabled the extensive use of manganese in ironmaking and steelmaking as a reducing, desulphurizing and dephosphorizing agent (Cardarelli, 2008; Gasik, 2013). Furthermore, manganese dissolves readily in acids with the evolution of hydrogen and the formation of manganous salts, an attribute that has been widely exploited in the production of a variety of manganese-based chemicals (Cardarelli, 2008).

Manganese is an essential nutrient for different microorganisms, plants and animals (Santos-Burgoa et al., 2001; Howe et al., 2004). Nevertheless, high concentrations of manganese and its compounds in the environment are undesirable as they are associated with toxicity in living organisms (Williams et al., 2012). Typically, manganese toxicity in humans is associated with a condition called manganism, whose symptoms include tremors, difficulty walking and facial muscle spasms (Williams et al., 2012). Manganese is also ubiquitous with the environment, but the major anthropogenic sources in the environment include municipal wastewater discharges, sewage sludge, mining and mineral processing, emissions from iron and steel production processes, combustion of fossil fuels and, to some extent, combustion of fuel additives (Howe et al., 2004). As a result, environmental pollution as a result of manganese contamination can largely be attributed to human activities.

5.4.2 Uses of Manganese and Its Alloys

5.4.2.1 Metallurgical Uses of Manganese as an Alloying Element

Manganese is an essential element in the production of iron and steel due to its desulphurizing, deoxidizing and alloying properties (Bristow et al., 2000; Cardarelli, 2008; Tangstad, 2013a). In general, manganese is added in steel in the form of manganese ferroalloys to improve the strength, hardenability and wear resistance properties of steel products (Akdogan and Eric, 1994). Particularly, manganese is extensively used in stainless steels in order to improve their hot ductility properties (Outokumpu, 2013). Basically, the manganese in steels acts as an austenite and carbide stabilizer and is undoubtedly one of the most prevalent alloying elements in various grades of steels (Bristow et al., 2000; Cardarelli, 2008; Outokumpu, 2013; Corathers, 2014). In particular, the effect of manganese on the ferrite/austenite balance varies with temperature, such that, at low temperature, manganese is an austenite stabilizer but tends to act as a ferrite stabilizer at higher temperatures (Outokumpu, 2013). Furthermore, manganese increases the solubility of nitrogen in steels and is thus used to obtain high nitrogen contents in duplex and austenitic stainless steels (Outokumpu, 2013).

The steel industry accounts for approximately 90% of the global demand for manganese, either in the form of ferromanganese or silicomanganese alloys, with the carbon steel market accounting for about 70% of manganese consumption within the steel industry itself (Zhang and Cheng, 2007a,b; Cardarelli, 2008; Tangstad, 2013a). Although manganese is not as powerful as nickel in stabilizing the austenite phase in steels, it is tactically used to reduce the cost of alloying elements used in the production of low-cost stainless steels, where it is applied to partially or entirely replace the nickel (Outokumpu, 2013). Finally, manganese is also used as an alloying element in the manufacture of important non-ferrous engineering materials such as aluminium and copper alloys (Cardarelli, 2008).

5.4.2.2 Uses of Manganese in Chemicals Industry

Manganese and its compounds have also found widespread applications in the chemicals industry, particularly in the manufacture of batteries, catalysts, water purifying agents, colourants and many other different chemicals (Post, 1999; Kholmogorov et al., 2000; Zhang and Cheng, 2007a,b; Cardarelli, 2008; Hwang et al., 2016). Manganese dioxide (MnO_2), in particular, is widely used as a cathodic depolarizer in dry cell batteries ($Zn/NH_4Cl/MnO_2$), alkaline batteries ($Zn/KOH/MnO_2$) and to some extent in lithium batteries (Li/MnO_2) (Zhang and Cheng, 2007a,b; Cardarelli, 2008). According to Cardarelli (2008), there are four different categories of the manganese dioxide feedstocks used in the manufacture of batteries, and these include, (1) conventional or natural manganese dioxide used in the production of dry cells; (2) activated manganese dioxide obtained from the roasting of high-grade oxidic manganese ores at 600°C and treating the roasted calcine with sulphuric acid; (3) chemical manganese dioxide, which is typically an alkali-activated metal manganite and (4) electrochemical manganese dioxide produced from the electrolysis of a manganous sulphate electrolyte. In addition, manganese-based chemicals such as potassium permanganate are powerful oxidizing agents with bacterial and algicidal properties and have since been extensively used in purifying drinking water and deodorizing pungent industrial waste discharges (Cardarelli, 2008).

5.4.3 Occurrence of Manganese

Manganese is ubiquitous in the environment and is the twelfth most abundant mineral in the earth's crust (Post, 1999; Steenkamp and Basson, 2013; Tangstad, 2013a). It is proposed that the crustal rocks contain about 0.1% Mn, and is second only to iron as the most common heavy metal (Post, 1999). Geochemically, manganese resembles iron in that the lower-valence forms are highly soluble, mobile and reactive, whereas the higher-valence forms are highly insoluble, immobile and both chemically and physically resistant (Corathers and Machamer, 2006). Furthermore, manganese in ores exist as a component of various minerals, mostly in the form of oxides, sulphides, carbonates and silicates (Olsen et al., 2007;

Steenkamp and Basson, 2013; Tangstad, 2013a), with the most commonly occurring manganese minerals being pyrolusite, rhodochrosite, rhodonite and hausmannite (Post, 1999; Howe et al., 2004; Chetty, 2010; Steenkamp and Basson, 2013; Tangstad, 2013a). Table 5.4 shows some of the typical manganese-bearing minerals commonly found in the Earth's crust (Chetty, 2010; Steenkamp and Basson, 2013; Tangstad, 2013a).

About 80% of the global land-based manganese reserves are found in South Africa (Steenkamp and Basson, 2013). Other countries with substantial reserves include Australia, China, Gabon and Ukraine (Corathers, 2014). Furthermore, significant amounts of seabed manganese reserves, in the form of manganese nodules or ferromanganese concretions containing up to 30–36 wt.% Mn, were also discovered and could form additional sources of manganese (Marjoram et al., 1981; Post, 1999; Calvert, 2004; Clark et al., Hein and Petersen 2013). Currently, the global production of manganese ores, based on both the gross weight and the contained weight, is driven by China and South Africa, followed by Australia, Brazil and Gabon (Corathers, 2014). Based on data available for 2012, China and South Africa were the leading global producers of ferromanganese and silicomanganese alloys (Corathers, 2014).

TABLE 5.4

Typical Categories and Composition of Manganese Minerals

Mineral	Composition	Mn Content (wt.%)
Oxides and Hydroxides		
Bixbyite	$(Mn,Fe)_2O_4$	55.6
Braunite	$3(Mn,Fe)_2O_3 \cdot MnSiO_3$	48.9–56.1
Braunite II	$7(Mn,Fe)_2O_3 \cdot CaSiO_3$	56.2
Hausmannite	$(Mn,Fe)_3O_4$	64.8
Jacobsite	Fe_2MnO_4	23.8
Manganite	γ-MnOOH	62.5
Pyrolusite	MnO_2	63.2
Vernadite	$MnO_2 \cdot H_2O$	44–52
Carbonates		
Mn-calcite	$Mn(Ca)(CO_3)_2$	<20–25
Kuohorite	$CaMn(CO_3)_2$	35.4
Rhodochrosite	$MnCO_3$	47.6
Oligonite	$(Fe,Mn)CO_3$	23–32
Silicates		
Rhodonite	$MnSiO_3$	42
Tephroite	Mn_2SiO_4	54.4
Andradite	$Ca_3Fe_2(SiO_4)_3$	–
Serpentine	$Mg_3Si_2O_5(OH)_4$	–

Sources: Chetty, D., A geometallurgical evaluation of the ores of the Northern Kalahari manganese deposits, South Africa, DPhil thesis, University of Johannesburg, Johannesburg, South Africa, 2010; Steenkamp, J.D. and Basson, J., *J. South. Afr. Inst. Min. Metal.*, 113, 667, 2013; Reprinted from *Handbook of Ferroalloys: Theory and Technology*, Tangstad, M., Manganese ferroalloys technology, Gasik, M. (ed.), Butterworth-Heinemann/Elsevier, Oxford, UK, p. 232. Copyright 2013a, with permission from Elsevier.

5.4.4 Beneficiation of Manganese Ores

Conventionally, manganese ores are characterized by their content of manganese, iron and various impurities, and can generally be categorized into (Corathers and Machamer, 2006; Olsen et al., 2007; Malayoglu, 2010; Tangstad, 2013a) (1) metallurgical (or high-grade) ores, typically containing 35–55 wt.% Mn and are used in the production of high-carbon ferromanganese and silicomanganese alloys, (2) ferruginous grade ores with 15–35 wt.% Mn but containing high levels of iron and (3) mangano-ferrous ores, which are typically iron ores with 5–10 wt.% Mn. On the other hand, chemical and battery-grade ores are often categorized by their MnO_2 content, which is typically in the range of 70–85 wt.% or 44–54 wt.% Mn (Malayoglu, 2010).

Typically, metallurgical-grade manganese ores require a Mn/Fe ratio greater than 7.5 by weight for the production of standard manganese ferroalloys with 78 wt.% Mn (Olsen et al., 2007; Tangstad, 2013a). Metallurgical-grade ores are typically produced from open-pit and underground operations using conventional mining techniques (Corathers and Machamer, 2006; Olsen et al., 2007; Tangstad, 2013a). After mining, the manganese ores are subjected to beneficiation processes to increase the manganese content and/or to remove undesirable impurities such as sulphur and phosphorus (Corathers and Machamer, 2006; Tangstad, 2013a). Some of the typical processes used in beneficiating manganese ores include crushing and screening, heavy media separation, jigging, flotation and high-intensity magnetic separation (Corathers and Machamer, 2006; Malayoglu, 2010; Tangstad, 2013a). The beneficiation of lower-grade manganese deposits has also been extensively evaluated. For example, several studies have focused on complementing the physical beneficiation techniques with chemical methods such as leaching, roasting and prereduction in order to upgrade the Mn/Fe ratio, particularly for lower-grade and more refractory manganese resources (Buckenham, 1961; Kanungo and Sant, 1981; Sharma, 1992; Corathers and Machamer, 2006; Malayoglu, 2010; Singh et al., 2011; Gao et al., 2012b; Makhula et al., 2013; Pereira et al., 2014; Zhou et al., 2015).

5.4.4.1 Agglomeration of Manganese Ore Fines

Generally, lumpy manganese ores are used as raw materials in the production of manganese ferroalloys (Olsen et al., 2007; Faria et al., 2012; Tangstad, 2013a; Eric, 2014). However, the high decrepitation behaviour of most manganese ores during heating is detrimental to the operation of conventional furnaces due to decreased charge permeability (Faria et al., 2012, 2014). As discussed in Section 5.2.2, the efficient operation of ferroalloy furnaces is predicated on the gas permeability of the solids-packed zone as a result of the close sizing of raw materials. In practice, fines in the raw materials are detrimental in the operation of submerged arc furnaces as they result in poor charge porosity, which in turn results in high power consumption per ton of ferroalloy, excessive fumes and dust losses and low furnace productivity (Yoshikoshi et al., 1984; Olsen et al., 2007; Tangstad et al., 2010; Faria et al., 2012; Tangstad, 2013a). Based on this, the standard practice among ferromanganese producers is to utilize a combination of lumpy ore and agglomerates.

Manganese ore fines are agglomerated into properly sized sinters, pellets and briquettes (Pienaar and Smith, 1992; Jorma et al., 2001; Olsen et al., 2007; Tangstad et al., 2010b; Faria et al., 2012; Tangstad, 2013a). Basically, the pelletization and briquetting have their own challenges, among which is the inability to develop enough strength in the agglomerated products due to the inherent endothermic dissociation reactions of MnO_2 and Mn_2O_3 during firing under conventional temperature conditions of 1150°C–1250°C (Faria et al., 2012; Tangstad, 2013a).

So far, the agglomeration processing by sintering is by far the most preferred technique, and has grown to become one of the leading technologies in the agglomeration of manganese ore fines (Pienaar and Smith, 1992; Jorma et al., 2001; Olsen et al., 2007; McDougall, 2013; Tangstad, 2013a; Eric, 2014). In general, the advantages of sintering the furnace feed include (Pienaar and Smith, 1992; Jorma et al., 2001; Tangstad, 2013a; European Commission, 2014) (1) greater utilization of agglomerated fine ore, which otherwise would have limited applications in conventional smelting; (2) reduced gas volumes, hence fewer chances of furnace eruptions when smelting sintered ore; (3) better porosity of the charge resulting in uniform gas distribution within the solids packed bed; (4) increased furnace availability and operating loads and (5) utilization of waste products such as coke breeze and return fines, thereby increasing the overall raw material and energy efficiency of the ferromanganese smelting plant.

During the sintering process, manganese ore and coke breeze are mixed with binders and return fines from dust abatement systems and are heated to about 1200°C in a sinter grate to form sintered manganese oxides bound in fayalite and silicate phases (Pienaar and Smith, 1992; Jorma et al., 2001; McDougall, 2013; Tangstad, 2013a). As discussed in Chapter 4, the emission of particulate dusts, volatile organic matter, and noxious gases is particularly problematic in the sintering process. However, the off-gases generated from the sintering process are usually de-dusted using electrostatic precipitators and fabric filters and recycled back into the process (European Commission, 2014). Further technical details on the sintering process and sinter product specifications in the production of manganese alloys were highlighted in detail by Olsen et al. (2007).

5.4.5 Production of Manganese Alloys

Manganese is used in the steel industry mostly in the form of different types of bulk manganese ferroalloys (Olsen et al., 2007; Tangstad, 2013a; Eric, 2014). In general, the different types of manganese ferroalloys are broadly categorized into (Olsen and Tangstad, 2004; Olsen et al., 2007; Tangstad, 2013a; Eric, 2014; European Commission, 2014) (1) ferromanganese alloys (FeMn), which are further subdivided into high-carbon ferromanganese alloys (HCFeMn; max. 7.5 wt.% C), medium-carbon ferromanganese alloys (MCFeMn; max. 2.5 wt.% C) and low-carbon ferromanganese alloys (LCFeMn; max. 0.75 wt.% C), (2) silicomanganese (SiMn) (3) ferrosilicomanganese (FeSiMn) alloys (max. 3.5 wt.% C), (4) metallic manganese and (5) nitrided manganese alloys (max. 0.2 wt.% C for MnN and max. 3.5 wt.% C for SiMnN). Among these, the high-carbon ferromanganese and ferrosilicomanganese alloys constitute the traditional form of manganese feedstock in steelmaking, with the medium- and low-carbon versions being used where the final carbon content in steel must be controlled (Tangstad, 2013a). On the other hand, the silicomanganese alloys are consumed mostly in silicon- and manganese-containing steels where the final composition of steel requires a combination of carbon, manganese, silicon and other trace elements (Tangstad, 2013a).

In general, manganese ferroalloys are commercially produced from the carbothermic reduction of manganese ores, either in the blast furnace or in submerged arc furnaces (Ding and Olsen, 1996; Vanderstaay et al., 2004; Olsen et al., 2007; Tangstad, 2013a; Eric, 2014). Due to inherent disadvantages of the blast furnace process, such as heavy reliance on metallurgical-grade coke as a thermal and reducing agent, the submerged arc furnace route has since become the most dominant route to produce the different grades of manganese ferroalloys (Olsen et al., 2007; Tangstad, 2013a; Eric, 2014). In practice, the submerged arc furnace process route offers several advantages such as higher overall yield of manganese from the ores, less carbon consumption, ability to utilize lower-quality reducing agents and the greater flexibility to produce different grades of manganese ferroalloys (Olsen et al., 2007).

Basically, manganese ferroalloys are produced by three distinct processes categorized by their slag chemistry and operating temperature regimes: (1) HCFeMn production using discard slag in the temperature range of 1400°C–1450°C, (2) SiMn production utilizing only ore as a source of manganese in the temperature range of 1500°C–1600°C and (3) HCFeMn production using a rich slag practice and the subsequent production of SiMn (commonly referred to as the duplex process) in the temperature range of 1400°C–1450°C (Holappa and Xiao, 2004; Olsen et al., 2007; Steenkamp and Basson, 2013; Tangstad, 2013a). In the HCFeMn–SiMn duplex process, the slag tapped from the HCFeMn production furnace contains about 30–50 wt.% MnO, and is then further reprocessed for the production of silicomanganese or manganese metal (Heo et al., 2015; Tangstad, 2013a,b). Because the HCFeMn and SiMn alloys constitute the bulk of manganese ferroalloys production, their production process and waste considerations will be discussed briefly in this section.

5.4.5.1 High-Carbon Ferromanganese Alloys

High-carbon ferromanganese alloys are produced by the carbothermic reduction of lumpy or sintered manganese ores in three-phase submerged arc furnaces (Olsen and Tangstad, 2004; Olsen et al., 2007; Tangstad, 2013a; Eric, 2014; European Commission, 2014). The typical alloy composition from a high-carbon ferromanganese process contains, on average, 72–82 wt.% Mn and 7 wt.% C, while the slag typically contains around 40 wt.% MnO (Tangstad, 2013a). Basically, the key design objectives in the operation of a ferromanganese furnace are to (Olsen et al., 2007) (1) operate on a stable and high load, (2) minimize coke and energy consumption, (3) produce metal and slag of required composition, (4) produce a high yield of manganese and (5) minimize the emission of greenhouse gases, noxious compounds and particulate dusts.

The submerged arc furnaces used in manganese alloy smelting can be subdivided into distinct thermochemical zones, namely, the upper and middle solid-state zones and the high-temperature liquid-state zone (Hayes, 2004; Yang et al., 2004; Ringdalen et al., 2010; Tangstad, 2013a; Eric, 2014). In the upper and middle zones, the main reactions taking place include the heating and prereduction in the solid state, where the manganese oxides are reduced to lower-state manganese oxides while the iron oxides are reduced to metallic iron (Akdogan and Eric, 1994; Vanderstaay et al., 2004; Ringdalen et al., 2010; Kalenga et al., 2013; Coetsee et al., 2014). The typical reactions taking place in the upper and middle zones are shown in Equations 5.22 through 5.27:

$$Mn_2O_{3(s)} + CO_{(g)} \rightarrow 2Mn_3O_{4(s)} + CO_{2(g)} \quad (5.22)$$

$$3Mn_2O_{3(s)} + CO_{(g)} \rightarrow 2Mn_3O_{4(s)} + CO_{2(g)} \quad (5.23)$$

$$Mn_3O_{4(s)} + CO_{(g)} \rightarrow 3MnO_{(s)} + CO_{2(g)} \quad (5.24)$$

$$Fe_2O_{3(s)} + CO_{(g)} \rightarrow 2FeO_{(s)} + CO_{2(g)} \quad (5.25)$$

$$FeO_{(s)} + CO_{(g)} \rightarrow Fe_{(s)} + CO_{2(g)} \quad (5.26)$$

$$C_{(s)} + CO_{2(g)} \rightarrow 2CO_{(g)} \quad (5.27)$$

The high-temperature zone of the furnace is typically in the temperature range of 1500°C–1550°C, and is characterized by the formation of liquid slag and the final reduction of manganese oxides. At temperatures above 1550°C, the reduction of silica from slag and the final reduction of the manganese oxides dissolved in slag also become dominant. Based on this supposition, Ostrovski and Swinbourne (2013) proposed that the liquid state reduction mechanism in ferromanganese smelting involved the direct reduction of MnO and/or the initial dissolution of (Mn,Ca,Mg)O components into slag followed by the reduction of the MnO component from the slag by solid carbon from coke and/or by the carbon dissolved in the alloy melt. Basically, the solubility of MnO in liquid slags depends on temperature and slag composition, that is, the solubility increases by increasing the process temperature and decreasing the slag basicity (Ostrovski and Swinbourne, 2013).

Depending on equilibrium process and temperature conditions, the reactions occurring in the liquid state are shown by Equations 5.28 through 5.37 (Ding and Olsen, 1996; Olsen and Tangstad, 2004; Olsen et al., 2007; Safarian et al., 2009; Ringdalen et al., 2010; Tangstad et al., 2010b; Kalenga et al., 2013; Tangstad, 2013a; Coetsee et al., 2014; Eric, 2014; Safarian and Kolbeinsen, 2015):

$$(MnO)_{(slag)} + C_{(s)} \rightarrow [Mn]_{(alloy)} + CO_{(g)} \tag{5.28}$$

$$(FeO)_{(slag)} + C_{(s)} \rightarrow [Fe]_{(alloy)} + CO_{(g)} \tag{5.29}$$

$$(MnO)_{(slag)} + [C]_{(alloy)} \rightarrow [Mn]_{(alloy)} + CO_{(g)} \tag{5.30}$$

$$7(MnO)_{(slag)} + 10C_{(s)} \rightarrow [Mn_7C_3]_{(alloy)} + 7CO_{(g)} \tag{5.31}$$

$$(SiO_2)_{slag} + 2C_{(s)} \rightarrow [Si]_{alloy} + 2CO_{(g)} \tag{5.32}$$

$$2(MnO)_{(slag)} + [Si]_{alloy} \rightarrow 2[Mn]_{(alloy)} + (SiO_2)_{slag} \tag{5.33}$$

$$(SiO_2)_{(slag)} + 2[Mn]_{alloy} \rightarrow [Si]_{(alloy)} + 2(MnO)_{slag} \tag{5.34}$$

$$(SiO_2)_{(slag)} + [Si]_{alloy} \rightarrow 2SiO_{(g)} \tag{5.35}$$

$$C_{(s)} \rightarrow [C]_{(alloy)} \tag{5.36}$$

$$[Mn]_{(alloy)} \rightarrow Mn_{(g)} \tag{5.37}$$

According to Safarian et al. (2009), the carbothermic reduction of FeO and MnO from slag takes place simultaneously. The typical products from the liquid slag reduction reactions consist of a carbon-saturated Mn–Fe–Si–C alloy and MnO–SiO_2–CaO–Al_2O_3–MgO slag in equilibrium with CO gas (Olsen and Tangstad, 2004; Coetsee et al., 2014). Typically, the liquid slag should have low viscosity for it to percolate easily through the coke-packed solids layer. Finally, the superheated liquid alloy collects in the hearth of the SAF and is either cast into ingots or granulated for shipment, while the supernatant slag floating on top of the liquid alloy is tapped into slag launders before being granulated and/or allowed to cool in the slag yard (Olsen et al., 2007; Tangstad, 2013a). In practice, the slag and metal separation in submerged arc furnace operation is enhanced by adjusting the viscosity of the slags. In essence, a high-viscosity slag results in the entrapment of metal droplets in slag, which is detrimental to metal yield and productivity. Further details on the thermodynamics and thermochemical aspects of manganese oxides smelting in the submerged arc furnaces were highlighted by several researchers (Olsen et al., 2007; Tangstad, 2013a; Eric, 2014), and thus will not be dealt with in this section.

In principle, the optimum operation of a ferromanganese furnace is every operator's dream, and basically achieved when the power consumption is low and the furnace is operated on a stable and high load (Olsen et al., 2007). In general, low power consumption and high tonnages can be achieved by operating under the following conditions: (1) utilizing highly oxidized manganese ores, (2) good prereduction in feed, (3) low slag/metal ratio due to high content of manganese in the ore and high yield of manganese to the metal and (4) good yield of manganese due to low MnO content in the slag by close control of the operating parameters such as slag basicity and process temperature.

5.4.5.2 Silicomanganese Alloys

Silicomanganese alloys are usually produced from the carbothermic reduction of oxide ores in the submerged arc furnaces (Olsen and Tangstad, 2004; Olsen et al., 2007; Tangstad, 2013a; Eric, 2014; European Commission, 2014), and are typically used as substitutes for ferromanganese and/or ferrosilicon alloys in steelmaking, and as raw materials for the production of medium- and low-carbon ferromanganese alloys as well as industrial-grade manganese metal (Eric, 2014). Depending on the quality of starting raw materials, the standard silicomanganese alloys typically contain 14–19 wt.% Si, while the special grades of these alloys can contain as high as 35 wt.% Si (Eric, 2014; European Commission, 2014). In general, the content of dissolved carbon in these alloys is controlled by the amount of dissolved silicon, that is, the solubility of carbon in SiMn alloys decreases with increasing the silicon content of the alloy (Eric, 2014).

There are basically three routes for the production of silicomanganese alloys, namely, (1) reduction of manganese ores and silica with coke and coal, (2) carbothermic reduction of MnO-rich slags from the production of HCFeMn alloys and quartzite blends and (3) reaction of standard ferromanganese alloys and quartzite blend with coke (Olsen et al., 2007; Tangstad, 2013a; Ostrovski and Swinbourne, 2013; Eric, 2014). However, the majority of the standard silicomanganese alloys are currently being produced from a blend of MnO-rich slag tapped from the HCFeMn process (approx. 35–40 wt.% MnO), manganese ores and sinter, quartz or quartzite, off-spec alloy remelts and coke and coal (Olsen and Tangstad, 2004; Olsen et al., 2007; Steenkamp and Basson, 2013; Tangstad, 2013a; Eric, 2014). According to Ostrovski and Swinbourne (2013), the typical charge in the production of silicomanganese alloys comprises of 1029 kg MnO, 105 kg FeO, 337 kg SiO_2, 191 kg Al_2O_3, 248 kg MgO and 79 kg Si sculls per ton of SiMn alloy with composition 67–70 wt.% manganese, 17–20 wt.% silicon and 1.5–2 wt.% carbon. In most cases, the production of SiMn alloys is integrated with the production of HCFeMn in such a way that the liquid slag from the latter process is reprocessed in the production of SiMn alloys (Olsen and Tangstad, 2004; Olsen et al., 2007; Steenkamp and Basson, 2013; Tangstad, 2013a; Eric, 2014).

The equipment and processes for the production of commercial silicomanganese alloys are generally similar to those of high-carbon ferromanganese alloys, with the only major difference being in the process temperatures (Olsen and Tangstad, 2004; Olsen et al., 2007; Steenkamp and Basson, 2013; Tangstad, 2013a; Eric, 2014). In general, higher process temperature conditions (1600°C–1650°C) are required in the silicomanganese alloy production in order to attain the desired silicon specification content in the alloy (Olsen and Tangstad, 2004; Olsen et al., 2007; Steenkamp and Basson, 2013; Tangstad, 2013a; Eric, 2014). As a result, operating using deeper electrode penetration is usually practised in order to attain the requisite high temperatures required in the production of silicomanganese alloys (Eric, 2014).

The chemical reactions taking place in the production of SiMn alloys are similar to those discussed under high-carbon ferromanganese process. In general, the equilibrium composition of MnO in SiMn slags depends on both the temperature and the silica content of the slag (Olsen and Tangstad, 2004). The efficacy of a silicomanganese process in terms of minimizing Mn loss to slag, either as metal inclusions or as MnO dissolved in slag, depends on the slag chemistry and temperature control (Olsen and Tangstad, 2004; Olsen et al., 2007; Tangstad, 2013a; Eric, 2014). In order to control the slag chemical properties, lime-dolomite ($CaCO_3 \cdot MgCO_3$) or olivine-based ($2MgO \cdot SiO_2$) fluxes are usually added as part of precalculated furnace charges (Olsen et al., 2007; Tangstad, 2013a).

5.4.5.3 Medium-Carbon Ferromanganese Alloys

Medium-carbon ferromanganese (MCFeMn) alloys typically contain 1–1.5 wt.% carbon and 75–85 wt.% manganese (Eric, 2014). These types of alloys are preferred in the production of low-carbon and other special grades of steel where the close control of carbon is essential (Tangstad, 2013a). Essentially, two process options exist in the commercial production of medium-carbon ferromanganese alloys, namely, (1) silicothermic reduction of manganese ore and MnO-rich slag and (2) manganese oxygen refining process, which involves the decarburization treatment of a high-carbon ferromanganese alloy with an argon-oxygen or nitrogen-oxygen gas mixtures (You et al., 1999; Tangstad, 2013a; Eric, 2014; European Commission, 2014). The two processes are discussed briefly in Sections 5.4.5.4 and 5.4.5.5, respectively.

5.4.5.4 Silicothermic Reduction

In the silicothermic reduction process, a high-grade slag or melt containing manganese ore and lime is reacted with SiMn containing 16–30 wt. Mn (Eric, 2014). The silicon in the alloy technically acts as the reductant, and in this case, the composition the SiMn alloy in terms of Si wt.% determines the final carbon concentration in the alloy (Tangstad, 2013a; Eric, 2014). According to Eric (2014), the equilibrium for the silicothermic reduction process is governed by the reaction:

$$[Si]_{(SiMn)} + 2(MnO)_{ore/slag} \rightarrow [Mn]_{(alloy)} + (SiO_2)_{slag} \tag{5.38}$$

Controlling the yield of the alloy is of critical importance in the silicothermic reduction process. Based on the reaction shown in Equation 5.34, the moderate additions of basic slag formers and fluxes, typically MgO due to the common use of MgO–C refractory lining in the smelting furnaces, favours the silicomanganese distribution in alloy by, especially by decreasing the activity of coefficient of SiO_2 in slag, thereby shifting the equilibrium to the right (Tangstad, 2013a; Eric, 2014). The silicothermic reduction of slag is exothermic, and the heat of reaction is normally sufficient to make the process autothermal and to cater for any heat losses in the process (Tangstad, 2013a). Furthermore, Tangstad (2013a) also proposed the use of ladles with stirring capabilities, such as shaking ladles, in improving the heat and mass transfer efficiency of the process.

Apparently, the scale of production of medium-carbon ferroalloys using the silicothermic reduction process is slowly diminishing globally (You et al., 1999; European Commission, 2014). According to You et al. (1999), this process is generally associated with a plethora of technical and economic problems such as high refractory consumption, generation of large amounts of slag and poor metal yields, among others. In addition, the European Commission (2014) also reported that the most important route in producing MCFeMn alloys within its member states is via the decarburization treatment of HCFeMn alloys.

5.4.5.5 Converter Decarburization Process

The converter decarburization process, also known as the manganese oxygen refining, involves decarburization treatment of HCFeMn alloys by blowing argon-oxygen or nitrogen-oxygen gas mixtures (You et al., 1999; Nelson et al., 2001; Olsen et al., 2007; Tangstad, 2013a; Eric, 2014). Just like in the decarburization treatment of steel, the process is very much autothermal, with the heat requirements being supplied by the oxidation reactions of manganese and carbon (Eric, 2014). The oxygen blown into the melt initially oxidizes part of the manganese and increases the temperature from about 1350°C to 1550°C. Thereafter, with increasing temperature, the carbon present in the HCFeMn melt also gets oxidized resulting in further increase in temperature from 1550°C to about 1750°C (You et al., 1999; Tangstad, 2013a; Eric, 2014; European Commission, 2014). Depending on the control of process parameters such as temperature, partial pressure of carbon monoxide in the furnace and the silicon reduction of slag in the final stages of the heat, the decarburization process is capable of producing alloys with carbon content in the range of 0.5–1.0 wt.% C (Tangstad, 2013a).

The manganese oxygen refining process is very similar to the decarburization treatment of liquid steel and HCFeCr alloys discussed in Chapter 4 and Section 5.3.4, respectively. Nevertheless, operating temperature conditions are key distinctive features that differentiate the manganese oxygen refining process to the oxygen refining process of liquid steel (Tangstad, 2013a; Eric, 2014). Basically, the manganese oxygen refining process requires higher final temperature of over 1750°C, 200°C to 300°C higher than the conventional 1550°C for the basic oxygen furnace refining of steel (Tangstad, 2013a; Eric, 2014). The higher operating temperature provides several challenges such as (Tangstad, 2013a; Eric, 2014) (1) more severe refractory attack, (2) challenges with casting a superheated alloy and supernatant slag, (3) higher vapour pressure of manganese resulting in the excessive evaporation and oxidation loss of manganese, and hence a high concentration of oxidized manganese compounds in the off-gas and (4) higher volume and temperature of the off-gas.

The economics of the manganese oxygen refining process is predicated on the final recovery of the manganese in the refined product (You et al., 1999; Tangstad, 2013a). Technically, a substantial amount of manganese loss occurs as a result of oxidation and evaporation during the refining stage (You et al., 1999; Tangstad, 2013a; Eric, 2014). During oxygen refining process, the manganese distribution among the process streams is as follows (You et al., 1999; Tangstad, 2013a; Eric, 2014): (1) manganese yields in medium-carbon ferroalloy, (2) oxidation loss to slag, (3) manganese loss by vaporization into off-gas and (4) other losses to sculls and splashes. In general, the oxidation loss of manganese to slag is controlled by slag composition, melt temperature and slag weight at the end of the blow (You et al., 1999). As a result, additional slag reduction by silicon is necessary at the end of the blow to reduce the MnO from slag and recover the oxidized manganese in the alloy (You et al., 1999; Tangstad, 2013a).

The converter off-gas is typically rich in oxidized manganese oxides, typically existing as Mn_3O_4 (Tangstad, 2013a). You et al. (1999) investigated the loss of manganese to the gas phase, as a function of evaporation rate and vapour pressure, during the oxygen refining of high-carbon ferromanganese melts in a 2 ton model converter. In their findings, the manganese content remained almost constant until when the carbon content of the melt decreased from a starting point of 6.8 wt.% to about 3 wt.%. Since the oxidation reaction of manganese is thermodynamically favoured over the decarburization reaction below the critical carbon content, the loss of manganese becomes more pronounced, thereby leading to increased oxidation loss of manganese (You et al., 1999).

5.4.6 Waste Generation in the Production of Manganese Ferroalloys

As discussed in Section 5.4.1, the anthropogenic sources of manganese in the environment are mostly due to human nature, particularly as waste emissions in the mining and beneficiation of manganese ores, manganese ferroalloys production processes, production of manganese chemicals, and emissions from the iron and steel production processes. According to the International Manganese Institute (2014), a significant amount of by-products and waste materials are produced in the production of manganese and its alloys, and these include (1) waste rock, overburden and tailings; (2) particulate matter and fugitive emissions from comminution and handling processes; (3) point-source particulate dust emissions at sinter plants, furnaces and refineries and (4) metallurgical slags from smelting and refining furnaces. Of particular interests are the process slags and particulate dust emissions which are produced in large amounts in the smelting of manganese ferroalloys.

So far, the broader processes involved in the production of various manganese alloys have been discussed in detail. In addition, the generation of waste from the mining and beneficiation processes was broadly discussed in Chapter 3. As a result, this section will focus on the generation, physicochemical characterization, and the potential valorization characteristics of metallurgical dusts, sludges and process slags produced in the production of the various manganese alloys.

5.4.6.1 Particulate Dust Emissions in Production of Manganese Ferroalloys

Typically, a significant amount of charge materials is converted to particulate dusts in the smelting and refining of ferromanganese and silicomanganese alloys in submerged arc furnaces (Shen et al., 2005; de Araujo et al., 2006; Shen et al., 2007a,b; Frías and Rodríguez, 2008; Kim et al., 2010; Lee et al., 2010; Singh and Tathavadker, 2011; Tangstad, 2013a; International Manganese Institute, 2014; Brakestad, 2015). In particular, particulate dusts from the SAFs are formed from the volatilized materials and entrained fines that are collected in the dry and wet scrubbing systems of the furnace off-gases (Shen et al., 2007b). Additional sources of particulate dusts in the manganese ferroalloy smelting include the tapping of alloy and slag, the alloy casting and crushing processes, and in the refining of ferromanganese alloys, particularly from the manganese oxygen refining process (You et al., 1999; Tangstad, 2013a; Gjønnes et al., 2011).

The problem of generation of the manganese-containing dusts in manganese alloy–producing furnaces is particularly exacerbated by the high vapour pressure of manganese (You et al., 1999; Tangstad, 2013a; Eric, 2014). In general, the particulate dusts collected from the furnace off-gas cleaning systems can contain up to 25 wt.% manganese, which is technically suitable for the production of manganese alloys (Shen et al., 2005;

Shen et al., 2007a,b; Kim et al., 2010; Tangstad, 2013a;). For example, Tangstad (2013a) proposed that the oxidized manganese collects as Mn_3O_4 in the gas cleaning systems of the SAF, and is normally agglomerated and recycled to the SAF to recover the contained manganese. As discussed in Chapter 2, the disposal of manganese-containing furnace dusts in landfills presents long-term environmental problems. Basically, this type of waste is classified as a toxic and hazardous waste in most jurisdictions, and as such, it has become imperative for the manganese ferroalloys industries to develop technologies and systems for the in-process recycling of these particulate dusts (Shen et al., 2005; de Araujo et al., 2006; Shen et al., 2007a,b; Frías and Rodríguez, 2008; Kim et al., 2010; Lee et al., 2010; Singh and Tathavadker, 2011; Tangstad, 2013a; International Manganese Institute, 2014).

5.4.6.2 Physicochemical Properties of Particulate Dusts

The characterization of particulate dusts from the production of ferromanganese alloys has been extensively studied with a view towards understanding health dust exposure levels at workplaces (Gunst et al., 2000; Gjønnes et al., 2011; Kero et al., 2015) and in developing effective processes to recover the manganese contained therein (Shen et al., 2005; Shen et al. 2007a,b; Brakestad, 2015). In principle, the physiochemical properties of particulate dusts vary extensively depending on the type of the process and the grade and type of the ferroalloy product produced. Generally, the chemical and morphological characterization of various dust samples collected from various processes and locations has revealed the following common characteristics for the furnace dusts: (1) manganese-containing compounds as MnO, Mn_3O_4, $MnFe_2O_4$, rhodochrosite ($MnCO_3$) and manganosite ((Ca,Mn) CO_3) (Shen et al., 2005; Gaal et al., 2010; Ravary et al., 2013; Brakestad, 2015); (2) intermetallic particles typically consisting of FeMn and SiMn entrained in dust particles; (3) complex oxides such as Fe_3O_4, ZnO, $MgFe_2O_4$, SiO_2 and MgO (Brakestad, 2015) and (4) carbonates, sulphides and sulphate compounds such as ZnS, CaS, K_2SO_4 and $MgCO_3$ (Brakestad, 2015). Furthermore, Gjønnes et al. (2011) characterized the dusts and aerosol emissions from the tapping of a high-carbon ferromanganese alloys as containing (1) particles dominated by MnO occurring as chain-like or compact agglomerates with minor contents of Mn_3O_4 and Fe_3O_4, (2) sheets or flakes of Mn and Fe oxides, (3) MnO and Mn_3O_4 dominating minor amounts of MnO_2 and Mn_2O_3, (4) occurrence of fibres and needles of MnO and Mn_3O_4 and (5) carbonaceous particles (mostly amorphous or poorly crystalline graphite) frequently encapsulating nanometre-sized particles of (Mn,Fe)O, Mn_3O_4 and Fe_3O_4 and to some extent intermetallic particles of MnSi and MnFe. In contrast, fine particulate emissions from the casting of SiMn alloys were dominated by MnSi phases existing as chain-like or more compact agglomerates, as well as individual particles (Gjønnes et al., 2011). Some ZnO needles as well as carbonaceous particles were also observed in the dust and aerosol samples (Gjønnes et al., 2011).

Although the chemical compositions of the particulate dusts tend to vary depending on process conditions, there is great potential for recovering high-value manganese products from these waste materials. Technically, the particulate dusts can economically be recycled into the process, particularly as feed in the sinter plant and/or as agglomerated furnace charges (Shen et al., 2005; Shen et al., 2007a,b; Tangstad, 2013a). The potential utilization of these dust materials as feed materials in the production of electrolytic grade manganese oxides was also investigated (de Araujo et al., 2006; Ghafarizadeh et al., 2011). The in-process recycling of manganese ferroalloy particulate dusts, particularly in the sinter plant, will be discussed briefly in Section 5.4.6.3.

5.4.6.3 In-Process Recycling of Ferromanganese Alloys Particulate Dusts

Particulate dusts from the processing of various grades of ferromanganese and silicomanganese alloys are typically rich in manganese oxides and other metal compounds. Since the disposal in landfills of particulate dusts is restricted due to environmental concerns, the most prudent approach would be to recycle the particulate dust material back into the process (Language et al., 2000; Shen et al., 2005, 2007a,b; Olsen et al., 2007; Gaal et al., 2010; Lee et al., 2010; Ghafarizadeh et al., 2011; McDougall, 2013; Ravary et al., 2013; Tangstad, 2013a).

Technically, particulate dusts from submerged arc furnace processing of FeMn and SiMn alloys can economically be agglomerated and recycled back into the process. In general, several laboratory and industrial scale tests have been conducted to test the potential of in-process recycling of, metal recovery from, and reuse of manganese ferroalloy dusts based the following approaches: (1) pelletizing and sintering processing (Language et al., 2000; Olsen et al., 2007; Shen et al., 2007a,b; Gaal et al., 2010), (2) solid-state reduction using rotary hearth technologies (Hansmann et al., 2007; Tateishi et al., 2008; Gao et al., 2012a; Degel et al., 2015), (3) direct furnace smelting to produce alloy-rich product (Jones et al., 1993; Denton et al., 2005; Kesseler and Erdmann, 2007; Gaal et al., 2010; Lee et al., 2010), (4) reductive leaching of dust to recover the manganese contained therein (Ghafarizadeh et al., 2011) and (5) using the dust as complementary cementing materials in blended cements (Frías and Rodríguez, 2008). As discussed in Chapter 4, the recycling of particulate dusts is particularly constrained by the in-process build-up of volatile metal species such as zinc and alkali metal compounds.

According to Shen et al. (2005, 2007a,b), the accumulation of zinc in furnace off-gas is one of the major factors affecting the recycling potential of the particulate dusts in the sintering process. The operating conditions in the sintering process are relatively oxidizing, and this results in the zinc remaining in the sinter as zinc oxide (Language et al., 2000; Shen et al. 2005, 2007a,b). Based on extensive sintering tests, the overall zinc removal was deemed to be difficult due to reoxidation of the zinc in the oxidizing atmosphere of the sintering process (Language et al., 2000; Shen et al., 2005, 2007a,b). Equations 5.39 through 5.41 show the most probable reactions that may occur thermodynamically at operating temperatures up to 1320°C (Language et al., 2000; Shen et al., 2007b):

$$Zn_{(v)} + CO_{2(g)} \rightarrow ZnO_{(s)} + CO_{(g)}; \quad K = \left(\frac{a_{ZnO}}{P^{e}_{Zn}}\right)\left(\frac{P_{CO}}{P_{CO_2}}\right) \tag{5.39}$$

$$Zn_{(v)} + \frac{1}{2}O_{2(g)} \rightarrow ZnO_{(s)} \tag{5.40}$$

$$Zn_{(v)} \rightarrow Zn_{(s,l)} \tag{5.41}$$

The extent of deposition of ZnO is dependent on temperature, the ratio of partial pressures of CO and CO_2 gases (P_{CO}/P_{CO2}) and the vapour pressure of zinc (P_{Zn}) (Language et al., 2000). When sintered product is eventually charged into the submerged arc furnace, the build-up of zinc in the process inevitably takes place (Language et al., 2000; Shen et al., 2007b). In this regard, Language et al. (2000) proposed the recirculation of zinc in the submerged arc furnace operation as involving the following mechanistic steps: (1) introduction of Zn-rich oxidic dust into the furnace as part of the charge; (2) descent of the Zn-rich charge to the hotter and more reducing regions of the furnace; (3) the reduction of ZnO and the eventual volatilization of the reduced Zn; (4) upward transport of the Zn (v) with the furnace off-gas, countercurrent to the descending furnace charge and (5) the condensation of $Zn_{(v)}$ to $Zn_{(l)}$ or $Zn_{(s)}$ and the eventual oxidation and heterogeneous condensation of the Zn to ZnO in the cooler and more oxidizing upper regions of the furnace charge. The recirculation and build-up of zinc in the SAF process results in increased chemical attack of the refractory materials and dust collection systems, and as such, the sustainable utilization of such particulate dusts should at least include a partial zinc removal step to reduce the zinc loading of the recycled materials (Language et al., 2000; Shen et al., 2005; Hansmann et al., 2007; Shen et al., 2007a,b; Gaal et al., 2010).

5.4.6.4 Recycling and Reuse of Alloy Fines

In the submerged arc furnace processing of FeMn and SiMn alloys, the alloy separates from the slag via density differentiation and is tapped from the furnace and is typically solidified into ingots (Dunkley and Norval, 2004; Olsen et al., 2007; de Faria et al., 2008; McDougall, 2013; Tangstad, 2013a; International Manganese Institute, 2014). Although the granulation processing of ferroalloys has gained traction in the recent years, mainly due to its ability to produce a homogenous and clean product which meets the increasing demands for the production of clean steels, the traditional approach in the post-tapping processing of FeMn and FeSi alloys involves solidification, crushing and sieving to produce alloy products with the right size specification demanded by customers (Dunkley and Norval, 2004; Olsen et al., 2007; de Faria et al., 2008; Singh and Tathavadker, 2011; McDougall, 2013; Tangstad, 2013a; Vesterberg et al., 2013; International Manganese Institute, 2014). However, the crushing processes inherently produce residual alloy fines, which essentially results in significant product losses and creates handling and health problems for producers and end users alike (de Faria et al., 2008; Singh and Tathavadker, 2011; Versterberg et al., 2013).

In general, the ferroalloys market imposes strict requirements in terms of the chemical and particle-size composition in order to optimize the alloy addition in secondary metallurgy processes where the automatic handling and feeding is commonly practised (Berg et al., 1999; Kutsin et al., 2015). Despite the metal fines and dusts generated in the crushing and screening processes being typically identical in composition to the lump products, the current practice is either to recycle them internally in the smelting furnaces and/or to sell them to customers at discounted prices (International Manganese Institute, 2014). As a result, finding the alternative methods of value beneficiation of these FeMn and SiMn fines is beneficial especially for the cost-sensitive ferromanganese and silicomanganese producers. As a result, extensive research has been conducted in an effort to increase the recycling and utilization potential of these metal fines and dusts. For example, Singh and Tathavadker (2011) proposed an agglomeration process using resin binders followed by curing to attain the right strength of the agglomerates to facilitate handling and charging in downstream processes. Furthermore, de Araujo et al. (2006) proposed the use of FeMn fines as a cost-effective substitute material to rhodochrosite in the production of electrolytic grade manganese.

5.4.6.5 Characteristics of Manganese Ferroalloy Slags

Significant amounts of process slags are produced as a by-product from the production of manganese ferroalloys (Olsen et al., 2007; Ostrovski and Swinbourne, 2013; Steenkamp and Basson, 2013; Tangstad, 2013a; Eric, 2014). Generally, slags produced in the production of generic manganese ferroalloys belong to the $MnO–Al_2O_3–CaO–MgO–SiO_2$ system (Warren et al., 1974; Kor, 1979; Cengizler and Eric, 1992; Olsen et al., 1995; Skjervheim and

Olsen, 1995; Holappa and Xiao, 2004; Zhao et al., 2005; Olsen et al., 2007; Groot et al., 2013; Kumar et al., 2013; Ostrovski and Swinbourne, 2013; Tangstad, 2013a; Eric, 2014). The control of process conditions such temperature and slag chemistry is important in controlling the partition of manganese between the alloy and $MnO–Al_2O_3–CaO–MgO–SiO_2$ slag phases. As a result, the effect of these parameters, and the thermodynamic considerations of manganese alloys in equilibrium with the slags were extensively investigated (Warren et al., 1974; Kor, 1979; Cengizler and Eric, 1992; Olsen et al., 1995; Swinbourne et al., 1995; Zaitsev and Mogutnov, 1995; Holappa and Xiao, 2004; Olsen et al., 2007; Ostrovski and Swinbourne, 2013; Eric, 2014). From a thermodynamic standpoint, the equilibrium conditions between the alloy and the $MnO–Al_2O_3–CaO–MgO–SiO_2$ slag phases control the recovery of Mn in the FeMn production process, with the most important factors being the MnO content and the basicity ratio of the slag (Holappa and Xiao, 2004). Ostrovski and Swinbourne (2013) further proposed a strong dependence of the equilibrium activity coefficient of MnO in slag on the composition the $MnO–Al_2O_3–CaO–MgO–SiO_2$ slags.

Depending on the process, typical slags in the production of ferromanganese alloys can be categorized as either high/rich slag or discard slag (Holappa and Xiao, 2004; Olsen et al., 2007; Steenkamp and Basson, 2013). Typically, the HCFeMn discard slag system typically consists of about 15 wt.% MnO, 24 wt.% SiO_2, 21 wt.% Al_2O_3, 35 wt.% CaO and 5 wt.% MgO (Olsen et al., 2007; Steenkamp and Basson, 2013). In contrast, the high-carbon ferromanganese alloy slags utilizing the rich slag process consist of about 40 wt.% MnO, 22–24 wt.% SiO_2, 12–13 wt.% Al_2O_3, 17 wt.% CaO and 6 wt.% MgO. Fundamentally, the HCFeMn slags rich in MnO are typically consumed in the production of SiMn alloys (Olsen et al., 2007; Ostrovski and Swinbourne, 2013; Steenkamp and Basson, 2013). Approximately, one ton of slag is produced in the rich slag process for every ton of high carbon ferromanganese alloy produced (Steelworld, 2012).

Based on FACTSage calculations, Ostrovski and Swinbourne (2013) proposed the chemical composition of final slags produced in the SiMn process in the temperature range of 1600°C–1650°C to be in the range of 6–12 wt.% MnO, 38–44 wt.% SiO_2, 10–25 wt.% Al_2O_3, 20–35 wt.% CaO and 5–15 wt.% MgO. The relatively low MnO in SiMn slag is as a result of the high process temperatures which, in essence, facilitate the fast reduction of MnO and SiO_2 from the slag according to Equations 5.28 through 5.37.

Typically, solidified ferromanganese slags contain both amorphous and glassy silicate phases (Frias et al., 2006; Steenkamp et al., 2011; Rong-jin et al., 2012; Groot et al., 2013; Kumar et al., 2013). The typical mineralogical phases in slow-cooled MnO-rich slags produced in the production of HCFeMn alloys include (Norval and Oberholster, 2011; Steenkamp et al., 2011; Groot et al., 2013) (1) gehlenite ($Ca_2Al(AlSiO_7)$), (2) glaucochroite ($CaMnSiO_4$), (3) manganosite ((Mn,Mg,Ca)O), (4) spinel ($(Mn,Mg)Al_2O_3$), (5) monticellite ($CaMgSiO_4$) and (6) quartz (SiO_2). The evolution of these phases was also confirmed by thermodynamic calculations by Zhao et al. (2005) based on the liquidus isotherms and phase equilibria of high-MgO ferromanganese and silicomanganese smelting slags in the temperature range of 1200°C–1400°C.

In contrast, the main mineralogical phases observed in the final SiMn slags include (Rai et al., 2002; Frias et al., 2006; Rong-jin et al., 2012; Kumar et al., 2013) (1) akermanite ($Ca_2MgSi_2O_7$), (2) dicalcium silicate (Ca_2SiO_4), (3) merwinite ($Ca_3MgSi_2O_8$), (4) gehlenite ($Ca_2Al_2SiO_7$), (5) diopside ($CaMgSi_2O_6$), (6) manganese oxides (MnO) and (7) wollastonite ($CaSiO_3$). In addition, Kumar et al. (2013) characterized a typical SiMn slag containing about 9.9 wt.% MnO using x-ray diffraction methods as consisting of 32.2 wt.% $Ca_2MgSi_2O_7$, 20.6 wt.% CaSiO3, 36.2 wt.% $CaMgSi_2O_6$ and 10.8 wt.% SiO_2. Furthermore, Rai et al. (2002) used x-ray diffraction methods to investigate the mineralogical phases of air-cooled MnO slags and concluded that forsterite (Mg_2SiO_4), Mn_2O_3, MnO and quartz (SiO_2) were the main common phases in the air-cooled low-MnO slags.

5.4.6.6 Recycling and Reuse of Ferromanganese Alloy Slags

Understanding the mineralogical and structural phases in ferromanganese slags is important in developing holistic metal recovery, and slag recycling and reuse techniques. In general, the link between slag structure properties such as mineralogy and crystallography and the functional properties such as grain size distribution of metallurgical slags is important in enhancing their valorization potential, thereby greatly improving the industrial ecology of the smelting operations (Durinck et al., 2008). Due to the irreplaceability of manganese in the production of high-value steel products, the demand for ferromanganese alloys is expected to remain stable in the long term (International Manganese Institute, 2014). As a result, the amount of slag materials produced in the production of ferromanganese alloys is also expected to increase, and as such, it is prudent for the ferromanganese alloys industry to take a long-term view on improving the valorization potential of these process slags. Because manganese-containing slags are typically classified as hazardous materials, there is pressure to reduce the amount of landfilled slags in order to mitigate the potential environmental impacts from the production of these alloys (International Manganese Institute, 2014).

According to Groot et al. (2013), over 20 million tonnes of ferromanganese slags has accumulated in slag dumps in South Africa, and a further 0.5 Mt are being added every year. To date, significant efforts have been directed in developing alternative processes and technologies, not only to recover the entrained manganese alloys before disposal, but also to develop the in-process recycling and reuse potential of the slags in alternative industries (Mynko et al., 1993; Rai et al., 2002; Frias et al., 2006; Norval and Oberholster, 2011; Rong-jin et al., 2012; Groot et al., 2013; Kumar et al., 2013; International Manganese Institute, 2014). According to the International Manganese Institute (2014), slags may directly be reprocessed within the manganese production process for their manganese content or can be used as a low-cost material for local building and road construction. Based on the global life cycle study of manganese alloys and manganese by-products, approximately 530 kg slag/tonnes Mn-alloy is sold to the construction industry and 360 kg slag/ton Mn-alloy is recovered in furnaces, while 140 kg slag/tonnes Mn-alloy is stockpiled (International Manganese Institute, 2014).

5.4.6.7 Recovery of Manganese from Ferromanganese Alloy Slags

As discussed in Section 5.4.6.5, two broad categories of slags are produced in the production of ferromanganese alloys, namely, HCFeMn discard slags and SiMn slags, with each typically containing up to 15 wt.% MnO and 6–12 wt.% MnO, respectively. Basically, the process slags can be recycled back in the SAF to recover the entrained alloys and/or alloys dissolved in the slags, or are consumed in the sinter plant as premelted fluxes (Kim et al., 2011; Norval and Oberholster, 2011). However, the direct in-process recycling of ferromanganese slags has several limitations such as the limitations on the amount of slag that can be recycled, and the danger of recycling the gangue components contained in the slag, which potentially increases the slag volumes and hence the energy consumption in the smelting processes (Kim et al., 2011; Norval and Oberholster, 2011).

In essence, several processes and techniques have been explored to maximize the recovery of metal values from ferromanganese slags (Shen and Forssberg, 2003; Durinck et al., 2008;

Semykina, 2010; Semykina et al., 2010; Kim et al., 2011; Norval and Oberholster, 2011; Groot et al., 2013; Shatokha et al., 2013; Baumgartner and Groot, 2014). The typical conventional approaches in the reprocessing of ferromanganese slags involve the use of mechanical and/or physical processing techniques such as crushing, grinding, magnetic separation and flotation techniques to recover entrained metallic manganese and iron components. The resultant barren components of slag from the metal recovery processes can thus be used as cement additives and/or as premelted flux additives in the sintering and smelting processes (Durinck et al., 2008; Semykina, 2010; Kim et al., 2011). For example, Kim et al. (2011) investigated a direct physical separation process for the upgrading of manganese from silicomanganese discard slag using magnetic separation techniques, and proposed a two-step process involving milling and magnetic separation to produce a manganese-rich slag product containing more than 20 wt.% Mn.

The hot stage processing of metallurgical slags to enhance metal recovery while developing the functional properties of the barren slag compatible with alternative reuse options is also a promising technology in the holistic valorization of slags (Durinck et al., 2008; Semykina, 2010; Semykina et al., 2010; Kim et al., 2011; Norval and Oberholster, 2011; Shatokha et al., 2013). Semykina (2010), Semykina et al. (2010) and Shatokha et al. (2013) proposed an innovative process based on the transformation of non-magnetic iron-bearing components in slag to magnetite by oxidation pretreatment followed by the selective recovery of the iron-bearing and non-iron-bearing slag constituents. According to these researchers, the reactions that take place during the oxidation of molten CaO–FeO–SiO_2 and CaO–FeO–SiO_2–MnO in the temperature range of 1350°C–1500°C and environment oxidizing enough to increase the valence of Fe and form stable ferrites can be depicted by Equations 5.42 through 5.44:

$$3(\mathrm{FeO})_{\mathrm{slag}} + \frac{1}{2}\mathrm{O}_{2(\mathrm{air})} \rightarrow (\mathrm{Fe_3O_4})_{\mathrm{solid}} \quad (5.42)$$

$$2(\mathrm{FeO})_{\mathrm{slag}} + \frac{1}{2}\mathrm{O}_{2(\mathrm{air})} \rightarrow (\mathrm{Fe_2O_3})_{\mathrm{solid}} \quad (5.43)$$

$$2(\mathrm{FeO})_{\mathrm{slag}} + (\mathrm{MnO})_{\mathrm{slag}} + \frac{1}{2}\mathrm{O}_{2(\mathrm{air})} \rightarrow (\mathrm{MnFe_2O_4})_{\mathrm{solid}} \quad (5.44)$$

The ferrites with high magnetic susceptibilities produced in the oxidation process are maintained during the solidification process to enable the separation of completely solidified slag by magnetic separation (Shatokha et al., 2013). Furthermore, $MnFe_2O_4$ ferrites have been also reported to possess magnetic properties that can potentially be exploited in their magnetic separation from the bulk nonmagnetic components of slags (Ding et al., 1995; Sugimoto, 1999; Li et al., 2004; Mahmoud, 2005). Potentially, the magnetic properties of manganese ferroalloy slags can be engineered at $MnFe_2O_4$ ferrite crystal level in order to enhance the potential for the magnetic separation of Mn-Fe-bearing components from the slags (Semykina, 2010).

5.4.6.8 Hydrometallurgical Processing of Ferromanganese Slags

5.4.6.8.1 Fundamentals of Manganese Oxide Leaching

Several efforts have been made to recover manganese from manganese ores and secondary resources using hydrometallurgical processes (Das et al., 1982; Veglio and Toro, 1994;

Zhang and Cheng, 2007a,b; Su et al., 2008; Ghafarizadeh et al., 2011; Groot et al., 2013; Baba et al., 2014; Yan and Qiu, 2014). Basically, the hydrometallurgical processing of manganese-bearing compounds involves a series of processing steps involving a chemical reductive step, purification of the Mn-rich leach liquors and the recovery of Mn from solutions using processes such as electrowinning, solvent extraction or precipitation (Das et al., 1982; Zhang and Cheng, 2007a,b; Su et al., 2008; Chow et al., 2010; Baba et al., 2014). Due to the refractory nature of most manganese oxides to conventional acid leaching, the efficacy of some of the reducing agents used in the reductive leaching of these materials was extensively investigated, and these include, among others, the following: (1) carbohydrate-based reductants (Veglio and Toro, 1994; Zhang and Cheng, 2007a), (2) aqueous sulphur dioxide or sulphurous acid (Petrie, 1995; Zhang and Cheng, 2007a; Su et al., 2008; Chow et al. 2010) and (3) aqueous ferrous sulphate (Das et al., 1982; Zhang and Cheng, 2007a; Baba et al., 2014).

5.4.6.8.2 Leaching and Recovery of Manganese from Ferromanganese Slags

To date extensive research on the leaching and recovery of manganese from ferromanganese slags has been reported by several researchers based on the use of organic acids (McIntosh and Baglin, 1992; Hariharan and Purnima, 2015) and inorganic acids (Sen Gupta and Dhananjayan, 1983; Groot et al., 2013; Baumgartner and Groot, 2014; Yan and Qiu, 2014) under normal and reductive leaching conditions. In essence, sulphuric acid is the most commonly preferred leaching reagent, mainly due to its relatively low cost and the ubiquity of downstream processes in the recovery of the manganese from the resulting manganese sulphate solutions (Sen Gupta and Dhananjayan, 1983; Zhang and Cheng, 2007a,b; Chow et al. 2010; Groot et al., 2013; Baumgartner and Groot, 2014; Yan and Qiu, 2014).

In particular, Groot et al. (2013) and Baumgartner and Groot (2014) investigated the sulphuric acid digestion of FeMn slags followed by water leaching of the resultant leach residue to produce high-grade manganese sulphate. The barren residue from the sulphuric acid leaching of FeMn slags was found to contain substantial amounts of amorphous silica and calcium sulphate, which have been empirically tested as important cementitious aggregates (Shi, 2002, 2004; Groot et al., 2013). Furthermore, Yan and Qiu (2014) investigated the possibility of recovering electronic grade manganese sulphate from ferromanganese slags by grinding, sulphuric acid leaching and $MnCO_3$ precipitation.

Typically, the hydrometallurgical processing of the manganese ferroalloy slags has its own challenges. Ferromanganese slags contain significant amounts of silicate phases which react to form soluble silicates when leached with sulphuric acid. The solubilized silicates tend to polymerize into gelatinous compounds, thereby hindering the solid–liquid separation (Sen Gupta and Dhananjayan, 1983; Groot et al., 2013; Baumgartner and Groot, 2014). Despite these challenges, the use of sulphuric acid to leach and recover the manganese from ferromanganese slags is a promising technology with potential industrial applications, especially if the increased intrinsic monetary value of the residue can subsidize the economics of the process (Groot et al., 2013; Baumgartner and Groot, 2014).

5.4.6.8.3 Granulation Processing of Ferromanganese Slags

The mineralogical and structural phases observed in FeMn and FeSi are controlled by the solidification and crystallization pathway after tapping from the furnace (Mynko et al., 1993; Durinck et al., 2008; Norval and Oberholster, 2011). In practice, ferromanganese slags can be solidified either by air-cooling in the slag yard or by granulation using water jets (Mynko et al., 1993; Norval and Oberholster, 2011). Fundamentally, the water granulation of slags is a proven technology to render metallurgical slags environmentally friendly and physically and chemically acceptable for use in the construction industry (Durinck et al., 2008; Norval and Oberholster, 2011). In particular, Norval and Oberholster (2011) studied the granulation process of reprocessed FeMn slags with a view of recovering the entrained FeMn alloy and reusing the granulated slag as a cement additive. In short, their findings can be summarized as (1) significant reduction in the MnO content from 21.8 wt.% in air-cooled slag to 6.1 wt.% in the reprocessed granulated slag, (2) the possibility of recovering the Mn in the CaMn-silicate phases of the granulated slags and (3) the solidification of granulated slag in glassy phases with properties compliant with construction industry standard for use as a cementitious material.

5.4.6.8.4 *Reuse of Ferromanganese Slags as Cementitious Additives*

In general, metallurgical slags have found widespread applications as cement additives in the construction industry (Rai et al., 2002; Horii et al., 2013; Shi, 2004; Durinck et al., 2008; Frías and Rodríguez, 2008; Kumar et al., 2008; Norval and Oberholster, 2011; Yildirim and Prezzi, 2011; Kumar et al., 2013; EUROSLAG, 2015). According to Durinck et al. (2008), the general application of metallurgical slags as cementitious systems is controlled by the interrelationships among the process parameters (such as composition and cooling path), slag structure (such as mineralogy, crystallography, phase distribution, porosity) and slag properties (such as strength and abrasivity). Generally, the trend is shifting towards granulation processing of metallurgical slags in order to increase the content of glassy and reactive silica phases while at the same time reducing the amount of disintegration-prone dicalcium silicate (C_2S) phases (Shi, 2004; Durinck et al., 2008). Based on the working knowledge of how the slag structure develops during solidification, significant progress has been made in transforming the metallurgical slags from being waste streams into a valuable by-products with potential applications to markets exogenous to the ferroalloys industry (Motz and Geisler, 2001; Durinck et al., 2008; Frías and Rodríguez, 2008; Kumar et al., 2008; Horii et al., 2013).

Extensive research on the use of manganese ferroalloy slags as construction materials has revealed the technical suitability of these materials as cementious and pozzolanic additives (Rai et al., 2002; Frias et al., 2006; Frías and Rodríguez, 2008; Norval and Oberholster, 2011; Kumar et al., 2013). For example, studies by Frias et al. (2006), based on a granulated SiMn slag consisting of about 43 wt.% SiO_2, 12 wt.% Al_2O_3, 25 wt.% CaO, 4 wt.% MgO and 10 wt.% MnO as the main components, have confirmed the suitability of SiMn-blended matrices in the manufacture of blended cements. These materials were evaluated based on standard specification requirements such as chemical composition, reactivity, volume stability, porosity and mechanical strength (Frias et al., 2006). The compliance to construction standards of granulated ferromanganese slags was also confirmed by Norval and Oberholster (2011) based on studies conducted on the remelting of the air-cooled slags to recover the FeMn alloy and granulation of the resultant barren slag.

Despite the extensive body of knowledge on the cementious properties of granulated blast furnace slags, the widespread adoption of manganese ferroalloy slags as cementious and pozzolanic materials in blended cements has been constrained, mainly due to the high content of entrained and/or dissolved manganese-bearing phases in the slag and the limited knowledge of these phases on the hydration and volume stability properties of the blended cements (Frias et al., 2006; Frías and Rodríguez, 2008). Notwithstanding these challenges, the use of manganese ferroalloy slags as a cementitious material has generated a lot of interest in the recent years, mainly due to increased environmental regulations and the protracted long-term growth in the production of manganese ferroalloys, which in effect has resulted in increased volume of slags to be disposed of in landfills (Rai et al., 2002; Frias et al., 2006; Frías and Rodríguez, 2008; Norval and Oberholster, 2011; Kumar et al., 2013).

5.4.6.8.5 Summary

The key aspects pertinent to the production of various grades of manganese alloys were highlighted in this section. Different parameters affecting the generation of, and amount of, waste streams in the production of manganese and its alloys were highlighted. In particular, the critical aspects affecting the generation of particulate dust emissions in the production of manganese alloys include the use of agglomerated raw materials and decrepitation behaviour of manganese-bearing furnace charge. In addition, the section highlighted the different strategies and best practices currently being adopted in mitigating the anthropogenic effects from the production of manganese, and its alloys were also highlighted.

Furthermore, the physicochemical and mineralogical properties of particulate dusts and slags were evaluated from a viewpoint of enhancing their valorization potential. Technically, manganese-bearing particulate dust can economically be recycled back into process, particularly through the sinter plant. A lot of potential also exists for the valorization of manganese-bearing slag by-products, particularly by granulation and engineering the solidification behaviour to enhance metal recovery and alternative use as additives in the manufacture of cementitious products.

5.5 Silicon and Its Alloys

5.5.1 Chemical and Physical Properties of Silicon

Silicon (Si) is a Group IV element with relative atomic number 28.085, electronic structure [Ne] $3s^2 3p^2$ and exhibits a face-centred diamond cubic crystal structure (Mitraŝinović, 2010; Greenwood and Earnshaw, 2012; Leroy and Rancoita, 2015; Los Alamos, n.d.). The melting and boiling points of pure silicon are 1410°C and 3265°C, respectively (Lenntech, n.d.; Los Alamos, n.d.). Typically, silicon has four valence electrons in its outer shell and tends to form covalent bonds as a result of electron sharing between atoms (Leroy and Rancoita, 2015). As a typical Group IV element, silicon has oxidation states between 1 and 4, with the quadrivalent state being the most stable (Tangstad, 2013b; Leroy and Rancoita, 2015). In addition, silicon is a metalloid with characteristics of both metals and non-metals (Mitraŝinović, 2010; Tangstad, 2013b; Leroy and Rancoita, 2015). In fact, silicon exhibits metallic properties such as thermal and electrical conductivity in the liquid state while predominantly exhibiting semiconductor properties in the solid state (Mitraŝinović, 2010; Tangstad, 2013b).

5.5.2 Occurrence of Silicon

Silicon is the second most abundant element after oxygen in the earth's crust (Schei et al., 1997; Laruelle et al., 2009; Tangstad, 2013b). In its natural form, silicon combines with oxygen to occur predominantly in the form of silica (SiO_2) and silicates (Schei et al., 1997; Tangstad, 2013b). Silicon is a major constituent of many rock-forming minerals such as quartz, clays, mica, feldspar, garnets, asbestos and talc.

Anthropogenically, most of the silicon is bound in the form of quartz and silicate minerals and is largely unavailable for uptake by organisms (Laruelle et al., 2009). However, anthropogenic cycle of silicon remains poorly understood (Laruelle et al., 2009). Furthermore, the benefits of silicon to plants and humans are also not clear (Guntzer et al., 2012; Martin, 2013). Guntzer et al. (2012) reported that silicon is not considered an essential element in agronomy based on the assumption that silicon is not technically limiting in soils. However, Martin (2013) contends that despite the specific biochemical and physiological functions of silicon in human beings being largely unknown, they are presumed to exist.

Contrary to bioavailable silicon forms, crystalline silica is widely documented as a potent respiratory hazard (Lenntech, n.d.). As discussed in Chapter 2, health problems such as lung cancer and silicosis are usually associated with occupational exposures to crystalline and non-crystalline forms of silica, particularly in the form of quartz, cristobalite and asbestos. Further details on the health effects of silicon and its compounds were covered in detail in Chapter 2.

5.5.3 Industrial and Commercial Applications of Silicon

Silicon, in its various forms, has found widespread industrial and specialist commercial applications. The industrial applications of silicon range from generic uses in bulk commodities such as concrete, clays and ceramics, to chemically modified systems such as soluble silicates, glasses and glazes (Ciftja et al., 2008; Siddique and Chahal, 2011a,b; Siddique, 2011; Roskill, 2011; Greenwood and Earnshaw, 2012). Other more advanced applications of silicon include the production of silicon-based chemicals, industrial polymers and semiconductors (Ciftja et al., 2008; Roskill, 2011; Siddique and Chahal, 2011a,b; Greenwood and Earnshaw, 2012). Silicon and its alloys also have important metallurgical applications such as alloying and deoxidizing functions in the production of ferrous and non-ferrous alloys (Turkdogan and Fruehan, 1998; Bristow et al., 2000; Suzuki et al., 2001; Outokumpu, 2013). In essence, most of the industrial and commercial applications of silicon have been attributed to its metalloid characteristics, that is, the ability to possess both metallic and non-metallic properties. The following sections briefly discuss some of the important applications of silicon and its compounds in the various industry sectors.

5.5.3.1 Metallurgical Applications of Silicon

5.5.3.1.1 Silicon as a Deoxidizing Agent in Steelmaking

Some of the critical iron and steel refining processes such as argon-oxygen decarburization refining are carried out under high temperatures and oxidizing conditions such that oxygen is bound to dissolve in the liquid steel (Patil et al., 1998; Turkdogan and Fruehan, 1998; Bristow et al., 2000; Suzuki et al., 2001; Outokumpu, 2013). However, dissolved and residual oxygen in steel has adverse effects on the cleanliness of steel, and eventually on the mechanical properties of solidified steel (Kor and Glaws, 1998; Patil et al., 1998; Turkdogan and Fruehan, 1998; Bristow et al., 2000; Zhang and Thomas, 2006; Outokumpu, 2013). As a result, the residual oxygen content in liquid steel is lowered during ladle deoxidation treatment using silicon and/or other reactive metal elements such as Ti, Mn, Ca and Al (Kor and Glaws, 1998; Turkdogan and Fruehan, 1998; Bristow et al., 2000; Zhang and Thomas, 2006; Outokumpu, 2013).

In the refining process of steel silicon is usually consumed in the form of various grades of silicon ferroalloys, particularly ferrosilicon, silicomanganese, silico-chromium, silicocalcium and silicon carbide alloys (Kor and Glaws, 1998; Turkdogan and Fruehan, 1998;

Suzuki et al., 2001). Furthermore, silicon in these various forms of silicon-based ferroalloys is also used in converter decarburization refining of high-carbon ferroalloys, particularly in the decarburization processes of HCFeCr and HCFeMn alloys where it is principally used to reduce the oxidized chromium and manganese from the slag (Rick, 2010; Gasik, 2013; Eric, 2014).

5.5.3.1.2 Silicon as an Alloying Element in Steels

Silicon is an integral alloying element in the production of certain grades of steels (Bristow et al., 2000; Outokumpu, 2013). In particular, silicon promotes the ferritic microstructure by solid solution hardening, and generally increases the strength of stainless steels (Bristow et al., 2000). Silicon is also added to increase the high-temperature oxidation resistance of stainless steels, particularly in high-temperature ferritic and austenitic grades (Outokumpu, 2013). In addition, silicon increases the resistance of steels to oxidation in strongly oxidizing solutions at lower temperatures (Bristow et al., 2000; Outokumpu, 2013). Silicon is also an integral alloying element in the production of spring steels, which typically contain about 1.7–2.2 wt.% Si. Another application of silicon is a principal alloying element in the production of electrical steels used in transformers, electric motor laminations, generators and relays. In these types of steels, silicon confers unique and beneficial electrical properties such as relatively high permeability, increased electrical resistivity and low hysteresis loss (Bristow et al., 2000).

5.5.3.1.3 Silicon as an Alloying Element in Aluminium Alloys

The recent years have seen an increased demand for versatile light materials possessing a combination of engineering properties such as plasticity, strength, ductility and corrosion resistance (Eskin and Kharakterova, 2001; Pedersen and Arnberg, 2001). In particular, aluminium-silicon (Al-Si) cast alloys have been proved to meet these requirements due to their low density, superior corrosion resistance, high specific strength and specific stiffness combined with castability (Eskin and Kharakterova, 2001; Pedersen and Arnberg, 2001; Ciftja et al., 2008; Prillhofer et al., 2014). As a result, aluminium-based alloys have attracted considerable attention as formidable construction and transportation materials, particularly in the automotive industry where manufacturers are focusing on designing lightweight vehicles with improved fuel efficiency and reduced exhaust emissions (Prillhofer et al., 2014).

Silicon has been proved to be an important alloying element in the production of the various grades of aluminium alloys (Eskin and Kharakterova, 2001; Pedersen and Arnberg, 2001; Warmuzek, 2004; Ciftja et al., 2008; Prillhofer et al., 2014). In general, heat treatable wrought alloys of the Al–Mg–Si–(Cu) system (6XXX series), typically containing about 0.5–1.0 wt.% silicon, represent one of the most important Al–Si alloys systems that meet the complex requirements of good formability with high strength, corrosion resistance, and weldability (Warmuzek, 2004; Ciftja et al., 2008; Prillhofer et al., 2014). Furthermore, silicon is also important in improving the castability of aluminium alloys (Di Sabatino and Arnberg, 2009). In essence, silicon is added to aluminium melts in order to improve the viscosity and fluidity of liquid aluminium during casting, thereby improving the mechanical properties of the Al-based commercial alloys (Ciftja et al., 2008; Di Sabatino and Arnberg, 2009). Further details on the mechanical and microstructural properties of the various Al–Si alloys are available in a comprehensive handbook published by Warmuzek (2004).

Based on the aforementioned unique properties of aluminium-silicon alloys, the long-term demand of silicon as an alloying element in aluminium lightweight alloys is expected to increase. Recent official data on the global flow of silicon into aluminium alloys are scarce, but Roskill (2011) estimated that about 45% of the silicon consumed globally was used in the manufacture of aluminium-silicon alloys.

5.5.3.1.4 Silicon Applications in Semiconductors

Silicon is an important semiconductor material that is extensively used in electronic devices such as transistors, printed circuit boards, integrated circuits and photovoltaic/solar panels (Lutz et al., 2011; Greenwood and Earnshaw, 2012; European Commission, n.d.). Generally, the electrical resistivity of silicon decreases with increasing temperature, and also with increasing the concentrations of doping elements such as boron, aluminium, gallium and phosphorus (Lutz et al., 2011; Greenwood and Earnshaw, 2012; European Commission, n.d.). As a result, controlling the composition of these elements is critical in the quality control of semiconductor-grade silicon.

Semiconductor-grade silicon is widely used in the manufacture of photovoltaic cells (Bathey and Cretella, 1982; Braga et al., 2008; Lynch, 2009; Pizzini, 2010; Ramos et al., 2015). According to Pizzini (2010), semiconductor-grade silicon (also referred to as solar-grade silicon [SoG-Si]), in its various configurations such as single crystal, multi-crystalline, amorphous and micro-crystalline silicon, supplies about 90% of the substrates used in the photovoltaic industry. Although silicon is generally consumed in small quantities within the photovoltaic industry, the onward trajectory in the demand for solar cells is most likely to persist as the world shifts from fossil fuels towards clean energy generated by harnessing solar energy (Bathey and Cretella, 1982; Braga et al., 2008; Ciftja et al., 2008; Lynch, 2009; Pizzini, 2010; Ramos et al., 2015).

Generally, semiconductor-grade silicon is manufactured by refining metallurgical-grade silicon (MG-Si) (Bathey and Cretella, 1982; Ikeda and Maeda, 1992; Morita and Miki, 2003; Braga et al., 2008; Ciftja et al., 2008; Lynch, 2009; Takiguchi and Morita, 2009; Pizzini, 2010; Ramos et al., 2015). MG-Si is produced by the high-temperature carbothermic reduction of silica, in the form of high-purity quartz, using charcoal or coke as a reductant (Bathey and Cretella, 1982; Braga et al., 2008; Ciftja et al., 2008; Takiguchi and Morita, 2009; Mitrašinović, 2010). The carbothermic reduction of quartz in the submerged arc furnace is further discussed in detail in Section 5.5.4.

Naturally, MG-Si contains impurities that are detrimental to semiconductor properties of silicon (Bathey and Cretella, 1982; Ikeda and Maeda, 1992; Morita and Miki, 2003; Istratov et al., 2006; Braga et al., 2008; Lynch, 2009; Takiguchi and Morita, 2009; Pizzini, 2010; Ramos et al., 2015). Typically, metal impurities have well-documented effects that are detrimental to solar cell efficiency (Istratov et al., 2006). Therefore, the silicon produced from the carbothermic reduction process requires a series of refining and purification steps to produce polycrystalline SoG-Si (Bathey and Cretella, 1982; Braga et al., 2008; Takiguchi and Morita, 2009). Basically, there are two major process options used in the production of SoG-Si from the refining of MG-Si, namely, the chemical gaseous and the metallurgical refining routes (Bathey and Cretella, 1982; Ikeda and Maeda, 1992; Morita and Miki, 2003; Braga et al., 2008; Ciftja et al., 2008; Lynch, 2009; Takiguchi and Morita, 2009; Pizzini, 2010; Ramos et al., 2015).

5.5.3.2 Chemical Refining Process of Metallurgical-Grade Silicon

Basically, the pure silicon suitable for use in solar cell and microelectronics industries is typically produced via sequential two steps: (1) reacting MG-Si with hydrochloric acid to form a range of chlorosilanes, including trichlorosilanes ($HSiCl_3$), followed by (2) the reduction of the chlorosilanes with hydrogen on pure silicon surfaces to produce polycrystalline Si (Braga et al., 2008; Ciftja et al., 2008; Takiguchi and Morita, 2009; Ranjan et al., 2011).

The typical chemical reactions taking place in the refining process are shown in Equations 5.45 and 5.46 (Braga et al., 2008; Ciftja et al., 2008; Takiguchi and Morita, 2009; Ranjan et al., 2011):

$$Si_{(s)} + 3HCl_{(g)} \rightarrow HSiCl_{3(g)} \tag{5.45}$$

$$HSiCl_{3(g)} + H_{2(g)} \rightarrow Si_{(s)} + SiCl_{4(g)} + 2HCl_{(g)} \tag{5.46}$$

The chemical refining process has two major advantages: (1) it is a proven technology with known thermodynamics and process parameters, and (2) it is capable of producing ultrapure silicon directly usable for growing either the single-crystalline Czochralski ingots or the multi-crystalline ingots using directional solidification methods (Braga et al., 2008; Pizzini, 2010; Ranjan et al., 2011). However, the main disadvantages of the chemical refining route include the high energy consumptions (with a minimum being 120 kWh/kg) and the potential environmental problems associated with toxic and corrosive reactants and products (Braga et al., 2008; Takiguchi and Morita, 2009; Pizzini, 2010; Ranjan et al., 2011).

5.5.3.3 Metallurgical Refining Process of Metallurgical-Grade Silicon

The metallurgical refining process involves obtaining SoG-Si directly from MG-Si (Ikeda and Maeda, 1992; Yuge et al., 2001; Khattak et al., 2002; Morita and Miki, 2003; Braga et al., 2008; Lynch, 2009; Pizzini, 2010; Ranjan et al., 2011; European Commission, n.d.). Due to the nature of the thermochemical reactions taking place in the carbothermic reduction of quartz in submerged arc furnaces, the final silicon metal product from the smelting process tends to contain a significant amount of impurities which are detrimental to the semiconductor properties of silicon (Bathey and Cretella, 1982; Yuge et al., 2001; Khattak et al., 2002; Istratov et al., 2006; Braga et al., 2008; Lynch, 2009; Pizzini, 2010). Typically, the impurities in MG-Si include metallic elements and intermetallic compounds such as aluminium, calcium, iron and titanium, as well as non-metallic elements and inclusions such as phosphorus, boron and to some extent carbon (Lynch, 2009; Pizzini, 2010). Technically, the presence of these impurity elements in the MG-Si presents challenges for its use in electronic and photovoltaic applications (Bathey and Cretella, 1982; Yuge et al., 2001; Khattak et al., 2002; Braga et al., 2008; Lynch, 2009; Pizzini, 2010).

Given the challenges associated with high impurity levels, several processes have been explored in effort to upgrade the purity of MG-Si to SoG-Si, and these include, inter alia, (1) acid leaching of pulverized MG-Si to dissolve metallic, intermetallic and non-metallic inclusions; (2) slag treatment of the silicon melt to dissolve the impurities and remove them as supernatant slag; (3) vacuum degassing of the silicon melt to remove the volatile impurities; (4) reactive plasma treatment of the melt to remove metallic elements such as Fe, Ca, Ti and Al, as well as non-metallic elements such as P, B and to some extent C; (5) purification of the silicon melt using gases or water vapour and (6) directional solidification of the molten silicon to segregate the impurities in the melt based on differential segregation coefficients (Bathey and Cretella, 1982; Yuge et al., 2001; Khattak et al., 2002; Morita and Miki, 2003; Istratov et al., 2006; Braga et al., 2008; Lynch, 2009; Pizzini, 2010; Martorano et al., 2011; Heuer, 2013). The detailed thermodynamic aspects in the refining process of MG-Si have been reported in several articles (Morita and Miki, 2003; Lynch, 2009; Pizzini, 2010; Heuer, 2013).

The energy consumption in the production of SoG-Si from MG-Si is reported to be considerably lower than that of the chemical refining processes (Braga et al., 2008; Pizzini, 2010;

Ranjan et al., 2011). As a result, the extensive research conducted to date has been focused on improving the efficiency of impurity removal from the MG-Si, and in consolidating the economies of scale to improve the overall economics of refining the silicon for solar cell applications. In particular, most these studies have focused on linking the efficiency of metallurgical production of the silicon with the requisite efficiencies in the refining processes (Bathey and Cretella, 1982; Yuge et al., 2001; Istratov et al., 2006; Braga et al., 2008; Lynch, 2009; Ranjan et al., 2011).

Further details on the production process and chemical composition of MG-Si are discussed in Section 5.5.4. Despite the focus of this chapter being on waste production and utilization in the ferroalloys industry, and silicon metal technically not being categorized as a ferroalloy, the current and future criticality of MG-Si in the production of semiconductor-grade silicon for solar cell applications warrants its discussion in this section.

5.5.3.4 Silicon Applications in Chemicals and Polymers

Silicon has profound applications in the chemicals and polymers industry, particularly in the manufacture of silicones (O'Lenick, 1999; Ciftja et al., 2008; Roskill, 2011; Colas, n.d.), synthetic silica (Ciftja et al., 2008) and silanes (Ciftja et al., 2008; Roskill, 2011). According to Roskill (2011), the manufacture of silicones second largest consumer global silicon metal produced globally, contributing about 35% in 2010. Due to the important and unique industrial applications of these functional silicon-based materials, the following sections briefly discuss the uses of silicon in the manufacture of some of these materials.

5.5.3.4.1 Silicones

Silicone compounds are unique functional polymeric materials with chemical structure based on alternate chains of silicon and oxygen atoms (Cornelius and Monroe, 2004; Lin et al., 2008).Technically, silicones are denoted by the chemical formula $[R_2SiO]_n$, where R is an organic group that can include the methyl, phenyl, alkenyl or hydrogen groups (O'Lenick, 1999; Lin et al., 2008; Colas, n.d.). Generally, silicones provide unique properties, such as impermeability to water, flexibility and resistance to chemical attack over a wide temperature range (O'Lenick, 1999). As a result, silicones are widely used in waterproofing treatment, mechanical seals, high-temperature grease and waxes, breast implants, explosives and pyrotechnics (Cornelius and Monroe, 2004). The unique functional properties and applications of silicones can be attributed to the strength and flexibility of the Si–O bond, its partial ionic character and the low interactive forces between the non-polar organic groups (O'Lenick, 1999; Lin et al., 2008; Colas, n.d.).

In general, the industrial manufacture of silicone compounds involves reacting silicon powders with methyl chloride (CH_3Cl) at high temperatures (250°C–350°C) and pressure (1–5 bars) over a copper catalyst (O'Lenick, 1999; Cornelius and Monroe, 2004; Lin et al., 2008; Colas, n.d.). In most cases, the silicon feedstock is produced from high-temperature carbothermic reduction of quartz in electric furnaces to produce MG-Si, which is further refined to meet the requisite chemical specifications (O'Lenick, 1999).

5.5.3.4.2 Uses of Crystalline and Synthetic Amorphous Silica

Generally, there are three forms of silicon dioxide, namely, crystalline silica, naturally occurring amorphous silica and synthetic amorphous silica (Greenwood and Earnshaw, 2012; Pölloth, 2012). Crystalline silica is usually naturally occurring oxide of silicon which exists in different forms of polymorphs, with the main ones being quartz, cristobalite and tridymite (Greenwood and Earnshaw, 2012). The most important applications of crystalline silica are in the manufacture of glass and ceramics, and also as construction materials and materials for piezoelectric applications (Greenwood and Earnshaw, 2012).

Synthetic amorphous silica, in the form of pyrogenic or fumed silica precipitated amorphous silica and gel or colloidal amorphous silica, represents an important category of synthetic functional silicon-based materials that have impacted the global demand for the silicon metal (Ciftja et al., 2008; Greenwood and Earnshaw, 2012; Pölloth, 2012). Basically, synthetic amorphous silica can be produced via two principal processes, namely, (1) the high-temperature hydrolysis of chlorosilanes ($SiCl_4$) formed from the chlorination of MG-Si under oxyhydrogen conditions for the pyrogenic synthetic amorphous silica and (2) the wet chemical synthesis route via the acidification of aqueous solutions of sodium silicate, for the precipitated, gel and colloidal forms (Greenwood and Earnshaw, 2012; Pölloth, 2012).

Generally, the various forms of synthetic silica compounds have found widespread applications in chemicals industry, particularly (1) as additives in silicone rubbers to increase the mechanical strength and elasticity of these elastomers (Greenwood and Earnshaw, 2012), (2) in the manufacture of pharmaceutical and cosmetic products (Pölloth, 2012) and (3) in the manufacture of coating systems and paints (Greenwood and Earnshaw, 2012). Other generic applications of synthetic silica products include use in abrasives and hard materials, and in the manufacture of refractory and ceramic materials for high-temperature applications (Greenwood and Earnshaw, 2012).

5.5.4 Production of Silicon and Its Alloys

Silicon and its alloys are commercially produced by the carbothermic reduction of silicon-bearing ores in submerged arc furnaces (Dosaj et al., 2005; Mitraŝinović, 2010; Tangstad, 2013b; Eric, 2014). Basically, two generic products of silicon products are produced commercially using the carbothermic reduction process, and these include silicon metal and silicon ferroalloys. The silicon ferroalloys are generally categorized based on their iron and silicon contents, as well as the maximum amount of allowable trace impurities (Schei et al., 1997; Corathers, 2003; Ciftja et al., 2008; Mitraŝinović, 2010; Tangstad, 2013b; Eric, 2014). The most common categories of silicon ferroalloys include ferrosilicon (FeSi), silicomanganese (SiMn), ferrosilicomanganese (FeSiMn), silico-chromium (SiCr), silico-calcium (SiCa) and silicon carbide (SiC) alloys (Dosaj et al., 2005; Tangstad, 2013b; European Commission, 2014). Among these, the ferrosilicon and ferrosilicomanganese alloys constitute the major feedstock for the silicon addition in the refining processes of steels (Bristow et al., 2000; Dosaj et al., 2005; Outokumpu, 2013; Tangstad, 2013b).

The production process of silicomanganese alloys was briefly discussed in Section 5.4.5. Furthermore, the production of silico-calcium and silico-chromium alloys is not discussed as these alloys currently contribute a minor proportion in the production of silicon alloys. As such, only the processes peculiar to the production of MG-Si metal and ferrosilicon alloys will be discussed in this section. Furthermore, the common industrial uses of silicon and its alloys were also discussed in Section 5.5.3, and as such, the following sections will briefly focus on the production of ferrosilicon alloys and MG-Si.

5.5.4.1 General Considerations in the Production of Silicon and Ferrosilicon Alloys

Generally, silicon and ferrosilicon alloys are produced by the carbothermic reduction of high-grade quartz and/or quartzite lumps in low-shaft three-phase submerged arc furnaces (Dosaj et al., 2005; Schei et al., 1997; Tangstaad, 2013b; Eric, 2014). The overall process is highly endothermic and the smelting reduction is accomplished by the conversion of electrical energy fed into the process through carbon electrodes into heat in the gas-filled cavities surrounding the electrode tips (Myrhaug et al., 2004; Saevarsdottir and Bakken, 2010; Tangstad, 2013b; Eric, 2014; Machulec and Bialik, 2016). In principle, the heating effect in SAFs comes from the energy transferred from the electric arc, which itself is considered a high-temperature plasma with temperatures up to 20,000K, to increase the temperature of the furnace feed up to 1800°C–2000°C (Tangstad, 2013b).

Depending on the process efficiency, about 2500–3000 kg quartz, 1200–1400 kg of low-ash coal and/or charcoal and 1500–3000 kg wood chips are required to produce 1 ton of silicon (Dosaj et al., 2005). In addition, the electric furnaces used in the production of silicon and ferrosilicon alloys can be open or semi-closed in order to facilitate charging and efficient charge distribution in the furnace through stoking (Tangstad, 2013b; European Commission, 2014). Because silicon alloys generally do not absorb carbon, the submerged arc furnaces utilize a carbon-based refractory material in the hearth and lower sidewalls (Eric, 2014). The carbothermic reaction taking in the high-temperature zone of the submerged arc furnace is shown in Equation 5.47 (Schei et al., 1997; Senapati et al., 2007; Tangstaad, 2013b; Vangskåsen and Tangstad, 2013b; Eric, 2014).

$$SiO_{2(s)} + 2C_{(s)} \rightarrow Si_{(l)} + 2CO_{(g)} \quad (T \sim 1800°C - 2000°C) \tag{5.47}$$

Mechanistically, the carbothermic reduction of silica can take place via intermediate steps as shown in Equations 5.48 and 5.49 (Videm, 1995; Eric, 2014):

$$2SiO_{2(s,l)} + 3C_{(s)} \rightarrow SiC_{(s)} + 2CO_{(g)} \tag{5.48}$$

$$SiO_{2(s,l)} + C_{(s)} \rightarrow SiO_{(g)} + CO_{(g)} \tag{5.49}$$

The silicon monoxide gas ($SiO_{(g)}$) produced in the high-temperature zone of the furnace further reacts with solid or gaseous carbon (Equations 5.50 and 5.51) and with the intermediate reaction product $SiC_{(s)}$ (Equation 5.52) as it ascends through the solids packed zone of the furnace charge (Videm, 1995; Schei et al., 1997; Tangstad et al., 2010a; Dal Martello et al., 2013; Tangstad, 2013b; Vangskåsen and Tangstad, 2013b; Eric, 2014):

$$SiO_{(g)} + 2C_{(s)} \rightarrow SiC_{(s)} + CO_{(g)} \tag{5.50}$$

$$3SiO_{(g)} + CO_{(g)} \rightarrow 2SiO_{2(s,l)} + SiC_{(s)} \tag{5.51}$$

$$SiO_{(g)} + SiC_{(s)} \rightarrow 2Si_{(l)} + CO_{(g)} \tag{5.52}$$

Essentially, Reaction 5.50 is controlled by the ability of the carbon materials to react with the SiO gas, and the reactivity of carbon reductants towards SiO gas is assumed to be critical to the performance of the process (Videm, 1995; Myrhaug et al., 2004). In practice, the unreduced $SiO_{(g)}$ that escapes from the high temperature reaction zone of the furnace is recovered by condensation in the lower-temperature zones of the furnace, and the condensate is usually deposited in the charge mix at temperatures below 1680°C, as shown in Equation 5.53 (Schei et al., 1997; Tangstad et al., 2010a; Dal Martello et al., 2013; Tangstad, 2013b; Vangskåsen and Tangstad, 2013b; Eric, 2014):

$$2SiO_{(g)} \rightarrow SiO_{(s,l)} + Si_{(s,l)} \tag{5.53}$$

The condensate behaviour of SiO gas is not only crucial in reducing the loss of silicon in the furnace off-gas (Schei et al., 1997; Tangstad et al., 2010a; Dal Martello et al., 2013; Tangstad, 2013b; Vangskåsen and Tangstad, 2013b; Eric, 2014) but has also been attributed to have a significant effect on the specific energy consumption of the process (Schei et al., 1997; Kamfjord et al., 2010; Kamfjord, 2012; Tangstad, 2013b; Eric, 2014). Technically, some of the SiO eventually leaves the furnace as off-gas, and is normally recovered as condensed silica fume in baghouse filters (Schei et al., 1997; Kamfjord et al., 2010; Dal Martello et al., 2013; Tangstad, 2013b). The characteristics of silica fume in submerged arc furnace dust and its recycling and/or reuse opportunities are further discussed in detail in Section 5.5.6.

5.5.4.2 Characteristics of Silicon-Bearing Raw Materials

The production of silicon and ferrosilicon alloys requires a strict control of the raw materials (Tangstad, 2013b). In general, the raw material requirements for an optimized silicon and ferrosilicon process include (Aasly et al., 2007; Schei et al. 2007; Tangstad, 2013b) (1) high silica content (not less than 97 wt.% SiO_2) and low Al_2O_3 (not more than 1.5 wt.% Al_2O_3); (2) low tramp element composition (i.e. trace amounts of tramp elements such as Al, Ti, B, P, Fe and Ca); (3) proper lump sizes (20–80 mm); (4) good mechanical strength, especially during handling, transportation and charging; (5) good thermal strength (disintegration and decrepitation upon heating) and the (6) softening, volume expansion and melting temperatures of quartz. The composition and characteristics of different quartz and quartzite minerals were discussed in detail by Tangstad (2013b).

In essence, the raw material parameters mentioned earlier have serious implications to the operation of the submerged arc furnace. Naturally, raw material chemistry is important in controlling the grade and yield of the product, particularly in the downstream refining and manufacturing processes such as the refining of silicon to semiconductor-grade silicon and the production of special grades of steels which are sensitive to the presence of tramp elements (Bathey and Cretella, 1982; Istratov et al., 2006; Braga et al., 2008; Lynch, 2009; Pizzini, 2010; Heuer, 2013). In addition, the control of the mechanical and thermal properties is important in controlling the dust generation during handling, charging and operation of the furnace (Tangstad, 2013b; Eric, 2014). Furthermore, fine materials in the electric furnace feed tend to segregate and, more importantly, deposit on the countercurrent charge thereby reducing the porosity of the solids packed bed and resulting in premature fusion and crusting of the furnace charge (Eric, 2014).

The softening and melting parameters essentially affect the properties of the solids packed zone of the SAF (Yang et al., 2004; Tangstad, 2013b). For example, if the quartz materials used begin to melt too high in the furnace, it can result in the formation of sticky solid mass which can inhibit the flow of gases through the charge, thereby resulting in the channelling of gases through the charge (Tangstad, 2013b). Basically, the channelling of gas through the charge bed not only affects the heat transfer between the countercurrent gases and the charge, but also results in high $SiO_{(g)}$ losses in the furnace off-gas which inadvertently results in low silicon yield (Tangstad, 2013b; Eric, 2014). The formation of liquid slag phases high up in the furnace is also detrimental to overall thermal balance of the process, particularly through the liquid phases affecting the electric current distribution in the furnace and also disturbing the arc formation and stability (Tangstad, 2013b).

Based on the aforementioned, the performance of the submerged arc furnace in the production of silicon and its alloys is sensitive to the properties and quality of silica minerals. Detailed discussions on the Si–O and Si–O–C thermodynamic systems and the physicochemical properties of silica are given in several articles (Aasly et al., 2007; Schei et al., 1997; Greenwood and Earnshaw, 2012; Tangstad, 2013b; Eric, 2014).

5.5.4.3 Characteristics of Carbon Reductant Materials

Carbonaceous materials in the form of charcoal, char, wood chips, petroleum coke, coal and metallurgical coke constitute the most common reductant materials used in the production of silicon and its alloys (Myrhaug et al., 2004; Elkin et al., 2010; Kim et al., 2013; Tangstad, 2013b; Kopeć and Machulec, 2016). Basically, coal is a naturally occurring fossil-based reductant, while the coke is produced from the pyrolysis and carbonization of metallurgical-grade coal under controlled conditions (Diez et al., 2002; Arnold, 2013; Tangstad, 2013b). Further details on the essentials of the coke-making process were discussed in Chapter 4.

Char constitutes carbonization products from wood, activated carbons or coal that do not pass through a plastic stage during pyrolysis and carbonization (Tangstad, 2013b). On the other hand, charcoal is derived from the controlled pyrolytic carbonization of organic cellulosic material products at moderate heating rates (Antal and Grønli, 2003; Tangstad, 2013b; Guiotoku et al., 2014). Petroleum coke is another important reductant widely used as a co-reductant in the production silicon and silicon alloys, and is produced as a by-product from the oil refining processes (Elkin et al., 2010; Tangstad, 2013b; Mekhtiev et al., 2014). Despite petroleum coke having desirable properties such as low ash and volatile matter content, it is mostly used in combination with charcoal due to its high cost and poor physical and chemical properties at high temperatures (Mekhtiev et al., 2014).

Basically, the selection of carbon reductant materials is governed by four fundamental considerations, namely, availability, process parameters, required product purity and environment factors (Elkin et al., 2010; Tangstad, 2013b; Li et al., 2015; Kopeć and Machulec, 2016). Based on the aforementioned fundamental considerations, the typical reducing agents used in the production of silicon and its alloys must meet the following criteria (Videm, 1995; Pistorius, 2002; Myrhaug et al., 2004; Sahajwalla et al., 2004; Lindstad et al., 2007; Myrvågnes, and Lindstad, 2007; Elkin et al., 2010; Tangstad, 2013b; Li et al., 2015; Kopeć and Machulec, 2016): (1) chemical composition and reactivity needed to achieve a high conversion rate and completeness of silica reduction, and hence high silicon yield; (2) physicochemical properties essential to achieve optimum charge permeability; (3) minimum impurity content (e.g. P, S, B, Hg, Pb, Na and K); (4) high-temperature reactivity to SiO gas and slag, particularly as the molten charge descends to the hearth during smelting and (5) must be cost effective and readily available. Essentially, the carbon materials are not only a source of carbon for the reduction of silica but also play an important role by acting as a gas filter in the upper zones in the working space of silicon and ferrosilicon furnaces (Kopeć and Machulec, 2016).

Technically, the reactivity of carbon reductants is considered to be a key control parameter in the selection of reductants in the silicon and ferrosilicon processes (Videm, 1995; Pistorius, 2002; Myrhaug et al., 2004; Sahajwalla et al., 2004; Lindstad et al., 2007; Myrvagnes and Lindstad, 2007; Elkin et al., 2010; Tangstad, 2013b; Li et al., 2015). According to these researchers, the reactivity of the reducing materials is categorized by their ability to react with SiO gas according to Equation 5.49. Extensive research on the reactivity of most carbonaceous materials to SiO gas under simulated experiments has revealed that the granulometric composition of the reductants is a key determinant factor (Videm, 1995; Myrhaug et al., 2004; Lindstad et al., 2007; Myrvågnes and Lindstad, 2007; Elkin et al., 2010; Tangstad, 2013b; Li et al., 2015). For example, several studies have explicitly highlighted the specific dependence of the SiO reactivity on the pore-size distribution of the reductant materials, which essentially was determined to be strongly correlated to the effective diffusivity of the SiO gas in the pores of the reductant materials (Raaness and Gray, 1995; Videm, 1995; Myrhaug et al., 2004; Lindstad et al., 2007; Myrvågnes and Lindstad, 2007; Kopeć and Machulec, 2016). The impact of the reactivity of carbon-based reductants with SiO gas on the key economic indicators and efficiency of the silicon and ferrosilicon processes is two-fold, vis-à-vis: (1) minimizing the loss of silicon in the form of SiO gas in furnace off-gas, thereby impacting the yield of silicon and (2) reducing the chemical and thermal energy losses from the process since the gaseous SiO is an energy carrier (Myrvagnes and Lindstad, 2007; Kamfjord et al., 2010; Kopeć and Machulec, 2016).

5.5.4.4 Composition of Industrial-Grade Silicon

The production process and applications of industrial MG-Si have been discussed in the proceeding sections. Typically, the silicon content in industrial grade silicon metal is typically in the range of 98.5 to 99.5 wt.% (Margaria et al., 1992; Dosaj et al., 2005; Tangstad, 2013b). Table 5.5 further summarizes the typical compositions of the different grades of metallurgical silicon (Tangstad, 2013b). According to Margaria et al. (1992), some of the impurity elements highlighted in the table tend to precipitate as grain boundary intermetallic compounds during the solidification of MG-Si. Based on scanning electron microscopy analyses, the typical intermetallic compounds that identified range from binary to quaternary compounds of silicon such as Si_2Ca, Si_2Al_2Ca, $Si_8Al_6Fe_4Ca$, Si_2FeTi, $Si_7Al_8Fe_5$, Si_2Al_3Fe and $Si_{2,4}Fe(Al)$ (Margaria et al., 1992). As discussed in Section 5.5.3.3, the presence of these impurity elements in the MG-Si presents challenges for its use in electronic and photovoltaic applications.

The thermochemical reactions in the submerged arc furnace process and the raw material regime make it difficult to control the level of impurities in industrial silicon to the levels demanded by semiconductor and electronic applications. The demand for low-impurity silicon feedstock for semiconductor applications has been highlighted in the preceding sections. Technically, the refining process of MG-Si reduces the level of the impurities shown in Table 5.5, from the wt.% level to the parts per million (ppm) order that is required in the semiconductor-grade silicon (Bathey and Cretella, 1982; Braga et al., 2008; Lynch, 2009; Pizzini, 2010; Ramos et al., 2015).

TABLE 5.5

Typical Composition of Metallurgical-Grade Silicon

	Typical Composition (Wt.%)				
Grade	**Si**	**Fe**	**Al**	**Ca**	**Fe + Al + Ca**
Si99	99	0.4	0.4	0.4	1
Si98	98	0.7	0.7	0.6	2
Si97	97	1	1.2	0.8	3
Si96	96	1.5	1.5	1.5	4

Source: Reprinted from *Handbook of Ferroalloys: Theory and Technology*, Tangstad, M., Ferrosilicon and silicon technology, Gasik, M. (ed.), Butterworth-Heinemann/Elsevier, Oxford, UK, p. 197. Copyright 2013, with permission from Elsevier.

5.5.4.5 Composition of Ferrosilicon Alloys

The raw materials and process requirements in the generic production of silicon and ferrosilicon alloys were discussed in Sections 5.5.4.2 and 5.5.4.3, respectively. As already discussed in the preceding sections, ferrosilicon is typically an alloy of iron and silicon (Eric, 2014). In addition, carbon may be present in the ferrosilicon alloy as well, mainly as dissolved carbon and SiC particles (Eric, 2014). Although, in principle, the silicon and ferrosilicon production processes are similar, the fundamental difference is that iron-bearing materials in the form of high-quality iron ore and scrap are added as part of the charge in the ferrosilicon alloy process (Dosaj et al., 2005; Tangstad, 2013b; Eric, 2014). In essence, silicon-based ferroalloys are produced by the carbothermic reduction of quartz and/or quartzite and iron oxide materials in semi-open or closed submerged arc furnaces (Dosaj et al., 2005; Tangstad, 2013b; Eric, 2014). Specific requirements of the reductant materials in the ferrosilicon alloy process include (Tangstad, 2013b) (1) high electrical resistance (to enable smelting of charge due to resistance heating), (2) high reactivity in relation to SiO_2 and SiO gas reduction and (3) constant moisture.

The thermodynamic conditions required for improved silicon yield in the production of ferrosilicon alloys are positively enhanced by the presence of iron in the process (Dosaj et al., 2005; Tangstad, 2013b; Eric, 2014). For example: (1) the presence of iron reduces the stability of SiC intermediate reaction product in the high-temperature zone of the furnace, thereby limiting the formation of SiC accretions in the process; (2) the iron lowers the thermodynamic partial pressure of SiO required for its complete reduction to Si, thereby reducing the losses of SiO to furnace off-gas; (3) iron reduces the activity of silicon due to the formation of Fe–Si solutions (e.g. the typical silicon recovery in ferrosilicon alloys is 90%–95% compared to less than 50% for silicon metal process) and (4) the presence of iron has a positive impact on the overall energy demand of the smelting process due to the solution of silicon in liquid iron having a favorable free energy change as well as an exothermic enthalpy change.

Once the smelting is completed, the molten ferrosilicon alloys are tapped from the furnace and solidified (Dosaj et al., 2005; Tangstad, 2013b; Eric, 2014). The most common methods of solidification include atomizing in an inert gas, air granulation, water granulation, casting on flat copper plates and multiple-layer casting (Tangstad, 2013b). It must be noted, however, that the casting process of ferrosilicon alloys requires special care and attention in order to control the particle size and element and phase distribution during the solidification process. Depending on the grade, the ferrosilicon alloys consumed in the steelmaking and foundry industries have variable silicon contents, ranging from 10 to 95 wt.% Si (Dosaj et al., 2005; Tangstad, 2013b; Eric, 2014). However, the most common commercial grades of ferrosilicon alloys are those containing 15, 45, 75 and 90 wt.% silicon (Eric, 2014).

5.5.5 Physicochemical Characteristics of Silicon and Ferrosilicon Slags

Technically, the ferrosilicon production can be considered a slag-free process and the amount of slag produced in the ferrosilicon process does not normally exceed 3–5 wt.% of the alloy mass (Dosaj et al., 2005; Tangstad, 2013b; Eric, 2014). Despite the small amounts produced, the supernatant slags in the production of ferrosilicon alloys are mostly heterogeneous silicate-based melts, consisting of 48–50 wt.% SiO_2, 20–25 wt.% Al_2O_3 and 15–18 wt.% CaO, as well as metallic inclusions of ferrosilicon alloys (Tangstad, 2013b). Depending on the composition of the melt, the main phases in solidified ferrosilicon process slags include anorthite ($CaO \cdot Al_2O_3 \cdot 2SiO_2$) and gehlenite ($2CaO \cdot Al_2O_3 \cdot SiO_2$).

The chemical and morphological properties and the alternative uses of solidified CaO–Al_2O_3–SiO_2–based slags were discussed in detail in Chapter 4. As already discussed, the CaO–Al_2O_3–SiO_2–based slags have found important applications in the production of cementitious and geopolymer materials as well as in the manufacture of ceramic materials. In essence, the chemical and morphological compositions of solidified ferrosilicon slags are similar to the CaO–Al_2O_3–SiO_2 type of slags produced from the blast furnace ironmaking process. Unlike blast furnace slags which are produced in large quantities and have found extensive uses outside the iron and steelmaking industry, the alternative uses of ferrosilicon slags have not been explored extensively to date, mainly due to the small quantities produced. As a result, developing alternative markets for the utilization of these types of slags can help in reducing the volumes of the slags that are disposed of in landfills, thereby improving the overall economics of the ferrosilicon production processes.

5.5.6 Physicochemical Characteristics of Silica Fume Dust

In general, the production of silicon and ferrosilicon alloys produces significant amounts of off-gas, particularly rich in fine silica particles. These silica-rich particulate dusts are generically referred to as silica fume (Schei et al., 1997; Dosaj et al., 2005; Silica Fume Association, 2005; Barati et al., 2011; Siddique and Khan, 2011; Siddique, 2011; Tangstad, 2013b; Vangskåsen and Tangstad, 2013). The chemical composition and morphological characteristics of silica fume vary significantly depending on the quality of raw materials, type of product produced, and overall efficiency of the process. For example, Tangstad (2013b) proposed the silica fume dust in the production of ferrosilicon alloys to be typically composed of 80–95 wt.% SiO_2, 1.5–9 wt.% Al_2O_3, 0.4–2.7 wt.% CaO, 1–5 wt.% MgO and less than 0.5 wt.% Fe_2O_3. In contrast, the silica fume produced from the silicon metal production furnace would predominantly consist of the various forms of amorphous silica. On average, a typical silicon furnace produces between 200 and 400 kg of silica fume dust per ton of silicon metal, representing about 8%–15% of the silica input into the process (Schei et al., 1997). The current global production of silica fume is estimated to be around 1.5 million metric tonnes per annum (Sakulich, 2011).

The thermochemical reactions taking place in the submerged arc furnace production of silicon and ferrosilicon alloys were discussed in the preceding sections. In addition to the entrainment of fine SiO_2 particles in the off gas, the formation of silica fume can also be attributed to the chemical reactions involving the $SiO_{(g)}$ intermediate reaction product. In this case, the SiO gas exits the furnace with other gases and is oxidized with infiltrate air and/or condenses to from fine particles of non-crystalline silica which are then collected with dust and other condensed phases in the gas cleaning system (Schei et al., 1997; Kamfjord et al., 2010; Barati et al., 2011; Siddique, 2011; Siddique and Khan, 2011; Tangstad, 2013b). Basically, the typical oxidation reactions leading to the formation of silica fume is shown in Equation 5.54 (Barati et al., 2011). In addition, the oxidation of SiO gas by CO_2 gas in the furnace freeboard where the oxygen potential is higher can also occur according to Equations 5.54 and 5.55:

$$SiO_{(g)} + \frac{1}{2}O_{2(g)} \rightarrow SiO_{2(\text{silica fume})} \quad (5.54)$$

$$SiO_{(g)} + CO_{2(g)} \rightarrow SiO_{2(\text{silica fume})} + CO_{(g)} \quad (5.55)$$

Fundamentally, the silica fume dust formed from these oxidation reactions is typically very fine; the average particle size is typically less than 5 μm and has high specific area

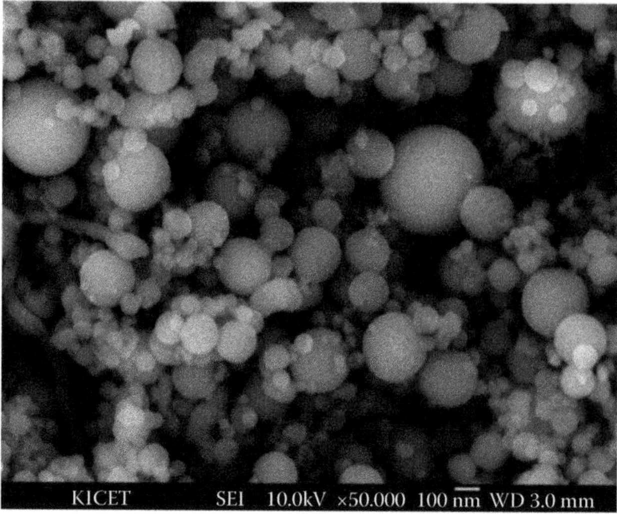

FIGURE 5.7
SEM micrographs of silica fume. (Reprinted from *Constr. Build. Mater.*, 21, Jo, B.W., Kim, C.H., Tae, G.H., and Park, J.B., Characteristics of cement mortar with nano-SiO_2 particles, 1351–1355. Copyright 2007, with permission from Elsevier.)

(~2000 m^2/g) and low bulk density (180–230 kg/m^3) (Jo et al., 2007; Tangstad, 2013b; Zhang et al., 2016). Figure 5.7 shows the scanning electron micrographs of a typical silica fume sample with particle size in the range of 100 nm (Jo et al., 2007).

The fine aggregation of silica fume is problematic in controlling the diffuse and fugitive emissions in the silicon and ferrosilicon production processes (Dosaj et al., 2005; Silica Fume Association, 2005; Tangstad, 2013b). In practice, the raw silica fume dust is usually densified in order to facilitate handling and mitigate risks associated with handling superfine materials (Silica Fume Association, 2005; Zhang et al., 2016). As discussed in Chapter 2, health problems are associated with exposure to these silica compounds, particularly the silica existing in the form of crystalline silica (Dosaj et al., 2005; Silica Fume Association, 2005; Kamfjord, 2012; Tangstad, 2013b). As discussed previously, prolonged exposure to crystalline silica dust results in serious health problems such as pulmonary fibrosis and other related chronic respiratory conditions.

5.5.6.1 Utilization of Silica Fume Waste Dusts

Extensive research has been explored to find alternative uses of silicon and ferrosilicon furnace dusts. Some of the areas of research exploring the valorization of silica fume dusts include (1) utilization in the construction industry (Chatterji et al., 1982; Buil et al., 1984; Mitchell et al., 1998; Chung, 2002; Silica Fume Association, 2005; Jo et al., 2007; Lothenbach et al., 2011; Siddique, 2011; Siddique and Chahal, 2011a,b; Siddique and Khan, 2011; Sanjuán et al., 2015; Zhang et al., 2016), (2) reuse as raw material in the manufacture of ceramic and composite materials (Ewais et al., 2009; Zhong et al., 2010; Suri et al., 2011; Khattab et al., 2012; Wang et al., 2012; Ji et al., 2014), (3) as raw material in the recovery of silicon by hydrometallurgical methods (Barati et al., 2011), (4) as source of agglomerated silica flux in metallurgical processes and (5) as raw material in the production of sodium silicate glasses (Tangstad, 2013b; Oxychem, n.d.). The following sections briefly discuss the potential applications of silica fume in ceramics and as supplementary cementitious materials for the construction industry.

5.5.6.2 Utilization of Silica Fume in Ceramic and Composite Materials

Several researchers have explored the use of waste silica fumes in the synthesis of ceramic and composite materials for high-temperature applications (Ewais et al., 2009; Zhong et al., 2010; Suri et al., 2011; Khattab et al., 2012; Wang et al., 2012; Ji et al., 2014). Furthermore, Table 5.6 summarizes some of the typical studies focused on the utilization of silica fumes in the production of engineered ceramic and composite materials.

5.5.6.3 Utilization of Silica Fume in Cementitious Systems

Extensive research has been conducted to explore the potential applications of silica fume as supplementary cementitious materials in the construction industry. As discussed in the preceding sections, silica fume mostly consists of ultra-fine particles of non-crystalline or amorphous silica. The applications of silica fume as supplementary cementitious materials in the blended cements are well documented (Chatterji et al., 1982; Buil et al., 1984; Mitchell et al., 1998; Chung, 2002; Silica Fume Association, 2005; Jo et al., 2007; Lothenbach et al., 2011; Siddique, 2011; Siddique and Chahal, 2011a,b; Siddique and Khan, 2011; ASTM C1240-15, 2015; Sanjuán et al., 2015; Zhang et al., 2016). Basically, some of the specific applications of silica fume in the construction industry include (Siddique, 2011; Siddique and Chahal, 2011a,b;

TABLE 5.6

Research Activities Focused on the Utilization of Silica Fume in Ceramics and Composite Materials

Process Description	Engineered Product Characteristics	Reference
Synthesis of carbon-free Si_3N_4/SiC nanopowders using integrated mechanical and thermal activation process at 1465°C under mild oxidizing conditions	High-temperature applications in turbine and automobile engine components due to good mechanical and thermal properties at elevated temperatures	Suri et al. (2011)
Synthesis of SiC nanopowder from silica fume at 1500°C via carbothermic reduction and mechano-chemical milling	Low-cost structural material in grinding tools, abrasive media, mechanical seals and blasting nozzles	Zhong et al. (2010)
Crystallization synthesis of nano-crystalline mullite whiskers using molten $Al_2(SO_4)_2$–Na_2SO_4 salts in the temperature range 750°C–900°C	Engineered high-temperature characteristics based on good mechanical strength, excellent thermal shock, high creep resistance, low thermal coefficient of expansion and good dielectric properties of mullite whiskers	Wang et al. (2012)
Synthesis of sintered cordierite-mullite nano-macro composites using silica fume, calcined ball clay and calcined alumina and magnesia at 1400°C	Improved mechanical and thermal properties for high-temperature applications	Khattab et al. (2012)
Synthesis of porous cordierite ceramic from silica fume, bauxite and talc using polymeric sponge replica technique	Potential engineering material for efficient particulate removal from hot gas streams due to porous thermostable and low thermal coefficient of expansion	Ewais et al. (2009)
Synthesis of Si_3N_4 powder with tunable α/β-Si_3N_4 content from carbothermal reduction of waste silica fume	Engineered high-temperature applications based on excellent thermal shock resistance, good mechanical properties and good ambient and high temperatures chemical stability of Si_3N_4	Ji et al. (2014)

Sanjuán et al., 2015) (1) production of high-performance concrete containing silica fume for highway bridges, parking decks, marine structures and bridge deck overlays; (2) high-strength concrete enhanced with silica fume for greater design flexibility; (3) silica fume shotcrete for use in rock stabilization, mine tunnel linings and rehabilitation of deteriorating bridges, marine columns and piles; (4) oil well grouting and (5) use in a variety of cementitious material repair products. In fact, these studies have proved the efficacy of silica fume as a pozzolanic admixture that is effective in enhancing the mechanical, chemical and physical properties of Portland cement blends (Chatterji et al., 1982; Buil et al., 1984; Mitchell et al., 1998; Chung, 2002; Silica Fume Association, 2005; Jo et al., 2007; Lothenbach et al., 2011; Siddique, 2011; Siddique and Chahal, 2011a,b; Siddique and Khan, 2011; ASTM C1240-15, 2015; Sanjuán et al., 2015; Zhang et al., 2016).

Silica fume is a very reactive pozzolanic material because of the extremely fine particles and very high amorphous silica content (Lothenbach et al., 2011; Siddique, 2011; Siddique and Chahal, 2011a,b; ASTM C1240-15, 2015; Zhang et al., 2016). In general, silica fume has been recognized as one of the most effective supplementary cementitious materials which can result in improved concrete durability through pozzolanic reactions with free lime, pore-size refinement and matrix densification as well as cement paste–aggregate interfacial improvements (Zhang et al., 2016). It is proposed that the effect of silica fume in concrete blends takes place via any one of the following mechanistic steps (Mitchell et al., 1998; Lothenbach et al., 2011; Siddique, 2011; Siddique and Chahal, 2011a,b; Siddique and Khan, 2011; Zhang et al., 2016): (1) pore-size refinement and matrix densification, (2) silica fume particles acting as nucleation sites for cement hydration thereby accelerating the process and (3) pozzolanic reactions occurring between the silica-rich silica fume particles and the cement hydration product portlandite.

The documented benefits of utilizing silica fume in cement blends include a substantial increase in compressive strength and durability of hardened concrete when added in optimum amounts (Lothenbach et al., 2011; Siddique and Chahal, 2011a,b; Siddique and Khan, 2011; Sanjuán et al., 2015). In particular, some of the specific advantages of using silica fume as supplementary cementitious materials include (Buil et al., 1984; Lothenbach et al., 2011; Siddique and Chahal, 2011a,b; Siddique and Khan, 2011; Sanjuán et al., 2015) (1) improved mechanical strength as a result of high early compressive strength, high tensile strength, high flexural strength and high bond strength and (2) higher durability as a result of sulphate resistance, higher electrical resistivity, low permeability to chloride and increased resistance to chemical attack, freeze thaw resistance, controlled expansion due to alkali–silica reaction and good fire resistance.

5.5.6.4 Summary

This section highlighted the key aspects pertinent in the production of silicon and ferrosilicon alloys. Furthermore, the different industrial uses of silicon and their effect in driving the future growth of global silicon production were highlighted. The critical raw material properties such as the physical and chemical properties of silicon-bearing raw materials and the reactivity of carbon reductants in the submerged arc furnace production of silicon and its alloys were highlighted. Although the processes of silicon and ferrosilicon alloy production are somewhat similar, the fundamental difference between these two processes is the presence of iron in the ferrosilicon alloy production.

The silicon and ferrosilicon processes can essentially be classified as slagless processes. This means that the volume of process slags produced in the production of silicon metal and ferrosilicon alloys is significantly lower that produced from the production of other ferroalloys. However, the these two processes produce significant amounts of particulate dusts. As such, this section focused mostly on the production and physicochemical properties of particulate dusts. The particulate dust produced in these processes is particularly rich in non-crystalline silica or silica fume. The alternative uses of silica fume as additives in cementitious and pozzolanic systems, and its utilization as a raw material in the production of ceramic materials for high-temperature applications, were also discussed in detail.

5.6 Concluding Remarks

This chapter discussed the ubiquitous industrial applications and production processes of various grades of bulk ferroalloys from their ores. In particular, the production of different ferrochromium, ferromanganese and ferrosilicon alloys in submerged arc furnaces was discussed. Although silicon is not technically a ferroalloy, the production processes of silicon metal were considered in this chapter, mainly due to the critical applications of silicon products in chemicals (silicon-based polymers), ferrous and non-ferrous alloys production, as well as in semiconductor and photovoltaic applications.

To date, the production of these bulk ferroalloys has been on the upward trend, critically buoyed by growth in special grades of steels and non-ferrous alloys for lightweight applications. In terms of waste production and utilization in the bulk ferroalloys industry, the factors considered in this chapter were broken down into raw materials effects, equipment design effects, process chemistry effects and valorization potential of the different waste streams. In detail, the key issues arising can be summarized as follows:

1. The need for raw materials quality control, particularly the engineered composition and physical characteristics of the various ores and the reactivity and physical characteristics of reductants. For example, the use of agglomerated fines and lumps with low decrepitation potential is particularly important in controlling the amount of particulate dust emissions. The effect of raw material melting and decrepitation behaviour in the submerged arc furnace processing of the different ferroalloys was also highlighted. Furthermore, the control of feed properties is essential in minimizing the amount and the physicochemical properties of the resulting slags, which essentially are responsible for increasing the alloy losses to slag.
2. The design of the furnace and ancillary equipment as well as the furnace operating procedures has been highlighted as critical in mitigating particulate dust emissions. In particular, the nature and quantity of particulate dust emissions is basically driven by fundamental factors such as type of furnace used (i.e. open, semi-closed or closed); type of process (i.e. continuous or batch processing); operating procedures such as operating at variable loads or with insufficient electrode submergence and the selection, design and maintenance practices of dust collection systems. Other critical factors in mitigating the particulate dust and alloy entrapment in slag include furnace atmosphere, charge of mineralogical composition and furnace equipment size. To date, the use of continuous processing and closed furnaces is more or less mandatory in mitigating the emission of particulate dusts in the production of these ferroalloys.
3. Process chemistry is fundamentally critical in controlling the loss of ferroalloys to slag. The fundamental effects of slag chemistry variables such as liquidus

temperature, slag viscosity and slag basicity on the partition of different alloy elements between the alloy bath and slag phases have been highlighted. For example, solidified slags in the production of ferrochromium alloys tend to contain significant amounts of chromium in the form of PACs, entrained ferrochromium alloy, entrained chromium carbides and entrained iron-chromium spinels, a phenomenon which is worsened by poor control of the physicochemical properties of the slags.

4. Various strategies applicable to increasing the valorization potential of ferroalloy wastes through metal recovery, recycling and reuse of the waste materials were highlighted. Most of the waste streams produced in the production of bulk ferroalloys are classified as hazardous and toxic materials, and as such the emphasis on reducing the amount of waste materials, recycling and reuse of the different waste streams formed the bulk of the material covered in this chapter. Furthermore, the emphasis of this chapter was to highlight the anthropogenic effects of the different waste streams, and also to highlight potential areas applicable to the valorization of the different waste streams.
5. In particular, the chapter discussed the different approaches applicable to the immobilization of heavy metal ions in stable slag phases, such as the spinel crystal structure in ferrochromium alloy slags, as a way of increasing the reuse potential of slags. In particular, future emphases on immobilizing the hazardous and toxic heavy metals in slag through engineered slag properties were fully discussed. In essence, slag engineering principles to meet both the metallurgical and environmental needs of the different ferroalloys production processes are imperative to developing solutions for the holistic management of bulk ferroalloy wastes, particularly process slags.

References

Aasly, K., Malvik, T. and Myrhaug, E.H. (2007). A review of previous work on important properties of quartz for FeSi and Si metal production. *Proceedings of the 11th International Ferroalloys Congress*, New Delhi, India, pp. 393–401.

Agarwal, S., Pal, J. and Ghosh, D. (2014). Development of chromite sinter from ultra-fine chromite ore by direct sintering. *ISIJ International* 54(3): 559–566.

Akdogan, G. and Eric, R.H. (1994). Carbothermic reduction behavior of wessel manganese ores. *Minerals Engineering* 7(5/6): 633–645.

Akyuzlu, M. and Eric, R.H. (1992). Slag-metal equilibrium in the smelting of high-carbon ferrochromium. *Journal of the Southern African Institute of Mining and Metallurgy* 92(4): 101–110.

Albertsson, G.J. (2011). Investigations of stabilization of Cr in spinel phase in chromium-containing slags. Licentiate thesis. Royal Institute of Technology, Stockholm, Sweden.

Albertsson, G.J. (2013). Abatement of chromium emissions from steelmaking slags-Cr stabilization by phase separation. Doctoral thesis. Royal Institute of Technology, Stockholm, Sweden.

Albertsson, G.J., Teng, L. and Björkman, B. (2014). Effect of basicity on chromium partition in CaO-MgO-SiO_2-Cr_2O_3 synthetic slag at 1873K. *Mineral Processing and Extractive Metallurgy (Transactions of the Institute of Mining and Metallurgy)* 132(2): 116–122.

Angadi, S.I., Rao, D.S., Prasad, A.R. and Rao, R.B. (2011). Recovery of ferrochrome values from flue dust generated in ferroalloy production: A case study. *Mineral Processing and Extractive Metallurgy (Transactions of the Institution of Mining and Metallurgy)* C120(1): 61–63.

Antal, M.J. and Grønli, M. (2003). The art, science and technology of charcoal production. *Industrial and Engineering Chemistry Research* (American Chemical Society) 42: 1619–1640.

Arnold, B.J. (2013). Coal formation. In: Osborne, D. (ed.), *The Coal Handbook: Towards Cleaner Production, Vol. 1: Coal Production*. Woodmead Publishing House, Cambridge, UK, pp. 31.

ASTM C1240-15 (2015). *Standard Specification for Silica Fume Used in Cementitious Mixtures*. ASTM International, West Conshohocken, PA. Available at www.astm.org.

Baba, A.A., Ibrahim, L., Adekola, F.A., Bale, R.B., Ghosh, M.K., Sheik, A.R., Pradhan, S.R., Ayanda, O.S. and Folorunsho, I.O. (2014). Hydrometallurgical processing of manganese ores: A review. *Journal of Minerals and Materials Characterization and Engineering* 2: 230–247.

Bai, Z.T., Zhang, Z.A., Guo, M., Hou, X.M. and Zhang, M. (2015). Magnetic separation and extraction of chrome from high carbon ferrochrome slag. *Materials Research Innovations* 19(S2): 113–188.

Barati, M., Sarder, S., McLean, A. and Roy, R. (2011). Recovery of silicon from silica fume. *Journal of Non-Crystalline Solids* 357: 18–23.

Barbieri, L., Leonelli, C., Manfredini, T., Pellacani, G.C. and Siligardi, C. (1994). Solubility, reactivity and nucleation effect of Cr_2O_3 in CaO-MgO-Al_2O_3-SiO_2 glassy system. *Journal of Materials Science* 29: 6273–6280.

Barnhart, J. (1997a). Occurrences, uses, and properties of chromium. *Regulatory Toxicology and Pharmacology* 26: S3–S7.

Barnhart, J. (1997b). Chromium chemistry and implications for the environmental fate and toxicity. *Journal of Soil Contamination* 6(6): 561–568.

Barozzi, S. (1997). Metal recovery from steel wastes by the INMETCO process. *Steel Times* 225: 191–192.

Basson, J. and Daavittila, J. (2013). High carbon ferrochrome technology. In: Gasik, M.M. (ed.), *Handbook of Ferroalloys: Theory and Technology*. Butterworth-Heinemann, Elsevier, Oxford, UK.

Batchelor, B. (2006). Overview of waste stabilization. *Waste Management* 26: 689–698.

Bathey, B.R. and Cretella, M.C. (1982). Solar-grade silicon: Review. *Journal of Materials Science* 17: 3077–3096.

Baumgartner, S.J. and Groot, D.R. (2014). The recovery of manganese products from ferromanganese slag using a hydrometallurgical route. *The Journal of the Southern African Institute of Mining and Metallurgy* 114: 331–340.

Bedinger, G.M., Corathers, L.A., Kuck, P.H., Papp, J.F., Polyak, D.E., Schnebele, E.K., Shedd, K.B. and Tuck, C.A. (2015). Ferroalloys (advance release). 2013 Minerals Yearbook. U.S. Geological Survey, Reston, VA. Available at http://minerals.usgs.gov/minerals/pubs/commodity/ferroalloys/myb1-2013-feall.pdf. Accessed 10 March 2016.

Berg, H., Laux, H., Johansen, S.T. and Klevan, O.S. (1999). Flow pattern and alloy dissolution during tapping of steel furnaces. *Ironmaking and Steelmaking* 26(2): 127–139.

Beskow, K., Rick, C.J. and Vesterberg, P. (2015). Refining ferroalloys with the CLU® process. *Proceedings of the 14th International Ferroalloys Congress*, Kiev, Ukraine, pp. 256–263.

Beukes, J.P., Dawson, N.F. and Van Zyl, P.G. (2010). Theoretical and practical aspects of Cr (VI) in the South African ferrochrome industry. *Journal of the Southern African Institute of Mining and Metallurgy* 110: 743–750.

Beukes, J.P. and Guest, R.N. (2001). Cr (VI) generation during milling: Technical note. *Minerals Engineering* 14(4): 423–426.

Beukes, J.P., Roos, H., Shoko, L., Van Zyl, P.G., Neomagas, H.W.J.P., Strydom, C.A. and Dawson, N.F. (2013). The use of thermochemical analysis to characterize Söderberg electrode paste raw materials. *Minerals Engineering* 46–47: 167–176.

Beukes, J.P., Van Zyl, P.G. and Ras, M. (2012). Treatment of Cr (VI)-containing wastes in the South African ferrochrome industry – A review of currently applied methods. *The Journal of the Southern African Institute of Mining and Metallurgy* 112: 347–352.

Bhonde, P.J., Ghodgaonkar, A.M. and Angal, R.D. (2007). Various techniques to produce low carbon ferrochrome. *Proceedings of the 11th International Ferroalloys Conference*, New Delhi, India, pp. 85–90.

Biermann, W., Cromarty, R.D. and Dawson, N.F. (2012). Economic modelling of a ferrochrome furnace. *The Journal of the Southern African Institute of Mining and Metallurgy* 112: 301–308.

Braga, A.F.B., Moreira, S.P., Zampieri, P.R., Bacchin, J.M.G. and Mei, P.R. (2008). New process for the production of solar-grade polycrystalline silicon: A review. *Solar Energy Materials and Solar Cells* 92: 418–424.

Brakestad, A.M. (2015). Characterization of workplace aerosols in the manganese alloy production by X-ray diffraction, inductively coupled plasma optical emission spectrometry, and anion chromatography. MSc thesis. Norwegian University of Life Sciences, Akershus, Norway.

Bristow, D.J., Carter, R.J., Frakes, H.E., Lewis, D.A., Mang, J.F., Satte, S. and Wardle, D. (2000). *Ferroalloys and Alloying Additives Online Handbook.* AMG Vanadium, Cambridge, OH. Available at http://www.metallurgvanadium.com/contents.html. Accessed 14 March 2016.

Buckenham, M.H. (1961). Beneficiation of manganese ores with particular reference to the treatment of a low grade ore from Viti Levu, Fiji. *New Zealand Journal of Geology and Geophysics* 4: 136–147.

Buil, M., Paillère, A.M. and Roussel, B. (1984). High strength mortars containing condensed silica fume. *Cement and Concrete Research* 14: 693–704.

Bulut, U., Ozverdi, A. and Erdem, M. (2009). Leaching behavior of pollutants in ferrochrome arc furnace dust and its stabilization/solidification using ferrous sulphate and Portland cement. *Journal of Hazardous Materials* 162: 893–898.

Calvert, J.B. (2004). Chromium and manganese. Available at http://mysite.du.edu/~jcalvert/phys/chromang.htm. Accessed 6 April 2016.

Cardarelli, F. (2008). *Materials Handbook: A Concise Desktop Reference*, 2nd edn. Springer Science and Business Media, London, UK.

Cengizler, H. and Eric, R.H. (1992). Thermodynamic activity of manganese oxide in ferromanganese slags, and the distribution of manganese between the metal and slag phases. *Proceedings of the International Ferroalloy Conference*, Johannesburg, South Africa, pp. 167–174.

Cha, W.Y., Miki, T., Sasaki, Y. and Hino, M. (2008). Temperature dependence of Ti deoxidation equilibria of liquid iron in coexistence with Ti_3O_5 and Ti_2O_3. *ISIJ International* 48(6): 729–738.

Chatterji, S., Thaulow, N. and Christensen, P. (1982). Pozzolanic activity of byproduct silica fume from ferrosilicon production. *Cement and Concrete Research* 12: 781–784.

Chen, Q.Y., Tyrer, M., Hills, C.D., Yang, X.M. and Carey, P. (2009). Immobilization of heavy metal in cement-based solidification/stabilization: A review. *Waste Management* 29: 390–403.

Chetty, D. (2010). A geometallurgical evaluation of the ores of the Northern Kalahari manganese deposits, South Africa. DPhil thesis. University of Johannesburg, Johannesburg, South Africa.

Chow, N., Nacu, A., Warkentin, D., Aksenov, I. and The, H. (2010). The recovery of manganese from low grade resources: Bench scale metallurgical test program completed. Report prepared for American Manganese Inc. KEMETCO Research Inc. Available at http://www.americanmanganeseinc.com/wp-content/uploads/2011/08/American-Manganese-Phase-II-August-19-2010-Final-Report-Internet-Version-V2.pdf. Accessed 1 May 2016.

Chung, D.D.L. (2002). Improving cement based materials by using silica fume: Review. *Journal of Materials Science* 37: 673–682.

Ciftja, A., Engh, T.A. and Tangstad, M. (2008). Refining and recycling of silicon: A review. Norwegian University of Science and Technology Report. Available at http://www.diva-portal.org/smash/get/diva2:123654/FULLTEXT01.pdf. Accessed 10 May 2016.

Coetsee, T., Ziestman, J. and Pistorius, C. (2014). Predicted effect of ore composition on slag formation in manganese ore reduction. *Mineral Processing and Extractive Metallurgy (Transactions of the Institution of Mining and Metallurgy)* 123C(3): 141–147.

Cohen, B., Cilliers, J.J. and Petrie, J.G. (1997). Optimization of solidification/stabilization treatment of ferroalloy waste products through factorial design. *Journal of Hazardous Materials* 54: 175–188.

Cohen, B. and Petrie, J.G. (1997). Containment of chromium and zinc in ferrochromium flue dust by cement-based solidification. *Canadian Metallurgical Quarterly* 36(4): 251–260.

Cohen, B. and Petrie, J.G. (2005). The application of batch extraction tests for the characterization of solidified ferroalloy waste product. *Journal of Environmental Management* 76: 221–229.

Colas, A. (n.d.). Silicones: Preparation, properties and performance. Dow Corning Life Sciences Report. Available at https://www.dowcorning.com/content/publishedlit/01-3077.pdf. Accessed 12 May 2016.

Conner, J.R. and Hoeffner, S.L. (1998). A critical review of stabilization/solidification technology. *Critical Review in Environmental Science and Technology* 28(4): 397–462.

Corathers, L.A. (2003). Silicon. *U.S. Geological Survey Minerals Yearbook*. Available at http://minerals.usgs.gov/minerals/pubs/commodity/silicon/silicmyb03.pdf. Accessed 15 May 2016.

Corathers, L.A. (2014). Manganese. (Advance Release). *2012 Minerals Yearbook*. U.S. Geological Survey. Available at http://minerals.usgs.gov/minerals/pubs/commodity/manganese/myb1-2012-manga.pdf. Accessed 6 April 2016.

Corathers, L.A., Gambogi, J., Kuck, P.H., Papp, J.F., Polyak, D.E. and Shedd, K.B. (2012). Ferroalloys (Advance release). *2010 Minerals Yearbook*. U.S. Geological Survey. Available at http://minerals.usgs.gov/minerals/pubs/commodity/ferroalloys/myb1-2010-feall.pdf. Accessed 10 March 2016.

Corathers, L.A. and Machamer, J.F. (2006). Chromite. In: Kogel, J.E., Trivedi, N.C., Barker, J.M. and Krukowski, S.T. (eds.), *Industrial Minerals and Rocks: Commodities, Markets and Uses*, 7th edn. Society for Mining, Metallurgy and Exploration, Englewood, CO.

Cornelius, D.J. and Monroe, C.M. (2004). Unique properties of silicone and fluorosilicone elastomers. *Polymer Engineering and Science* 25(8): 467–473.

Cox, X.B., Linton, R.W. and Butler, F.E. (1985). Determination of chromium speciation in environmental particles. Multitechnique study of ferrochrome smelter dust. *Environmental Science and Technology* 19: 345–352.

Dal Martello, E., Tranell, G., Ostrovski, O., Zhang, G., Raanes, O., Larsen, R.B., Tang, K. and Koshy, P. (2013). Trace elements in the Si furnace: Part II: Analysis of condensate in carbothermal reduction of quartz. *Metallurgical and Materials Transactions* 44B: 244–251.

Darcovich, K., Jonasson, K.A. and Capes, C.E. (1997). Developments in the control of fine particulate materials: Invited review. *Advanced Powder Technology* 8(3): 179–215.

Das, B., Mohanty, J.K., Reddy, P.S.R. and Ansari, M.I. (1997). Characterization and beneficiation studies of charge chrome slag. *Scandinavian Journal of Metallurgy* 26(4): 153–157.

Das, S.C., Sahoo, P.K. and Rao, P.K. (1982). Extraction of manganese from low grade manganese ores by $FeSO_4$ leaching. *Hydrometallurgy* 8: 35–47.

De Araujo, J.A.M., Das Mercês Reis de Castro, M. and De Freitas Cunha Lins, V. (2006). Reuse of furnace fines of ferroalloy in the production of electrolytic manganese production. *Hydrometallurgy* 84: 204–210.

De Faria, G.L., Reis, E.L. da Silva Araujo, F.G., Vieira, C.B., von Kruger, F.L. and Jannotti Jr., N. (2008). Characterization of manganese alloy residues for the recycling of FeSiMn and high carbon FeMn fines. *Materials Research* 11(4): 405–408.

Degel, R., Fröhling, C., Hansmann, T., Kappes, H. and Barozzi, S. (2015). Zero waste concept in steel production. *METEC & 2ND ESTAD*, Dusseldorf, Germany, pp. 1–9.

Denton, G.M., Barcza, N.A., Scott, P.D. and Fulton, T. (2005). EAF stainless steel dust processing. *John Floyd International Symposium on Sustainable Developments in Metals Processing*, Melbourne, Australia, pp. 273–284.

Department of Water Affairs and Forestry (1998). Minimum requirements for waste disposal by landfill, 2nd edn. Pretoria, South Africa. Available at https://www.dwa.gov.za/dir_wqm/docs/Pol_Landfill.PDF. Accessed 25 May 2016.

Deqing, Z., Jian, L., Jian, P. and Aoping, H. (2008). Sintering behavior of chromite fines and the consolidation mechanism. *International Journal of Mineral Processing* 86: 58–67.

Dhal, B., Thatoi, H.N., Das, N.N. and Pandey, B.D. (2013). Chemical and microbial remediation of hexavalent chromium from contaminated solid and mining/metallurgical solid waste: A review. *Journal of Hazardous Materials* 250–251: 272–291.

Diez, M.A., Alvarez, R. and Barriocanal, C. (2002). Coal for metallurgical coke production: Predictions of coke quality and future requirements for coke making. *International Journal of Coal Geology* 50: 389–412.

Ding, J., McCormick, P.G. and Street, R. (1995). Formation of spinel Mn-ferrite during mechanical alloying. *Journal of Magnetism and Magnetic Material* 171: 309–314.

Ding, W. and Olsen, S.E. (1996). Reaction equilibria in the production of manganese ferroalloys. *Metallurgical and Materials Transactions* 27B: 5–17.

Di Sabatino, M. and Arnberg, L. (2009). Castability of aluminium alloys. *Transactions of the Indian Institute of Metals* 62(4–5): 321–325.

Dosaj, V., Kroupa, M. and Bittar, R. (2005). Silicon and silicon alloys, chemical and metallurgical. In: Kirk-Othmer (ed.), *Kirk-Othmer Encyclopedia of Chemical Technology*. John Wiley & Sons, NY.

Downing, J.H. (1975). Smelting chrome ore. *Geochemica et Cosmochimica Acta* 39: 854–856.

Downing, J.H. (1985). Basis for producing ferroalloys in an electric smelting furnace. In: Taylor, R. and Custer, C.C. (eds.), *Electric Furnace Steelmaking*. Iron and Steel Making Society, Book Crafters, Chelsea, New York.

Downing, J.H. (2013). Manganese processing. *Encyclopedia Britannica*. Available at http://global.britannica.com/technology/manganese-processing. Accessed 6 April 2016.

Dunkley, J.J. and Norval, D. (2004). Atomization of ferroalloys. *Proceedings of the 10th International Ferroalloys Congress*, Cape Town, South Africa, pp. 541–547.

Du Preez, S.P., Beukes, J.P. and Van Zyl, P.G. (2015). Cr (VI) generation during flaring of CO-rich off-gas from closed ferrochromium submerged arc furnaces. *Metallurgical and Materials Transactions* 46B: 1002–1010.

Durinck, D., Engstrom, F., Arnout, S., Huelens, J., Jones, P.T., Bjorkman, B., Blanpain, B. and Wollants, P. (2008). Hot stage processing of metallurgical slags: Review. *Resources, Conservation and Recycling* 52: 1121–1131.

Durinck, D., Jones, P.T., Blainpain, B. and Wollants, P. (2007). Slag solidification modeling using the Scheil-Gulliver assumptions. *Journal of American Ceramic Society* 90(4): 1177–1185.

Eksteen, J.J., Frank, S.J. and Reuter, M.A. (2002). Dynamic structures in variance based data reconciliation adjustments for a chromite smelting furnace. *Minerals Engineering* 15: 931–943.

Elkin, K.S., Fedorov, N.I., Sporykhin, V.S. and Cherevko, A.E. (2010). Production of high quality ferrosilicon using petroleum coke. *Steel in Translation* 40(11): 983–984.

Environmental Protection Agency (EPA) (1995). Metallurgical industry. In: *Complication of Air Pollutant Emissions Factors: Stationary Point and Area Sources*, (450AP425ED), Vol. 1, 5th edn. Office of Air Quality and Planning Standards, Research Triangle Park, NC 12: 4–1.

Eric, R.H. (2014). Production of ferroalloys. In: Seetharaman, S., McLean, A., Guthrie, R. and Seetharaman, S. (eds.), *Treatise on Process Metallurgy, Vol. 3: Industrial Processes, Part A*. Elsevier, Oxford, UK.

Eric, R.H. and Akyuzlu, M. (1990). Slag-metal equilibria in the system Fe-Cr-Si-C-Ca-Mg-Al-O. Mintek Report No. M405. Johannesburg, South Africa.

Eric, R.H. and Demir, O. (2014). Dissolution of chromite in slags. *Mineral Processing and Extractive (Transactions of Institute of Mining and Metallurgy C)* 123(1): 2–9.

Eskin, D.G. and Kharakterova, M.L. (2001). The effect of silicon and copper on the precipitation hardening of sheets of a 6XXX-series alloy. *Materiali in Tehnologije* 35(1–2): 5–8.

European Commission (2014). Ferroalloys. Best Available Techniques (BAT) Reference Document for the Non-ferrous Metals Industries. Industrial Emissions Directive 2010/75/EU (Integrated Pollution Prevention and Control). Available at http://eippcb.jrc.ec.europa.eu/reference/BREF/NFM_Final_Draft_10_2014.pdf. Accessed 10 March 2016.

European Commission (n.d.). Critical raw materials profiles. Report on Critical Raw Materials for the EU. Available at http://ec.europa.eu/growth/sectors/raw-materials/specific-interest/critical/index_en.htm. Accessed 15 May 2016.

EUROSLAG (2015). Slag: A high grade product out of a high quality controlled industry. The European Association Representing Metallurgical Slag Producers and Processors. http://www.euroslag.com/products/properties/. Accessed 15 December 2015.

Ewais, E.M.M., Ahmed, Y.M.Z. and Ameen, A.M.M. (2009). Preparation of porous cordierite ceramic using a silica secondary resource (silica fumes) for dust filtration purposes. *Journal of Ceramic Processing Research* 10(6): 721–728.

Faria, G.L., Jannotti, N. and Da Silva Araujo, F.G. (2012). Decrepitation behavior of manganese lump ores. *International Journal of Mineral Processing* 102–103: 150–155.

Faria, G.L., Jannotti, N. and Da Silva Araújo, F.G. (2014). Particle disintegration of an important Brazilian manganese lump ore. *REM Revista Escola de Minas Ouro Preto* 67(1): 55–60.

Forsbarcka, L. and Holappa, L. (2004). Viscosity of SiO_2-CaO-CrOx slags in contact with metallic chromium and application of the Iida model. *Seventh International Conference on Molten Slags, Fluxes and Salts*, Cape Town, South Africa, pp. 129–136.

Frías, M. and Rodríguez, C. (2008). Effect of incorporating ferroalloy industry wastes as complimentary cementing materials on the properties of blended cement matrices. *Cement and Concrete Composites* 30: 212–219.

Frias, M., Sáchez de Rojas, M.I., Santamaría, J. and Rodríguez, C. (2006). Recycling of silicomanganese slag as pozzolanic material in Portland cements: Basic and engineering properties. *Cement and Concrete Research* 36: 487–491.

Gaal, S., Tangstad, M. and Ravary, B. (2010). Recycling of waste materials from the production of FeMn and SiMn. *Proceedings of the 12th International Ferroalloys Congress*, Helsinki, Finland, pp. 81–87.

Gao, J.T., Li, S.Q., Zhang, Y.L. and Chen, P.Y. (2012a). Experimental study on the solid state recovery of metallic resources from EAF dust. *Ironmaking and Steelmaking* 39(6): 446–453.

Gao, Y., Olivas-Martinez, M., Sohn, H.Y., Kim, H.G. and Kim, C.W. (2012b). Upgrading of low grade manganese ore by selective reduction of iron oxide and magnetic separation. *Metallurgical and Materials Transactions* 43B: 1465–1475.

Gasik, M.M. (2013). Basics of ferroalloys. In: Gasik, M.M. (ed.), *Handbook of Ferroalloys: Theory and Technology*. Butterworth-Heinemann, Elsevier, Oxford, UK.

Ghafarizadeh, B., Rashchi, F. and Vahidi, E. (2011). Recovery of manganese from electric arc furnace dust of ferromanganese production units by reductive leaching. *Minerals Engineering* 24: 174–176.

Gjønnes, K., Skogstad, A., Hetland, A., Ellingsen, D.G., Thomassen, Y. and Weinnruch, S. (2011). Characterization of workplace aerosols in the manganese alloy production industry by electron microscopy. *Analytical and Bioanalytical Chemistry* 399: 1011–1020.

Glasser, F.P. (1997). Fundamental aspects of cement solidification and stabilization. *Journal of Hazardous Materials* 52(2–3): 151–170.

Greenwood, N.N. and Earshaw, A. (2012). *Chemistry of the Elements*, 2 edn. Elsevier-Butterworth-Heinemann, Burlington, VT, p. 345.

Groot, D.R., Kazadi, D.M., Pollmann, H., de Villiers, J.P.R., Redtmann, T. and Steenkamp, J. (2013). The recovery of manganese and generation of a valuable residue from ferromanganese slags by a hydrometallurgical route. *Proceedings of the 13th International Ferroalloys Congress*, Almaty, Kazakhstan, pp. 1051–1059.

Gu, F. and Wills, B.A. (1988). Chromite – Mineralogy and processing. *Minerals Engineering* 1(3): 235–240.

Guézennec, A. G., Huber, J. C., Patisson, F., Sessiecq, P., Birat, J.P. and Ablitzer, D. (2005). Dust formation in electric arc furnace: Birth of the particles. *Powder Technology* 157: 2–11.

Guiotoku, M., Rambo, C.R. and Hotza, D. (2014). Charcoal produced from cellulosic raw materials by microwave-assisted hydrothermal carbonization. *Journal of Thermal Analysis and Calorimetry* 117: 269–275.

Gunst, S., Weinbruch, S., Wentzel, M., Ortner, H.M., Skogstad, A., Hetland, S. and Thomassen, Y. (2000). Chemical composition of individual aerosol particles in workplace air during the production of manganese alloys. *Journal of Environmental Monitoring* 2: 65–71.

Guntzer, F., Keller, C. and Meunier, J.D. (2012). Benefits of plant silicon for crops: A review. *Agronomy for Sustainable Development* 32: 201–213.

Haga, T., Katoh, K. and Ibaraki, T. (2012). Technical developments for saving natural resources and increasing material recycling. Nippon Steel Report 101, pp. 196–202.

Hansmann, T. Fontana, P., Chiappero, A., Both, I. and Roth, J.L. (2007). Paul Wurth technologies for the optimum recycling of steelmaking residues. *Third International Steel Conference on New Developments in Metallurgical Process Technologies*, METEC InSteelCon, Düsseldorf, Germany, pp. 1044–1051.

Harbison-Walker (2005). *Handbook of Refractory Practice*. Harbison-Walker Refractory Company, Pittsburgh, PA.

Hariharan, A.V.L.N.S.H. and Purnima, K.C. (2015). Leaching and extraction of manganese from ferromanganese slag. *International Journal of Chemical Science and Technology* 5(2): 405–408.

Harman, C.N. and Rao, N.S.S. (2013). A process for the recovery of chromium and iron oxide in high carbon ferrochrome slag to obtain chromium and iron in the form of a saleable metal. *Proceedings of the 13th International Ferroalloys Congress*, Almaty, Kazakhstan, pp. 103–108.

Hayes, P.C. (2004). Aspects of SAF smelting of ferrochrome. *Proceedings of the 10th International Ferroalloys Congress*, Cape Town, South Africa, pp. 1–14.

Hayhurst, A. (1974). Crystallization processes in a high carbon ferrochromium slag. *Proceedings of the First International Ferroalloys Congress*, Johannesburg, South Africa, pp. 187–193.

Hein, J.R. and Petersen, S. (2013). The geology of manganese nodules. In: Baker, E. and Beaudoin, Y. (eds.), *Deep Sea Minerals: Manganese Nodules, A Physical, Biological, Environmental and Technical Review*, Vol. 1B. Secretariat of the Pacific Community, Suva, Fiji Islands.

Heo, J.H., Chung, Y.S. and Park, J.H. (2015). Effect of CaF_2 addition on the silicothermic reduction of MnO in ferromanganese slag. *Metallurgical and Materials Transactions* 46B: 1154–1161.

Heuer, M. (2013). Metallurgical grade and metallurgically refined silicon for photovoltaics. In: Willeke, G. and Weber, E.R. (eds.), *Advances in Photovoltaics Part 2*, Semiconductors and Semimetals, Vol. 89. Elsevier, San Diego, CA.

Hino, M., Higuchi, K., Nagasaka, T. and Banya, S. (1998). Thermodynamic estimation on the reduction behavior of iron-chromium ore with carbon. *Metallurgical and Materials Transactions* 29B: 351–360.

Holappa, L. (2010). Towards sustainability in ferroalloy production. *Journal of the Southern African Institute of Mining and Metallurgy* 110: 703–710.

Holappa, L. and Louhenkilpi, S. (2013). On the role of ferroalloys in steelmaking. *Thirteenth International Ferroalloys Congress: Efficient Technologies in Ferroalloys Industry*, Almaty, Kazakhstan, pp. 1083–1090.

Holappa, L. and Xiao, Y. (2004). Slags in ferroalloys production-review of present knowledge. *The Journal of Southern African Institute of Mining and Metallurgy* 104: 429–438.

Horii, K., Kitano, Y., Tsutsumi, N. and Kato, T. (2013). Processing and reusing technologies for steelmaking slags. Nippon Steel Report 104, Tokyo, pp. 123–129.

Howe, P.D., Malcom, H.M. and Dobson, S. (2004). Manganese and its compounds: Environmental aspects. Concise International Chemical Assessment Document 63, World Health Organization, Geneva, Switzerland.

Hwang, S., Jo, S.H., Kim, J., Shin, M.C., Chun, H.H., Park, H. and Lee, H. (2016). Catalytic activity of MnO_x/TiO_2 catalysts synthesized with different manganese precursors for the selective catalytic reduction of nitrogen oxides. *Reaction Kinetics Mechanism and Catalysis* 117: 583–591.

Ikeda, T. and Maeda, M. (1992). Purification of metallurgical grade silicon for solar grade silicon by electron beam button melting. *ISIJ International* 32(5): 635–642.

International Chromium Development Association (2011a). Discover chromium: Uses of chromium. Available at http://icdacr.com/index.php?option=com_content&view=article&id=141&Itemid=347&lang=en. Accessed 14 March 2016.

International Chromium Development Association (2011b). Discover chromium: Geology and mineralogy. Available at http://www.icdacr.com/index.php?option=com_content&view=article&id=157:geology-and-mineralogy&catid=38&Itemid=341&lang=en. Accessed 14 March 2016.

International Chromium Development Association (2011c). Discover chromium: Mining. Available at http://icdacr.com/index.php?option=com_content&view=article&id=141&Itemid=347&lang=en. Accessed 14 March 2016.

International Chromium Development Association (2011d). Discover chromium: Ore processing. Available at http://icdacr.com/index.php?option=com_content&view=article&id=141&Itemid=347&lang=en. Accessed 14 March 2016.

International Chromium Development Association (n.d.). Stainless steel. http://icdacr.com/index.php?option=com_content&view=article&id=137&Itemid=343&lang=en. Accessed 16 March 2016.

International Manganese Institute (2014). Lifecycle assessment of global manganese alloy production. Paris, France. Available at http://www.manganese.org/images/uploads/pdf/H337630-0000-07-124-003_-_Final_Manganese_Alloy_LCA_Report.pdf. Accessed 15 April 2016.

Istratov, A.A., Buonassisi, T., Pickett, M.D., Huer, M. and Weber, E.R. (2006). Control of metal impurities in "dirty" multi-crystalline silicon for solar cells. *Materials Science Engineering* B134: 282–286.

Jahanshahi, S., Sun, S. and Zhang, L. (2004). Recent developments in the physicochemical characterization and modelling of ferroalloy slag systems. *The Journal of the Southern African Institute of Mining and Metallurgy* 104: 529–540.

Jansson, Å., Brabie, V., Fabo, E. and Jansson, S. (2002). Slag formation and its role in ferrochromium production. *Scandinavian Journal of Metallurgy* 31(5): 314–320.

Jelkina, G., Teng, L., Björkman, B. and Seetharaman, S. (2012). Effect of low oxygen partial pressure on the chromium partition in CaO-MgO-SiO_2-Al_2O_3 synthetic slag at elevated temperatures. *Ninth International Conference on Molten Slags, Fluxes and Salts*, Beijing, China, pp. 46–59.

Ji, H., Huang, Z., Chen, K., Li, W., Gao, Y., Fang, M., Liu, Y. and Wu, X. (2014). Synthesis of Si_3N_4 powder with tunable α/β-Si_3N_4 content from waste silica fume using carbothermal reduction nitridation. *Powder Technology* 252: 51–55.

Jo, B.W., Kim, C.H., Tae, G.H. and Park, J.B. (2007). Characteristics of cement mortar with nano-SiO_2 particles. *Construction and Building Materials*, 21: 1351–1355.

Johnson, C.A. (2004). Cement stabilization of heavy metal-containing wastes. In: *Energy, Waste, and the Environment: A Geochemical Perspective*, Geological Society Special Publications 236 Giere, R. and Stille, P(eds.), pp. 595–606.

Jones, R.T., Curr, T.R. and Barcza, N.A. (1993). Developments in plasma furnace technology. *High Intensity Pyrometallurgy, Institution of Mining and Metallurgy*, London : 1–18, Available at http://www.mintek.co.za/Pyromet/Files/PlasmaDev-OCR.pdf. Accessed 20 April 2016.

Jorma, D., Helge, K., Päivi, O. and Riku, S. (2001). Sintered manganese ore and its use in ferromanganese production. *Proceedings of the Ninth International Congress on Ferroalloys*, Quebec, Montreal, Canada, pp. 212–223.

Kalenga, M., Xiaowei, P. and Tangstad, M. (2013). Manganese alloys production: Impact of chemical compositions of raw materials on energy and materials balance. *Proceedings of the 13th International Ferroalloys Congress*, Almaty, Kazakhstan, pp. 647–654.

Kamfjord, N.E. (2012). Mass and energy balances of the silicon process: Improved emissions standards. Doctoral thesis. Faculty of Natural Sciences and Technology, Norwegian University of Science and Technology, Trondheim, Norway.

Kamfjord, N.E., Myrhaug, E.H., Tveit, H. and Wittgens, B. (2010). Energy balance of 45MW (ferro)-silicon submerged arc furnace. *Proceedings of the 12th International Ferroalloys Congress*, Helsinki, Finland, pp. 729–738.

Kanungo, S.B. and Sant, B.R. (1981). Dephosphorization of phosphorus-rich manganese ores by selective leaching with dilute hydrochloric acid. *International Journal of Mineral Processing* 8: 359–375.

Kapure, G., Kari, C., Rao, S.M.M. and Raju, K.S. (2007). Use of chemical energy in submerged arc furnaces to produce ferrochrome: Prospects and limitations. *Proceedings of the 11th International Ferroalloys Conference*, New Delhi, India.

Karbowniczek, M., Gladysz, J. and Slezak, W. (2012). Current situation on the production market of FeMn and FeCr. *Journal of Achievements in Materials and Manufacturing Engineering* 55(2): 870–875.

Kato, M. and Minowa, S. (1969). Viscosity measurements of molten slag-properties of slag at elevated temperatures: Part 1. *Transactions of the Iron and Steel Institute of Japan* 9: 31–38.

Keene, B.J. (1995). *Slag Atlas*, 2nd edn. Verlag Stahleisen GmbH, Dusseldorf, Germany.

Kero, I., Naess, M.K. and Tranell, G. (2015). Particle size distributions of particulate emissions from the ferroalloy industry evaluated by electrical low pressure impactor. *Journal of Occupational and Environmental Hygiene* 12: 37–44.

Kesseler, K. and Erdmann, R. (2007). OxyCup® slag – A new product for demanding markets. Techforum 1. ThyssenKrupp AG, Düsseldorf, Germany, pp. 23–30.

Khattab, R.M., El-Rafei, A.M. and Zawrah, M.F. (2012). In situ formation of sintered cordierite-mullite nano-macro composites by utilizing waste silica fume. *Materials Research Bulletin* 47: 2662–2667.

Khattak, C.P., Joice, D.B. and Schmid, F. (2002). A simple process to remove boron from metallurgical grade silicon. *Solar Energy Materials and Solar Cells* 74: 77–89.

Kholmogorov, A.G., Zhyzhaev, A.M., Konomov, U.S., Moiseeva, G.A. and Paskov, G.L. (2000). The production of manganese dioxide from manganese ores of some deposits of the Siberian regions of Russia. *Hydrometallurgy* 56: 1–11.

Kilau, H.W. and Shah, I.D. (1984). Chromium-bearing waste slags: Evaluation of leachability when exposed to simulated acid precipitation. In: Jackson, L.P., Rohlik, A.R. and Conway, R.A. (eds.), *Symposium on Hazardous and Industrial Wastes Management: Third Symposium American Society for Testing and Materials (ASTM STP1984)*, Philadelphia, PA, pp. 61–80.

Kim, B.S., Jeong, S.B., Jeong, M.H. and Ryu, J.W. (2011). Upgrading of manganese from waste silico-manganese slag by a mechanical separation process. *Materials Transactions* 52(8): 1705–1708.

Kim, B.S., Lee, J.C., Jeong, S.B., Lee, H.I. and Kim, C.W. (2010). Kinetics of the volatilization removal of zinc from manganese dust. *Materials Transactions* 51(7): 1313–1318.

Kim, V., Tolymbekov, M., Kim, S., Ulyeva, G. and Kudarinov, S. (2013). Carbon reductant for silicon metal production. *Proceedings of the 13th International Ferroalloys Congress*, Almaty, Kazakhstan, pp. 519–525.

Kindness, A., Macias, A. and Glasser, F.P. (1994). Immobilization of chromium in cement matrices. *Waste Management* 14(1): 3–11.

Koleli, N. and Demir, A. (2016). Chromite (Chapter 11). In: Prasad, M.N.V. and Shih, K. (eds.), *Environmental Materials and Waste: Resource Recovery and Pollution Prevention*. Elsevier, Amsterdam, the Netherlands, pp. 252.

Kopeć, G. and Machulec, B. (2016). Selection of carbon reducers for the ferrosilicon smelting process. *Solid State Phenomena* 246: 256–259.

Kor, G.J.W. (1979). Equilibria between liquid Mn-Si alloys and MnO-SiO_2-CaO-MgO slags. *Metallurgical Transactions* 10B: 367–374.

Kor, G.J.W. and Glaws, P.C. (1998). Ladle refining and vacuum degassing. In: Fruehan, R.J. (ed.), *The Making, Shaping and Treatment of Steels: Steelmaking and Refining*, Vol. 2. The AISE Steel Foundation, Pittsburg, PA.

Kühn, M. and Mudersbach, D. (2004). Treatment of liquid EAF slag from stainless steelmaking to produce environmental friendly construction materials. *Proceedings of the Second International Conference in Iron and Steelmaking (SCANMET II)* Lulea, Sweden, p. 369.

Kumar, P.H., Srivastava, A., Kumar, V. and Kumar-Singh, V. (2014). Implementation of industrial waste ferrochrome slag in conventional and low cement castables: Effect of calcined alumina. *Journal of Asian Ceramic Studies* 2: 371–379.

Kumar, S., Garcia-Trinanes, P., Teixeira-Pinto, A. and Bao, M. (2013). Development of alkali activated cement from mechanically activated silico-manganese slag. *Cement and Concrete Composites* 40: 7–13.

Kumar, S., Kumar, R., Bandopadhyay, A., Alex, T.C., Kumar, B.R., Das, S.K. and Mehrotra, S.P. (2008). Mechanical properties of granulated blast furnace slag and its effect on the properties and structure of Portland slag cement. *Concrete and Concrete Composites* 30: 679–685.

Kutsin, V.S., Olshansky, V.I., Filippov, I.Yu., Chumakov, A.A. and Ganotsky, V.G. (2015). The development and introduction of resource saving technology of non-standard manganese ferroalloy fines recycling. *Proceedings of the 14th International Ferroalloys Congress*, Kiev, Ukraine, pp. 745–749.

Laforest, G. and Duchesne, J. (2006). Characterization and leachability of electric arc furnace dust made from remelting of stainless steel. *Journal of Hazardous Materials* B135: 156–164.

Language, P., Schimdt, F.O. and Van Blerk, J. (2000). Recycling of ferrochrome furnace bag filter dust at Samancor Chrome: A success story. *The Clean Air Journal* 10(5): 15–22.

Laruelle, G.G. et al. (2009). Anthropogenic perturbations of the silicon cycle at the global scale: Key role of the land-ocean transition. *Global Biogeochemical Cycles* 23: 1–17.

Lee, K.J., Min, D.S. Park, C.S., Park, Y.K., Jo, H.C. and Hong, S.H. (2010). High purity Mn metal from Mn oxide dust produced during FeMn refining process. *Proceedings of the 12th Ferroalloys Congress*, Helsinki, Finland, p. 197–205.

Lenntech (n.d.). Manganese. Available at: http://www.lenntech.com/periodic/elements/mn.htm. Accessed 6 April 2016.

Leont'ev, L.I., Smirnov, L.A., Zhuchkov, V.I., Zhdanov, A.V., Dashevskii, V. Ya. and Kuriva, S.A. (2016). Status and prospects of ferroalloys production in the Russian Federation. *Metallurgist* 59(11–12): 1001–1006.

Leroy, C. and Rancoita, P.G. (2015). *Principles of Adiation Interaction in Matter and Detection*, 4th edn. World Scientific, Singapore.

Li, F., Tangstad, M. and Solheim, I. (2015). Quartz and carbon black pellets for silicon production. *Proceedings of the 14th International Ferroalloys Congress*, Kiev, Ukraine, pp. 390–401.

Li, J.L., Xu, A.J., He, D.F., Yang, Q.X. and Tian, N.Y. (2013). Effect of FeO on the formation of spinel phases and chromium distribution in ther CaO-SiO_2-MgO-Al_2O_3-Cr_2O_3 system. *International Journal of Minerals, Metallurgy and Materials* 20(3): 253–258.

Li, Y., Jiang, J. and Zhao, J. (2004). X-ray diffraction and Mossbauer studies of phase transformation in manganese ferrite prepared by combustion synthesis method. *Materials Chemistry and Physics* 87: 91–95.

Liao, C., Tang, Y., Liu, C., Shih, K. and Li, F. (2016). Double barrier mechanism of chromium immobilization: A quantitative study of crystallization and leachability. *Journal of Hazardous Materials* 311: 246–253.

Lin, S.B., Durfee, L.D., Knott, A.A. and Schalau, G.K.II (2008). Silicone pressure sensitive adhesives. In: Benedek, I. and Feldstein, M.M. (eds.), *Technology of Pressure Sensitive Adhesives and Products: Handbook of Pressure Sensitive Adhesives and Products*. CRC Press, Boca Raton, FL.

Lind, B.B., Fällman, A.M. and Larsson, L.B. (2001). Environmental impact of ferrochrome slag in road construction. *Waste Management* 21: 255–264.

Lindstad, T., Gaal, S., Hansen, S. and Prytz, S. (2007). Improved SINTEF SiO-reactivity test. *Proceedings of the 11th International Ferroalloys Congress*, New Delhi, India, pp. 414–423.

Liu, Y., Jiang, M.F., Xu, L.X. and Wang, D.Y. (2012). Mathematical modelling of refining of stainless steel in smelting reduction converter using chromium ore. *ISIJ International* 52(3): 394–401.

Los Alamos National Laboratory (n.d.). Chromium: Periodic Table of Elements. Available at http://periodic.lanl.gov/24.shtml. Accessed 14 March 2016.

Lothenbach, B., Scrivener, K. and Hooton, R.D. (2011). Supplementary cementitious materials. *Cement and Concrete Research* 41: 1244–1256.

Lutz, J., Schlangenotto, H., Scheurmann, U. and De Doncker, R. (2011). *Semiconductor Power Devices: Physics, Characteristics and Reliability*. Springer-Verlag, Berlin, Germany.

Lynch, D. (2009). Winning the global race for solar silicon. *Journal of Materials* 61(11): 41–48.

Ma, G. and Garbers-Craig, A.M. (2006a). A review on the characteristics, formation mechanism and treatment processes of Cr (IV)-containing pyrometallurgical wastes. *Journal of the Southern African Institute of Mining and Metallurgy* 106: 753–763.

Ma, G. and Garbers-Craig, A.M. (2006b). Cr (IV) containing electric arc furnace dusts and filter cake from stainless steel waste treatment plant. Part 1: Characteristics and microstructure. *Ironmaking and Steelmaking* 33(3): 229–237.

Ma, G. and Garbers-Craig, A.M. (2006c). Cr (IV) containing electric arc furnace dusts and filter cake from stainless steel waste treatment plant. Part 2: Formation mechanisms and leachability. *Ironmaking and Steelmaking* 33(3): 238–244.

Ma, G. and Garbers-Craig, A.M. (2009). Stabilization of Cr (IV) in stainless steel plant dust through sintering using silica-rich clay. *Journal of Hazardous Materials* 169: 210–216.

Machulec, B. and Bialik, W. (2016). Comparison of the physicochemical model of ferrosilicon smelting process with results observations of the process under industrial conditions. *Archives of Metallurgy and Materials* 61(1): 265–270.

Maeda, M., Sano, N. and Matsushita, Y. (1981). Chromium recovery from chromium-containing slags. *Conservation and Recycling* 4(3): 137–144.

Mahmoud, M.H. (2005). Low temperature Mossbauer study of gallium substitution for iron in manganese-ferrite. *Solid State Ionics* 176: 1333–1336.

Makhula, M., Bada, S. and Afolabi, A. (2013). Evaluation of reduction roasting and magnetic separation for upgrading Mn/Fe ratio of fine ferromanganese. *International Journal of Mining Science and Technology* 23: 537–541.

Malayoglu, U. (2010). Study on the gravity processing of manganese ores. *Asian Journal of Chemistry* 22(4): 3292–3298.

Malviya, R. and Chaudhary, R. (2006). Factors affecting hazardous waste solidification/stabilization: A review. *Journal of Hazardous Materials* B137: 267–276.

Margaria, T., Anglezio, J.C. and Servant, C. (1992). Intermetallic compounds in metallurgical silicon. *Proceedings of the Sixth International Ferroalloys Congress*, Johannesburg, South Africa, pp. 209–214.

Marjoram, T., Cameron, H., Ford, G., Garner, A. and Gibbons, M. (1981). Manganese nodules and marine technology. *Resource Policy* 7: 45–57.

Martin, K.R. (2013). Silicon: The health benefits of a metalloid. *Metal Ions in Life Sciences* 13: 451–473.

Martorano, M.A., Ferreira Neto, J.B., Oliveira, T.S. and Tsubaki, T.O. (2011). Refining of metallurgical grade silicon by directional solidification. *Materials Science and Engineering* B176: 217–226.

Mashanyare, H.P. and Guest, R.N. (1997). The recovery of ferrochrome from slag at Zimasco. *Minerals Engineering* 10(11): 1253–1258.

McClelland, J.M. and Metius, G.E. (2003). Recycling ferrous and nonferrous waste streams with FASTMET. *Journal of Materials* 55: 30–34.

McDougall, I. (2013). Basics of ferroalloys. In: Gasik, M.M. (ed.), *Handbook of Ferroalloys: Theory and Technology*. Butterworth-Heinemann, Elsevier, Oxford, UK.

McGrath, S.P. (1995). Chromium and nickel. In: Alloway, B.J. (ed.), *Heavy Metals in Soils*, 2nd edn. Blackie Academic & Professional, Glasgow, UK, p. 155.

McGuire, M.F. (2008). Austenitic stainless steels. In: *Stainless Steels for Design Engineers*. ASM International, Materials Park, OH. Available at www.asminternational.org. Accessed 20 May 2016.

McIntosh, S.N. and Baglin, E.G. (1992). Recovery of manganese from steel plant slag by carbamate leaching. Report of Investigations 9400. U.S. Department of Interior, Bureau of Mines, Washington DC.

Mekhtiev, A.D., Tolymbekov, M.J., Kim, A.V., Zholdubaeva, Z.H.D., Issin, D.K., Issagulov, A.Z. and Issin, B.D. (2014). *Metabk* 53(4): 563–566.

Mitchell, D.R.G., Hinczak, I. and Day, R.A. (1998). Interaction of silica fume with calcium hydroxide solutions and hydrated cement pastes. *Cement and Concrete Research* 28(11): 1571–1584.

Mitraŝinović, A. (2010). Characterization of the Cu-Si system and the utilization of metallurgical techniques in silicon refining for solar cell applications. Doctoral thesis. University of Toronto, Toronto, Ontario, Canada.

Morita, K. and Miki, T. (2003). Thermodynamics of solar-grade silicon refining. *Intermetallics* 11: 1111–1117.

Mosier, D.L., Singer, D.A., Moring, B.C. and Galloway, J.P. (2012). Podiform chromite deposits – Database, grade and tonnage models. U.S. Geological Survey Scientific Investigations Report 2012–5157, pp. 1–45.

Mostafaee, S. (2011). A study of the EAF high chromium stainless steelmaking slags characteristics and foamability. PhD thesis. Royal Institute of Technology, Stockholm, Sweden.

Motz, H. and Geiseler, J. (2001). Products of steel slag: An opportunity to save natural resources. *Waste Management* 21: 285–293.

Muan, A. (1984). Slag-metal equilibria involving chromium as a component. *MINTEK 50*, Sandton, South Africa, pp. 897–904.

Murthy, Y.R., Tripathy, S.K. and Kumar, C.R. (2011). Chrome ore beneficiation challenges and opportunities: A review. *Minerals Engineering* 24: 375–380.

Mynko, N.I., Gubarev, A.V. and Nevedomskii, V.A. (1993). Crystallization mechanism and characteristics of silicomanganese slag melts used to make glass ceramics: Review. *Glass and Ceramics* 50(3): 103–107.

Myrhaug, E.H., Tuset, J.K. and Tveit, H. (2004). Reaction mechanism of charcoal and coke in the silicon process. *Proceedings of the 10th International Ferroalloys Congress*, Cape Town, South Africa, pp. 108–121.

Myrvågnes, V. and Lindstad, T. (2007). The importance of coal and coke properties in the production of high silicon alloys. *Proceedings of the 11th International Ferroalloys Congress*, New Delhi, India, pp. 402–413.

Nafziger, R.H. (1982). A review of the deposits and beneficiation of lower-grade chromite. *Journal of the Southern African Institute of Mining and Metallurgy* 28: 205–226.

Nakamoto, M., Forsbacka, L. and Holappa, L. (2007). Assessment of viscosity of slags in ferrochromium process. *Eleventh International Ferroalloys Congress*, New Delhi, India, pp. 159–164.

Nandy, B., Chaudhury, M.K., Paul, J. and Bhattacharjee, D. (2009). Sintering characteristics of Indian chrome ore fines. *Metallurgical and Materials Transactions* 40B: 662–675.

Nelson, L.R. (2014). Evolution of the mega-scale in ferroalloy electric furnace smelting. In: Mackey, P.J., Grimsey, E.J., Jones, R.T. and Brooks, G.A. (eds.), *Celebrating the Megascale: Proceedings of the Extraction and Processing Division. Symposium on Pyrometallurgy in Honor of David G.C. Robertson*. The Minerals, Metals and Materials Society, San Diego, pp. 39–68.

Nelson, L.R., Bouwer, P.H.F. and Buenk, N.B. (2001). Mineralogical and thermodynamic investigation of CLU converter lining failures in commercial production of medium carbon ferromanganese. *Proceedings of the Ninth International Ferroalloys Congress*, Quebec, Montreal, Canada, pp. 360–376.

Niemelä, P. and Kauppi, M. (2007). Production, characteristics and use of ferrochromium slags. *Proceedings of the 11th International Ferroalloys Congress*, New Delhi, India, pp. 171–179.

Niemelä, P., Krogerus, H. and Oikarinen, P. (2004). Formation, characterization and utilization of CO-gas formed in ferrochrome smelting. *Proceedings of the 10th Ferroalloys Congress*, Cape Town, South Africa, pp. 68–77.

Norval, D. and Oberholster, R.E. (2011). Further processing and granulation of slags with entrained metal and high metal compound content, with specific reference to present manganese slag dumps as well as additions due to daily production. In: Jones, R.T. and den Hoed, P. (eds.), *Southern African Pyrometallurgy*. Southern African Institute of Mining and Metallurgy, Johannesburg, South Africa, pp. 129–143.

O'Lenick Jr., J. (1999). Basic silicone chemistry – A review. Silicone Spectator. Available at http://www.scientificspectator.com/documents/silicone%20spectator/Silicone_Spectator_January_2009.pdf. Accessed 12 May 2016.

Oliveria, H. (2012). Chromium as an environmental pollutant: Insights on induced toxicity. *Journal of Botany* 2012: 1–8.

Olsen, S.E., Ding, W., Kossyreva, O.A. and Tangstad, M. (1995). Equilibrium in production of high carbon ferromanganese. *Proceedings of the International Ferroalloys Congress*, Trondheim, Norway, pp. 591–600.

Olsen, S.E. and Tangstad, M. (2004). Silicomanganese production: Process understanding. *Proceedings of the 10th International Ferroalloys Congress*, Cape Town, South Africa, pp. 231–238.

Olsen, S.E., Tangstad, M. and Lindstad, T. (2007). *Production of Manganese Ferroalloys*. SINTEF and Tapir Akademisk Forlag, Trondheim, Norway.

Onuma, K. and Tohara, T. (1983). Effect of chromium on phase relations in the join forsterite-anorthite-diopside in air at 1atm. *Contributions to Mineralogy and Petrology* 84: 174–181.

Ostrovski, O. and Swinbourne, D. (2013). Slags in production of manganese alloys. *Steel Research International* 84(7): 680–686.

Outokumpu (2013). *Handbook of Stainless Steel*. Outokumpu Stainless AB. Sweden, Europe. Available at www.outokumpu.com. Accessed 10 December 2015.

Owada, S. and Harada, T. (1985). Grindability and magnetic properties of chromites. *The Journal of the Mining and Metallurgical Institute of Japan* 101(1174): 781–786.

Oxychem (n.d.). *The Oxychem Sodium Silicate Handbook*. Available at http://www.oxy.com/ourbusinesses/chemicals/products/documents/silicates/silicate.pdf. Accessed 20 May 2016.

Panda, C.R., Mishra, K.K., Panda, K.C., Nayak, B.D. and Nayak, B.B. (2013). Environmental and technical assessment of ferrochrome slag as concrete aggregate material. *Construction and Building Materials* 49: 262–271.

Pande, M.M., Guo, M., Devisscher, S. and Blanpain, B. (2012). Influence of ferroalloy impurities and ferroalloy addition sequence on ultra-low carbon (ULC) steel cleanliness after RH treatment. *Ironmaking and Steelmaking* 39(7): 519–529.

Pande, M.M., Guo, M., Guo, X., Geysen, D., Devisscher, S., Blainpain, B. and Wollants, P. (2010). Ferroalloy quality and steel cleanliness. *Ironmaking and Steelmaking* 37(7): 502–511.

Papp, J.F. and Lipin, B.R. (2001). Chromium and its alloys. In: *Kirk-Othmer Encyclopedia of Chemical Technology*, Vol. 6. John Wiley & Sons.

Papp, J.F. and Lipin, B.R. (2006). Chromite. In: Kogel, J.E., Trivedi, N.C., Barker, J.M. and Krukowski, S.T. (eds.), *Industrial Minerals and Rocks: Commodities, Markets and Uses*, 7th edn. Society for Mining, Metallurgy and Exploration, Littleton, CO.

Patil, B.V., Chan, A.H. and Choulet, R.J. (1998). Refining of stainless steels. In: Fruehan, R.J. (ed.), *The Making, Shaping and Treatment of Steels: Steelmaking and Refining*, Vol. 2. The AISE Steel Foundation, Pittsburg, PA.

Pedersen, L. and Arnberg, L. (2001). The effect of solution heat treatment and quenching rates on mechanical properties and microstructures in AlSiMg foundry alloys. *Metallurgical Transactions* 32A: 525–532.

Pereira, M.J., Lima, M.M.F. and Lima, R.M.F. (2014). Calcination and characterization studies of a Brazilian manganese ore tailing. *International Journal of Mineral processing* 131: 26–30.

Person, R.A. (1971). Control of emissions from ferroalloy furnace processing. *Journal of Metals* 23: 17–29.

Petavratzi, E., Kingman, S. and Lowndes, I. (2005). Particulates from mining operations: A review of sources, effects and regulations. *Minerals Engineering* 18: 1183–1199.

Petersen, J., Stewart, M. and Petrie, J. (2000). Management of ferroalloy wastes. In: Warhurst, A. and Noronha, L. (eds.), *Environmental Policy in Mining: Corporate Strategy and Planning for Closure*. CRC Press, Boca Raton, FL.

Petrie, L.M. (1995). Molecular interpretation for SO_2 dissolution kinetics of pyrolusite, manganite and hematite. *Applied Geochemistry* 10: 253–267.

Pienaar, P.C. and Smith, W.F.P. (1992). A case study of the production of high grade manganese sinter from low grade Mamatwan manganese ore. *Proceedings of the Sixth International of Ferroalloys Congress*, Johannesburg, South Africa, pp. 131–138.

Pillay, K., von Blottnitz, H. and Petersen, J. (2003). Ageing of chromium (III)-bearing slag and its relation to the atmospheric oxidation of solid chromium (III) oxide in the presence of calcium oxide. *Chemosphere* 52: 1771–1779.

Pistorius, P.C. (2002). Reductant selection in ferroalloy production: The case for the importance of dissolution in the metal. *The Journal of the Southern African Institute of Mining and Metallurgy* 102: 33–36.

Pizzini, S. (2010). Towards solar grade silicon: Challenges and benefits for low cost photovoltaics. *Solar Energy Materials and Solar Cells* 94: 1528–1533.

Pölloth, C.F. (2012). The toxicological mode of action and the safety of synthetic amorphous silica: A review. *Toxicology* 294: 61–79.

Post, J.E. (1999). Manganese oxide minerals: Crystal structures and economic and environmental significance. *Proceedings of the National Academy of Sciences* 96: 3447–3454.

Pretorius, E.B. and Nunnington, R.C. (2002). Stainless steel slag fundamentals: From furnace to tundish. *Ironmaking and Steelmaking* 29(2): 133–139.

Prillhofer, R., Rank, G., Berneder, J., Antrekowitsch, H., Uggowitzer, P.J. and Pogatscher, S. (2014). Property criteria for automotive Al-Mg-Si sheet alloys. *Materials* 7: 5047–5068.

Raaness, O. and Gray, R. (1995). Coal in the production of silicon rich alloys. *Proceedings of the Seventh International Ferroalloys Congress*, Trondheim, Norway, pp. 201–220.

Rai, D., Eary, L.E. and Zachara, J.M. (1989). Environmental chemistry of chromium. *The Science of the Total Environment* 86: 15–23.

Rai, A., Prabakar, J., Raju, C.B. and Morchalle, R.K. (2002). Metallurgical slag as a component in blended cement. *Construction and Building Materials* 16: 489–494.

Ramos, A., Filtvedt, W.O., Lindholm, D., Ramachandran, P.A. and Rodríguez, A. (2015). Deposition reactors for solar grade silicon: A comparative thermal analysis of a Siemens reactor and a fluidized reactor. *Journal of Crystal growth* 431: 1–9.

Ranjan, S., Balaji, S., Panella, A. and Ydstie, B.E. (2011). Silicon solar cell production: Review. *Computers and Chemical Engineering* 35: 1439–1453.

Rao, D.S., Angadi, S.I., Muduli, S.D. and Nayak, B.D. (2010). Valuable waste: Recovery of chromite values from ferrochrome industry flue dust. *AT Mineral Processing English Edition* 51(5): 2–6.

Ravary, B., Hunsbedt, L. and Kristensen, O. (2013). Progress in recycling sludge from off-gas cleaning of manganese alloy furnaces. *Proceedings of the 13th International Ferroalloys Congress*, Almaty, Kazakhstan, pp. 1023–1027.

Rennie, M.S., Howat, D.D. and Jochens, P.R. (1972). The effects of chromium oxide, iron oxide and calcium oxide on the liquidus temperatures, viscosities and electrical conductivities of slags in the system MgO-Al_2O_3-SiO. *Journal of the Southern African Institute of Mining and Metallurgy* (August): 1–9.

Reuter, M., Xiao, Y. and Boin, U. (2004). Recycling and environmental issues of metallurgical slags and salt fluxes. *VII International Conference on Molten Slags, Fluxes and Salts*, The Southern African Institute of Mining and Metallurgy, Johannesburg, pp. 349–356.

Richard, F.C. and Bourg, A.C.M. (1991). Aqueous geochemistry of chromium: A review. *Water Research* 25(7): 807–816.

Rick, C.J. (2010). Refining of charge chrome: A study of some products and applications. *The Journal of Southern African Institute of Mining and Metallurgy* 110: 751–757.

Riekkola-Vanhanen, M. (1999). Finnish expert report on best available techniques in ferrochromium production. Finnish Environment Institute, Helsinki, Finland. Available at https://helda.helsinki.fi/bitstream/handle/10138/40531/FE_314.pdf?sequence=1. Accessed 12 May 2016.

Ringdalen, E., Gaal, S., Tangstad, M. and Ostrovski, O. (2010). Ore melting and reduction in silicomanganese production. *Metallurgical and Materials Transactions* 41B: 1220–1229.

Rodríguez-Pinero, M., Fernández Pereira, C., Ruiz de Elvira Francoy, C. and Vale Parapar, J.F. (1998). Stabilization of a chromium-containing solid waste: Immobilization of hexavalent chromium. *Journal of Air and Waste Management Association* 48(11): 1093–1099.

Rong-jin, L., Qing-jun, D., Ping, C. and Guang-yao, Y. (2012). Durability of concrete made with manganese slag as supplementary cementitious materials. *Journal of Shanghai Jiaotong University* 17(3): 345–349.

Roskill (2011). Silicon and ferrosilicon: Global industry markets and outlook. 2011 Roskill Report, UK. Available at https://roskill.com/wp/wp-content/uploads/2014/11/new-silicon-and-ferrosilicon-report.attachment1.pdf. Accessed 20 April 2016.

Royal Society of Chemistry (2012). Chromium. Periodic Table. Available at http://www.rsc.org/periodic-table/element/24/chromium. Accessed 14 March 2016.

Saevarsdottir, G. and Bakken, J.A. (2010). Current distribution in submerged arc furnaces for silicon metal/ferrosilicon production. *Proceedings of the 12th International Ferroalloys Congress*, Helsinki, Finland, pp. 717–728.

Safarian, J. and Kolbeinsen, L. (2015). Microscopic study of carbon surfaces interacting with high carbon ferromanganese slag. *Metallurgical and Materials Transactions* 46B: 125–134.

Safarian, J., Kolbeinsen, L., Tangstad, M. and Tranell, G. (2009). Kinetics and mechanism of the simultaneous carbothermic reduction of FeO and MnO from high carbon ferromanganese slag. *Metallurgical and Materials Transactions* 40B: 929–939.

Sahajwalla, V., Dubikova, M. and Khanna, R. (2004). Reductant characterisation and selection: Implications for ferroalloys processing. *Proceedings of the 10th International Ferroalloys Congress*, Cape Town, South Africa, pp. 351–362.

Sakulich, A.R. (2011). Reinforced geopolymer composites for enhanced material greenness and durability. *Sustainable Cities and Society* 1(4): 195–210.

Samada, Y., Miki, T. and Hino, M. (2011). Prevention of chromium elution from stainless steel slag into seawater. *ISIJ International* 51(5): 728–732.

Sanjuán, M.A., Argiz, C.M., Gálvez, J.C. and Moragues, A. (2015). Effect of silica fume fineness on the improvement of Portland cement strength performance. *Construction and Building Materials* 96: 55–64.

Sano, N. (2004). Reduction of chromium oxide in stainless steel slags. *Proceedings of 10th International Ferroalloys Congress*, Cape Town, South Africa, pp. 670–677.

Sano, N., Lu, W.K. and Riboud, P.V. (1997). *Advanced Physical Chemistry for Process Metallurgy*, 1st edn. Academic Press, San Diego, CA

Santos-Burgoa, C., Rios, C., Mercado, L.A., Arechiga-Serrano, R., Cano-Valle, F., Eden-Wynter, R.A., Texcalac-Sangrador, J.L., Villa-Barragan, J.P., Rodriguez-Agudelo, Y. and Montes, S. (2001). Exposure to manganese: Health effects on the general population, a pilot study in Central Mexico. *Environmental Research* A85: 90–104.

Schei, A., Tuset, J.K. and Tveit, H. (1997). *Production of High Silicon Alloys*. Tapir Forlag, Trondheim, Norway.

Sedumedi, H.N., Mandiwana, K.L., Ngobeni, P. and Panichev, N. (2009). Speciation of Cr (VI) in environmental samples in the vicinity of the ferrochrome smelter. *Journal of Hazardous Materials* 172(2–3): 1686–1689.

Semykina, A. (2010). Recovery of iron and manganese values from metallurgical slags by the oxidation route. Doctoral thesis. Royal Institute of Technology, Stockholm, Sweden.

Semykina, A., Shatokha, V., Iawse, M. and Seetharaman, S. (2010). Kinetics of oxidation of divalent iron to trivalent state in liquid FeO-CaO-SiO_2 slags. *Metallurgical Transactions* 41B: 1230–1239.

Senapati, D., Uma Maheswar, E.V.S. and Ray, C.R. (2007). Ferrosilicon operation at IMFA: A critical analysis. *Proceedings of the 11th International Ferroalloys Congress*, New Delhi, India, pp. 371–380.

Sen Gupta, P.L. and Dhananjavan, N. (1983). Utilization of ferro-manganese slag for production of manganese sulphate and electrolytic metal/dioxide. *Proceedings of Seminar on Problems and Prospects of Ferroalloys Industries in India*, Jamshedpur, India, pp. 285–289.

Sen Gupta, S. (2015). Indian ferroalloy scenario – A review. *A Steelworld* (March), pp. 41–47.

Shanker, A.K., Cervantes, C., Loza-Tavera, H. and Avudainayagam, S. (2005). Chromium toxicity in plants. *Environmental Engineering* 31: 739–753.

Sharma, T. (1992). Physico-chemical processing of low grade manganese ore. *International Journal of Mineral Processing* 35: 191–203.

Shatokha, V., Semykina, A., Nakano, J., Sridhar, S. and Setharaman, S. (2013). A study on transformation of some transition metal oxides in molten steelmaking slag to magnetically susceptible compounds. *Journal of Mining and Metallurgy, Section B: Metallurgy* 49(2): 169–174.

Shen, H. and Forssberg, E. (2003). An overview of recovery of metals from slags. *Waste Management* 23: 933–949.

Shen, H., Forssberg, E. and Nordstrom, U. (2004). Physicochemical and mineralogical properties of stainless steel slags oriented to metal recovery. *Resources, Conservation and Recycling* 40: 245–271.

Shen, R., Zhang, G., Dell'amico, M., Brown, P. and Ostrovski, O. (2005). Characterization of manganese furnace dust and zinc balance in production of manganese alloys. *ISIJ International* 45(9): 1248–1254.

Shen, R., Zhang, G., Dell'amico, M., Brown, P. and Ostrovski, O. (2007a). Feasibility study of recycling of manganese furnace dust. *Proceedings of the 11th Ferroalloys International Conference*, New Delhi, India, pp. 507–519.

Shen, R., Zhang, G., Dell'amico, M., Brown, P. and Ostrovski, O. (2007b). Sintering pot test of manganese ore with addition of manganese furnace dust. *ISIJ International* 47(2): 234–239.

Shen, H. and Forssberg, E. (2003b). An overview of recovery of metals from slags. *Waste Management* 23: 933–949.

Shi, C. (2002). Characteristics and cementitious properties of ladle slag fines from steel production. *Cement and Concrete Research* 32: 459–462.

Shi, C. (2004). Steel slag: Its production, processing, characteristics, and cementitious properties. *Journal of Materials in Civil Engineering*. 16(3): 230–236.

Shibata, E., Egawa, S. and Nakamura, T. (2002). Reduction behavior of chromium oxide in molten slag using aluminium, ferrosilicon and graphite. *ISIJ International* 42(6): 609–613.

Shoko, L., Beukes, J.P. and Strydom, C.A. (2013). Determining the baking isotherm temperature of Söderberg electrodes and associated structural changes. *Minerals Engineering* 49: 33–39.

Siddique, R. (2011). Utilization of silica fume in concrete: Review of hardened properties. *Resources, Conservation and Recycling* 55: 923–932.

Siddique, R. and Chahal, N. (2011a). Use of silicon and ferrosilicon industry byproducts (silica fume) in cement paste and mortar. *Resources, Conservation and Recycling* 55(8): 739–744.

Siddique, R. and Chahal, N. (2011b). Utilization of silica fume in concrete: Review of durability properties. *Resources, Conservation and Recycling* 57: 30–35.

Siddique, R. and Khan, M.I. (2011). *Supplementary Cementing Materials*, Engineering Materials. Springer-Verlag, Berlin, Germany, pp. 67–119.

Silica Fume Association (2005). Silica fume user's manual. Available at www.silicafume.org. Accessed 18 May 2016.

Simbi, D.J. and Tsomondo, B.M.C. (2002). Kinetics of chromite ore reduction from MgO-CaO-SiO_2-FeO-Cr_2O_3-Al_2O_3 slag system by carbon dissolved in high carbon ferrochromium alloy bath. *Ironmaking and Steelmaking* 29(1): 22–28.

Singh, V., Ghosh, T.K., Ramamurthy, Y. and Tathavadkar, V. (2011). Beneficiation and agglomeration process to utilize low-grade ferruginous manganese ore fines. *International Journal of Mineral Processing* 99: 84–86.

Singh, V. and Tathavadker, V. (2011). Development of agglomeration process to utilize the ferromanganese fines in steel making process. *ISIJ International* 51(1): 59–62.

Singh, V., Tathavadkar, V., Mohan Rao, S. and Kumar, S. (2008). Estimating effect of chrome ore granulometry on sintered pellet properties. *Ironmaking and Steelmaking* 35(1): 27–32.

Skjervheim, T.A. and Olsen, S.E. (1995). The rate and mechanism for reduction of manganese oxide from silicate slags. *Proceedings of the Seventh International Ferroalloys Congress*, Trondheim, Norway, pp. 631–639.

Soykan, O., Eric, R.H. and King, R.P. (1991). The reduction mechanism of a natural chromite at 1416°C. *Metallurgical Transactions* 22B: 53–63.

Sripriya, R. and Murty, V.G.K. (2005). Recovery of metal from slag/mixed metal generated in ferroalloy plants – A case study. *International Journal of Mineral Processing* 75: 123–134.

Stasinakis, A.S., Thomaidis, N.S., Mamais, D., Karivali, M. and Lekkas, T.D. (2003). Chromium species behavior in the activated sludge process. *Chemosphere* 52(6): 1059–1067.

Steelworld (2012). Ferroalloy sector hit by high costs and imports. Steelworld, Mumbai, India, pp. 48–50. Available at http://steelworld.com/newsletter/jan12/infocus0112.pdf. Accessed 20 April 2016.

Steenkamp, J.D. and Basson, J. (2013). The manganese ferroalloys industry in Southern Africa. *The Journal of the Southern African Institute of Mining and Metallurgy* 113: 667–676.

Steenkamp, J.D., Tangstad, M. and Pistorius, P.C. (2011). Thermal conductivity of solidified manganese-bearing slags: A preliminary review. In: Jones, R.T. and den Hoed, P. (eds.), *Southern African Pyrometallurgy*. Southern African Institute of Mining and Metallurgy, Johannesburg, South Africa, pp. 327–343.

Stegemann, J.A., Roy, A., Caldwell, R.J., Schilling, P.J. and Tittsworth, R. (2000). Understanding environmental leachability of electric arc furnace dust. *Journal of Environmental Engineering* 126(2): 112–120.

Su, H., Wen, Y., Wang, F., Sun, Y. and Tong, Z. (2008). Reductive leaching of manganese from low-grade manganese ore in H_2SO_4 using cane molasses as reductant. *Hydrometallurgy* 93(3–4): 136–139.

Sugimoto, M. (1999). The past, present and future of ferrites. *Journal of American Ceramic Society* 82(2): 269–280.

Suri, J., Shaw, L.L. and Zawrah, M.F. (2011). Synthesis of carbon-free Si_3N_4/SiC nano powders using silica fume. *Ceramics International* 37: 3477–3487.

Suzuki, K., Ban-ya, S. and Hino, M. (2001). Deoxidation equilibria of chromium stainless steel with Si at the temperatures from 1823 to 1923K. *ISIJ International* 41(8): 813–817.

Swinbourne, D.R., Rankin, W.J. and Eric, R.H. (1995). The effect of alumina in slag on manganese and silicon distribution in silicomanganese smelting. *Metallurgical Transactions* 26B: 59–65.

Takiguchi, H. and Morita, K. (2009). Sustainability of silicon feedstock for low-carbon society. *Sustainability Science* 4: 117–131.

Tangstad, M. (2013a). Manganese ferroalloys technology. In: Gasik, M. (ed.), *Handbook of Ferroalloys: Theory and Technology*. Butterworth-Heinemann, Elsevier, Oxford, U.K.

Tangstad, M. (2013b). Ferrosilicon and silicon technology. In: Gasik, M. (ed.), *Handbook of Ferroalloys: Theory and Technology*. Butterworth-Heinemann, Elsevier, Oxford, UK.

Tangstad, M., Ksiazek, M., Andersen, V. and Ringdalen, E. (2010a). Small scale laboratory experiments simulating an industrial silicon furnace. *Proceedings of the 12th International Ferroalloys Congress*, Helsinki, Finland, pp. 661–669.

Tangstad, M., Leroy, D. and Ringdalen, E. (2010b). Behavior of agglomerates in ferromanganese production. *Proceedings of the 12th International Ferroalloys Congress*, Helsinki, Finland, pp. 457–466.

Tanskanen, P. and Makkonen, H. (2006). Design of slag mineralogy and petrology: Examples of useful methods for slag composition and property design. *Global Slag Magazine* 5: 16–20.

Tateishi, M., Fujimoto, H., Harada, T. and Sugitatsu, H. (2008). Development of EAF dust recycling and melting technology using the coal-based FASTMELT process. MIDREX RHF Technologies Report, Charlotte, NC, pp. 9–15.

Total Materia. (2008). Production of stainless steel: Part 2. http://www.totalmateria.com/page.aspx?ID=CheckArticle&site=KTS&NM=220. Accessed 4 December 2015.

Trussell, S. and Spence, R.D. (1994). A review of solidification/stabilization interferences. *Waste Management* 14(6): 507–519.

Turkdogan, E.T. and Fruehan, R.J. (1998). Fundamentals of iron and steel making. In: Fruehan, R.J. (ed.), *The Making, Shaping and Treatment of Steels: Steelmaking and Refining*, Vol. 2. The AISE Steel Foundation, Pittsburg, PA.

US Geological Survey (2015). Chromium. Available at http://minerals.usgs.gov/minerals/pubs/commodity/chromium/mcs-2015-chrom.pdf. Accessed 15 March 2016.

Uslu, E. and Eric, R.H. (1991). The reduction of chromite in liquid iron-chromium-carbon alloys. *Journal of the Southern African Institute of Mining and Metallurgy* 91(11): 397–409.

Vanderstaay, E.C., Swinbourne, D.R. and Monteiro, M. (2004). A computational thermodynamics model of submerged arc electric furnace for ferromanganese smelting. *Minerals Processing and Extractive Metallurgy (Transactions of the Institute of Mining and Metallurgy)* 113: C38–C44.

Vangskåsen, J. and Tangstad, M. (2013). Condensate in metallurgical silicon process-Reaction mechanism. *Proceedings of the 13th International Ferroalloys Congress*, Almaty, Kazakhstan, pp. 283–290.

Van Reenen, J.H., Thiele, H. and Bergman, C. (2004). The recovery of chrome and manganese alloy fines from slag. *Proceedings of the 10th International Ferroalloy Congress*, Cape Town, South Africa, pp. 548–554.

Veglio, F. and Toro, L. (1994). Reductive leaching of a concentrate manganese dioxide ore in acid solution: Stoichiometry and preliminary kinetic analysis. *International Journal of Mineral Processing* 40: 257–272.

Vesterberg, P., Beskow, K. and Rick, C.J. (2013). Granulation of ferroalloys: Results from industrial operations and comparative study on fines generation. *Proceedings of the 13th International Ferroalloys Congress*, Almaty, Kazakhstan, pp. 65–72.

Videm, T. (1995). Reaction rate of the reduction materials for the (ferro) silicon process. *Proceedings of the Seventh International Ferroalloys Congress*, Trondheim, Norway, pp. 221–230.

Visser, J. and Barret, W. (1992). An evaluation of process alternatives for the reclamation of ferrochrome from slag. *Proceedings of the Sixth International Ferroalloy Congress*, Johannesburg, South Africa, pp. 107–112.

von Schéele, J. (2004). Oxyfines™ technology for the re-melting of fines, dust and sludge. *Tenth International Ferroalloys Congress, INFACON X: Transformation through Technology*, Cape Town, South Africa, pp. 678–686.

Wang, H., Wang, X., Lang, Y. and Chu, S. (2013). New process for sintering chromite concentrate with continuous-strand sinter machine. *Proceedings of the 13th International Ferroalloys Congress*, Almaty, Kazakhstan, pp. 57–64.

Wang, W., Li, H.W., Lai, K.R. and Du, K.H. (2012). Preparation and characterization of mullite whiskers from silica fume by using a low temperature molten salt method. *Journal of Alloys and Compounds* 510: 92–96.

Warmuzek, M. (2004). Introduction to aluminium-silicon casting alloys. In: *Aluminium-Silicon Casting Alloys: Atlas of Microfractographs (#06993G)*. ASM International, Materials Park, OH. Available at www.asminternational.org. Accessed 20 April 2016.

Warren, G.F., Jochens, P.R. and Howat, D.D. (1974). Liquidus temperatures and the activities of manganese (II) oxide in slags associated with the production of high carbon ferromanganese alloys. *Proceedings of the First International Congress on Ferroalloys*, Johannesburg, South Africa, pp. 175–185.

Weber, P. and Eric, R.H. (2006). The reduction of chromite in the presence of silica flux. *Minerals Engineering* 19: 318–324.

Wiles, C.C. (1987). A review of solidification/stabilization technology. *Journal of Hazardous Materials* 14: 5–21.

Williams, M., Todd, G.D., Roney, N., Crawford, J. and Coles, C. (2012). Toxicological profile for manganese. Agency for Toxic Substances and Disease Registry. U.S. Department of Health and Human Sciences, Atlanta, GA. Available at http://www.atsdr.cdc.gov/toxprofiles/tp151.pdf. Accessed 6 April 2016.

Wu, X.R., Zhu, B.X., Li, L.S. and Lu, H.H. (2014). Crystallization kinetics of $CaO\text{-}MgO\text{-}SiO_2\text{-}Al_2O_3\text{-}Fe_2O_3$ slag containing Cr_2O_3. In: Pal, M. (ed.), *International Conference on Material Science and Materials Engineering*. DeStech Publications, Pennsylvania, USA, pp. 482–487.

Xiao, Y. and Holappa, L. (1993). Determination of activities in slags containing chromium oxides. *ISIJ International* 33(1): 66–74.

Yan, S. and Qiu, Y. (2014). Preparation of electronic grade manganese sulfate from leaching solution of ferromanganese slag. *Transactions of Nonferrous Metals Society of China* 24: 3716–3721.

Yang, Y., Xiao, Y. and Reuter, M.A. (2004). Analysis of transport phenomena in submerged arc furnace for ferrochrome production. *Proceedings of the 10th International Ferroalloys Congress*, Cape Town, South Africa. Available at http://www.pyrometallurgy.co.za/InfaconX/. Accessed 6 April 2016.

Yildirim, I.Z. and Prezzi, M. (2011). Chemical, mineralogical and morphological properties of steel slag. *Advances in Civil Engineering* 2011: 1–13.

Yoshikoshi, H., Takeuchi, O., Miyashita, T., Kuwana, T. and Kishikawa, K. (1984). Development of composite cold pellet for silicomanganese production. *Transactions of Iron and Steel Institute of Japan* 24: 492–497.

You, B.D., Lee, B.W. and Pak, J.J. (1999). Manganese loss during oxygen refining of high carbon ferromanganese melts. *Metals and Materials* 5(5): 497–502.

Yuge, N., Abe, M., Hanazawa, K., Baba, H., Nakamura, N., Kato, Y., Sakaguchi, Y., Hiwasa, S. and Aratani, F. (2001). Purification of metallurgical grade silicon up to solar grade. *Progress in photovoltaics: Research and Applications* 9: 203–209.

Zaitsev, A.I. and Mogutnov, B.M. (1995). Thermodynamic properties and phase equilibria in the $MnO\text{-}SiO_2$ system. *Journal of Materials Chemistry (Royal Society of Chemistry)* 5: 1063–1073.

Zaldat, G.I., Kamyshnikov, V.V., Suvorov, S.A., Volkov, V.S., Koshkin, G.A., Yurman, M.G., Ostrovskii, Y.I. and Naryzhnyi, V.D. (1984). Fused refractory materials based on magnesia slags. *Refractories and Industrial Ceramics* 3: 125–130.

Zelić, J. (2005). Properties of concrete pavements prepared with ferrochromium slag as concrete aggregate. *Cement and Concrete Research* 35: 2340–2349.

Zhang, W. and Cheng, C.Y. (2007a). Manganese metallurgy review. Part 1: Leaching of ores/secondary materials and recovery of electrolytic/chemical manganese dioxide. *Hydrometallurgy* 89(3–4): 137–159.

Zhang, W. and Cheng, C.Y. (2007b). Manganese metallurgy review. Part 2: Manganese separation and recovery from solution. *Hydrometallurgy* 89(3–4): 160–177.

Zhang, L. and Thomas, B. (2006). State of the art in the control of inclusions during ingot casting. *Metallurgical and Materials Transactions* 37B: 733–761.

Zhang, Z., Zhang, B. and Yan, P. (2016). Comparative study of raw and densified silica fume in the paste, mortar and concrete. *Construction and Building Materials* 105: 82–93.

Zhao, B., Jak, E. and Hayes, P.C. (2005). Phase equilibria in high MgO ferromanganese and silicomanganese smelting slags. *ISIJ International* 45(7): 1019–1026.

Zhong, Y., Shaw, L.L., Manjarres, M. and Zawrah, M.F. (2010). Synthesis of silicon carbide nanopowder using silica fume. *Journal of the American Ceramic Society* 93(10): 3159–3167.

Zhou, F., Chen, T., Yan, C., Liang, H., Chen, T., Li, D. and Wang, Q. (2015). The flotation of low-grade manganese ore using a novel linoleate hydroxamic acid. *Colloids and Surfaces A: Physiochemical and Engineering Aspects* 466: 1–9.

Zhou, Q., Milestone, N.B. and Hayes, M. (2006). An alternative to Portland cement for waste management encapsulation: The calcium sulphonate cement system. *Journal of Hazardous Materials* 136: 120–129.

6

Base Metals Waste Production and Utilization

6.1 Introduction

Base metals broadly refer to a category of non-ferrous metal commodities typically consisting of copper, nickel, cobalt, lead, zinc and mercury (Craig and Vaughan, 1990; Jones, 1999; Cole and Ferron, 2002; Crundwell et al., 2011; Schlesinger et al., 2011). Although the platinum group metals (PGMs) (also referred to as platinum group elements (PGE)) are technically classified as precious metals, they are mostly associated in nature with base metal minerals such as copper-nickel-cobalt-iron sulphide minerals. Furthermore, the primary production of PGMs is closely associated with the production processes of copper, nickel and cobalt metals, hence their inclusion in this chapter (Jones, 1999; Cramer, 2001; Cole and Ferron, 2002; Crundwell et al., 2011).

Basically, base metals have irreplaceable industrial and commercial applications, and their consumption is directly correlated with the global economic growth (IMF, 2015). In general, base metals have ranging applications, such as in electrical and electronics industry for copper and in stainless steels and super alloys, shape memory, magnetic materials and chemicals industry for nickel and cobalt (Davies, 2000; Total Materia, 2001, 2002; CDI, 2006a,b,c; BGS, 2009; Farzin-Nia and Yoneyama, 2009; Outokumpu, 2013). On the other hand, PGMs have irreplaceable industrial uses such as in the manufacture of autocatalysts, catalysts in chemicals industry and the production of fuel cells devices (Crundwell et al., 2011).

Despite the ubiquity of the base metals and PGMs in various applications, the production processes of these metals inherently produce by-products and waste streams with the potential to cause significant health effects and environmental pollution. As a result, Chapter 6 discusses the primary production processes of base metals (nickel, copper and cobalt) and PGMs from sulphide ores, as well as the production of waste streams from the respective processes. Furthermore, this chapter discusses the potential processes and techniques used in the reduction, recycling and utilization of the waste streams in line with the global trends of valorization and inertization of metallurgical wastes.

6.2 Pyrometallurgical Extraction of Copper

6.2.1 Chemical and Physical Properties

Copper is a non-polymorphous transition metal with a face-centred cubic lattice, atomic mass 63.546 and electron structure $[Ar]3d^{10}4s^1$ (Drexler et al., 1992; BGS, 2007; Konečná and Fintová, 2012; CDA, n.d.). The melting point and density of pure copper are 1083°C and 8900 kg/m^3, respectively (Konečná and Fintová, 2012; CDA, n.d.). In its elemental form, copper exhibits diamagnetic properties and has high electrical conductivities, which are only second to silver (Drexler et al., 1992; Davies, 2001; CDA, n.d.). In fact the electrical conductivity of pure copper is usually used as a basic standard in evaluating and characterizing the electrical conductivity of metals and alloys (Davies, 2001; Konečná and Fintová, 2012).

6.2.2 Industrial Uses of Copper and Copper Alloys

Copper is one of the most versatile and widely used engineering and commercial materials, ranking third behind only iron and steel and aluminium (Davies, 2001; Total Materia, 2009; IMF, 2015). In fact, the applications and use of copper and copper alloys (i.e. metal compounds with copper as principal component) date back to many millennia (Davies, 2001; Total Materia, 2009; CDA, n.d.). Copper and its alloys have found widespread industrial applications due to their excellent electrical and thermal conductivities, corrosion resistance, ease of fabrication and good strength and fatigue resistance (Davies, 2001). Typically, copper is widely used in the construction industry as an alloy element in both ferrous and non-ferrous metallurgical industries in industrial machinery and automotive industries in electronic, electrical and sustainable energy applications and in a variety of industrial chemicals and consumer products (Davies, 2001; CDA, n.d.). The characteristic applications of copper and copper alloys include in power transmission lines, architectural applications, cooking utensils, electrical wiring, high conductivity wires, heat exchangers and tubing, plumbing and in water-cooled copper crucibles (Total Materia, 2009). In fact, the growth in the copper industry has been intimately linked to the increasing use of electricity, with electrical applications being one of copper's principal uses due to its ability to conduct electricity and be drawn into wire without breaking (Geoscience Australia, 2015).

Despite the current global recession in the commodities market, the long-term growth and global demand of copper and copper alloys is expected to remain buoyant (ICSG, 2013, 2016; IMF, 2015). For example, the global refined copper production was about 19 million metric tonnes (Mt) in 2010 and is projected to increase to 23 Mt by 2017 (ICSG, 2013, 2016). Figure 6.1 shows the actual and projected data for mine production, refined copper production and the usage of refined copper products in the period 1960–2017 (ICSG, 2013, 2016).

6.2.3 Occurrence of Copper

Copper is a non-ferrous metal with average concentration in the earth's crust of about 50 parts per million (BGS, 2007). In general, copper is commonly found in the earth's crust as copper-iron sulphide (Cu-Fe-S) minerals and, to a lesser extent, as native copper and oxidized minerals such as carbonates, oxides, hydroxyl silicates and sulphates that are typically formed from the weathering of copper-iron sulphide minerals (EPA, 1994; BGS, 2007; Schlesinger et al., 2011; Konečná and Fintová, 2012; Geoscience Australia, 2015). Scrap copper, particularly post-consumer scrap, is also another important source in the production of copper and copper alloys. Further details on the production of copper from secondary resources such as post-consumer scrap were covered in Chapter 9.

Copper minerals, some of which are summarized in Table 6.1, are found worldwide in a variety of geological environments depending on the predominant rock-forming processes that took place (EPA, 1994; BGS, 2007). In general, the copper deposits can be grouped into five broad categories, namely (EPA, 1994; BGS, 2007): (1) porphyry and related copper deposits, predominantly occurring as disseminated mineral deposit in magmatic and volcanic regions, and contributing to the bulk of global copper production (actually contributing to about 50%–60% of world copper production); (2) sediment-hosted deposits, mainly

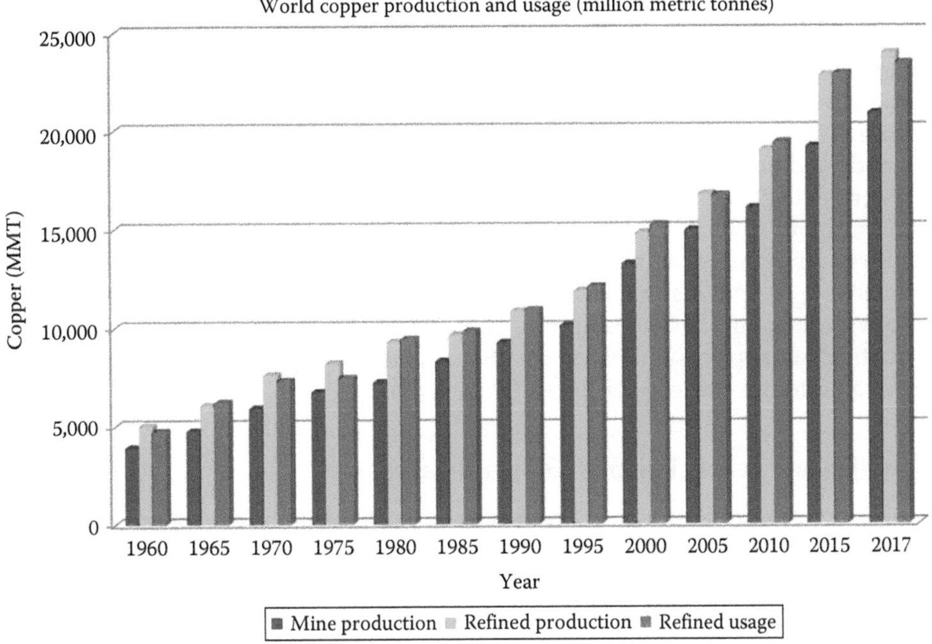

FIGURE 6.1
Actual and projected data for the mine production, refined copper production and the usage of refined copper products. (Adapted from International Copper Study Group (ICSG), The world copper fact book 2013, 2013, Available at http://copperalliance.org/wordpress/wp-content/uploads/2012/01/2013-World-Copper-Factbook.pdf, Accessed 11 June 2016; International Copper Study Group (ICSG), Copper market forecast 2016/2017, Press Release, 2016, Available at http://www.icsg.org/index.php/component/jdownloads/finish/113-forecast-press-release/2128-2016-03-icsg-forecast-press-release?Itemid=0, Accessed 11 June 2016.)

TABLE 6.1
Common Copper Minerals in Economically Exploitable Deposits

Mineral	Chemical Formula	Theoretical (wt.% Cu)
Native copper	Cu	100
Cuprite	Cu_2O	88.9
Chalcocite	Cu_2S	79.9
Covellite	CuS	66.4
Bornite	Cu_5FeS_4	63.3
Malachite	$Cu_2CO_3(OH)_2$	57.5
Azurite	$2CuCO_3 \cdot Cu(OH)_2$	55.3
Antlerite	$Cu_3SO_4(OH)_4$	53.7
Enargite	Cu_3AsS_4	49.0
Chrysocolla	$CuSiO_3 \cdot 2H_2O$	36.2
Chalcopyrite	$CuFeS_2$	34.6

Source: Adapted from Schlesinger, M.E. et al., *Extractive Metallurgy of Copper*, 5th edn., Elsevier, Oxford, UK, 2011, p. 19. With permission.

consisting of disseminations of fine-grained sulphides in a stratiform sedimentary and metasedimentary rocks and are the second most important source of copper (contributing to about 20% of global copper production); (3) volcanogenic massive sulphide deposits, formed as a result of the discharge of metal-rich hydrothermal fluids; (4) veins and replacement bodies associated with metamorphic rocks and (5) deposits associated with ultramafic, mafic and ultrabasic, and carbonate rocks. In general, the different types of copper deposits present both challenges and opportunities in the subsequent metal extraction processes. For example, the large size of porphyry copper deposits, their relatively low grade and intimate association with fractured host rock, alteration and mineralization make them amenable to opencast mining methods (BGS, 2007). Furthermore, the polymetallic nature of volcanogenic massive sulphide deposits is always advantageous during periods of fluctuating metal prices and is especially so when the recovery of coexisting minerals can act as a buffer for low copper prices (BGS, 2007).

6.2.4 Mining and Beneficiation of Copper-Bearing Ores

Generally, copper-bearing minerals are extracted from the copper-bearing ore body using three principal techniques (BGS, 2007): (1) open-pit mining, (2) underground mining and (3) in situ leaching. The applicability, merits and demerits of the different mining methods and processes were discussed in Chapter 3 and will not be discussed further in this section.

After the mining process, the run-of-mine ore is conveyed to the beneficiation plant. Typically, the beneficiation processes include crushing and grinding followed by chemical and/or physical processing and separation stages in order to upgrade the copper content into a concentrate (EPA, 1994; BGS, 2007; Schlesinger et al., 2011). Basically, the pure metal copper is further recovered from the resulting copper concentrates using two techniques (EPA, 1994; BGS, 2007; Schlesinger et al., 2011): (1) pyrometallurgical processes, including roasting, smelting, converting and electrolytic refining and (2) hydrometallurgical processes, including leaching, solvent extraction and electrowinning. Although these two processes are fundamentally different, they are somewhat complementary based on the fact that they can be installed in parallel, and the sulphur dioxide from the smelting and converting processes can be used to produce the sulphuric acid that is required for leaching in hydrometallurgical processes (BGS, 2007; Schlesinger et al., 2011).

In general, hydrometallurgical and pyrometallurgical processes are commonly applied for the extraction of copper from oxidized ores and the copper-bearing sulphide ores, respectively (BGS, 2007; Schlesinger et al., 2011). About 80% of the world's copper originates in Cu-Fe-S ores, and these are largely not amenable to extraction by conventional hydrometallurgical processes (Schlesinger et al., 2011). With the exceptions of the option of sulphide roasting and hydrometallurgical processing and/or pressure leaching, the vast majority of the sulphide copper concentrates are predominantly being processed via the pyrometallurgical route (BGS, 2007; Schlesinger et al., 2011). In fact, the pyrometallurgical extraction process is theoretically capable of producing metallic copper directly from the sulphide concentrates by oxidizing the sulphides to elemental copper (Schlesinger et al., 2011).

The fundamentals of metal extraction from ores using hydrometallurgical processes were covered in detail in Chapter 7 and will not be discussed further in this section. As such, this chapter is limited to recovery of metallic copper from the sulphide ores and will discuss the typical unit processes such as the concentration of copper sulphide minerals by flotation, smelting of copper sulphide ores to produce high-grade mattes, converting of high-grade copper mattes to produce blister copper and the final refining of the blister copper to produce high-purity copper.

6.2.5 Concentration of Copper-Bearing Minerals by Flotation Process

Basically, the first step in the recovery of copper from run-of-mine copper-bearing ores involves the comminution processing by successive crushing and grinding to liberate the copper-bearing minerals from gangue (BGS, 2007; Schlesinger et al., 2011). When the desired degree of liberation is achieved, the copper-bearing sulphide minerals are concentrated by physicochemical processes such as flotation processing. Basically, flotation is the physical separation process used to selectively separate the copper-iron sulphide (Cu-Fe-S) minerals from gangue based on the surface and physicochemical characteristics of the ore. The separation process takes place in flotation banks where special reagents are added to alter the hydrophobicity properties of the sulphide minerals, causing them to float on rising bubbles and be recovered as concentrate while the gangue minerals remain largely unfloated (EPA, 1994; BGS, 2007; Schlesinger et al., 2011; European Commission, 2014a). The concentrate produced from the flotation process typically contains about 25–35 wt.% Cu from upgrading the run-of-mine ore containing Cu in the range 0.5–3 wt.%. The flotation concentrate is then dewatered, dried and conveyed to the smelting units for further processing (BGS, 2007; Schlesinger et al., 2011).

6.2.6 Production of Copper from Copper-Iron Sulphide Concentrates

6.2.6.1 Fundamentals Considerations

The production of copper from copper-iron sulphide concentrates by pyrometallugical processing involves a number of processing steps such as roasting, smelting, converting, refining and electrorefining. Figure 6.2 summarizes some of the generic steps in the pyrometallurgical extraction of copper from copper sulphide concentrates (EPA, 1994; BGS, 2007; Schlesinger et al., 2011; European Commission, 2014a). Further details on the respective unit processes are discussed in the following sections.

6.2.6.2 Smelting of Copper Sulphide Concentrates

The smelting stage involves the melting and oxidation of copper sulphide concentrates to produce a copper-enriched molten sulphide matte phase (35–68 wt.% Cu), molten oxide slag containing as little entrained or dissolved copper as possible, and SO_2-bearing gases (BGS, 2007; Schlesinger et al., 2011; European Commission, 2014a). Basically there are two processes used in the smelting of copper-bearing concentrates, namely, flash smelting and bath smelting (European Commission, 2014a). The flash smelting process is currently the most dominant one, and uses oxygen-enriched air to result in an autothermal smelting operation (Schlesinger et al., 2011; European Commission, 2014a). Basically, silica fluxes are usually added as part of furnace charge in order to control the slag properties such as liquidus temperature and fluidity (Schlesinger et al., 2011). Controlling the physicochemical properties of slag is crucial in promoting the matte–slag immiscibility properties, thereby reducing the copper losses by entrainment and/or dissolution of the copper in slag (Schlesinger et al., 2011). The typical reactions taking place in the smelting process to produce copper-rich matte and the formation of molten slag phases are shown in Equations 6.1 and 6.2 (Schlesinger et al., 2011):

$$2CuFeS_{2(s)} + \frac{13}{4}O_{2(g)} \rightarrow Cu_2S - 0.5FeS_{(l)} + \frac{3}{2}FeO_{(s)} + \frac{5}{2}SO_{2(g)} \quad \text{(Molten matte)} \tag{6.1}$$

$$2FeO_{(s)} + SiO_{2(s)} \rightarrow Fe_2SiO_{4(l)} \quad \text{(Molten slag)} \tag{6.2}$$

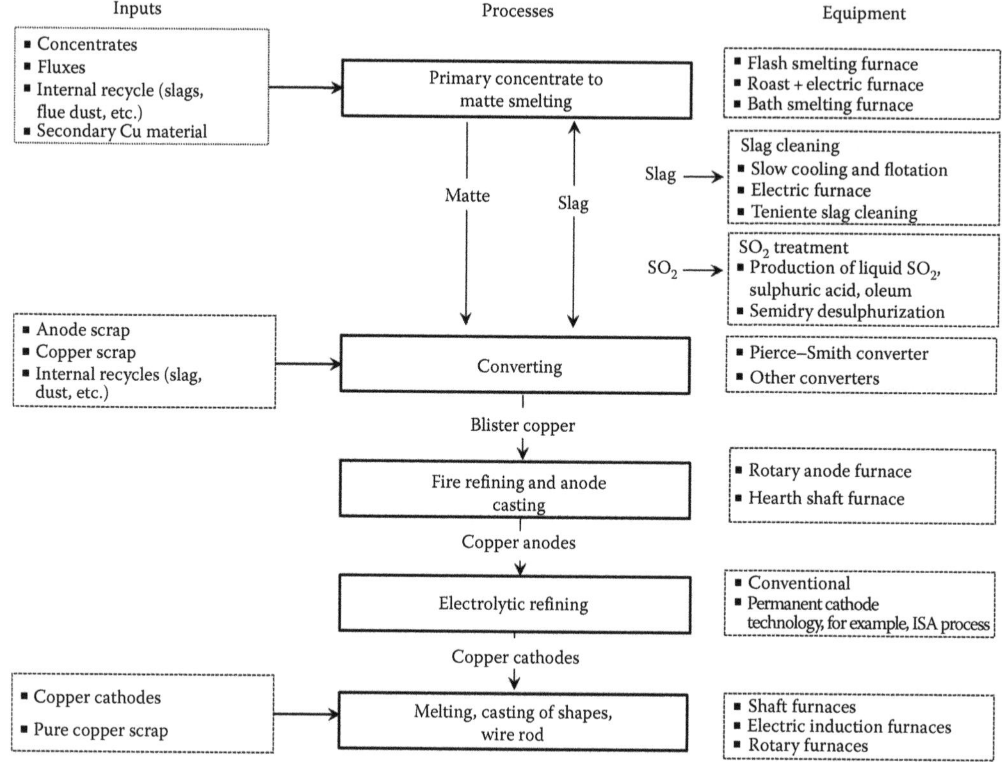

FIGURE 6.2
Unit processes in the primary production of copper from copper-bearing sulphide concentrates. (From European Commission, 2014a.)

Once the smelting and slag formation is complete, the matte and supernatant slag are separately tapped from the furnace for further refining (in the converting furnaces) and entrained metal recovery, respectively.

6.2.6.3 Converting of Copper Mattes

The converting process involves blowing oxygen-enriched air through the molten matte in order to oxidize the iron and sulphur produce a crude molten copper product (97–99 wt.% Cu) (BGS, 2007; Schlesinger et al., 2011; European Commission, 2014a). Silica fluxes are usually added to combine with the oxidized iron and form slag, and the heat generated from the oxidation reactions is sufficient to make the process autothermal (BGS, 2007; Schlesinger et al., 2011; European Commission, 2014a). Basically, the converting process takes place in two sequential steps: (1) the oxidation of FeS and formation of slag (Equation 6.3) and (2) the copper-forming stage from oxidation of Cu_2S in matte (Equation 6.4).

$$2FeS_{(l)} + 3O_{2(g)} + SiO_{2(s)} \rightarrow Fe_2SiO_{4(l)} + 2SO_{2(g)}; \quad \Delta H^0 < 0 \tag{6.3}$$

$$Cu_2S_{(l)} + 3O_{2(g)} + SiO_{2(s)} \rightarrow 2Cu_{(l)} + SO_{2(g)}; \quad \Delta H^0 < 0 \tag{6.4}$$

The use of basic CaO-based fluxes to form calcium ferrite slags is also gaining traction in the recent years, mainly due to the need to control the precipitation of magnetite (Fe_3O_4) by chemically binding it as calcium ferrite phases in the slag.

In practice, the sulphur dioxide in the converter off-gas is captured and is combined with that produced in the smelting furnaces to produce sulphuric acid (Schlesinger et al., 2011).

6.2.6.4 Fire Refining of Blister Copper

Basically, fire refining is a further purification step applied in the refining of blister copper produced from the converting stage (European Commission, 2014a). The molten blister copper is mixed with fluxes and charged into the anode furnace, and air is blown into the molten metal to oxidize impurities and to remove any remaining traces of sulphur (BGS, 2007; European Commission, 2014a). This oxidation stage is followed by the reduction stage, which entails subjecting the metal bath to reducing conditions to partially remove the dissolved oxygen and reduce the copper oxides that may have been formed as a result of the oxidation reactions (BGS, 2007; European Commission, 2014a). Finally, the fire-refined metal is tapped from the anode furnace, cast into anodes and is conveyed to the electrorefining plant (BGS, 2007; Schlesinger et al., 2011; European Commission, 2014a).

6.2.6.5 Electrorefining of Copper Anodes

Copper electrorefining basically entails electrochemically dissolving the copper from impure anodes into a $CuSO_4$–H_2SO_4–H_2O electrolyte, followed by the electrochemical plating of pure copper from the electrolyte onto copper and/or stainless steel cathodes (Schlesinger et al., 2011). Basically, some metal impurities such as nickel (Ni), iron (Fe), arsenic (As), bismuth (Bi) and antimony (Sb) are dissolved into the electrolyte, while the more noble metal impurities such as precious metals (e.g. gold, silver and PGMs), selenium (Se) and tellurium (Te) remain undissolved to form anode sludge that settles at the bottom of the electrolytic cells. The quality of electrorefined copper is determined by the level of impurities co-deposited on the cathodes, which in turn is controlled by the level of residual impurities in the anode as well as the type of concentrate and/or secondary material feedstock (European Commission, 2014a). Table 6.2 highlights the typical changes in the levels of impurities before and after the copper electrorefining process (European Commission, 2014a).

Basically, the anode slimes constitute a valuable source of precious metals and other minor elements, and are periodically removed from the electrolytic tanks and treated for metal recovery. Furthermore, the efficiency of the electrochemical cells tends to decrease as the concentration of dissolved metal elements in the electrolyte increases, and at some point, the electrolyte is bled from the system, neutralized, and is then treated for metal recovery before disposal.

TABLE 6.2

Impurity Levels in the Electrorefining Process of Copper

Element	Content in Anode (g/ton)	Content in Cathode (g/ton)
Silver	600–700	9–10
Selenium	50–510	<0.5
Tellurium	20–130	<0.5
Arsenic	700–760	<1.0
Antimony	330–700	<1.0
Bismuth	60	<0.5
Lead	990–5000	<1.0
Nickel	1000–5000	<3.0

Source: European Commission, 2014a.

6.3 Pyrometallurgical Processing of Nickel, Cobalt and Platinum Group Metals

6.3.1 Chemical and Physical Properties of Nickel, Cobalt and Platinum Group Metals

6.3.1.1 Chemical and Physical Properties of Nickel

Nickel is a transition metal element with a face-centred cubic lattice across the entire solid temperature range, atomic number 58.71 and electronic configuration $[Ar]3d^8 4s^2$ (Davies, 2000). The melting point and density of nickel are 1453°C and 8902 kg/m^3, respectively (Davies, 2000; BGS, 2008). In its solid state, nickel is a lustrous, silvery-white metal, which is hard, ductile and malleable, and can take a high polish (BGS, 2008). In addition, the thermal and electrical conductivities of nickel are fairly low, and it can easily be magnetized (Davies, 2000; BGS, 2008). In other words, nickel is strongly ferromagnetic at ambient temperatures, and as a result, special nickel-containing alloys have found widespread applications in the manufacture of different grades of magnets, ranging from high-permeability and soft magnetic alloys to high-coercivity, permanent magnet alloys (Davies, 2000). Other important properties peculiar to elemental nickel include resistance to oxidation and to corrosion attack by alkalis, high-temperature strength properties and high alloyability properties with other metals (Davies, 2000; Total Materia, 2001; BGS, 2008). Furthermore, nickel has typical metallic properties, which means it can easily be rolled, drawn into wire, forged and polished (Davies, 2000; Total Materia, 2001).

6.3.1.2 Chemical and Physical Properties of Cobalt

Cobalt is a transition metal element with atomic number 58.93, electronic structure $[Ar]3d^7 4s^2$, density 8850 kg/m^3, and falls between iron and nickel in the Periodic Table (Davies, 2000; Total Materia, 2002; CDI, 2006a; BGS, 2009). Cobalt exhibits a hexagonal close-packed structure at temperatures below 417°C, while exhibiting a face-centred cubic structure between 417°C and its melting point of 1493°C (CDI, 2006a). In its solid state, cobalt is a hard ferromagnetic, lustrous, greyish-silver and brittle metal (CDI, 2006a; BGS, 2009). Just like nickel, cobalt has the following peculiar attributes (Davies, 2000; Total Materia, 2002; CDI, 2006b,c; BGS, 2009): (1) ability to retain its strength at high temperatures; (2) fairly low thermal and electrical conductivities; (3) high ability to form alloys with many other metals, wherein it is able to impart strength properties at high temperatures; (4) ability to maintain its magnetic properties at high temperatures (up to 1121°C, which is higher than any other metal) and (5) good chemical and corrosion resistance properties. The special physical and chemical attributes of cobalt have enabled its widespread uses in applications in engineered materials that require unique properties such as magnetic properties, corrosion resistance, wear resistance and/or strength at elevated temperatures (Davies, 2000; Total Materia, 2002; CDI, 2006b,c; BGS, 2009).

6.3.1.3 Chemical and Physical Properties of Platinum Group Metals

The PGMs are a family of six greyish to silver-white metals with close physical and chemical properties (Hartley, 1991; Jones, 2004; BGS, 2009b; Crundwell et al., 2011; Johnson Matthey, n.d.). The chemical and physical properties of the elements belonging to the PGMs family include (Hartley, 1991; Jones, 2004; BGS, 2009b; Crundwell et al., 2011) (1) platinum (Pt) – atomic number 78, electronic structure $[Xe]4f^{14}d^9 6s^1$, density 21.45 g/cm^3 and melting point

1769°C; (2) palladium (Pd) – atomic number 46, electronic structure $[Kr]4d^{10}5s^0$, density 12.02 g/cm^3 and melting point 1554°C; (3) rhodium (Rh) – atomic number 45, electronic structure $[Kr]4d^85s^1$, density 12.41 g/cm^3 and melting point 1960°C; (4) iridium (Ir) – atomic number 77, electronic structure $[Xe]4f^{14}d^76s^2$, density 22.65 g/cm^3 and melting point 2443°C; (5) ruthenium (Ru) – atomic number 44, electronic structure $[Kr]4d^75s^1$, density 12.45 g/cm^3 and melting point 2310°C and (6) osmium (Os) – atomic number 76, electronic structure $[Xe]4f^{14}d^66s^2$, density 22.61 g/cm^3 and melting point 3050°C. Basically, the chemical properties of platinum group elements are closely similar, but the aforementioned physical properties tend to vary significantly from element to element (BGS, 2009b). For example, platinum, iridium and osmium are the densest known metals, being significantly denser than gold (19.3 g/cm^3) (BGS, 2009b). Furthermore, platinum and palladium tend to be soft, ductile and resistant to oxidation and high-temperature corrosion (Johnson Matthey, n.d.). On the other hand, rhodium and iridium are more difficult to work than platinum and palladium, and ruthenium and osmium are hard, brittle and almost unworkable in the solid state. In addition, ruthernium and osmium are also characterized by poor oxidation resistance (BGS, 2009b; Johnson Matthey, n.d.).

6.3.2 Industrial and Commercial Uses of Nickel, Cobalt and Platinum Group Metals

6.3.2.1 Nickel and Cobalt

Nickel and cobalt are indispensable non-ferrous metals in many industrial and commercial applications such as in the chemical and metallurgical industries. Basically, the different industrial and commercial applications of nickel and cobalt are summarized in Tables 6.3 and 6.4, respectively. As highlighted in Tables 6.3 and 6.4, the metallurgical uses of nickel and cobalt are in the production of alloys, ranging from the production of generic low alloy and stainless steels to high-specialty alloys such as nickel-base and cobalt-base superalloys, magnetic alloys and shape memory alloys (Davies, 2000; Total Materia, 2001, 2002; CDI, 2006b,c; Campbell, 2008; Farzin-Nia and Yoneyama, 2009; Li et al., 2010; Fernandes et al., 2011; Outokumpu, 2013).

In the chemical industries, nickel and cobalt are commonly consumed in the production of batteries and energy storage systems, in metal finishing processes (e.g. electroplating) and in the manufacture of a wide variety of paints and pigments (Davies, 2000; CDI, 2006a; BGS, 2009). Due to their important applications in these industry sectors, nickel and cobalt are actually considered as critical and strategic minerals in most economic jurisdictions (Shedd, 2002; European Commission, 2014b).

6.3.2.2 Industrial and Commercial Uses of Platinum Group Metals

The industrial and commercial applications of PGMs are predicated on their unique properties discussed earlier, and these include (Crundwell et al., 2011) (1) the ability to catalyse chemical reactions, (2) the ability to resist corrosion, (3) ease with which they can be worked, (4) visual appearance and appeal and (5) the high conductivities, densities and melting points. In general, the platinum group elements have high melting points, are chemically inert to a wide variety of substances across a wide temperature range, and have excellent catalytic properties (Jones, 2004; Hartley, 1991; BGS, 2009b; Crundwell et al., 2011; Mpinga, et al., 2015; Johnson Matthey, n.d.). The key commercial and industrial applications driving the global demand of platinum group metals include in (BGS, 2009b; Johnson Matthey, n.d.): (1) autocatalysts (Pt, Pd and Rh) to convert noxious emissions such as carbon monoxide, oxides of nitrogen and hydrocarbons from car exhaust systems to harmless and non-toxic gaseous emissions; (2) the manufacture of jewellery; (3) as catalysts in chemical industry, such as in the manufacture of nitric

TABLE 6.3

Major Industrial and Commercial Applications of Nickel

Field	Industrial Applications	References
Sustainable energy applications	Component in various battery systems and hydrogen storage applications. Typical examples include use in rechargeable Ni–Cd batteries containing Ni electrodes, Ni metal hydride batteries.	Davies (2000)
Stainless steels, e.g. Cr–Ni and Cr–Ni–Mo grades, high performance austenitic steels	Nickel is an indispensable alloying element in different grades of steels. It imparts corrosion and mechanical properties. The typical beneficial effects of nickel addition include the following: 1. Austenitic stainless steels: Ni stabilizes the austenite structure, promotes repassivation and corrosion resistance. It also increases ductility and toughness. 2. Precipitation hardening steels: Ni forms intermetallic compounds that are used to increase strength. 3. Martensitic grades: Improves weldability, especially when combined with low carbon levels. Different grades of steels and stainless steels have found widespread uses in medical applications, chemical and petrochemical industries, automotive industry, in other applications which require a combination of attributes such as high strength, corrosion resistance, weldability and formability.	Davies (2000); Outokumpu (2013); Total Materia (2001); Campbell (2008)
Nickel-base alloys	High strength and excellent heat and corrosion resistant properties; typically can withstand highly corrosive environments, high temperatures, high stresses (creep and fatigue resistance) and a combination of these factors. For example, (1) Ni–Cr alloys (Cr > 15 wt.%) designed for oxidation and corrosion resistance at temperatures exceeding 760°C, and (2) low expansion alloys, designed to have very low thermal expansion, constant modulus of elasticity and high strength. Typical applications of nickel-base alloys are in harsh environments at high temperatures, for example, aircraft gas turbines, aerospace applications, steam turbine power plants, nuclear power systems, chemical and petrochemical industries.	Davies (2000); Total Materia (2001); Campbell (2008)
Electrical resistance alloys (resistance and resistance heating alloys)	Typical resistance alloys containing nickel include Cu–Ni alloys (2–45 wt.% Ni), Ni–Cr–Al alloys (35–95 wt.% Ni), Ni–Cr–Fe alloys (35–60 wt.%) and Ni–Cr–Si alloys (70–80 wt.% Ni). Resistance alloys are typically used in instruments and control equipment to measure and regulate electrical characteristics. Resistance heating alloys: Ni–Cr alloys (65–50 wt.% Ni and 1.5 wt.% Si) and Ni–Cr–Fe alloys (35–70 wt.% Ni, 1.5 wt.% Si and 1.0 wt.% Nb). Resistance heating alloys are typically used in furnaces and appliances to generate heat.	Davies (2000); Total Materia (2001); Campbell (2008)
Soft magnetic alloys	High-nickel alloys (~79 wt.% Ni, 4–5 wt.% Mo and balance Fe) possess special magnetic properties, for example, high initial permeability and low saturation induction. Typically used in the manufacture of different grades of magnets ranging from high-permeability and soft magnetic alloys to high-coercivity, permanent magnet alloys.	Davies (2000); Total Materia (2001); Campbell (2008)
Shape memory alloys	Metallic materials with ability return to their previously defined shape when subjected to appropriate heating schedule, for example, Ni–Ti alloys (50Ni–50Ti) for medical applications (e.g. orthodontic wires and constructive surgery applications).	Fernandes et al. (2011); Li et al. (2010); Total Materia (2001); Campbell (2008); Farzin-Nia and Yoneyama (2009)
Other uses	Electroless plating (e.g. $Co(NH_2SO_3 \cdot 3H_2O)$) to provide a thin protective layer for corrosion resistance of metal pieces and components. Catalyst a wide range of chemical industries, for example, in the hydrogenation of vegetable oils, reforming hydrocarbons, and production of fertilizers, pesticides and fungicides.	BGS (2008)

TABLE 6.4

Major Industrial and Commercial Applications of Cobalt

Field	Industrial and Commercial Applications	References
Chemicals	Widespread applications in the chemicals industry, particularly in the manufacture of paints and pigments (e.g. $CoSO_4 \cdot 7H_2O$, Co_2S_3, $CoWO_4$) and cobalt catalysts, mostly for use in the petrochemical industries (e.g. mono sulphide, CoS).	Davies (2000); BGS (2009); CDI (2006a)
Batteries and electronics	Important component in rechargeable battery technologies such as Ni–Cd, nickel metal-hydride and lithium ion batteries. Electronic applications include cobalt alloyed with silicon to produce a metal silicate with low resistivity, high thermal stability and good bonding properties for use as electrical connectors in integrated circuits.	BGS (2009)
Stainless steels, e.g. martensitic steels	Alloying element in martensitic steels to increase the hardness and tempering resistance, especially at higher temperatures.	Davies (2000); Campbell (2008); BGS (2008); Outokumpu (2013)
Cobalt-base alloys	Co-base wear resistant alloys: Wear-resistant, corrosion-resistant and heat-resistant alloys for wear-resistant components and/or applications. Typical examples include Stellite® alloys containing 25–30 wt.% Cr, max. 1 wt.% Mo, 2–5 wt.% W, 0.25–3.3 wt.% C, max. 3 wt.% Fe, max. 3 wt.% Ni, max. 2 wt.% Si, max. 1 wt.% Mn and balance Co. The strength arises from the solid-solution strengthening effects of chromium, tungsten and molybdenum, as well as the formation of carbides. The corrosion resistance of cobalt-base alloys is imparted by chromium. Co-base high temperature alloys: High temperature strength, resistance to thermal fatigue and oxidation resistance, and have applications in aircraft gas turbines, aerospace applications, steam turbine power plants and nuclear power systems. Typically contain 20–23 wt.% Cr, 7–15 wt.% W, 10–22 wt.% Ni, 3 wt.% Fe (max.), 0.1–0.6 wt.% C and balance Co. Cobalt base-corrosion resistance alloys: Resistance to aqueous corrosion for applications in chemical and petrochemical industries. Typical composition include 20–25 wt.% Cr, 2 wt.% W, 5–10 wt.% Mo, 9–35 wt.% Ni, 3 wt.% Fe (max.), 0.8 wt.% C (max.), 0.1 wt.% N (max.) and balance Co.	Davies (2000); Total Materia (2002); CDI (2006b); Campbell (2008); BGS (2009)
Magnetic alloys	Hard ferromagnetic metal capable of maintaining its magnetic properties at high temperatures. Can be alloyed with other metals (e.g. SmCo, FeCrCo and AlNiCo alloys) to produce specialized magnetic components. Cobalt-based magnetic materials have applications in electrical motors and generators, automotive sensors and in computer hard discs.	Davies (2000); Total Materia (2002); CDI (2006c); BGS (2008)

acid and silicones where addition of Pt compounds to the silicone mixture catalyses the cross-linking process that results in engineered silicone product, and in petrochemical refining processes; (4) fuel cell devices that generate electricity from electrochemical reactions such as proton exchange membrane fuel cells using platinum-based electrodes; (5) manufacture of electrical and electronics components; (6) the manufacture of special types of glass; (7) special alloys for medical and biomedical applications and (8) assets for investment. To date, the use of platinum group metals in the production of catalytic converters and fuel cells is by far the fastest growing segment driving the demand and consumption of these metals (Jones, 2004; Hartley, 1991; BGS, 2009b; Crundwell et al., 2011; Mpinga, et al., 2015; Johnson Matthey, n.d.).

6.3.3 Occurrence of Nickel, Cobalt and Platinum Group Metals

6.3.3.1 Occurrence of Nickel

Nickel is generally produced from sulphide minerals, and from laterite and saprolite ores (BGS, 2008; Crundwell et al., 2011; European Commission, 2014a). In general, the most common mode of occurrence of some of the important nickel-bearing minerals found in economic deposits include (BGS, 2008; Crundwell et al., 2011) (1) sulphide-hosted minerals such as pentlandite ($(Fe,Ni)_9S_8$), which is the most important nickel-bearing sulphide mineral in economic deposits, occurring together with pyrrhotite ($Fe_{1-x}S_x$), millerite (NiS), chalcopyrite ($CuFeS_2$) and pyrite (FeS_2) in mafic and ultramafic (iron and magnesium rich) igneous rocks; (2) laterites related to ultramafic rocks, and these include garnierite ($(NiMg)_3Si_2O_5(OH)_4$ and nickel-ferrous limonite ((Fe,Ni)O(OH)) (BGS, 2008); (3) hydrothermal replacement of pentlandite in mafic intrusions typically containing minerals such as niccolite (NiAs) and (4) hydrothermal veins hosting sulphide minerals such as siegenite ($(Ni,Co)_3S_4$). Although laterites constitute about 70% of the nickel contained in land-based deposits, their exploitation only contributes to about 40% of the world nickel production (BGS, 2008; Crundwell et al., 2011).

6.3.3.2 Occurrence of Cobalt

Cobalt ores exist mainly as mixed sulphides, either in copper or nickel oxides/sulphide mixtures (CDI, 2006a; BGS, 2009; Crundwell et al., 2011; European Commission, 2014a). The most common cobalt minerals found in economic deposits include (BGS, 2009) (1) erythrite ($Co_3(AsO_4)_2 \cdot 8H_2O$), (2) skutterudite ($(Co,Ni)As_3$), (3) cobaltite (CoAsS), (4) carrollite ($Cu(Co,Ni)_2S_4$), (5) linnaeite ($Co^{2+}Co_2^{3+}S_4$) and (6) asbolite ($(Ni,Co)_{2-x}Mn^{4+}(O,OH)_4 \cdot nH_2O$). In general, cobalt can easily substitute for transition metals in many mineral and chemical compounds, and is commonly found in the place of iron and nickel as they share similar properties (particularly through nickel and iron replacement in pentlandite $(Ni,Fe,Co)_9S_8$) (BGS, 2009; Crundwell et al., 2011). As a result, cobalt is mostly produced as a by-product from the production of other base metals, mainly in the production of nickel and copper. And as such, the production of cobalt is largely discussed together under the production processes of nickel (BGS, 2009).

6.3.3.3 Occurrence of Platinum Group Metals

Platinum group elements (PGEs) are very rare in the earth's crust, and their scarcity technically qualifies them to be classified as precious metals (Cole and Ferron, 2002; Jones, 2005; BGS, 2009b; Crundwell et al., 2011). In nature, platinum group elements are mostly associated either with copper-nickel metal sulphide minerals in magmatic rocks or with the PGE-bearing minerals (Cramer, 2001; Cole and Ferron, 2002; Jones, 2005; BGS, 2009b; Crundwell et al., 2011). The deposits containing hosting the platinum group elements can further be into (Cramer, 2001; Cole and Ferron, 2002; BGS, 2009b) (1) PGE-dominant deposits, wherein the platinum group elements constitute the main economic components, with the other base metals (Ni, Cu and Co) being recovered as minor by-products; (2) nickel-copper dominant deposits, mostly dominated by nickel mineralizations and copper, cobalt and PGEs being produced as by-products and (3) other deposits such as laterites and placers and Alaskan or Ural–Alaskan type of deposits.

In general, the mineralogy of PGMs in these various deposits can be classified into (Cole and Ferron, 2002; BGS, 2009b; Crundwell et al., 2011) (1) small (10–50 μm) platinum group mineral grains such as braggite ((Pt,Pd)S), isoferroplatinum (Pt_3Fe) and laurite ($(Ru,Ir,Os)S_2$), associated with mineral grains of nickel and copper sulphides; (2) discrete atoms of PGEs in pentlandite ($(Fe,Ni,Co)_9S_8$); (3) arsenides, for example, sperrylite (PtS_2) and stillwaterite (Pd_8As_3); (4) antimonides, for example, genkinite ($(Pt,Pd,Rh)_4Sb_3$) and sudburite ((Pd,Ni) Sb); (5) bismuthides, for example, froodite ($PdBi_2$) and (6) tellurides, for example, merenskyite ($(Pd,Pt)(Te,Bi)_2$) and michenerite (Pd(Bi,Sb)Te).

The primary production of platinum group elements from the different types of deposits is dominated by South Africa, Russia, Zimbabwe, Canada and the United States (Jones, 2004, 2005; BGS, 2009b; Crundwell et al., 2011; Mpinga et al., 2015). In fact, the major proportion of platinum and palladium (about 58%) produced from primary sources globally originates from the Merensky, Upper Group Two (UG-2) and Platreef deposits within the Bushveld Complex in South Africa (Jones, 2004, 2005; Nel, 2004; Mpinga et al., 2015).

6.3.4 Primary Production Processes of Nickel, Cobalt and Platinum Group Metals

6.3.4.1 Fundamentals of Nickel Production

As discussed in the preceding sections, the nickel-bearing ores in economic deposits occur predominantly either as lateritic and saprolite oxides or as sulphide-dominant minerals. Fundamentally, the extraction processes for nickel from such ores tend to be different, and the different generic processes are depicted in Figures 6.3 and 6.4, respectively. Figure 6.5 further highlights the generic process flowsheet for the production of nickel from sulphide concentrates using flash smelting and converting furnaces (European Commission, 2014a).

Some aspects on the production of nickel from oxidized nickel-bearing ores by pyrometallurgical processes (ferronickel smelting) were discussed briefly in Chapter 5. Furthermore, the fundamentals of hydrometallurgical extraction of nickel from nickel oxide ores were covered in Chapter 7. As a result, this section will focus on the extraction of nickel from sulphide minerals using base metal smelting and converting processes.

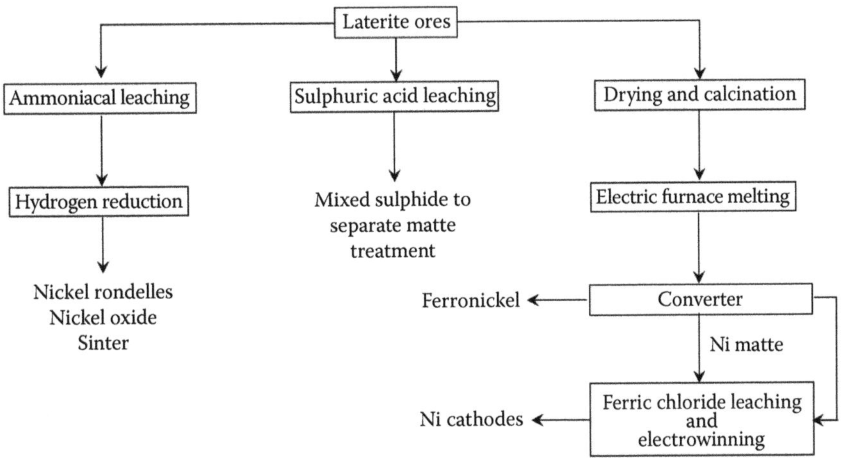

FIGURE 6.3
Generic flowsheet for the production of nickel from laterite ores. (From European Commission, 2014a.)

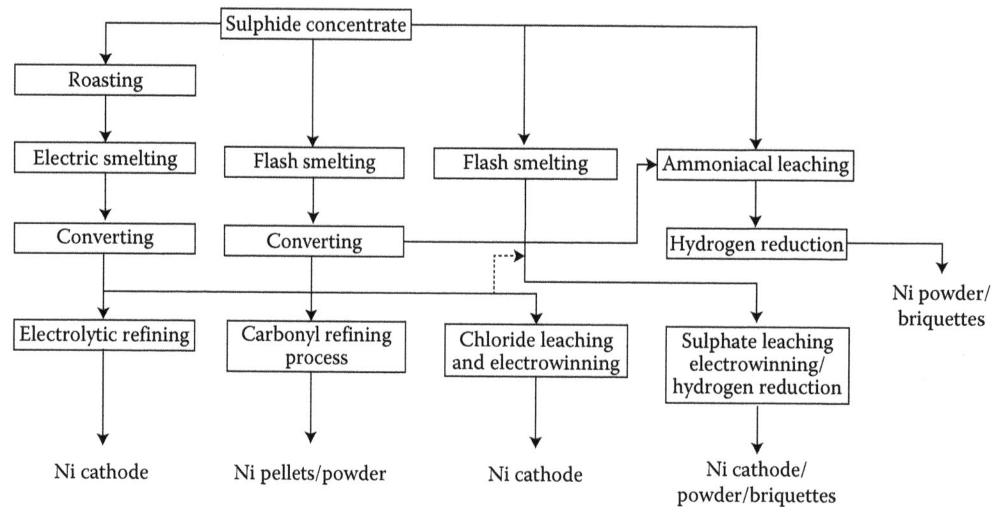

FIGURE 6.4
Generic flowsheet for the production of nickel from sulphide ores. (From European Commission, 2014a.)

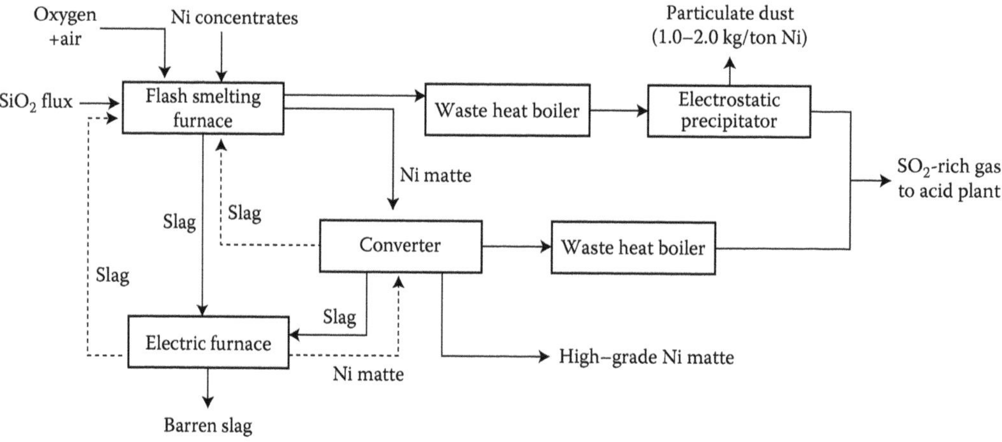

FIGURE 6.5
Generic flowsheet for flash smelting and converting processing of nickel- and cobalt-sulphide concentrates. (From European Commission, 2014a.)

6.3.4.2 *Fundamentals of Cobalt Production*

As discussed in Section 6.3.3, cobalt ores are mainly mixed sulphides, coexisting in either copper and/or nickel oxides/sulphide mixtures such as pentlandite, $(Ni,Fe,Co)_9S_8$, carrollite $(Cu(Co,Ni)_2S_4$ and asbolite $((Ni,Co)_{2-x}Mn^{4+}(O,OH)_4 \cdot nH_2O)$ (BGS, 2009). This means that cobalt is predominantly produced as a by-product from the production of other base metals, particularly in the production of nickel, and to some extent copper and PGMs. As such, the technical aspects in the production cobalt will generally be discussed under the production processes of nickel in the following sections (BGS, 2008, 2009; Crundwell et al., 2011; European Commission, 2014a).

6.3.4.3 Fundamentals of Platinum Group Metals Production

Basically, the production of platinum group metals consists of the following steps (Cole and Ferron, 2002; Crundwell et al., 2011): (1) selective mining of ore rich in platinum group metals; (2) crushing and grinding of the ore to liberate the sulphide mineral grains containing PGMs, nickel, copper and iron from the host rock; (3) concentrating the platinum group elements in the ore into a concentrate predominantly consisting of nickel-copper-iron sulphides (Ni-Cu-Fe-S) and rich in platinum group metals; (4) smelting the base metal (Ni-Cu-Fe-S) and PGM enriched flotation concentrates into base- and PGMs enriched matte; (5) converting the base metal and PGMs- enriched smelter mattes by blowing oxygen to oxidize the iron and sulfur in matte, thereby increasing the content of the desired metal elements in the converter matte; (6) separation of the platinum group elements in the converter matte from the base metals, either by magnetic concentration or by leaching, to produce a platinum group metal-rich product and (7) chemical refining the platinum group metal concentrate into high-purity (excess of 99.9%) and separate individual platinum group metals. Figures 6.6 and 6.7 show the

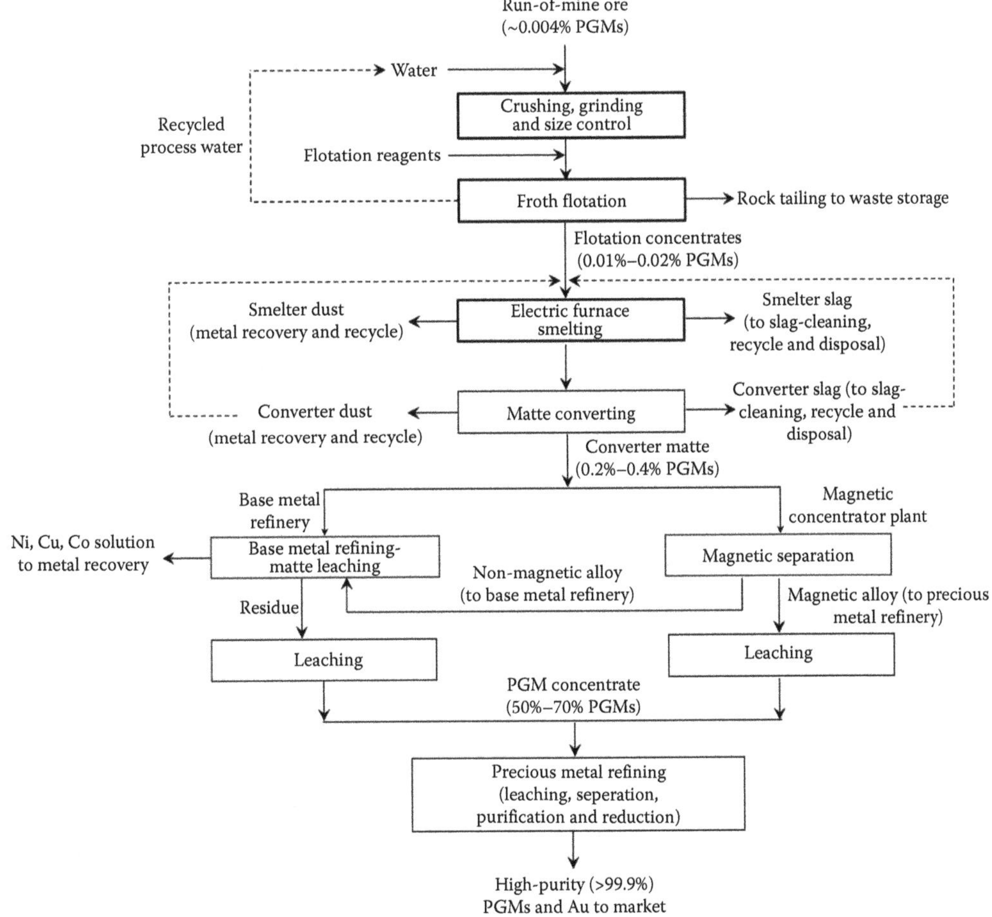

FIGURE 6.6
General flowsheet for the production of pure platinum group metals from PGE-dominant deposits. (Reprinted from *Extractive Metallurgy of Nickel, Cobalt and Platinum-Group Metals*, Crundwell, F.K., Moats, M.S., Ramachandran, V., Robinson, T.G. and Davenport, W.G., p. 412. Copyright 2011, with permission from Elsevier.)

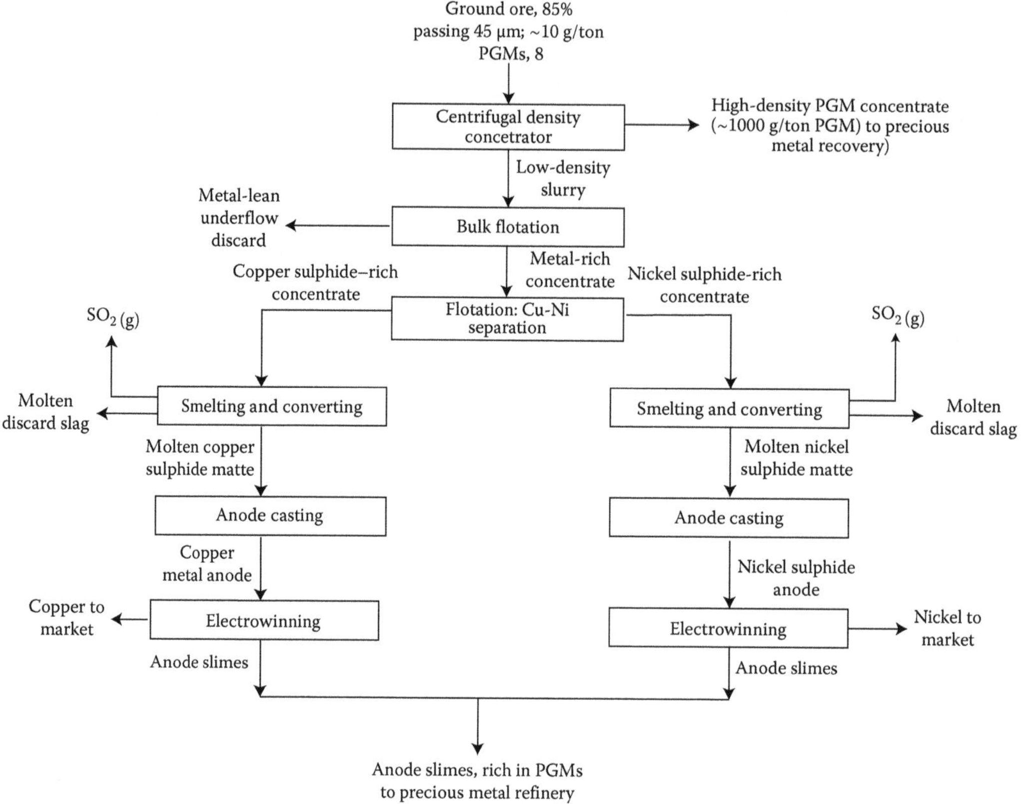

FIGURE 6.7
General flowsheet for producing PGEs from Noril'sk Complex's nickel-copper–dominant deposits. (Reprinted from *Extractive Metallurgy of Nickel, Cobalt and Platinum-Group Metals*, Crundwell, F.K., Moats, M.S., Ramachandran, V., Robinson, T.G. and Davenport, W.G., p. 430. Copyright 2011, with permission from Elsevier.)

generalized process flowsheet for the production of pure metals from PGE–dominant and from nickel-copper–dominant deposits, respectively (Crundwell et al., 2011).

6.3.4.4 Beneficiation of Base Metal and Platinum Group Metal Sulphide Minerals

Sulphide minerals containing nickel, cobalt and platinum group elements can easily be upgraded using the conventional flotation processes to produce concentrates particularly enriched in these elements (Jones, 1999; Cramer, 2001; Cole and Ferron, 2002; BGS, 2008; Crundwell et al., 2011; European Commission, 2014a). Basically, the typical beneficiation steps entail (Cole and Ferron, 2002; Crundwell et al., 2011) (1) comminution (crushing and grinding) of the ores to liberate the nickel- and cobalt-bearing minerals from other minerals and gangue, (2) froth flotation to separate the nickel- and cobalt-bearing sulphide minerals from the gangue and other sulphide minerals (e.g. pyrrhotite) and (3) froth flotation to separate the nickel- and cobalt-bearing minerals (and containing platinum group metals) from copper-bearing minerals. Basically, the fundamental processes peculiar to the concentration of these sulphide minerals are similar to those discussed under the production of copper, and therefore will not be discussed further in this section.

6.3.4.5 Smelting and Converting of Nickel, Cobalt and Platinum Group Metal Concentrates

In general, the pyrometallurgical smelting of nickel-, cobalt- and PGM-bearing sulphide concentrates can be carried out via three basic processes, namely (Toscano and Utigard, 2003; Warner et al., 2006; Warner et al., 2007; Crundwell et al., 2011; European Commission, 2014a): (1) partial roasting of sulphide concentrates followed smelting in electric furnaces, (2) conventional flash or direct electric furnace smelting of the sulphide concentrates to produce nickel-enriched matte and (3) the direct Outokumpu nickel process. Among these processes, the roasting-electric furnace smelting and the conventional flash smelting of these base metal bearing sulphide concentrates are discussed further in Sections 6.3.4.5.1 and 6.3.4.5.2.

6.3.4.5.1 Roasting–Electric Furnace Smelting of Nickel- and Cobalt-Bearing Concentrates

Basically, the roasting–electric furnace processing, predominantly applicable for nickel- and cobalt-bearing sulphide concentrates, involves the following steps (Warner et al., 2006; Crundwell et al., 2011; European Commission, 2014a): (1) partially oxidizing the nickel-bearing concentrates in fluidized-bed roasters or any other roasting equipment to produce a calcine and (2) melting of the SiO_2-fluxed calcine in electric furnaces to form nickel-cobalt–rich molten matte and nickel-cobalt–lean molten slag. When the melting process is complete, the matte and supernatant slag are tapped from the furnace, and the nickel-enriched matte is then taken for converting and recovery of the nickel, copper and cobalt.

6.3.4.5.2 Flash Smelting of Nickel and Cobalt Sulphide Concentrates

Basically, the objective of direct smelting of sulphide concentrates is to produce a nickel- and cobalt-enriched matte by (Warner et al., 2007; Crundwell et al., 2011) (1) oxidation of sulphur from the sulphide concentrates, (2) removal of iron as molten fayalite slag by oxidation and silica-fluxing reactions and (3) removal of gangue components in the concentrates by dissolving them in molten iron silicate slags. Technically, the flash smelting process entails continuously blowing oxygen, air, sulphide concentrates and silica flux in a flash furnace at around 1300°C to produce three product streams (Crundwell et al., 2011): (1) nickel-cobalt–enriched molten sulphide matte, (2) molten slag lean in nickel and other base metals and (3) hot and dust-laden furnace off-gas containing about 20–50 vol.% SO_2. Due to autothermal conditions, the flash smelting process has low electrical energy and fossil fuel consumption compared to roasting and electric furnace process. However, the oxidizing conditions results in increased oxidation loss and hence the entrainment of nickel and cobalt in slag (Warner et al., 2007; Crundwell et al., 2011; European Commission, 2014b). As such, the slags from conventional flash smelting furnaces are retreated in electric slag cleaning furnaces to recover the metal values from slag as shown in Figure 6.5.

6.3.4.6 Converting of Sulphide Mattes

In general, the molten matte tapped from either the electric or flash smelting processes typically contains about (1) 15–40 wt.% Ni and 20–40 wt.% Fe and 20–25 wt.% S for nickel-cobalt smelter matte, and (2) 40 wt.% Fe, 27 wt.% S and 0.1 wt.% PGMs for PGM smelter matte (Warner et al., 2006, 2007; Crundwell et al., 2011). Basically, the converting process involves transferring the molten matte to a converting furnace (e.g. Pierce–Smith or Ausmelt) for the final oxidation of iron and sulphur by blowing air or oxygen-enriched air through the molten bath. The converting process produces an even higher-grade matte containing about 40–70 wt.% Ni, 0.5–4 wt.% Fe and 21–24 wt.% S for nickel converter mattes and about 0.6–3 wt.% Fe, 20 wt.% S, 0.25 wt.% PGMs converter mattes, with the remainder being Cu and Co (Toscano and Utigard, 2003; Warner et al., 2006, 2007; Crundwell et al., 2011). In the converting process, the oxidized iron is removed as iron silicate phases in the molten slag according to the following reactions (Crundwell et al., 2011; Schlesinger et al., 2011):

$$FeS_{(l)} + \frac{3}{2}O_{2(g)} \rightarrow FeO_{(l)} + SO_{2(g)} \tag{6.5}$$

$$2FeO_{(l)} + SiO_{2(s)} \rightarrow Fe_2SiO_{4(l)} \tag{6.6}$$

The molten fayalite slags are periodically tapped from the furnace during the converting cycle, and the fresh matte is added after each slag tap to ensure enough build-up of nickel and other metals in the high-grade matte (Crundwell et al., 2011). Finally, the high-grade nickel-cobalt matte is tapped from the converter and is either cast into ingots or is granulated (Crundwell et al., 2011; European Commission, 2014a).

There are two basic principles governing the converting process of nickel-cobalt sulphide mattes, vis-à-vis (Crundwell et al., 2011): (1) iron oxidizes preferentially to nickel, copper, cobalt and platinum group metals and (2) the slag produced by the oxidation of iron is immiscible with the unoxidized matte to collect as supernatant layer above the denser matte layer. In this regard, the control of slag properties and bath operating conditions is critical in mitigating the entrainment and dissolution of the value metal elements in slag, as well as the reducing loss of these metals by entrainment in the converter off-gases (Crundwell et al., 2011). The typical bath control and operating philosophies in converting operations include (Crundwell et al., 2011) (1) high operating temperatures, for example, around 1275°C for nickel and cobalt; (2) steady input of small, evenly sized particles of silica fluxes; (3) deep tuyeres placement in the matte to avoid over-oxidizing the slag; (4) vigorous mixing to enhance the reaction between matte, oxygen and fluxes and (5) the use of reactive fluxes to facilitate the early slag formation.

Despite the conceited efforts to control the dissolution and entrainment of base metals in converter slags, the loss of these metals to slag is inevitable due to the oxidizing nature and chemistry of converting operations (Toscano and Utigard, 2003). As a result, converter slags typically contain significantly higher amounts of dissolved and/or entrained metal elements than the slags produced from the primary smelting processes. As shown in Figure 6.5, the converter slag is typically treated in separate electric furnaces for metal recovery or is recycled as pre-melted fluxes in flash smelting furnaces (Toscano and Utigard, 2003; European Commission, 2014).

6.3.4.7 Refining of High-Grade Nickel and Cobalt Matte

The high-grade nickel and cobalt mattes tapped from the converter furnaces are further treated in order to recover and refine the base metals and platinum group metals contained (European Commission, 2014a). In general, the typical matte-refining processes and techniques peculiar to nickel and cobalt include (1) chloride leaching of matte followed by electro winning, (2) sulphate-based atmospheric leaching followed by electrowinning and/or hydrogen reduction, (3) ammonia pressure leach and hydrogen reduction,

(4) ferric chloride leaching and (5) refining using the carbonyl process (European Commission, 2014a). The majority of these processes incorporate solvent extraction stages to remove impurities such as iron, calcium and zinc and to separate the nickel and cobalt prior to electro winning or transformation (European Commission, 2014a).

6.3.4.8 Sulphide Smelting Peculiar to Platinum Group Metal Concentrates

Basically, the dried platinum group metal concentrates are smelted in electric furnaces (typically six-in-line submerged-arc electric furnaces) to produce PGEs-rich molten matte and PGEs-lean slags (Jones, 1999, 2005; Cole and Ferron, 2002; Nel, 2004; Crundwell et al., 2011). As opposed to nickel bearing sulphide concentrates which can be smelted autothermally using conventional flash smelting furnaces, the use of autothermal flash smelting in the processing of PGM concentrates is inherently limited by (Jones, 1999, 2005; Cole and Ferron, 2002; Nel, 2004; Crundwell et al., 2011) (1) insufficient sulphide content in concentrates and (2) the high liquidus temperatures of slags (up to 1450°C–1600°C) due to the presence of chromium and magnesium oxides in the pyroxenite and chromitite gangue. The high slag liquidus temperatures are particularly evident in the smelting of the UG-2 flotation concentrates that typically contain considerable amounts (in the range 7–10 times that of Merensky concentrate) of chromite ($FeCr_2O_4$) (Jones, 1999, 2005; Cramer, 2001; Cole and Ferron, 2002; Nel, 2004).

The smelting of chromite-rich PGM concentrates is particularly problematic from energy and process control point of view. Depending on the chromium content of the concentrates and furnace operating conditions, the slags become saturated with chromium-rich spinel ($(Fe,Mg)Cr_2O_4$). Basically, these spinels have a very high melting point, which is around 1600°C, and as such tend to cause serious process and operational problems such as (Jones, 1999, 2005; Cole and Ferron, 2002; Nel, 2004; Crundwell et al., 2011) (1) inefficient settling of matte caused by the spinel accumulation at the slag–matte interface; (2) increase in the effective slag viscosity, thereby causing emulsification of slag and matte; (3) spinel freezing in the lower temperature zones of the furnace, thereby reducing the effective working volume; (4) matte contamination with $FeCr_2O_4$, thereby increasing the chromium content of the converter slag and (5) increase in liquidus temperatures of slag, resulting in excessive superheating of matte and excessive refractory corrosion attack. In order to reduce these operational problems caused by the high chromite content, most operators in South Africa tend to use blended feed materials, such as Merensky and high Cr_2O_3 UG-2 blend, in smelting furnaces (Jones, 1999).

6.3.4.9 Converting and Refining of PGM Mattes

As discussed in the preceding sections, the converting process entails oxidizing the iron and sulphur in the matte using oxygen-enriched air to produce a molten matte product enriched in platinum group metals and lean in iron and sulphur. Currently, the PGMs industry typically uses the horizontal and cylindrical shaped Peirce-Smith and the Ausmelt continuous converters in the converting process of PGMs mattes (Jones, 1999; Crundwell et al., 2011). According to Jones (1999), silica fluxes are added to the converter to flux the iron oxide that is formed by the oxidation of the iron, and to form an iron silicate slag with the approximate composition of fayalite ($2FeO{\cdot}SiO_2$). Basically, the turbulent conditions in the converter results not only in the entrainment of converter matte in the slag, but also in the dissolution of the valuable elements in slag (Jones, 1999, 2005; Cole and Ferron, 2002;

Crundwell et al., 2011). Technically, the converter slags require a retreatment process in order to recover the valuable elements from the slag.

The production of pure platinum group metals from PGE-dominant and nickel-copper-dominant deposits were highlighted in Figures 6.5 and 6.6., respectively. As shown in the figures, the high-grade PGMs and nickel-cobalt mattes tapped from the converting furnaces are further treated in order to recover the individuals metals contained (Cole and Ferron, 2002; Crundwell et al., 2011; European Commission, 2014a). Basically, there are fundamentally two options currently being adopted in the refining of molten matte produced from the recovery of PGMs from PGE-dominant ores, that is, (1) slow cooling and magnetic concentration, followed by the base metal recovery from the non-magnetic components and precious metal recovery from the magnetic components, and (2) whole-matte leaching in a base metal refinery (Cole and Ferron, 2002; Crundwell et al., 2011). Further details of the refining processes of PGM-mattes were highlighted by Crundwell et al. (2011), and therefore will not be discussed further in this section.

6.4 Waste Production and Utilization in the Base Metal Production Processes

6.4.1 General Considerations

As discussed in the preceding sections, base metals and platinum group metals are indispensable metals in various industrial, chemical and household applications. Due to their irreplaceable applications, the long-term demand of these metals and alloy products is expected to remain positive (Warner et al., 2006, 2007; Crundwell et al., 2011; ICSG, 2013, 2016). While the technologies and processes used in the production of base and platinum group metals from raw ore, sulphide concentrates, and secondary and/or recycled resources have matured, the demand to reduce the anthropogenic effects in their production processes is increasingly becoming a global issue. As a result, this section discusses some of the processes and techniques applicable to the reduction, recycling and utilization of these base metal production wastes.

Significant amounts of waste materials are produced from the key process steps used in the production of base metals and platinum group metal products, and these include (1) particulate dusts and sludges emitted from smelting, converting and refining operations that are particularly contaminated with volatile and potentially toxic metal elements such as arsenic, antimony, lead and bismuth (Azawa and Azakami, 1969; Steele et al., 1995; Morales et al., 2010; Shibayama et al., 2010; Crundwell et al., 2011; Schlesinger et al., 2011; Montenegro et al., 2013; European Commission, 2014a; Guo et al., 2015); (2) smelting and converting process slags containing both dissolved and entrained metal species (Banda et al., 2002; Toscano and Utigard, 2003; Nel, 2004; Rudnik et al., 2009; Shibayama et al., 2010; Schlesinger et al., 2011; Crundwell et al., 2011; European Commission, 2014a); (3) gaseous emissions, particularly rich in sulphur dioxide and organometallic compounds from the contaminants inherent in secondary materials (European Commission, 2014a); (4) sludges from base and precious metals refining and separation containing precious metals but also contaminated with potentially toxic metal elements such as selenium, zinc and tellurium (Crundwell et al., 2011; Schlesinger et al., 2011; European

Commission, 2014a); (5) spent solutions containing dissolved base metal elements (Crundwell et al., 2011; Schlesinger et al., 2011; European Commission, 2014a) and (6) metal scrap and sludges from the production and fabrication of base metal alloys and/or components (European Commission, 2014a). In general, the amount and characteristics of waste streams produced from the primary production of base metals and their alloys largely depend on the desired metal content and the content of other deleterious metals (such as As, Se, Hg, Sb, Bi and Zn) in the concentrate and/or secondary materials used (European Commission, 2014a). Table 6.5 summarizes some of the characteristics of some of the waste streams produced in the primary production of base metals. Furthermore, Figure 6.8 illustrates the typical material flow in the primary production processes of copper. In fact, the material flow highlighted can also be generalized to other base metals as well (European Commission, 2014a).

The following sections discuss in detail the different types of waste streams generated in the production processes of base metals and their alloys. Furthermore, the characteristics of the typical waste streams produced in the primary production processes of base metals, and their valorization processes and techniques are summarized in Tables 6.5 and 6.6, respectively.

6.4.2 Smelter and Converter Particulate Dusts

As discussed in the preceding sections, base metal smelting, converting and refining operations generate substantial amounts of particulate dust emissions (Azawa and Azakami, 1969; Steele et al., 1995; Morales et al., 2010; Shibayama et al., 2010; Schlesinger et al., 2011; Montenegro et al., 2013; European Commission, 2014a; González-Castanedo et al., 2014; Guo et al., 2015). To date, the formation of particulate dusts in the primary production processes of base metals has been extensively documented (Azawa and Azakami, 1969; Montenegro et al., 2013). For example, Azawa and Azakami (1969) proposed the removal of impurities in copper smelting process to involve two fundamental mechanisms, that is oxidation followed by slagging and volatilization, with the volatilization mechanism being responsible for the removal of impurity elements such as lead, antimony, bismuth and arsenic from the molten matte bath. Furthermore, Montenegro et al. (2013) proposed the generation of these particulate dusts to involve volatilization and/or entrainment of condensate matter and fine particles from the semi-melted concentrate transported with the furnace off-gas as a result of volatilization and entrainment of condensate matter and fine particles from the semi-melted concentrate transported with the furnace off-gas. Basically, the amount and form of the particulate dusts produced in the smelting and converting processes varies greatly depending on the type of reactor, the chemical composition of the feed materials and the process operating conditions (Yazawa and Azakami, 1969; Montenegro et al., 2013).

The direct and diffuse dust emissions from the primary production processes have serious economic and anthropogenic implications due to the entrainment of valuable metal elements such as Ni, Co, Cu and Fe, and the presence of volatile and toxic metal elements such as Zn, As, Sb, Pb and Hg, respectively (Shibayama et al., 2010; Montenegro et al., 2013; European Commission, 2014a). Furthermore, the disposal of untreated particulate dusts from the primary production processes of these metals presents serious challenges to the environment, largely due to the presence of potentially toxic, easily solubilized and mobile metal species such as Zn, As, Sb, Pb and Bi (Kim et al., 1995; Steele et al., 1995; Morales et al., 2010; Shibayama et al., 2010; Montenegro et al., 2013; European Commission, 2014a; González-Castanedo et al., 2014; Guo et al., 2015).

TABLE 6.5

Characteristics of Waste Streams in the Primary Production of Base Metals

Waste	Characteristics Source	Characteristic Composition	References
Smelting, converting and refining particulate dusts	Entrainment of fine and semi-molten particles in furnace in off-gases volatilization and entrainment of condensate matter in furnace off-gas. Course and fine particulate dusts are recovered from furnace off-gas cleaning systems (e.g. electrostatic precipitators, baghouses, dry and wet scrubbers, etc.).	Typical metal elements contained include toxic metal species such as Zn, As, Sb, Pb, Cd and Hg, as well as valuable metal species such as Cu, Ni, Fe, Co and Ca. Various primary smelter dust samples were characterized as typically containing As_2O_3, $FeAsO_4 \cdot 2H_2O$, $CuSO_4$, $ZnSO_4$, $PbHAsO_4$, NiS, $CuFeS_2$, $CaSO_4 \cdot 2H_2O$, SiO_2, $(Na,K)AlSi_3O_8$, PbS, $PbSO_4$, $Pb_5(ASO_4)_3OH$, Sb_2O_3, Cu_2O and ZnO.	Kim et al. (1995); Morales et al. (2010); Shibayama et al. (2010); Crundwell et al. (2011); Montenegro et al. (2013); Guo et al. (2015)
Gaseous emissions	SO_2 emissions from primary smelting formed as a result of oxidation of sulphides in concentrate. SO_2 emissions from converter refining are as a result of oxidation of sulphur in mattes. Organic compounds (including PCDD/Fs and mixed halogenated dioxins) formed from combustion of organic additives, scrap contaminants, as well as de novo synthesis during cooling.	The amount of SO_2 produced is typically affected by raw material composition and design and configuration off-gas systems. The emission of PCDD/Fs and mixed halogenated dioxins is largely determined by the nature and amount of raw material additives, as well as operational abatement techniques such as after burning, activated carbon adsorption and controlled gas handling and cooling.	Crundwell et al. (2011); Schlesinger et al. (2011); European Commission (2014a); González-Castanedo et al. (2014)
Smelting and converting slags	By-product from fluxing reactions of SiO_2 fluxes with gangue components in concentrates and oxidized metals in mattes. The chemical, textural and mineralogical characteristics are largely controlled by the metallurgical properties (e.g. slag chemical composition, liquidus temperature, furnace conditions, feed material properties as well as the cooling path).	Typical phases in smelting slags include magnetite, fayalite, glassy phases with high silica and low iron content, entrained sulphides such as bornite (Cu_5FeS_4), chalcopyrite ($CuFeS_2$), chalcocite (Cu_2S), oxidized Cu, Ni and Co and entrained oxides and metallic phases of copper, nickel, cobalt and iron. PGM smelter slags, particularly from processes using chromite-rich concentrates, contain significant amounts of $(Fe,Mg)Cr_2O_4$ spinel. Typical phases in converter and refining slags include oxidized Cu, Ni and Co, and entrained metallic copper, nickel, cobalt and iron.	Jones (1999, 2005); Cole and Ferron (2002); Nel (2004); Shen and Forssberg (2003); Toscano and Utigard (2003); Baghalha et al. (2007); Li et al. (2008); Das et al. (2010); Schlesinger et al. (2011); Crundwell et al. (2011); Khorasanipour and Esmailzadeh (2016)

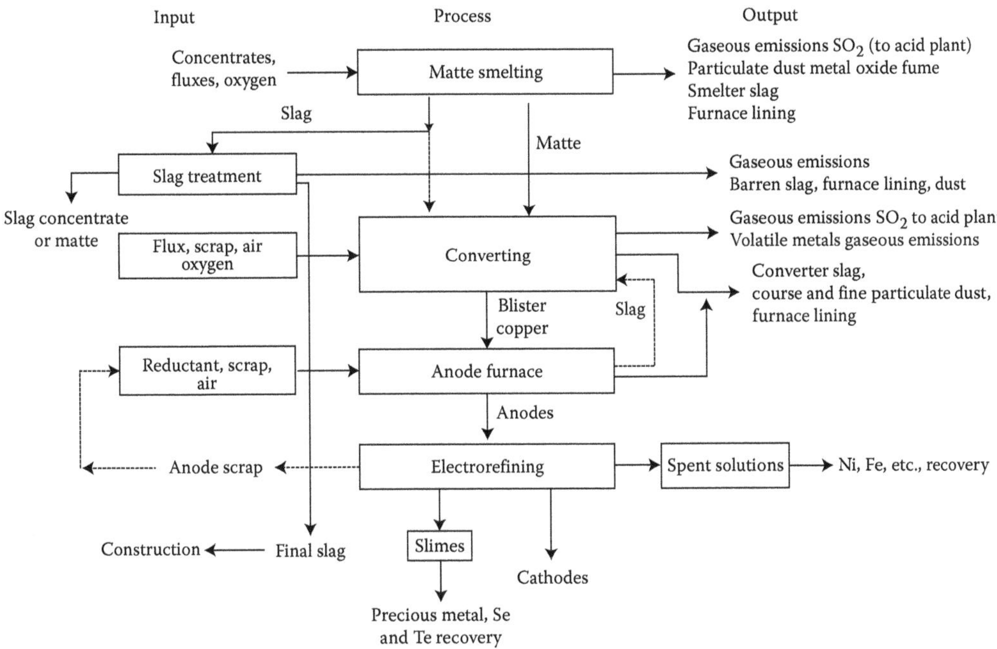

FIGURE 6.8
Material flow in the primary production of copper. (From European Commission, 2014a.)

As previously discussed, the chemical and morphological characteristics of particulate dusts from smelting, converting and refining processes vary significantly according to the composition of concentrates and secondary materials, as well as the type and configuration of reactor equipment (Kim et al., 1995; Morales et al., 2010; Shibayama et al., 2010; Montenegro et al., 2013; Guo et al., 2015). Table 6.5 summarizes some of the typical chemical and morphological attributes of particulate dusts emitted in the primary production processes of base metals.

6.4.3 Recycling and Treatment of Base Metal Smelter Particulate Dusts

As discussed earlier, particulate dusts from primary production processes of base metals contain both valuable metals and potentially toxic metal species. The presence of the potentially toxic metal species particularly presents challenges to the handling, recycling and disposal of these particulate dusts. Furthermore, there is need to reduce the amount of disposed dusts in line with increased environmental demands. As a result, it is imperative to increase the in-process recycling of these materials, increase the amount of dusts that are retreated for metal recovery and to develop viable methods to stabilize the toxic metal species before disposal. In view of the aforementioned demands, several processes and techniques have been proposed with regard to the holistic management of smelter dusts, vis-à-vis (Kim et al., 1995; Steele et al., 1995; Morales et al., 2010; Shibayama et al., 2010; Montenegro et al., 2013; European Commission, 2014a; Samuelsson and Björkman, 2014; Guo et al., 2015; Ha et al., 2015): (1) internal recycling as secondary material in the smelting furnace, (2) treatment for metal recovery using pyrometallurgical and/or hydrometallurgical techniques and (3) stabilization and/or solidification before disposal. The various processes and techniques used in the recycling, metal recovery and treatment of particulate dusts from primary production processes are summarized in Table 6.6.

TABLE 6.6

Potential Valorization Processes and Techniques for Waste Streams in Primary Production of Base Metals

Waste Stream	Characteristics	Valorization Processes and Techniques	References
Smelting, converting and refining particulate dusts	Typically rich in entrained valuable metal elements (e.g. Cu, Ni, Co and Fe) as well as potentially toxic metal species (e.g. Zn, As, Bi, Sb, Pb and Hg). The granulometric characteristics vary according to the type and configuration of furnace off-gas cleaning systems.	Particulate dusts can be internally recycled (particularly converting and refining) back into the primary smelting or to secondary furnaces (References 1–5). However, potential challenges to internal recycling include the recirculation of volatile metal species. As a result, other options include adopting a separate metal recovery smelting unit or removal of volatile metal species before charging into the smelting units. Several treatment have been explored to recover metal species in dusts, such as Bi, Zn, As and Pb (References 5–8), germanium (Reference 9) using hydrometallurgical and/or combination of hydrometallurgical and pyrometallurgical process (Reference 10). Controlled disposal after stabilization and/or solidification processes using cementitious systems have also been reported (discussed in Chapters 4 and 5).	[1]European Commission (2014b); [2]Samuelsson and Björkman (2014); [3]Schlesinger et al. (2011); [4]Montenegro et al. (2013); [5]Ha et al. (2015); [6]Steele et al. (1995); [7]Morales et al. (2010); [8]Guo et al. (2015); [9]Font et al. (2011); [10]Shibayama et al. (2010)
Gaseous emissions	Rich in SO_2 from oxidation reactions, metal plumes from the volatilization and/or entrainment of volatile metal species and VOCs and PCDD/Fs from incomplete combustion and cooling, and from de novo synthesis reactions.	The SO_2 rich off-gas from gas handling and cleaning systems is cleaned and used in the production of sulphuric acid. Otherwise, the furnace off-gases can be desulphurized and recirculated into the plant to provide energy for drying of concentrates. Typical mitigation control of VOCs and PCDD/Fs emissions include measures such as controlled gas handling and cooling, effective dust removal, activated carbon adsorption and raw material sorting.	Schlesinger et al. (2011); Crundwell et al. (2011); European Commission (2014a)
Smelting, refining and converting slags	Mostly fayalitic and silicate glassy phases, magnetite, entrained sulphides and entrained metallic elements. The chemical, textural and mineralogical aspects key to the valorization potential of slags. The mineralogical characteristics further affect the metal recovery from, and the valorization potential of, copper making slags.	Primary smelting slags usually undergo slag treatment in slag cleaning furnaces before disposal or reused as construction materials. Due to lower contents of deleterious impurities, converting and refining slags are typically recycled into the smelting furnaces as premelted fluxes. Typical valorization options include (1) slag treatment for metal recovery in slag cleaning furnaces; (2) metal recovery using techniques such as flotation, magnetic separation and gravity separation; (3) metal recovery by pyro-and/or hydrometallurgical techniques; (4) utilization of slags as construction materials and (5) stabilization and/or solidification in cementitious systems. The presence of chromium rich spinel, $(Fe,Mg)Cr_2O_4$, presents challenges to slag-cleaning metal recovery, recycling due to in-process build-up of spinel and to direct disposal without stabilizing the chromium.	Jones et al. (1996); Jones (1999, 2005); Cole and Ferron (2002); Nel (2004); Shen and Forssberg (2003); Gorai et al. (2003); Nel (2004); Toscano and Utigard (2003); Rudnik et al. (2009); Das et al. (2010); Schlesinger et al. (2011); Crundwell et al. (2011); Sánchez and Sudbury (2013); Kim et al. (2013); European Commission (2014a); Samuelsson and Björkman (2014); Khorasanipour and Esmailzadeh (2016).

6.4.4 Gaseous Emissions

Gaseous emissions from primary production processes of base metals mostly consist of (European Commission, 2014a; González-Castanedo et al., 2014) (1) sulphur dioxide from the oxidation of sulphide concentrates and sulphur dissolved in mattes; (2) direct and diffuse smelter plumes, particularly rich in volatile metals and metalloids, such as As, Sb, Hg, Cd and Pb and (3) volatile organic carbon compounds, mostly in the form of polychlorinated dibenzodioxins/furans, formed as a result of poor combustion of oil and plastic in the feed materials as well as the de novo synthesis due to poor cooling conditions of off-gases. The characteristics of gaseous emissions from primary and secondary production processes were also summarized in Table 6.5.

6.4.5 Physicochemical Characterization of Base Metal Slags

Process slags constitute a major by-product produced in the primary production of base metals and platinum group metals (Crundwell et al., 2011; Schlesinger et al., 2011). Technically, slags play an important metallurgical role in these processes. SiO_2 fluxes are deliberately charged together with feed materials in order to engineer the desired physicochemical slag properties (Toscano and Utigard, 2003; Crundwell et al., 2011; Schlesinger et al., 2011). Despite the ubiquity of slags in these processes, large volumes of base metal slags contaminated with potentially toxic metal species are produced in these operations (Arslan and Arslan, 2002; Banda et al., 2002; Shen and Forssberg, 2003; Sánchez et al., 2004; Rudnik et al., 2009; Das et al., 2010; Schlesinger et al., 2011; Kim et al., 2013; European Commission, 2014a; Khorasanipour and Esmailzadeh, 2016).

Depending on feed and process conditions, the composition of base metal slags is typically in the range (1) 0.04–1.2 wt.% Ni, 0.21–0.7 wt.% Co and 0.3–3.7 wt.% Cu for copper smelter slags (Shen and Forssberg, 2003; Li et al., 2008); (2) 2.87–4.80 wt.% Ni, 0.77–1.59 wt.% Co and 0.17–1.4 wt.% Cu for converter slags (Baghalha et al., 2007; Li et al., 2008) and (3) about 1.1–2.3 wt.% Ni and 0.1–0.3 wt.% Co for laterite smelting slags (Agatzini-Leonardou and Zafiratos, 2004; Li et al., 2008). Basically, the mechanical entrainment and chemical dissolution of metals and metal oxides in the slags is largely controlled by the slag physicochemical properties such as slag liquidus temperature, slag basicity, slag viscosity and settling time (Toscano and Utigard, 2003; Crundwell et al., 2011; Schlesinger et al., 2011). For example, base metals such as nickel and cobalt also tend to exist in the form of oxides in base metal slags, thereby increasing the loss of these metals due to mechanical entrainment and chemical dissolution (Toscano and Utigard, 2003).

The physicochemical characterization of base metal slags plays an important role, not only in understanding the behaviour of slags in the smelting, converting and refining processes (Crundwell et al., 2011; Schlesinger et al., 2011) but also in providing comprehensive knowledge on the anthropogenic behaviour of, and metal reclamation from, such slags (Shen and Forssberg, 2003; Li et al., 2008; Sánchez and Sudbury, 2013; Khorasanipour and Esmailzadeh, 2016). As a result, the characterization of metal slags based on the chemical, textural and mineralogical properties becomes critical in developing potential metal reclamation and slag valorization techniques (Shen and Forssberg, 2003; Khorasanipour and Esmailzadeh, 2016). As shown in Figure 6.9, the solidified base metal slags are estimated to exist mostly as non-crystalline complex oxide and sulphide phases predominantly associated with fayalite- and silicate-glass phases, as well as

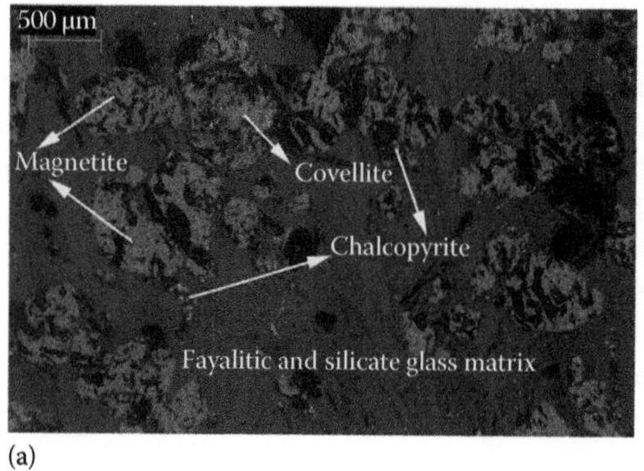

(a)

(b)

FIGURE 6.9
Mineralogical phases in solidified copper slags. Micrograph (a) shows chalcopyrite, covellite (CuS) and magnetite in the fayalite matrix, and Micrograph (b) shows chalcocite, native copper and magnetite in the fayalite and silicate glass matrix. (Reprinted from *J. Geochem. Explor.*, 166, Khorasanipour, M. and Esmailzadeh, E., Environmental characterization of Sarcheshmeh Cu-smelting slag, Kerman, Iran: Application of geochemistry, mineralogy and single extraction methods, 7. Copyright 2016, with permission from Elsevier.)

some entrained sulphides, oxides and metallic phases (Shen and Forssberg, 2003; Das et al., 2010; Schlesinger et al., 2011; Khorasanipour and Esmailzadeh, 2016). In addition, the typical mineralogical characteristics of typical base metal slags are further summarized in Table 6.5.

As discussed in the preceding sections, the solidified process slags from the smelting of UG-2 platinum group metal concentrates contain significant amounts of chromium-containing spinels ($(Fe,Mg)Cr_2O_4$). The typical chromite contents of converter slags from various smelters in South Africa are highlighted in Table 6.7 (Jones, 1999). The presence of chromium-bearing phases in these slags technically limits not only the in-process recycling of converter slag due to internal build-up of the chromium, but also presents challenges to the direct disposal of slags without the prior stabilization of the chromium species in the slag. The presence of chromium-bearing phases in the slag is also of particular concern to the environment due to the potential generation of toxicity, and mobility in soil and water of Cr(VI) in the solidified slags (Jones, 1999, 2005; Nel, 2004; Crundwell et al., 2011).

TABLE 6.7
Chemical Analysis of Chromite-Rich Converter Slags from South African Smelters

	Composition (wt.%)										
Source	**Al_2O_3**	**CaO**	**Co**	**Cr_2O_3**	**Cu**	**MgO**	**FeO**	**Ni**	**S**	**SiO_2**	**Total**
Amplats: Waterval	0.7	0.4	0.45	0.4	1.17	1.1	63	2.25	2.4	27	99
Impala	1.8	0.3	0.43	1.4	1.06	0.71	64	1.90	1.0	27	100
Lonmin	0.7	0.5	0.39	1.4	0.94	0.78	65	1.43	1.7	28	100
Northam	1.3	0.7	0.4	0.36	1.37	0.82	64	2.18	–	27	98

Source: Jones, R.T., *S. Afr. J. Sci.*, 95, 530, 1999, Available at http://www.mintek.co.za/Pyromet/Files/Platinum.pdf, Accessed 17 June 2016. With permission.

6.4.6 Recycling, Metal Recovery and Utilization of Base Metal Slags

As previously discussed, the amount and characteristics of slags produced from the primary production processes of base metals vary significantly depending on the type and quality of raw materials used (Shen and Forssberg, 2003). For example, it is estimated that the production of 1 ton of copper produces about 2–3 tonnes of slags (Shen and Forssberg, 2003; Das et al., 2010; Khorasanipour and Esmailzadeh, 2016). Technically, the large volumes of slags produced in the production base metals presents challenges in terms of their disposal and valorization. As discussed in the preceding sections, primary base metal slags are generally estimated to contain significant amounts of valuable and recoverable metals, and thereby essentially making them potential secondary sources of these metals (Shen and Forssberg, 2003; Kim et al., 2013).

In general, the uncontrolled disposal of base metal slags can result in environmental problems, and the anthropogenic effects of such slags are especially compounded by the presence of potentially toxic trace metal species (Khorasanipour and Esmailzadeh, 2016). Nevertheless, several options, technologies and processes have been developed to date in order to minimize the environmental footprint of primary base metal slags, and these include (Arslan and Arslan, 2002; Gorai et al., 2003; Shen and Forssberg, 2003; Sánchez et al., 2004; Li et al., 2008; Rudnik et al., 2009; Das et al., 2010; Crundwell et al., 2011; Schlesinger et al., 2011; Kim et al., 2013; Sánchez and Sudbury, 2013; European Commission, 2014a; Samuelsson and Björkman, 2014; Khorasanipour and Esmailzadeh, 2016) (1) internal recycle as pre-melted fluxes, (2) slag cleaning for metal recovery by smelting reduction in electrical furnaces, (3) metal recovery by physical beneficiation methods, (4) metal recovery by hydrometallurgical or pyrometallurgical techniques, (5) utilization of base metal slags as construction materials and (6) stabilization and/or solidification of base metal and platinum group metal slags in cementitious systems, particularly for chromium-bearing slags. Some of the potential processes and techniques used in the valorization of primary base metals process slags were summarized in Table 6.6.

6.5 Concluding Remarks

The technical and process aspects in the primary production of metals were discussed in this chapter. Basically, the production processes of base metals were categorized according to the type of primary ore, and the chapter focused particularly on the extraction of base metals from sulphide concentrates. The typical processes discussed in this chapter include the smelting, converting and refining of (1) copper from copper–sulphide dominant ores; (2) nickel and cobalt from nickel sulphide dominant ores and (3) the recovery of platinum group elements from both the platinum group metal–dominant and from nickel-cobalt sulphide–dominant ores.

The typical waste streams in the selected primary processes were discussed in line with the global trend towards waste reduction, recycling and utilization. Basically, the key categories of metallurgical wastes produced from the primary production of base and platinum group metals include (1) particulate dusts from emission abatement systems, particularly contaminated with volatile and potentially toxic metals species such as Zn, As, Pb, Cd, Sb, Hg and Bi; (2) process slags from the smelting, converting and refining processes and (3) process sludges from the final refining and metal purification processes. Furthermore, this chapter discussed the typical processes and techniques applicable in the potential valorization of the highlighted waste streams.

References

Agatzini-Leonardou, S. and Zafiratos, I.G. (2004). Beneficiation of a Greek serpentine nickeliferrous ore: Part II. Sulphuric acid heap and agitation leaching. *Hydrometallurgy* 74: 267–275.

Arslan, C. and Arslan, F. (2002). Recovery of copper, cobalt, and zinc from copper smelter and converter slags. *Hydrometallurgy* 67: 1–7.

Baghalha, M., Papangelakis, V.G. and Curlook, W. (2007). Factors affecting leachability of Ni/Co/Cu slags at high temperature. *Hydrometallurgy* 85(1): 42–52.

Banda, W., Morgan, N. and Eksteen, J.J. (2002). The role of slag modifiers on the selective recovery of cobalt and copper from waste smelter slag. *Minerals Engineering* 15: 899–907.

British Geological Survey (BGS) (2007). Mineral profile-Copper. Centre for Sustainable Mineral Development. Available at http://www.bgs.ac.uk/mineralsUK/statistics/mineralProfiles.html. Accessed 10 June 2016.

British Geological Survey (BGS) (2008). Mineral profile-Nickel. Centre for Sustainable Mineral Development. Available at http://www.bgs.ac.uk/mineralsUK/statistics/mineralProfiles.html. Accessed 10 June 2016.

British Geological Survey (BGS) (2009a). Mineral profile-Cobalt. Centre for Sustainable Mineral Development. Available at http://www.bgs.ac.uk/mineralsUK/statistics/mineralProfiles.html. Accessed 10 June 2016.

British Geological Survey (BGS) (2009b). Mineral profile-Platinum. Centre for Sustainable Mineral Development. Available at http://www.bgs.ac.uk/mineralsUK/statistics/mineral Profiles.html. Accessed 13 June 2016.

Campbell, F.C. (2008). Nickel and cobalt. In: *Elements of Metallurgy and Engineering Alloys*. Campbell, F.C. (ed.) ASM International, OH.

Cobalt Development Institute (CDI) (2006a). Cobalt facts – Chemicals. Available at http://www.thecdi.com/cdi/images/documents/facts/COBALT_FACTS-Chemicals.pdf. Accessed 12 June 2016.

Cobalt Development Institute (CDI) (2006b). Cobalt facts – Metallurgical uses. http://www.thecdi.com/cdi/images/documents/facts/COBALT_FACTS-Metallurgical_%20uses.pdf. Accessed 12 June 2016.

Cobalt Development Institute (CDI) (2006c). Cobalt facts – Magnetic alloys. http://www.thecdi.com/cdi/images/documents/facts/COBALT_FACTS-Magnetic_Alloys.pdf. Accessed 12 June 2016.

Cole, S. and Ferron, C.J. (2002). A review of the beneficiation and extractive metallurgy of the platinum group elements, highlighting recent process innovations. Technical Paper 2002-02. SGS Minerals Services, SGS Global, pp. 1–43.

Copper Development Association (CDA) (n.d.) General, atomic and crystallographic properties and features of copper. Available at http://www.copper.org/resources/properties/ atomic_properties.html. Accessed 11 June 2016.

Craig, J.R. and Vaughan, D.J. (1990). Compositional and textural variations of the major iron and base metal sulphide minerals. In: Gray, P.M.J., Bowyer, G.J., Castle, J.F., Vaughan, D.J. and Warner, N.A. (eds.), *Sulphide Deposits – Their Origin and Processing*. The Institution of Mining and Metallurgy, London, UK, pp. 1.

Cramer, L.A. (2001). The extractive metallurgy of South Africa's platinum ores. *The Journal of Materials* 53(10): 14–18.

Crundwell, F.K., Moats, M.S., Ramachandran, V., Robinson, T.G. and Davenport, W.G. (2011). *Extractive Metallurgy of Nickel, Cobalt and Platinum-Group Metals*. Elsevier, Oxford, UK.

Das, B., Mishra, B.K., Angadi, S., Pradhan, S.K., Prakash, S. and Mohanty, J. (2010). Characterization and recovery of copper values from discarded slag. *Waste Management and Research* 28: 561–567.

Davies, J.R. (2000). *Nickel, Cobalt, and Their Alloys. ASM Specialty Handbook*. Davis, J.R. (ed.) ASM International, Ohio, Available at asminternational.org. Accessed 20 April 2016.

Davies, J.R. (2001). *Copper and Copper Alloys. ASM Specialty Handbook.* ASM International, Cleveland, OH, p. 1.

Drexler, E.S., Reed, R.P. and Simon, N.J. (1992). Properties of copper and copper alloys at cryogenic temperatures. NIST MN-177. National Institute of Standards and Technology, Gaithersburg, MD.

Environmental Protection Agency (EPA) (1994). Copper. Extraction and beneficiation of ores and minerals. Technical Resource Document Volume 4. EPA 530-R-94-031. Washington, DC.

European Commission (2014a). Processes to produce copper and its alloys from primary and secondary raw materials. Best Available Techniques (BAT) Reference Document for the Non-Ferrous Metal Industries. Final Draft. Industrial Emissions Directive 2010/75/EU (Integrated Pollution Prevention and Control).

European Commission (2014b). Report on ad hoc working group on defining critical raw materials. Report on Critical Raw Materials for the EU. Available at http://ec.europa.eu/growth/sectors/raw-materials/specific-interest/critical/index_en.htm. Accessed 13 June 2016.

Farzin-Nia, F. and Yoneyama, T. (2009). Orthodontic devices using Ti-Ni shape memory alloys. In: Yoneyama, T. and Miyazaki, S. (eds.), *Shape Memory Alloys for Biomedical Applications.* Woodhead Publishing House, Cambridge, UK, p. 257.

Fernandes, D.J., Peres, R.V., Mendes, A.M. and Elias, C.N. (2011). Understanding the shape-memory alloys used in orthodontics: Review. *Dentistry – International Scholarly Research Network* 11: 1–6.

Font, O., Moreno, N., Aixa, G., Querol, X. and Navia, R. (2011). Copper smelting flue dust: A potential source for germanium. *Revista Sociedad Espanola de Mineralogica* (September): 87–88.

Geoscience Australia (2015). Copper fact sheet. Australian Atlas of Minerals Resources, Mines and Processing Centers. Commonwealth of Australian. Available at http://www.australianminesatlas.gov.au/education/fact_sheets/copper.html. Accessed 10 June 2016.

González-Castanedo, Y., Moreno, T., Fernández-Camacho, R., de la Campa, S., Alastuey, A., Querol, X. and de la Rosa, J. (2014). Size distribution and chemical composition of particulate matter stack emissions in an around a copper smelter. *Atmospheric Environment* 98: 271–282.

Gorai, B., Jana, R.K. and Premchand (2003). Characteristics and utilization of copper slag: A review. *Resources, Conservation and Recycling* 39: 299–313.

Guo, X., Shi, J., Tian, Q. and Li, D. (2015). Separation and recovery of arsenic from arsenic-bearing dust. *Journal of Environmental Chemical Engineering* 3: 2236–2242.

Ha, T.K., Kwon, B.K., Park, K.S. and Mohapatra, D. (2015). Selective leaching and recovery of bismuth as Bi_2O_3 from copper smelter converter dust. *Separation and Purification Technology* 142: 116–122.

Hartley, F.R. (1991). The occurrence, extraction, properties and uses of platinum group metals. In: Hartley, F.R. (ed.), *Chemistry of the Platinum Group Metals: Recent Developments.* Elsevier, Amsterdam, the Netherlands, p. 9.

International Copper Study Group (ICSG) (2013). The world copper factbook 2013. Available at http://copperalliance.org/wordpress/wp-content/uploads/2012/01/2013-World-Copper-Factbook.pdf. Accessed 11 June 2016.

International Copper Study Group (ICSG) (2016). Copper market forecast 2016/2017. Press Release. Available at http://www.icsg.org/index.php/component/jdownloads/finish/113-forecast-press-release/2128-2016-03-icsg-forecast-press-release?Itemid=0. Accessed 11 June 2016.

International Monetary Fund (IMF) (2015). Commodity Market Developments and Forecasts, with a Focus on Metals in the World Economy. Commodity Special Feature. Available at https://www.imf.org/external/np/res/commod/pdf/WEOSpecialOCT15.pdf. Accessed 17 June 2016.

Johnson Matthey (n.d.). About PGMs. Available at http://www.platinum.matthey.com/about-pgm. Accessed 15 June 2016.

Jones, R.T. (1999). Platinum smelting in South Africa. South African Journal of Sciences 95, pp. 525–534. Available at http://www.mintek.co.za/Pyromet/Files/Platinum.pdf. Accessed 17 June 2016.

Jones, R.T. (2004). JOM world nonferrous smelter survey, part II: Platinum Group Metals. *Journal of Minerals, Metals and Materials Society* (December): 59–63.

Jones, R.T. (2005). An overview of Southern African PGM smelting. *Nickel, Cobalt 2005: Challenges in Extraction and Production. Proceedings of the 44th Annual Conference of Metallurgists,* Calgary, Alberta, Canada, pp. 147–178.

Jones, R.T., Hayman, D.A. and Denton, G.M. (1996). Recovery of cobalt, nickel, and copper from slags using DC-arc furnace technology. Mintek Paper 8360. Randburg, South Africa. Available at http://www.mintek.co.za/Pyromet/Cobalt/Cobalt.htm. Accessed 15 June 2016.

Khorasanipour, M. and Esmailzadeh, E. (2016). Environmental characterization of Sarcheshmeh Cu-smelting slag, Kerman, Iran: Application of geochemistry, mineralogy and single extraction methods. *Journal of Geochemical Exploration* 166: 1–17.

Kim, B.S., Jo, S.K., Shin, D., Lee, J.C. and Jeong, S.B. (2013). A physicochemical separation process for upgrading iron from waste copper slags. *International Journal of Mineral Processing* 124: 124–127.

Kim, J.Y., Lajoie, S. and Godbehere, P. (1995). Characterization of copper smelter dusts and its effect on metal recovery. In: Rao, S.R., Amaratunga, L.M. and Richards, G.G. (eds.), *Waste Processing and Recycling in Mineral and Metallurgical Industries II: Proceedings of International Waste Processing and Recycling in Mineral and Metallurgical Industries II*. Canadian Institute of Mining, Metallurgy and Petroleum, Montreal, Quebec, Canada, pp. 221–234.

Konečná, R. and Fintová, S. (2012). Copper and copper alloys: Casting, classification, and characteristic microstructure. In: Collini, L. (ed.), *Copper Alloys: Early Applications and Current Performance–Enhancing Processes*. InTech, Rijeka, Croatia. Available at: http://www.intechopen.com/books/copper-alloys-early-applications-and-current-performance-enhancing-processes/copper-and-copper-alloys-casting-classification-and-characteristics-. Accessed 10 June 2016.

Li, Q., Zeng, Y. and Tang, X. (2010). The applications and research progresses of nickel-titanium shape memory alloy in reconstructive surgery. *Australasian Physical and Engineering Science in Medicine* 33: 129–136.

Li, Y., Perediriy, I. and Papangelakis, V.G. (2008). Cleaning of waste smelter slags and recovery of valuable metals by pressure oxidative leaching. *Journal of Hazardous Materials* 152(2,1): 607–615.

Montenegro, V., Sano, H. and Fujisawa, T. (2013). Recirculation of high arsenic content copper smelting dust to smelting and converting processes. *Minerals Engineering* 49: 184–189.

Morales, A., Cruells, M., Roca, A. and Bergó, R. (2010). Treatment of copper flash smelter flue dust for copper and zinc extraction and arsenic stabilization. *Hydrometallurgy* 105: 148–1154.

Mpinga, C.N., Eksteen, J.J., Aldrich, C. and Dyer, L. (2015). Direct leach approaches to platinum group metal (PGM) ores and concentrates: A review. *Minerals Engineering* 78: 93–113.

Nel, J. (2004). Melting of platinum group metal concentrates in South Africa. *The Journal of the Southern African Institute of Mining and Metallurgy* (August): 423–428.

Outokumpu (2013). *Handbook of Stainless Steel*. Outokumpu Stainless AB, Avesta, Sweden. Available at www.outokumpu.com. Accessed 10 December 2015

Rudnik, E., Burzyńska, L. and Gumowska, W. (2009). Hydrometallurgical recovery of copper and cobalt from reduction-roasted copper converter slag. *Minerals Engineering* 22: 88–95.

Samuelsson, C. and Björkman, B. (2014). Copper recycling. In: Worrell, E. and Reuter, M.A. (eds.), *Handbook of Recycling: State of the Art for Practitioners, Analysts, and Scientists*. Elsevier, Amsterdam, the Netherlands, pp. 85.

Sánchez, M., Parada, F., Parra, R., Marquez, F., Jara, R., Carrasco, J.C. and Palacios, J. (2004). Management of copper pyrometallurgical slags: Giving additional value to copper mining industry. *Proceedings of the VII International Conference on Molten Slags, Fluxes and Salts*. The Southern African Institute of Mining and Metallurgy, Johannesburg, pp. 543–550.

Sánchez, M. and Sudbury, M. (2013). Physicochemical characterization of copper slag and alternatives of friendly environmental management. *Journal of Mining and Metallurgy, Section B: Metallurgy* 149(2): 161–168.

Schlesinger, M.E., King, M.J., Sole, K.C. and Davenport, W.G. (2011). *Extractive Metallurgy of Copper*, 5th edn. Elsevier, Oxford, UK.

Shedd, K.B. (2002). Cobalt. US Geological Survey, Reston, VA. Available at http://minerals.usgs.gov/minerals/pubs/commodity/cobalt/cobalmyb02.pdf. Accessed 11 June 2016.

Shen, H. and Forssberg, E. (2003). An overview of recovery of metals from slags. *Waste Management* 23: 933–949.

Shibayama, A., Takasaki, Y., William, T., Yamatodani, A., Higuchi, Y., Sunagawa, S. and Ono, E. (2010). Treatment of smelting residue for arsenic removal and recovery of copper using pyro-hydrometallurgical process. *Journal of Hazardous Materials* 181: 1016–1023.

Steele, D.K., Gritton, K.S. and Odekirk, S.B. (1995). Treatment of copper smelting and refining wastes. Bureau of Mines. RI 9522. Report of Investigations/1994. US Department of Interior, Washington, DC, pp. 1–13.

Toscano, P. and Utigard, T.A. (2003). Nickel, copper, and cobalt slag losses during converting. *Metallurgical Transactions* 34B: 121–125.

Total Materia (2001). Nickel and nickel alloys. Available at http://www.totalmateria.com/Article9.htm. Accessed 11 June 2016.

Total Materia (2002). Cobalt and cobalt alloys. Available at http://www.totalmateria.com/Article54.htm. Accessed 11 June 2016.

Total Materia (2009). Classification and properties of copper alloys. Available at http://www.totalmateria.com/page.aspx?ID=CheckArticle&site=ktn&NM=216. Accessed 11 June 2016.

Yazawa, A. and Azakami, T. (1969). Thermodynamics of removing impurities during copper smelting. Canadian Metallurgical Quarterly. *The Canadian Journal of Metallurgy and Materials Science* 8(3): 257–261.

Warner, A.E.M., Díaz, C.M., Dalvi, A.D., Mackey, P.J., Tarasov, A.V. and Jones, R.T. (2006). JOM world nonferrous smelter survey Part III: Nickel: Laterite. *Journal of Materials* 58(4): 11–20.

Warner, A.E.M., Díaz, C.M., Dalvi, A.D., Mackey, P.J., Tarasov, A.V. and Jones, R.T. (2007). JOM world nonferrous smelter survey Part IV: Nickel: Sulphide. *Journal of Materials* 51(4): 58–72.

7

Hydrometallurgical Waste Production and Utilization

7.1 Introduction

Hydrometallurgical processing involves the use of aqueous media, such as acids, alkalis and organic solvents, in the extraction of metals from metal-bearing resources. These processes consist of a classic number of steps commonly known as unit processes. These include (1) leaching for the purpose of metal extraction into solution, (2) purification of the leach liqour to remove impurities, (3) concentration of the leach liquor to upgrade the metal content and finally, (4) recovery of the solid metal in different forms to be sold to the end market. The initial leaching step of getting the metal into solution is common for most of the material processed via a hydrometallurgical route. This is then followed by techniques such as solvent extraction (SX), ion exchange, adsorption, precipitation, chemical and gaseous reduction and electroreduction. The techniques generate two product streams: one containing value materials and the second a potential recycle in the process or a discharge stream.

Most of the waste streams generated in hydrometallurgical production of, for example base metals, precious metals and alumina have in the past been dumped on the plant site and occasionally used for landfilling purposes. However, the rapid growth of industrialization and the subsequent demand for space for plant expansion has led to a decrease in the available land for landfilling of large quantity of waste. Correspondingly, the associated disposal cost have also skyrocketed. Another major challenge faced by mining countries is the degradation of environmental quality and adverse effects on the human health and the ecosystem due to unrestricted discharge of process waste solids and solutions from the hydrometallurgical plants. In response to this challenge, many governments have introduced stringent environmental regulations and policies in order to encourage more environmentally friendly processes for metal extraction and environmental protection. These regulations and policies have thus, stimulated the need for the development of new technologies for minimization, valorization and recycling of industrial waste.

This chapter first gives an overview of the hydrometallurgical processing of minerals and metals. This is followed by an insight into the management and treatment of the associated waste materials produced with special attention being paid to resource recovery and the reutilization of the streams in other industries. The chapter is structured as follows: An overview of the steps and/or process units involved in the hydrometallurgical processing of minerals and metals is given first. This is followed by a discussion of the hydrometallurgical flowsheets for the processing of copper, zinc, aluminium, gold, platinum group metals (PGMs), uranium and rare earth elements. The nature of the process waste generated, for example, chemical and physical properties, will also be considered as this plays a role in identifying options for their alternative utilization. The treatment of the process residue and its potential application and reutilization in other industries will be given.

7.2 Leaching

Leaching or dissolution is the first process step involved in the hydrometallurgical extraction of metals from ores or any other metal-bearing material. This step involves contact of the value-containing mineral or any metal-bearing material with different chemical reactants known as the leaching solution or the lixiviant. The dissolution process can involve different mechanisms such as simple chemical reactions (acid–base), complexation, redox, dissolved organic matter and microbiological activities. The dissolution process can only take place under appropriate conditions, and most metals are extractable at very low pH and high potentials or high pH and moderate potential. However, the choice of the leaching solution usually depends on the type of materials to be leached, the solubility of the material in the leach solution, the cost of and ease of access to the specific reagent and its regeneration capabilities. The lixiviants can be water, acids, alkalis, salts or, as in modern processes, a biological solution containing microorganisms. Water is the cheapest of all reagents but has limited use since not many minerals are water soluble. It is mainly used for dissolving naturally occurring sodium and potassium salts such as carbonates, chlorides, sulphates and nitrates. It is also used for the leaching of calcines obtained after sulphatizing and chloridizing roasting.

In ideal situations, leaching gives rise to two end products: a solid residue that is in most cases devoid of valuable metals and is destined for the tailings dumps, and the other, a metal-laden solution that advances to the next stage of processing. In other situations, however, the solid residue can contain some valuable metals that, if unrecovered, can be lost to the environment. On the other hand, the residue can also contain some toxic components that, if left untreated, would have a negative impact on the environment. These toxic components will therefore need to be removed before disposal. Thus, modern developments in the hydrometallurgical flowsheet design consider the processing of waste streams in order to recover valuable metals if any and to reduce the level of toxicity of the final discharge stream.

The subsequent subsections discuss various types of leaching media for the dissolution of metals from metal bearing materials.

7.2.1 Acidic and Alkaline Leaching

Acidic leaching is conducted under low pHs. Inorganic acids such as sulphuric acid, nitric acid, hydrochloric acid and hydrofluoric acid are commonly used in most industrial applications. The acids can be used in dilute form as in the leaching of a number of oxide ores, in combination with an oxidizing agent, for example, in the leaching of uranium and sulphide ores. They can also be used in combination with a reducing agent as in the leaching of manganese ores or heterogeneous copper–cobalt oxide ores. Organic acids such as oxalic acid, citric acid and acetic acid have also been applied as leaching reagents although not on a wide scale due to their prohibitive costs and sometimes more complex process flowsheets involved. In spite of the fact that inorganic acids can be highly corrosive, especially when used at high concentrations, sulphuric acid has been popular in most applications as it has the advantage of being cheap and readily available. In addition, it allows for easier downstream metal purification and recovery processes such as SX, ion exchange, electrowinning (EW) and the possibilities for regenerating it during the EW stage.

Acidic leaching with highly concentrated solutions generally produces residues that require treatment for neutralization before disposal into the environment. Neutralization of the residues before discharge likely adds to the disposal costs. Some residues also contain heavy metals that can ultimately leach into the underground water when exposed to conditions favouring metal dissolution. In this context, inorganic acids such as nitric, hydrochloric and sulphuric acid as well as organic acids such as citric and oxalic acid have been used for the extraction of heavy metals from sludge wastes before disposal (Wong and Henry, 1983; Marchioretto et al., 2002; Dacera and Babel, 2006).

Alkaline leaching is done under high pH and is mostly applied when significant quantities of acid-consuming gangue minerals usually in oxide form are present in the feed material. In this case, acid leaching is not economic. Typical leaching reagents that are commonly used include sodium hydroxide as in the leaching of bauxite ores, a mixture of sodium carbonate and sodium bicarbonate applied in uranium leaching and ammonium hydroxide used in the dissolution of reduced lateritic ores containing cobalt and nickel. Alkaline leaching has also been very popular in the development of process routes for the extraction of metals from secondary waste resource streams such as coal fly ash and red mud due to the high residual value of oxides in such material (Thakur and Sant, 1983a; Padilla and Sohn, 1985; Paramguru et al., 2005).

7.2.2 Leaching with Metal Salts

Leaching with metal salts involves the use of reagents such as ferric chloride, ferric sulphate and cupric chloride in the dissolution of base metal sulphides and cyanide salts in the extraction of precious metals such as gold and silver from ores. These reagents can be applied on their own but are usually used in combination with another reagent. A typical example is the use of ferric sulphate as an oxidant in the presence of sulphuric acid for the leaching of primary copper sulphides such as chalcopyrite (Dutrizac and MacDonalds, 1974; Smalley and Davis, 2000; Ferron, 2003). The leaching is based on the oxidation of the iron sulphide component of the material to sulphate accompanied by the reduction of ferric to ferrous ions. The ferric ion can be regenerated by oxidizing the produced ferrous ions in air or in the presence of bacteria and then recycled in the process (Gericke and Pinches, 1999; Watling, 2006). The use of sodium cyanide as a complexing agent in the presence of oxygen for the leaching of gold ores is another well established process. Oxygen is very important in the leaching of gold using cyanide, as the rate of gold dissolution in cyanide is usually directly proportional to the amount of oxygen present in the system (Marsden and House, 2006).

7.2.3 Bioleaching

Bioleaching is the ability of certain bacteria and archaea, found in nature, to catalyse the oxidation of metal sulphides and iron minerals from ores and concentrates. For example, microorganisms are currently used in commercial leaching operations for processing the ores of copper, nickel, cobalt, zinc and uranium and in the biooxidation of gold ores and the desulphurization of coal. In general, this process occurs in acidic medium at pH values between 1.5 and 3 where the metal ions can easily be mobilized in an aqueous system.

In the case where microorganisms catalyse the oxidation of iron sulphides, ferric sulphate and sulphuric acid are generated. Ferric sulphate, which is a powerful oxidizing agent, then oxidizes the sulphide minerals, for example, copper sulphide. The copper contained in the mineral is then leached by the sulphuric acid produced. The technique can also be applied to lean and so far unusable resources for the recovery of valuable metals (Mishra and Young-Ha, 2010).

Bioleaching is an efficient and cost-effective alternative to conventional chemical and physical methods because of its low demand for energy. In addition it generates minimal waste by-products such that it can be perceived as a green technology (Mishra and Young-Ha, 2010). The bacterial species that are of importance under the acidic leaching conditions are the *Acidithiobacillus* group such as *Thiobacillus ferrooxidans* and *Thiobacillus thiooxidans*. Other bacteria are also able to oxidize sulphur and sulphides, but they are more stable at higher pH values at which most metal ions do not remain in solution. In the recent past, extreme mesophiles and thermophiles (bacteria that work under extremely high temperatures) have been tested for the processing of waste-bearing materials. These bacteria have shown significant advantages such as the ease of dissolution of refractory metal matrixes at higher temperatures and the increased overall reaction kinetics. At elevated temperatures, some archaeal genera including *Acidianus*, *Metallosphaera* and *Sulfolobus*, all belonging to the order *Sulfolobales*, appear to be the most important organisms (Rodríguez et al., 2003; Vilcáez et al., 2008).

A lot of research in the past has focused on the use of microorganisms for the recovery of valuable metals from waste such as spent petroleum catalysts (Mishra et al., 2007; Bosio et al., 2008; Beolchini et al., 2010), electronic waste (Brandl et al., 2001; Cui and Zhang, 2008; Yang et al., 2009), lithium battery wastes (Zhao et al., 2008; Xin et al., 2009) , sewage sludge (Wong and Henry, 1983; Pathak et al., 2009) , fly ash and incineration waste (Seidel et al., 2001; Xu and Ting, 2009). These wastes mostly contain Ni, V, Mo, Co, Cu, Pb, Zn and Cr. The *Acidithiobacillus* groups of bacteria are the key genus that has been used to leach out most of these metals. However, these waste samples lack the key energy sources for the *Acidithiobacillus* group such as iron and sulphur required for bacteria metabolism; thus, both ferrous sulphate and elemental sulphur need to be added externally for the bacterial activity. This can be an expensive procedure unless a natural and an inexpensive source of sulphur or iron is readily available (Simate et al., 2010).

Since the microorganisms used for leaching tend to flourish under highly acidic conditions, the alkaline nature of some of the waste samples such as coal fly ash and bauxite residues (due to the high oxide material present) also tends to reduce the bacterial growth, as well as the leaching efficiency of metals. Therefore, bioleaching alkaline waste material has to be managed through the proper adaptation of the bacteria so that the bacteria can be able to retain their viability during the bioleaching process and enhance the metal dissolution process (Seidel et al., 2001). These mentioned challenges have resulted in significant limitations in the commercial applications of bacteria during the processing of waste material.

7.3 Purification and Concentration Processes

The metal extraction processes discussed in section 7.2 are in most cases followed by solution concentration and purification before final solid metal recovery. There are a number of unit operations that can be used to upgrade the concentration of metal ions and to remove the impurities that would ultimately lower the final purity of the product. Processes such as precipitation, crystallization, SX, ion exchange, adsorption and biosorption have been employed in most hydrometallurgical plants. The choice of the purification and concentration process is largely influenced by the type and quantities of impurities in the process solution and the final purity target level in the resultant product or discharge streams. The next sections discuss some of the processes used to purify and concentrate solutions. These processes can also be used to remove toxic metals and unwanted organic components from waste solutions before discharge into the environment.

7.3.1 Chemical Precipitation Processes

Chemical precipitation that includes hydroxide and sulphide precipitation is by far the most widely used process in the industry because of its simplicity, cost-effectiveness and ease of operation (Ku and Jung, 2001). Chemical precipitation can be used to recover metals from pregnant leach solutions and also as a remediation technique for the treatment of process waste solutions before disposal.

All chemical precipitation processes operate under the same fundamental chemical principles: the ionic metals are converted into an insoluble form (particle) by a chemical reaction between the soluble metal compounds and the precipitating reagent. Most often, an alkaline reagent is used to raise the solution pH so as to lower the solubilities of the metallic components. Equation 7.1 shows the conceptual mechanism of metal removal by the use of an alkaline precipitant:

$$M^{2+}(aq) + 2OH^{-}(aq) = M(OH)_2(s) \tag{7.1}$$

where

M^{2+} and OH^- represent the dissolved metal ions and the precipitant respectively.

$M(OH)_2$ is the insoluble metal hydroxide generated

The precipitates formed can be separated from the aqueous solution by sedimentation or filtration. The barren solution stream is then decanted and appropriately discharged or reused. The effectiveness of a chemical precipitation process is dependent on factors such as the type and concentration of ionic metals present in solution, the type of precipitant used, the pH of the solution and the presence of other constituents that may inhibit the precipitation reactions (Neutralac, 2016).

The most widely used chemical precipitation technique is hydroxide precipitation due to its relative simplicity, low cost and ease of pH control (Huisman et al., 2006). Indeed, hydroxide precipitation has been used in a number of well-established hydrometallurgical processes such as the Bayer process for alumina recovery, removal of iron from zinc sulphate leach solutions and separation of cobalt and nickel from sulphate leach solutions. It has also been used in the treatment of wastewater streams to remove heavy metals and for the treatment of acidic mine water for neutralization and heavy metal removal. Figure 7.1 shows the basic hydroxide precipitation process.

In order for the hydroxide precipitation process to work effectively, enough hydroxide (OH^-) ions must be supplied to raise the solution pH so that the dissolved metals can form insoluble metal hydroxides and settle out of the water. A variety of chemicals such as lime, caustic soda and limestone can be used to precipitate metal hydroxides from leach solutions and wastewater streams. A flocculant is added to the mixed slurry that is fed into a solid/liquid separation step. In most processes, a clarifier is used for solid/liquid separation,

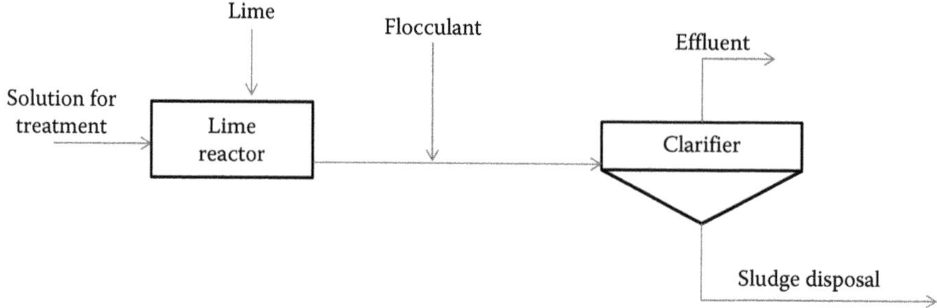

FIGURE 7.1
The basic hydroxide precipitation process. (From Aube, B., The science of treating acid mine drainage and smelter effluents, 2004, Available at http://www.infomine.com/publications/docs/Aube.pdf, Accessed June 2016.)

although a settling pond can also be used. Although widely used, hydroxide precipitation processes tend to generate large volumes of sludge, which can present dewatering and disposal problems (Kongsricharoern and Polprasert, 1996).

Sulphide precipitation is another effective process for the treatment of pregnant leach and effluent streams. Sulphide precipitation can be effected using either solid (FeS, CaS), aqueous (Na_2S, NaHS, NH_4S) or gaseous sulphide sources (H_2S). There is also the possibility of using the degeneration reaction of sodium thiosulphate ($Na_2S_2O_3$) as a source of sulphide for metal precipitation. Furthermore, industrial effluents that contain high sulphate and metal concentrations can be treated using a combination of bacterial sulphate reduction to generate sulphide, followed by removal of the metals as metal sulphide precipitates (Boonstra and Buisman, 2003; Adams et al., 2008) (Figure 7.2). The distinct advantage of this technology is that it can be used for a profitable operation in cases where the solution concentrations are too low for solvent extraction–electrowinning (SX-EW).

Although hydroxide precipitation is widely used in the industry for metal removal, a number of investigations have shown that sulphide precipitation processes remove heavy metals from industrial solutions much more effectively (Bhattacharyya et al., 1979;

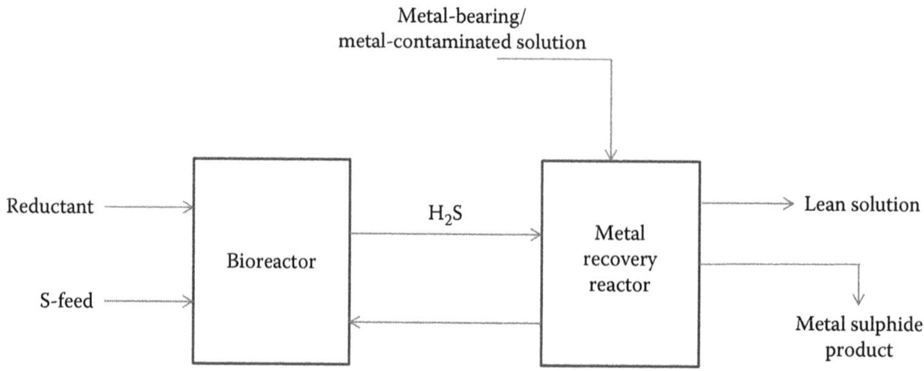

FIGURE 7.2
Process diagram for metal recovery using biogenic H_2S. (From Boonstra, J. and Buisman, C.J.N., Biotechnology for sustainable hydrometallurgical extraction, in Young, C., Alfantazi, A., Anderson, C., James, A., Dreisinger, D. and Harris, B., eds., *Electrometallurgy and Environmental Hydrometallurgy*, Vol. 2, 2003, pp. 1105–1119, Wiley Online Library.)

Peters et al., 1985; Lewis, 2010). Some advantages of sulphide precipitation include the lower solubility of metal sulphide precipitates, potential for selective metal removal, fast reaction rates, better settling properties and potential for reuse of sulphide precipitates by smelting (Lewis, 2010). However, the application of sulphide precipitation, especially in the treatment of process waste streams still remains very limited due to a number of reasons. The costs of sulphide reagents are relatively higher than those used in hydroxide precipitation (MEND, 1994). In addition, the sulphide reagents used tend to produce hydrogen sulphide fumes when contacted with acidic wastes. However, this can be prevented by maintaining the pH of the solution between 8 and 9.5, though it may require ventilation of the treatment tanks (MEND, 1994).

Precipitation processes are of great importance in relation to the treatment of waste solutions and certain impurity problems occurring in leach solution such as iron and arsenic. However, the stability of the products of such precipitation processes, in relation to safe disposal, is increasingly coming under scrutiny. Sulphide sludge, for example, is more prone to oxidation, resulting in resolubilization of the metals as sulphates. Thus, the sludge should be stored carefully, recycled for resource recovery or disposed of in an environmentally acceptable manner. As sludge disposal costs can be high, development of processes that result in the minimal quantities of high-density sludge is paramount. In addition, generation of a sludge that can be used in other industrial applications would result in a more sustainable approach to the environmental challenges faced in the metals extraction industry.

The Bayer process for alumina production, iron removal in zinc, titanium production and removal of arsenic from copper solutions are some of the few typical examples where precipitation is used as a means of solution purification. However, the processes also result in the production of waste material that requires consideration in disposal or other alternative applications.

7.3.2 Solvent Extraction

Solvent extraction (SX) of metals is based on the use of an organic solution to extract metal ions from an aqueous solution (Ritcey and Ashbrook, 1984; Vladimir, 2011). In hydrometallurgy, the aqueous phase from the leaching process contains the metal that is to be concentrated into the organic phase. Ideally, there are three basic stages in SX: extraction, scrubbing and stripping. Figure 7.3 shows the typical flowsheet for a SX process.

During extraction, the aqueous phase is brought into contact with the organic phase by thorough mixing. By mixing the aqueous and organic phases, the metal ions bind to the extracting agent and are, thus, transferred to the organic phase, a process known as loading. Thereafter, the organic and aqueous phases are allowed to disengage, and then the two phases are separated. The products of the extraction stage are the organic solvent loaded with the desired metals and the metal depleted aqueous phase known as the raffinate. The raffinate is either treated so as to recover other metals of interest or recycled in a process that can make use of its highly acidic nature. The washing stage may also be necessary for processes where there is considerable entrainment of the organic phase into the aqueous phase or vice versa.

During extraction, there is sometimes co-extraction of unwanted impurities, and these are usually removed from the loaded organic phase using a suitable aqueous solution in a process known as scrubbing. Once the metal ions have been extracted into the organic phase and separated from all the impurity metal ions, they must be recovered back into the aqueous phase in a process known as stripping. Stripping uses a suitable aqueous reagent,

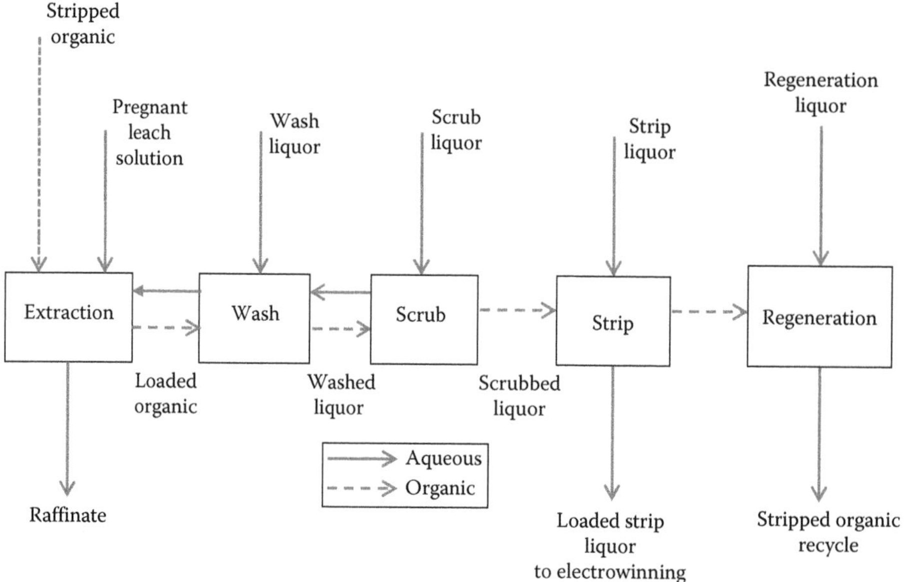

FIGURE 7.3
A generalized SX circuit. (From Aguilar, M. and Cortina, J.L., *Solvent Extraction and Liquid Membranes*, CRC Press, London, UK, 2008.)

to produce a concentrated aqueous phase, known as the loaded strip liquor. After stripping, the organic phase is regenerated so as to convert it back to a state suitable for recycling to the metal extraction stage.

There are different types of extraction reagents used in SX, and the choice of each extractant is largely influenced by the metals under consideration. Selectivity is the key requirement and the aim is to choose an extractant that would extract a specific metal ion into the organic phase while leaving the other impurity elements in the aqueous or raffinate phase. The extraction reagents are generally classified into three broad groups: (1) acidic, extract metals in cationic form, for example, dialkyl phosphoric acids; (2) basic, extract metals in anionic form, for example, amines and (3) neutral, extract metals as solvates, for example, tributyl-n-phosphate (TBP) (Ritcey and Ashbrook, 1984; Vladimir, 2011).

Most extractants tend to have a high viscosity, and they are, therefore, unsuitable for use without alteration due to the poor phase separation of the organic and aqueous solutions after mixing. Furthermore, extractants are sold at high concentrations, and if used as they are, would result in the unnecessary recycling of unreacted extractant and possible high operational reagent losses and costs. The addition of aliphatic or aromatic hydrocarbons to the extractant helps to reduce the viscosity and the amount of the extractant used. These hydrocarbons are known as diluents (Ritcey and Ashbrook, 1984; Vladimir, 2011).

SX has three features that have led to its widespread application in the metallurgical industry. First, it is selective, therefore, very useful in metal separation processes. Second, it can be operated continuously in closed loop with very little reagent make-up requirements, and finally, it can generate a marketable product. SX has become well established in the processing, purification and upgrading of the concentration of metal ions such as copper, zinc, cobalt, nickel, PGMs and rare earth elements. It has also found widespread application in the recovery of metals from less-concentrated solutions such as process waste streams.

7.3.3 Ion Exchange

Ion exchange is a process applied in the purification, concentration and decontamination of aqueous and other ion containing solutions using solid polymeric or mineralic ion exchangers. It is a reversible chemical reaction where an ion from a solution is exchanged for a similarly charged ion attached to an immobile solid particle (resin). The ions taken up from the solution have a greater preference than the ones existing on the resins, and thus, the existing ions are released into the solution as the preferred ones are taken up by the resin (Wang et al., 2009). As previously stated, the process is reversible and the ion exchanger can be regenerated.

The solid ion exchange particles used are either naturally occurring inorganic zeolites or synthetically produced organic resins (Wang et al., 2009). They can be either cation exchangers that exchange positively charged ions (cations) or anion exchangers that exchange negatively charged ions (anions). Ion exchangers can have binding preferences for certain ions or classes of ions or non-selective preferences depending on their chemical structure. (Ćurkovic et al., 1997).

Ion exchange particles or resins have obvious applications in the field of hydrometallurgy; however, high specificity usually also means a lower degree of reversibility. For many specific ion exchangers, the economics of regenerating the resins outweighs the high degree of specificity achieved in the separation process, thus limiting their commercial applications.

In the most popular form of ion exchange application, the metal-bearing solution is passed through a bed of solid organic resins, wherein the sorption of the metal ions on the resin takes place by ion exchange reactions (Damons, 2001). The most common type of contactor used in such an ion exchange process is the fixed-bed column in which a static bed of resin is placed in a vertical column and the solution is pumped down through the bed. Since the metal ion is usually distributed between the two phases, a single separation process is generally insufficient to obtain the desired separation; thus, multiple contacting must be applied.

The extraction of PGMs, uranium, copper, cobalt and nickel from dilute leach and lean process solutions are some of the most studied of all applications of ion exchange processes. Ion exchange has also been used in nuclear reprocessing and the treatment of radioactive waste. A very important example is the plutonium–uranium redox extraction (PUREX) process (Irish and Reas, 1957). The process uses SX and ion exchange techniques to reprocess spent nuclear fuel, in order to extract primarily uranium and plutonium, independent of each other, from other constituents. Ion exchange is also an essential component in the in-situ leaching of uranium. In-situ leaching of uranium results in the generation of very low uranium ion concentration through process wells. The extracted uranium solution is then filtered through the ion exchange resin beads. Uranium-loaded resins are then transported to a processing plant, where tri-uranium octoxide (U_3O_8) is separated from the resin beads and yellowcake is produced. The resin beads can then be regenerated and returned to the ion exchange facility where they are reused.

Ion exchange is frequently employed in conjunction with other liquid treatment processes, usually as a final or polishing step. For example, conventional filtration can be used to remove particulate crud and the filtrate then treated by ion exchange. In other instances, the total dissolved solids may be reduced by chemical precipitation and the clarified liquid can then be polished by ion exchange. Ion exchange resins have also been used in conjunction with SX in the recovery of metals from solutions, for example, in uranium processing at the Rössing Uranium Mine in Namibia and the Vaal River South Uranium Plant in South Africa (Sole et al., 2005; Lunt et al., 2007). The readers' attention is drawn to Figure 7.15 which shows the production of uranium. The application of SX after resin ion exchange as a second stage of purification helps to obtain very-high-grade products.

Ion exchange is suitable for the treatment of dilute solutions because it allows metals to be removed without releasing the extractant back into the solution. However, the major limitation in the use of ion exchange for processing inorganic effluent is primarily the high-cost requirements for appropriate pretreatment systems such as filtration. This is because the resin matrix gets easily fouled by organics and other solids in the solution thus, prior filtration of the solution is necessary.

Spent ion-exchange materials are considered to be a special type of waste that poses problems in the selection of their disposal options. This is particularly true for the resins arising from the nuclear industry where solutions containing radioactive material are treated. The disposal and storage of the spent ion-exchange material has proved to be problematic over the years as they require special approach and precautions during their conditioning in order to meet the acceptance criteria for disposal (Gan et al., 2008; Braet et al., 2012). Thus, in order to guarantee sustainable disposal techniques, it is crucial to investigate more effective ways to correctly dispose spent ion-exchange materials. A viable spent ion-disposal method would be one that produces an intermediate stable product that may be used further. Therefore, the proper collection and conversion of spent ion-exchange wastes into materials that can be used in other industrial applications is a topic of vast interest to researchers.

7.3.4 Adsorption Processes

Adsorption is a natural process by which molecules of a dissolved compound collect on and adhere to the surface of a solid material. It is defined as the mass transfer of a substance (adsorbate) from liquid onto the solid surface (adsorbent) and becomes attached by physical and/or chemical interactions (Kurniawan et al., 2006). Thus, adsorption can be either physical (physisorption) or chemical (chemisorption) in nature and frequently involves both (Cheremishinoff and Morresi, 1978). In chemisorption, the transfer of electrons is significant and equivalent to the formation of a chemical bond between the sorbate and the solid surface. In physisorption, the interactive forces are relatively weak (Sing et al., 1985). In addition, physisorption involves attraction by electrical charge differences between the adsorbent and the adsorbate.

Adsorption is recognized as an effective and economic method for recovery of metals from process solutions. The adsorption process offers flexibility in design and operation and in many cases will produce high-quality treated effluent. In addition, because adsorption is sometimes reversible, adsorbents can be regenerated by a suitable desorption process (Fu and Wang, 2011). Studies and practical experience have shown that adsorption capacity depends on the physical and chemical characteristics of both the adsorbent and adsorbate, concentration of the adsorbate in liquid solution, experimental conditions such as temperature and solution pH and the amount of time the adsorbate is in contact with the adsorbent (Cheremishinoff and Morresi, 1978; Kurniawan et al., 2006).

Various adsorbents such as activated carbon (AC) and activated alumina have been developed and applied for the removal of metal ions from leach and process waste streams

(Richardson and Harker, 2002). The commonly applied adsorbent used in process solutions is AC. AC is an excellent adsorbent which has proven to be very effective in the removal of both inorganic and organic compounds from process solutions. This is because carbon is particularly adept at removing low levels of dissolved ions (Free, 2013). Although any carbon material can be used to make AC commercial AC is however, manufactured from only a few carbon sources such as wood, peat, coal, oil products, nut shells and pits (Davidson et al., 1968; Marsden and House, 2006).

AC has been widely used for the adsorption of metals from process solutions because of its well-developed porous structure, large active surface area and good mechanical properties. It has a high specific surface area ranging between 500 and 1500 m^2/g and a wide range of surface functional groups (Yin et al., 2007). Its utilization for metal adsorption greatly relies upon surface acidity and special surface functionality (Marsden and House, 2006), where the removal mechanisms may comprise of ion exchange, basal plane–cation interaction and coordination to functional groups (Marsden and House, 2006).

7.3.5 Biosorption Processes

Microorganisms and plant cell walls can be used to recover valuable metals from process streams using biosorption and bioaccumulation techniques in a process known as bioremediation (Atkinson et al., 1998; Volesky, 2001; Abia et al., 2002, 2003; Ahalya et al., 2003; Beolchini et al., 2010). The biological ion exchange resin is used like a conventional ion exchange resin involving a typical ion-exchange process where the metal ion is exchanged for a counter-ion attached to the biomass.

Biosorption is based on binding capacities of various biological materials such as living or dead microorganisms or plants (Volesky, 2001; Abia et al., 2002). The process involves a solid phase (sorbent or biosorbent, biological material) and a liquid phase containing the dissolved species to be sorbed (adsorbate, metal) (Ahalya et al., 2003). Due to the high affinity of the biosorbent for the adsorbate species, the latter is attracted and bound to the biosorbent by different mechanisms. The process continues until equilibrium is established between the solid-bound adsorbate species and the portion remaining in the solution. The degree of adsorbent affinity for the adsorbate determines its distribution between the solid and liquid phases.

The biosorption processes have many advantages over the conventional methods in that they are highly selective, more efficient, easy to operate and result in the minimal production of chemical and/or biological sludge. They also allow for regeneration of biosorbent and are cost effective for the treatment of large volumes of effluent streams containing low concentrations of heavy metals (Bhattacharyya et al., 1979; Sud et al., 2008). The biosorption techniques have the unique capabilities of being able to concentrate and reduce the levels of metal ions to environmentally acceptable limits in an economically and environmentally friendly manner (Volesky, 2001). Metals such as copper, iron, aluminium, cobalt, nickel, chromium, cadmium, lead, zinc, silver, gold and platinum can be recovered from solutions and waste streams using such techniques. The cost-effective nature of the biosorption process in removing most metals ions and, in particular, heavy metal ions places this technology in an interesting position of being able to open and successfully penetrate the suppressed potentially huge environmental clean-up markets.

Agricultural waste is one of the rich sources for low-cost adsorbents besides industrial by-products or natural materials. The use of agrowastes in a pure or chemically modified form for the remediation of contaminated aqueous solutions and industrial effluents has attracted considerable attention (Randall et al., 1974; Ho et al., 1995; Low et al., 1995; Gardea-Torresdey et al., 1998; Quek et al., 1998; Abia et al., 2002, 2003; Ndlovu et al., 2013; Simate and Ndlovu, 2015). Most of the research has focused on the use of algae, seaweed, alfalfa, sago waste, banana pith, sunflower and cassava waste. This is because large amounts of excess plant biomass that possess little economic value are produced by the agri-industry creating serious disposal problems. These materials are readily available, cheap, biodegradable and sludge free, involve small initial cost and land investment and, thus, create a significant renewable resource for sustainable bioremediation and biodegradation processes.

7.4 Electrochemical Processes

Electrochemical processes in aqueous solutions can be used either for recovery of a metal from solution or for refining an impure metal. The processes utilize a potential (applied or spontaneous) to drive electrochemical reactions in the desired direction. Electrodes (cathode and anode) must be connected electrically through an external circuit, and an electrolyte or ion-conducting medium must exist between the anode and the cathode.

Electrochemical processes are divided into galvanic and electrolytic processes. In a galvanic process, the reduction and oxidation reactions occur spontaneously without the aid of an external power supply. In electrolytic processes, chemical reactions are promoted and controlled by means of an applied external potential. The chemical reactions that do not occur 'spontaneously' are forced to take place through the application of an external electrical current. Thus, electrical power supply from the external circuit controls the conditions and extent at which electrons are transferred from the anode to the cathode. Electrolytic processes are divided into two, namely, electrowinning and electrorefining. In both electrorefining and EW, the main cathodic reaction is the reduction of the metal ions in solution to metallic form, while the main anodic reaction is the dissolution of the impure metal (electrorefining) or oxygen evolution at an inert electrode (EW). The next sections give a brief overview of the electrochemical processes used in hydrometallurgical processes.

7.4.1 Cementation Processes

Cementation is a process that involves the reduction of metal ions from a solution to form elemental metal. To obtain the reduction of a metal from the aqueous solution, the potential of the reducing metal system must be more negative than the potential of the metal system undergoing reduction, that is, have a tendency to donate electrons. The process can be considered as a galvanic reaction, and in general, for solutions in which the metal is present as simple aquo-cations, inspection of a table of standard electrode potentials will indicate the general feasibility of reduction.

Cementation has been used for the recovery of valuable metal ions from wastewaters and dilute solutions. The process has also been used in the removal of impurities from electrolytes prior to EW. A typical example in this case is in the production of zinc through the roast–leach–electrowin (RLE) process where zinc dust is used to remove impurities such as copper, cadmium, cobalt and nickel before EW (Jackson, 1986). High-purity zinc is produced through this process. Copper cementation by iron is also one of the well-known processes in hydrometallurgy (Jackson, 1986). This can be applied by considering the standard electrode potentials of the two systems:

$$Cu^{2+} + 2e = Cu \quad E^0 = +0.337 \text{ V} \tag{7.2}$$

$$Fe^{2+} + 2e = Fe \quad E^0 = -0.440 \text{ V} \tag{7.3}$$

The iron system has a more negative potential and iron will thus dissolve anodically, while copper ion has a more positive potential and will therefore, undergo reduction. The overall process is summarized in the reactions below.

$$\text{Anodic reaction: } Fe = Fe^{2+} + 2e \tag{7.4}$$

$$\text{Cathodic reaction: } Cu^{2+} + 2e = Cu \tag{7.5}$$

$$\text{Overall reaction: } Fe + Cu^{2+} = Fe^{2+} + Cu \quad E^0 = +0.777 \text{ V} \tag{7.6}$$

This process was practised significantly in the past for the production of copper prior to the advent of SX. It is also practised in the removal of copper from wastewater solutions.

7.4.2 Electrowinning

Electrowinning (EW) is the process of recovering solid metals from a solution (electrolyte) containing dissolved metal ions using an applied electrical potential. The electrolyte is derived from upstream leaching, or purification and concentration processes. The EW process takes place in an electrolytic cell and utilizes an applied external electrical power source to supply the potential and current to drive the electrochemical reactions in the desired direction. Reduction of the metal ions to metallic form occurs at the cathode whereas the anodes used are usually inert and generally support a gaseous evolution reaction.

EW has proved itself to be a cost-effective and technologically sound alternative for the recovery of metals such as gold, silver, cadmium, zinc, nickel and copper from plating wastes and rinse waters. Historically, EW needed high concentrations of metals to operate efficiently, but new technologies and process routes have drastically lowered the minimum metal concentration of the feed stream. A typical example is the integration of ion exchange or SX processes with EW so as to concentrate the solution before metal recovery as applied in most copper hydrometallurgical extraction processes.

7.4.3 Electrorefining

In electrorefining an impure metal is used as the anode in the electrolytic cell. Conditions are set up in the cell such that the metal at the anode is dissolved and the metal ions so formed migrate to the cathode where the metal is deposited in the same way as in EW. As dissolution occurs at the anode, impurities are released either as solid phases or as aqueous species. Any metal in the impure anode that is more noble than the metal being refined does not go into the solution as ions but stays as a metal and falls to the bottom of the cell forming what is known as 'anode sludge'. This phenomenon is very common in the electrorefining of copper. The anode sludge formed can be further treated to recover noble, but precious, metals. Metals less noble than copper (like zinc, nickel, cobalt) also dissolve and form ions along with copper into the bulk electrolyte. Being the noblest element in the solution, copper will deposit preferentially as its discharge potential is reached first. The elements that are less noble than copper, however, will not deposit at the cathode, provided their concentrations are not too high. Thus, under the right conditions, most of the impurities present in the anode are not deposited on the cathode with the metal.

7.5 Base Metals Production

Base metals are more abundant in nature and therefore far cheaper than precious metals such as gold, silver and platinum. Base metals include aluminium, copper, lead, nickel, tin and zinc and are widely used in commercial and industrial applications. The subsequent sections will look at the production of selected base metals, namely, copper, nickel, zinc and aluminium. Some of the waste generated from the production of these metals and the potential value that can be realized from the waste material is also looked at.

7.5.1 Production of Copper

The treatment of copper feed material in a hydrometallurgical plant generally involves a process of acid leaching followed by purification to remove impurities using processes such as precipitation and SX, with the final generation of a solid metal through the EW process.

The leaching followed by SX and EW processes has been widely adopted for the hydrometallurgical processing of copper ores. Figure 7.4 shows the basic process flowsheet for oxide or non-refractory sulphide material.

The process starts with the dissolution of oxide or non-refractory sulphide material in a dilute aqueous solution of sulphuric acid in the presence of an oxidant with ferric sulphate being the oxidant commonly used. This is followed by SX for purification and/or solution

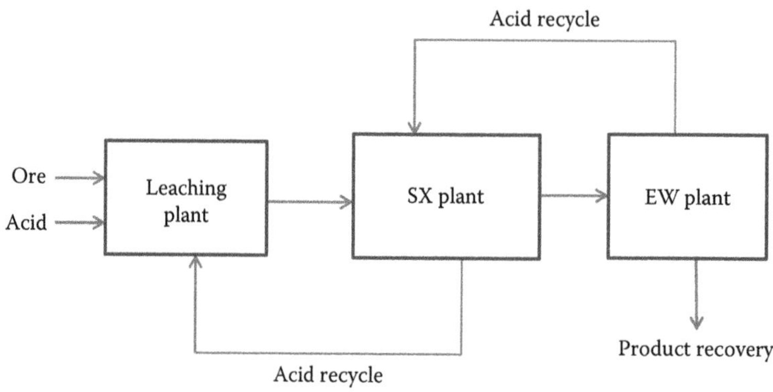

FIGURE 7.4
Basic process flowsheet for copper production from oxide and non-refractory sulphide material.

concentration upgrade before EW. The process ends with the generation of a solid metal through the EW process.

It must be noted, however, that several other techniques are used to extract copper from the sulphide ores; some ores are roasted or calcined before leaching. For example, Zambia and the DRC leach the roaster calcine to extract Cu and Co from Co–Cu–Fe–S concentrate. Bacteria that exist naturally within sulphide mine water can also facilitate the leaching process of sulphide ores or waste material in a process known as bioleaching. For primary copper sulphide ores such as chalcopyrite, a more intensive leaching approach is generally applied, such as (1) leaching under high temperatures and pressure (Placer process), (2) leaching under highly oxidizing conditions in the presence of a catalyst (CESL), (3) fine grinding the material before leaching (Activox), (4) the use of low-grade coal as an additive (Dynatec process) and (5) leaching in the presence of microorganisms (bioleaching) (Dreisinger, 2006).

After copper has been extracted from the ore as indicated earlier, it can be recovered from leach solutions through precipitation or by SX. The reduction and precipitation of copper from dilute sulphuric acid leach solutions by iron is the oldest hydrometallurgical method known for copper recovery. In practice, iron scrap or sponge iron with a large surface area is used. The precipitate, cement copper, is impure copper, containing from about 60 to >90 wt.% Cu. Further treatment of the cement copper is usually performed via a pyrometallurgical process (Jackson, 1986).

Due to the high costs and inefficiencies associated with the iron precipitation process, SX has, instead, become very popular and in essence largely replaced the precipitation process in the recent past. The copper SX consists of contacting the pregnant aqueous leach liquor with a copper-specific organic liquid extractant in a mixer–settler contactor. The Cu-loaded organic extractant is then separated from the aqueous solution (raffinate). The raffinate or barren solution that is highly acidic is sent back to the leaching stage. During stripping, the Cu loaded on the organic extractant is transferred back into the aqueous phase by contacting the organic solution with a concentrated sulphuric acid solution. The now-stripped organic extractant is returned to the SX stage. The enriched electrolyte is sent to EW where metallic copper is removed by electrodeposition, while the spent electrolyte that is highly acidic is returned to the stripping stage where it is utilized as acid make-up. The basic process overview of the SX circuit has been shown in Figure 7.3.

Because of its dependence on sulphuric acid, the SX/EW process tends to be integrated with conventional smelting, smelter production of sulphur dioxide for acid production (see Figure 7.5) or the manufacture of sulphuric acid from sulphur or pyrite. However, it is also applicable in locations where acid is not readily available through purchase of

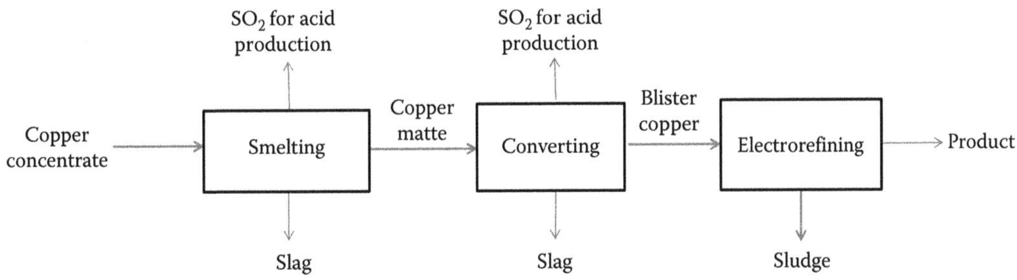

FIGURE 7.5
The basic process for the production of copper from sulphide material via the pyrometallurgical route.

sulphuric acid or the on-site production of the acid through pressure leaching of sulphidic material such as in the Sepon copper process in Laos (Dreisinger, 2006).

The importance of the leach SX-EW process is that it offers the opportunity to recover copper from an entirely different set of ores and mining by-products than is possible by smelting. For example, oxidized copper-cobalt ores in the DRC are leached in the presence of sulphuric acid and a reducing agent to dissolve both copper and cobalt (Stuurman et al., 2014). The solution is treated by SX/EW process to recover copper, while cobalt is recovered through precipitation.

An analysis of the hydrometallurgical processing circuit for copper does not reveal a lot of opportunities for by-product recovery or process residue utilization. The SX/EW process itself has very little environmental impact because its liquid streams are very easily contained. There is no effluent as all impurities are returned to the site where they originated, and the sulphuric acid is eventually neutralized by the limestone in the ore body or waste dump where it is deposited as calcium sulphate (gypsum) – a very insoluble substance (Dresher, 2001). In the copper production by SX/EW, the potential residue that can be generated is from the SX circuit where crud can be formed.

7.5.1.1 Crud Formation, Treatment and Recycle

Crud is the solid-stabilized emulsion formed by mixing an aqueous phase containing fine suspended solids with an organic extractant phase. It commonly accumulates at the aqueous/organic interface in the settlers during extraction stages. According to Ritcey and Ashbrook (1984), factors such as ore type, solution composition, solvent composition, presence of other organic constituents and design and type of agitation can adversely affect the chemical and physical operation of the SX circuit and result in crud formation. Materials such as clays, jarosites, gypsum, quartz and micas have all been linked to crud formation in copper SX (Readett and Miller, 1995; Virnig, 2007).

Crud can result in increased entrainment in the electrolyte, causing contamination of electrolyte if allowed to build up in the settlers. The entrainment into the extraction section of the SX process can result in the loss of copper to the raffinate (Mukutuma et al., 2009).

Crud contains typically 50% organic by weight or more; thus, recovery of the organic minimizes organic losses and results in cost savings as the organic can be recycled into the extraction circuit (Mukutuma et al., 2009). Since the crud formed usually differs from one plant to another or in another circuit in the same plant depending on the conditions that caused its formation, the treatment scheme adapted for the recovery of the solvent will also be different.

Ritcey and Ashbrook (1984) reviewed the problem of crud formation in SX. They highlighted a number of methods that can be used to prevent or minimize the formation of crud such as the addition of surfactants to prevent precipitation, breaking down the cruds by extreme agitation such as cycling through a centrifugal pump, followed by settling and skimming, filtration, acidification or neutralization to an alkaline condition. Crud arising from colloidal silica can be treated via coagulation followed by centrifuge separation and filtration. These methods are based on the fact that in the water treatment industry, silica in solution is often removed by adsorption or by coagulation followed by filtration. Crud treatment such as centrifuging or clay application is commonly used to recover the high-value organic extractant (Bishop et al., 1999; Hartmann and Corbella, 2007). The use of fluoride to decompose silicate scale is also well known (Fletcher and Gage, 1985).

7.5.1.2 Treatment of Anode Slimes from Copper Electrolytic Refining

In the conventional pyrometallurgical production of copper, the copper-bearing ore is enriched in order to obtain the copper concentrate from which, a copper anode of purity 97%–99% is produced. Figure 7.5 shows the basic process for the production of copper via the pyrometallurgical route.

In this process, copper ore is mined, crushed, ground, concentrated, smelted and then refined. The process is applied to ores that contain copper minerals in the sulphide form such as chalcocite (Cu_2S), chalcopyrite ($CuFeS_2$) and covellite (CuS). In the smelting operation, the concentrate from the flotation process is fed to a smelter together with oxygen. In the smelter, copper and iron sulphide concentrates are oxidized at high temperature resulting in the production of impure molten metallic copper, a mixture of molten oxides (iron oxide, silica, alumina, calcium oxide and magnesium oxide, etc.) and the emission of gaseous sulphur dioxide. The gaseous sulphide is used in the production of sulphuric acid while there is also numerous potential application of the generated slag; readers are referred to Chapter 6 for more details.

Following the smelting process, the matte copper is converted to produce blister copper then purified by electrolytic purification to produce cathodic copper with a marketable purity of 99.99% copper, while the iron oxide is disposed of as slag.

As previously discussed (under electrolytic processes), in the electrorefining process, the copper anode dissolves, enters the electrolyte as copper ions and deposits onto the cathode under an applied direct electric current. During dissolution, any metal in the impure copper anode that is more noble than copper does not go into solution as ions but stays as a metal and falls to the bottom of the cell forming what is known as an 'anode sludge/slime'. Figure 7.6 shows a typical copper electrorefining cell with the anode slime deposit.

These slimes are generated at a rate of 3–5 kg/ton of anode processed and are a big source of valuable metals (Hait et al., 2009). The anode slime contains precious metals (silver, gold and platinum), selenides and tellurides of copper and silver, lead sulphate, stannic oxide hydrate and complex compounds of arsenic, antimony and bismuth. Table 7.1 shows the typical composition of the anode sludge.

The recovery of these valuable metals through the treatment of the anode slime is of great importance in a copper refinery process. Methods of treating the anode slime to

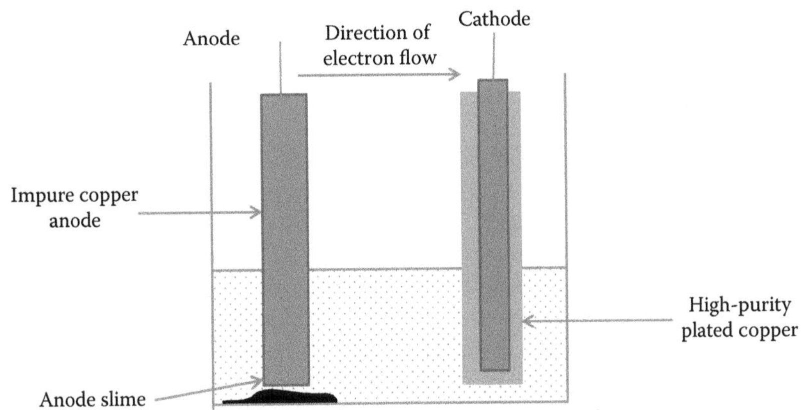

FIGURE 7.6
Copper electrorefining with anode slime deposit.

TABLE 7.1
Distribution of Constituents of Copper Anode Slime

Element	Distribution (wt.%)
Copper	20–50
Nickel	0.5–5
Lead	5–10
Arsenic	0.5–5
Antimony	0.5–5
Bismuth	0.5–2
Tellurium	1–4
Selenium	1.5–28.4
Silver	≅25
Gold	≅4
Platinum	Trace–1.5

Sources: Lu, D. et al., *Trans. Nonferrous Met. Soc. China*, 25, 1307, 2015; Ally, M.R. et al., Economical recovery of by-products in the mining industry, A report produced by the Engineering Science and Technology Division, Nuclear Science and Technology Division and the Environmental Sciences Division, US Department of Energy, 2001; Young, R.S., *Talanta*, 43, 125, 1975.

recover the valuable metals include processes whereby these metals are extracted one at a time under specific conditions, with the PGMs being recovered last. Owing to the fact that the compositions of the slime and the local conditions vary greatly from one refinery to another, a number of different processes have been developed and applied for the treatment of anode slimes. Although these methods may vary depending on the composition of the anode slime, some features are however, common and may be summarized as follows:

- Leaching with dilute sulphuric acid to oxidize copper
- Recovery of selenium and tellurium by pyrometallurgical or hydrometallurgical methods
- Removal of unwanted elements and production of silver alloy
- Separation of precious metals (silver and gold) and platinum metals

In general, existing anode slime treatment processes have been designed to remove copper first and then to recover the precious metals in the form of an alloy known as doré metal (Schloen and Elkin, 1950; Scott, 1990; Heimala et al., 1975). The doré metal is a silver–gold mixture bearing small quantities of copper and the PGMs present in the slime. According to Heimala et al. (1975), the treatment of anode slimes involves first the removal of copper and nickel if present, followed by selenium and tellurium thereafter. Impurities such as selenium and tellurium constitute useful by-products. The doré metal is then obtained by smelting the residual slime generated after removal of the above components, that is, copper, selenium and tellurium. Pure silver is then obtained by electrolysis of the doré metal. The anode slime generated from silver electrolysis is then used to cast anodes for gold refining. In the gold electrolysis, pure gold is obtained on the cathode, and the Pt and Pd dissolved in the electrolyte can be separated from it.

In a process developed by Lindroos et al. (2004) and with a patent filed by Outotec, the doré smelting stage is eliminated from the process through the introduction of a sulphatizing process. In this process, the copper anode slime is leached under atmospheric pressure using sulphuric acid in the presence of oxygen. This helps to dissolve the copper whilst keeping the precious metals in the sludge. The slime is then calcined in two steps. The first step is for separating selenium and removing it as selenium oxide. The second step is for sulphatizing silver and any residual copper. The sulphatized silver and copper is separated by leaching into a neutral aqueous solution. The gold and platinum metals from the residual sludge can then be recovered.

In another process developed by Wang et al. (1980) three major unit operations, leaching, SX and reduction, are involved in the treatment of anode slimes for the recovery of valuable metals. The decopperized anode slime is first leached with nitric acid at an elevated temperature to obtain a leach liquor high in silver, selenium and tellurium. Silver in the nitric acid leach liquor is recovered in the form of silver chloride. Thereafter, the solution containing selenium, tellurium, copper and other impurities is denitrated and chlorinated by a SX technique to separate the tellurium from the other impurities. Selenium and tellurium are then recovered individually by passing sulphur dioxide through the selenium-containing and tellurium-containing solutions. The nitric acid leach residue is treated with aqua regia to leach gold and other impurities at elevated temperatures. The gold-containing solution is treated for gold recovery by use of a SX technique followed by reduction.

Outotec (2015) has developed a process flowsheet for the recovery of precious metals from the anode slime produced in the electrorefining plant as shown in Figure 7.7.

The first step in the anode slime handling is to remove the copper and nickel present in the slime either by atmospheric or by pressure leaching depending on the slime composition. A pyrometallurgical step for the recovery of selenium either from the off-gases produced in the Kaldo furnace or from the selenium roasting furnace then follows. In the selenium roasting stage, copper-free anode slime is dried and selenium is evaporated as selenium dioxide gas using oxygen and SO_2 gas as reagents. The selenium is recovered from the off-gases into an aqueous solution. Smelting, reduction and refining either in a Kaldo furnace or in a tilting rotating oxy-fuel (TROF) converter produces doré metal. The TROF converter takes care of the smelting, reduction and refining steps in the same

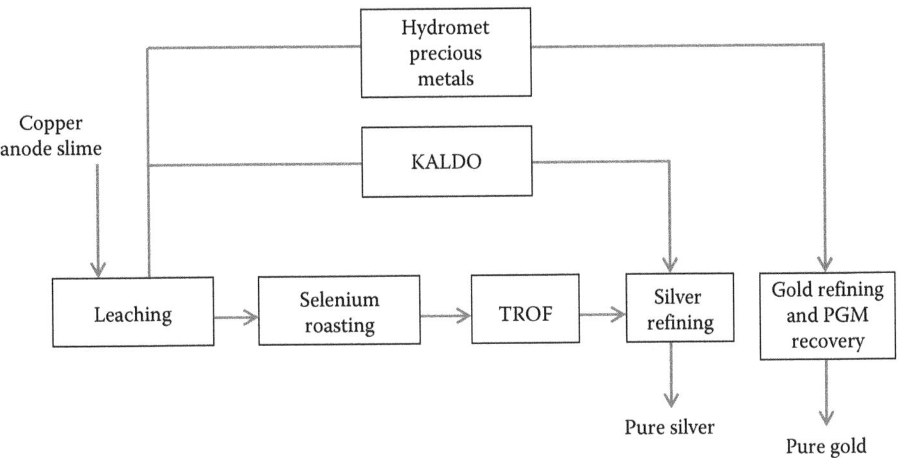

FIGURE 7.7
Outotec process flowsheet for the recovery of metals from the anode slime. (From Outotec, 2015, Available at http://www.outotec.com/en/Products--services/Non-ferrous-metals-processing/Precious-metals/Precious-metals/, Accessed December 2015.)

furnace, which is energy efficient. Silver is electrolytically refined from doré anodes. Gold and PGM metals are leached and precipitated from gold mud produced in silver electrorefining. The end products of the process include selenium, silver, gold and PGMs. Tellurium can be recovered as copper telluride and lead can be recovered as lead sulphate using the hydrometallurgical process.

7.5.2 Nickel and Cobalt Production

Cobalt is usually not mined alone and tends to be produced as a by-product of nickel and copper production activities. Nickel can also be produced during the production of copper from sulphide ores. The process for the production of cobalt and nickel is usually similar to that of copper production as it follows the same route of leaching, SX or reductive precipitation with hydrogen gas. The process residue arises mostly from the crud during SX which can be treated in a similar manner as described in Section 7.5.1.

Nickel can also be produced from the processing of nickel laterites. These materials are processed by ammonium carbonate technology (Caron) acid leaching, high-pressure acid leaching and bacteria leaching to extract the nickel. These processes generate a solid residue with a complex structure that is physically characterized by its black colour, semi-metallic lustre, fine particle sizes and predominantly magnetic nature with notable amounts of metals like Ni (0.25%) and Co (0.09%) that are partially oxidized (Cabrera et al., 2011). The process also generates some iron hydroxide precipitates in the form of goethite and hematite; however, current metallurgical processes usually focus on the recovery of Ni and/or Co from the residues.

Microbiological processes such as bioleaching have been popular as an alternative to treat this kind of residue due to the associated low cost, clean technology and low energy requirement (Coto et al., 2007, 2008; Bosio et al., 2008; Hernández et al., 2009). The recovery of metal species from lateritic material by microorganisms is mainly based on the use of organic acids as leaching agents, and these are produced by filamentous fungi. However, the use of fungi suffers from certain limitations such as the need to add organic compounds as a source of carbon and energy to culture the fungi, which increases the cost of future commercial processes, and the formation of chelates, which hinders the recovery of dissolved metals and decreases the amount of leached metals (Cabrera et al., 2011).

The starvation acid leach technology (SALT) has also been proposed for recovering nickel and/or cobalt and/or by-product metals from a lean ore, such as a nickel laterite ore or a Caron plant tailing material (Dreisinger and Clucas, 2013). This method involves subjecting the material to a prolonged acid leach with application of starvation levels of acid. An alkali is added to the leach slurry to neutralize the residual acidity in the slurry and to precipitate metals such as iron, aluminium and chromium as metal hydroxides. The leached ore/tailing residue may be separated from the iron-/aluminium-/chromium-free solution to recover nickel and cobalt, for example, by a process of mixed hydroxide precipitation, mixed sulphide precipitation, SX/EW or SX and hydrogen reduction. The final solutions and/or washed-ore (or tailing) residue is subjected to environmental treatment prior to disposal.

Laterite residues have also been suggested as potential material in the production of zeolites that can be used for CO_2 capture from flue gas streams (Liu et al., 2014). The zeolite is prepared from a mixture of laterite residue and bauxite using an alkali fusion process followed by hydrothermal treatment. The product showed similar surface areas to that of commercial ones. The adsorption property of the zeolite was further found to be comparable to the commercial ones in terms of CO_2 and N_2 adsorption equilibrium capacities.

7.5.3 Production of Zinc

Traditionally, zinc has been produced from sulphide materials using two methods – the RLE process and the direct pressure leaching process with EW used for final metal recovery. Recently, a novel process for the production of zinc from silicate material has been implemented at the Skorpion Zinc plant in Namibia (Cole and Sole, 2003; Gnoinski et al., 2005). This process involves the leaching of zinc silicate/oxide material in dilute sulphuric acid followed by neutralization to remove silica and some of the iron and aluminium, and then SX to upgrade the zinc concentration and remove the rest of the impurities. The last stage is EW to produce super high grade zinc.

In general, the RLE process has enjoyed a lot of commercial success in the production of zinc. The first stage in the process involves the roasting of zinc sulphide concentrate in which the zinc occurs mostly as sphalerite, ((Zn, Fe) S) in order to generate a calcine that can be easily leached in the hydrometallurgical plant. During roasting, sulphur dioxide is also generated. The sulphur dioxide released is captured and used to produce sulphuric acid that is subsequently used in the leaching process. During roasting, sphalerite is not only transformed to the simple calcine, ZnO, but also to zinc ferrite, which has an ideal formula of $ZnO \cdot Fe_2O_3$. The following equations show the reactions that take place during roasting.

$$2ZnS + 3O_2 = 2ZnO + SO_2 \tag{7.7}$$

$$4FeS_2 + 11O_2 = 2Fe_2O_3 + 8SO_2 \tag{7.8}$$

$$ZnO + Fe_2O_3 = ZnO \cdot Fe_2O_3 \tag{7.9}$$

$$2ZnO + 2SO_2 + O_2 = 2ZnSO_4 \tag{7.10}$$

The compound $ZnO \cdot Fe_2O_3$ is not leached in the primary leaching step that happens under slightly neutral pHs but reports to the neutral leach residue instead. As a result, this residue tends to contain large amounts of zinc. To improve zinc recovery, it has been found that zinc ferrite could be dissolved in a highly acidic leaching solution containing more than 100 g/L H_2SO_4 at temperatures ranging from 90°C to 95°C and at a residence time of 2–4 h (Gupta, 2003). However, this procedure not only dissolves the zinc but also the associated iron. Any iron precipitated in the neutral leaching step is, therefore, also redissolved to give a zinc-rich process solution typically containing between 15 and 30 g/L of iron that is mostly in the ferric form.

Recently direct leaching of concentrates at elevated temperatures and pressures and bacterial leaching have become established. These processes have the potential to become serious contenders in the future hydrometallurgical production of zinc (Peters, 1976; Sandstrom and Peterson, 1997; Konishi et al., 1998; Dew et al., 1999; Fowler and Crundwell 1999; Petersen and Dixon, 2007). In these processes, zinc ferrites do not form, but significant amounts of iron that will require an efficient iron removal are still solubilized. The challenge in the removal and subsequent disposal of iron has been a major difficulty for the industry for many decades and was directly responsible for low overall zinc recoveries. Therefore, the efficient removal of iron from the leach liquors is a critical step in the hydrometallurgical extraction of zinc from its ores. This is primarily because the EW of zinc from aqueous solutions can be negatively affected by the presence of even a trace amount of impurities such as iron. Therefore, consideration must be given to the iron removal and purification stages to ensure that the zinc-bearing solution is amenable to producing high-quality zinc cathode in the tankhouse. Many existing electrolytic plants and all new plants have embedded one or more of the iron removal processes into their leaching circuit (Meyer et al., 1996; Hearne and Haegele, 1998; Claassen et al., 2002). The introduction of the iron removal processes such as jarosite, goethite or haematite, depending on preference and economics, has allowed the precipitation of iron in an easily filterable form, thereby enhancing the overall zinc recovery.

7.5.3.1 Jarosite Production and Utilization

Figure 7.8 shows the typical flowsheet for the removal of iron as jarosite in a zinc production plant.

The bulk of the calcine enters the plant into the neutral leaching reactors where it is mixed with a return solution from the jarosite process and spent electrolyte from EW. About 80% of the soluble zinc is dissolved in the neutral leaching reactors. The neutral leach liquor is separated from the solid residues and passes to the purification stages to remove elements such as copper, nickel, cobalt and cadmium through cementation with zinc dust prior to EW.

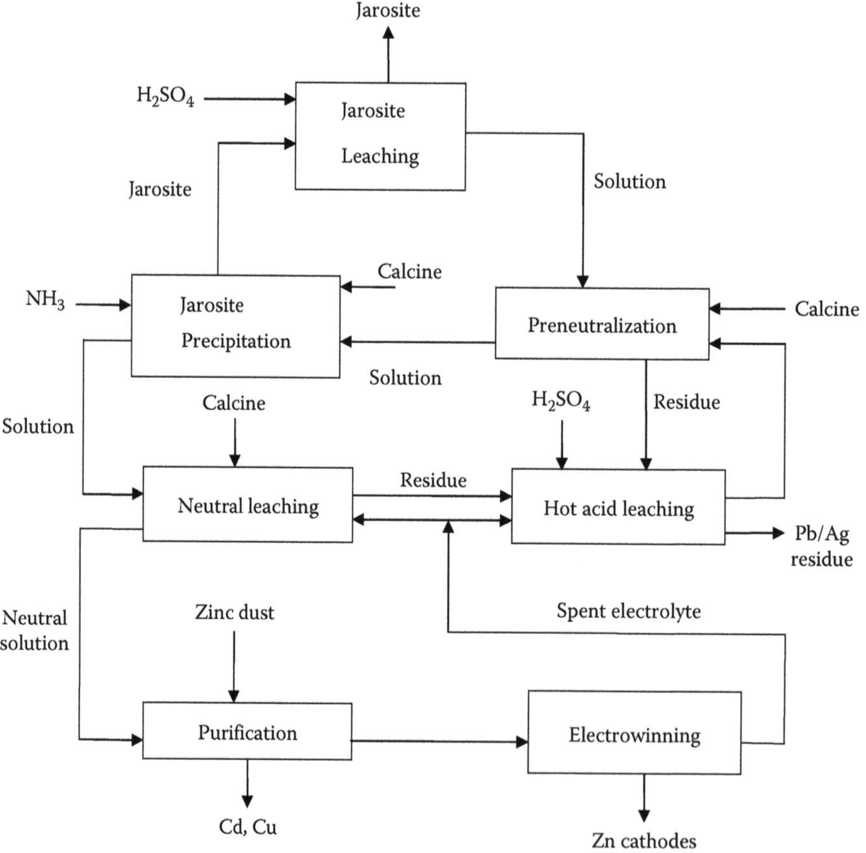

FIGURE 7.8
The basic jarosite process used in zinc production. (From Gupta, C.K., *Chemical Metallurgy: Principles and Practice*, Wiley-VCH Verlag GmbH & Co. KGaA, Weinheim, Germany, 2003, 831pp.)

The solid residues from the neutral leaching stage are transferred to the hot acid leach stage where they are leached at 95°C with spent electrolyte containing about 50 g/L zinc and 150–200 g/L make up sulphuric acid (Gupta, 2003; van Dyk, 2006). Under these conditions, the zinc ferrites break down, and most of the zinc and iron are leached out together with much of the copper and cadmium in the residues. The solution leaving the hot acid leaching is partially neutralized by adding zinc calcine. Only the zinc oxide in this calcine is dissolved and so the ferrite residue is returned to the hot acid leaching stage. The neutralized solution is then ready for jarosite precipitation.

The precipitation of iron as jarosite (i.e. one of a group of basic ferric sulphates) occurs from acidic sulphate solutions as follows:

$$M^+ + 3Fe^{3+} + SO_4^{2-} + 6H_2O \rightarrow MFe_3(SO_4)_2(OH)_6 + 6H^+ \tag{7.11}$$

where M stands for the added monovalent cation such as NH_4^+ or Na^+.

The process generates hydrogen ions and the solution is therefore, neutralized by adding zinc calcine (ZnO) to maintain the necessary pH levels. Jarosite is precipitated at 95°C and a pH of 1.5, that is, in a relatively acidic environment. The kinetics of jarosite formation is slow at temperatures below 60°C but is rapid at temperatures around 80°C–95°C, which is used in commercial zinc plants (Wang et al., 1985). The jarosite residue contains the zinc ferrite from the calcine added as a neutralizing agent during jarosite precipitation. Jarosite is immune to acid attack, and therefore, undissolved zinc roast combined with jarosite can be readily removed by washing with acid (Claassen et al., 2002; van Dyk, 2006). This releases ferritic zinc into the solution while the jarosite remains largely unaffected. The solution from jarosite leaching is returned to preneutralization and the acid-washed jarosite is discarded or recycled.

The main advantage of precipitating jarosite compounds is the high precipitation efficiency and ease of filtration. In addition, it allows the process to be carried out at lower temperatures than other iron removal processes and in slightly acidic media that avoids the expensive neutralization step. The jarosite produced however, has a low iron content resulting in huge amounts being produced in order to significantly reduce the iron content in the zinc sulphate solution.

7.5.3.2 Potential Utilization of Jarosite

The large amount of jarosite generated in the zinc production process suggests a need for the complete utilization of this waste material in order to save on disposal costs and space. The bulk of the world production of jarosite is generated mainly in Spain, Holland, Canada, France, Australia, Yugoslavia, Korea, Mexico, Norway, Finland, Germany, Argentina, Belgium and Japan (Arslan and Arslan, 2003). Research on recycling of this residue is also established in these countries.

The major component of jarosite is oxides of iron, sulphur and zinc. It contains approximately 30%–50% Fe (which explains the interest in its use as a source of iron), approximately 30% SO_3, and about 10% ZnO. The other constituents are aluminium, cadmium, calcium, cobalt, copper, lead, magnesium, manganese and nickel, and each constituent is present below 7% (Pappu et al., 2006a).

The presence of toxic materials in jarosite poses a serious problem for its disposal. The toxicity in jarosite is mainly due to the presence of metals such as lead, cadmium, arsenic and chromium, some of which are present in an environmentally mobile form (Pappu et al., 2006a,b). Safe management of the hazardous nature of jarosite is of paramount importance. As a result it has become a global concern to find a socio, techno-economic, environmentally friendly solution to maintain a clean and green environment.

Stabilization or the production of solidified products is usually the first step in the treatment of jarosite waste (or residue). Stabilization inhibits the transportation of pollutant elements into the surrounding environment and improves the physical characteristics of jarosite; thereby further reducing leaching of contaminants and toxic metals. Because jarosite is produced under very acidic conditions, the sludge generated is very acidic and requires neutralization prior to disposal or alternative reuse. The addition of a neutralizing agent helps to transform most of the metals into oxides, which are immobile under neutral conditions. Jarosite has been converted into a stable material for disposal in the Jarofix process by an addition of 2% lime and 10% cement (Seyer et al., 2001). Cement and lime enable the longterm physical and chemical stabilization of the jarosite residue. The addition of cement and lime, however, results in high costs for the process. This is an important factor when considering the utilization of this process for jarosite reuse.

Jarosite should be considered as a potential resource, which can be recycled in a technically and economically feasible and environmentally friendly manner to generate products for other potential applications. Various applications of jarosite residues have been proposed such as use in the production of construction material, catalysts, pigments, gypsum and refractories (Sanjeev et al., 1999; Rathore et al., 2014). The physicochemical characteristics of jarosite indicate that there is also potential utilization of a jarosite–sand mixture for building materials like bricks, blocks, cement, tiles, composites, glass and ceramics (Pelino, 2000; Pappu et al., 2006a,b).

Construction industry: Stabilized products can be produced from jarosite for use in construction industry. In this regard, a lot of researchers have focused on the use of waste from other industries, that is, cross waste recycling, in order to generate useful products and at the same time play a significant role in minimizing the amount of waste to be disposed of across different industries. According to Mymrin et al. (2005), jarosite wastes can be bound or neutralized with dump ferrous slag (DFS), alkaline coal fly ash, highly alkaline Al-surface cleaning waste, etc. to make new construction material. An activator (also an industrial waste) and optional addition of Portland cement (PC) or CaO can also be used in the process. These small additions of lime or PC to the mixtures can accelerate the materials' strengthening process. The use of DFS is of much more crucial concern for many countries, due to the large amounts of DFS produced and the large areas needed to store such slags. Pappu et al. (2010) studied the recycling potential of jarosite waste and coal combustion residues generated in large quantities in thermal power plants for the production of composite products that met the quality standard for use in construction applications. The mechanical properties of these composite materials such as high compressive strength, water absorption, shrinkage and density make them highly recommendable for use as construction materials for road and airfield runways, levee cores, industrial and municipal dumps and multistorey building foundations, in tile and in brick production.

Jarosite has also been used in the development of alternative and lightweight building blocks as a substitute to the existing fired clay bricks (Pappu et al., 2006a,b, 2007; Council of Scientific & Industrial Research, 2014). There is a possibility of using jarosite as a substitute for natural gypsum in cement production

Orru et al. (1999) proposed a technique for the treatment of goethite waste that consists of blending the goethite waste with suitable amounts of reducing agents (aluminium or aluminium silicon) and ferric oxide. The technique included igniting the resulting mixture so that a self-propagating reaction in the form of combustion wave rapidly could travel through the mixture without requiring additional energy, thus converting the reactants into products where hazardous and toxic species are incorporated. Reactants were converted into two solids, (1) an amorphous product where toxic species are immobilized and could be used in the glass-ceramic industry and (2) other zinc-containing products that could be recycled back into the zinc production plant.

7.5.3.5 Hematite Production and Utilization

The hematite process for the removal of iron from zinc sulphate leach solutions is practised by Iijima Zinc Refinery in Japan (Okada and Yamazaki, 1984; Gupta 2003). A flowsheet of the process is shown in Figure 7.10.

The hydrolysis of ferrous sulphate to hematite is represented by the following reaction:

$$2FeSO_4 + \frac{1}{2}O_2 + 2H_2O \rightarrow Fe_2O_3 + 2H_2SO_4 \quad (7.13)$$

The reaction takes place under an oxidizing atmosphere at high pressure (2 MPa) and high temperatures (200°C).

One major advantage of the process is the high iron content of the hematite residue produced, about 60% in practice which leads to the smaller amounts of iron in the final disposal residues. For instance a given quantity of iron, when precipitated in the form of hematite, will weigh less than half of that same quantity of iron when precipitated in the form of jarosite (Claassen et al., 2002; Gupta, 2003). As a result, the problem related with the storage of residues is also minimized.

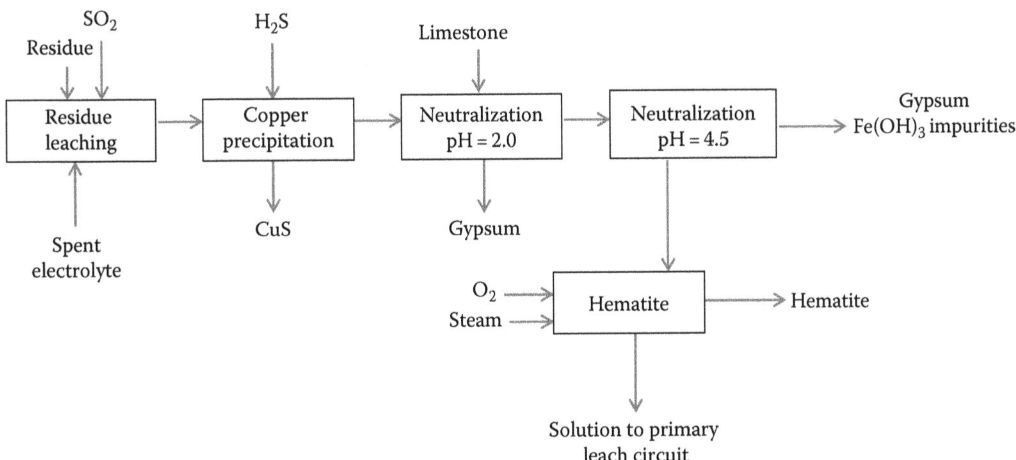

FIGURE 7.10
The hematite process. (From Gupta, C.K., *Chemical Metallurgy: Principles and Practice*, Wiley-VCH Verlag GmbH & Co. KGaA, Weinheim, Germany, 2003, 831pp.).

One marked feature of the removal of iron from solutions as hematite is the possibility of using it in the cement industry, in the iron industries and/or as a pigment. Thus, the hematite process enables iron-containing residue from zinc production to be converted into a useful by-product and to be disposed of without environmental problems. However, the cost and complexity of the hematite-processing technology restricts its widespread application (Ullmann's, 1993). In addition, the hematite so produced is not acceptable for iron and steel production since small amounts of zinc in the blast furnace charge can cause refractory problems and contribute to large quantities of flue dust produced (Zheng and Kozinsk, 1997).

7.5.3.6 Summary

There is a tremendous potential for recycling and using solid waste from the zinc hydrometallurgical plants as construction and building materials. The new and alternative construction and building materials developed using the waste material have ample scope for reducing, to an extent, the costs of traditional building materials. However, to effectively utilize the waste as a raw material, filler, binder and additive in developing alternative building materials, glass ceramics, etc., a detailed physicochemical, engineering, thermal, mineralogical and morphological characterization of the material needs to be undertaken.

7.5.4 The Bayer Process for Alumina Production

Aluminium is virtually in all economic segments, but its principal uses have been developed in major industries such as transportation, construction, electrical, containers, packing and mechanical equipment. The traditional source of aluminium is bauxite. Bauxite is a heterogeneous material that contains aluminium oxides and hydroxides in levels between 30% and 65% (measured as aluminium oxide). Current estimates of known world bauxite reserves are some 30 billion tonnes with indications of unproven reserves being much higher. Bauxite mined in many countries including Australia, Brazil, China, Ghana, Greece, Guinea, Guyana, Hungary, India, Indonesia, Jamaica, Sierra Leone, Suriname, Venezuela and Vietnam. The Bayer process shown in Figure 7.11 is the principal industrial means of refining high-grade or medium-grade bauxite to produce alumina, an intermediate in aluminium production. Although other processes have since been investigated for the recovery of alumina from bauxite, the traditional Bayer process still predominates.

In the Bayer process, bauxite is washed with a hot solution of sodium hydroxide, NaOH, at 175°C (a process called digestion). The digestion process removes the hydrated alumina from other insoluble oxides by reacting it with sodium hydroxide according to the following reaction:

$$Al_2O_3 + 2OH^- + 3H_2O \rightarrow 2[Al(OH)_4^-] \tag{7.14}$$

The other components of bauxite do not dissolve and are filtered from the solution in a thickener during the clarification stage resulting in a mixture of solid impurities known as red mud or bauxite residue. The hydroxide solution is cooled and seeded with small crystals of alumina tri-hydrate that promote the precipitation of the dissolved aluminium hydroxide as a white, fluffy solid. The hydroxide is then calcined at 1050°C, decomposing to alumina and giving off water vapour in the process:

$$2Al(OH)_3 \rightarrow Al_2O_3 + 3H_2O \tag{7.15}$$

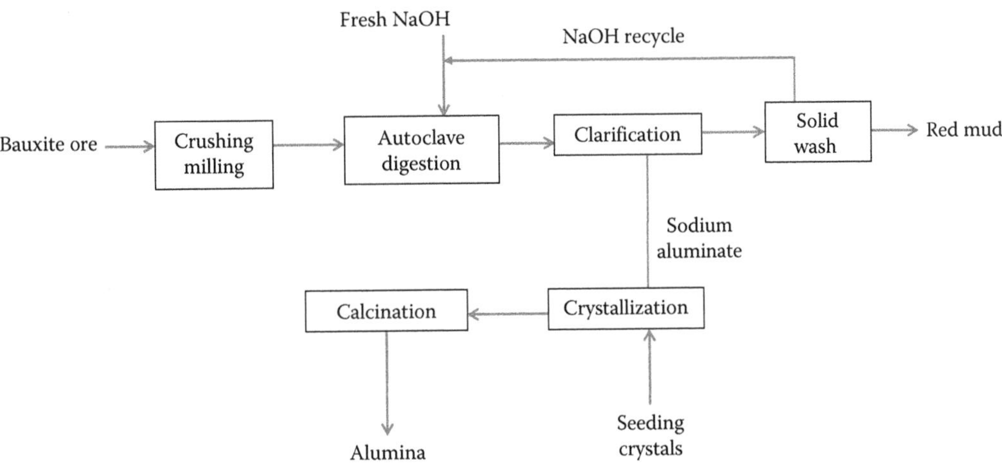

FIGURE 7.11
The Bayer process for the production of alumina.

It should be noted that the Bayer process does not produce pure aluminium metal; it is the first step in the overall production process. The large amount of the alumina generated in the Bayer process is subsequently sent to a smelter where the aluminium metal is produced in the Hall–Héroult process. A detailed discussion on the production of aluminium from the alumina is given in a section on base metals smelting presented in Chapter 6. This chapter focuses on the residue that is generated in the alumina production step, which is a pure hydrometallurgical process.

The amount of bauxite residue produced by an alumina producing plant is primarily dependent on the sources of the bauxite and on the extraction conditions used at the plant. The most important factors are the aluminium content of the bauxite, the type of aluminium oxide/hydroxide present (e.g. gibbsite, boehmite or diaspore) and the temperature and pressure conditions used for the extraction. For every ton of alumina produced, the red mud generated can vary from 0.3 to as high as 2.5 tonnes, though typically it lies between 0.7 and 2 tonnes per ton of alumina produced. According to the World Aluminium Report (2015), aluminium mining leaves behind a staggering 120 million metric tonnes (Mt) per year of the salty, highly alkaline, heavy-metal–laden residue material. Because of the chemical nature and the mineralogical species present in the red mud, this solid material causes a negative impact on the environment, and thus, proper disposal of red mud presents a huge challenge where alumina industries are installed.

Current technology allows the red mud to be stored harmlessly for now, although there is the possibility of leaks and floods, as seen in Western Hungary (Winkler, 2014). In October 2010, Hungary witnessed a major industrial disaster when the reservoir wall of the Ajka red mud depository failed, spilling one million cubic metres of highly caustic bauxite residue. The collapse, which was the single largest environmental release of red mud in history, led to 10 deaths and over 120 injuries, polluting hectares of agricultural land and the river systems (Winkler, 2014). As a result of this disaster, there are now extremely strict conditions on the development of new areas for the disposal of this waste. There is therefore little doubt that there will be no alumina producer who can afford to maintain giant storage pits for these residues in the long-term future.

The aluminium industry has long tried to find ways to recycle the environmentally problematic red mud, and research has been undertaken by various entities on the ways to contain and utilize the red mud. However, since the bauxite composition differs greatly depending on the deposit, there is no single established bauxite residue processing technology that can be applied at all the alumina refineries in the world.

7.5.4.1 Characterization of the Red Mud

The red mud, given its name due to the colour of the original bauxite ore and the iron oxide it contains (up to 60% of the mass of the red mud), is largely a mixture of solid alkaline waste (pH ranging from 10 to 13) due to the sodium hydroxide used in the process.

The red mud, according to the production process of the aluminium, can be divided into Bayer process red mud, sintering process red mud and combined-process red mud (Liu et al., 2007, 2014). The red mud from the Bayer process gives a significantly higher iron and alumina content compared to that obtained from the sintering process. The alkalinity of the material is strong enough to kill plant and animal life and to cause burns and damage to airways if the fumes are breathed. It is believed that the main damages in the Ajka Western Hungary disaster arose from the high pH of the mud slurry, which was responsible for both severe chemical burns to humans and animals as well as killing specimens in the rivers and contaminating soils (Liu et al., 2014). In order to safely dispose of red mud, it is crucial to wash, dewater and neutralize before it is discharged. The first step of washing the residue helps to recover valuable caustic soda and dissolved alumina. The caustic soda is recycled back into the digestion process, reducing production costs and, in turn, lowering the alkalinity of the residue.

Red mud is an extremely fine particle material with an average particle size less than 100 μm depending on the grain size of the primary minerals (Liu and Li, 2015). Due to its fine-grained nature, the material is hazardous to human health, causing respiratory and cardiovascular problems, and is highly mobile in aqueous environments.

The composition and properties of red mud, especially the trace minerals, may vary within wide ranges due to the bauxite origin and technology utilized in the alumina processing plants. The most abundant gangue minerals in bauxite are free silica, for example, quartz, bound silicon oxide (in the form of aluminosilicates, commonly kaolinite), iron oxides (hematite, magnetite, hydrohematite, goethite, limonite) and titanium in the form of rutile, anatase or brucite, of which anatase is the commonest (World Aluminium Report, 2015). These gangue minerals report to the red mud. The composition of the red mud is therefore a mixture of compounds originally present in the parent mineral, bauxite and of compounds formed or introduced during the Bayer process. Not infrequently, the variations are notable within one and the same deposit (Liu and Li, 2015). Table 7.2 shows the typical composition of red mud, and the chemical analysis indicates that it is mainly composed of the oxides of Si, Al, Fe, Na, Ca and Ti.

Like most ores and soils, bauxite can also contain trace quantities of metals such as arsenic, beryllium, cadmium, chromium, lead, manganese, mercury, nickel and naturally occurring radioactive materials such as thorium and uranium. Most of these trace elements remain with the residue after extraction of the alumina as they are not present in sufficient amounts for their extraction to be economic, or their extraction simply does not align with the company's strategy.

7.5.4.2 Utilization of Red Mud

The utilization of red mud as a secondary resource can be useful to reduce management cost. The key focus in the application of red mud should be its bulk utilization with

TABLE 7.2

Distribution of Mineral Phases in Red Mud

Component	wt.%	Source Point
Al_2O_3	10–20	Un-extracted aluminium oxide
Fe_2O_3	30–60	Impurity present in bauxite gives the red colour of the mud
SiO_2	3–50	Present as sodium- or calcium-alumino-silicate
TiO_2	Trace–25	Impurity present in bauxite
Na_2O	2–10	Added as NaOH for leaching, responsible for the high (alkaline) pH and the chemical burns
CaO	2–8	Added in digestion process to enhance alumina extraction and minimize soda losses in red mud

Source: World Aluminium Report, Bauxite residue management: Best practice, 2015, Available at www.bauxite.world-aluminium.org, Accessed November 2015.

minimal pretreatment or processing steps on the red mud itself due to the large quantities generated. A complete utilization of red mud is a more economically feasible solution to achieve zero waste and fully utilize the resources hidden in the waste. The classification as hazardous or nonhazardous is also important to the economics and the possible exclusion in certain applications (World Aluminium Report, 2015). Further points to consider would be the consumption or utilization of the red mud near the point source or plant site to avoid transportation costs and handling difficulties. Transportation costs for delivering red mud to a processing plant and other technical difficulties normally diminish the possibility of its application (Klauber et al., 2009, 2011; Liu and Naidu, 2014).

The various uses of red mud include metallurgical ones in the production of iron and steel, titania and alumina and the recovery of minor metallic constituents such as rare earth elements (Liu and Li, 2015). It has also been used in the building and construction industry (bricks, lightweight aggregates, roofing and flooring tiles, cements, etc.), in catalysis, ceramics (pottery, sanitary ware, special tiles and glasses, glazes, ferrites) and other miscellaneous direct uses, for example, in wastewater treatment, as a filler and as a fertilizer in agricultural industries (Klauber et al., 2009, 2011; Samal et al., 2013; Liu and Naidu, 2014).

The subsequent sections discuss some of the research that has been undertaken with regard to the potential utilization of the red mud in different industries.

7.5.4.3 Metallurgical Utilization through Metal Recovery

Bauxite residue contains a number of valuable elements and several interesting minor or trace elements. Table 7.3 shows a summary of some of the processes that have been used for the recovery of major metals from red mud and the related references.

According to Liu and Li (2015), the options of recovery techniques should be made on the basis of economic feasibility and practicability rather than the recovery efficiency. As seen from Table 7.3, a lot of research in the past has focused on the recovery of iron, titanium, rare earth elements and any residual aluminium oxide, although the economics of most of the proposed processes are still not favourable. This is because single metal recovery is not always economical, with multiple recoveries of valuable metals needing to be undertaken in order to make the process economically feasible. Thus, the only way in which processing of bauxite residue for the production of major elements such as iron, alumina and titanium could possibly be viable is through the production of several metals in a single integrated process.

TABLE 7.3

Summary of Some of the Processes for Extraction of Metals from Red Mud

Metals	Process	Advantages	Disadvantages	Reference
Fe	Magnetic separation	Low energy input	Low efficiency due to fine particle size and coexistence of iron minerals with other minerals	
Fe	Hydrometallurgy Pyrometallurgy	Improved efficiency 1. A product with reduced metallic Fe or pig iron amenable to steelmaking is possible 2. Additional values can be added such as recovery of alumina, titanium, alkali, etc.	Introducing secondary pollutants 1. High energy input 2. Needs pre-removal of alumina and alkali oxides if red mud is mixed into blast furnace 3. Pure solid iron rich product is difficult to obtain due to the association between iron minerals and other oxides	Hammond et al. (2013)
Al	Hydrometallurgy	1. High efficiency 2. Green process to use bacterial leaching	Usage of organic acids involves high costs Bioleaching processes usually take a longer time	Vachon et al. (1994)
Ti	Incorporated with iron recovery by pyrometallurgy	Can be additional value to iron recovery	High energy input	
	Hydrometallurgy	High Efficiency	Introduces acid and solvent waste	
REEs	Various technique	1. Additional value to the existing process 2. Recovery to generate economic income	1. Generally complex process 2. Other chemicals needed	

Source: Liu, Y. and Naidu, R., *Waste Manag.*, 34, 2662, 2014.

7.5.4.4 Recovery of Iron

Since iron, present as oxides/oxyhydroxides, is usually the largest component, iron recovery has attracted major attention (Kumar and Premchand, 1998). The iron content can be wide ranging (4.5 up to 60%) depending on the bauxite used and the refining processes employed. According to Zhu et al. (2012), the recovery of the iron is possible for red mud containing >51% Fe, which is competitive with iron ore in its Fe content. The recycling process of iron from red mud can be divided into techniques such as magnetic recovery, pyrometallurgical processing and hydrometallurgical extraction through leaching methods.

Magnetic separation techniques: Since hematite and goethite are the major constituents of red mud, magnetic separation has been the most employed basic method for iron recovery from red mud. In this process, the iron minerals are separated from gangue minerals by high-gradient magnetic separation due to the magnetic susceptibility difference between iron/iron oxides and gangue minerals. Direct magnetic separation, under high- and low-intensity magnetic separation conditions has been utilized (Xiang et al., 2001; Jamieson et al., 2006; Li et al., 2011; Peng and Huang, 2011). Li et al. (2011) used a high-gradient superconducting magnetic separation (HGSMS) system to separate the extremely fine red mud particles (<100 μm) into the high- and low-iron content fractions. The HGSMS system is reported to be more efficient in separating fine magnetic particles in liquid suspension. The resulting magnetic product from these processes can then be used as a feed material for iron production or as a pigment for pottery or ceramic (Jamieson et al., 2006; Hammond et al., 2013). It should be noted that aluminosilicate residues, after the recovery of iron, are up to 60% of the total amount of red mud, and they also face the problem of storage and secondary pollution. This residual non-magnetic portion can be used in building materials or added back into the Bayer process to try and recover some of the aluminium.

Pyrometallurgical techniques: Pyrometallurgical techniques such as solid-state carbothermic reduction and smelting in a blast furnace have also been used for iron recovery. Direct reduction of red mud is possible for obtaining pig iron after magnetic separation. Guo et al. (2013) produced iron nuggets of about 96.52% Fe by directly reducing the carbon-bearing pellets of red mud and coal at 1400°C for 30 min. The nuggets so obtained are low in Si and Mn but have a high content of S and P. Direct reduction of Fe from Indian red mud using thermal plasma technology by smelting in a DC extended arc thermal plasma reactor with addition of fluxes and graphite has been shown by Jayasankar et al. (2012) and Zhang et al. (2013). In addition, the entire Indian alumina industry is exploring ways to produce pig and cast irons and alumina-rich slag by plasma smelting from the bauxite residue (Indian Bureau of Mines, 2014).

The 'Elgai' process has also been proposed for the recovery of iron from red mud (Kumar and Premchand, 1998). The process involves mixing red mud, charcoal and soda in a rod mill and pelletizing, roasting the pellets at 700°C–925°C, then leaching the roasted materials with water, followed by magnetic separation of residue for iron recovery. Treatment of the residue with carbon monoxide and hydrogen at <350°C, then addition of calcium chloride and heating to 530°C–600°C followed by treatment with water allowed the iron, titania and silica to be separated (World Aluminium Report, 2015). According to Hammond et al. (2013), a new concept of using red mud directly for iron making/smelting gives a lot of promise. However, the impurities existing in red mud, such as Al, S, P, Na, Ca, Mg and Si, hinder its realization in practice as the alumina content can cause problems for slag fluidity and the high alkalinity is not compatible with the refractory and alkali accumulation.

Hydrometallurgical techniques: Although various efforts have been made for improved iron recovery using hydrometallurgical techniques, the general iron recovery and reaction rates tend to be low, and there is also a tendency of introduction of secondary waste. Sulphuric acid has been investigated for dissolution of iron from red mud (Uzun and Gulfen, 2007; Debadatta and Pramanik, 2013). The work by Debadatta and Pramanik (2013), gave iron recoveries of about 47%. Yu et al. (2012) treated red mud using oxalic acid to form an Fe(III) oxalate solution. The Fe (III) oxalate solution obtained was then irradiated with UV light to produce Fe (II) oxalate precipitate with >90% transformation. Hydrothermal processes have also been used to recover both iron and aluminium (Uzun and Gulfen, 2007). In this work, red mud was calcined at different temperatures (200°C–1000°C), followed by leaching using sulphuric acid solution. The dissolution of iron was found to be faster than that of aluminium, giving 97.46% iron and 64.40% aluminium.

The bioleaching process for the recovery of metals from iron oxide– and sulphide-containing minerals is a well-established technique. In general, bioleaching involves contacting the feedstock material with a bacterial microorganism culture in order to oxidize the metal species, rendering them soluble in inorganic lixiviants. These microorganisms are mostly indigenously associated with several oxide mineral deposits and secrete exopolysaccharides, several proteins, enzymes and organic acids such as citric, acetic, formic and oxalic acids formed during their metabolism of carbohydrate substrates. The organic acids usually have the dual effect of increasing metal dissolution by lowering the pH and increasing the load of soluble metals by complexation/chelating into soluble organic–metallic complexes (Vasan et al., 2001). A wide range of bacteria that can produce organic acids for extracting iron from an insoluble phase are used for this purpose, with many of them being aerobic. Although bioleaching processes have been investigated and applied in several fields for iron removal from kaolin (He et al., 2011; Aghaie et al., 2012) and from bauxite (Anand et al., 1996; Papassiopi et al., 2010), no such efforts have been made for the recovery of iron from red mud. This is primarily due to the high pH of the red mud mixture, which is not favourable for bacterial growth. In addition, bioleaching suffers the disadvantages of slow kinetics and poor extraction. Therefore, despite the known advantages such as low capital and energy costs, no commercial applications have been reported for this extraction route.

7.5.4.5 Recovery of Aluminium and Titanium

Thakur and Sant (1983a,b) and Paramguru et al. (2005) have reviewed processes for the recovery of iron, aluminium and titanium, individually and in combination. The key objectives of the recovery of multiple major metals are to simultaneously maximize the value created and minimize the residual solids produced. Iron, aluminium and titanium have been extracted from Jamaican bauxite residues by calcining the residue with sodium carbonate and then dissolving out the resultant sodium aluminate with water. The iron is recovered by magnetic separation after conversion to a ferromagnetic state by a reduction process (World Aluminium Report, 2015). The bauxite residues have also been treated with carbon monoxide and hydrogen at temperatures <350°C, followed by the addition of calcium chloride and heating to 530°C–600°C. Subsequent treatment of the calcined product with water allowed the iron, titania and silica to be separated. In Romania, some 60–70,000 tonnes/year of bauxite residue from the Tulcea plant are used locally, in steel production as a titaniferous material that is fed into the furnaces to reduce the ilmenite consumption needed for the protection of the hearth refractories against erosion (World Aluminium Report, 2015).

7.5.4.6 Recovery of Rare Earth Elements

In recent times, demand for rare earths elements (REEs) has grown exponentially, owing predominantly to advances in modern technologies that utilize these elements. Although REEs are not scarce, few places exist with enough concentrations to mine profitably. Furthermore, they are difficult to isolate in purified form and require advanced technology to extract. REEs consist of 17 elements from the lanthanide group (15 metals) plus scandium and yttrium. Typically, REEs are key components of computers, liquid crystal displays and cathode ray tube monitors, hybrid automobiles, wind power turbines, magnets, television sets, energy-efficient light bulbs, sensors, GPS technologies, CD and DVD drives, digital cameras, most optic lenses, most communication devices and satellites. Traces of REEs have been found in red mud residues, especially those of lower iron content (Chuan-sheng and Dong-yan, 2012). More than 90% of the trace metal value in bauxite residue can be attributed to the presence of REEs, namely, scandium (Binnemans et al., 2013).

Extraction of minor elements from bauxite residue can only make sense if the process is economically viable. It is for this reason that attention has been largely focused on scandium, partly because it is relatively concentrated in some bauxite and also because it has been stated to have high commercial value in the international market (Tsakanika et al., 2004). The bauxite residue plant in Greece has relatively high scandium content, which is favourable in comparison to the naturally occurring scandium resources. Research by Ochsenkühn-Petropoulou et al. (2002) on the Greek bauxite residue based on a nitric acid leach followed by an anion exchange separation of the scandium was taken to pilot stage. Xue et al. (2010) explored the use of sulphuric acid for the extraction of scandium, and about 80% recovery was achieved. Wang et al. (2011) gave a review on the recovery of scandium from various resources, while Zhang et al. (2005) studied the extraction of scandium from bauxite residue using concentrated HCl at around 50°C, and about 80% recovery was achieved. Although HCl is the popular reagent of choice in the present technologies applied for the leaching of the REEs from the residue, it is highly acidic and corrosive in nature, necessitating high equipment cost for the extraction since most of the equipment should be treated by preservatives (Liu et al., 2014).

Jamaica is one of the largest exporters of bauxite in the world behind countries such as Australia, China, India, Brazil and Guinea. The concentration of REEs found in Jamaica's red mud deposits is significantly greater than what is known about other red mud sites around the world. Table 7.4 shows the typical REE composition of the Jamaican bauxite residue.

A number of studies designed to mitigate the negative effects of red mud have been undertaken by the bauxite companies that operate in Jamaica over the years. There was considerable interest in extracting the REEs from Jamaican bauxite residues during the 1980s, but when the market prices fell, most of the ventures were brought on halt. The extraction routes were typically based on acid extraction to leave most of the iron and titanium behind, followed by selective SX of the resulting liquor. In 1991, Fulford et al. developed a process for the extraction of REEs from Jamaican bauxite residue using sulphur

TABLE 7.4

REE Composition of Jamaican Bauxite Residue

Element	Composition (mg/kg)
Scandium	135
Lanthanum	500
Cerium	650
Neodymium	250
Samarium	65
Europium	15
Terbium	10
Ytterbium	30
Lutetium	5
Tantalum	10

Source: World Aluminium Report, Bauxite residue management: Best practice, 2013, Available at www.bauxite.world-aluminium.org, Accessed November 2015.

dioxide, whereby REEs were selectively dissolved while leaving iron and titanium substantially undissolved.

In the recent past, Japan's Nippon Light Metal Company Limited approached Jamaica with a proposal to use its capacity and technology to extract REES from the local red mud. In 2013, the company's researchers confirmed high concentrations of REEs in Jamaica's red mud (Table 7.4) and indicated that those elements could be extracted efficiently (World Aluminium Report, 2015). The company then entered into a Memorandum of Understanding with the Jamaica Bauxite Institute for the establishment of the pilot project to determine the commercial scope of the project. Since then, it has done chemical research and successfully extracted some REEs. If the pilot project is a success, Nippon Light Metal Company Limited hopes to extract 1500 metric tonnes of rare earth oxides annually.

In late 2012, Orbite Aluminae Inc. of Canada and Veolia Environmental Services, a France-based multinational specializing in waste management, also signed a joint agreement for the treatment and recycling of red mud in order to recover REEs and other components using a hydrochloric acid extraction route (World Aluminium Report, 2015).

Although both these mentioned ventures are facing challenges due to the low commodity prices prevailing in the current market, they however signify a very important development in the utilization of red mud as a potential source of REEs. This is also supported by the concerted efforts and activities in China in reclaiming these elements from bauxite residue, with several small-scale plants currently in operation. RUSAL, one of the world's largest alumina producers, with the support of the Russian Ministry of Science, is also currently testing large-scale bauxite residue REEs recovery programmes as well as for steel and concrete manufacturing (Mann, 2013). Russia has also investigated the treatment of bauxite residues with methods similar to those used for low-grade uranium ores, using direct extraction methods such as mineral acid leaching together with ion exchange separation of radioactive and valuable components. Using a resin-in-pulp method in sulphuric acid, they demonstrated the feasibility of recovering titanium, scandium, uranium and thorium from the residue produced at an alumina plant in Ukraine. This plant processes bauxite residue from a number of sources including Brazil, Guyana, Australia, Guinea and India (Pawlek, 2008).

Considering the revived interest on REEs due to supply concerns, novel research in alternative sources is set to continue. Recovery of REEs from discarded consumer products and other hardware is being heavily researched in Japan, Europe, the United States and elsewhere. The recovery of these elements from bauxite residue could form a valuable part of this scheme. The advantage over traditional rare earths mining and processing methods is that processing of bauxite residue averts the problem of low concentration in a given mineral which tends to make the extraction non-commercially viable. However, although a lot of novel research has been documented in the recent past, application at commercial scale has been limited because of economic or engineering reasons.

7.5.4.7 Utilization of Red Mud in Building and Construction Materials

Considerable research has been done on the utilization of red mud as a raw material for the production of a range of building and construction materials. The utilization of red mud in the construction industry includes the generation of materials such as bricks, stabilized blocks, lightweight aggregates and low-density foamed products.

When dewatered, compacted and mixed with a suitable binder, bauxite residue makes a good road-building material and has been used to construct haul roads. In the south of France, bauxite residue termed Bauxaline, produced from Gardanne alumina refinery, has been used in the construction of several roads and platforms (World Aluminium Report, 2015). In Australia, Alcoa has used bauxite residues to construct the Perth to Bunbury Highway, which opened in 2009. A successful pilot project of a road embankment construction using Greek bauxite residue has been carried out by the Highway Engineering Laboratory of the Aristotle University of Thessaloniki, Greece (Fontini, 2008). These options of bauxite red mud usage are an attractive option with a high potential for large-volume use.

The presence of 4%–5% of alkali in red mud provides good fluxing action resulting in good plasticity and better bonding in the bricks. The red mud can also be applied in the cement industry for making cements, special cements, additive to cements and mortars, concrete construction and repair of roads, pavements, etc. Since there is less pozzolanically active substance in the Bayer red mud, it is not very feasible to use it for construction materials directly. In general, only 3%–5% red mud can be mixed with other raw materials (Liu et al., 2009). Although this does not allow for the full utilization of the red mud considering the huge amount of production, it still goes a long way to mitigating the environmental impact of the waste material.

7.5.4.7.1 Brick Making

Red mud has been popularly applied in the manufacture of bricks. The composition of the red mud gives bricks a pale brown, orange or golden yellow colour depending on the composition of the raw material and the firing temperatures. There are two distinct approaches to brick and block manufacture from the fine fraction of bauxite residue depending on whether the material is kiln fired or not. Conventional clay bricks are fired at temperatures ranging from 1000°C to 1500°C to achieve the required strength. Unfired bricks are made by including cement or other inorganic or organic binders. The most popular agents are calcium-containing compounds, most often quicklime, limestone, gypsum or organics (Thakur and Sant, 1983a). The Institute of Shandong Aluminum Company and the Institute of Chinese Great Wall Aluminum Company separately achieved the production of non-steam–cured and non-fired brick using red mud and fly ash as raw materials (Sutar et al., 2014). Recently, the Indian Company has found that by adding siliceous materials, the alumina red mud can be effectively utilized in the production of bricks made by either hand moulding or extrusion (Samal et al., 2013). The so-produced bricks have properties similar to commercial building bricks.

The mechanical properties of structural brick and clay products made with red mud are highly dependent on three different starting points in terms of the raw material used. In cases where the red mud is used with no additives or extractions from the as-received state, a firing temperature range between 1000°C and 1100°C tends to be used (Thakur and Sant, 1983b). This method generally gives higher-porosity and lower-strength values compared to other methods. The use of mineral additives such as sand, calcium carbonate or fly ash, slag, volcanic ash, refractory clay, brick clay, amorphous silicic acid or boric acid to replace part of the red mud in the brick making process, has also been undertaken. In such cases, a firing temperature range of between 950°C and 1400°C is applied and considerably higher-strength and lower-porosity values are achieved (Thakur and Sant, 1983b). The third approach to manufacturing construction materials from red mud requires the extraction of Na_2O from the raw material before it is processed. This treatment enhances the resulting mechanical properties and gives lower water absorption ratios. However, adding and extracting minerals adds to the cost of the material, making it less desirable for use in brick manufacturing. For this reason, there is need for more research on the use of the red mud without additives or extractions.

Although bricks made from residue have been shown to be technically sound, there has been a limited uptake in the commercialization and industrial utilization of bricks. Issues such as the perception that bauxite residue bricks would be deployed in poor areas only and the uncertainties about their long-term stability, leaching and ionizing radiation properties have hindered their widespread utilization.

The use of bauxite residue in the production of bricks has been reported in countries such as Australia, China, India, Greece, Turkey and Jamaica (McCarthy et al., 1992; Kavas, 2006; Samal et al., 2013). In China, it is reported that approximately 10% of red mud produced is recycled for further metal extraction or utilized as a raw material for brick production (Peng et al., 2005). In Australia, bricks made from bauxite residue from Alcoa's alumina processing plant at Kwinana, Western Australia, have been used to build homes in the South West of Western Australia, according to a report in The West Australian, 1 February 2002. According to the report, the use of waste bauxite residue was tested in a building in the early 1980s. However, the Health Department rejected the building after tests registered radioactivity readings that bordered on the maximum acceptable radiation exposure levels for 19 h a day. The residue contained radioactive thorium and uranium.

In the 1990s, a project to build a sports pavilion using bricks made predominantly from bauxite residue from Ewarton was set up by the Jamaica Bauxite Institute and the Jamaican Building Research Institute (Pinnock, 1991). Bricks made by a silicate-bonded system and red mud pozzolanic cement gave good characteristics, and the building is still in use today. The trace radioactive material remaining in bauxite residue gave rise to some concerns in Jamaica, but an investigation showed that using 100% bauxite residue gave a dose equivalent to just over 2 mSv/year and was judged to be acceptable (Pinnock, 1991).

7.5.4.7.2 Cement

Bauxite is commonly used directly in cement making as a source of alumina and iron for the production of both Portland and calcium aluminate cements. However, raw material cost is a factor limiting increased usage of bauxite in cement applications; hence, if bauxite residues could be supplied at a lower cost instead, then their utilization rates could be increased. The key components of cost are processing and transportation, and the application would be more feasible and cost effective if the cement-making process was undertaken at residue point source. The residue is usually chosen as a source of alumina and iron in preference to other industrial by-products because of its low silica content. The aluminium and iron contents provide valuable benefits to the cement in terms of strength and setting characteristics of the cement, although the presence of sodium ions can be a problem. The residue can be mixed with additives such as gypsum and bauxite (Singh et al., 1996, 1997) in the preparation of special cements. The largest current usages of bauxite residues in cement production, with an estimated 400,000 ton/year, are recorded for Greece, Ukraine, Georgia, Moldova and Belorussia (World Aluminium Report, 2015).

Iron-rich, special setting cements with improved strength (when compared to Portland cement) have been made with levels of up to 50% bauxite residue from Renukoot, India, together with bauxite and gypsum (World Aluminium Report, 2015). In Japan, plant scale tests were initiated in 2003 to confirm the possibility of using dry bauxite residues in cement production (Japan Aluminum Organization, 2004). Residual bauxite was introduced as a raw material along with other raw materials such as lime, clay, silica, iron source, etc. The tests confirmed that residue-added cement, as well as mortar and concrete made from this cement, met the Japanese Industrial Standards. Aluminium producer Alcoa also has a process to carbonate red mud using CO_2 from industrial gas streams. The resulting 'red sand' is used to make cement and is also used in road construction.

7.5.4.8 ***Glass-Ceramics and Pigments***

The red mud residue can be used as a major component of the ceramic mix, as a minor pigment component or as a glazing agent (Puskas, 1983). The use of residue in this area has been covered in reviews by Klauber et al. (2011), Paramguru et al. (2005) and Thakur and Sant (1974). The high iron content and finely divided state of the red mud have attracted an interest as a pigment in a wide variety of materials. A wide range of formulations have been developed for various applications, including pots, household fittings, domestic and refractory tiles and building blocks, in which the bauxite residue may be used in combination with other by-products, such as fly ash and slags (Klauber et al., 2011). Additions to bricks of a few percent of high iron oxide bauxite residue have been done to reduce the cost of the raw materials and provide a uniform red colouration to the bricks.

The presence of silica, soda and calcium in bauxite residue is also beneficial in the formation of vitreous glazes. Up to 37% of bauxite residue addition is possible in the production of glazes in the manufacture of a range of domestic and industrial porcelain products, due to the formation of a range of sodium and calcium silicates and aluminosilicates (Klauber et al., 2011). Although the simpler technical route suggests that these areas are potentially attractive in developing economies, it does not mean that energy efficiency, product performance or product safety will necessarily follow. Pigment application is limited in tonnage, and any ceramics applications of residue as pigment or filler would likely employ only a minor tonnage (Yalcin and Sevinc, 2000).

7.5.4.9 ***Other Applications***

Catalyst: The use of red mud as a catalyst can be a good alternative to the existing commercial catalysts. Its properties such as the iron content in the form of Fe_2O_3, TiO_2 content, high surface area, sintering resistance, resistance to poisoning and low cost make it an attractive potential catalyst for many reactions. Work has been done confirming the ability of bauxite residue to act as a catalyst to suppress adverse coke formation in the processing of heavy hydrocarbonaceous feedstocks in petroleum refining. It has been employed as a catalyst for hydrogenation, hydrodechlorination and hydrocarbon oxidation (Sushil and Batra, 2008; Balakrishnan et al., 2009; Karimi et al., 2010; Oliveira et al., 2011).

The developments of bauxite residue as a catalyst on a commercial scale will, however, depend on finding applications in which it can be cost competitive compared to other alternatives. The performance of unmodified residue is generally poor compared to straight iron oxides or commercial catalysts (Sushil and Batra, 2008). Most applications require some prior treatment or activation of the residue to improve catalytic efficiencies. The activation procedures include size reduction, heat treatment, sulphidization and even acid addition to dissolve the iron and other minerals that are then precipitate in a more active form. However, the presence of sodium and calcium oxides causes problems of catalyst sintering at high temperatures. Thus, the red mud will not generally be able to compete on technical performance, so implementation would depend on low unit cost outweighing the disadvantage in unit performance for each particular application (Klauber et al., 2011).

Adsorbent in water treatment: In the past decades, much effort has been expended to utilize the components of red mud for bulk applications, especially in water treatment (Gupta et al., 2001; Gupta and Sharma, 2002; Brunori et al., 2005; Cengeloglu et al., 2007; Wang et al. 2008). However, it should be noted that red mud contains many metal ions and organics, which are released under certain conditions. A leaching test should be conducted before the mud is considered for water treatment applications. The red mud has also found application as a low-cost adsorbent for removal of pollutants from aqueous solutions. Two characteristics of bauxite residues contribute to this sequestration potential: (1) high alkalinity, which favours hydrolysis and precipitation of metals as hydroxides and carbonates, and (2) large concentration of iron, aluminium and titanium oxides (including hydroxides and oxyhydroxides), which provides surface sites for sorption reactions of metals and metalloids. Commonly used coagulants are Fe^{3+} and Al^{3+} compounds. Due to high contents of Fe and Al in red mud, this waste is thus, believed to be a promising material for coagulant production.

Australia-based industrial waste management firm Virotec has developed a process to neutralize red mud with copious amounts of seawater or brine (Ritter, 2015). The process uses a technology known as Bauxsol® and is generally used to remediate mining sites. The Bauxsol mixture is also used as filler to make bricks and as a sorbent to trap metals and phosphorus in wastewater (Clark et al., 2004). Similarly in Italy, residues from the Eurallumina plant were neutralized with seawater to give a material with good heavy metal absorption properties (World Aluminium Report, 2014). In some formulations, the bauxite residue was mixed with fly ash, which improved the absorption of arsenic. In Korea, pellets made by heat treatment of mixtures of bauxite residue, polypropylene, sodium metasilicate, magnesium chloride and fly ash at 600°C were shown to have good heavy-metal absorption properties, especially for lead, copper and cadmium (World Aluminium Report, 2014).

7.5.4.10 Summary

The ability to fully utilize all the components in the red mud can lead to increasingly sustainable red mud utilization. The performance of the residue in any particular application must however, be competitive with the alternatives in relation to quality, cost and risk. For example, residue sand as a building material must be competitive with existing resources of mined virgin sand; extraction of iron from residue must compete with established iron ore resources; otherwise there will be no driving force to substitute the traditional source with the residue material.

From the information presented in this section, the largest potential uses of red mud are seen in cement production, iron recovery, landfill restoration, road construction and building materials, most probably not as the major component but as a sizeable fraction to ensure that all fears regarding radioactivity are allayed. Although some of these applications have been commercialized, matching the annually rising tonnage with possible commercial applications continues to be a major challenge. Most of the applications are not capable of utilizing significant tonnages of bauxite residue. Other applications, such as REEs recovery might be economically very attractive, but have no impact on reducing the large volumes of bauxite residue created annually unless the recovery of iron, alumina and titanium is carried out as part of an integrated process. The ultimate goal should be to cleverly segregate the different materials within the red mud and utilize each component to its highest value, whether that is metal recovery, functional materials for higher-value applications, construction, chemicals industry and commodity usage. This approach on the use of red mud resources has a scope for creating new industrial synergies that can enhance the sustainability of the aluminium/alumina industries and many economies.

7.6 Precious Metals Production

Precious metals such as gold and PGMs exist in many industrial products and have wide commercial applications. Gold, silver and palladium are frequently used for electroplating of connectors and contacts because they have excellent corrosion resistance and high electrical conductivity. In addition, silver and gold are used in hybrid links and solders. Such metals also exist in vehicles exhaust catalysts. These precious metals can also be used for decorative purposes and in jewellery. The next section looks at the production of gold and PGMs as well as the potential value recovery from the waste material generated during the production process.

7.6.1 Production of Gold

Cyanide leaching has been the industry standard for gold processing for more than 100 years.

This process involves the dissolution of gold-containing ores in dilute cyanide solution in the presence of lime and oxygen. Other complexing agents such as thiourea and halides such as chlorides have been used to extract gold from ores, but these are not generally cost effective and also present environmental and health concerns (Marsden and House, 2006). Cyanide complexes are more stable and effective and do not require additional aggressive chemicals to affect gold recovery.

Once the gold has been dissolved in the cyanide, it is then extracted from the pregnant leach solution. The common processes for recovery of gold from the leach solution include carbon adsorption, ion exchange, Merrill Crowe and EW. Traditionally, the Merrill Crowe process was used to precipitate gold from a cyanide solution by using zinc dust. Nowadays, the use of AC has proven to be very popular and successful.

Activated carbon has a remarkable affinity for gold and, under the right conditions, can become loaded with gold up to a level of more than 30% of its mass (typically 1000–4000 g/ton) (Marsden and House, 2006). The major advantage of carbon-in-pulp recovery over Merrill–Crowe recovery is the elimination of the solid–liquid separation unit step. After the loading step, the loaded carbon is removed from the slurry, and the adsorbed gold is stripped out at high temperature and pressure using sodium hydroxide and cyanide solutions to form a high-value electrolyte solution. Gold bullion is then recovered from the electrolyte by EW. In recent years, some operations have started replacing carbon with resin as the phase that gold is adsorbed on. The process used in resin absorption is similar to carbon adsorption, but synthetic spherical resin particles replace the grains of AC (Fleming and Cromberge, 1984; Kotze et al., 1993; Green et al., 2002; Pilśniak-Rabiega and Trochimczuk, 2014).

In order to ensure good leaching kinetics and high overall gold recovery, it is always necessary to add more cyanide than will be consumed in the leaching process. This excess cyanide reports to the tailings as uncomplexed or free cyanide (Fleming, 2003). In the recent past, rising costs associated with cyanide consumption, strict environmental regulations, the need to ensure sustainable operations and the growing negative perception of the general public for the gold mining industry due to several highly publicized cyanide spills have led to an increased interest in technologies that can be used to recover and recycle a portion of cyanide. The cyanide recovery brings a number of important benefits such as (Fleming, 2003)

- Reduced detoxification costs for tailings.
- Cost-effective recycling of cyanide that reduces the overall yearly requirements associated with purchase and delivery of new cyanide. However, the cost of recovering and recycling cyanide from tailings should generally be lower than the cost of purchasing new cyanide.
- Increased revenue as a result of the creation of saleable by-products such as copper and zinc sulphides.
- Improved compliance with environmental regulations.

The next section looks at some of the techniques that have been developed to recover and recycle cyanide from leach tailings.

7.6.1.1 Recovery and Recycling of Cyanide from Gold Tailings

The Sulphidization–Acidification–Recycle–Thickening (SART) process (Figure 7.12) is one of the techniques that have been developed to treat gold plant tailings that contain high concentrations of copper cyanide (Adams et al., 2008; Hedjazi and Monhemius, 2014).

The cyanide associated with the copper cyanide complexes is released by Sodium hydrogen sulphide (NaHS) dosing to precipitate copper and zinc and convert cyanide to hydrogen cyanide (HCN) gas under weak acidic conditions, thus allowing it to be recycled back to the leaching process as free cyanide. However, the chemical sulphide (NAHS) can be replaced by biogenically produced hydrogen sulphide (H_2S), which has the added advantage of lowering the acid demand for copper and zinc cyanide treatment. In addition, the overall process cost is lower as the cost of production of H_2S through the bioprocess is much lower. Metals such as copper and zinc are converted to high-grade sulphides, the solids are thickened and filtered and the liquor is neutralized and recycled to the leaching

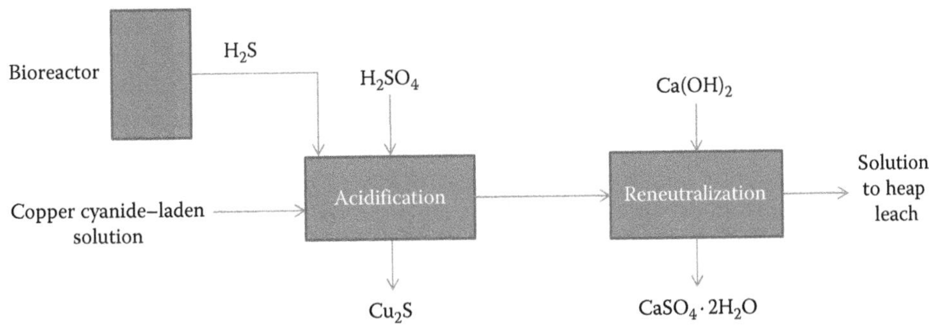

FIGURE 7.12
The SART process for copper recovery and recycle of cyanide-bearing solution to heap leach, with biogenic sulphide replacing chemical sulphide. (From Adams, M. et al., *Miner. Eng.*, 21(6), 509, 2008.)

step (Fleming, 2003; Adams et al., 2008). Cyanide recycling allows the leaching circuit to be operated at higher cyanide levels thus, maximizing the leaching efficiency and minimizing the copper and zinc deportment to the gold EW stage.

One other significant advantage of the process is that revenue can be generated from the sale of high-grade copper or zinc sulphide precipitates. This revenue can offset the operating costs of the cyanide recovery plant in most cases, and hence, the technology has the potential to transform uneconomic ore bodies into viable mines (Fleming, 2010).

The Hannah process, on the other hand, has been used in situations where direct acidification of the tailings could result in high operating costs due to the presence of high acid consumers (Fleming, 2010). This process uses strong resin technology to extract free cyanide radicals, as well as metal cyanide complexes, from gold tailings. Normally, conventional commercial strong-base resins are used in this application.

The Metallgesellschaft Natural Resources (MNR) process is suitable for tailings containing copper cyanide complexes (Stewart and Kappes, 2011). In the MNR process, sulphide ions are added to the cyanide solution and the acid. The process gives very high recovery rates, up to 100% of copper-complexed cyanide. The copper is recovered as a saleable sulphide precipitate. The limitation is that this is a relatively new technology that has not yet been extensively proven.

The Acidification, Volatilization and Reneutralization (AVR) was the first process used to recover cyanide from gold tailings and, more recently, pulps (Fleming, 2003; Stewart and Kappes, 2011). During cyanide leaching, pH is always carefully controlled and maintained at a high level to avoid the production of highly toxic HCN. The AVR process involves acidification of the gold plant tailings with sulphuric acid, to lower the pH from ~10 to less than 7, usually to pH 3 to 5. During this acidification process, free cyanide and weak complexed metal cyanide are converted to HCN, which is then volatilized in a sealed system by passing a stream of air bubbles through the solution or tailings pulp. The air/HCN gas stream is then treated using lime slurry. This converts the HCN back to free cyanide ions for recycling. The AVR process is the only process involving HCN generation that has been applied successfully in the mining industry, with high rates of cyanide recovery and a wealth of experience gained from several operating plants around the world.

Most commercial processes are effective for the destruction of cyanide, but not all have showed an affinity to thiocyanate destruction. The biological degradation of thiocyanate and cyanide in effluents provides an alternative treatment to the more traditional processes. The Activated Sludge Tailings and Effluent Remediation (ASTER™)

process can effectively treat solutions containing high levels of cyanide and thiocyanate using microorganisms that include bacteria, actinomycetes and fungi (van Zyl et al., 2011, 2015). The degradation process relies on a consortium of aerobic microorganism that utilizes cyanide, thiocyanate and certain cyanide metal complexes as a nitrogen and carbon source. Some of the major advantages of the process involve a robust technology suited to remote locations, low operating costs and low levels of skill required for process monitoring. To date, two commercial ASTER™ plants are in operation, one at the Barberton Mines Limited Consort plant in South Africa and the other at the Nordgold's Suzdal plant in Kazakhstan.

Another potentially very attractive alternative approach to precipitation and acidification processes is the use of membrane technology (EC, 2009). The membrane can be applied where the efficacy of cyanide use is hindered due to the presence of copper and other metals such as zinc and silver in the gold ore. The presence of these metals causes an increase in the consumption of cyanide, a lowering of the gold recovery efficiency and also poses a heightened environmental management issue for the tailings. This approach involves the integration of the industry's well-tested and tried membrane and EW technologies allowing for the recovery of metallic copper and the simultaneous liberation of free cyanide from the copper cyanide complexes. The process consists of three parts (EC, 2009):

- A solids removal step to provide a clean liquor for subsequent processing.
- A membrane step that concentrates the copper cyanide complexes. This step also recovers a portion of the free cyanide.
- A metal recovery unit that deposits the copper electrolytically, thereby liberating a portion of the WAD-cyanide as free cyanide.

This process has a number of benefits such as lower reagents consumption compared to the traditional cyanide destruction processes. The cyanide that would otherwise be lost to the circuit is recovered from the tailings and reused, reducing the cyanide inventory on site, and also the costs of purchasing the cyanide and the cyanide destruction. Further to this, copper metal can be recovered as a saleable by-product. The process has environmental benefits; the amount of cyanide and copper in the tailings stream is reduced significantly prior to cyanide destruction or disposal of the waste to tailings storage facilities. This process may be installed in the tailings circuit prior to discharge to the tailings pond or in the returned water circuit recovered from the tailings dam (EC, 2009).

7.6.2 Production of Platinum Group Metals

About 58% of world PGMs production occurs in South Africa whilst Russia accounts for a further 26%; most of this as a co-product of nickel mining. Nearly all of the rest come from Zimbabwe, Canada and the United States. Figure 7.13 shows an overview of the PGM production process.

In the process, the ore is first crushed, then milled into fine particles. This is followed by froth flotation resulting in the production of a concentrate that is dried and smelted in an electric furnace at temperatures over 1500°C. A matte containing the valuable metals is transferred to converters where iron and sulphur are removed. The PGMs are then separated from the base metals (nickel, copper and cobalt), and refined to a high level of purity using a combination of SX, distillation and ion exchange techniques. After the metal

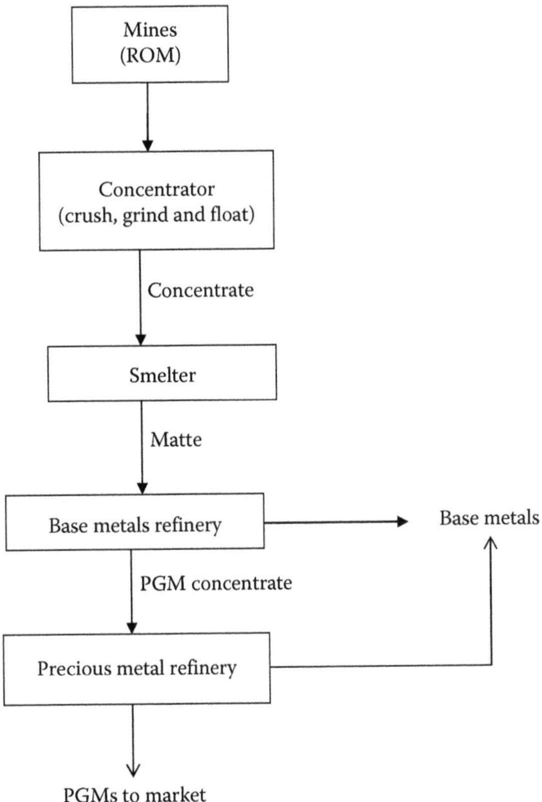

FIGURE 7.13
Overview of the PGM production process: from mine to market.

refining, however, there may be small amounts of PGMs that remain in solution at levels that are not economic to refine and are thus, lost to effluent streams. Likewise, the widespread use of noble metals such as PGMs as either heterogeneous or homogeneous catalysts for chemical processes generates substantial amounts of waste solutions or streams of various compositions. The discussion on the recovery of PGM from secondary sources such as spent catalysts is contained in Chapter 9. It must be noted, however, that the economical use of PGMs based catalysts almost always dependent on the efficient recovery of the catalyst, whether by recycling the catalysts themselves or by the efficient recovery and refining of the PGMs.

7.6.2.1 Recovery of PGMs from Refining Waste

Effluent that contains low concentrations of metals presents a problem in terms of both metal cost, particularly in the case of PGMs and the environment, in the case of base or heavy metals. All metal used in industry has a value, and this is particularly true of precious metals; hence, such losses can accumulate over time, which can be a significant cost to the company. Furthermore, because of limited availability from natural deposits, the improved recycling procedures for less-concentrated streams of PGMs from end-of-life process and low-level residual process solutions are needed to ensure sustainable supplies of these metals. The greater efficiency in the refining and recycling of precious metals reduces the need for obtaining new metals from mining and replaces processing steps in the conventional production processes that consume excess energy and pollute the environment. Therefore, a wide range of techniques such as precipitation, cementation, ion exchange and adsorption have been studied for the recovery and separation of platinum from dilute processes and waste solutions (Guibal et al., 1999; Amos et al., 2000; Saktas and Morcali, 2011; Izatt et al., 2012; Oke et al., 2014). Precipitation and cementation techniques tend to result in high inefficiencies, that is, they generally do not ensure complete purification; therefore, a further treatment is inevitably necessary. In addition, these processes tend to generate waste streams that carry significantly more contaminants than the original waste streams.

Ion exchange and adsorption technology can resolve difficulties associated with the recovery of precious metals from low-grade and waste streams by minimizing waste generation and improving the overall economics of the process. Processes involving adsorption have found widespread usage due to their associated cost-effectiveness and high efficiency. A wide range of adsorbents such as biomass and AC have been investigated (Das, 2010; Saktas and Morcali, 2011). Ion exchange can reduce the concentration of precious metals in the waste streams to extremely low levels (μg/L). It also allows relatively high upgrade ratios from the solution onto the resin without major water balance concerns (Yahorava and Kotze, 2014).

Smopex® and molecular recognition technology (MRT) are some of the recent innovations that are widely used as an option for recovery of valuable metals in the precious metals industry (Amos et al., 2000; Izatt et al., 2012). Johnson and Matthey developed the novel fibrous metal scavenger known as Smopex, which can selectively recover precious metal dissolved in solution, down to extremely low levels. The fibres can also be used to clean up base and heavy metals from waste streams, particularly in areas where legislation dictates limits on the metal content of the final discharge solution (Frankham and Kauppinen, 2010). The fibres consist of a polypropylene or viscose backbone (making them suitable for use in both organic and aqueous streams), grafted with specific functional groups that can selectively remove PGMs from the solution. These fibres are also thermally stable up to at least 120°C, so they can be used at elevated temperatures, where metal recovery is often easier.

MRT uses specially designed ligands that are useful for separating, recovering or purifying ions (Izatt et al., 2009, 2011). The ligands form coordinate chemical bonds with the ions to be separated. Large-scale MRT separation systems used in the refining of precious metals incorporate SuperLig® solid-phase particles such as silica gel or polymer substrates to which the selective ligand has been chemically attached. The process involves packing SuperLig beads into fixed-bed columns followed by passing the solution through the column while the target species is removed selectively from the solution.

While conventional separation technologies such as ion exchange, SX and precipitation generally utilize the differences between ions based only on a single parameter (e.g. charge, solubility, size), MRT differs from these processes in that it utilizes combinations of ion size, chemistry and geometry to selectively recognize and bind to specific ions in solutions (Izatt et al., 2012). The high selectivity of the process results in the MRT being able to bind ions present in highly acidic or basic conditions at very low levels even in solutions of high concentrations of competing ions.

MRT exhibits very rapid kinetics, high selectivity, efficiency of separation, simple elution chemistry and the ability to bring the concentration of the metal ions to extremely low levels. Furthermore, the SuperLig products applied to MRT have proven selective metal separation properties that are effective at low concentration levels (mg/L to µg/L) where traditional separation techniques are ineffective (Izatt et al., 2011). These advantages translate to lower capital and operating costs compared to conventional technologies (Izatt et al., 2009).

Several applications of MRT especially for residual and waste stream treatment are proven such as (Izatt et al., 2012):

- Recovery of platinum, palladium and rhodium from spent automobile catalytic converters. The Tanaka Kinkinzoku Kogyo, a leading Japanese precious smelter, uses SuperLig technology to purify precious metals in one of the world's largest automobile catalytic converter recycling process.
- Recovery of gold and copper from gold cyanide discharge streams, and recovery and recycle of the cyanide.
- Treatment of base-metals refinery waste streams and sludge.
- Recovery of iron, aluminium, zinc, copper and manganese from acid mine drainage and waste streams.

7.7 Strategic and Critical Metals

7.7.1 Production of Uranium

Uranium ore deposits are found in a number of countries with the largest resources being located in Australia, Canada, Kazakhstan, Brazil and South Africa. In the hydrometallurgical process for the production of uranium, the uranium ore is leached in a solution in a process commonly known as milling. The resulting liquor is then processed using a number of hydrometallurgical separation and concentration techniques to produce a uranium oxide (U_3O_8) concentrate, commonly known as a yellowcake. It should therefore be noted that the main product of the uranium hydrometallurgical processing plant is not pure elemental uranium but rather a uranium oxide concentrate. The uranium oxide is then packaged and sent for further processing before it can be used as fuel for a nuclear reactor. Figure 7.14 shows a flowsheet overview of the production of uranium oxide concentrate (Rossing Uranium, 2016).

The hydrometallurgical process for the production of the uranium oxide involves a number of steps such as leaching, ion exchange, SX and precipitation.

Leaching: Uranium can be extracted into solution by means of alkaline or acidic conditions in the presence of an oxidant. Uranium naturally occurs in the tetravalent (U^{4+}) and hexavalent (U^{6+}) states. However, the U^{4+} has a very low solubility in acidic and alkaline solutions; thus, U^{4+} must be oxidized to U^{6+} state, which has a higher solubility. Hence, the need for the maintenance of proper oxidizing conditions in order to achieve high uranium extraction.

The choice between the acidic or alkaline route depends on the process economics and these are largely influenced by the nature of the ore (Weil, 2012). Acidic leaching has the advantage of being more effective with difficult ores, requiring lower temperatures and leaching times compared to alkaline leaching. In the acidic leaching process, commonly applied in the Australian mines, the uranium content of the ore is dissolved in sulphuric acid in the presence of an oxidizing agent such as ferric sulphate. Other commonly used

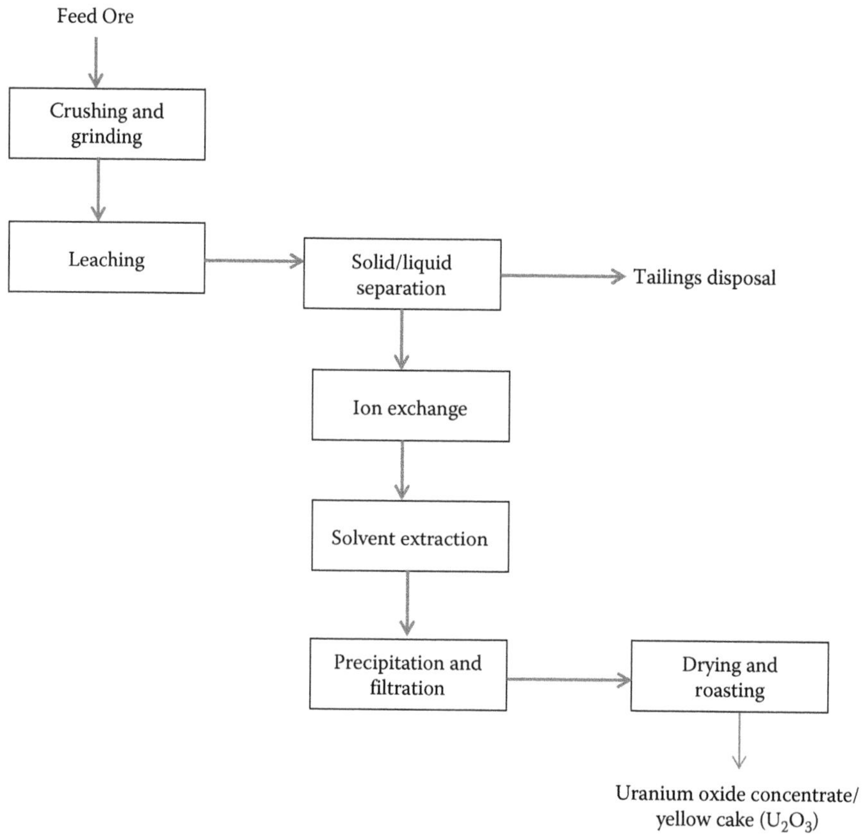

FIGURE 7.14
Simplified flowsheet for the processing of uranium ore to produce a uranium oxide concentrate. (From Rössing Uranium, Rössing Uranium's production of uranium oxide and the nuclear fuel cycle, 2016, Available at http://www.rossing.com/uranium_production.htm, Accessed September 2016.)

oxidizing agents include manganese dioxide and sodium chlorate. The end product from the leaching process is usually a highly soluble uranyl sulphate complex $[UO_2(SO_4)_3]^{4-}$.

Alkaline leaching on the other hand is applied when significant quantities of acid-consuming minerals are present, and therefore, acid leaching is not economic. Alkaline processing is commonly applied in the United States due to the presence of significant quantities of acid-consuming minerals such as gypsum and limestone in the host aquifers. Alkaline solutions tend to be more selective to uranium minerals, which means the solution tends to contain fewer impurities (Kweto et al., 2012; Weil, 2012). Consequently, the uranium oxide can be directly precipitated without purification. Furthermore, the solution is less corrosive and can be recycled without increasing impurity concentrations in the plant (Weil, 2012). Common alkaline leaching solutions are mixtures of sodium carbonate and sodium bicarbonate. Just as in acidic leaching, an oxidant has to be applied. This can be achieved by simply introducing oxygen or by bubbling air into the solution. The end product for the alkaline leaching process is usually a highly soluble uranyl tricarbonate complex, $UO_2(CO_3)_3^{4}$.

Purification and concentration: After leaching, the pulp is taken to the solid–liquid separation section. The uranium-bearing solution is sent to the purification and upgrading stages (ion exchange and SX), while the slime is washed, mixed with sand and pumped to the tailings area.

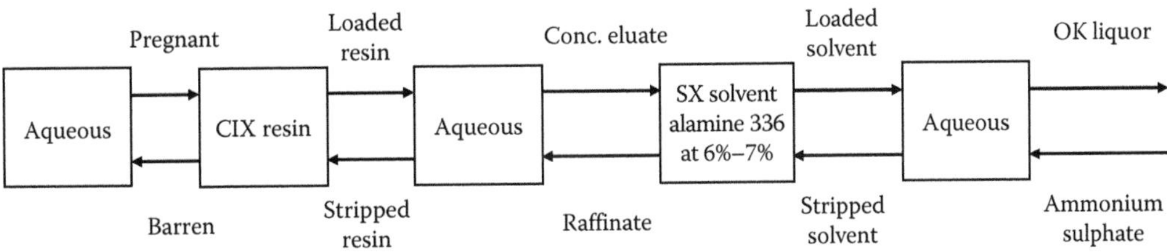

FIGURE 7.15
An integrated ion exchange and SX uranium production plant.

SX is used to treat the clarified acid liquor and has also been used after ion exchange as a second stage of purification in order to obtain very high-grade products (Figure 7.15).

The clear uranium-bearing solution is sent to the continuous ion exchange (CIX) plant where uranium ions are adsorbed onto the resin beads and preferentially extracted from the solution. The resins prefer ions with the smallest solvated volume, which in general increases with an increase in the atomic number. Relative affinity for anionic complexes of uranium in dilute solutions is therefore high. For the Rossing Uranium plant in Namibia (Lewis and Kesler, 1980; Cole and Sole, 2003; Sole et al., 2005) that uses a sulphuric acidic leach operation, the loaded resin from the CIX plant is sent to the elution stage where an acid wash is used to get the uranium ions back into solution. The product of the CIX plant is a uranyl sulphate solution. On the other hand, for the Langer Heinrich operation that uses a carbonate leach, the loaded resin is eluted with sodium chloride, producing a uranyl carbonate solution. The resin is regenerated with sodium bicarbonate solution before returning to the loading cycle.

The resulting eluate from the CIX section is further purified and concentrated in the SX section. During SX, the eluate from the ion exchange plant is mixed with an organic solution. The choice of the extractant depends on the lixiviant used in the leaching process, that is, alkaline or acidic. The common extraction systems used for the recovery of uranium from sulphate leach liquors are tertiary amines such as Alamine 336 or dialkyorganophosphorus acids such as D2EHPA. D2EHPA is used to selectively extract uranium from vanadium and iron (III) (Lewis and Kesler, 1980; Boydell and Viljoen, 1982; Lloyd, 1983; Nicol et al., 1987). Tri-n-butyl phosphate (TBP) is usually added to the organic phase to prevent the formation of a third phase.

However, amine extractants are now more widely used as they achieve higher purity than the organophosphorus systems due to the greater selectivity of amines for uranium. They also have lower extractant losses due to their lower aqueous-phase stability (Nicol et al., 1987). Uranium can be stripped from the loaded organic phase using a variety of reagents including NaCl, $(NH_4)_2SO_4$ and Na_2CO_3. The choice of the stripping reagent is usually determined by economic factors. The Rossing Uranium plant uses $(NH_4)_2SO_4$ as the stripping medium and the product is a uranyl sulphate (Lewis and Kesler, 1980; Sole et al., 2005).

Precipitation and yellowcake production: The stripped uranium liquor from the SX section is then sent to the precipitation plant where it is mixed with gaseous ammonia raising the pH and resulting in the precipitation of ammonium diuranate, which is then thickened to yellow slurry. The slurry is thickened further through a filtration process resulting in a yellow paste known as a yellowcake. This is then dried and roasted to drive off the ammonia leaving uranium oxide.

In situ leaching of uranium has been explored and commercialized by a number of uranium producing companies in the recent past. The chemistry of these processes is very similar to the extraction methods described earlier, except that the leaching process takes places underground and the leach solution is simply pumped into the ore zone containing the uranium ore. An alkaline or acidic leach process can also be utilized. The resulting leach liquor is then pumped to the surface for further processing. This process has the advantage of minimizing solid waste disposal as only the leach solution is pumped to the surface while the solid residue remains mostly in situ. In addition, most of the orebody's radioactivity remains well underground, and hence there is a minimal increase in radon release and no ore dust (Weil, 2012). However, the leaching chemicals used can cause long-term and potentially irreversible changes to groundwater quality if the processes are not well engineered and monitored. The use of acidic solutions mobilizes high levels of heavy metals such as cadmium, strontium, lead and chromium. Alkaline solutions tend to mobilize only a few heavy metals such as selenium and molybdenum. Thus, for these operations, remediation and restoration of the groundwater should be of paramount importance in the original design of the mining operations. The ore zone aquifer must be restored to pre-leaching conditions, and groundwater restoration is a very protracted and troublesome process, which is not yet completely understood (Diehl, 2011).

Uranium minerals are always associated with radioactive elements such as radium and radon in the ore, which are a result of the radioactive decay of uranium over a few million years. Uranium itself is barely radioactive; however, the ore from which it is mined, especially if it is of very high grade, needs to be handled with some care due to the presence of the radioactive components.

The hydrometallurgical process for the production of uranium oxide suggests that the residue/tailings in the plant are mainly generated from the solid–liquid separation or clarification steps that follow the leaching section. This residue ranges from slimes to coarse sand. A major portion of the radionuclides present in the ore remain in the tailings after uranium has been extracted. These can have a long-term impact on the environment, owing to the long half-lives and the ready availability of some of the toxic radionuclides (Weil, 2012). The spent ion exchange resins from the CIX plant are another source of residue that needs to be considered for safe disposal.

These waste materials are dumped in special ponds or piles, where they are normally covered by some metres of clay and topsoil with enough rock to resist erosion. This is to reduce both gamma radiation levels and radon emanation rates to levels near those normally experienced in the region of the orebody, and for a vegetation cover to be established. The largest of such piles in the United States and Canada contains up to 30 million tonnes of solid material. In Saxony, Germany, the Helmsdorf pile near Zwickau contains 50 million tonnes, and in Thuringia, the Culmitzsch pile near Seelingstädt contains 86 million tonnes of solids (Diehl, 2011).

Looking at current literature on uranium recycling and recovery, it can be noted that research has largely focused on the reprocessing of spent nuclear fuel (and not uranium production tailings) to recover fissile and fertile materials (IAEA, 2007; WNA, 2014). The commercial recovery and purification of uranium from spent fuel is performed worldwide using the PUREX process or modified versions of it (WNA, 2014). This process separates uranium and plutonium very effectively. The process involves dissolving the fuel elements in concentrated nitric acid. Chemical separation of uranium and plutonium is then undertaken by SX steps.

The PUREX process and the other various modifications such as the UREX and COEX processes are all based on the use of nitric acid in the leaching process (WNA, 2014). In 2015, Stassen and Suthiram (2015), however, investigated the potential use of ammonium carbonate–based process in the recovery of uranium from a residue generated by an alkaline dissolution process used in the production of the medical isotope 99Mo. Their test works showed that complete recovery of uranium from the residue could be achieved with three successive ammonium carbonate/peroxide leaches. The motivation for this approach is that a carbonate dissolution process offers advantages in that there is minimal metal corrosion and NOx gas emissions. There is also reduced release of volatile fission products during dissolution (Stassen and Suthiram, 2015). The uranium carbonate leach solution can then be converted to a HNO_3 solution for final U purification, using a steam-stripping precipitation process described by Kweto et al. (2012).

Field site phytoremediation has also been carried out at a former uranium mining area in East Germany where dump leaching of U and radionuclides was carried out in the past (Willscher et al., 2013). This treatment method involves extraction and/or immobilization of the contaminants which contributes to the improvement of erosion control, soil chemical ecological biodiversity and the chemical quality of the soil (Miller, 1996; Maegher, 2000). In the test plant experiments by Willscher et al. (2013), triticale, *Helianthus annuus* and *Brassica juncea* were used to investigate the influence of biological additives and soil amendment strategies on biomass production and plant tolerance to heavy metals. The microbial and soil amendment strategies resulted in a reduction of the concentration of contaminants in the seepage waters, as well as of the seepage water rates and loads. The tests also showed that the harvested plant biomass from phytoremediation of the test field site could be subsequently utilized for the winning of bioenergy.

7.7.2 Production of Rare Earth Elements

Despite their name, REEs are relatively plentiful in the earth's crust. However, because of their geochemical properties, REEs are typically dispersed and not often found concentrated in economically exploitable ore deposits (Haxel et al., 2006). It was the very scarcity of these minerals that led to the term 'rare earth elements'. Most production of REEs is from monazite and bastnasite ores. The REEs are extracted from these ores by sulphuric and hydrochloric acid digestion. Thereafter, processes such as fractional crystallization and precipitation, SX, ion exchange and reduction are used to recover the metals from the leach solutions.

It must be noted, however, that the production of REEs is often associated with environmental issues. For example, mining, leaching, pre-concentration and the numerous separation stages are needed to achieve the degree of purity required in certain applications, sometimes as high as 99.999% (US Department of Energy, 2009, 2011). This often leads to high amounts of secondary waste that have a negative impact on the environment. Once all processing has been undertaken, approximately 92% of the entire product mix has no commercial value and, therefore, needs to be disposed of (US Department of Energy, 2009, 2011). This, as previously indicated, is due to the fact that REEs are typically dispersed and not often found concentrated; hence, the large amount of gangue produced. While this waste mix is often considered hazardous, containing slightly radioactive thorium and uranium and other toxic substances, it is usually dumped into a tailings pond where, because it is still in liquid form, it is able to leach into groundwater. In 2012, however, a rare earth mining firm, Molycorp Inc., announced that it is attempting to harvest REEs in an environmentally friendly manner (Currie, 2015). The company claims that instead of dumping tailings, its process enables the firm to reuse the water by pressing it out of the tailings. The final product is no longer a contaminated liquid but rather a paste that can be added to cement and laid out at a disposal site.

The recovery of rare earth elements from secondary resources is discussed in more detail in Chapter 9 where end-of-life products and electronic waste are considered.

7.8 Final Remarks

Hydrometallurgical processes generate a significant amount of waste through different metal extraction processes. While most of these hydrometallurgical operations are inherently limited in time, the environmental legacies they leave behind can last for a large number of years. Although research into recycling of such materials for metal recovery and the generation of other valuable products has been widespread, large-scale production plants dedicated solely to the processing and utilization of the waste material have not been completely successful. Most of the processes are highly capital intensive, and success would rely on a consistent and uniform supply of waste and a market for the products that are competitive with similar products from primary raw materials. As a result, only a few commercial applications exist, and these are largely dedicated to large-volume, high value waste that require minimal processing before utilization in a secondary or alternative industry. A more detailed analysis of the challenges and future outlook in terms of the value recovery from metal extraction and processing waste will be given in the concluding chapter; Chapter 10.

References

Abia, A.A., Horsfall, M.Jr. and Didi, O. (2002). Studies on the use of agricultural by-products for the removal of trace metals from aqueous solution. *Journal of Applied and Science Environment* 6: 89–95.

Abia, A.A., Horsfall, M.Jr. and Didi, O. (2003). The use of chemically modified and unmodified cassava waste for the removal of Cd, Cu and Zn ions from aqueous solution. *Bioresource Technology* 90: 345–348.

Adams, M., Lawrence, R. and Bratty, M. (2008). Biogenic sulphide for cyanide recycle and copper recovery in gold-copper ore processing. *Minerals Engineering* 21(6): 509–517.

Aghaie, E., Pazouki, M., Hosseini, M.R. and Ranjbar, M. (2012). Kinetic modeling of the bioleaching process of iron removal from kaolin. *Applied Clay Science* 65–66: 43–47.

Aguilar, M. and Cortina, J.L. (2008). *Solvent Extraction and Liquid Membranes*. CRC Press, London, UK.

Ahalya, N., Ramachandra, T.V. and Kanamadi, R.D. (2003). Biosorption of heavy metals. *Journal of Chemistry and Environment* 7: 71–79.

Ally, M.R., Berry, J.B., Dole, L.R., Ferrada, J.J. and Van Dyke, J.W. (2001). Economical recovery of by-products in the mining industry. A report produced by the Engineering Science and Technology Division, Nuclear Science and Technology Division and the Environmental Sciences Division, US Department of Energy. Oak Ridge National Laboratory, Oak Ridge, Tennessee. 39 pages.

Amos, G., Hopkins, W., Izatt, S.R., Bruening, R.L., Dale, J.B. and Krakowiak, K.E. (2000). Extraction, recovery, and recycling of metals from effluents, electrolytes, and product streams using molecular recognition technology. Plenary Lecture, *V International Conference on Clean Technologies for the Mining Industry*, Santiago, Chile, May 9–13, 2000. In: Sánchez, M.A., Vergara, F. and Castro, S.H. (eds.), *Environmental Improvements in Mineral Processing and Extractive Metallurgy*, Vol. 2. University of Concepción, Concepción, Chile, pp. 1–16.

Anand, P., Modak, J.M. and Natarajan, K.A. (1996). Biobeneficiation of bauxite using Bacillus polymyxa: Calcium and iron removal. *International Journal of Mineral Processing* 48: 51–60.
Arslan, C. and Arslan, F. (2003). Thermochemical review of jarosite and goethite stability regions at 25 and 95°C. *Turkish Journal of Engineering and Environmental Sciences* 27: 45–52.
Atkinson, B.W., Bux, F. and Kasan, H.C. (1998). Considerations for application of biosorption technology to remediate metal-contaminated industrial effluents. *Water SA* 24(2): 129–136.
Aube, B. (2004). The science of treating acid mine drainage and smelter effluents. Available at http://www.infomine.com/publications/docs/Aube.pdf. Accessed June 2016.
Balakrishnan, M., Batra, V.S., Hargreaves, J.S.J., Monaghan, A., Pulford, I.D. and Rico, J.L. (2009). Hydrogen production from methane in the presence of red mud – Making mud magnetic. *Green Chemistry* 11(1): 42–47.
Beolchini, F., Fonti, V., Ferella, F. and Veglio, F. (2010). Metal recovery from spent refinery catalysts by means of biotechnological strategies. *Journal of Hazardous Materials* 178: 529–534.
Bhattacharyya, D., Jumawan, A.B. and Grieves, R.B. (1979). Separation of toxic heavy metals by sulphide precipitation. *Separation Science and Technology* 14: 441–452.
Binnemans, K., Pontikes, Y., Jones, P.T., Van Gerven, T. and Blanpain, B. (2013). Recovery of rare earths from industrial waste residues: A concise review. *Proceedings of the Third International Slag Valorisation Symposium: The Transition to Sustainable Materials Management, 2013,* Leuven, Belgium, pp. 191–205.
Bishop, M.D. et al. (1999). Investigation of evaporative losses in solvent extraction circuits. In: Young, S.K., Dreisinger, D.B., Hackl, R.P. and Dixon, D.G. (eds.), *Copper*, Phoenix, AZ, 10–13 October 1999. Canadian Institute of Mining, Metallurgy and Petroleum, Montreal, Quebec, Canada, pp. 277–289.
Boonstra, J. and Buisman, C.J.N. (2003). Biotechnology for sustainable hydrometallurgical extraction. In: Young, C., Alfantazi, A., Anderson, C., James, A., Dreisinger, D. and Harris, B. (eds.), *Electrometallurgy and Environmental Hydrometallurgy*, Vol. 2, pp. 1105–1119. TMS, Warrandale PA. Wiley Online Library
Bosio, V., Viera, M. and Donati, E. (2008). Integrated bacterial process for the treatment of a spent nickel catalyst. *Journal of Hazardous Materials* 154: 804–810.
Boydell, D.W. and Viljoen, E.B. (1982). Uranium processing in South Africa from 1961 to 1981. In: Glen, H.W. (ed.), *Proceedings of the 12th Council of Mining and Metallurgical Institutions Congress*. South African Institute of Mining and Metallurgy, Johannesburg, South Africa, pp. 575–582.
Braet, J., Charpentier, D., Centner, B. and Vanderperre, S. (2012). Radioactive spent ion-exchange resins conditioning by the hot supercompaction process at Tihange NPP – Early Experience – 12200. WM 2012 Conference, 26 February–1 March 2012, Phoenix, Arizona. pp. 1–10.
Brandl, H., Bosshard, R. and Wegmann, M. (2001). Computer-munching microbes: Metal leaching from electronic scrap by bacteria and fungi. *Hydrometallurgy* 59: 319–326.
Brunori, C., Cremisini, C., Massanisso, P., Pinto, V. and Torricelli, L. (2005). Reuse of a treated red mud bauxite waste: Studies on environmental compatibility. *Journal of Hazardous Materials* 117(1): 55–63.
Cabrera, G., Gómez, J.M., Hernández, I., Coto, O. and Cantero, D. (2011). Different strategies for recovering metals from CARON process residue. *Journal of Hazardous Materials* 189: 836–842.
Cengeloglu, Y., Tor, A., Gulsin, A., Mustafa, E. and Sait, G. (2007). Removal of boron from aqueous solution by using neutralized red mud. *Journal of Hazardous Materials* 142(1–2): 412–417.
Cheremisinoff, N.P. and Moressi, A.C. (1978). Carbon adsorption applications. In: Cheremisinoff, N.P. and Ellerbusch, F. (eds.), *Carbon Adsorption Handbook*. Ann Arbor Science, Ann Arbor, MI, pp. 1–53.
Chinnam, R.K., Francis, A.A., Will, J., Bernardo, E. and Boccaccini, A.R. (2013). Review. Functional glasses and glass-ceramics derived from iron rich waste and combination of industrial residues. *Journal of Non-Crystalline Solids* 365: 63–74.
Chuan-sheng, W. and Dong-yan, L. (2012). Mineral phase and physical properties of red mud calcined at different temperatures. *Journal of Nanomaterials* 2012: 6.

Claassen, J.O., Meyerb, E.H.O., Rennieb, J. and Sandenbergh, R.F. (2002). Iron precipitation from zinc-rich solutions: Defining the Zincor Process. *Hydrometallurgy* 67: 87–108.

Clark, M., McConchie, D., Berry, J., Caldicott, W., Davies-McConchie, F. and Castro, J. (2004). Bauxsol™ technology to treat acid and metals: Applications in the coal industry. *National Meeting of the American Society of Mining and Reclamation and The 25th West Virginia Surface Mine Drainage Task Force*, 18–24 April 2004. ASMR, Lexington, KY, pp. 293–332.

Cole, P.M. and Sole, K. (2003). Zinc solvent extraction in the process industries. *Mineral Processing and Extractive Metallurgy Review* 24: 91–137.

Coto, O., Galizia, F., González, E., Hernández, I., Marrero, J. and Donati, E. (2007). Cobalt and nickel recoveries from lateritic tailings by organic and inorganic bioacids. *Advances in Materials Research* 20–21: 107–110.

Coto, O., Galizia, F., Hernández, I., Marrero, J. and Donati, E. (2008). Cobalt and nickel recoveries from lateritic tailings by organic and inorganic bioacids. *Hydrometallurgy* 94: 18–22.

Council of Scientific & Industrial Research (2014). A composition for preparing nonhazardous building blocks and process for preparation thereof. Patent 259781, 27 March 2014.

Cui, J. and Zhang, L. (2008). Metallurgical recovery of metals from electronic waste: A review. *Journal of Hazardous Materials* 158: 228–256.

Ćurkovic, L., Cerjan-Stefanović, Š. and Filipan, T. (1997). Metal ion exchange by natural and modified zeolites. *Water Resources* 31: 1379–1382.

Currie, A. (2015). Rare earth green waste disposal: Here to stay. Available at http://investingnews.com/daily/resource-investing/critical-metals-investing/rare-earth-investing/rare-earth-green-waste-disposal-here-to-stay/. Accessed December 2015.

Dacera, D. and Babel, S. (2006). Use of citric acid for heavy metals extraction from contaminated sewage sludge for land application. *Water Science and Technology* 54(9): 129–135.

Damons, R.E. (2001). Development of an aspen model for the treatment of acid mine drainage. Thesis. Master of Technology, The Cape Technikon, Cape Town, South Africa.

Das, G. (2010). Recovery of precious metals through biosorption – A review. *Hydrometallurgy* 103: 180–189.

Davidson, H.W., Wiggs, P.K., Churchouse, A.H., Maggs, F.A. and Bradley, R.S. (1968). *Manufactured Carbon*. Pergamon, Oxford, UK.

Debadatta, D. and Pramanik, K. (2013). A study on chemical leaching of iron from red mud using sulphuric acid. *Research Journal of Chemistry and Environment* 17: 50–56.

Dew, D.W., van Buuren, C., McEwan, K. and Bowker, C. (1999). Bioleaching of base metal sulphide concentrates: A comparison of mesophilic and thermophilic bacterial cultures. In: Amils, R. and Ballester, A. (eds.), *Biohydrometallurgy and the Environment toward the Mining of the 21st Century: Part A. IBS '99*. Elsevier, Amsterdam, the Netherlands, pp. 229–238.

Diehl, P. (2011). Uranium mining and milling wastes: An introduction. Available at http://www.wise-uranium.org/uwai.html. Accessed 15 February 2016.

Dreisinger, D. (2006). Copper leaching from primary sulfides: Options for biological and chemical extraction of copper. *Hydrometallurgy* 83: 10–20.

Dreisinger, D. and Clucas, J. (2013). The starved acid leaching technology (SALT) for recovery of nickel and cobalt from saprolites and Caron plant residues. *2013 TMS Annual Meeting & Exhibition*, San Antonio, TX, 3–7 March 2013.

Dresher, W.H. (2001). How hydrometallurgy and the SX/EW process made copper the "green" metal. Available at http://www.copper.org/publications/newsletters/innovations/2001/08/hydrometallurgy.html. Accessed November 2015.

Dutrizac, J.E. and MacDonald, R.J. (1974). Ferric ion as a leaching medium. *Minerals Science and Engineering* 6(2): 59–100.

EC (European Commission) (2009). Reference document on best available techniques for the management of tailings and waste rock in mining activities. Available at http://eippcb.jrc.ec.europa.eu/reference/BREF/mmr_adopted_0109.pdf. Accessed 21 February 2016.

Ferron, C.J. (2003). Leaching of secondary copper minerals using regenerated ferric sulphate. SGS Minerals Services Technical Paper 2003-14. SGS Minerals Services, Lakefield, Ontario Canada.

Fleming, C.A. (2003). The economic and environmental case for recovering cyanide from gold plant tailings. SGS Minerals Services Technical Paper 2003-02. SGS Minerals Services, Lakefield, Ontario Canada.

Fleming, C.A. (2010). Cyanide management in the gold industry. *Mining Environmental Management* July: 26–27.

Fleming, C.A. and Cromberge, G. (1984). The extraction of gold from cyanide solutions by strong- and weak-base anion-exchange resins. *Journal of the Southern African Institute of Mining and Metallurgy* 84(5): 125–137.

Fletcher, A.W. and Gage, R.C. (1985). Dealing with a siliceous crud problem in solvent extraction. *Hydrometallurgy* 15: 5–9.

Fontini, K. (2008). An innovative geotechnical application of bauxite residue. *Electronic Journal of Geotechnical Engineering* 13/G: 1–9.

Fowler, T.A. and Crundwell, F.K. (1999). Leaching of zinc sulphide by Thiobacillus ferrooxidans: Bacterial oxidation of sulphur product layer increases the rate of zinc sulphide dissolution at high concentrations of ferrous iron. *Applied and Environmental Microbiology* 65(12): 5285–5292.

Frankham, J. and Kauppinen, P. (2010). The use of metal scavengers for recovery of precious, base and heavy metals from waste streams. *Platinum Metals Review* 54(3): 200–202.

Free, L.M. (2013). *Hydrometallurgy: Fundamentals and Applications.* John Wiley & Sons Inc., Hoboken, NJ.

Fu, F. and Wang, Q. (2011). Removal of heavy metal ions from wastewaters: A review. *Journal of Environmental Management* 92: 407–418.

Gan, X., Lin, M., Bao, L., Gao, Y. and Zhang, Z. (2008). Investigation of the immobilization of the radioactive ion exchange resins into specific cement using the mixture response surface approach. *Journal of Nuclear Science and Technology* 45: 1084–1090.

Gardea-Torresdey, J.L., Gonzalez, J.H., Tiemann, K.J., Rodriguez, O. and Gamez, G. (1998). Phytofiltration of hazardous cadmium, chromium, lead and zinc ions by biomass of *Medicago sativa* (alfalfa). *Journal of Hazardous Materials* 57: 29–39.

Gericke, M. and Pinches, A. (1999). Bioleaching of copper sulphide concentrate using extreme thermophilic bacteria. *Minerals Engineering* 12: 893–904.

Gnoinski, J., Bachmann, T. and Holtzhausen, S. (2005). Skorpion Zinc: Defining the cutting edge of zinc refining. *Proceedings of the International Symposium on Lead and Zinc Processing*, Koto, Japan, pp. 1315–1325.

Green, B.R., Kotze, M.H. and Wyethe, J. (2002). Developments in ion exchange: The mintek perspective (industrial insight). *Journal of the Minerals, Metals and Materials Society* 54(10): 37–43.

Guibal, E., Vincent, T., Larkin, A. and Tobin, J.M. (1999). Chitosan sorbents for platinum recovery from dilute solutions. *Journal of Industrial & Engineering Chemistry Research* 38(10): 4011–4022.

Guo, Y.H., Gao, J.J., Xu, H.J., Zhao, K. and Shi, X.F. (2013). Nuggets production by direct reduction of high iron red mud. *Journal of Iron and Steel Research, International* 20: 24–27.

Gupta, C.K. (2003). *Chemical Metallurgy: Principles and Practice.* Wiley-VCH Verlag GmbH & Co. KGaA, Weinheim, Germany, 831pp.

Gupta, V.K., Gupta, M. and Sharma, S. (2001). Process development for the removal of lead and chromium from aqueous solutions using red mud – An aluminium industry waste. *Water Research* 35(5): 1125–1134.

Gupta, V.K. and Sharma, S. (2002). Removal of cadmium and zinc from aqueous solutions using red mud. *Environmental Science & Technology* 36(16): 3612–3617.

Hait, J., Jana, R.K. and Sanyal, S.K. (2009). Processing of copper electrorefining anode slime: A review. *Transactions of the Institution of Mining and Metallurgy, Section C* 118: 240–252.

Hammond, K., Apelian, B.M.D. and Blanpain, B. (2013). CR3 communication: Red mud – A resource or a waste? *Journal of the Minerals, Metalals and Materials Society* 65: 340–341.

Hartmann, T. and Corbella, J. (2007). Tailor made crud treatment with 3 phase separating centrifuge. In: Riveros, P.A., Dixon, D.G., Dreisinger, D.B. and Collins, M.J. (eds.), *Copper*, Toronto, Ontario, Canada, 26–29 August 2007. Canadian Institute of Mining Metallurgy and Petroleum, Montreal, Quebec, Canada.

Haxel, G., Hedrick, J. and Orris, J. (2006). *Rare Earth Elements – Critical Resources for High Technology*. United States Geological Survey, Reston, VA. USGS Fact Sheet: 087-02. (PDF). Retrieved 13 March 2012.

He, Q.X., Huang, X.C. and Chen, Z.L. (2011). Influence of organic acids, complexing agents and heavy metals on the bioleaching of iron from kaolin using Fe(III)-reducing bacteria. *Applied Clay Science* 51: 478–483.

Hearne, T.M. and Haegele, R. (1998). Hydrometallurgical recovery of zinc from sulphide ores and concentrates. In: Dutrizac, J.E., Gonzalez, J.A., Bolton, G.L. and Hancock, P. (eds.), *Zinc and Lead Processing*. The Metallurgical Society of CIM, Montreal, Quebec, Canada, pp. 765–780.

Hedjazi, F. and Monhemius, A.J. (2014). Copper–gold ore processing with ion exchange and SART technology. *Minerals Engineering* 64: 120–125.

Heimala, S.O., Hyvarinen, J.O.V. and Tiitinen, H.A. (1975). Hydrometallurgical process for the recovery of valuable components from the anode slime produced in the electrolytical refining of copper. Patent US 4002544 A.

Hernández, I., Galizia, F., Coto, O. and Donati, E. (2009). Improvement in metal recovery from laterite tailings by bioleaching. *Advances in Materials Research* 71–73: 489–492.

Ho, Y.S., Wase, D.A.J. and Forster, C.F. (1995). Batch nickel removal from aqueous solution by sphagnum moss peat. *Water Research* 29: 1327–1332.

Huisman, J., Schouten, G. and Schultz, C. (2006). Biologically produced sulphide for purification of process streams, effluent treatment and recovery of metals in the metal and mining industry. *Hydrometallurgy* 83(1–4): 106–113.

IAEA (2007). Management of reprocessed uranium: Current status and future prospects IAEA TECDOC 1529 (2007), International Atomic Energy Agency, Vienna, Austria. Available at http://www-pub.iaea.org/MTCD/publications/PDF/te_1529_web.pdf. Accessed February 2016.

Indian Bureau of Mines. January 2014. *Indian Minerals Yearbook 2012 (Part II: Metals & Alloys)*, 51st edn. Indian Bureau of Mines, Ministry of Mines, Government of India, Nagpur, India.

Irish, E.R. and Reas, W.H. (1957). The Purex Process – A solvent extraction reprocessing method for irradiated uranium. Unclassified AEC Research and Development Report, USA Atomic Energy Commission, pp. 1–25.

Izatt, S.R., Bruening, R.L. and Izatt, N.E. (2012). Some applications of molecular recognition technology (MRT) to the mining industry. In: Wang, S., Dutrizac, J.E., Free, M.L., Hwang, J.Y. and Kim, D. (eds.), *T.T. Chen Honorary Symposium on Hydrometallurgy, Electrometallurgy and Materials Characterisation*. TMS, Warrendale, PA, pp. 51–63.

Izatt, S.R., Bruening, R.L., Izatt, N.E. and Dale, J.B. (2009). The application of molecular recognition technology (MRT) in the nuclear power cycle: From uranium mining and refining to power plant waste separation and recovery, as well as element analysis and isotope purification. *9075 WM2009 Conference*, Phoenix, AZ, 1–5 March 2009.

Izatt, S.R., Izatt, N.E. and Bruening, R.L. (2011). Review of selective separations of cobalt, uranium, zinc, nickel and associated contaminants from various process streams. *The Southern African Institute of Mining and Metallurgy, Sixth Southern African Base Metals Conference 2011*, 18–21 July 2011, Palaborwa, South Africa. pp. 221–235.

Jackson, D.A. (1976). (330 W. Diversey Parkway, Chicago, IL, 60657), Electrolytic zinc refining process including production of by-products from jarosite residues. Jackson, David A., assignee. United States Patent 3,937,658, 02 October 1976.

Jackson, E. (1986). *Hydrometallurgical Extraction and Reclamation*. Ellis Horwood Limited, Halsted Press, Chichester, UK.

Jamieson, E., Jones, A., Cooling, D. and Stockton, N. (2006). Magnetic separation of red sand to produce value. *Minerals Engineering* 19: 1603–1605.

Japan Aluminum Organization (2004). A field survey for the environmental impact assessment of disposal at sea of bauxite residue, International Maritime Organisation (IMO), Scientific Group (SG) – 27th Meeting, March 2004.

Jayasankar, K., Ray, P.K., Chaubey, A.K., Padhi, A., Satapathy, B.K. and Mukherjee, P.S. (2012). Production of pig iron from red mud waste fines using thermal plasma technology. *International Journal of Minerals, Metallurgy, and Materials* 19: 679–684.

Karimi, E., Briens, C., Berruti, F., Moloodi, S., Tzanetakis, T. and Thomson, M.J. (2010). Red mud as a catalyst for the upgrading of hemp-seed pyrolysis bio-oil. *Energy and Fuels* 24(12): 6586–6600.

Katsioti, M. (2005). Use of jarosite/alunite precipitate as a substitute for gypsum in Portland cement. *Cement and Concrete Composites* 27(1): 3–9.

Kavas, T. (2006). Use of boron waste as a fluxing agent in production of red mud brick. *Building and Environment* 41(12): 1779–1783.

Klauber, C., Gräfe, M. and Power, G. (2009). Review of Bauxite re-use options, CSIRO Document DMR-3609, CSIRO, pp. 1–79.

Klauber, C., Gräfe, M. and Power, G. (2011). Bauxite residue issues: II. Options for residue utilization. *Hydrometallurgy* 108: 11–32.

Kongsricharoern, N. and Polprasert, C. (1996). Chromium removal by a bipolar electrochemical precipitation process. *Water Science and Technology* 34(9): 109–116.

Konishi, Y., Nishibura, H. and Asai, S. (1998). Bioleaching of sphalerite by the acidophilic thermophile *Acidianus brierleyi*. *Hydrometallurgy* 47: 339–352.

Kotze, M.H., Fleming, C.A., Green, B.R., Steinbach, G. and Virnig, M.J. (1993). Progress in development of the Minix gold selective strong-base resin. In: Brent Hiskey, J. and Warren, G. (eds.), *Hydrometallurgy – Fundamentals, Technology and Innovation*. Society for Mining, Metallurgy and Exploration, Colorado, pp. 395–401.

Ku, Y. and Jung, I.L. (2001). Photo catalytic reduction of Cr(VI) in aqueous solutions by UV irradiation with the presence of titanium dioxide. *Water Research* 35: 135–142.

Kumar, R.S. and Premchand, J.P. (1998). Utilization of iron values of red mud for metallurgical applications. In: Bandopadhyay, A., Goswami, N.G. and Rao, P.R. (eds). *Environmental Waste Management*. National Metallurgical Laboratory, Jamshedpur, India. pp 108–119.

Kunda, W. and Veltman, H. (1979). Decomposition of jarosite. *Metallurgical Transactions B* 10B: 439–446.

Kurniawan, T.A., Chan, G.Y.S., Lo, W.H. and Babel, S. (2006). Physico–chemical treatment techniques for wastewater laden with heavy metals. *Journal of Chemical Engineering* 118: 83–98.

Kweto, B., Groot, D.R., Stassen, L., Suthiram, J. and Zeevaart, J.R. (2012). The use of ammonium carbonate as lixiviant in uranium leaching. *Proceedings of Uranium Sessions at Alta*, Perth, Western Australia, Australia, 31 May to 1 June 2012, pp. 139–149.

Lewis, A.E. (2010). Review of metal sulphide precipitation. *Hydrometallurgy* 104: 222–234.

Lewis, I.E. and Kesler, S. (1980). A case study – Optimising the performance of the uranium SX plant at Rossing Uranium Limited in Namibia. *Proceedings International Solvent Extraction Conference ISEC'80*, Vol. 3, Association des Ingenieurs Sortis de l'Universite de Liege, Liege, Belgium, Paper 80-146.

Li, Y., Wang, J., Wang, X., Wang, B. and Luan, Z. (2011). Feasibility study of iron mineral separation from red mud by high gradient superconducting magnetic separation. *Physica C: Superconductivity* 471: 91–96.

Lindroos, L., Virtanen, H. and Järvinen, O. (2004). Method for processing anode sludge. Patent US 7,731,777 B2.

Liu, W., Chen, X., Li, W., Yu, Y. and Yan, K. (2014). Environmental assessment, management and utilization of red mud in China. *Journal of Cleaner Production* 84: 606–610.

Liu, W., Yang, J. and Xiao, B. (2009). Application of Bayer red mud for iron recovery and building material production from aluminosilicate residues. *Journal of Hazardous Materials* 161: 474–478.

Liu, Y., Lin, C. and Wu, Y. (2007). Characterization of red mud derived from a combined Bayer Process and bauxite calcination method. *Journal of Hazardous Materials* 146: 255–261.

Liu, Y. and Naidu, R. (2014). Hidden values in bauxite residue (red mud): Recovery of metals. *Waste Management* 34: 2662–2673.

Liu, Z. and Li, H. (2015). Metallurgical process for valuable elements recovery from red mud – A review. *Hydrometallurgy* 155: 29–43.

Lloyd, P.J.D. (1983). Commercial processes for uranium from ore. In: Lo, T.C., Baird, M.H.I. and Hanson, C. (eds.), *Handbook of Solvent Extraction*. John Wiley & Sons, New York, pp. 763–782.

Low, K.S., Lee, C.K. and Leo, A.C. (1995). Removal of metals from electroplating wastes using banana pith. *Bioresource Technology* 51: 227–231.

Lu, D., Chang, Y.-F., Yang, H.-Y. and Xie, F. (2015). Sequential removal of selenium and tellurium from copper anode slime with high nickel content. *Transactions of the Nonferrous Metals Society of China* 25(2015): 1307–1314.

Lunt, L., Boshoff, P., Boylett, M. and El-Ansary, Z. (2007). Uranium extraction: The key process drivers. *The Journal of the Southern African Institute of Mining and Metallurgy* 107: 419–426.

Maegher, R. (2000). Phytoremediation of toxic elemental and organic pollutants. *Current Opinion in Plant Biology* 3: 153–162.

Mann, V. (2013). Red Mud: A resource for the future. *Metal Bulletin* 7 May 2013.

Marabini, A.M., Plescia, P., Maccari, D., Burragato, F. and Pelino, M. (1998). New materials from industrial and mining wastes: Glass-ceramics and glass- and rock-wool fibre. *International Journal of Mineral Processing* 53: 121–134.

Marchioretto, M., Bruning, H., Loan, N.T.P. and Rulkens, H. (2002). Heavy metals extraction from anaerobically digested sludge. *Water Science and Technology* 46(10): 1–8.

Marsden, I.C. and House, J.O. (2006). *The Chemistry of Gold Extraction*, 2nd edn. Society for Mining, Metallurgy, and Exploration, Littleton, CO.

McCarthy, S., Armour-Brown, V.S., Iyer, V.S., Desu, S.B., Kander, R.G. and Vaseashta, A. (1992). Utilization of Jamaica bauxite tailings as a building material and its socio-economic considerations. In: Glenister, D.J. (ed.), *International Bauxite Tailings Workshop*. Australian Bauxite and Alumina Producers, 2–6 November Perth, Western Australia, Australia, pp. 366–376.

MEND (1994). The Mine Environment Neutral Drainage (MEND) Program. Acid Mine Drainage – Status of Chemical Treatment and Sludge Management Practices, MEND Report 3.32.1. Richmond Hill, Ontario, Canada.

Meyer, E.H.O., Howard, G., Heagele, R. and Beck, R.D. (1996). Iron control and removal at the Zinc Corporation of South Africa. In: Dutrizac, J.E. and Harris, G.B. (eds.), *Iron Control and Disposal*. Canadian Institute of Mining, Metallurgy and Petroleum, Montreal, Quebec, Canada, pp. 163–182.

Miller, R.R. (1996). *Phytoremediation, Technology Overview Report*. Ground-Water Remediation Technologies Analysis Center, Pittsburgh, PA, 80pp.

Mishra, D., Kim, D.J., Ralph, D.E., Ahn, J.G. and Rhee, Y.H. (2007). Bioleaching of vanadium rich spent refinery catalysts using sulfur oxidizing lithotrophs. *Hydrometallurgy* 88: 202–209.

Mishra, D. and Young-Ha, R. (2010). Current research trends of microbiological leaching for metal recovery from industrial wastes. In: Mendez-Vilas, A. (ed.), *Current Research Technology and Education Topics in Applied Microbiology and Microbial Technology*, Formatex Research Centre, Badajoz, Spain, pp. 1289–1296.

Mukutuma, A., Schwarz, N., Chisakuta, G., Mbao, B. and Feather, A. (2009). A case study on the operation of a Flottweg Tricanter® centrifuge for solvent-extraction crud treatment at Bwana Mkubwa, Ndola, Zambia. *The Fourth Southern African Conference on Base Metals*, 23–27 July 2007, Swakopmund, Namibia. The Southern African Institute of Mining and Metallurgy, pp. 393–404.

Mymrin, V., Ponte, H. and Impinnisi, P. (2005). Potential application of acid jarosite wastes as the main component of construction materials. *Construction and Building Materials* 19(2): 141–146.

Ndlovu, S., Simate, G.S., Mchibwa, K.A., Iyuke, S.E. and Giaveno, A. (2014). Characterization of precipitates formed from the forced hydrolysis of bioleach liquors under different pH conditions. *Journal of Industrial and Engineering Chemistry* 20: 3578–3583.

Ndlovu, S., Simate, G.S., Seepe, L., Shemi, A., Sibanda, V. and Van Dyk, L. (2013). The removal of Co^{2+}, V^{3+} and Cr^{3+} from waste effluents using cassava waste. *SAICHE Journal* 18: 51–69.

Neutralac (2016). Waste water and sludge treatment. Available at http://www.neutralac.com/wastewater_precipitation.html. Accessed 10 January 2016.

Nicol, M.J., Fleming, C.A. and Preston, J.S. (1987). Application to extractive metallurgy. In: Wilkinson, G., Gillard, R.D. and McCleverty, J.A. (eds.), *Comprehensive Coordination Chemistry: The Synthesis, Reactions, Properties and Application of Coordination Compounds*, Vol. 6. Permagon, Oxford, UK, pp. 779–842.

Ochsenkühn-Petropoulou, M., Hatzilyberis, K., Mendrinos, L. and Salmas, C. (2002). Pilot plant investigation of the leaching process for the recovery of scandium from red mud. *Industrial and Engineering Chemistry Research* 41(23): 5794–5801.

Okada, S. and Yamazaki, N. (1984). Zinc residue treatment at Iijima refinery – Energy consumption in hematite process and research on iron utilisation. *113th TMS-AIME Annual Meeting*, Los Angeles, CA, paper A84-51.

Oke, D., Ndlovu, S. and Sibanda, V. (2014). Removal of platinum group metals from dilute process streams: Identification of influential factors using doe approach. *Journal of Environmental Chemical Engineering* 2: 1061–1069.

Oliveira, A.A.S., Tristão, J.C., Ardisson, J.D., Dias, A. and Lago, R.M. (2011). Production of nanostructured magnetic composites based on Fe^0 nuclei coated with carbon nanofibers and nanotubes from red mud waste and ethanol. *Applied Catalysis B: Environmental* 105(1–2): 163–170.

Orru, R., Sannia, M., Cincotti, A. and Cao, G. (1999). Treatment and recycling of zinc hydrometallurgical wastes by self-propagating reactions. *Chemical Engineering Science* 54: 3053–3061.

Outotec (2015). Recovery of precious metals from anode slimes. Available at http://www.outotec.com/en/Products--services/Non-ferrous-metals-processing/Precious-metals/Precious-metals/. Accessed 16 December 2015.

Padilla, R. and Sohn, H.Y. (1985). Sodium aluminate leaching and de-silication in lime–soda sinter process for alumina from coal wastes. *Metallurgical and Materials Transactions B: Process Metallurgy and Materials Processing Science* 16(4): 707–713.

Papassiopi, N., Vaxevanidou, K. and Paspaliaris, I. (2010). Effectiveness of iron reducing bacteria for the removal of iron from bauxite ores. *Minerals Engineering* 23: 25–31.

Pappu, A., Saxena, M. and Asolekar, S.R. (2006a). Hazardous jarosite use in developing nonhazardous product for engineering application. *Journal of Hazardous Materials* 137(3): 1589–1599.

Pappu, A., Saxena, M. and Asolekar, S.R. (2006b). Jarosite characteristics and its utilisation potentials. *Science of the Total Environment* 359(1): 232–243.

Pappu, A., Saxena, M. and Asolekar, S.R. (2007). Solid wastes generation in India and their recycling potential in building materials. *Building and Environment* 42(6): 2311–2320.

Pappu, A., Saxena, M. and Asolekar, S.R. (2010). Recycling hazardous jarosite waste using coal combustion residues. *Materials Characterization* 61: 1342–1355.

Paramguru, R., Rath, P. and Misra, V. (2005). Trends in red mud utilization – A review. *Mineral Processing and Extractive Metallurgy Review* 26(1): 1–29.

Pathak, A., Dastidar, M.G. and Sreekrishnan, T.R. (2009). Bioleaching of heavy metals from sewage sludge by indigenous iron-oxidizing microorganisms using ammonium ferrous sulfate and ferrous sulfate as energy sources: A comparative study. *Journal of Hazardous Materials* 171: 273–278.

Pawlek, R.P. (2008). *Alumina Refineries and Producers of the World*. Aluminium-Verlag GmbH, Duesseldorf, Germany.

Pelino, M. (2000). Recycling of zinc-hydrometallurgy wastes in glass and glass ceramic materials. *Waste Management* 20: 561–568.

Pelino, M., Cantalini, C., Abbruzzese, C. and Plescia, P. (1996). Treatment and recycling of goethite waste arising from the hydrometallurgy of zinc. *Hydrometallurgy* 40: 25–36.

Pelino, M., Cantalini, C., Boattini, P.P., Abbruzzese, C., Rincon, J.M. and Garcia Hernandez, J.E. (1994). Glass-ceramic materials obtained by recycling goethite industrial wastes. *Resources, Conservation and Recycling* 10: 171–176.

Peng, F., Liang, K., Shao, H. and Hu, A. (2005). Nano-crystal glass-ceramics obtained by crystallization of vitrified red mud. *Chemosphere* 59(6): 899–903.

Peng, X. and Huang, G. (2011). Method for recovering iron concentrates from alumina red mud. Google Patents, Patent number CN 101648159 A.

Peters, E. (1976). Direct Leaching of sulfides: Chemistry and applications. *Metallurgical Transactions B* 7(4): 505–517.

Peters, R.W., Ku, Y. and Bhattacharyya, D. (1985). Evaluation of recent treatment techniques for removal of heavy metals from industrial wastewaters. *AICHE Symposium Series, Separation of Heavy Metals and Other Contaminants* 81(243): 165–203.

Petersen, J. and Dixon, D.G. (2007). Modelling zinc heap bioleaching. *Hydrometallurgy* 85: 127–143.

Pilśniak-Rabiega, M. and Trochimczuk, A.W. (2014). Selective recovery of gold on functionalized resins. *Hydrometallurgy* 146: 111–118.

Pinnock, W.R. (1991). Measurement of radioactivity in Jamaican building materials and gamma dose equivalents in a prototype red mud house. *Journal of Health Physics* 61(5): 647–651.

Pisciella, P., Crisucci, S., Karamanov, A. and Pelino, M. (2001). Chemical durability of glasses obtained by vitrification of industrial wastes. *Waste Management* 21: 1–9.

Piskunof, V.M., Matveev, A.F. and Yaroslavtev, A.S. (1988). Utilizing iron residues from zinc production in USSR. *Journal of Meteorology* 40: 36–39.

Puskas, F. (1983). US Patent 4,368,273.

Quek, S.Y., Wase, D.A.J. and Forster, C.F. (1998). The use of sago waste for the sorption of lead and copper. *Water SA* 24: 251–256.

Randall, J.M., Reuter, F.C. and Waiss, A.C. (1974). Removal of cupric ions from solution by contact with peanut skins. *Journal of Applied Polymer Science* 19: 156–171.

Rathore, N., Patil, M.P. and Dohare, D. (2014). Utilization of jarosite generated from lead-zinc smelter for various applications: A review. *International Journal of Civil Engineering and Technology* 5(11): 192–200.

Readett, D.J. and Miller, G.M. (1995). The impact of silica on solvent extraction: Girilambone Copper Company case study. In: Cooper, W.C., Dreisinger, D.B., Dutrizac, J.E., Hein, H. and Ugarte, G. (eds.), *Copper*, Santiago, Chile, 26–29 November 1995. Canadian Institute of Mining Metallurgy and Petroleum, Montreal, Quebec, Canada, pp. 679–690.

Richardson, J.F. and Harker, J.H. (2002). *Chemical Engineering*, 5th edn. Butterworth-Heinemann, Oxford, UK.

Rincon, J.M., Callejas, P. and Romero, M. (1992). Recycling of goethite wastes, originated from hydrometallurgical plants for production of glass-ceramic materials. E.C. Contract No. MA2R-CT90-0007, II Semestral Rep., CORDIS, European Commission.

Ritcey, G.M. and Ashbrook, A.W. (1984). *Solvent Extraction. Principles and Applications to Process Metallurgy. Part I*. Elsevier, Amsterdam, the Netherlands.

Ritter, S.K. (2015). Making most of the red mud. *Chemical and Engineering News* 92(8). Available at http://cen.acs.org/articles/92/i8/Making-Red-Mud.html. Accessed 22 January 2016. pp. 33–35.

Rodríguez, Y., Ballester, A., Blázquez, M.L., González, F. and Muñoz, J.A. (2003). New information on the chalcopyrite bioleaching mechanism at low and high temperature. *Hydrometallurgy* 71: 47–56.

Ropenack, A., Bohmer, W. and Grimm, H. (1995). Method of reprocessing jarosite containing residues. Ruhr-Zink GmbH, assignee. United States Patent 5,453,253, 6 September 1995.

Rössing Uranium (2016). Rössing Uranium's production of uranium oxide and the nuclear fuel cycle. Available at http://www.rossing.com/uranium_production.htm. Accessed 20 September 2016.

Saktas, S. and Morcali, M.H. (2011). Platinum recovery from dilute platinum solutions using activated carbon. *Transactions of Nonferrous Metals Society of China* 21: 2554–2558.

Samal, S., Ray, A.K. and Bandopadhyay, A. (2013). Proposal for resources, utilization and processes of red mud in India – A review. *International Journal of Mineral Processing* 118: 43–55.

Sandstrom, A. and Peterson, S. (1997). Bioleaching of a complex sulphide ore with moderate thermophilic and extreme thermophilic microorganisms. *Hydrometallurgy* 46: 181.

Sanjeev, B., Amritphale, S.S. and Chandra, N. (1999). Utilisation of toxic solid waste generated in a nonferrous metal industry for making frit/glazing material and its application for glazing of ceramics. In: Rao, P.R., Kumar, R., Srikanth, S. and Goswami, N.G. (eds.), *Proceedings, Nonferrous Extractive Metallurgy in the New Millennium*, Jamshedpur, India. NML, pp. 375–384.

Schloen, J.H. and Elkin, E.M. (1950). Treatment of electrolytic copper refinery slimes. *Journal of Metals* 42(8): 764–777.

Scott, J. (1990). Electrometallurgy of Copper Refinery Anode Slimes. *Metallurgical Transactions B* 21B: 629–635.

Seidel, A., Zimmels, Y. and Armon, R. (2001). Mechanism of bioleaching of coal fly ash by *Thiobacillus thiooxidans*. *Chemical Engineering Journal* 83: 123–130.

Seyer, S., Chen, T.T. and Dutrizac, J.E. (2001). Jarofix: Addressing iron disposal in the zinc industry. *JOM* December: 32–35.

Simate, G.S. and Ndlovu, S. (2015). The removal of heavy metals in a packed bed column using cassava waste. *Journal of Industrial and Engineering Chemistry* 21: 635–643.

Simate, G.S., Ndlovu, S. and Walubita, L.F. (2010). The fungal and chemolithotrophic leaching of nickel laterites – Challenges and opportunities. *Hydrometallurgy* 103: 150–157.

Sing, K.S.W., Everett, D.H., Haul, R.A.W., Moscou, L., Pierotti, R.A., Rouquerol, J. and Siemieniewska, T. (1985). Reporting physisorption data for gas/solid systems with special reference to the determination of surface area and porosity. *Pure and Application Chemistry* 57: 603–619.

Singh, M., Upadhayay, S. and Prasad, P. (1996). Preparation of special cements from red mud. *Waste Management* 16(8): 665–670.

Singh, M., Upadhayay, S. and Prasad, P. (1997). Preparation of iron rich cements using red mud. *Cement and Concrete Research* 27(7): 1037–1046.

Smalley, N. and Davis, G. (2000). Operation of the Las Cruces ferric sulphate leach pilot plant. *Minerals Engineering* 13(6): 559–608.

Sole, K.C., Feather, A.M. and Cole, P.M. (2005). Solvent extraction in Southern Africa: An update of some recent hydrometallurgical developments. *Hydrometallurgy* 78: 52–78.

Stassen, L. and Suthiram, J. (2015). Initial development of an alkaline process for recovery of uranium from ^{99}Mo production process waste residue. *Journal of Radioanalytical and Nuclear Chemistry* 305: 41–50.

Stewart, M. and Kappes, D. (2011). SART for copper control in cyanide heap leaching. *International Conference, Percolation Leaching: The Status Globally and in Southern Africa, Symposium Series S69*, The Southern Africa Institute of Mining and Metallurgy, 7–9 November 2011, Muldersdrift, South Africa, pp. 145–164.

Stuurman, S., Ndlovu, S. and Sibanda, V. (2014). Comparing the extent of the dissolution of copper-cobalt ores from the DRC region. *Journal of the Southern African Institute of Mining and Metallurgy* 114: 347–353.

Sud, D., Mahajan, G. and Kaur, M.P. (2008). Agricultural waste material as potential adsorbent for sequestering heavy metal ions from aqueous solutions. *Bioresource Technology* 99: 6017–6027.

Sushil, S. and Batra, V.S. (2008). Catalytic applications of red mud, an aluminium industry waste: A review. *Applied Catalysis B: Environmental* 81(1–2): 64–77.

Sutar, H., Mishra, S.C., Sahoo, S.K., Chakraverty, A.P. and Maharana, H.S. (2014). Progress of Red Mud utilization: An overview. *American Chemical Science Journal* 4(3): 255–279.

Thakur, R.S. and Sant, B.R. (1974). Utilization of red mud. *Journal of Scientific and Industrial Research* 33(8): 408–416.

Thakur, R.S. and Sant, B.R. (1983a). Utilization of red mud-part I. *Journal of Scientific and Industrial Research* 42(1): 87–108.

Thakur, R.S. and Sant, B.R. (1983b). Utilization of red mud. 2. Recovery of alkali, iron, aluminum, titanium, and other constituents and the pollution problems. *Journal of Scientific and Industrial Research* 42(8): 456–469.

Tsakanika, L., Ochsenkuhn-Petropoulou, M. and Mendrinos, L. (2004). Investigation of the separation of scandium and rare earth elements from red mud by use of reversed-phase HPLC. *Analytical and Bioanalytical Chemistry* 379(5–6): 796–802.

Ullmann's (1993). *Ullmann's Encyclopedia of Industrial Chemistry*, 5th edn. VCH Publishers Inc., Weinheim, Germany.

US Department of Energy (2009). Lamps standards final rule TSD Appendix 3C (Rare Earth Phosphor Market). Available at http://www1.eere.energy.gov/buildings/appliance_standards/residential/pdfs/app_3c_lamps_standards_final_tsd.pdf. Accessed 18 November 2015.

US Department of Energy (2011). Critical materials strategy. Available at http://energy.gov/sites/prod/files/DOE_CMS2011_FINAL_Full.pdf. Accessed 18 November 2015.

Uzun, D. and Gulfen, M. (2007). Dissolution kinetics of iron and aluminium from red mud in sulphuric acid solution. *Indian Journal of Chemical Technology* 14: 263–268.

Vachon, P., Tyagi, R.D., Auclair, J.C. and Wilkinson, K.J. (1994). Chemical and biological leaching of aluminum from red mud. *Environmental Science & Technology* 28: 26–30.

van Dyk, J.P. (2006). An overview of the Zincor Process. In: Jones, R.T. (ed.), *Southern African Pyrometallurgy 2006*, 5–8 March 2006. South African Institute of Mining and Metallurgy, Johannesburg, South Africa, pp. 273–282.

van Zyl, A.W., Harrison, S.T.L. and van Hille, R.P. (2011). Biodegradation of thiocyanate by a mixed microbial population. *Proceedings of the 11th International Mine Water Association Congress*, 4–11 September, Aachen Germany, pp. 119–123.

van Zyl, A.W., Huddy, R., Harrison, S.T.L. and van Hille, R.P. (2015). Evaluation of the ASTER™ process in the presence of suspended solids. *Minerals Engineering* 76: 72–80.

Vasan, S.S., Jayant Modak, M. and Natarajan, K.A. (2001). Some recent advances in the bioprocess of bauxite. *International Journal of Mineral Processing* 62: 173–186.

Vilcáez, J., Suto, K. and Inoue, C. (2008). Response of thermophiles to the simultaneous addition of sulfur and ferric ion to enhance the bioleaching of chalcopyrite. *Minerals Engineering* 21: 1063–1074.

Virnig, M.J. (2007). Crud formation: Field studies and fundamental studies. In: Riveros, P.A., Dixon, D.G., Dreisinger, D.B., and Collins, M.J. (eds.), *Copper*, Toronto, Quebec, Canada, 26–29 August 2007. Canadian Institute of Mining Metallurgy and Petroleum, Montreal, Quebec, Canada, pp. 291–303.

Vladimir, S.K.I. (2011). *Solvent Extraction: Classical and Novel Approaches*. Elsevier, Amsterdam, The Netherlands, 576pp.

Volesky, B. (2001). Detoxification of metal – bearing effluents: Biosorption for the next century. *Hydrometallurgy* 59: 203–216.

Vu, H., Jandová, J. and Hron, T. (2010). Recovery of pigment-quality magnetite from jarosite precipitates. *Hydrometallurgy* 101: 1–6.

Wang, J., Xu, F., Xie, W., Mei, Z., Zhang, Q., Cai, J. and Cai, W. (2009). The enhanced adsorption of dibenzothiophene onto cerium/nickel-exchanged zeolite Y. *Journal of Hazardous Materials* 163: 538–543.

Wang, Q.K., Ma, R.J. and Tan, Z.Z. (1985). The jarosite process – Kinetic study. In: Tozawa, K. (ed.), *Zinc '85: International Symposium on Extractive Metallurgy of Zinc*, Tokyo 1985. Mining and Metallurgical Institute of Japan, Tokyo, Japan, pp. 675–690.

Wang, S., Ang, H.M. and Tade, W.O. (2008). Novel applications of red mud as coagulant, adsorbent and catalyst for environmentally benign processes. *Chemosphere* 72: 1621–1635.

Wang, W., Pranolo, Y. and Cheng, C.Y. (2011). Metallurgical processes for scandium recovery from various resources: A review. *Hydrometallurgy* 108: 100–108.

Wang, W.K., Hoh, Y., Chuang, W. and Shaw, I. (1980). Hydrometallurgical process for recovering precious metals from anode slime. Patent US 4,293,332 A.

Watling, H.R. (2006). The bioleaching of sulphide minerals with emphasis on copper sulphides – A review. *Hydrometallurgy* 84: 81–108.

Weil, B. (2012). Uranium mining and extraction from ore. Available at http://large.stanford.edu/courses/2012/ph241/weil2/. Accessed 10 February 2016.

Willscher, S., Mirgorodsky, D., Jablonski, L., Ollivier, D., Merten, D., Büchel, G., Wittig, J. and Werner, P. (2013). Field scale phytoremediation experiments on a heavy metal and uranium contaminated site, and further utilization of the plant residues. *Hydrometallurgy* 131–132: 46–53.

Winkler, D. (2014). Collembolan response to red mud pollution in Western Hungary. *Applied Soil Ecology* 83: 219–229.

WNA (World Nuclear Association) (2014). Processing of used nuclear fuel. Available at http://www.world-nuclear.org/info/Nuclear-Fuel-Cycle/Fuel-Recycling/Processing-of-Used-Nuclear-Fuel/. Accessed 15 January 2016.

Wong, L.T.K. and Henry, J.G. (1983). Bacterial leaching of heavy metals from anaerobically digested sewage sludge. *Water Pollution Research Journal of Canada* 18: 151–162.

World Aluminium Report (2013). Bauxite residue management: Best practice. Available at www.bauxite.world-aluminium.org. Accessed 10 November 2015.

World Aluminium Report (2014). Bauxite residue management: Best practice. Available at www.aluminium.org.au/_literature_175772/Bauxite_Residue_Management_-_Best_Practice. Accessed 20 November 2015.

World Aluminium Report (2015). Bauxite residue management: Best practice. Available at www.bauxite.world-aluminium.org. Accessed 4 November 2015.

Xiang, Q., Liang, X., Schlesinger, M.E. and Watson, J.L. (2001). Low-temperature reduction of ferric iron in red mud. *Light Metals: Proceedings of Sessions, TMS Annual Meeting, Warrendale, Pennsylvania*, New Orleans, LA, pp. 157–162.

Xin, B., Zhang, D., Zhang, X., Xia, Y., Wu, F., Chen, S. and Li, L. (2009). Bioleaching mechanism of Co and Li from spent lithium-ion battery by the mixed culture of acidophilic sulfur-oxidizing and iron-oxidizing bacteria. *Bioresource Technology* 100: 6163–6169.

Xu, T.-J. and Ting, Y.-P. (2009). Fungal bioleaching of incineration fly ash: Metal extraction and modeling growth kinetics. *Enzyme and Microbial Technology* 44: 323–328.

Xue, A., Chen, X. and Tang, X. (2010). The technological study and leaching kinetics of scandium from red mud. *Nonferrous Metals Extractive Metallurgy* 2: 51–53.

Yahorava, V. and Kotze, M. (2014). Ion exchange technology for the efficient recovery of precious metals from waste and low-grade streams. *Journal of the Southern African Institute of Mining and Metallurgy* 114(2): 173–181.

Yalcin, N. and Sevinc, V. (2000). Utilization of bauxite waste in ceramic glazes. *Ceramics International* 26(5): 485–493.

Yang, T., Xu, Z., Wen, J. and Yang, L. (2009). Factors influencing bioleaching copper from waste printed circuit board by *Acidithiobacillus ferrooxidans*. *Hydrometallurgy* 97: 29–32.

Yin, Y.C., Aroua, M.K. and Wan Daud, W.M. (2007). Review of modifications of activated carbon for enhancing contaminant uptakes from aqueous solutions. *Separation and Purification Technology* 52: 403–415.

Young, R.S. (1975). The analysis of copper refinery slimes. *Talanta* 43: 125–130.

Yu, Z.-L., Shi, Z.-X., Chen, Y.-M., Niu, Y.-J., Wang, Y.-X. and Wan, P.-Y. (2012). Red-mud treatment using oxalic acid by UV irradiation assistance. *Transactions of the Nonferrous Metals Society of China* 22: 456–460.

Zhang, J., Deng, Z. and Xu, T. (2005). Experimental investigation on leaching metals from red mud. *Light Metals* 2005(2): 13–15.

Zhang, J., Liu, J., Zhang, J., Wang, L., Huang, S. and Li, B. (2013). A process for the production of direct reduced iron from red mud. Google Patents, Patent number CN 102559979 B.

Zhao, L., Yang, D. and Zhu, N.-W. (2008). Bioleaching of spent Ni-Cd batteries by continuous flow system: Effect of hydraulic retention time and process load. *Journal of Hazardous Materials* 160: 648–654.

Zheng, G.-H. and Kozinsk, J.A. (1997). Solid waste remediation in the metallurgical industry: Application and environmental impact. *Environmental Progress & Sustainable Energy* 15(4): 283–292.

Zhu, D.Q., Chun, T.J., Pan, J. and He, Z. (2012). Recovery of iron from high-iron red mud by reduction roasting with adding sodium salt. *Journal of Iron and Steel Research, International* 19: 1–5.

8

Metal Manufacturing and Finishing Waste Production and Utilization

8.1 Introduction

The production of iron and steel and ferroalloy products were discussed in Chapters 4 and 5, respectively. As discussed in Chapters 4 and 5, significant amounts of liquid steel and ferroalloys produced globally are consumed in the production of value-added steel products. The two main processes involved in the production of iron and steel products include smelting and refining. Once the appropriate quality of the liquid steel has been achieved in the two process streams, the liquid steel is then cast and solidified into semi-finished products, such as billets, blooms or slabs, for subsequent rolling in the finishing mills (Vijayaram, 2012; Remus et al., 2013; Kozak and Dzierzawski, n.d.).

The continuous casting process is commonly used in the solidification process of steel, mainly due to its high productivity, high product quality control and cost efficiencies as a result of its ability to solidify large volumes of liquid metal into simple shapes for subsequent processing (Kaushish, 2010; Vijayaram, 2012; Remus et al., 2013; Kozak and Dzierzawski, n.d.). Ingot casting processes can also be adopted, particularly in foundries, where the smelting and refining processes are done batchwise to match the capacity of downstream manufacturing and finishing processes (Okumura, 1994; Thomas, 2001).

The role of metal manufacturing and finishing processes is to transform the intermediate products from the solidification processes into value-added manufactured and finished goods and components such as machine components, machinery, instruments and tools, which can directly be used by end users (Roto, 1998). Some of the common metal manufacturing processes include metal forming (either hot or cold forming), rolling and extrusion, machining and forging (Kaushish, 2010). Typically, metal forming processes essentially involve using an externally applied force to plastically deform a metal product into various desired shapes and sizes (Kaushish, 2010).

Generally, metal finishing processes refer to a generic group of surface coating techniques adopted to enhance the surface finish of steel products and impart properties such as corrosion and chemical resistance, biocompatibility, reflectivity and appearance, hardness, electrical conductivity, wear resistance and other engineering or mechanical properties (Roto, 1998; Lindsay, 2007; Regel-Rosocka, 2009; Rögener et al., 2012). Technically, the metal finishing industry involve two distinct operations, that is, plating (i.e. electroplating and electroless plating) and surface treatment (i.e. chemical and electrochemical conversion, case hardening, metallic coating and chemical coating) (EPA, 1997).

Despite the criticality of the metal forming, finishing and manufacturing processes, they processes are associated with the production of toxic and hazardous by-products. This chapter discusses some of the technical aspects of the typical metal forming, finishing and manufacturing processes; the different waste products produced and the possible techniques and opportunities in reducing, recycling and reusing the waste materials.

8.2 Metal Forming Processes

The continuous casting process produces standard intermediate products such as slabs, billets and blooms, and these intermediate products require further processing through the various metal forming processes to produce finished products such as plates, sheets, rods, tubes and structural sections (Rentz et al., 1999). In general, metal forming processes utilize the application of direct uniaxial forces to plastically deform the metal into desired near-net shapes and sizes, and are usually associated with high productivity and high dimensional accuracies (Kaushish, 2010).

The metal forming processes can be broadly categorized into (Kaushish, 2010) (1) direct compression type processes, such as forging and rolling; (2) indirect compression processes, such as deep drawing and extrusion and (3) bending processes. Essentially, the metal forming processes can further be subdivided into two main categories (Kaushish, 2010; Klocke, 2014): (1) hot forming, where the deformation is conducted above the recrystallization temperature but below the melting point of the metal and (2) cold forming, essentially carried out under ambient conditions or below the recrystallization temperature of the metal. Hot forming processes are most commonly applied for the primary solid state shaping processes such as forging, rolling, extruding, especially where larger reduction in the size without cracking of the specimen is required (Kaushish, 2010). In practice, cold forming would be more ideal over hot forming practices, mainly because (1) there are no energy costs for preheating the metal work piece, (2) there are no material losses due to scale formation and no finishing treatment process required to remove the oxide scales and (3) there are no dimensional faults due to shrinkage during the cooling of the hot-worked work piece (Klocke, 2014). However, the main disadvantages of cold forming processes include, inter alia, the requirement for higher force and limited formability of materials at ambient conditions. As a result, the hot forming processes are usually adopted industrially to specifically prevent the overload of forming machines and fracture of materials to be processed (Klocke, 2014).

In essence, cold and hot rolling processes constitute some of the most extensively used mechanical forming processes in the production of a variety of steel products such as flat-rolled steel sheets and plates, long products, pipes and tubes (Rentz et al., 1999; Chen and Yuen, 2003; Li and Celis, 2003; Boljanovic, 2010; Kaushish, 2010; Klocke, 2014). As a result, the following sections discuss in detail the technical and process aspects of the cold and hot rolling processes. In this case, the generic process considerations, typical waste produced, as well as the waste valorization opportunities in the rolling and treatment processes of low alloy and stainless steels will be discussed.

8.2.1 General Considerations of the Rolling Process

In general, the rolling process utilizes a two-dimensional deformation process using compressive forces between two rolls, where the material to be rolled is drawn into the revolving roll gap by means of friction to reduce the thickness or change the cross-sectional area of the as-cast material (Boljanovic, 2010; Klocke, 2014). As already discussed, the rolling process can be classified into two broad categories, namely, (1) the cold rolling process conducted at ambient temperature conditions, and (2) the hot rolling process usually conducted above the recrystallization temperature of the material to be processed (Rentz et al., 1999; Li and Celis, 2003; Boljanovic, 2010; Klocke, 2014).

8.2.2 Cold Rolling Process

The cold rolling process is usually conducted under ambient temperature conditions or at temperatures way below the recrystallization temperature of the material (Rentz et al., 1999; Boljanovic, 2010; Klocke, 2014). As shown in Figure 8.1, the cold rolling process of steel includes sub-operations such as surface cleaning or pickling, cold rolling, heat treatment and finishing operations (Rentz et al., 1999).

In general, the cold rolling process has several advantages, namely, (1) greater tolerances and better surface finish due to the absence of work piece cooling and surface oxidation, (2) the ability to roll thin products, (3) the ability to control the grain size properties of the

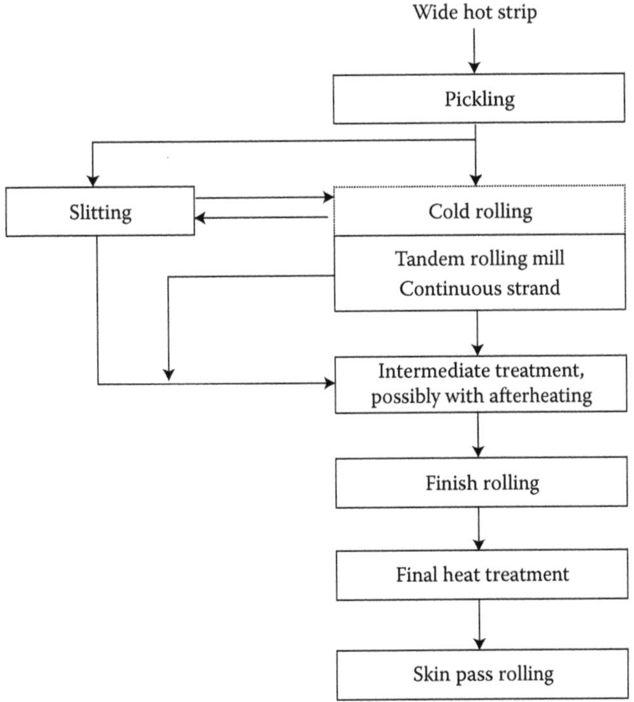

FIGURE 8.1
Process flow for cold strip rolling. (From Rentz, O. et al., Report on best available techniques (BAT) in the German Ferrous Metals Processing Industry, Final draft, Deutsch-Französiches Institut für Umweltforschung, 1999, p. 137, Available at https://www.umweltbundesamt.de/sites/default/files/medien/publikation/long/2490.pdf, Accessed 2 June 2016. With permission.)

rolled materials before annealing and (4) the ease of lubrication during the rolling process (Boljanovic, 2010). However, the rolled products usually require homogenization treatment since the cold rolling process tends to result in the hardening effect of steel products (Boljanovic, 2010). Furthermore, cold forming require higher force and, thus there is limited formability of materials.

8.2.3 Hot Rolling Process

Hot rolling is usually conducted above the recrystallization temperature of the material (Boljanovic, 2010; Kaushish, 2010; Klocke, 2014). The temperature range for the plastic deformation of most metallic materials is given as $0.6T_M \leq T \leq 0.75T_M$, where T_M and T are the melting point for that particular metal and the working temperatures (K), respectively (Boljanovic, 2010). Depending on the type of steel processed, the typical operating temperatures are usually in the range 1070°C–1260°C (Boljanovic, 2010). The hot rolling of steel products is usually preferred in most industrial applications due to several reasons (Rentz et al., 1999): (1) the process is conducted at high temperatures where plastic deformation of the material is higher, hence improved formability is achieved and less force is required, (2) it requires a reduced number and frequency of rolling steps and hence, reduced processing time and (3) it has beneficial effects on the microstructure and properties of the steel.

The hot-rolling process comprises of several major steps, and these include, inter alia, (1) surface preparation, (2) slab reheating, (3) descaling, (4) hot rolling and (5) finishing (Rentz et al., 1999; European Commission, 2001; Chen and Yuen, 2003; Basabe and Szpunar, 2008). The as-cast material is charged either at ambient temperatures or into reheating furnaces and heated to the required rolling temperatures (usually 600°C–800°C) (Rentz et al., 1999). In most cases, in order to supply the metal with the right surface and internal temperature profiles, industrial reheating furnaces consist of three zones, namely, the preheating, heating and homogenization zones (Rentz et al., 1999).

The reheating furnaces are essentially refractory-lined and are equipped with burners which combust a mixture of natural gas and preheated oxygen-enriched air to heat the steel to its discharge and rolling temperatures (Rentz et al., 1999; European Commission, 2001). Table 8.1 shows the input–output material flows in a typical reheating furnace (Rentz et al., 1999).

TABLE 8.1

Input–Output Material Flows in a Typical Reheating Furnace

Input		Output	
Slabs, blooms, billets	1010–1030 kg/t	Slabs, blooms, billets	1000 kg/t at about 1150°C
Energy (varies with charging temperature 20°C–1000°C)	0.10–1.5 GJ/t	Emissions to air	Dependent on fuel input
Fuels	Gas (e.g. natural, Blast furnace, COREX)	NOx	25–500 g/t
	Oil	SOx	0.02–900 g/t
	Fuel-mix	Dust	0.2–30 g/t
Cooling water	Closed loop	Scale	5–28 kg/t

Source: Rentz, O. et al.,. Report on best available techniques (BAT) in the German Ferrous Metals Processing Industry, Final draft, Deutsch-Französiches Institut für Umweltforschung, 1999, p. 147, Available at https://www.umweltbundesamt.de/sites/default/files/medien/publikation/long/2490.pdf, Accessed 2 June 2016. With permission.

8.2.4 Formation of Adherent Oxide Scale during the Reheating Process

The reheating furnaces are usually heated by combustion of natural gas to the desired temperatures, and as a result, the generation of oxide scales on the surfaces of steel is inevitable because of the oxidizing atmosphere within the reheat furnaces (Rentz et al., 1999; European Commission, 2001; Chen and Yuen, 2003; Basabe and Szpunar, 2008). Therefore, when the steel work piece is discharged from the reheat furnace, it is usually coated with an adherent oxide scale, and the impact of such oxide scale in hot rolling operations is two-fold (Krzyzanowski et al., 2010): (1) by forming the interface between the tool and workpiece, it determines the friction and heat transfer during the thermomechanical processing, and (2) it affects the quality of the formed product by forming intergrowths onto the rolled surface. To some extent, the heat insulating effect, and the effect of the interfacial friction as a result of the oxide scale, is considered to have beneficial effects in the hot working of steel (Luong and Heijkoop, 1981; Wei et al., 2009). For example, it is proposed that the rolling force and the friction conditions depend on the thickness and microstructure of the oxide scale (Luong and Heijkoop, 1981; Wei et al., 2009; Krzyzanowski et al., 2010).

In general, the oxide scale represents a loss of metal during the hot forming processes (Luong and Heijkoop, 1981). In practice, the initial adherent oxide layer, or primary layer formed during reheating processes is usually removed by the use of hydraulic descalers positioned near the exit of the reheating furnaces (Luong and Heijkoop, 1981; Rentz et al., 1999; Chen and Yuen, 2003; Li and Celis, 2003; Basabe and Szpunar, 2008; Wei et al., 2009). Secondary oxide scales are also formed during the course of the rolling process, particularly during and after the removal of the primary oxide layer; and are usually removed either by using high-pressure water jets after each pass or by using a second hydraulic descaler before the rolled products are conveyed into finishing rolling mill (Chen and Yuen, 2003). Further formation of oxide scales on the surface of rolled steel can also take place during and after the finishing rolling stage, as well as during cooling, and coiling of hot-formed products (Chen and Yuen, 2003).

8.2.5 Adherent Oxide Scale Formation during the Heat Treatment of Formed Products

Metal forming operations induce some internal energy and microstructure deformation within the metal substrate (Li and Celis, 2003). Cold working, in particular, leads to anisotropy and increased stiffness and strength in metals (Boljanovic, 2010). In most industrial processes, metal forming processes are combined with heat treatment processes, such as annealing and normalizing, in order to release the work hardening and internal stresses induced by the deformation processes (Digges et al., 1966; Brooks, 1996). Basically, the heat treatment processes involve subjecting a metal to a definite time–temperature cycle, and may be divided into three steps (Digges et al., 1966; Brooks, 1996; Rentz et al., 1999): (1) heating to a requisite temperature, (2) holding or soaking at heat treatment temperatures to ensure the uniformity of temperature within the metal body and (3) cooling the heat-treated work piece to ambient temperature. Typical heat treatment processes applicable to steels include annealing, tempering, normalizing and hardening (Digges et al., 1966; Brooks, 1996; Rentz et al., 1999).

The annealing of cold-worked metals is usually done to soften the metal by heating below its recrystallization temperature (Brooks, 1996). The annealing temperatures depend on the eutectoid composition of the steel to be heat treated, and are typically around 650°C and 1100°C–1180°C followed by slow cooling for low-alloy and austenitic stainless steels, respectively (Li and Celis, 2003). During the annealing process, oxide scales are formed on the surface of steel as a result of the thermal gradients and oxidative furnace conditions (Li and Celis, 2003; Basabe and Szpunar, 2008). The structure and composition of the adherent oxide layer depends on the annealing conditions such as temperature, oxygen partial pressure and the alloy composition of the metal (Zielinski and Kurzydlowski, 2000; Cheng et al., 2003; Li and Celis, 2003; Green and Higginson, 2011). The physicochemical characteristics of the oxide layer formed during hot working and heat treatment vary depending on the composition and type of steel being processed, and these are addressed further in Section 8.2.6. In this context, the oxide materials originating from the descaling processes, together with the metal cutting and rejects from other manufacturing processes, are broadly referred to as mill scale. The recycling and utilization opportunities of mill scale are discussed later in the Section 8.2.7.

8.2.6 Characteristics of Adherent Oxide Layers

In general, the characteristics and chemical composition of the oxide scales formed on the surface of steel during hot working and heat treatment processes largely depend on (1) alloy composition of the steel, (2) oxygen partial pressure in the reheat and heat treatment furnace atmosphere, (3) reheat and heat treatment temperature conditions and (4) presence of residual elements within the steel (Chen and Yuen, 2003; Cheng et al., 2003; Li and Celis, 2003; Basabe and Szpunar, 2008; Wei et al., 2009; Green and Higginson, 2011). In this section, the characteristics of adherent oxide scale are discussed based on two broad categories of steel products, namely, low-alloy carbon steels and stainless and high-alloy steels.

Low-alloy steels: In general, the oxide scales formed during the hot working of low-alloy steels consist of oxides of iron such as wustite (FeO), hematite (α-Fe_2O_3) and magnetite (Fe_3O_4) (Atkinson and Kolarik, 2001; Chen and Yuen, 2003; Cho and Lee, 2008; Umadevi et al., 2009). The morphology of the oxide scale formed in the high temperature oxidation of iron and carbon steels was characterized in studies by Chen and Yuen (2003). In this case, oxide scale formed from the oxidation of iron in the temperature range 700°C–1250°C comprised of a thin outermost hematite layer, a thin intermediate magnetite layer and a thick inner wustite layer (Chen and Yuen, 2003). Chen and Yuen (2003) proposed that the oxide scale structure differed at different locations across the width of hot-rolled strip, and the morphology was affected mainly by the starting cooling temperature, cooling rate and oxygen availability.

Stainless steels and high-alloy steels: As discussed in Chapter 4, stainless steels typically contain between 10 and 30 wt.% chromium. In general, stainless steels are grouped within the chromia-forming alloys, wherein the Cr_2O_3 layer protects the bulk of the metal from oxidation, particularly at temperatures of up to 627°C (Wei et al., 2009). In essence, the unique properties and applications of the different grades of stainless steels are largely controlled by the presence and integrity of this protective Cr_2O_3 layer (Sánchez et al., 2008; Wei et al., 2009; Green and Higginson, 2011; Xia et al., 2015).

According to Green and Higginson (2011), the growth of the Cr_2O_3 layer is largely dependent on factors such as (1) the substrate's microstructure and chemical composition, (2) the degree of disorder in the substrate, (3) the homogeneity of the surface finish, (4) the atmospheric conditions and (5) temperature conditions. The interactive effect of these factors on the oxidation behaviour of stainless steels is rather complex, but the effect of microstructure and grain boundary size on the strength and oxidation resistance of stainless steels is commonly acknowledged, with the fine grained stainless steels having been observed to possess superior mechanical properties and oxidation resistance (Green and Higginson, 2011).

Based on the knowledge of subatomic-level interactions in high-chromium steels proposed by Green and Higginson (2011), a great deal of effort has been channelled towards understanding the effects of hot forming, heat treatment and surface preparation processes have on the characteristics and morphology of the oxide scale and protective chromium oxide layers (Cheng et al., 2003; Li and Celis, 2003; Sánchez et al., 2008; Green and Higginson, 2011; Xia et al., 2015). In general, the chemical composition of the protective oxide layer formed during the hot forming and heat treatment of selected grades of stainless steels has been proposed to consist mainly of (1) Cr_2O_3, Fe_2O_3 and $(Fe,Cr)_3O_4$ spinel for austenitic stainless steels (Li and Celis, 2003), (2) Fe_3O_4, Fe_2O_3 and $(Fe,Cr)_3O_4$ for martensitic (P-91) steels (Sánchez et al., 2008; Xia et al., 2015), (3) Cr_2O_3 and $(Fe,Cr)_2O_4$ spinel for AISI 430 steels (Sánchez et al., 2008) and (4) Fe_2O_3, Fe_3O_4, $FeCr_2O_4$, Cr_2O_3, CrO_3, $FeCrO_3$ and MnO_2 on duplex stainless steels (Donik, 2011).

Based on the findings from studies highlighted earlier, the oxide scales on stainless steels are naturally enriched in chromium, mainly due to the high chromium content in stainless steels and also due to the high affinity of chromium to oxygen (Li and Celis, 2003; Li et al., 2005; Sánchez et al., 2008; Xia et al., 2015). In addition, the preferential oxidation of chromium at hot forming temperatures tends to result in the formation of a continuous chromium-depleted layer between the oxide scale film and the base metal (Scott and Wei, 1989; Li and Celis, 2003; Li et al., 2005; Sánchez et al., 2008; Xia et al., 2015). Furthermore, Li and Celis (2003) proposed that the Cr_2O_3 layer tends to be fairly adhesive and protective, while the iron oxide films tend to be quite permeable due to their loose and porous structure. As such, the morphology of these oxide scales and the chromium-depleted layers tend to significantly influence the efficiency of the scale removal, surface treatment and pickling processes (Li and Celis, 2003; Li et al., 2005).

8.2.7 Recycling and Utilization of Surface Oxide Scale Material

The mechanism of and conditions for the formation of the adherent oxide scale, its chemical and morphological characteristics and its removal from steel surfaces in the hot forming processing of different grades of steels were discussed so far. The descaled oxide materials, together with other forms of oxide wastes generated in the casting and metal forming processes, and metal cuttings and rejects from metal processing steps constitute a valuable resource for iron and alloy elements (Cho and Lee, 2008; Umadevi et al., 2009; Farahat et al., 2010; Bagatini et al., 2011; Martin et al., 2012). In industry, it is standard practice to categorize such a combination of material wastes from the heat treatment, metal forming and other manufacturing processes as mill scale (Cho and Lee, 2008; Umadevi et al., 2009; Farahat et al., 2010; Bagatini et al., 2011; Martin et al., 2012). In fact, mill scale is a generic and broad term given to the solid by-products produced during the continuous casting, reheating and forming processes of steel that predominantly contain elemental iron and oxides of iron such as wustite (FeO), hematite (α-Fe_2O_3) and magnetite (Fe_3O_4), as well as oxides of the various alloying elements (Atkinson and Kolarik, 2001; Cho and Lee, 2008; Umadevi et al., 2009; Farahat et al., 2010; Bagatini et al., 2011; Martin et al., 2012).

It is estimated that over 13.5 million tonnes of mill scale wastes were being produced annually, translating to about 2% of the global steel production (Cho and Lee, 2008; Bagatini et al., 2011; Umadevi et al., 2009). Depending on the process from which it is generated and the nature of the product, the weight of mill scale can vary from 20 to 50 kg/t of hot-rolled product (Martin et al., 2012). The total iron content is normally in the range of 70%, with traces of non-ferrous metals and alkaline compounds, as well as contamination from lubricants, oils and greases from the equipment and lubricants associated with the rolling operations (Umadevi et al., 2009; Bagatini et al., 2011; Martin et al., 2012).

The chemical composition and characteristics of mill scale makes it suitable for direct internal recycling in the steel plant (Atkinson and Kolarik, 2001; Poulalion, 2001; Cho and Lee, 2008; Umadevi et al., 2009; Bagatini et al., 2011; Martin et al., 2012; Meynerts, et al., 2013). In fact, about 90% of the mill scale produced globally is recycled directly within the iron and steel making process, either as pre-reduced briquetted charge for the blast furnace and/or basic oxygen furnace or as partial raw material additives in the sinter plant (Poulalion, 2001; Martin et al., 2012; Meynerts, et al., 2013). However, it is reported that full-scale recycling of mill scale is still difficult for mini-mill operators whose operations are solely based on the electric arc furnace with no pre-reduction facilities (Cho and Lee, 2008; Bagatini et al., 2011). In order to circumvent the capacity constraints of recycling mill scale in mini-mill operations, several investigations also focused on the utilization of mill scale to produce sponge iron using carbothermic reduction techniques (Salak and Kubelik, 1974; Farahat et al., 2010; Martin et al., 2012). These studies have shown that, using mill scale, it is possible to produce a highly metallized iron product with a chemical composition similar to low carbon steel for direct melting in electric arc furnaces (Salak and Kubelik, 1974; Farahat et al., 2010; Martin et al., 2012).

The main challenges to the direct recycling of mill scale arise as a result of mill scale contamination with lubricants, oils and greases (Atkinson and Kolarik, 2001). The lubricant, oil and grease contaminants are particularly detrimental in the control of gaseous emissions in steel plants, mainly due to their propensity to result in the formation of toxic polychlorinated-p-dibenzodioxins and dibenzofurans (PCDD/Fs) (Buekens et al., 2001; Kasai et al., 2001; Fisher et al., 2004; Ooi and Lu, 2011; Remus et al., 2013). As discussed in Chapter 4, the production of these PCDD/Fs and other organometallic compounds occur as a result of high temperature combustion reactions and volatilization of organic compounds in feed materials and recycled oily mill scale additives (Buekens et al., 2001; Kasai et al., 2001; Fisher et al., 2004; Ooi and Lu, 2011; Remus et al., 2013). Nevertheless, mill scale is still considered a valuable source of high grade alloy-rich oxides by most steel producers (Atkinson and Kolarik, 2001; Poulalion, 2001; Gavrilovski et al., 2011; Martin et al., 2012; Meynerts, et al., 2013).

8.3 Pickling of Formed and Heat-Treated Steel Products

8.3.1 Overview of the Pickling Process

The aesthetics, properties and quality of stainless steel steels are based on a defined, homogenous and corrosion-resistant surface finish (Rögener et al., 2012). The generation of a homogenous and passivated surface finish on stainless steels by the removal of the adherent oxide and chromium-depleted layers is done through a pickling process (Li and Celis, 2003; Li et al., 2005; Rögener et al., 2012; Devi et al., 2014). In general, metal pickling processes typically consist of mechanical, electrochemical and/or chemical operations used to remove adherent oxide scales, rust, dust and the chromium-depleted layers that form on the surface of steels during metal forming and heat treatment processes (Li et al., 2005).

In order to modify the oxide scale film and make the subsequent pickling steps more efficient, the oxide descaling step is usually carried out using mechanical scale breaking and/or shot blasting techniques (European Commission, 2001; Li and Celis, 2003; Li et al., 2005). The descaling step is usually followed by preliminary pickling processes, such as chemical pickling, anodic or cathodic electrochemical pickling in acidic or neutral electrolytes and salt pickling (Li and Celis, 2003; Li et al., 2005). According to Li and Celis (2003), the main reasons for adopting preliminary pickling processes are two-fold: (1) to electrochemically and chemically alter the oxide scale, thereby increasing its solubility in pickling solutions, and (2) to oxidize metallic ions and loosen the oxide scale, thereby enhancing the effectiveness of the final pickling operations.

8.3.2 Preliminary Pickling Processes

Technically, chemical acid pickling is essentially used to reduce or oxidize the oxides in the adherent oxide scale in order to increase its solubility in the pickling solutions (Li and Celis, 2003). Typically, the chemical acid pickling process uses various mixtures of inorganic acids such as hydrochloric acid (HCl) and sulphuric acid (H_2SO_4) or mixtures of sulphuric acid and sodium chloride (NaCl) (Li and Celis, 2003). The use of nitric acid has been extensively adopted as well and mainly plays the role of oxidizing the metal oxides in the oxide scale to higher oxidation states without destroying the corrosion-resistant qualities of stainless steels (AISI, 1982; European Commission, 2001; Li and Celis, 2003).

Basically, the electrochemical pickling processes can be classified into anodic, cathodic and alternating current pickling operations (Li and Celis, 2003; Li et al., 2005, 2008). In essence, electrochemical pickling processes are aimed at either electrochemically altering the oxides in the oxide scale layer or the oxides beneath the oxide scale layer so as to enhance the effectiveness of the final chemical pickling operations (Li et al., 2008). The application of the various electrochemical pickling processes to various grades of stainless steels has been discussed in detail in several studies (AISI, 1982; Li and Celis, 2003; Li et al., 2005, 2008).

Other variations of electrochemical pickling processes include salt bath and electrolytic pickling (AISI, 1982; Li and Celis, 2003; Li et al., 2008). Essentially, salt bath pickling can be performed in a reducing molten bath, oxidizing molten or aqueous bath or electrolytic molten bath (AISI, 1982; Li and Celis, 2003; Li et al., 2008). The efficiency of the salt bath pickling process is usually high, and the process is capable of producing a smooth and uniform surface finish, as well as decreasing the duration of the pickling process (Li and Celis, 2003; Li et al., 2008). However, the main disadvantages of the salt bath processes are the associated high costs and challenges associated with handling molten salts at high temperatures (Li et al., 2008).

The electrolytic pickling process using neutral Na_2SO_4 at 65°C–85°C is widely used in combination with the HNO_3-HF-mixed acid pickling to provide the fast removal of oxide scale (Braun, 1980; Hildén et al., 2000; Li and Celis, 2003; Li et al., 2008). In fact, this technique has widely been adopted in modern pickling operations because of the following advantages (Braun, 1980; Hildén et al., 2000; Li and Celis, 2003; Li et al., 2008): (1) the ability to obtain good surface finish, (2) low metal loss, (3) easy control of the pickling rate, (4) improved in-service life of the plant due to lower operating temperatures and less harsh conditions, (5) low operating and maintenance costs, (6) fewer environmental problems associated with effluents and (7) more hygienic working conditions for operators.

8.3.3 Final Pickling Processes

The highly acidic HNO_3-HF mixtures are commonly used as final pickling solutions for the industrial pickling of many grades of stainless steels (Li and Celis, 2003; Rögener et al., 2012). The main function of the mixed acid pickling solution is to remove the oxide scale and the chromium-depleted layers by dissolution, as well as passivating the newly formed surface to form a homogenous and corrosion-resistant surface finish (Li and Celis, 2003). The mechanism of the pickling process from the mixed acid solutions can be described as involving either of the following steps: (1) HNO_3, being a strong oxidant, dissolves and removes the oxides and the chromium-depleted layer and (2) the HF then acts as a regenerator of H^+ ions, complexing agent for the Fe^{3+} and Cr^{3+} ions, a redox potential stabilizer of solutions and effective de-passivator of the newly formed surface film (Li and Celis, 2003).

8.3.4 Waste Generation in Pickling Process

The various pickling processes discussed earlier generate solid waste in the form of scrap, scale dusts and pickling tank sludge, and liquid wastes in the form of spent pickling solutions and wastewater from the neutralization process (European Commission, 2001; Twidwell and Dahnke, 2001; Li et al., 2005; Rögener et al., 2012; Devi et al., 2014). For example, the acid regeneration process can generate solid waste in the range of 0.05–15 kg of sludges on a dry basis per ton of steel, while the solid sludge from a stainless steel pickling process can contain 55–66 wt.% Fe, 5–10 wt.% Cr and 3–5 wt.% Ni (European Commission, 2001). In general, the chemical constituents and amount of metal ions contained in the solid and liquid pickling wastes in the processing of stainless steels are largely influenced by factors such as (1) the composition and characteristics of the oxide scale and chromium-depleted layer, (2) the pickling duration and temperature conditions and (3) the type and concentration of the pickling solutions (Li and Celis, 2003). Due to the nature and composition of the oxide scale, the dewatered sludges formed from the neutralization process are particularly rich in valuable elements such as iron, chromium, nickel, molybdenum and cobalt (Rögener et al., 2012).

The process steps involved in the generation of waste streams in a typical stainless steel pickling process is shown in Figure 8.2 (Rögener et al., 2012; Devi et al., 2014). Figures 8.3 through 8.5 further depict the process steps and material flow in the hydrochloric acid, sulphuric acid and mixed acid (HF-HNO_3) pickling operations, respectively (European Commission, 2001). In practice, the spent acid effluent is neutralized using basic chemical compounds such as slacked lime or caustic soda, resulting in the precipitation of metal ions as hydroxides or oxides (Regel-Rosocka, 2009; Rögener et al., 2012). According to Rögener et al. (2012), the dewatered sludges from pickling wastes produced by German manufacturers typically contain metal compounds in the range of 18 wt.% Fe, 40 wt.% Cr, and 2.5 wt.% Ni. Due to the presence of toxic heavy metal ions, particularly the potential of generating hexavalent chromium species, the pickling process wastes are typically categorized as controlled and hazardous waste materials which must be treated before disposal and/or recycled for metal recovery (Atkinson and Kolarik, 2001; Rögener et al., 2012; Remus et al., 2013).

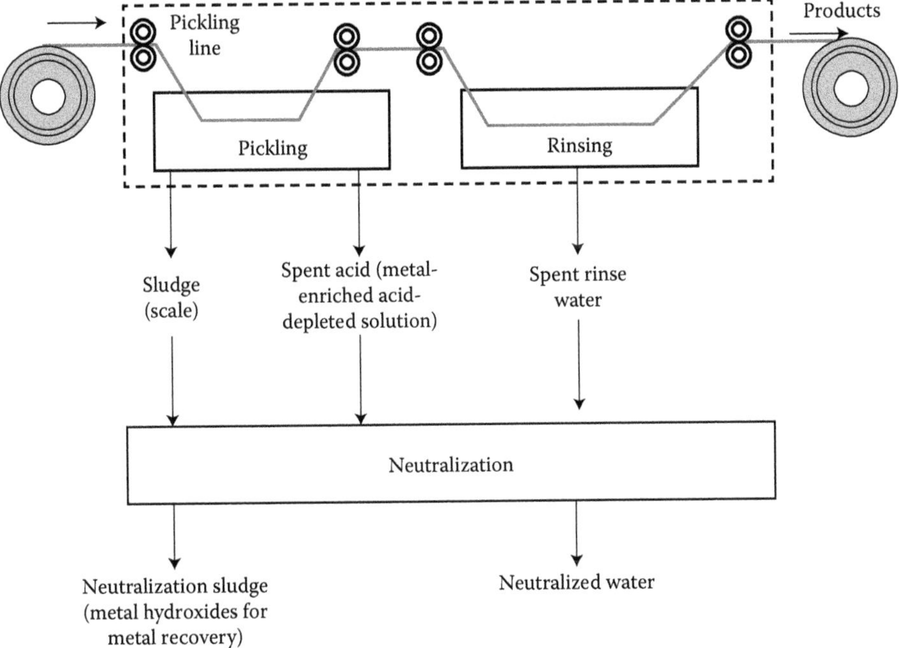

FIGURE 8.2
Generation of waste streams in pickling lines. (Adapted from *Resourc. Conserv. Recycl.*, 60, Rögener, F., Sartor, M., Bàn, A., Buchloh, D. and Reichardt, T., Metal recovery from spent stainless steel pickling solutions, 73. Copyright 2012, with permission from Elsevier.)

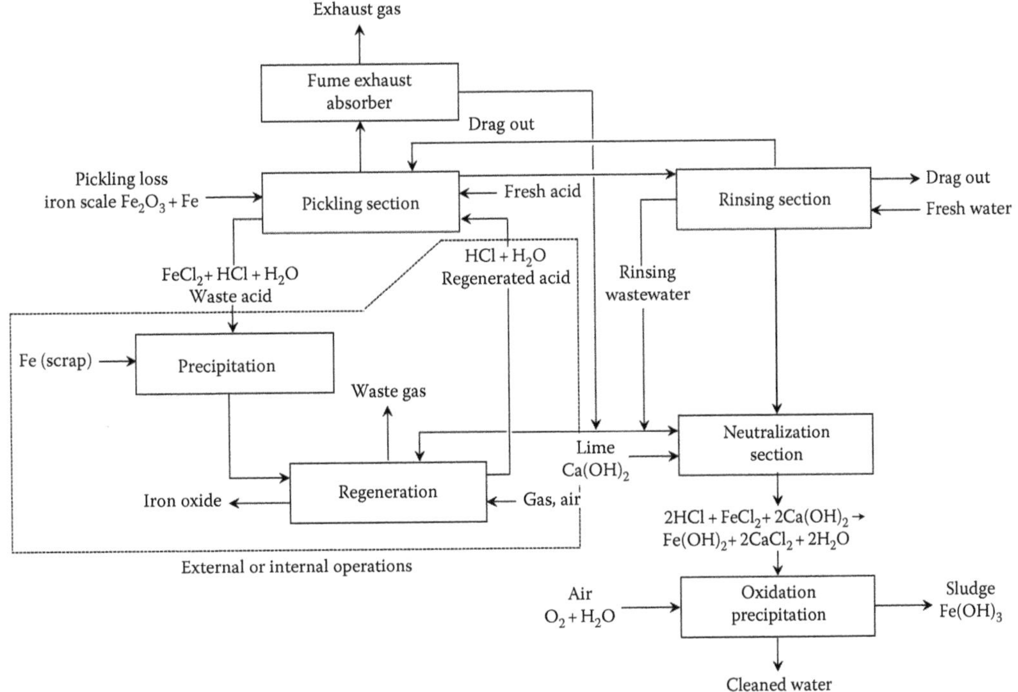

FIGURE 8.3
Process steps and material flow in the hydrochloric acid pickling line. (From European Commission, 2001.)

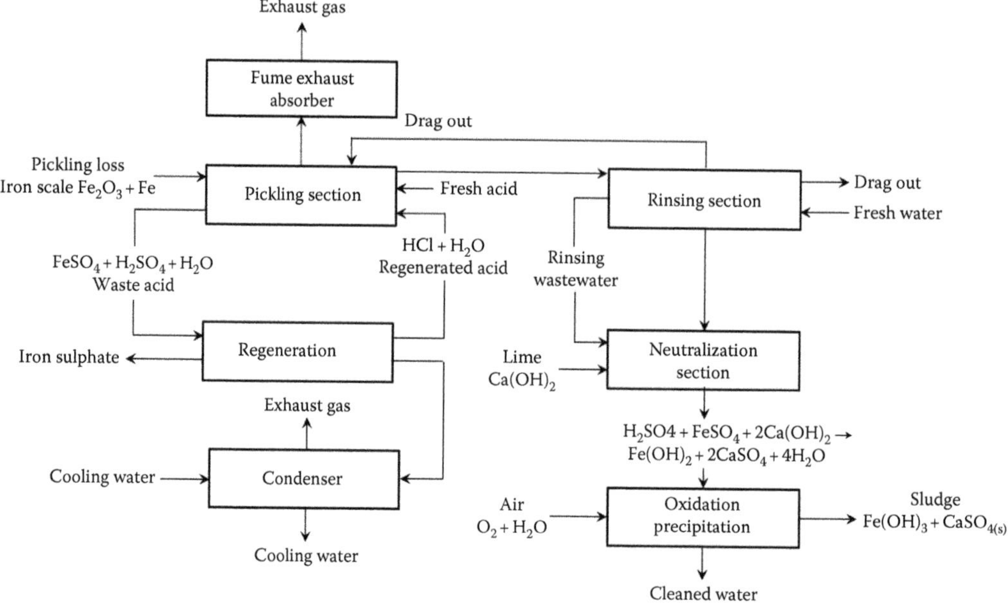

FIGURE 8.4
Process steps and material flow in the sulphuric acid pickling line. (From European Commission, 2001.)

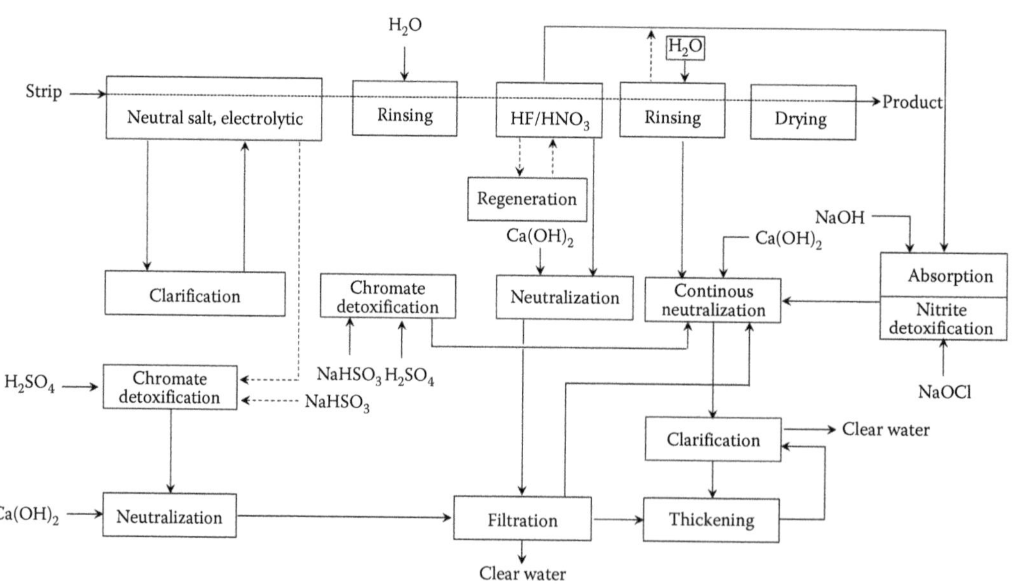

FIGURE 8.5
Process steps and material flow in the mixed ($HF\text{-}HNO_3$) acid pickling line. (From European Commission, 2001.)

8.4 Metal Finishing Processes

Generally, the final finishing processes include surface coating techniques, such as electroplating, galvanizing, and painting, and coating, that are adopted to enhance the surface finish of steel products and impart desirable properties such as corrosion and chemical resistance, biocompatibility, reflectivity and appearance, hardness, electrical conductivity, wear resistance, and other engineering or mechanical properties (EPA, 1984, 1997; Sharma, 1985; Lindsay, 2007; Devi et al., 2014). Technically, the metal finishing industry classifies the metal finishing operations into two main categories (EPA, 1997): (1) plating processes (i.e. electroplating and electroless plating) and (2) surface treatment processes (i.e. chemical and electrochemical conversion, case hardening, metallic coating and chemical coating).

8.4.1 Electroplating and Surface Treatment Processes

Electroplating: Electroplating refers to the generic electrochemical finishing processes that involve applying metallic coatings by passing an electric current through a solution containing dissolved metal ions and the metal object to be plated (EPA, 1997). In other words, electroplating is a process by which a thin layer of metal is cathodically deposited on an electrically conductive metal or substrate material so as to achieve the desired surface characteristics (Sharma, 1985). In practice, electroplating results in the production of a thin surface coating of one metal upon another by cathodic electrodeposition, and the process is largely controlled by process parameters such as voltage and current, bath temperature, residence time and the purity of bath solutions (EPA, 1984, 1997; Sharma, 1985). In the electroplating process, metal ions supplied by the dissolution of metal from anodes are reduced on the workpiece to be plated (which acts as the cathode) (EPA, 1984). In addition to the metal plating ions, electroplating baths also contain metal salts, organometallic and organic additives to induce grain refinement, uniformity of plating and brightness of the plated surfaces (EPA, 1984).

Electroless plating: Electroless plating involves the chemical deposition of a metal coating onto an object using chemical reactions rather than electric current (EPA, 1984, 1997). In most cases, complexing agents and buffers are used to hold the metal ions in solution and maintain bath stability, respectively (EPA, 1984, 1997). A typical example of electroless plating is the plating and electroforming of nickel from nickel ions in solution using chemical reductants such as sodium hypophosphite, sodium borohydride and dimethylamine borane (Parkinson, 2001).

Chemical and electrochemical conversion: According to the EPA (1997), the chemical and electrochemical conversion treatments are used to deposit a protective and/or decorative coating on metal surfaces, and include processes such as (1) chromating treatment using hexavalent chromium within a certain pH range to deposit an adherent protective film on metal surfaces (EPA, 1997; Lindsay, 2007); (2) phosphating treatment, which is usually used to prepare the metal surfaces for further processes such as painting (EPA, 1997); (3) anodizing treatment, where metal workpiece is immersed in a solution containing metal salts and/or acids where electrochemical reactions occur to form a non-porous metal oxide layer that provides good corrosion resistance (EPA, 1997; Lindsay, 2007); (4) passivation treatment, where the metal workpiece is immersed in acid solution, usually nitric acid or mixture of nitric acid and dichromate solution, to enhance the chromium fraction in the passive film (EPA, 1997; Wegrelius and Sjödén, 2004) and (5) metal colouring treatment, which involves treating the metal workpiece to impart a decorative and colourful finish (EPA, 1984, 1997).

Chromium plating processes: Due to the sensitivity surrounding the anthropogenic and health effects of chromium compounds, particularly hexavalent chromium, the different chromium plating processes warrant further discussion in this section. Technically, as elaborated by EPA (1997), the chromium plating and anodizing operations include (1) hard chromium electroplating, where a relatively thick layer of chromium is deposited directly on the substrate metal to provide a surface with wear resistance, low coefficient of friction, hardness and corrosion resistance; (2) decorative chromium plating applied to metals and plastics and (3) chromic acid anodizing, particularly used on components subjected to high stress and corrosion (such as aircraft and architectural structures), and trivalent chromium plating, particularly developed to circumvent the toxicity problems associated with decorative hexavalent chromium plating baths.

8.4.2 Galvanizing Process

Generally, metal plating or galvanized coatings using zinc and its alloys are primarily applied to iron and steel products to protect the base metal from corrosion (EPA, 1984; Mansur et al., 2008; Sciscenko et al., 2016). The process of metal plating using zinc and its alloys is termed zinc plating or hot-dip galvanizing and involves the deposition of a layer of zinc on the substrate metal by immersing metal substrates in molten zinc at temperatures around 450°C–460°C (Mansur et al., 2008). Essentially, galvanizing provides corrosion resistance and decoration.

In essence, the galvanizing process involves a sequence of pretreatment steps (such as washing, degreasing and oxide removal using hydrochloric acid media), coating and surface conversion baths (Mansur et al., 2008; Hackbarth et al., 2016). The molten metal plating bath typically contains zinc compounds such as acid sulphates and chlorides and alkaline non-cyanide and cyanide solutions, as well as conducting salts, buffers and surfactant additives (EPA, 1984, 1997; Sciscenko et al., 2016).

8.4.3 Waste Generation from Electroplating and Galvanizing Coating Processes

Chromium-based plating processes: Typical pretreatment processes in hard chromium plating processes include pretreatment (e.g. polishing, grinding and degreasing), alkaline cleaning, acid dipping, chromic acid anodizing and chromium electroplating (EPA, 1997). Typically, the degreasing pretreatment step consists of either dipping the metal to be coated in organic solvents or using organic vapours from organic solvents to remove the surface grease (EPA, 1997). In addition to high levels of both the hexavalent and trivalent chromium species in dewatered neutralization sludges, other waste emissions produced from chromium plating processes would include the chromium-contaminated wastewaters, spent electroplating bath solutions, mists generated as a result of gas evolution from electrochemical reactions and organometallic sludges and vapours emitted from the pretreatment processes (EPA, 1997). The processes and methods to mitigate the emissions, as well as the recycling and utilization opportunities, of these toxic waste materials are discussed briefly in Sections 8.4.4 and 8.4.5.

Galvanizing coating process: The galvanic treatment process results in the generation of potentially toxic waste streams in the form of wastewater, neutralization sludge and mists contaminated with heavy metal cations, principally zinc, iron, chromium and nickel (EPA, 1997; Twidwell and Dahnke, 2001; Mansur et al., 2008; Hackbarth et al., 2016). As discussed in the preceding sections, the presence of these heavy metal cations have serious health and environmental implications, and as such, must be treated for metal recovery or inertized before disposal.

8.4.4 Treatment of Electroplating and Galvanizing Waste Streams

In general, the treatment of waste streams produced in the chromium plating and galvanizing processes may involve any of the following steps (Twidwell and Dahnke, 2001; Mansur et al., 2008; Hackbarth et al., 2016): (1) reduction of Cr (VI) to Cr (III) under controlled pH conditions using ferric chloride ($FeCl_2$), sodium sulphides (Na_2S) or sodium bisulphite ($NaHSO_3$); (2) oxidation of Fe (II) to Fe (III) using hydrogen peroxide (H_2O_2); (3) precipitation of metals through the addition of basic oxides and/or hydroxides in the pH range 7.0–8.5; (4) coagulation and flocculation of suspended solids and (5) sludge thickening and dewatering. As a result of a series of neutralization and redox reactions stated earlier, the dewatered sludge is typically rich in metal ions and compounds, mostly as oxides and hydroxides, and is thus treated for metal recovery. Despite the efforts to recover most of the heavy metals as solid sludge using the conventional treatment processes, the complete removal of heavy metals from process water is usually not 100% efficient, and as a result, the wastewater requires further treatment and quality control before it is disposed into the environment (Mansur et al., 2008; Bhatnagar and Sillanpää, 2010).

8.4.5 Waste Recycling and Utilization from Pickling and Metal Finishing Processes

The generation of waste streams from the pickling and metal finishing processes was discussed in detail in the preceding sections. Due to the similarities in the nature and composition of the waste streams, the treatment of waste streams from these two processes, in terms of metal recovery from neutralization sludges, is broadly discussed in this section. The typical and emerging techniques in the treatment and regeneration of spent pickling solutions are also discussed.

8.4.5.1 Metal Recovery from Pickling and Plating Sludges

Notwithstanding the environmentally toxic characteristics of solid and liquid pickling and surface treatment wastes, these waste materials also provide an opportunity to recover the valuable metals from the metal-rich dewatered neutralization sludges (Twidwell and Dehnke, 2001; Rögener et al., 2012). In general, the incentives for the recovery and recycling of metals from pickling and plating sludges include (Brooks, 1985; EPA, 2002) (1) maximizing the reuse of strategic metals with significant economic value, (2) minimizing the anthropogenic effects from uncontrolled emissions and disposal and (3) minimizing the costs of hazardous waste disposal. In fact, the recovery of metals as hydroxides or oxides from the dewatered neutralization sludges can contribute to the sustainable utilization of these pickling sludges. Several studies have successfully developed solutions to recover the valuable metals from pickling and plating waste streams using various processes such as (1) electrochemical deposition of metal cations from solution (Rögener et al., 2012), (2) chemical recovery processes (such as precipitation, crystallization, ion exchange and reduction) (Brooks, 1986; Twidwell and Dahnke, 2001; EPA, 2002) and (3) high temperature carbothermic reduction of dewatered neutralization sludges (Ma et al., 2005; Matinde et al., 2009; Li et al., 2014).

8.4.5.2 Treatment and Regeneration of Spent Pickling Solutions

The efficiency of the pickling process typically decreases with increasing amount of dissolved metal ions in the bath (Regel-Rosocka, 2009). In most cases, the pickling solutions are considered spent when the acid concentration in them decreases by 75%–85%, which is usually accompanied by significant metal content increases of up to 150–250 g/dm^3 (Regel-Rosocka, 2009). Naturally, spent pickling solutions contain both significant amounts of toxic heavy metal ions and residual free acid (Regel-Rosocka, 2009; Rögener et al., 2012). Despite the spent liquor being relatively dilute solutions of low-cost heavy metals, the recovery of metal ions and the regeneration of free acid from the spent pickling solutions are recommended, mostly from an environmental and economic point of view (Brooks, 1986; Regel-Rosocka, 2009; Rögener et al., 2012; Devi et al., 2014; Nleya et al., 2016). The typical methods and processes applied in the regeneration of spent pickling solutions (SPS) have been discussed in detail by several researchers. In particular, the merits and demerits of the traditional techniques of treating spent pickling solutions are summarized in Table 8.2 (Regel-Rosocka, 2009).

8.5 Foundry Processes

8.5.1 Introduction

The different unit processes in the integrated iron and steel process were discussed in detail in Chapter 4. Basically, Chapter 4 covered the production of bulk molten steel from ores and metal scrap, the refining processes of liquid steel and the solidification and casting processes. The integrated iron and steel plants operate on huge economies of scale, require strict control of raw materials and, as a result, tend to be specific to certain geographic locations. In contrast, foundries and mini-mills are ubiquitous in terms of scale, raw materials flexibility and investment costs, and hence play an important role in meeting the requirements for near-net shape ferrous and non-ferrous components, as well as enabling the circular economy of steel through the recycling of metallic scrap.

This section discusses in detail the role of foundry processes in the melting of scrap, pig iron, ingots and alloys and the near-net casting of ferrous and non-ferrous products. Furthermore, the typical unit steps, particularly in ferrous foundries, the categories and types of waste streams produced, as well as the valorization opportunities of the waste materials, are also discussed.

TABLE 8.2

Merits and Demerits of Typical Methods for Treating Spent Pickling Solutions (SPS)

Method	Type of SPS	Merits	Demerits
Fluidized bed pyro-hydrolysis at 800°C	HCl, Fe	Able to recover HCl from condensed gas and Fe as iron oxide granules Capable to treat high volumes High efficiency of acid recovery	Energy intensive process Limited capabilities on solutions containing Zn (II) ions above 0.5 g/dm^3, which tends to accrete in installation walls and contaminate iron oxides
Spray roasting at 450°C	HCl, Fe, $ZnHNO_3$, HF, Fe (III)	Effective for large amounts of spent solutions Reduced wastewater volumes and sludge Operating costs recovered by value of regenerated acid and recovered metal oxides Widely adopted in industry	Limited by Zn (II) concentrations High operating costs, energy consumption and requires complex installation High release of NOx
Precipitation/ neutralization	HCl, Fe, $ZnHNO_3$, HF	Low operating costs Neutral sludge instead of acidic wastewater Simple technique and equipment Applicable to small plants	Large consumption of chemicals Hazardous precipitation with high chloride content Metals forming hazardous sludges/precipitates Expensive storage of sludge High nitrogen content above regulated amount
Evaporation	HNO_3, HF	Recovery of HNO_3, HF, hence reduced fresh acid consumption Small changes in content of fluoride and nitrate in acid mixture No nitrates emission in wastewater Adoptable in industry	High investment and operating costs High energy consumption
Retardation/ion exchange	HCl, Fe, $ZnHNO_3$, HF, Fe (III)	Effective retention of Zn in the resin High selectivity to Zn and Fe separation Low operational costs Simple equipment and space	Production of high volumes of waste Production of diluted solutions of metal salts High consumption of fresh water
Crystallization	HF, HNO_3, Fe, Cr, NiH_2SO_4, Fe	Low cost and reduction of waste disposal Recovery of total amount of metal in recyclable form Recirculation of HNO_3/HF mixture to the pickling process, hence reduced consumption of fresh acid	Bleeding out of acid containing high nickel concentration Risk of scale formation in the crystallizer Possible emissions to air Increased consumption of energy

(Continued)

TABLE 8.2 (*Continued*)

Merits and Demerits of Typical Methods for Treating Spent Pickling Solutions (SPS)

Method	Type of SPS	Merits	Demerits
Solvent extraction	HCl, Zn, Fe	Tributyl phosphate (TBP) effective for wide range of Zn concentration in feed Good selectivity of Zn over Fe (II) with TBP Acid extractants permit Zn extraction up to 100 g/dm^3 after stripping High throughput with compact equipment	Organic impurities in aqueous phase and extractant loss Adsorptive pretreatment necessary to remove impurities Co-extraction of Fe (III) with Zn High treatment costs at high Zn concentrations
	HNO_3, HF, Fe (III)	High flexibility on Zn concentration in feed Closed circuit continuous process Clean valuable product solutions	Problems with phase separation after stripping Diluted strip solution requires further treatment
	H_2SO_4, Fe (III)	Recycling of the recovered acid to pickling Ability to manage large volumes of solutions with high content of toxic solutes	

Source: Adapted from *J. Hazard. Mater.*, 177, Regel-Rosocka, M., A review on methods of regeneration of spent pickling solutions from steel industry, 60. Copyright 2009, with permission from Elsevier.

8.5.2 Process Description of Foundry Processes

Foundries play an important role in the manufacture of value-added ferrous and non-ferrous products (European Commission, 2005). Technically, foundry processes are used to melt ferrous and non-ferrous metals and alloys, and reshape them into products at or near their finished shape by pouring and solidifying molten metal or alloy in a mould (European Commission, 2005). In other words, foundry processes produce metal castings and components by the near-net shape solidification of molten metal in moulds, allowing it to cool and solidify, and then separating the solidified product from the mould and core materials in the shake-out process (Oman, 1998).

The foundry industry consists of a wide range of installations of different sizes and highly differentiated capabilities, as well as diverse products range (European Commission, 2005). Basically, foundries can be classified into three broad categories (Bawa, 2004): (1) jobbing foundries, which are basically based on job orders; (2) production foundries, with a high degree of mechanization, which produce castings on a large scale and (3) captive industry foundries, which basically form an integral part of internal manufacturing value chain to produce castings for further processing within the organizational system boundaries.

Furthermore, the foundry sector can also be broadly categorized according to the type of metal, vis-à-vis, ferrous and non-ferrous foundries (European Commission, 2005). In addition, the ferrous foundry sector can further be categorized into two broad groups based on the type of alloy being produced, that is, into (1) cast iron foundries (for producing different grades of cast iron components) and (2) cast steel foundries (carbon steels, low-alloy steels and stainless steels). On the other hand, the non-ferrous foundries can be subdivided according to the type of metal or alloy processed (Bawa, 2004; European Commission, 2005): (1) aluminium and its alloys such as those containing silicon, copper and/or magnesium, (2) copper and its alloys, (3) zinc and its alloys, (4) super alloys (such as nickel or cobalt-based alloys in combination with chromium, iron, manganese, molybdenum, etc.), (5) magnesium and its alloys and (6) titanium and its alloys. Tables 8.3 and 8.4 summarize the different attributes and respective foundry processes for the selected types of ferrous and non-ferrous alloys.

The typical process steps in foundry operations include (1) melting and metal treatment (melting shop); (2) mould/core preparation (moulding shop); (3) casting of the molten metal into a mould, cooling for solidification and removing the casting from the mould (casting shop) and (4) cleaning and finishing of the solidified castings (finishing shop) (European Commission, 2005). The generic steps and the material flows in a typical foundry process are depicted in Figures 8.6 and 8.7, respectively.

The following sections discuss some of the key foundry processing steps, and the different types of waste streams generated from such processes. Due to the sheer volumes and diversity of ferrous metals foundry products, particular emphasis is given to the processes and material flows in the cast iron and steel foundry operations.

8.5.3 Generation of Waste Streams in the Foundry Melting Process

8.5.3.1 General Fundamentals

The melting and refining of metallic charge is the initial step in preparing the liquid metal for casting in foundry operations (Katz and Tiwari, 1998; European Commission, 2005). Typically, the steps in the melting and refining processes of cast iron and cast steel comprise the following steps (EPA, 1995): (1) scrap preparation, (2) furnace charging, (3) melting, (4) backcharging (i.e. addition of more metal and alloys), (5) refining by single (oxidizing) slag or double (oxidizing and reducing) slagging operations, (6) oxygen lancing (i.e. the injection of oxygen into the steel bath to adjust the chemistry of the metal and speed up the melting process) and (7) tapping the metal into ladles or directly into mould. In general, wide ranges of melting furnaces are employed by the foundry industry, and these include cupola furnace melting, induction furnace melting, and the classical electric arc furnace melting (Davies, 1996; Katz and Tiwari, 1998; European Commission, 2005).

8.5.3.2 Waste Generation in the Melting Process

The different melting processes and the waste streams produced in the respective processes are discussed in Table 8.5. Furthermore, Figures 8.8 and 8.9 highlight the respective flowsheets in the production of cast iron and cast steel foundry products. Generally, the physicochemical characteristics and chemical composition of the different waste streams produced in the melting processes differ from operation to operation and are largely controlled by the type and quality of raw materials (particularly the metallic scrap), grade and type of alloy produced and the type of the melting process. Technical discussions on the smelting and refining processes of different grades of ferrous alloys, particularly using the electric arc furnace, were covered in Chapter 4, and as such, are not discussed further in this section.

TABLE 8.3

Typical Foundry Processes of Selected Ferrous and Superalloy Castings

Alloy	Properties and Uses	Melting and Casting Processes	References
Cast iron	These are basically Fe-C alloys (2.4–4 wt.% C) with low melting point and good castability properties. Cast irons are classified according to (1) amount of carbon into lamellar iron (carbon in form of flakes), nodular iron (spheroidal carbon form), and compact graphite iron (carbon in bonded form), (2) according to material properties into grey iron, ductile iron, and malleable iron or (3) according to alloying elements (alloy cast irons) such as Ni, Cr, Mo, Cu, and Mn. Typical example of alloy cast irons is high-alloy white cast iron with alloy content above 4 wt.%.	Cast iron is commonly melted using cupola furnaces, induction (coreless) furnaces or in rotary furnace. Casting of cast irons is commonly done in green sand mould with resin-bonded cores. In some cases, chemically bonded sand moulding and die casting are also practised for iron castings that are required in smaller numbers and/or intricate shapes and smooth finish. The generic process flow sheet for the cupola melting process is shown in Figure 8.5.	Davies (1996), Katz and Tiwari (1998), European Commission (2005), Boljanovic (2010)
Cast steels	Steels constitute one of the most versatile and widely applied engineering materials. Technically, cast steels contain other alloying elements such as Cr, Ni, Mo and Mn, and the carbon content does not usually exceeding 2 wt.% (exceptions may exist for some Cr-cast steels). Categories of cast steels include low-alloy steels (typically less than 5 wt.% of combined alloy elements such as Cr, Ni, Mn and Mo), high-alloy steels (typically more than 5 wt.% combined alloying elements) and special steels (these are specifically produced with engineered properties such as higher strength, better resistance to corrosion and fatigue wear).	Cast steels are usually melted in basic refractory-lined electric arc furnaces (shown in Figure 8.6) or in induction (coreless) furnaces. Due to the high-alloy content, the molten cast steel may require refining (such as decarburization, desiliconization, dephosphorization, and desulphurization), deoxidation and degassing treatment. The casting is usually done in sand and ceramic moulds. However, the low fluidity of cast steels limits the design of thin sections. Heat treatment of solidified cast steel components is usually required to homogenize the microstructures and obtain isotropic properties.	European Commission (2005), Boljanovic (2010)
Super alloys	Typically Ni, Ni–Fe and Co based alloys with Cr, Ti, W, Al additions. Ni-base alloys typically contain over 50 wt.% Ni and less than 10 wt.% Fe. Co-base superalloys contain 40–70 wt.% Co and 7–15 wt.% W.Super alloys are commonly used for high temperature applications (e.g. aircraft and gas turbines) and/or in severe corrosive environments, in cryogenic temperature appliances and in biomedical applications (e.g. orthopaedic and dental prosthesis).	Typical melting processes include vacuum induction furnace melting to reduce the content of interstitial gases. Some alloys such as Ni–Fe and Co-based alloys can be directly melted in EAFs by classical methods similar to stainless steels.Casting processes include investment casting (in ceramic moulds) up to very precise dimensions. Directional casting applied to aircraft gas turbine manifolds to eliminate grain boundaries and increases strength. Hot isostatic pressing also applied to eliminate internal porosity in large castings.	European Commission (2005), Maurer (2012), Boesch (2012), Boljanovic (2010)

TABLE 8.4
Typical Foundry Processes for Selected Non-Ferrous Alloys

Alloy	Uses	Melting Processes	Casting Processes	References
Aluminium	Light weight material with applications in automotive, aircraft and architectural industries	Ingot melting in reverberatory furnaces Induction furnace melting Crucible furnace melting	Pressure die casting Low pressure die casting Gravity casting Sand casting	European Commission (2005), Davies (1993), Boljanovic (2010)
Magnesium	Light-weight material with applications in aerospace and automotive industries	Induction melting Resistance furnaces Gas-heated furnaces	Pressure die casting Gravity casting Sand casting	Kainer (2003), European Commission (2005), Boljanovic (2010)
Copper	Electrical and thermal conductor applications Marine engineering applications due to corrosion resistance	Reverberatory furnaces Induction furnaces Crucible (tilting or stationary furnaces)	Sand casting Die casting Centrifugal molding Investment casting	Davies (2001), Konečná and Fintová (2001), European Commission (2005), Boljanovic (2010)
Titanium alloys	Light weight material applications in aerospace, biomedical and chemical industries	Vacuum arc remelting	Investment casting Hot isostatic pressing	Donachie (2000)

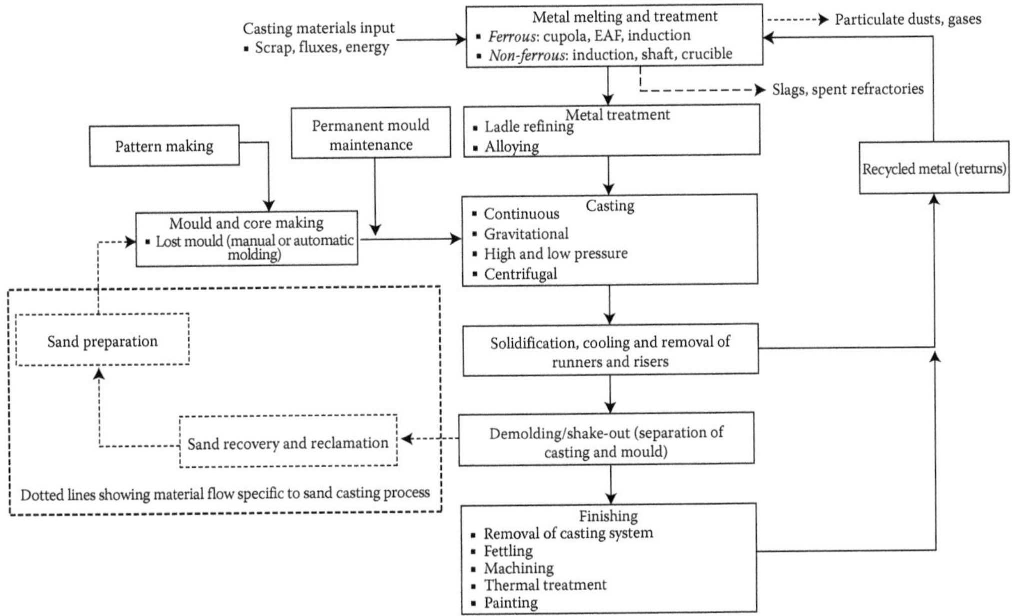

FIGURE 8.6
Schematic representation of typical steps in a foundry process. (From European Commission, Integrated pollution prevention and control: Reference document on the best available techniques in the smitheries and foundries industry, 2005, p. iii, Available at http://eippcb.jrc.ec.europa.eu/reference/BREF/sf_bref_0505.pdf, Accessed 6 June 2016. With permission.)

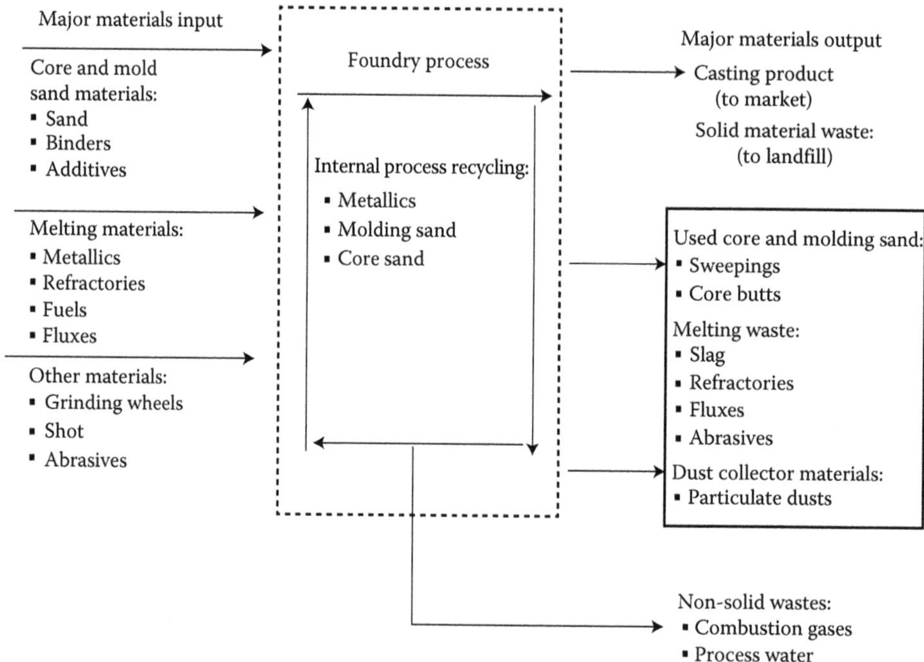

FIGURE 8.7
Typical material flow in a generic foundry operation. (From Oman, D.E., *J. Air Pollut. Control Assoc.*, 38, 933, 1998. With permission.)

8.5.4 Generation of Waste Streams in the Moulding Process

8.5.4.1 Fundamental Considerations

Basically, the shape and dimensions of the manufactured casting products are determined by the shape and dimensions of the mould (Blair and Stevens, 1995). The mould is produced by shaping a refractory material to form a cavity of desired shape and dimensions such that molten metal can be poured into the cavity (ECS, 2006). In essence, the mould must retain its shape until the liquid metal has solidified and the casting is removed and, as a result, must meet the following minimum requirements (Blair and Stevens, 1995; ECS, 2006): (1) must be strong enough to sustain the weight of the molten metal, (2) must be constructed in such a way as to permit the flow of gases formed within the mould or mould cavity to escape, (3) must be capable to resist the erosive and corrosive action of liquid metal, (4) must be collapsible to allow the metal to contract without undue restraint during solidification, (5) must be able to strip away from the casting after the casting has sufficiently cooled and (6) must be cost-effective due to shear volumes of castings and large amounts of refractory materials consumed.

According to ECS (2006), the versatility and selection of metal casting is controlled by several factors, and these include (1) the required surface quality, (2) the required dimensional accuracy, (3) the type of pattern/core box equipment, (4) the cost of making the moulds and (5) the impact of the selected process on the design of the casting. Basically, moulding processes can be broken down into four general categories (EPA, 1992; Davies, 1996; European Commission, 2005; ECS, 2006): (1) sand casting process, (2) permanent mould processes, (3) ceramic processes and (4) rapid prototyping. Out of these four

TABLE 8.5

Typical Waste Streams in Foundry Processes

Unit	Process Description	Description of Typical Waste Streams	References
Cupola furnace melting	Traditionally the most predominant melting furnace in ferrous foundries. Essentially, a cupola furnace is a fixed bed cylindrical shaft furnace, in which alternate layers of metal scrap, foundry coke, fluxes and ferroalloys are charged from the top. Cupolas are refractory lined (mostly with fireclay bricks) vessels and the melting takes place by heat transfer as a result of the counter-current gas–metal flow. Combustion of coke using oxygen-enriched air takes place in the tuyeres zones of the shaft. Hot blast cupola process uses preheated air, as opposed to cold blast process. Supernatant slag and liquid metal are tapped separately from the tap holes. Generally, they have low energy efficiency as a result of incomplete combustion and transfer of sensible heat.	The cupola melting unit generates typical waste streams such as particulate dust and gaseous emissions, slags, slag and metal spillages and reverts. Dust generation occurs from the breakage and mechanical abrasion of feed constituents during charging, entrainment of particulate fines, and from volatilization and condensation of heavy metals species such as Zn, Cr, Pb and Cd. Gaseous emissions are a result of incomplete combustion of carbon materials, contaminants in scrap contaminants such as oil and greases. The tapping process is also a major candidate of particulate dust and gaseous emissions. Liquid skim slag and metal spillages during tapping are also major waste by-products produced.	EPA (1992), Davies (1996), Katz and Tiwari (1998), Oman (1998), European Commission (2005)
Induction furnace melting	Induction furnaces are used in foundry operations to melt both ferrous and non-ferrous metals by passing an alternating current through water cooled-coils surrounding the refractory lining of the furnace. The electric current flow generates a magnetic field which interacts with the metal charge thereby generating an induced current in the metal. The electrical resistance of the metal produces the heat required to melt the metal. Basically, there is no contact between the charge and the energy carrier, hence the versatility of induction melting to different types of alloys.	Data on broad emissions form induction furnace melting process is largely available. However, particulate dust and gaseous emissions are mostly determined by the cleanliness of the metallic scrap charged into the process. For example, charge cleanliness can be affected by rust, dirt, foundry sand, paints, galvanized metal, which in essence results in the emissions of particulate dusts as well as organometallic fumes.	EPA (1992), Blair and Stevens (1995), Davies (1996), Oman (1998), European Commission (2005)
Electric arc furnace (EAF) melting	Electric arc furnaces are commonly used in the large-scale intermittent foundry scrap-based melting of steels, and are basically limited for the high efficiency melting purposes. The furnaces can either be lined with an acidic refractory (SiO_2-based) or basic refractory (MgO-C based). Further details on the operation and physicochemical aspects of EAF processes were discussed in Chapter 4.	Basically, the EAF process generates waste emissions during the charging, melting, refining, and tapping processes. Typical waste streams include particulate dusts, slags, gaseous emissions, as well as spent refractory materials. The mechanisms of generation, physicochemical characteristics and abatement approaches of primary and secondary waste stream emissions in EAF processes were discussed in Chapter 4.	EPA (1992), Blair and Stevens (1995), Davies (1996), Oman (1998), European Commission (2005)

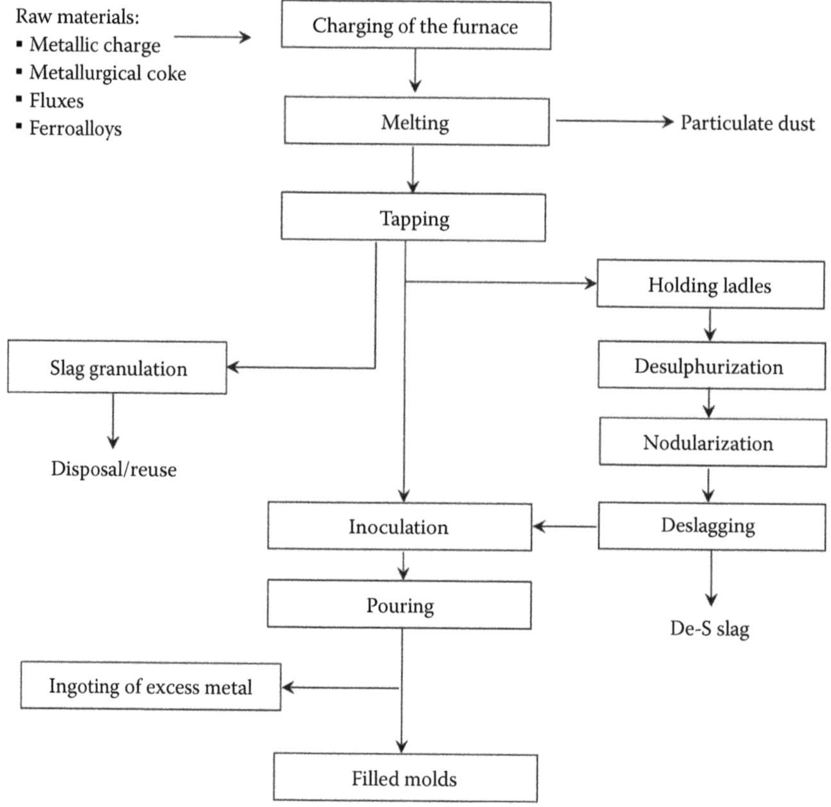

FIGURE 8.8
Process flowsheet for cupola furnace melting. (From European Commission, Integrated pollution prevention and control: Reference document on the best available techniques in the smitheries and foundries industry, 2005, p. 16, Available at http://eippcb.jrc.ec.europa.eu/reference/BREF/sf_bref_0505.pdf, Accessed 6 June 2016. With permission.)

categories, only the sand casting process will be discussed. This process is of interest mostly due to its scale of application and the bulk nature of the waste streams produced.

8.5.4.2 Sand Moulding Processes

Moulds used in the sand casting process consist of particulate refractory material that is bonded together to hold its shape during metal pouring (EPA, 1992). In fact, sand moulding is one of the most versatile metal forming processes due to its flexible design in terms of size, shape and product quality (Davies, 1996). In addition, the sand moulding processes are broadly classified according to the way in which the sand is bonded, and the typical classifications include (Davies, 1996; European Commission, 2005; ECS, 2006) (1) inorganic bonded systems, such as green sand moulding, dry sand moulding, sodium silicate/carbon dioxide systems and phosphate-bonded systems and (2) organically or chemically bonded systems utilizing unique chemical binders and catalyst to cure and harden the mould.

Basically, the next step after the production of moulds is to produce mould cores. Basically, mould cores shape the interior surfaces of a casting that cannot be shaped by

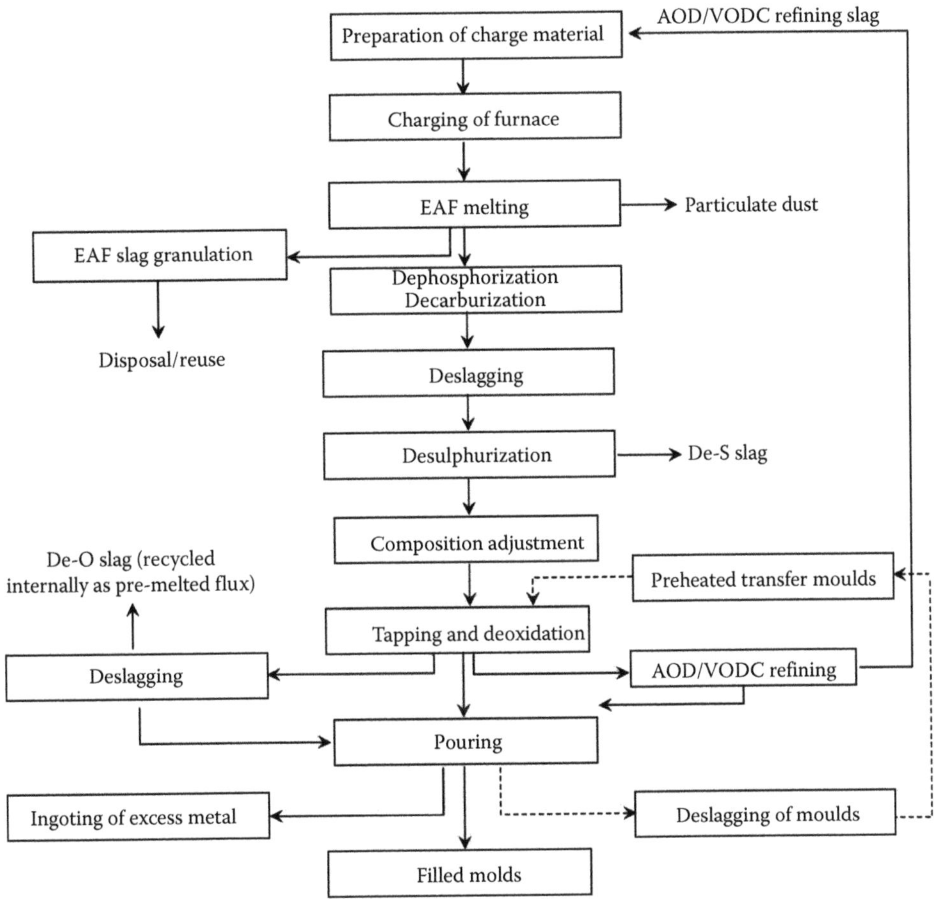

FIGURE 8.9
Process flowsheet for steel foundry with a basic refractory-lined EAF furnace. (From European Commission, Integrated pollution prevention and control: Reference document on the best available techniques in the smitheries and foundries industry, 2005, p. 18, Available at http://eippcb.jrc.ec.europa.eu/reference/BREF/sf_bref_0505.pdf, Accessed 6 June 2016. With permission.)

the mould cavity surface (EPA, 1992). Mould cores are made of sand (typically silica, olivine and zirconia) and a binder, and must be strong enough to be inserted into a mould without losing dimensional accuracy (EPA, 1992; European Commission, 2005). Different binder materials are used to hold the individual grains of sand together, and the choice of binder technology used depends on factors such as the size of the casting, production rate, type of metal poured, and the shake-out properties (EPA, 1992; European Commission, 2005). The binder technology commonly used for mould cores can broadly be categorized into (1) oil binders (combination of vegetable or animal oils and petrochemicals); (2) synthetic resin binders (such as phenolic, phenol-formaldehyde, urea-formaldehyde, phenolic-isocyanate, and alkyd isocyanate resin) and (3) chemical resin binders (such as furan acid catalysed no-bake binders, phenolic acid catalysed no-bake binders, ester-cured alkaline no-bake binders, silicate/ester-catalysed no-bake binders and phenolic urethane no-bake binders) (EPA, 1992; European Commission, 2005).

TABLE 8.6

Characteristic, and Recycling and Utilization of Different Ferrous Foundry Waste Streams

Waste Stream	Characteristics	Reduction, Recycling and Reuse	References
Particulate dust from melting operations	The particulate dusts from the melting process are particularly rich in toxic heavy metals species such as Zn, Cr, Pb and Cd. Gaseous emissions are typically rich in volatile organic carbons, SO_X, NO_X and a plethora of organometallic fumes.	In most, the particulate dusts from the melting units are agglomerated and recycled back into the process. However, the in-process recycling presents challenges of the in-process build-up of volatile tramp elements. Several techniques and processes applicable to the recycling, metal recovery and stabilization of metallurgical dusts were discussed in Chapter 4. Furthermore, the use of baghouse wet scrubbing systems to capture the fine particulate dusts and organometallic fumes is of critical importance to managing the foundry particulate dusts and gaseous emissions.	EPA (1992), Oman (1998), European Commission (2005), Jezierski and Janerka (2011)
Process slags	The composition is mostly controlled by the type and quality of raw materials, and the type and grade of alloys being produced. Generally, foundry slags from melting and holding furnaces are as a result of materials in metallic charge (such as recycled sand-encrusted gates and risers from the casting process or rust-and dirt-encrusted metallic scrap), oxidation of elements in metal bath (e.g. Si, Mn, Mg and Al), fluxes added to react with the metal oxides in slag, and refractory materials as a result of the slag metal-refractory interactions.In cast steel production, process slags typically contain entrained and oxidized alloy elements such as chromium and nickel (discussed in detail in Chapter 4).	In most cases, primary smelting slags usually undergo slag treatment in slag cleaning furnaces before disposal and/or are reused as construction materials. Refining slags can easily be recycled as pre-melted fluxes in primary melting units. In general, typical valorization options include: (1) slag treatment for metal recovery in slag cleaning furnaces, (2) metal recovery using physical beneficiation techniques, (3) metal recovery by pyro-and/or hydrometallurgical techniques, (4) utilization of slags as construction materials and (5) stabilization and/or solidification in cementitious systems.	EPA (1992), European Commission (2005), Yuvaraj et al. (2015), Naro and Williams (2015), Sharma et al. (2016)
Sand molding wastes	The bulk of wastes include spent foundry sand from molding and core making operations and dust and sludge from dust abatement systems of various processes.The diffuse emission of gaseous emissions contaminated with volatile organometallic compounds and polyaromatic hydrocarbons (PAH) is particularly problematic from occupational health point of view.	Green sand is regenerated and reutilized internally as molding sand. Core butts can also be crushed and recycled as molding sand. In essence, a combination of mechanical and thermal reclamation processes are widely being adopted. Completely degraded foundry sand can be reused exogenous to the foundry plant, particularly supplementary materials in civil engineering (e.g. asphalt concrete, brick making, Portland cement and in self-compacting backfill materials). Other opportunities to utilize spent foundry sand include reuse as pneumatically injected powdered slag foamers in EAF.	EPA (1992), Oman (1998), European Commission (2005), Ghosh (2013), CWC (1996), Tanaka (1987), Lahl (1992), Jezierski and Janerka (2011)

8.5.4.3 Waste Generation in the Moulding Process

In general, the typical waste streams produced in the moulding and core making processes include (EPA, 1992; European Commission, 2005) (1) spent foundry sand from moulding and core making operations; (2) particulate dust and sludges from dust abatement systems of sand preparation, moulding and shake-out processes; (3) investment castings and waxes and (4) volatile organic compounds emissions as a result of thermal decomposition of organic binders during metal pouring. The different characteristics of recycling and utilization schemes of the waste streams from the moulding process are summarized in Table 8.6.

In practice, green sand can be regenerated and reutilized internally as moulding sand (Tanaka, 1987; EPA, 1992; CWC, 1996; Oman, 1998; European Commission, 2005). In addition, mould core butts can also be crushed and recycled as moulding sand (EPA, 1992; European Commission, 2005). Both the green sand and the core butts can be reclaimed using mechanical attrition methods such as vibration and fluidization (Ghosh, 2013). However, the chemically bonded foundry sands tend to undergo thermal and chemical transformations during service, and as such, such thermochemical may present challenges to the amenability of the mould sand to mechanical reclamation processes (Ghosh, 2013). As a result, thermal reclamation processing is gaining traction as a viable option to reclaim the chemically bonded foundry sands, and essentially involves the heating of mechanically reclaimed sand at temperatures around 800°C in specially designed furnaces (Ghosh, 2013).

Eventually, the foundry sand may degrade completely due to repeated and cyclic reclamation treatments (Tanaka, 1987; Lahl, 1992; CWC, 1996). Thus, the continual reclamation of foundry sand might eventually impact the quality of moulds, for example, the sand may become too fine from repeated cyclic thermal and mechanical stresses and as a result, has to be flushed out of the system (Tanaka, 1987; Lahl, 1992; CWC, 1996). Ideally, such materials can be reused in industries external to the foundry plants, particularly supplementary materials in civil engineering (e.g. asphalt concrete, brick making, Portland cement and in self-compacting backfill) (Tanaka, 1987; CWC, 1996).

8.5.5 Recycling and Utilization of Foundry Process Wastes

Ferrous foundries are major consumers of recycled waste materials and play a critical role in a circular economy (Oman, 1998). In as much as they consume externally generated waste materials such as post-consumer scrap in their melting processes, foundries also play a major role in the internal recycling and utilization of waste streams produced in their processes (Oman, 1998; European Commission, 2005; Jezierski and Janerka, 2011). The internal and external recycling and utilization of different foundry waste streams are summarized in Table 8.6.

8.6 Concluding Remarks

This chapter discussed the technical and process aspects, waste production, and the waste valorization opportunities in the selected metal forming and finishing processes. Typical processes covered in detail include the hot forming (rolling process), heat treatment, surface preparation (pickling process) and electroplating and surface coating processes.

Furthermore, the chapter provided a detailed discussion on the different ferrous and non-ferrous foundry processes used in the casting of near-net components. The different types of waste streams, their valorization opportunities and the contribution of foundry operations to a circular economy were discussed. Based on the key areas highlighted in this chapter, a lot of opportunities can be explored in order to reduce the amount of waste produced, recycle the waste streams, as well as utilize waste streams in industries that are external to the metal manufacturing, finishing and foundry processes.

References

American Institute of Iron and Steel (AISI) (1982). *Cleaning and Descaling of Stainless Steel*. Designers' Handbook Series. AISI Committee of Stainless Steel Producers, Washington, DC, p. 27.

Atkinson, M. and Kolarik, R. (2001). *Steel Industry Technology Roadmap*. The American Iron and Steel Institute, Washington, DC.

Bagatini, M.C., Zymla, V., Osorio, E. and Vilela, A.C. (2011). Characterization and reduction of mill scale. *ISIJ International* 51(7): 1072–1079.

Basabe, V.V. and Szpunar, J.A. (2008). Effect of O_2 in heating atmosphere on hydraulic descaling in hot rolling of low carbon steel. *ISIJ International* 48(4): 467–474.

Bawa, H.S. (2004). Foundry. In: *Manufacturing Processes II*. Tata McGraw Hill, New Delhi, India.

Bhatnagar, A. and Sillanpää, M. (2010). Utilization of agro-industrial and municipal waste materials as potential adsorbents for water treatment: A review. *Chemical Engineering Journal* 157: 277–296.

Blair, M. and Steven, T.I. (eds.), (1995). Melting. *Steel Castings Handbook*, 6th edn. Steel Founders Society of America and ASM International, Ohio, pp. 14.1–14.10.

Boesch, W. (2012). Super alloys. In: Tien, J.K. and Caulfield, T. (eds.), *Superalloys, Super Composites and Super Ceramics*, Vol. 1. Materials Science and Technology, Academic Press Inc. Boston.

Boljanovic, V. (2010). *Metal Shaping Processes: Casting and Molding, Particulate Processing, Deformation Processes and Metal Removal*. Industrial Press Inc., New York.

Braun, E. (1980). How to improve pickling of stainless steel strip. *Iron and Steel Engineer* 57(4): 79–81.

Brooks, C.R. (1996). *Principles of the Heat Treatment of Plain Carbon and Low Alloy Steels*. ASM International, Ohio, pp. 235–282.

Brooks, C.S. (1985). Metal recovery from waste sludges. *Proceedings of the 39th Purdue Industrial Waste Conference Purdue University*, pp. 545–553.

Brooks, C.S. (1986). Metal recovery from industrial wastes. *Journal of Metals* 38: 50–57.

Buekens, A., Stieglitz, L., Hell, K., Huang, H. and Segers, P. (2001). Dioxins from thermal and metallurgical processes: Recent studies for the iron and steel industry. *Chemosphere* 42: 729–735.

Chen, R.Y. and Yuen, W.Y.D. (2003). Review of the high temperature oxidation of iron and carbon steels I air or oxygen. *Oxidation of Metals* 59(5/6): 433–468.

Cheng, S.Y., Juan, C.T., Kuan, S.L. and Tsai, W.T. (2003). Effect of annealing atmosphere on scale formation and pickling performance of 410SS. *Oxidation of Metals* 60(5/6): 409–425.

Cho, S.Y. and Lee, J.H. (2008). Metal recovery from stainless steel mill scale by microwave heating. *Metals and Materials International* 14(2): 193–196.

CWC (1996). Beneficial reuse of spent foundry sand. Report submitted prepared for the Recycling Technology Assistance Partnership (IBP-95-1). Available at http://www.cwc.org/industry/ibp951fs.pdf. Accessed 9 June 2016.

Davies, J.R. (1993). Fabrication and finishing of aluminium alloys. In: *Aluminium and Aluminium Alloys. ASM Specialty Handbook*. ASM International, Ohio, pp. 199–451. Available at www.asminternational.org. Accessed 2 June 2016.

Davies, J.R. (1996). *Cast Irons. ASM Specialty Handbook*. ASM International, Ohio, p. 133. Available at www.asminternational.org. Accessed 2 June 2016.

Davies, J.R. (2001). *Copper and Its Alloys. ASM Specialty Handbook.* ASM International, Ohio, p. 171. Available at www.asminternational.org. Accessed 2 June 2016.

Devi, A., Singhal, A., Gupta, R. and Panzade, P. (2014). A study on treatment methods of spent pickling liquor generated by pickling process of steel. *Clean Technologies and Environmental Policy* 16: 1515–1527.

Digges, T.G., Rosenberg, S.J. and Geil, G.W. (1966). Heat treatment and properties of iron and steel. US National Bureau of Standards Monograph 88. Available at http://www.hnsa.org/wp-content/uploads/2014/07/heat-treatment-steel.pdf. Accessed 2 June 2016.

Donachie Jr., M.J. (2000). Castings. In: *Titanium: A Technical Guide*, 2nd edn. ASM Specialty Handbook. ASM International, Ohio. Available at www.asminternational.org, p. 39. Accessed 2 June 2016.

Donik, Č. (2011). Surface characterization and pickling characteristics of the oxide scale on duplex stainless steel. *Materials and Technologies* 45(4): 329–333.

Engineered Casting Solutions (ECS) (2006). Guide to casting and molding processes. Casting Source Directory. Available at http://www.afsinc.org/files/methods.pdf. Accessed 7 June 2016.

Environmental Protection Agency (EPA) (1984). Guidance manual for electroplating and metal finishing pre-treatment standards. EPA-440/1-84/091g. U.S. Environmental Protection Agency, Washington, DC.

Environmental Protection Agency (EPA) (1992). Guides to pollution prevention. Metal Casting and Heat Treating Industry. EPA/625/R-92/009. U.S. Environmental Protection Agency, Washington, DC.

Environmental Protection Agency (EPA) (1995). Metallurgical industry. *Complication of Air Pollutant Emissions Factors*, Vol. I. Stationary Point and Area Sources (450AP425ED), 5th edn. Office of Air Quality and Planning Standards, Research Triangle Park, NC, p. 12.4-1.

Environmental Protection Agency (EPA) (1997). Pollution prevention for the metal finishing industry: A manual for pollution prevention technical assistance providers. EPA/742/B-97/005. U.S. Environmental Protection Agency, Washington, DC.

Environmental Protection Agency (EPA) (2002). Development document for final effluent limitations guidelines and standards for the iron and steel manufacturing point source category. EPA-821-R-02-004. U.S. Environmental Protection Agency, Washington, DC.

European Commission. (2001). Integrated pollution prevention and control: Reference document on the best available techniques in the ferrous metals processing industry. Available at http://eippcb.jrc.ec.europa.eu/reference/BREF/fmp_bref_1201.pdf. Accessed 20 April 2016.

European Commission (2005). Integrated pollution prevention and control: Reference document on the best available techniques in the smitheries and foundries industry. Available at http://eippcb.jrc.ec.europa.eu/reference/BREF/sf_bref_0505.pdf. Accessed 6 June 2016.

Farahat, R., Eissa, M., Megahed, G. and Baraka, A. (2010). Reduction of mill scale generated by steel processing. *Steel Grips* 8: 88–92.

Fisher, R., Anderson, D.R., Wilson, D.T., Aries, E., Hemfrey, D. and Fray, T.A.T. (2004). Effect of chloride on the formation of PCCD/Fs and WHO-12 PCBs in iron ore sintering. *Organohalogen Compounds* 66: 1116–1123.

Gavrilovski, M., Kamberovic, A., Filipovic, M. and Majinski, N. (2011). Optimization of integrated steel plant recycling: Fines-grain remains and by-products synergy. *Strojarstvo* 53 (5): 359–365.

Ghosh, A. (2013). Modern sand reclamation technologies for economy, environment friendliness and energy efficiency. *Transactions of the 61st Indian Foundry Congress*, Kolkata, India. Available at http://www.foundryinfo-india.org/images/pdf/2A5.pdf. Accessed 9 June 2016, pp. 1–5.

Green, G. and Higginson, R.L. (2011). Oxide scale formation and morphology on austenitic stainless steel at medium annealing temperatures. *Materials at High Temperatures* 28(4): 269–273.

Hackbarth, F.V., Maass, D., Augusto, A., Souza, U. De, Vilar, V.J.P. and Guelli, S.M.A. (2016). Removal of hexavalent chromium from electroplating wastewaters using marine macroalga *Pelvetia canaliculata* as natural electron donor. *Chemical Engineering Journal* 290: 477–489.

Hildén, J.M.K., Virtanen, J.V.A. and Ruoppa, R.L.K. (2000). Mechanism of electrolytic pickling of stainless steels in neutral sodium sulphate solution. *Materials and Corrosion* 51: 728–739.

Jezierski, J. and Janerka, K. (2011). Waste utilization in foundries and metallurgical plants. In: Kumar, S. (eds.), *Integrated Waste Management*, Vol. 1. INTECH. Available at http://www.intechopen.com/books/ integrated-waste-management-volume-i/solid-waste-utilization-in-foundries-and-metallurgical-plants. Accessed 9 June 2016.

Kainer, K.U. (2003). The current state of technology and potential for further development of magnesium applications. In: Kainer, K.U. (eds.), *Magnesium Alloys and Technology*. Wiley-VCH GmbH & Co, Weinheim, Germany, pp. 1–22.

Kasai, E., Aono, T., Tomita, Y., Takasaki, M., Shiraishi, N. and Kitano, S. (2001). Macroscopic behaviour of dioxins in the iron ore sintering plants. *ISIJ International* 41(1): 86–92.

Katz, S. and Tiwari, B. (1998). A critical overview of liquid metals processing in the foundry. In: Katz, S. and Landfield, C.F. (eds.), *Foundry Processes: Chemistry and Physics*. Plenum Press, New York, pp. 1.

Kaushish, J.P. (2010). Metal forming: Hot and cold working and press working. *Manufacturing Processes*, 2nd edn. PHI Learning (Pvt) Ltd, New Delhi, India, p. 624.

Klocke, F. (2014). *Forming. Manufacturing Processes 4*, RWTH Edition. Springer, Aachen, Germany, pp. 12–15.

Konečná, R. and Fintová, S. (2001). Copper and copper alloys: Casting, classification and characteristic microstructures. In: Collini, L. (eds.), *Copper Alloys – Early Applications and Current Performance-Enhancing Processes*. INTECH. Available at http://www.intechopen.com/books/copper-alloys-early-applications-and-current-performance-enhancing-processes/copper-and-copper-alloys-casting-classification-and-characteristics-. Accessed 6 June 2016

Kozak, B. and Dzierzawski, J. (n.d.). Continuous casting of steel: Basic principles. Available at https://www.steel.org/making-steel/how-its-made/processes/processes-info/continuous-casting-of-steel---basic-principles.aspx?siteLocation=88e232e1-d52b-4048-9b8a-f687fbd5cdcb. Accessed 25 February 2016.

Krzyzanowski, M., Beynon, J.H. and Farrugia, D.C.J. (2010). *Oxide Scale Behavior in High Temperature Metal Processing*. Wiley-VCH Verlag GmbH, Weinheim, Germany, pp. 1–19.

Lahl, U. (1992). Recycling of waste foundry sands. *The Science of Total Environment* 114: 185–193.

Li, L.F., Caenen, P., Daerden, M., Vaes, D., Meers, G., Dhondt, C. and Celis, J.P. (2005). Mechanism of single and multiple steps pickling of 304 stainless steel in acid electrolytes. *Corrosion Science* 47: 1307–1324.

Li, L.F., Caenen, P. and Jiang, M.F. (2008). Electrolytic pickling of the oxide layer on hot-rolled 304 stainless steel in sodium sulphate. *Corrosion Science* 50: 2824–2830.

Li, L.F. and Celis, J.P. (2003). Pickling of austenitic stainless steels—A review. *Canadian Metallurgical Quarterly* 42(3): 365–376.

Li, X.M., Xie, G., Hojamberdiev, M., Cui, Y.R. and Zhao, J.X. (2014). Characterization and recycling of nickel-and chromium contained in pickling sludge generated in production of stainless steel. *Journal of Central South University* 21: 3241–3246.

Lindsay, T.C. (2007). Metal finishing and electroplating. In: Kutz, M. (ed.), *Environmentally Conscious Manufacturing*. Wiley Series in Environmentally Conscious Engineering. John Wiley & Sons, New Jersey, p. 123.

Luong, L.H.S. and Heijkoop, T. (1981). The influence of scale on friction in hot metal working. *Wear* 71: 93–102.

Ma, P., Lindblom, B. and Björkman, B. (2005). Experimental studies on solid-state reduction of pickling sludge generated in the stainless steel production. *Scandinavian Journal of Metallurgy* 34(1): 31–40.

Mansur, M.B., Ferreira Rocha, S.D., Magalhaes, F.S. and dos Santos Benedetto, J. (2008). Selective extraction of zinc(II) over iron(II) from spent hydrochloric acid pickling effluents by liquid-liquid extraction. *Journal of Hazardous Materials* 150: 669–678.

Martin, M.I., Lopez, F.A. and Torralba, J.M. (2012). Production of sponge iron powder by reduction of rolling mill scale. *Ironmaking and Steelmaking* 39(3): 155–162.

Matinde, E., Hino, M., Todoroki, H. and Kobayashi, Y. (2009). Method for reducing valuable metal raw material. Japanese Patent (JP 2009-209411A). Available at http://ip.com/patapp/JP2009209411A. Accessed 6 June 2016.

Maurer, G.E. (2012). Primary and secondary melt processing – Superalloys. In: Tien, J.K. and Caulfield, T. (eds.), *Superalloys, Supercomposites and Superceramics*. Materials Science Technology. Elsevier, p. 49.

Meynerts, U., Seabra da Rocha, S.H. and Maurer, M. (2013). Mill scale briquetting. US Patent US 2013/0192421 A1.

Naro, R.L. and Williams, D.C. (2015). Controlling slag defects, insoluble build-up in melting and holding furnaces. Foundry Management and Technology. Available at http://foundrymag.com/meltpour/controlling-slag-defects-insoluble-buildup-melting-and-holding-furnaces. Accessed 18 June 2016.

Nleya, Y., Simate, G.S. and Ndlovu, S. (2016). Sustainability assessment of the recovery and utilization of acid from acid mine drainage. *Journal of Cleaner Production* 113: 17–27.

Okumura, H. (1994). Recent trends and future prospects of continuous casting technology. *Nippon Steel Technical Report* 61: 9–14.

Oman, D.E. (1998). Hazardous waste minimization: Part IV waste minimization in the foundry industry. *Journal of the Air Pollution Control Association* 38: 932–940.

Ooi, T.C. and Lu, L. (2011). Formation and mitigation of PCCD/Fs in iron ore sintering. *Chemosphere* 85: 291–298.

Parkinson, R. (2001). Nickel plating and electroforming: Essential industries for today and the future. NiDI Technical Report Series 10 088. Nickel Development Institute. Available at https://www.nickelinstitute.org/~/media/Files/TechnicalLiterature/PlatingandElectroforming_EssentialIndustriesforTodayandtheFuture_10088_.ashx. Accessed 6 June 2016.

Poulalion, A. (2001). Process of recycling mill scale of alloyed steel in an electric arc furnace into a ferro-silicon product. European Patent, EP 1122319A1.

Regel-Rosocka, M. (2009). A review on methods of regeneration of spent pickling solutions from steel industry. *Journal of Hazardous Materials* 177: 57–69.

Remus, R., Mononet, M.A.A., Roudier, S. and Sanch, L.D. (2013). Best available techniques (BAT) reference document for iron and steel production. Industrial Emissions Directive 2010/75/EU. European Commission JRC Reference Report, Seville, Spain. Available at http://eippcb.jrc.ec.europa.eu/reference/BREF/IS_Adopted_03_2012.pdf. Accessed 15 December 2015

Rentz, O., Jochum, R. and Schultmann, F. (1999). Report on best available techniques (BAT) in the German Ferrous Metals Processing Industry. Final draft. Deutsch-Französiches Institut für Umweltforschung, Karlsruhe, Germany. Available at https://www.umweltbundesamt.de/sites/default/files/medien/publikation/long/2490.pdf. Accessed 2 June 2016.

Rögener, F., Sartor, M., Bàn, A., Buchloh, D. and Reichardt, T. (2012). Metal recovery from spent stainless steel pickling solutions. *Resources, Conservation and Recycling* 60: 72–77.

Roto, P. (1998). Smelting and refining. In: Stellman, J.M. (ed.), *Encyclopedia of Occupational Health and Safety: Chemical, Industries and Occupations*, Vol. III, 4th edn. International Labour Office, Geneva, Switzerland, p. 82.2.

Salak, A. and Kubelik, J. (1974). Some properties of reduced iron powder manufactured from mill scale. *Soviet Powder Metallurgy and Metal Ceramics* 13(3): 251–254.

Sánchez, L., Hierro, M.P. and Pérez, F.J. (2008). Effect of chromium content on the oxidation behavior of ferritic steels for applications in steam atmospheres at high temperatures. *Oxidation of Metals* 71: 173–186.

Sciscenko, I., Pedre, I., Hunt, A., Bogo, H. and González, G.A. (2016). Determination of a typical additive in zinc electroplating baths. *Microchemical Journal* 127: 226–230.

Scott, F.H. and Wei, F.I. (1989). High temperature oxidation of commercial austenitic stainless steels. *Materials Science and Technology* 5(11): 1140–1147.

Sharma, B.K. (1985). *Electrochemistry*, 1st edn. Goel Publishing House, Meerut, India, p. E154.

Sharma, D., Sharma, S. and Goyal, A. (2016). Utilization of waste foundry slag and alccofine for developing high strength concrete. *International Journal of Electrochemical Science* 11: 3190–3205.

Tanaka, T. (1987). Regeneration of waste mold sand. *Conservation and Recycling* 10(4): 253–263.

Thomas, B.G. (2001). Continuous casting. In: Yu, K.O. (ed.), *Modelling for Casting and Solidification*, 1st edn. Marcel Dekker Inc., New York: pp. 499.

Twidwell, L.G. and Dahnke, D.R. (2001). Treatment of metal finishing sludge for detoxification and metal value. *The European Journal of Mineral Processing and Environmental Protection* 1(2): 76–88.

Umadevi, T., Sampath Kumar, M.G., Mahapatra, P.C., Mohan Babu, T. and Ranjan, M. (2009). Recycling of steel plant mill scale via iron ore pelletization process. *Ironmaking and Steelmaking* 39(6): 409–415.

Vijayaram, T.R. (2012). Continuous casting technology for ferrous and non-ferrous foundries. Metalworld News Digest, Mumbai, India, pp. 40–44.

Wegrelius, L. and Sjoden, B. (2004). Passivation treatment of stainless steel. Outokumpu Stainless. Available at http://www.outokumpu.com/sitecollectiondocuments/passivation-treatment-of-stainless-steel-acom.pdf. Accessed 6 June 2016.

Wei, D.B., Huang, J.X., Zhang, A.W., Tieu, A.K., Shi, X., Jiao, S.H. and Qu, X.Y. (2009). Study on the oxidation of stainless steels 304 and 304L in humid air and the friction during hot rolling. *Wear* 67: 1741–1745.

Xia, Z.X., Zhang, C., Huang, X.F., Liu, W.B. and Yang, Z.G. (2015). Improve oxidation resistance at high temperature by nanocrystalline surface layer. *Scientific Reports* 5(13027): 1–7.

Yuvaraj, T., Palanivel, M., Vigneswar, S., Bhoopathy, R. and Gnanasekaran, V. (2015). Environmental feasibility in utilization of foundry solid waste (slag) for M20 concrete mix proportions. *Journal of Environmental Science, Toxicology and Food Technology* 9(1): 16–23.

Zielinski, W. and Kurzydlowski, K.J. (2000). TEM studies of the oxide scales formed on type 316 stainless steel during annealing at 600°C in a vacuum and air. *Scripta Materialia* 43: 33–37.

9

Post-Consumer Waste Production and Utilization

9.1 Introduction

The recycling of post-consumer wastes is becoming increasingly important in today's world due to public awareness, legislative promotion and imposition, economic benefits and the availability of appropriate technologies (Baeyens et al., 2010). So, what is recycling of post-consumer waste? What do these terms mean? Typically, raw materials are processed so as to manufacture products for consumption. Any leftovers and/or obsolete products are then discarded. This linear process, from extraction of raw materials to production, then consumption and disposal, has created a waste crisis. To decrease this one-way flow of resources to overburdened waste disposal facilities, materials that are no longer needed or wanted can be remanufactured. This is what is termed 'recycling'. According to the Master Recycler Program (2015), recycling is a dynamic process that restores the life cycle of a material. On the other hand, post-consumer material is defined as a waste material generated by households or by commercial, industrial and institutional facilities in their role as end users of the product which can no longer be used for its initial purpose (Leppo, 2009).

Another term that needs to be contextualized is 'recyclability'. Although widely discussed, there is considerable confusion about its meaning. However, recyclability is both a technological and an economic issue (Field et al., 1994a,b). On the technological side, recyclability requires the existence of methods that can be used to extract the constituent materials from an obsolete product. On the economic side, recyclability depends upon the existence of a market for these extracts. Furthermore, there must be a balance between the cost of employing the extraction technology and the quality of the extract such that the recycler has an economic incentive to undertake the recycling.

This chapter deals with the recyclable constituents of post-consumer metallic materials. The chapter starts by giving a snapshot of various international legislation and policies governing a selected number of post-consumer wastes. It then unpacks the major metallic recyclables, their reuse potential and recycling techniques used. This is followed by a discussion of a selected number of critical post-consumer sectors.

9.2 Legislation and Policies Governing Post-Consumer Wastes

In general, waste management laws govern the transport, treatment, storage and disposal of waste (Qasioun, 2015; Shantgiri, 2015). These laws are designed to minimize or eliminate the uncontrolled disposal of waste materials into the environment in a manner that may cause ecological or biological harm and include laws designed to reduce

TABLE 9.1

Regulations Addressing End-of-Life Vehicles

	EU	USA	China	Japan
Law	Directive 2000/53/EC on end-of-life vehicles	Automobile Recycling Study Act of 1991 (HR 3369)	End-of-Life Vehicle Recycling Regulation (i.e. Statute 307) and the subsequent policies	Law on Recycling of End-of-Life Vehicles
Enforcement year	2000	1991	2001	2005
Comment	It aims to control the generation and disposal of wastes from automobiles and to enhance environment-consciousness among parties involved in ELV treatment, through the promotion of reuse, recycling and collection of ELVs and their components.	This act was incorporated into a set of amendments to the Resource Conservation and Recovery Act of 1976 (RCRA), but it failed to gain approval. Since then, regulatory activities pertaining to end-of-life vehicles in the United States have been limited to the state level.	The most relevant of the policies in this chapter is 'The Automotive Products Recycling Technology Policy' of 2006'. This is aimed at establishing the end-of-life vehicle recycling system, including developing and promoting scrapped automotive products.	The purpose of this law is to ensure that waste from end-of-life vehicles is properly disposed of and that resources are used effectively through reduction of the amount of waste from end-of-life vehicles, and recycling and sufficiently using the recycled parts of such end-of-life vehicles.
References	EU (2005) and Sakai et al. (2014)	Jody et al. (2010)	Li et al. (2014)	GEC (2012)

the generation of waste and promote or mandate waste recycling (Qasioun, 2015; Shantgiri, 2015). At present, there is no single agreed-upon legislation and/or policy governing post-consumer wastes. Nevertheless, there are discrete laws in which post-consumer wastes fall, and ideally, the 3Rs – reduce, reuse, recycle – are the general strategies for dealing with generated waste. However, it must be noted that advocates of a sustainable solid waste management system argue that waste avoidance and reduction should be given more priority before the 3Rs (Antonio, 2010). Tables 9.1 through 9.3 show some of the key legislations governing different types of waste in selected countries. It must be noted, however, that various countries have developed different approaches for managing waste that differ both in the legislative scope and effectiveness (Kumar and Singh, 2013).

9.3 Recyclable Metals

As already stated, the recycling of metallic post-consumer wastes is becoming increasingly important. In fact, the importance of metal recycling is indisputable because of increasingly stringent environmental regulations (Bell et al., 2003). It is argued that the remelting of recycled metallic materials saves a lot of the energy required to manufacture pure metals from ores. This, in turn, significantly reduces pollution and greenhouse gas emissions from mining, ore refining and melting. For example, 95% of the energy used to make aluminium

TABLE 9.2
Regulations Addressing E-Waste

	EU	USA	China	Japan
Law	Directive 2012/19/ EU on Waste Electrical and Electronic Equipment	–	Regulation on Management of the Recycling and Disposal of Waste Electrical and Electronic Equipment	Home Appliance Recycling Law
Enforcement year	2014	–	2011	2008
Comment	This directive follows the repealed Directive 2002/96/ EC (the 'old Waste Electrical and Electronic Equipment Directive'). The key purpose of the directive is to contribute to sustainable production and consumption by, as a first priority, the prevention of waste electrical and electronic equipment and, in addition, by the reuse, recycling and other forms of recovery of such wastes.	There is no specific legislation at the federal level targeting the management of waste electrical and electronic equipment. Therefore, individual states have developed and implemented their own policies, reflecting local conditions and each state's specific political dynamics. For example, at least 20 states including New York city municipality have legislation about waste electrical and electronic equipment, and have instituted recovery programs.	This is a pivotal piece of national legislation for e-waste management. The regulation stipulates that e-waste should be collected through multiple channels and recycled by licensed recycling enterprises. The regulation established a special fund to subsidize the formal collection and recycling of e-waste. Producers and importers of electronic products are required to contribute to this fund.	The law which was revised in 2008 deals with four classes of items: refrigerators, washing machines, TVs and air conditioning units. The law has been criticized for excluding certain types of waste electrical and electronic equipment, such as mobile phones. Personal computers are covered by the Law of Promotion of Effective Utilization of Resources (LPUR), which came into effect in 2001.
References	European Union (2012)	Veit and Bernardes (2015)	Kumar and Singh (2013)	Bo and Yamamoto (2010) and Veit and Bernardes (2015)

from bauxite ore is saved by using recycled material (EPA, 1990). However, recycling is not an absolute solution to the demand for finished metal products, though it is useful in the case of materials whose future availability from primary sources is declining sharply. Nevertheless, 100% recycling is practically impossible. Moreover, recovery of metals from secondary and/or post-consumer materials is complicated where metals are used in alloys or, for example, in electronic equipment where recovery yields a mixed product with impurities that are difficult to remove; hence, secondary and/or post-consumer materials cannot fully supply the total demand for metals due to such technical constraints (Verhoef et al., 2004).

The targeted recyclables discussed in this chapter include aluminium, copper, lead, zinc, tin and steel. For each recyclable, the chapter describes its properties first, followed by its primary sources. The chapter then discusses its uses and/or the consumer products that contain the recyclable. Finally, the processes that may be used to recover the post-consumer waste are discussed in detail.

TABLE 9.3

Regulations Addressing Waste such as Spent Catalysts

	EU	USA	China	Japan
Law	Directive 2008/98/EC on waste and repealing certain Directives (Waste Framework Directive)	Hazardous waste regulation of the EPA under the Resource Conservation and Recovery Act (RCRA)	Law on the Prevention and Control of Environmental Pollution by Solid Waste	Law for Promotion of Effective Utilization of Resources
Enforcement year	2008	1999	1996	2001
Comment	This directive establishes the legislative framework for the handling of waste in the EU member states. Therefore, the regeneration (off-site), recovery or disposal of spent catalysts are subject to the Directive. The waste directive lays down all definitions on waste, by-products, end-of-waste criteria, waste hierarchy, operations and – as a very important part – recovery versus disposal.	In general, the EPA regulates household, industrial and manufacturing solid and hazardous wastes under RCRA. The EPA originally added spent hydrotreating and hydrorefining catalysts to the list of RCRA hazardous wastes because they were shown to pose unacceptable risk to human health and the environment when mismanaged, and can at times exhibit pyrophoric properties. However, in 2015, the EPA has revised the generator-controlled exclusion to allow spent petroleum catalysts to be eligible for that exclusion, and is also allowing spent petroleum catalysts to be eligible for the verified recycler exclusion.	The purpose of the law is to prevent and control environmental pollution by solid waste, safeguard human health, maintain the ecological safety and promote the sustainable development of the economy and society. It sets forth provisions relating to solid waste management infrastructure and solid waste collection, storage, transportation and treatment. The three principles of solid waste treatment (waste reduction, decontamination and recycling) are also incorporated into the law.	The purpose of this law is to comprehensively promote reduction of waste, reuse of parts and recycling of used products as raw materials. The law requires that businesses take measures to apply the 3Rs (reduction, reuse and recycling) under specific criteria (ministerial ordinances) with respect to 10 designated and/or specified industries and 69 product items (covering about 50% of municipal and industrial waste).
References	European Union (2008) and ECMA (2012)	EPA (2015)	McElwee (2011) and Dongping (2013)	METI (2007)

9.3.1 Aluminium

9.3.1.1 Chemical and Physical Properties of Aluminium

Aluminium, symbolized as Al, is a shiny, silvery white coloured metal. It has an atomic number of 13 and atomic mass of 26.98. Its electronic configuration is [Ne] $3s^2 3p^1$. Aluminium has a specific gravity of 2.7, which makes it one-third lighter than iron or brass (Davis, 1999;

TABLE 9.4
Summary of Some Properties of Aluminium

Property	Value
Density (solid)	2699 kg/m^3 (theoretical density based on lattice spacing; 2697–2699 kg/m^3 (polycrystalline materials)
Density (liquid)	2357 kg/m^3 at 973 K; 2304 kg/m^3 at 1173 K
Coefficient of expansion	23 × 10^{-6}/K at 293 K
Thermal conductivity	2.37 W/cm K at 298 K
Volume resistivity	2.655 × 10^{-8} Ω m
Magnetic susceptibility	16 × 10^{-3}/m^3 g/atom at 298 K
Surface tension	868 dyne/cm at the melting point
Viscosity	0.012 poise at the melting point
Melting point	933.5 K
Boiling point	2767 K
Heat of fusion	397 J/g
Heat of vapourization	1.08 × 10^{-4} J/g K
Heat capacity	0.90 J/g K

Sources: Hatch, J.E., *Aluminum: Properties and Physical Metallurgy*, ASM International, Cleveland, OH, 1984; Davis, J.R., *Corrosion of Aluminum and Aluminum Alloys*, ASM International, Cleveland, OH, 1999.

Chatterjee, 2007). Its strength to weight ratio is higher than steel, a property that allows the design and construction of strong, lightweight structures that are particularly good for anything that moves – space vehicles and aircrafts, as well as all types of land and waterborne vehicles (Davis, 1999). Table 9.4 is a summary of some of the important properties of aluminium (Davis, 1999), and Table 9.5 shows some of its advantages that include lightweight, corrosion resistance, high thermal and electrical conductivity, good machinability and excellent response to a variety of finishing processes (Capus, 2010).

9.3.1.2 Occurrence

Aluminium is the third most abundant element after oxygen and silicon and the most abundant metallic element comprising 8% of the earth's crust (Earnshaw and Greenwood, 1997; Chatterjee, 2007). Despite its abundance, aluminium usually occurs in clays, soil and rocks that cannot be utilized for its extraction (Chatterjee, 2007). At present, bauxite is the only principal ore used for large-scale production of aluminium (Authier-Martin et al., 2001; Chatterjee, 2007). The bauxites used for the production of aluminium typically contain 35%–60% total aluminium oxide (Davis, 1999). However, bauxite is not a distinct mineral, but a generic term for a number of colloidal aluminium hydroxides like trihydrate gibbsite ($Al_2O_3 \cdot 3H_2O$), alpha-type monohydrate boehmite ($Al_2O_3 \cdot H_2O$) and the beta-type monohydrate diaspore ($Al_2O_3 \cdot H_2O$) (Chatterjee, 2007). Although both boehmite and diaspore have the same chemical composition, boehmite is a secondary mineral formed due to dehydration of gibbsite, is not suitable for metal recovery and is softer than diaspore, which is a product of metamorphism (Chatterjee, 2007). Besides these compounds, bauxite also contains iron oxide (which gives it a reddish-brown colour), as well as silicates (clay and quartz) and titanium oxide (Davis, 1999).

A report by Bray (2012) estimated bauxite resources to be 55–75 billion tonnes in the following ratios: Africa (32%), Oceania (23%), South America and the Caribbean (21%), Asia (18%) and elsewhere (6%). Though there are significant bauxite reserves worldwide, most of it is too difficult to mine without major improvement in the infrastructure (Davies, 2014).

TABLE 9.5
Advantages of Aluminium

Property	Comment
Light weight	Aluminium is one of the lightest available commercial metals with a density approximately one third that of steel or copper or brass. Its high strength-to-weight ratio makes it particularly important to transportation industries allowing increased payloads and fuel savings.
Excellent corrosion resistance	Aluminium has excellent resistance to corrosion due to the thin layer of aluminium oxide that forms on the surface of aluminium when it is exposed to air.
Strong	Although tensile strength of pure aluminium is not high, mechanical properties can be markedly increased by the addition of alloying elements and tempering.
Strong at low temperatures	Whereas steel becomes brittle at low temperatures, aluminium increases in tensile strength and retains excellent toughness.
Easy to work	Aluminium can be easily fabricated into various forms such as foil, sheets, geometric shapes, rod, tube and wire. It also displays excellent machinability and plasticity ideal for bending, cutting, spinning, roll forming, hammering, forging and drawing. Aluminium can be turned, milled or bored readily, using the correct tools.
Good heat conductor	Aluminium is about three times as thermally-conductive as steel. This characteristic is important in heat-exchange applications (whether heating or cooling).
High reflectivity	Aluminium is an excellent reflector of radiant energy through the entire range of wave lengths, from ultra-violet through the visible spectrum to infra-red and heat waves, as well as electromagnetic waves such as radio and radar. Aluminium also has a light reflectivity of over 80% which has led to its wide use in lighting fixtures.
Good electrical conductor	Aluminium is one of the two common metals having electrical conductivity high enough for use as an electrical conductor. The conductivity of electrical-conductor grade (alloy 1350) is about 62% that of the International Annealed Copper Standard. However, aluminium is only a third the weight of copper, which means it conducts about twice as much electricity as copper of the same weight.
Non-toxic	It is this characteristic which enables the metal to be used in cooking utensils without any harmful effect on the body. Aluminium with its smooth surface is easily cleaned, promoting a hygienic environment for food processing.
Easy to recycle	Due to a low melting temperature, it is economically recyclable, requiring only about 5% the energy required for smelting. It is an ideal material in this age of energy and resource saving.

Sources: Capus, J.M., *Metal Powders: Global Survey of Production, Applications and Markets 2001–2010*, Elsevier, Oxford, UK, 2010; Anand, S.H. et al., *Int. J. Emerg. Technol. Comput. Sci. Electron.*, 21(3), 91, 2016.

However, based on a study by Menzie et al. (2010), the capacity of bauxite mines worldwide was expected to increase to 270 million metric tonnes (Mt) by 2015 from 183 Mt in 2006, or by almost 48%. Subsequently, aluminium production capacity based upon the estimated capacity of bauxite mines worldwide was expected to reach 61 Mt in 2015 compared with 45.3 Mt in 2006, which is an increase of 35%, or almost 3.4% per year (Menzie et al., 2010).

Coal fly ash is also an important source of pre-mined alumina (Al_2O_3) (Torma, 1983; Matjie et al., 2005). Coal fly ash, formed as a result of coal combustion in coal-fired power plants, typically contains about 26%–31% alumina, possibly second only to bauxite in alumina content (Shemi, 2013). Some coal fly ashes have even much higher alumina content

TABLE 9.6

Typical Chemical Compositions of Bauxite and Coal Fly Ash

Component	Bauxite (wt.%)	Coal Fly Ash (wt.%)[a]
SiO_2	<0.5–10	46–60
Al_2O_3	30–60	26–31
Fe_2O_3	1–30	4–6
TiO_2	<0.5–10	1.3–1.7
CaO	0.1–2.0	3–11
P_2O_5	0.02–1.0	0.3–1.1

Sources: Shemi, A., Extraction of aluminium from coal fly ash using two-step acid leach process, MSc (Eng) dissertation, University of the Witwatersrand, Johannesburg, South Africa, 2013; Authier-Martin, M. et al., *J. Miner. Met. Mater. Soc.*, 53(12), 36, 2001; Shemi, A. et al., *Miner. Eng.*, 34, 30, 2012.

[a] Coal fly ash from Eskom power plants in South Africa.

in the range of 40%–45% (Guang-hui et al., 2010). Table 9.6 compares the chemical composition of bauxite and coal fly ash. As can be seen from the table, coal fly ash has higher silica, lower ferric oxide and within-range alumina content compared to bauxite. Despite the high silica content, the current metallurgical processes can be used to extract a significant amount of alumina present in coal fly ash (Shcherban et al., 1995; Nayak and Chitta, 2009; Shemi, 2013). Research studies have also shown that some materials, such as clays, alunite, anorthosite and oil shales, are technically feasible sources of alumina (Bray, 2012). Essentially, these nonbauxitic materials could satisfy the demand for alumina although they would require new plants using different technologies.

9.3.1.3 Uses of Aluminium

Today, aluminium is the leading non-ferrous metal in use (Constellium, 2014). Although there are many thousand applications of aluminium worldwide (>3000), its industrial use is limited to nine sectors, that is, utensils, electrical, consumer durables, transportation, building and construction, canning and packaging, chemical processing, defence equipment and electronics (Chatterjee, 2007). The nine sectors are listed and briefly discussed in Table 9.7.

9.3.1.4 Post-Consumer Recycling Processes

Aluminium is the most common recycled material (Yglesias, 2014). The sources of secondary aluminium can be divided into two main categories, that is, the old scrap and the new scrap (European Aluminium Association, 2002). The old scrap comes from products that have reached the end of their useful life (e.g. end-of-life vehicles). This category can also be subdivided into (1) intermediate return products, products with an estimated lifetime of less than one year (e.g. packaging, lithographic plates), and (2) installed base, products with a longer life time of more than a year (e.g. passenger cars). New scrap represents material that is generated in the process steps from semi-finished product to the end-product fabrication (e.g. scrap from automotive part manufacturer). This type of scrap is discussed in more detail in Chapter 8. However, new scrap is normally recycled within the aluminium industry or may end up in the open scrap market (European Aluminium Association, 2002).

TABLE 9.7

Summary of the Industrial Applications of Aluminium

Sector	Comment
Utensils	Easy workability, attractive colour, lightness, strong affinity for oxygen and non-toxicity make aluminium a cheaper substitute of copper in this use. The thin coating of oxide formed on its surface renders it corrosion-resistant.
Electrical	The use of aluminium predominates in most conductor applications. The use of aluminium rather than competing materials is based on a combination of low cost, high electrical conductivity, adequate mechanical strength, low specific gravity and excellent resistance to corrosion. Furthermore, its high electrical conductivity, and ductility when hot make it a good and cheap substitute for copper in transmission lines. However, it is an inferior substitute, because of its lower conductivity for the same diameter. Besides, lower tensile strength makes it more susceptible to snapping, and lower-than-copper melting point makes it unsafe in case of short circuit. To compensate for the lower conductivity, thick wires are used. Although this itself does not make the wire heavy due to its low specific gravity, its low tensile strength and consequent susceptibility to snapping requires the use of a core of steel wire around which the aluminium wire is stranded (aluminium conductor with steel-reinforced centre), and this makes the cable heavy with a tendency to sag.
Consumer durables	Light weight, strength, excellent appearance, ability to take organic or inorganic colour coatings, corrosion resistance, adaptability to all forms of fabrication and low cost of fabrication are the reasons for the broad usage of aluminium in household electrical appliances and furniture. Light weight is an important characteristic in vacuum cleaners, electric irons, portable dishwashers, food processors, refrigerators and blenders. Light weight, low maintenance, corrosion resistance, durability and attractive appearance are the principal advantages of aluminium in furniture.
Transportation	The second biggest use of aluminium is in transportation. Strength combined with light weight, natural and coated colours, corrosion resistance and recyclability are the primary criteria for using aluminium in parts of the bodies of cars, trucks, vans, etc. in wheel rims, in bumpers, in engine parts and in door panels of railway coaches. In addition, it is particularly suited for aeroplanes. A modern aeroplane contains about 80% aluminium by weight. For example, a Boeing 747 contains about 75 tonnes of aluminium.
Building and construction	Strength combined with light weight, natural and coated colours, corrosion resistance and easy workability into any shape – strips, bars, angles, corrugated sheets, partition panels, perforated strips, door and window panels – make aluminium a good material in this sector. When used in roofing sheets, high heat reflectance and low emissivity add to its suitability. Aluminium is also used for painting other surface.
Canning and packaging	The most extensive and fastest growing market for aluminium is packaging. Easy workability, colour, lightness combined with strength and corrosion resistance are the main criteria. Products include household wrap, flexible packaging and food containers, bottle caps, collapsible tubes and beverage and food cans. Furthermore, the food and drug industries use aluminium extensively because it is non-toxic, non-adsorptive and splinter-proof.
Chemical processing	Storage tanks and pipes for chemicals, paints, etc. use aluminium because mainly of its corrosion resistance which can be enhanced by increasing the thickness of the oxide coating, if necessary.
Defence equipment	Primarily due to its lightness combined with strength, aluminium is used in mortar barrel mountings, missiles, etc.
Electronic components	Aluminium of very high purity (>99.99%) is used for making capacitor foils and other electronic components.

Source: Chatterjee, K.K., *Uses of metals and metallic minerals*, New Age International, New Delhi, India, 2007.

Generally, most aluminium-bearing scrap is recycled through a smelting process using either a reverberatory or rotary furnace in the presence of halide fluxes (Shamsuddin, 1986; Hwang et al., 2006). Although not all the aluminium scrap (e.g. used beverage cans, cables and wires) follow the same charge preparation route, in general, the initial steps involved in aluminium scrap recycling before smelting include collection, sorting, shredding and/or sizing and media separation (Muchová and Eder, 2010a).

9.3.2 Copper

9.3.2.1 Chemical and Physical Properties of Copper

Copper is a chemical element with the symbol Cu. Copper, atomic number 29 with an atomic weight of 63.54, exhibits a face-centred cubic crystal structure (Reddy et al., 2015). It is a transitional element, and being a noble metal, it has inherent properties similar to those of silver and gold. It is a soft (2.5–3.0 on Mohs scale of hardness), malleable and ductile metal with very high thermal and electrical conductivities (Saravanan and Rani, 2012). A freshly exposed surface of pure copper has a reddish-orange colour. The density of copper is 8.89 g/cm^3, and its melting point is 1083°C. However, all of these properties and characteristics are significantly modified when copper is alloyed with other metals (NERC, 2007; MatWeb, 2015; Nielsen, 2015). Table 9.8 lists the common physical properties of copper.

From the chemistry point of view, copper is a moderately active metal that occurs in all plants and animals, as it is an essential element for all known living organisms (NERC, 2007). It dissolves in most acids and alkalis. Like other metals, copper forms oxides whenever it is exposed to the air. However, an important chemical property of copper is the way it reacts with oxygen, that is, it creates verdigris that cakes onto the copper

TABLE 9.8

Physical Properties of Copper

Property	Value
Density	8920 kg/m^3
Melting point	1083°C
Boiling point	2595°C
Latent heat of fusion	205 J/g
Linear coefficient of thermal expansion	16.8×10^{-6}/°C
Specific heat (thermal capacity)	0.386 J/g at 20°C; 0.393 J/g at 100°C
Thermal conductivity	3.94 W/cm °C at 20°C; 3.85 W/cm °C at 100°C
Electrical resistivity (volume)	1.7241–1.70 μΩ cm (annealed at 20°C); 1.78 μΩ cm (fully cold worked at 20°C)
Modulus of elasticity (tension)	118,000 MPa (annealed at 20°C)
Modulus of rigidity (torsion)	44,000 MPa (annealed at 20°C)

Sources: Chapman, D., High conductivity copper for electrical engineering, 1998, Available at http://copperalliance.org.uk/docs/librariesprovider5/pub-122---high-conductivity-copper-for-electrical-engineering/pub-122-hicon-copper-for-electrical-engineering.pdf?sfvrsn=2, Accessed October 2016; Li, M. and Zinkle, S.J., Physical and mechanical properties of copper and copper alloys, in: Konings, R.J.M. (ed.), *Comprehensive Nuclear Materials*, Vol. 4, Elsevier, Amsterdam, the Netherlands, 2012, pp. 667–690; Nielsen, W.D., Metallurgy of copper-base alloys, 2015, Available at http://www.copper.org/resources/properties/703_5//, Accessed December 2015.

(Klazema, 2014). Verdigris is a thin layer of cuprous oxide (Cu_2O) that develops on the surface of copper and copper alloys during prolonged exposure to moist air (Leones, 2009). Unlike iron, verdigris is a protective layer that prevents any copper material from corroding.

9.3.2.2 Occurrence

Copper deposits are widely distributed in many parts of the world in free state and in chemical combinations with a number of elements such as iron, sulphur, carbon and oxygen (NERC, 2007). Ideally, it occurs in native form (i.e. as nuggets of the free metal), in oxide ores (e.g. cuprite, Cu_2O; malachite, $Cu_2CO_3(OH)_2$) and in sulphide ores (e.g. chalcocite, CuS_2; chalcopyrite, $CuFeS_2$ or $Cu_2S \cdot Fe_2S_3$ and bornite, Cu_3FeS_3) (O'Leary, 2000). The average concentration of copper in the earth's crust is about 50 ppm, and the average exploitable grade for a copper deposit is 0.4%, which equates to a concentration factor of 80 based on average crustal abundance (NERC, 2007). More than 150 copper minerals have been identified, although only a small fraction shown in Table 9.9 is of economic importance (NERC, 2007; Lusty and Hannis, 2009). Currently, however, the most common source of copper ore is chalcopyrite, which accounts for about 50% of copper production (NERC, 2007; Lusty and Hannis, 2009).

One important property, particularly for this chapter, is that copper is recyclable. Copper can be recycled without any loss of quality. Actually, 40% of the world's demand is met by recycled copper.

9.3.2.3 Uses of Copper

Copper ranks third after iron and aluminium as one of the most used metals. It is an excellent electrical conductor and is also a good thermal conductor. Actually, most of its uses are based on these properties. For example, about a quarter of all copper produced is used in

TABLE 9.9

Common Copper Minerals Found in Economic Deposits

Mineral Name	Chemical Formula	Maximum Copper Content (wt.%)
Native copper	Cu	100
Chalcocite	Cu_2S	79.9
Cuprite	Cu_2O	88.8
Covellite	CuS	66.4
Bornite	Cu_5FeS_4	63.3
Chalcopyrite	$CuFeS_2$	34.6
Malachite	$Cu_2CO_3(OH)_2$	57.5
Azurite	$2CuCO_3{\bullet}Cu(OH)_2$	55.3
Antlerite	$Cu_3SO_4(OH)_4$	53.7
Enargite	Cu_3AsS_4	49.0
Chrysocolla	$CuSiO_3{\bullet}2H_2O$	36.2

Sources: Natural Environment Research Council (NERC), Copper, 2007, Available file:///C:/Users/a0009328/Downloads/comm_profile_copper%20(2).pdf, Accessed December 2015; Lusty, P.A.J. and Hannis, S.D., Commodity profile: Copper, 2009, Available at http://nora.nerc.ac.uk/7977/1/OR09041.pdf, Accessed October 2015.

TABLE 9.10

Summary of the Industrial Applications of Copper

Usage	Comment
Corrosion resistant	Copper is low in the reactivity series and thus, it does not tend to corrode. This is important in copper's use for pipes, electrical cables, saucepans and radiators. It is also well suited for decorative use. For example, jewellery, statues and parts of buildings can be made from copper, brass or bronze and remain attractive for thousands of years. Furthermore, its long life span is illustrated by its extensive use in architecture, particularly roofing, where it can survive for more than a hundred years.
Electrical	Copper is the best electrical conductor after silver and is widely used in the production of energy-efficient power circuits. Copper wires allow electric current to flow without much loss of energy. This is why copper wires are used in mains cables in houses and underground (although overhead cables tend be aluminium because it is less dense). However, copper is the best choice where size rather than weight is important. Thick copper strip is used for lightning conductors on tall buildings like church spires. The cable has to be thick so that it can carry a large current without melting. Electron tubes used in televisions and computer monitors, audio and video amplification and microwave ovens depend on copper for their internal components.
Construction	Copper is used in many forms in buildings, including wire, plumbing pipes and fittings, electrical outlets, switches and locks. Copper roofing is highly rated for its corrosion resistance and architectural characteristics.
Transportation	Copper is commonly used for radiators, brakes and wiring in motor vehicles.
Consumer and general products	Copper is commonly used in household products. Most silver-plated cutlery has a copper–zinc–nickel alloy base. Copper is used in many other domestic applications including cooking pans, lighting, clocks and for decorative purposes.
Catalytic compounds	Copper is a vital element that can act as a catalyst. For example, it speeds up the reaction between zinc and dilute sulphuric acid. It is found in some enzymes, one of which is involved in respiration.

Sources: Alavudeen, A. et al., *A Textbook of Engineering Materials and Metallurgy*, Firewall Media, New Delhi, India, 2006; Copper Development Association (CDA), Copper: Properties and applications, 2016, Available at http://copperalliance.org.uk/education-and-careers/education-resources/copper-properties-and-applications, Accessed October 2016.

electrical applications (NERC, 2007). However, many of its applications also rely on one or more of its other properties. For example, copper would not make very good water and gas pipes if it was highly reactive. Table 9.10 highlights the important industrial uses of copper.

9.3.2.4 Post-Consumer Recycling Processes

A significant quantity of copper is produced from secondary resources (Shamsuddin, 1986) comprising copper-containing products, landfills and waste or slag dumps from existing mining operations (Giurco, 2005). This chapter is only concerned with copper-containing products or post-consumer materials.

Copper scrap is commonly divided into four groups based on copper content as shown in Table 9.11 (Davenport et al., 2002). The method of production of secondary copper

TABLE 9.11

Classification of Copper Scraps

Category	Copper Content (%)	Example
No. 1 Scrap	>99	Copper wire, heavy scrap
No. 2 Scrap	88–99	Auto radiators, wires and cables
Low grade scrap	10–88	Printed circuit boards, electronics
Alloy scrap	65–80	Yellow and red brass

Source: Reprinted from *Extractive Metallurgy of Copper*, Davenport, W.G., King, M., Schlesinger, M., and Biswas, A. Copyright 2002, with permission from Elsevier.

depends on the quality of scrap (Shamsuddin, 1986), but unlike aluminium, the chemical and physical properties of copper permit a wide variety of pyrometallurgical and hydrometallurgical process alternatives for recovery of scrap (EIC Corporation, 1979). Before use, copper scrap may (or may not) require some upgrading by picking and sorting processes, sweating furnaces, wire burners and wire choppers (Goonan, 2004).

Generally, like aluminium, recycling techniques for copper scrap are based almost exclusively on pyrometallurgical processes. High-grade scrap containing more than 40% copper is melted and refined in reverberatory or rotary furnaces (Shamsuddin, 1986). However, if well sorted, No. 1 scraps and alloys can be melted and transformed into a finished product without any other processing (Bonnin et al., 2015), whereas No. 2 scrap has to be refined. Low-grade scrap is normally compacted and smelted in a blast furnace along with return converter slag, flux and reductant to produce molten black copper (Shamsuddin, 1986). In other words, low-grade scrap is processed much like copper ore, by being crushed and smelted in dedicated secondary blast furnaces (Giurco, 2005). Thereafter, it is converted and refined before being transformed into finished products (Bonnin et al., 2015).

Several efforts have also been made to develop hydrometallurgical processes for recycling copper metal scrap by either a leach-electrowinning approach or by direct electrorefining of scrap (Shamsuddin, 1986). For example, cupric chloride dissolved in concentrated brine has been used to leach brass scrap in the absence of air to produce a solution of cuprous chloride and zinc chloride (EIC Corporation, 1979; Shamsuddin, 1986). The proposed process sequence requires zinc removal by precipitation with caustic or soda ash, followed by electrolytic recovery of copper. Several scrap recovery processes have also been proposed based on the use of complexing agents (Koyama et al., 2006). Majima et al. (1993) reported the selective leaching of copper from motor scraps using cupric ammine solutions. Nigo et al. (1993) also studied the recovery of copper and iron from motor scraps by either spray leaching or upstream penetration leaching using cupric ammine solutions. Zhou et al. (1995a,b) studied the removal of copper from ferrous scrap by an ammonia leaching method. Several scrap recovery processes have also been proposed based on the use of an ammonia/ammonium carbonate lixiviant (EIC Corporation, 1979).

9.3.3 Lead

9.3.3.1 Chemical and Physical Properties of Lead

Elemental lead (Pb) is a heavy, soft and bluish metal that occurs naturally in the earth's crust. In comparison to the two most abundant metals in the earth's crust, aluminium (8%) and iron (5%), lead is a relatively uncommon metal (Abadin et al., 2007). It exists in three oxidation states: 0 (the metal), +2 and +4 (Abadin et al., 2007). However, in the environment, lead primarily exists in the +2 oxidation state. The +4 state is only formed under extremely oxidizing conditions, and inorganic compounds with an oxidation sate of +4 are not found under ordinary environmental conditions. In the ambient air, lead exists primarily as lead vapours, very fine lead particles and organic halogens such as lead bromide and lead chloride (Thompson, 2011). The most common sources in the atmosphere are gasoline additives, non-ferrous smelting plants and battery and ammunition manufacturing.

In general, lead possesses the physical properties of metals, that is, it is a conductor of electricity and heat (though not as good a conductor as some other metals, such as copper and aluminium), has a metallic lustre, albeit a dull one, and has high density. Compared with most other metals, it has a very low melting point of 327°C. However, the low melting point coupled with its extreme malleability allows easy casting, shaping and joining of lead articles and also influences some of the mechanical properties of lead (Thornton et al., 2001).

The problem with lead is that once it is mined, processed and introduced into the environment, it becomes a health hazard forever (Chambers, 2003). In particular, there exists extensive evidence that even at low dosages, lead may contribute to mental retardation and learning disabilities in children under the age of seven years (EPA, 1998). Currently, there is no technology that can destroy lead or render it permanently harmless (Chambers, 2003). Furthermore, compared with other metals, lead has extremely low strength, exacerbated by its creep and fatigue behaviour (Thornton et al., 2001). Thus, lead is unsuitable for applications that require even moderate strength. Actually, lead is rarely used in its pure form, but is alloyed with other metals so as to increase its strength.

9.3.3.2 Occurrence

Lead occurs naturally in the earth's crust at concentrations of about 8–20 mg/kg (Kimbrough and Krouskas, 2012). However, it is rarely found naturally as a metal, but it is usually found combined with other elements such as zinc, silver, copper and gold (Abadin et al., 2007). Essentially, lead is a coproduct of zinc mining or a by-product of copper and/or gold and silver mining (BCS Inc., 2002). Nevertheless, lead ore is mined worldwide, though three-quarters of the world output comes from only six countries: China, Australia, the United States, Peru, Canada and Mexico (Thornton et al., 2001). Actually, China and Australia represent about 30% and 22% of global mining production, respectively (UNEP, 2010). The most important lead-containing ore is galena (PbS) followed by anglesite ($PbSO_4$) and cerussite ($PbCO_3$) (Abadin et al., 2007; Soni and Suar, 2015), which are formed from the weathering of galena (Abadin et al., 2007). However, as stated already, lead-rich minerals most often occur together with other metals, and about two-thirds of worldwide lead output is obtained from mixed lead–zinc ores (UNEP, 2010). The world's reserves of lead are estimated at 8.7×10^7 tonnes, with over 40% located in Australia (Statista, 2015).

9.3.3.3 Uses of Lead

Lead is used and traded globally as a metal in various products (UNEP, 2010). In worldwide metal use, it only ranks behind iron, copper, aluminium and zinc (Howe, 1981; Jameson et al., 2003). The extensive usage of lead may be derived from its low melting point (327°C), high density (11.4 g/cm^3), malleability, chemical stability and resistance to acid corrosion (Howe, 1981; Jameson et al., 2003). The largest industrial use of lead today is for the production of lead-acid storage batteries which are largely used in the automobile and general industry (Jameson et al., 2003; UNEP, 2010). The very high density of lead also lends it to some quite different applications, such as shielding against sound, vibrations and radiation, for example, as protection for users of computer and TV screens (Thornton et al., 2001). For these purposes, lead is used in metallic form or as lead compounds in lead glasses. Other uses of lead include the production of lead alloys, the use in soldering materials and bearing metals for machinery and in the manufacture of corrosion and acid-resistant materials used in the building industry (Jameson et al., 2003; Abadin et al., 2007). Tables 9.12 and 9.13 are summaries of the main uses of lead and lead compounds, respectively.

9.3.3.4 Post-Consumer Recycling Processes

Lead has one of the highest recycling rates in the world, and unlike most materials, lead can be recycled indefinitely without any reduction in quality, making it ideal for a circular economy (International Lead Association, 2014). The sources of secondary lead include the following: (1) lead-acid batteries; (2) lead sheet, piping and cable sheathing; (3) end-of-life vehicles; (4) electrical and electronic products and (5) cathode ray tubes (Thornton et al., 2001). Among these sources, lead-acid batteries are by far the most important raw material of the secondary industry's feed stock (EIC Corporation, 1979; Shamsuddin, 1986; Vest, 2002; Abadin et al., 2007; Knežević et al., 2010). In fact, lead-acid batteries are the most recycled of all the consumer products (Gaines, 2014). Therefore, recycling of lead-acid batteries that is illustrated in Figure 9.1 will be the only focus of this section.

TABLE 9.12

Summary of the Main Uses (Current and Previous) of Lead

Usage	Comment
Lead-acid storage batteries	The major use worldwide, primarily as a starter battery in motor vehicles, but also as traction batteries for zero-emission electric vehicles and to provide emergency backup power supply, mostly for computer and telecommunication systems. Alternatives are under development for some applications, though at present these could not replace lead at comparable cost, or for technical reasons.
Constructional uses: pipe and sheet	Lead piping is now a minor application, as it is no longer used for domestic water supplies because of concerns that lead slowly dissolves in soft water and may pose a risk to health, and because of improvements in alternative materials. Lead sheet is widely used on roofs for flashings and weather-proofings, and is often used for complete roofs on both historic and modern buildings.
Cable sheathing	Lead sheaths are used to protect underwater and some underground power cables. However, this application is now minor.
Radiation screening	Lead is the most effective of the commonly available materials for screening from x-rays and some other types of radiation. It is widely used in hospitals as part of x-ray equipment, and also in nuclear power stations.
Miscellaneous products	Lead is widely used in shot and other munitions. Lead is also used extensively in weighting applications.
Lead alloys	Lead-tin solder is widely used, particularly by the electronics industry. Very minor applications are in bearings and ornamental ware (pewter) – though alternative materials are now generally used. Small additions of lead are made to some steels, brasses and bronzes so as to improve machinability.

Source: Thornton, I. et al., *Lead: The Facts*, IC Consultants Limited, London, UK, 2001.

TABLE 9.13
Summary of the Main Uses (Current and Previous) of Lead Compounds

Usage	Comment
Batteries	This is a major use of lead oxide. Lead dioxide is pasted on to the battery grids, and is the active material in the electrochemical reaction.
Pigments and other paint additives	Lead compounds were widely used until a few decades ago. They have been replaced in certain applications following concerns about potential impacts to human health. Leaded paints are still used in specialized outdoor applications as coatings for commercial vehicles and other industrial applications because of excellent rust-proofing properties. Lead dryers are still used in alkyd-based air drying paints as very efficient and cost effective through dryers.
Glasses and glazes	Lead additions improve the appearance and cutting properties of crystal glass. Small additions are also made to optical and electrical glass. The major application of leaded glass is in television screens and computer monitors, to protect viewers from the harmful x-rays generated by these appliances. Lead containing glazes are used for some pottery, tiles and tableware.
Functional ceramics	Lead titanates/zirconates are used in the electronics industry in various functions.
Additions to PVC	Small additions of organic lead compounds to some grades of PVC improve durability and heat resistance, both in manufacture and in service. This is a significant market for lead compounds.
Leaded petrol	Lead compounds were universally added to petrol to improve its efficiency at low cost. This has been the major source of lead emissions to the environment. It is now being phased out almost universally because of concerns about health impacts.

Source: Thornton, I. et al., *Lead: The Facts*, IC Consultants Limited, London, UK, 2001.

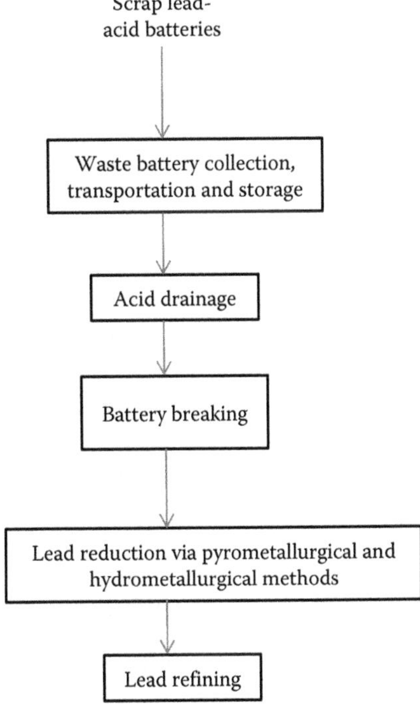

FIGURE 9.1
Generalized lead-acid battery recycling process.

9.3.3.4.1 Pre-Recycling Steps

Before reaching the recycling plant, used batteries must be collected, transported and stored with proper care, in order to avoid adverse health effects and environmental contamination (UNEP, 2003).

9.3.3.4.2 Acid Drainage

Irrespective of the recycling technology used, the batteries must always be drained before they enter the recycling process, since the acidic electrolyte produces several complications in the lead fusion–reduction process (UNEP, 2003). The acid electrolyte must be treated. The mining and metallurgical industry that uses acid in various leaching operations is the possible customer for the recycled acid (Vest, 2002). The remaining battery sludge is neutralized with lime. After passing through a filter press, the filter cake may be charged together with the fine fraction into the melting and reduction furnace.

9.3.3.4.3 Battery Breaking

After the electrolyte has been drained, batteries may or may not be broken, depending on the specific recycling process adopted. Battery-breaking methods differ from one another in process details and evolve as new technology becomes available. The suitability of each one for a given lead recovery plant depends on several specific factors such as local economy, quantity of raw materials and the demands of the smelting facility. Nevertheless, every effort should be made to eliminate the use of manual battery breaking because of the health and safety risks that are associated with the practice (UNEP, 2003). Figure 9.2 is an illustration of the modern battery-breaking process.

9.3.3.4.4 Lead Reduction

The battery scrap obtained from the breaking process is a mixture of several substances: metallic lead, lead oxide (PbO), lead sulphate ($PbSO_4$) and other metals such as calcium, copper, antimony, arsenic, tin and sometimes silver (UNEP, 2003). In order to isolate the metallic lead from this mixture, two methods may be applied: pyrometallurgical processes, also known as fusion–reduction methods, and hydrometallurgical processes, or electrolytic methods. It is also possible to combine the two and use a hybrid process.

Pyrometallurgical Methods This is the technique that is mainly used to treat waste lead-acid batteries (Knežević et al., 2010). The objective of the pyrometallurgical methods, or fusion–reduction methods, is to chemically reduce all metallic compounds to their metallic, or reduced forms by means of heating and providing adequate fluxing and reducing substances (UNEP, 2003). Prior to smelting, some methods may be employed to desulphurize the lead sulphate by reacting it with a mixture of sodium carbonate (Na_2CO_3) and sodium hydroxide (NaOH), thus converting it to lead oxide (PbO). Other desulphurizing agents include iron oxide (Fe_2O_3) and limestone ($CaCO_3$). This procedure reduces the amount of slag formation, and also, depending on the smelting method, the amount of sulphur dioxide (SO_2) released into the air is minimized. However, other methods simply add controlled amounts of lead sulphate and desulphurizing agent directly into the furnace.

The metallic fraction and the lead compounds derived from the desulphurization and neutralization processes are then added to the furnace and smelted with fluxing and reducing agents. The necessary heat is provided by several sources depending on the specific method, for example, oil, gas, coke and electricity. There are also several different vessels

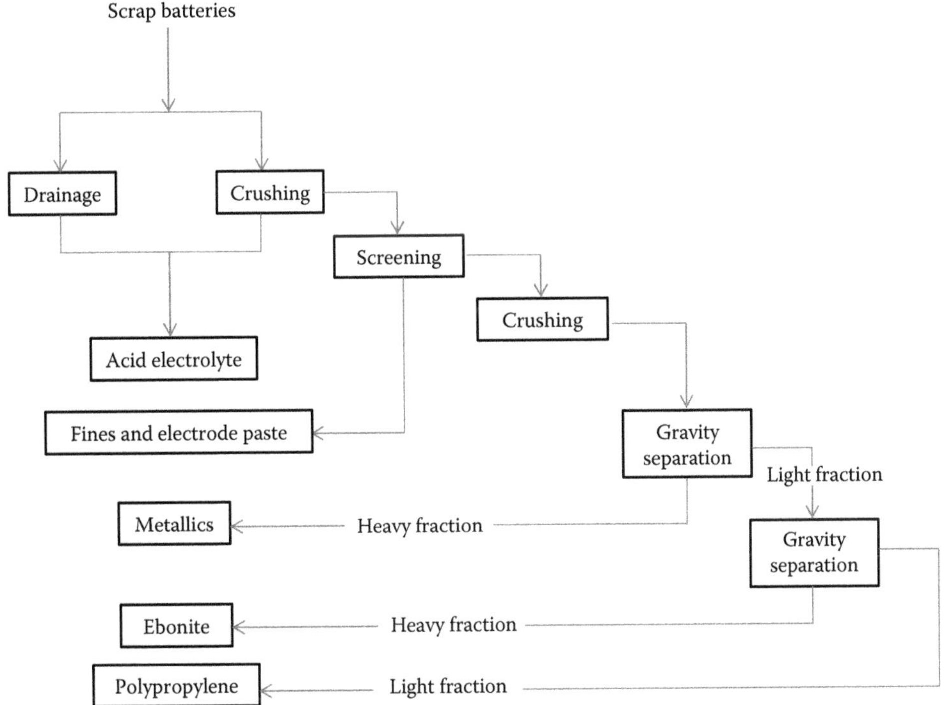

FIGURE 9.2
Battery breaking process. (From UNEP, Technical guidelines for the environmentally sound management of waste lead-acid batteries, 2003, Available at http://www.basel.int/Portals/4/Basel%20Convention/docs/pub/techguid/tech-wasteacid.pdf, Accessed January 2016.)

in which the smelting process may be carried out: rotary furnace, reverberatory furnace and blast or electric furnace, rotary kiln, etc. The choice of method depends, again, on several factors that include local economics, planned amount of recycling, etc.

The fluxing agents, which melt at a temperature below the lead compounds melting temperature, are added not only to reduce the lead smelting temperature, but also to provide a liquid solvent, which traps several unwanted compounds during the smelting and reducing processes. As the flux starts to be contaminated with all sorts of impurities from the smelting process, the formation of slag also starts. The physical and chemical properties of the slag, which are important characteristics to be considered in a later treatment, are entirely dependent on the chemical composition of the flux that is used. Reducing agents, which are usually carbon-based compounds such as coke, coal fines or other natural carbon source, are added with the purpose of reducing the lead oxide and lead hydroxide ($Pb(OH)_2$) to metallic lead.

During the process, the melted metallic lead accumulates at the bottom of the vessel and is normally contaminated with other metals of economic value. Therefore, the lead bullion must undergo a refining process before pure lead of marketable qualities can be recovered from it (UNEP, 2003; Ecofys, 2009; European Commission, 2009).

Hydrometallurgical Methods The objective of hydrometallurgical methods, or electrolytic methods, is to electrically and selectively reduce all lead compounds to metallic lead (UNEP, 2003). Ideally, in the electrolytic process, all lead compounds are converted to lead of the oxidation state +2 (i.e. Pb^{2+} or plumbous lead), which is then electrolytically reduced to produce metallic lead. Although it may be costly when considered as an isolated plant, this process is advantageous when linked to a low-temperature smelting plant since, with the appropriate separation of raw materials, it is a technological solution to overcome the lead refining processes.

9.3.3.4.5 Lead Refining

The objective of the refining process is to remove almost all the copper, antimony, arsenic and tin, since the soft lead standard does not allow more than 10 g per ton of these metals (UNEP, 2003). Two methods used for lead refining include hydrometallurgical methods, which are described in Section 9.3.3.4.4, and pyrometallurgical or thermal processes.

Pyrometallurgical refining is performed in liquid phase at temperatures higher than 327°C (lead melting point), but less than 650°C (lead boiling point). In this method, specific reagents are added to the molten lead at appropriate temperatures so as to remove the unwanted metals in a specific order as they are added selectively (UNEP, 2003). This process occurs in five steps: (1) removal of antimony, tin and arsenic; (2) removal of precious metals by Parkes process, in which zinc combines with gold and silver to form an insoluble intermetallic at operating temperatures; (3) vacuum removal of zinc; (4) removal of bismuth by the Kroll–Betterton process, in which calcium and magnesium are added to form an insoluble compound with the bismuth that is skimmed from the kettle and (5) removal of the remaining traces of metal impurities by adding NaOH and $NaNO_3$ (EPA, 1995; UNEP, 2003).

The final refined lead, from 99.990% to 99.999% pure, is typically cast into 45 kg pigs for shipment (EPA, 1995).

9.3.4 Zinc

9.3.4.1 Chemical and Physical Properties of Zinc

Zinc is a metallic element with atomic symbol Zn. It is the twenty-fourth most abundant element in the earth's crust, has atomic number 30, an atomic weight of 65.37 (Leon, 2006; Kaur et al., 2014) and a density of 7.133 g/cm^3 (Leon, 2006). Pure zinc is a lustrous, bluish-white metal that burns in air with a bluish-green flame (Roney et al., 2005). It is brittle (i.e. shatters easily) at room temperature, but malleable (i.e. easily bends) between 100°C and 150°C (Leon, 2006). At these temperatures, zinc also becomes ductile, which means it can be drawn into fine wires and threads (Leon, 2006). Zinc is stable in dry air, but reacts with moist air to form a film of zinc oxide or basic carbonate that covers the surface of the metal and, thus, retarding further corrosion (Roney et al., 2005). Compared to other metals, zinc has a low melting point (420°C). If it is heated above the melting point, zinc boils and turns into a gas at 907°C.

9.3.4.2 Occurrence

Zinc is a naturally occurring element found in the earth's crust with a distribution ranging from 20 to 200 ppm (Goodwin, 1998). Typical soil concentrations range from 5 to 770 ppm of zinc, with an average concentration of about 60 ppm (Hogan, 2007). Oceans contain 30 ppb zinc, and air has 0.1–4.0 µg of zinc per cubic metre (Hogan, 2007). Despite being widely distributed, zinc metal is never found as a free element in nature because of its reactivity (Lloyd and Showak, 1984). However, zinc minerals are generally associated with other metal minerals, the most common associations in ores being zinc–lead, lead–zinc, zinc–copper, copper–zinc, zinc–silver or zinc only (BCS Inc., 2002). Zinc is also combined with sulphur in a mineral called zinc blende or sphalerite (ZnS) (BCS Inc., 2002), which provides about 90% of zinc production (Goodwin, 1998; BCS Inc., 2002; Roney et al., 2005). Sphalerite contains 60%–62% Zn, and the largest deposits of sphalerite are found in Australia, Canada and the United States (Hogan, 2007). Other zinc-containing minerals (non-sulphide zinc) include hemimorphite, hydrozincite, calamine, franklinite, smithsonite, willemite and zincite (BCS Inc., 2002; Boni, 2003; Boni and Large, 2003).

9.3.4.3 Uses of Zinc

Zinc and its compounds have many important industrial and domestic applications (Leon, 2006), and it has the third highest usage rate among non-ferrous metals after aluminium and copper (Villanueva et al., 2010). The main use of zinc that accounts for about half the world production is galvanization (Reilly, 2002; Chandler, 2006). In other words, most zinc metal is used as a protective coating of other metals such as iron and steel (Roney et al., 2005). Zinc metal itself is also used extensively in the construction of buildings, roofing and rain water systems, as well as for cladding (Reilly, 2002). It is also found in many manufactured items, such as transistors, circuit boards, photocopiers, dry cell batteries and many others.

Due to lack of strength in zinc metal, it is frequently alloyed with other metals (e.g. aluminium, copper, titanium and magnesium) to impart a range of properties (Roney et al., 2005). When zinc metal is the primary component of the alloy, it is called a 'zinc-base' alloy, which is primarily used for casting and wrought applications. Other important applications of zinc alloys are in die-casting, construction and in other alloys (e.g. brass and bronze), which may be found in electrical components of many household goods.

Zinc compounds (e.g. zinc oxide) have also found widespread applications in the chemical and pharmaceutical industries, such as in the manufacture of paints, pigments, rubber and plastics (Reilly, 2002; Roney et al., 2005). They are also found in cosmetics, medicine, nutritional supplements and a variety of household consumables (Lloyd, 1984; Lloyd and Showak, 1984; Reilly, 2002; Roney et al., 2005). Other uses of zinc compounds include smoke bombs for crowd dispersal (e.g. zinc chloride), fertilizers, sprays, and animal feed (e.g. zinc sulphate), phosphor for watches and TV screens (e.g. zinc sulphide), wood preservative, mordant for antiseptics, a water-proofing agent (e.g. zinc acetate), dental materials (e.g. zinc phosphate), electroplating and gold extraction (e.g. zinc cyanide), gas sensors, varistors and transducers (e.g. zinc oxide) and an adsorbent in surgical dressings (e.g. zinc hydroxide) (Lewis, 1997; Goodwin, 1998; Wang and Muhammed, 1999; O'Neil et al., 2001; Roney et al., 2005).

9.3.4.4 Post-Consumer Recycling Processes

Except in certain specialized applications, zinc is not itself used widely as a metal in products, and therefore, zinc does not usually occur as a separate waste in isolation (OECD, 1995). For example, one-fourth of the total zinc produced is used as a protective coating that has a limited scope for recovery (Shamsuddin, 1986). Moreover, zinc-coated steel and other zinc products are very durable and, therefore, very slow in entering the recycling circuit (Villanueva et al., 2010). For instance, the life of zinc products can range from 10 to 15 years in household and car appliances, and up to 100 years for zinc sheets in roof protection.

A significant source of zinc for recycling is old and new brass scraps (OECD, 1995). The motor industry where a number of automobile parts are made from die-cast zinc alloys that contain a high percentage of zinc is also a significant source. Many other uses for zinc are dispersive, thus making it unavailable for direct recycling. However, zinc contained in metal-bearing wastes processed for the recovery of other metals increasingly provides additional sources of zinc for recovery. To a much lesser extent, zinc-containing residues from the chemical industry and those from chemical leaching processes for stripping zinc from coated steel are also possible sources of zinc for recovery, but are not widely used.

Technology for the recovery of zinc from scrap materials is mature, relatively simple and not highly sensitive to the grade of the scrap processed. Generally, there are three different operations used to recover secondary zinc from scrap: pretreatment, smelting (or electrolytical) and refining. These operations are discussed below in detail.

9.3.4.4.1 Pretreatment

When the concentration of zinc in materials for recycling is comparatively low (less than ~35%), preprocess must be carried out on the waste to increase its zinc content to at least 55%–60% so that zinc recovery becomes economically viable (OECD, 1995). Generally, scrap pretreatment includes (1) sorting, (2) cleaning, (3) crushing and screening and (4) sweating (EPA, 1995).

In the sorting operation, zinc scrap is manually separated according to zinc content and any subsequent processing requirements (EPA, 1995). Cleaning removes foreign materials so as to improve product quality and recovery efficiency. Crushing facilitates the ability to separate the zinc from the contaminants. Screening and pneumatic classification concentrates the zinc metal for further processing. A sweating furnace (rotary, reverberatory or muffle furnace) slowly heats the scrap containing zinc and other metals to approximately 364°C. This temperature is sufficient to melt zinc, but is still below the melting point of the remaining metals. Molten zinc collects at the bottom of the sweat furnace and is subsequently recovered. The remaining scrap metal is cooled and removed to be sold to other secondary processors.

9.3.4.4.2 Smelting

Afterwards, depending on the type and quality of scrap, the relatively pure zinc scrap is charged into a furnace such as kettle, crucible, reverberatory and electric induction furnace, where it is slowly heated up until the melting point of zinc is reached (EPA, 1995; Villanueva et al., 2010). Flux is used in these furnaces to trap impurities from the molten zinc. Facilitated by agitation, flux and impurities float to the surface of the melt as dross and are skimmed from the surface. Leaching with sodium carbonate solution converts dross (identified as $FeZn_{13}$) and skimming to zinc oxide, which can be reduced to zinc metal (EPA, 1995; Villanueva et al., 2010). Zinc oxide is transformed into zinc metal through a retort reduction process using coke as a reducing agent. Carbon monoxide produced by the partial oxidation of coke reduces the zinc oxide to metal and carbon dioxide, and the zinc vapour is recovered by condensation (EPA, 1995).

The remaining molten zinc from the melting process in the furnaces may be poured into moulds or transferred to the refining stage in a molten state. Zinc alloys, which are much stronger than unalloyed zinc, are usually produced during sweating or melting processes (EPA, 1995; Villanueva et al., 2010).

9.3.4.4.3 *Refining*

Refining processes remove further impurities from the zinc scrap. Molten zinc is heated until it vapourizes. The vapour is then condensed and recovered in several forms depending on temperature, recovery time, presence or absence of oxygen and the equipment used (Villanueva et al., 2010). The final products from the refining processes include zinc ingots, dust, zinc oxide and zinc alloys.

9.3.4.4.4 *Hydrometallurgical Techniques*

While the pyrometallurgical method discussed earlier is applicable to scrap material containing preponderantly metallic zinc, it is not efficient in processing zinc alloy materials (EIC Corporation, 1979). Hydrometallurgical techniques could offer a more attractive route to the recovery of zinc from such materials. Furthermore, hydrometallurgical methods have proved to be suitable for producing high-purity end products including zinc (Virolainen, 2013), which would be suitable for die-casting or alloying to brass rather than being generally restricted to use in galvanizing (EIC Corporation, 1979). Generally, sulphuric acid solutions have been found to be effective lixiviants in the proposed hydrometallurgical routes for recovery of zinc from scrap (Sharma and Row, 1985; Jha et al., 2001). The leach solution so obtained during leaching could be purified by conventional zinc cementation, solvent extraction or ion exchange process. The zinc metal or salt could be produced from the purified solution by electrowinning or crystallization (Jha et al., 2001).

9.3.5 Tin

9.3.5.1 Chemical and Physical Properties of Tin

Tin has the atomic symbol Sn, the atomic number 50, and an atomic mass of 118.71 (Westrum and Thomassen, 2002; Howe and Watts, 2005). It is a silver-white, malleable and ductile metal (Harper et al., 2005). Ideally, pure tin exists in two allotropic crystalline forms: grey tin (alpha form) and white tin (beta form) (Howe and Watts, 2005). At low temperatures (at about 18°C and below), the grey tin changes to white tin. The melting point (232°C) of pure tin is low compared with those of the common structural metals, whereas the boiling point (2602°C) exceeds that of most metals except tungsten and the platinum group (Westrum and Thomassen, 2002). Therefore, loss by volatilization during melting and alloying with other metals is insignificant. Only small quantities of some metals can be dissolved in pure liquid tin near its melting point, but intermetallic compounds are freely formed, some of which are of metallurgical importance.

The +2 (stannous) and +4 (stannic) oxidation states of tin, also known as tin (II) and tin (IV), respectively, are both reasonably stable and interconverted by moderately active reagents (Howe and Watts, 2005). Due to its amphoteric nature, tin reacts with strong acids and strong bases, but remains relatively resistant to neutral solutions (Westrum and Thomassen, 2002; Howe and Watts, 2005; Gitlitz and Moran, 2006).

Industrially important tin compounds can be categorized as inorganic (those without a tin–carbon bond) and organic (those having a tin–carbon bond) (Lide, 2003; Harper et al., 2005). The most commercially significant inorganic tin compounds include tin (II) and tin (IV) chlorides, tin (II) and tin (IV) oxide, potassium and sodium stannates, tin (II) fluoride, tin (II) difluoroborate and tin (II) pyrophosphate (Howe and Watts, 2005).

A summary of the physical and chemical properties of tin and that of some of the inorganic compounds of tin are given in Table 9.14.

TABLE 9.14
Physical and Chemical Properties of Tin and Some Inorganic Tin Compounds

Compound	Melting Point (°C)	Boiling Point (°C)	Density (g/cm³)	Solubility in Water
Sn	232	2606	5.77[a]7.27[b]	Insoluble
$SnBr_4$	31	205	3.34	Soluble
$SnCl_2$	247	623	3.90	Soluble
$SnCl_4$	213	850	4.57	Soluble
SnI_2	320	714	5.28	Slightly soluble
SnI_4	143	364.5	4.46	Soluble
SnO	1080		6.45	Insoluble
SnO_2	1630	1900	6.85	Insoluble
$Sn_2P_2O_7$	Decomposes at 400°C		4.01	Insoluble
SnS	880	1210	5.08	Insoluble
$SnSO_4$	Decomposes at >4378°C		4.15	Reacts with SO_2

Source: Lide, D.R., *CRC Handbook of Chemistry and Physics*, CRC Press, Boca Raton, FL, 2003.
[a] Grey tin, cubic crystalline
[b] White tin, silvery tetragonal crystalline form, stable above 13°C.

9.3.5.2 Occurrence

Tin occurs naturally in the earth's crust, with an average concentration of approximately 2–3 mg/kg (Moore, 1991; Howe and Watts, 2005; Brimer, 2011) and is present in the form of nine different minerals from two types of deposits (Gitlitz and Moran, 2006; van Nostrand, 2006). However, total tin concentrations in soil can range from <1 to 200 mg/kg, but levels of 1000 mg/kg may occur in areas of high tin deposits (Howe and Watts, 2005). Actually, some ore deposits may contain up to 50,000 mg/kg as tin. Tin is a borderline metal; thus, it has equal preference for complexation with sulphur and oxygen/nitrogen (Moore, 1991). Nevertheless, the only commercially significant mineral ore for tin is cassiterite (SnO_2), which is harder, heavier and more chemically resistant than the granite in which it typically forms (Ecclestone, 2014). The more economically valuable deposits of cassiterite are heavily concentrated in bands and layers of varying thickness, in Malaysia, Thailand, Indonesia and the People's Republic of China (Gitlitz and Moran, 2006). The other ores of tin are complex sulphides such as stannite (Cu_2FeSnS_4), teallite ($PbSnS_2$), canfieldite (Ag_8SnS_6) and cylindrite ($PbSn_4FeSb_2S_{14}$) (Beliles, 1994; Howe and Watts, 2005).

9.3.5.3 Uses of Tin

An important property of tin is its ability to form alloys with other metals (Westrum and Thomassen, 2002; Howe and Watts, 2005). Tin alloys cover a wide range of compositions and many applications (Howe and Watts, 2005). Actually, the major use for tin, which accounts for 34% of the production, is in the manufacture of solder alloys that are mainly used in the electrical and electronic industry. For example, common solder, an alloy of 63% tin and lead, is mainly used in the electrical industry; lead-free tin solders containing up to 5% silver or antimony are used at higher temperatures. Other tin alloys that are widely employed in electrical/electronic industry and for general industrial applications include those containing lead, antimony, silver, zinc or indium, babbit; babbit (containing mainly copper, antimony, tin and lead); Wood's metal (50% bismuth, 25% lead, 12.5% tin and 12.5% cadmium); brasses and bronzes (essentially tin–copper alloys); pewter (0%–95% tin plus

TABLE 9.15

Uses of Some Inorganic Compounds of Tin

Compound	Application
$SnCl_2$	Preparation of glass and plastic for metallizing, metallized glazing and electronic components on a plastic base, as a soldering flux, as a mordant in dyeing, and in the manufacture of tin chemicals, colour pigments and sensitized paper; food preservative and colour retention agent; nuclear medicine as an essential component in diagnostic agents used to visualize blood, heart, lung, kidney, and bone; reducing agent in manufacturing ceramics, glass and inks.
$SnCl_4$	Dehydrating agent in organic synthesis, in the production of organotin compounds, in the production of tin (IV) oxide films on glass, as a mordant in the dyeing of silks, in the manufacture of blueprint and other sensitized paper and as an antistatic agent in synthetic fibre and stabilizer for plastics.
SnO	Reducing agent in the preparation of stannous salts, and in the preparation of gold-tin and copper-tin ruby glass.
SnO_4	Polishing of glass and enamels, in the manufacture of milk-coloured ruby and alabaster glass and enamels, as a mordant in printing and dyeing of fabrics, and in fingernail polish.
SnF_2	An ingredient of caries-preventing toothpaste.
$Sn_2P_2O_7$	An ingredient in caries-preventing toothpaste.
Sn^{2+} ions	Sn^{2+} ions possess antibacterial activity, whereas Sn^{4+} ions do not.

Sources: Westrum, B. and Thomassen, Y., The Nordic expert group for criteria documentation of health risks from chemicals and the Dutch Expert Committee on Occupational Standards. 130. Tin and inorganic tin compounds, Arbete och Halsa, 10, 2002, pp. 1–48; Howe, P. and Watts, P., Tin and inorganic tin compounds, World Health Organization, Geneva, Switzerland, 2005.

1%–8% bismuth and 0.5%–3% copper) and dental amalgams (silver–tin–mercury alloys) (Bulten and Meinema, 1991; Howe and Watts, 2005).

The physicochemical properties of tin also make it useful in the prevention of corrosion and other chemical reactions with various metals (Moore, 1991; Westrum and Thomassen, 2002). For example, tinplating is widely used as a protective coating in food and beverage packaging, foil and wires (Moore, 1991; Brimer, 2011). The uses of some of the inorganic compounds of tin are listed in Table 9.15.

9.3.5.4 Post-Consumer Recycling Processes

The most important post-consumer materials for secondary tin include tin-containing products such as tin cans and electronic equipment, which have been discarded after use (Villanueva et al., 2010). Ideally, the most important raw material for secondary tin is the scrap generated directly by the tinplate process (Kékesi et al., 2000; Sokic et al., 2006). These wastes are generated by both industrial and domestic users (Villanueva et al., 2010), and the common sources of tinplate scrap are the canning factories and the magnetic separators of communal waste incinerating plants (Kékesi et al., 2000). Another secondary source of tin is bronze (an alloy of tin and copper) and is mostly remelted into ingots or processed in copper refineries (Rao, 2011).

Due to the detrimental effect of tin as an impurity, direct recycling of tinned steel scrap for steelmaking is usually not possible (Sokic et al., 2006). Therefore, the production of high-grade steels requires the preliminary detinning of scrap before remelting. On the other hand, the high commercial value of tin also warrants effort for its extraction from secondary raw materials. It must be noted, however, that most of the revenue is generated from the sale of the upgraded steel scrap rather than from the recovered tin (Rao, 2011).

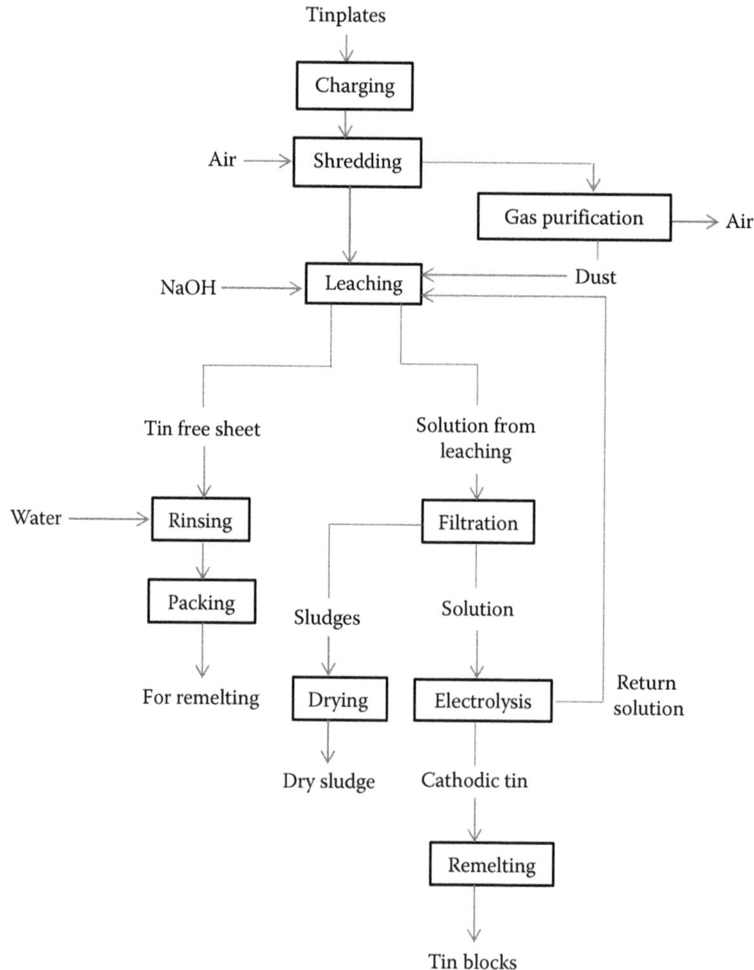

FIGURE 9.3
Technological procedure for processing waste tin plates and metallic packages. (From Sokic, M. et al., *Acta Metall Slovaca*, 12, 354, 2006.)

The following two procedures are widely used for the recovery of tin from tin-containing metallic waste: (1) pretreatment and (2) extraction. Some of these procedures shown in Figure 9.3 are discussed later in detail.

9.3.5.5 Pretreatment

The initial treatment of the metallic waste is aimed at removing organic and undesirable metallic impurities. Pretreatment usually includes the following operations: metallic waste collection and storage, rinsing, sorting, shredding, degreasing, varnish removal and final washing.

Waste collection and storage: Proper collection of the metallic waste and its temporary storage before being taken to a recycling facility are of utmost importance. Usually, collection sites and/or centres are setup where waste is received and classified.

Rinsing: The aim of rinsing is to remove unwanted impurities such as food residues and paper labels from the cans, for example. The rinsing of cans is carried out in rinsing drums. Shredding machines may also be used to rinse and shred the metallic waste simultaneously.

Sorting of the waste: The rinsed tin waste is transferred to the sorting belt (conveyor) where stones and other metallic wastes such as iron, aluminium, zinc and copper are separated. Magnetic separation is used to remove ferrous metals from the tin waste. It is important that all the metallic impurities (e.g. aluminium, zinc, galvanized iron, copper and brass) are removed at this stage because their presence affects the tinplating solutions' life, the plating efficiency and techno-economical indexes of the electrolysis process during the metal extraction procedure.

Shredding of the waste: After the sorting stage, tin waste is shredded and/or reduced in size using shredding machines.

Varnish removal: Varnish is usually removed by heating the metallic waste in air at 240°C–270°C.

Degreasing and final washing: The last operation in the metallic waste pretreatment includes degreasing and final washing. Degreasing is performed in metallic baskets or drums or on conveyors with either 2%–3% sodium carbonate solution at 80°C or 1% sodium hydroxide solution.

9.3.5.6 Extraction

The widely used methods for tin recovery from the waste tinplates and tin-bearing metallic packages in recent past include chlorination, alkaline leaching and electrolytic dissolution and deposition (Sokic et al., 2006). However, as stated already, most of the revenue is generated from the sale of the upgraded steel scrap rather than from the recovered tin (Rao, 2011).

9.3.5.6.1 Chlorination Method

The chlorination technique is based on the reaction between tin and gaseous chlorine-forming tin chloride ($SnCl_4$). The absence of moisture and organic compounds in the starting material are prerequisites to application of the chlorination process. The presence of moisture results in iron dissolution, thus contaminating tin chloride. On the other hand, organic materials react with tin resulting in a decreased degree of recovery. Furthermore, since the chlorination process generates a substantial amount of heat that favours the formation of iron chloride, the process is only economical for large-scale recycling.

Tin metal can be produced from tin chloride ($SnCl_4$) by two methods: (1) cementation with less noble metal than tin (e.g. aluminium, zinc), and (2) electrolysis process with insoluble anodes producing tin metal containing 99.8% tin and 0.1%–0.25% lead, with iron and zinc as trace elements. The major disadvantage of the chlorine process is that the removal of tin as volatile tin chloride needs relatively large and sophisticated equipment and is accompanied by health and corrosion hazards (Kékesi et al., 2000).

Besides the production of tin chloride, another product of the chlorination process is the upgraded steel scrap (i.e. remaining processed metallic waste), which is used as a secondary raw material in steelmaking plants.

9.3.5.6.2 Alkaline Leaching

The fact that tin is soluble and iron is insoluble in alkaline solutions (or caustic solutions) makes it possible to process waste tinplates by alkaline solutions (Sokic et al., 2006). In other words, selective chemical dissolution of the tin layer from the surface of steel plates can be achieved in caustic solutions, in the presence of an oxidizing agent (Kékesi et al., 2000).

Only the leaching of tin from tinplate using hot sodium hydroxide (NaOH) will be discussed here since it is the only one that is probably still commercially viable today. The dissolution of tin in hot NaOH solutions is a heterogeneous reaction, where the transport of material, adsorption and desorption steps can also play an important part. The overall reaction given by Equation 9.1 is principally electrochemical in nature.

$$Sn + 2NaOH \rightarrow Na_2SnO_2 + H_2 \tag{9.1}$$

Reaction (9.1) occurs very slowly because of the high hydrogen overvoltage on tin. In the presence of oxygen (or any other oxidant), the dissolution of tin becomes more intensive and forms sodium stannate as given in the following:

$$Sn + 2NaOH + O_2 \rightarrow Na_2SnO_3 + H_2O \tag{9.2}$$

The following methods are used to recover tin from the resulting stannate liquor: (1) electrolysis process with insoluble anodes; (2) precipitation with CO_2, $NaHCO_3$ or Ca $(OH)_2$ and (3) treatment with acids such as H_2SO_4.

9.3.5.6.3 Electrolytic Dissolution and Deposition

The caustic dissolution reaction discussed earlier is accelerated if it is simultaneously performed with electrolysis (Rao, 2011). The essence of the process is to separately leach tin from the waste tinplates using NaOH solution containing a suitable oxidant and, thereafter, to electrolytically deposit tin onto insoluble cathodes (Sokic et al., 2006). This electrolytic precipitation process is a widely applied procedure for tin extraction and recovery from the waste tinplates and other industrial scraps (Shamsuddin, 1986; Sokic et al., 2006; Rao, 2011). Ideally, due to its high electrical conductivity and passivity towards iron, the sodium hydroxide solution serves as the electrolyte. The metallic materials containing tin and/or the waste tinplates are used as soluble anodes in the electrolytic cell at a temperature of about 80–85°C. Tin is dissolved from tin-containing material by anodic reaction and electrolytically deposited at the cathode in the form of a porous, spongy metal. A spongy tin deposit formed on the cathodes is manually removed, compacted, melted and cast into ingots. Thereafter, the recovered tin with a purity in the range of 95%–97% (the main contaminants being lead and iron) is sold to tin refiners (Rao, 2011).

9.3.6 Steel

9.3.6.1 Introduction

Steel is a generic term for a large family of iron–carbon alloys (Huyett, 2004). Fundamentally, steel is a crystalline alloy of iron, carbon and several other elements (Séblin et al., 2014). These alloys vary both in the way they are made and in the proportions of the materials added to the iron (Sharif, 2008; Keerthiprasad et al., 2014). However, as discussed in Chapter 4, the principal raw materials used in steelmaking are iron ore, carbon and limestone. These materials are converted in a blast furnace into a product known as 'pig iron'. Pig iron is composed of about 93% iron, 3%–5% carbon and various amounts of other elements (Talbert and Keeton, 1996). Pig iron is comparatively weak and brittle; thus, it has a limited use (Talbert and Keeton, 1996). Therefore, about 90% of pig iron (as well as iron and steel scrap) is refined by removing undesirable elements and adding desirable elements in predetermined amounts (Talbert and Keeton, 1996; Huyett, 2004). Thus the properties of steel are linked to the chemical composition, processing path and resulting microstructure of the material (Bramfitt, 1998). Generally, an increase in the carbon content from 0.01% to 1.5% increases the strength and hardness of steel, but an increase beyond 1.5% causes appreciable reduction in the ductility and malleability of the steel (Sharif, 2008; Cabrera et al., 2014). Actually, when there is more than 2% carbon, the material comes into the category of 'cast irons' that are characterized as having dominant iron content, low melting point and good cast ability (Krauss, 2015). Historically, however, cast irons were brittle, a characteristic that differentiated them from steels with good combinations of strength and ductility. It must be noted that the basis for this historical differentiation is challenged by modern technology: good foundry practice produces nodular and austempered ductile cast irons with good combinations of strength and toughness (Krauss, 2015).

9.3.6.2 Types of Steel

Steel classification is important so as to understand what types are used in certain applications and which types are used for others (Capudean, 2003). Though any rigid classification of steel is impossible (Dossett and Boyer, 2006), there are fundamentally two main kinds of steels: plain carbon steels and alloy steels (Meyrick and Wagoner, 2001; Krauss, 2015). Plain carbon steels, in addition to iron and carbon, contain manganese and a variety of residual elements (Meyrick and Wagoner, 2001) and account for 90% of total steel production (Bell, 2015). Ideally, residual elements are either present within the raw materials used in the production process, for example, iron ore and scrap steel additions, or they were added in the production process for a specific purpose, for example, deoxidization by means of silicon or aluminium (Meyrick and Wagoner, 2001). According to the American Iron and Steel Institute, a plain carbon steel is defined as an alloy of iron and carbon that contains specified amounts of manganese below a maximum amount of 1.65% weight, less than 0.6% weight silicon, less than 0.6% weight copper and that does not have any specified minimum content of any other deliberately added alloying element (Meyrick and Wagoner, 2001). Plain carbon steels can be divided further into three main groups depending on the carbon content: low-carbon steel, also known as mild steel, medium-carbon steel and high-carbon steel (Uzorh, 2013; Bell, 2015).

If one or more elements (usually metals) are intentionally added to produce properties that are not found in plain carbon steel, then the product is an alloy steel (Meyrick and Wagoner, 2001). Some of the properties that may be enhanced by alloying include hardenability, machinability, strength through heat treatment, strength as manufactured, corrosion resistance, wear resistance and retention of hardness and strength at high temperatures (Outokumpu, 2013). Alloy steels can also be subdivided into many groups according to chemistry (e.g. standard low-alloy steels), amount of alloying additives (e.g. unalloyed to medium-alloy to high-alloy steels), applications (e.g. tool steels) or particular properties (e.g. stainless steels) or even as a consequence of a particular heat-treatment (e.g. maraging steel), etc. (Janke et al., 2000; Meyrick and Wagoner, 2001).

9.3.6.3 Uses of Steel

Steel has desirable properties such as great strength, uniformity, lightweight and ease of use that makes it the material of choice in numerous applications (Assakkaf, 2002). Indeed, steel is ubiquitous in applications that directly affect the quality of our lives (Meyrick and

TABLE 9.16
Plain Carbon Steels and Their Uses

Category of Carbon Steel	% by Weight of Carbon in Steel	Application
Low carbon	0.05–0.20	Automobile bodies, buildings, pipes, chains, rivets, nails.
	0.20–0.30	Gears, shafts, bolts, forgings, bridges, buildings.
Medium carbon	0.30–0.40	Connecting rods, crank pins, axels, drop forgings.
	0.40–0.50	Car axels, crankshafts, rails, boilers, auger bits, screwdrivers.
	0.50–0.60	Hammers, sledges.
High carbon	0.60–0.70	Stamping and pressing dies, drop forging dies, drop forgings, screwdriver, blacksmith hammers, table knives, set screws.
	0.70–0.80	Punches, cold chisels, hammers, sledges, shear blades, table knives, drop forging dies, anvil faces, wrenches, vise jaws, band saws, crowbars, lathe centres, rivet sets.
	0.80–0.90	Punches, rivet sets, large taps, threading dies, drop-forging dies, shear blades, table knives, saws, hammers, cold chisels, woodworking chisels, rock drills, axes, springs.
	0.90–1.00	Taps, small punches, threading dies, needles, knives, springs machinist hammers, screwdrivers, drills, milling cutlers, axes reamers, rock drills, chisels, lath centres, hacksaw blades.
	1.00–1.10	Axes, chisels, small taps, hand reamers, lath centres, mandrels, threading dies, milling cutters, springs, turning and planning tools, knives, drills.
	1.10–1.20	Milling cutters, reamers, woodworking tools, saws, knives, ball bearing, cold cutting dies, threading dies, taps, twist drills, pipe cutters, lath centres, hatchets, turning and planning tools.
	1.20–1.30	Turning and planning tools, twist drills, scythes, files, circular cutters, engravers tools, surgical cutlery, saws for cutting metals, tools for turning brass and wood, reamers.
	1.30–1.40	Small twist drills, razors, small engravers tools, surgical instruments, knives, boring tools, wire drawing dies, tools for turning hard metals, files, woodworking chisels.
	1.40–1.50	Razors, saws for cutting steel, wire drawing dies, fine cutters.

Sources: Florida Artist Black Smith (FABA), 1989, The clinker breaker. Available at http://blacksmithing.org/CB-Archive/1989/1989-02-cb.pdf, Accessed October 2016; Helsel, L. and Liu, P., *Industrial Materials*, Goodheart-Willcox Publisher, Chicago, IL, 2008; Faieza, A.A. et al., *Int. J. Adv. Res. Sci. Eng.*, 4(1), 14, 2015.

Wagoner, 2001). Table 9.16 shows carbon compositions of plain carbon steels and their typical applications, while typical applications of some selected alloy steels are listed in Table 9.17.

9.3.6.4 Post-Consumer Recycling Processes

Steel scrap is used as a source of iron in both the induction and electric arc furnaces (Christopher, 2011). However, the scrap proportions in the feed of the two types of furnaces are quite different in each case, with scrap being the main raw material in electric arc furnaces (Janke et al., 2000, 2006; Morici et al., 2013). Steel scrap encompasses a variety of different materials and qualities (Morici et al., 2013), and generally, steel scrap can be divided into three categories: (1) home scrap (plant scrap), (2) process scrap (prompt scrap) and (3) obsolete scrap (capital scrap) (Janke et al., 2000).

TABLE 9.17

Typical Applications of Selected Alloy Steels

Type of Alloy Steel	Main Chemical Composition	Application
Manganese steels	0.30% C, 1.75% Mn	High strength bolts, bicycle frames
Chromium steels	0.30% C, 0.80% Mn, 0.80% Cr	Steering parts
	0.60% C, 0.88% Mn, 0.80% Cr	Spring wheels
Chromium–molybdenum steels	0.30% C, 0.50% Mn, 0.95% Cr, 0.20% Mo	Pressure vessels, aircraft structural parts, auto axles, steering knuckles
	0.40% C, 0.88% Mn, 0.95% Cr, 0.20% Mo	
Chromium–vanadium steels	0.50% C, 0.80% Mn, 0.95% Cr, 0.15% Vn	Valves and springs
Nickel–chromium–molybdenum steels	0.40% C, 0.88% Mn, 0.50% Cr, 0.20% Mo, 0.55% Ni	Auto springs, small machine axles, shafts
Silicon steels	0.60% C, 0.88% Mn, 2.0% Si	Leaf springs

Sources: Yang, Z.G. et al., Materials properties database for selection of high-temperature alloys and concepts of alloy design for SOFC applications, 2002, Available at http://www.pnl.gov/main/publications/external/technical_reports/PNNL-14116.pdf, Accessed October 2016; Huyett, G.L., *Engineering Handbook: Technical Information*, Huyett Expressway, Minneapolis, MN, 2004; Poweleit, D. and Monroe, R., Steel alloys, Steel Founders' Society of America, Crystal Lake, IL, 2006; Helsel, L. and Liu, P., *Industrial Materials*, Goodheart-Willcox Publisher, Chicago, IL, 2008.

Home scrap (or new scrap) is generated in steel mills during the production of steel. It is relatively pure and its chemical composition is known, so it can be recycled easily. Process scrap, also falling in the category of new scrap, is generated in the manufacturing of products made from steel. This scrap occurs during the production of both industrial and consumer end products. Process scrap is available for recycling in a relatively short time after its generation.

Obsolete scrap consists of iron or steel products discarded after the end of their service life. It is often mixed or coated with other materials, such as copper, zinc, tin, glass and plastics. For this reason, the content of tramp elements in obsolete scrap is usually high. Moreover, the chemical composition of obsolete scrap fluctuates widely depending on its origin and degree of processing.

Another way of classifying steel scrap is to classify it according to the products in which the steel was used before it became a waste (Muchová and Eder, 2010b). The main steel scrap sources in this sense are automobiles, ships, aeroplanes, railroads, construction buildings, machinery, white goods, packaging, electric and electronic equipment, etc.

Although it may take very long for some steel products to become scrap, all steel can potentially be recycled (Villanueva et al., 2010). In general, iron and steel scrap recycling involves size reduction, cleaning, sorting, shredding, pressing, shearing and crushing, possibly also cryogenic processes and final melting at the steelworks (Muchová and Eder, 2010b).

9.3.6.4.1 Pretreatment

Using a variety of technologies, scrap dealers collect and process scrap into a physical form and chemical composition that steel mill furnaces can handle (Villanueva et al., 2010).

At the scrap pretreatment plant, different types of metals are separated and then shredded into fist-size pieces (Villanueva et al., 2010). The fist-size pieces of various materials are then sorted out into ferrous and non-ferrous materials by means of magnetic separation (Muchová and Eder, 2010b; Villanueva et al., 2010). After magnetic separation, the non-ferrous metals, stainless steel scrap and non-metal fractions are further separated by using combinations of density and eddy current separators (Muchová and Eder, 2010b). For example, fluids with different densities may be used to separate the light metals from the heavy metals. The heavy metals that usually contain mixtures of copper, zinc, lead and stainless steel may be manually hand-picked or sorted by sensors. On the other hand, eddy current separators exploit the electrical conductivity of non-magnetic metals. This is achieved by passing a magnetic current through the feed stream and using repulsive forces interacting between the magnetic field and the eddy currents in the metals (Javaid and Essadiqi, 2003). The simplest application of the process is the inclined ramp separator. The process uses a series of magnets on a sloped plate covered with a non-magnetic sliding surface such as stainless steel. When a feed of mixed materials is fed down the ramp, non-metallic items slide straight down, while metals are deflected sideways by the interaction of the magnetic field with the induced eddy current. The two streams are then collected separately.

At the end of the pretreatment process, steel components can be (1) recycled into the induction and electric arc furnace steelmaking processes to make new steel for different applications, and (2) reused, for example, by dismantling components from scrap vehicles and fitting them to another car (Villanueva et al., 2010).

9.3.6.4.2 Scrap Purification Methods

Despite the fact that steel is one of the easiest materials to recover from waste streams, the recycling of metal alloy is more complicated (Villanueva et al., 2010). For example, many tramp elements (i.e. elements used as alloying additions in certain steel grades, elements used as coating material for steel products or elements in pure state coexisting with pieces of steel scrap) found with and/or dissolved in steel (e.g. copper, tin, antimony and lead) are not oxidized in the presence of iron due to their low affinity for oxygen (Janke et al., 2000; Savov et al., 2003). For example, studies on the decopperization of molten steel have not achieved remarkable progress for a long time because the decomposition pressure of copper oxides is higher than that of iron oxides (Jianjun et al., 2009). This means that these elements (e.g. copper, tin, antimony and lead) cannot be removed from a steel scrap melt by a common pyrometallurgical process, as is the case with silicon, manganese and aluminium that are easily oxidized and dissolved in slag. Nevertheless, contaminating elements must be removed from steel scrap because they affect the mechanical strength and resistance to corrosion of the steel products (Savov et al., 2003).

As a result of low affinity for oxygen, tramp elements may be removed by pretreating scrap at lower temperatures while it remains in the solid state. Pretreatment of scrap in solid state often has the advantage that the tramp elements are present in pure state, either mingled with the ferrous portion of the scrap or existing at scrap surfaces, a fact that should facilitate their removal.

A summary of scrap purification methods (detinning, dezincing and decopperization) as discussed by Janke et al. (2000) and Savov et al. (2003) is outlined in the following discussion.

9.3.6.4.2.1 Detinning The main source of tin in steel is the recycling of post-consumer tinplate packaging. The tinplate scrap is detinned electrolytically, and the electrolytic detinning of tinplate scrap has been a commercialized process for a long time. First, tinplate scraps are prepared by pressing them into bundles with a density of about 1.5 ton/m^3. The bundles that serve as anodes in the electrolytic process are immersed in a caustic soda bath at a temperature of about 80–85°C. Tin is then deposited on a steel cathode as a sponge material that is then scraped off, pressed into large piles and sold to the tin industry. After detinning, the remaining tinplate scrap (ideally a clean steel scrap) would contain as low as 0.02% of residual tin.

It must be noted, however, that electrolytic detinning is suitable for prompt scrap, but is problematic for obsolete scrap.

9.3.6.4.2.2 Dezincing The main source of zinc contamination in steel scrap is recycling of zinc-coated steel. Zinc is also used as an alloying element in brass; thus, steel scrap that is mixed with brass is expected to be contaminated with zinc. Zinc is also present in the form of zinc compounds in rubber, ceramics and paints, which can be an additional source of zinc in steel scrap.

Due to its high vapour pressure (70 bar at 1600°C), most of the zinc evaporates during the steelmaking process. Although the removal of zinc at the scrap smelting stage is not problematic, it is proposed that dezincing at a scrap pretreatment stage would avoid the problems associated with recycling large amounts of galvanized scrap.

A continuous process for the electrolytic dezincing of process scrap from automotive industry involves immersing galvanized scrap into a hot caustic solution where zinc dissolves while steel remains unaffected. After leaving the dissolution reactor, the dezinced scrap is washed and compacted. The zinc-enriched solution is circulated to electrolytic cells where the zinc is recovered electrolytically by deposition onto cathode plates.

Several other methods of dezincing the steel scraps have also been investigated. These include thermal treatment, treatment with Cl_2–O_2 gas mixtures and mechanical post-treatment after thermal treatment.

9.3.6.4.2.3 Decopperization The main source of copper in steel is the obsolete scrap that is obtained from discarded cars. Copper in old cars is present mainly in the form of wires, electric motors and cooling elements. Copper is also introduced into steel melts by the smelting of scrap that originates from steel grades containing an increased amount of copper.

Unlike tin and zinc, copper cannot be removed from scrap-based melts by a conventional electrolytic refining method. Nevertheless, several approaches to reduce the copper content of steel have since been proposed. Carefully selected methods suggested by Savov et al. (2003) for copper removal from steel scrap are outlined in the following details.

1. *Smelting of copper parts from steel*: In this method, the difference in the melting temperature between the impurity elements (e.g. copper whose melting temperature is 1083°C) and steel (1520°C) is exploited. The liquid copper flows off, thus separating the premelted copper parts from the solid steel. The low separation rate caused by copper losses (e.g. flow of copper into cavities) and steel losses caused by oxidation are the major disadvantages of this technique.
2. *Non-ferrous metal baths*: This technique is based on the fact that the solubility of copper in a bath of lead or aluminium metals is higher than that of steel. Despite a considerable degree of copper removal, the high lead consumption makes this method unacceptable. For example, copper content in a ton of steel scrap can be reduced from 0.3% to 0.1%, but a ton of lead would also be required. On the other hand, the solubility of copper in an aluminium bath is 65% at 730°C, and research has also shown that 80% of copper can be removed from the steel scrap at 750°C in just about 20 min.
3. *Sulphur-containing slag melts*: This method illustrated by Equation 9.3 is based on the fact that copper sulphide is more stable than iron sulphide at temperatures above 600°C. In other words, the chemical affinity of copper for sulphur is stronger than that of iron such that copper is transferred into the sulphide slag (Jianjun et al., 2009).

$$2Cu + FeS \rightarrow Cu_2S + Fe \tag{9.3}$$

The standard Gibbs free energy (ΔG^0) and equilibrium constant (K) of Equation 9.3 are 19.814 kJ/mol and 4.284, respectively. This shows that the ferrous sulphide can be used as an effective decopperization agent to make the copper transfer from the molten steel into the slag. Sodium sulphide is another sulphur-containing slag that has also been used successfully at a laboratory scale.

As a result of the low copper distribution coefficient, the consumption of sulphur-containing slag is usually high. For example, a ton of steel requires about 100 kg of slag. Another disadvantage of this method is increased sulphur content in steel.

4. *Treatment of solid steel scrap using chlorine-containing gases*: A chlorine-containing gas mixture may be used to remove copper as shown in Equations 9.4 and 9.5.

$$Cu + 2HCl \rightarrow CuCl_2 + H_2 \tag{9.4}$$

$$Cu + Cl_2 \rightarrow CuCl_2 \tag{9.5}$$

The main advantage of this technique is that copper and other tramp metals such as lead and tin can be removed simultaneously. The presence of toxic treatment gases and loss of steel through the oxidation of iron are some of the disadvantages of the method.

5. *Vacuum distillation*: This method is based on the distinct differences in vapour pressure between copper and steel. Both elementary copper and copper compounds can vapourize; therefore, they can be removed from steel baths by vacuum distillation. In other words, the removal of copper from steel melts under reduced pressure is possible.
6. *Dilution of contaminated scrap*: The effect of the properties of copper can be minimized by the addition of alloying elements such as nickel and aluminium. For example, research studies have shown that the addition of nickel to steel baths in a proportion of 1:1 or 2:1 can neutralize the effects of the properties of copper. However, though dilution of the steel scrap by the addition of the copper-free materials represents an 'apparent' purification method, it is not acceptable as a 'real' long-term metallurgical solution to the problem.
7. *Filtration of steel melts through ceramic filters*: Research has shown that copper can be adsorbed on the surface of certain ceramic materials (e.g. Al_2O_3–ZrO ceramics), thus enabling its removal from steel melts.

9.4 Critical Post-Consumer Sectors

In the context of this section, a selected number of sectors are termed 'critical sectors' because of their importance in providing metals that are susceptible to future scarcity (i.e. the sectors are secondary sources for critical metals). It must be noted, however, that 'scarcity' is not defined solely in geological terms, but also in a socio-economic context (Schmidt, 2012). In other words, geological scarcity, such as is frequently discussed in public, is less crucial here. Instead, the relevant issue is whether the metals are available on the market in sufficient quantities and quality.

Nevertheless, the criteria used to differentiate critical metals from ordinary metals vary from study to study. Ayres and Peiró (2013) argue that metals are classed as critical if (because) they are geologically scarce, subject to potential supply constraints, costly and needed for an economically important purpose where substitution is difficult because of their special or unique properties. Of these characteristics, according to Ayres and Peiró (2013), the potential for supply constraints seems to be pre-eminent. By contrast, ordinary metals are generally not scarce and typically have many uses and many substitutes. Buchert et al. (2009) uses three criteria to differentiate critical metals from ordinary metals. The first criterion is growth in demand (rapid demand growth and moderate demand growth). Second, risks in supply. This implies (1) physical scarcity (i.e. less reserves compared to annual demand), (2) temporary scarcity (i.e. there is time lag between production and demand) and (3) structural or technical scarcity (i.e. the metal is just a minor product in a coupled production, and inefficiencies occur in the mining process, production and manufacturing). The third criterion is concerned with recycling restrictions, which may mean that the concerned metal has (1) a high scale of dissipative applications, (2) physical and/or chemical limitations for recycling, (3) no suitable recycling technologies and/or recycling infrastructures and (4) no price incentives for recycling.

Figure 9.4 shows the criticality of metals based on supply risk and economic importance as defined by the European Union (EU). It can be seen from the figure that the following metals (in alphabetical order) that lie on the topmost right-hand corner are critical: antimony, beryllium, cobalt, fluorspar, gallium, germanium, graphite, indium, magnesium, niobium, PGMs, rare earths, tantalum and tungsten (European Commission, 2010). Each of these metals is used in at least one of the following sectors: (1) electrical and electronic equipment (e.g. tantalum in electronic capacitors), (2) catalysts (e.g. platinum as catalysts in chemical industries), (3) automobiles (e.g. germanium for night-vision systems in automobiles) and (4) household equipment (e.g. cobalt in batteries for mobile phones). Therefore, all these products become a great source of critical metals when they reach the end of life.

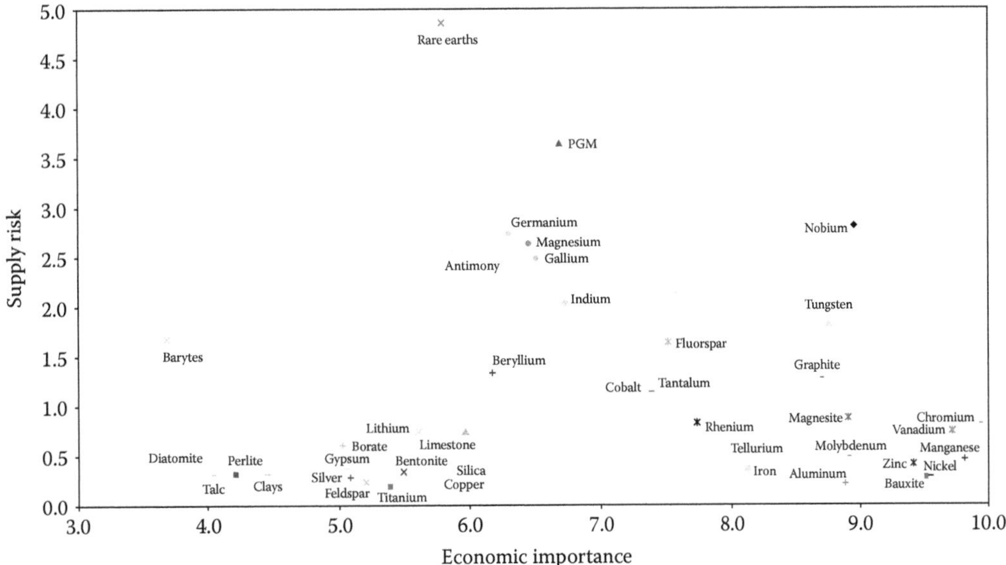

FIGURE 9.4
Criticality of metals based on supply risk and economic importance. (From European Commission, Critical raw materials for the EU, 2010, Available at http://www.euromines.org/files/what-we-do/sustainable-development-issues/2010-report-critical-rawmaterials-eu.pdf, Accessed January 2016.)

9.4.1 Electronic Waste

9.4.1.1 Introduction

The worldwide use of information and communications technology equipment and other electrical and electronic equipments is growing; thus, there is a growing amount of equipment that become waste after their usage time has elapsed (Balde et al., 2015). Indeed, this type of waste, which makes up 5% of all municipal solid waste worldwide, is now the fastest growing component of the municipal solid waste (Greenpeace, 2015). Furthermore, this waste material is diverse and complex, in terms of materials and components makeup, as well as the original equipment's manufacturing processes (Cui and Forssberg, 2003). Yet, at the moment, there is no standard definition of electronic and/or electrical waste (e-waste). The Organization for Economic Cooperation and Development defines e-waste as any appliance using an electric power supply that has reached its end of life (UNEP, 2007; Lundgren, 2012; Iqbal et al., 2015), while Directive 2008/98/EC of the EU defines it as 'electrical or electronic equipment which is waste, including all components, sub-assemblies and consumables which are part of the product at the time of discarding' (European Union, 2012). Another term that is also used alongside the term e-waste is waste electrical and electronic equipment (WEEE) (Iqbal et al., 2015). Nevertheless, there is no universal agreed framework among stakeholders (government, academic, developed countries, developing countries, etc.) pertaining to the definition and/or classification of e-waste. The differences in definitions and scope of what constitutes e-waste have the potential to create disparities in both the quantification of e-waste generation and the identification of e-waste flows (Lundgren, 2012). Therefore, in order to understand the dynamics of this complex waste stream, a framework for classification is needed that could ultimately provide guidance to both developed and developing countries to classify and measure e-waste more consistently (Balde et al., 2015).

In view of the lack of a framework, the United Nations University (UNU) has developed criteria for classifying e-waste based on similar function, comparable material composition (in terms of hazardous substances and valuable materials) and related end-of-life attributes (Wang et al., 2012). This classification developed by the UNU is referred to as the UNU-KEYS (Wang et al., 2012; Balde et al., 2015). Ideally, the UNU classification system is connected to other classifications, such as the 10 categories in the WEEE Directive, the 6 categories in the recast of the WEEE Directive, the WEEE Forum Key Figures and the harmonized combined nomenclature that is used by customs organizations in Europe (CWIT, 2014). As already stated, the UNU-KEYS clusters appliances according to functionality and end-of-life characteristics (CWIT, 2014), which makes the system very useful for compiling e-waste statistics (Balde et al., 2015).

The full list of the UNU-KEYS is presented in Table 9.18. As can be seen from the table, the 54 categories can be grouped into 10 primary categories, according to the original EU-WEEE Directive. The classification can also be linked to the new reporting categories for the recast of the WEEE-Directive, which will come into effect in 2017 in the EU (Balde et al., 2015).

9.4.1.2 Post-Consumer Recycling Processes

E-waste is chemically and physically distinct from other types of industrial and municipal wastes and includes both valuable metals like copper, iron, aluminium and precious metals, as well as various hazardous substances (flame retardants, lead, mercury, arsenic, etc.) (Sepúlveda et al., 2010; Iqbal et al., 2015), which require special handling and recycling

TABLE 9.18

Detailed Description of the UNU Product Classification and Its Correlation to Other E-Waste Classifications

UNU KEY	Description	Recast WEEE Directive	Old EU-WEEE Directive
0001	Central heating (household installed)	Large equipment	Out of scope
0002	Photovoltaic panels (including converters)	Large equipment	Out of scope
0101	Professional heating and ventilation (excluding cooling equipment)	Large equipment	01
0102	Dishwashers	Large equipment	01
0103	Kitchen (e.g. large furnaces, ovens, cooking equipment)	Large equipment	01
0104	Washing machines (including combined dryers)	Large equipment	01
0105	Dryers (wash dryers, centrifuges)	Large equipment	01
0106	Household heating and ventilation (e.g. hoods, ventilators, space heaters)	Large equipment	01
0108	Fridges (including combi-fridges)	Cooling and freezing	01
0109	Freezers	Cooling and freezing	01
0111	Air conditioners (household installed and portable)	Cooling and freezing	01
0112	Other cooling (e.g. dehumidifiers, heat pump dryers)	Cooling and freezing	01
0113	Professional cooling (e.g. large air conditioners, cooling displays)	Cooling and freezing	01
0114	Microwaves (including combined, excl. grills)	Small equipment	01
0201	Other small household (e.g. small ventilators, irons, clocks, adapters)	Small equipment	02
0202	Food (e.g. toaster, grills, food processing, frying pans)	Small equipment	02
0203	Hot water (e.g. coffee, tea, water cookers)	Small equipment	02
0204	Vacuum cleaners (excl. professional)	Small equipment	02
0205	Personal care (e.g. tooth brushes, hair dryers, razors)	Small equipment	02
0301	Small IT (e.g. routers, mice, keyboards, external drives and accessories)	Small IT	03
0302	Desktop PCs (excluding monitors, accessories)	Small IT	03
0303	Laptops (including tablets)	Screens	03
0304	Printers (e.g. scanners, multifunctionals, faxes)	Small IT	03
0305	Telecom (e.g. [cordless] phones, answering machines)	Small IT	03
0306	Mobile phones (including smartphones, pagers)	Small IT	03
0307	Professional IT (e.g. servers, routers, data storage, copiers)	Large equipment	03
0308	Cathode ray tube monitors	Screens	03
0309	Flat display panel monitors (LCD, LED)	Screens	03
0401	Small consumer electronics (e.g. headphones, remote controls)	Small equipment	04
0402	Portable audio and video (e.g. MP3, e-readers, car navigation)	Small equipment	04
0403	Music instruments, Radio, HiFi (including audio sets)	Small equipment	04
0404	Video (e.g. Video recorders, DVD, Blue Ray, set-top boxes)	Small equipment	04
0405	Speakers	Small equipment	04
0406	Cameras (e.g. camcorders, photo and digital still cameras)	Small equipment	04

(Continued)

TABLE 9.18 **(*Continued*)**

Detailed Description of the UNU Product Classification and Its Correlation to Other E-Waste Classifications

UNU KEY	Description	Recast WEEE Directive	Old EU-WEEE Directive
0407	Cathode ray tube TVs	Screens	04
0408	Flat display panel TVs (LCD, LED, Plasma)	Screens	04
0501	Lamps (e.g. pocket, Christmas, excl. LED and incandescent)	Lamps	05
0502	Compact fluorescent lamps (including retrofit and non-retrofit)	Lamps	05
0503	Straight tube fluorescent Lamps	Lamps	05
0504	Special lamps (e.g. professional mercury, high and low pressure sodium)	Lamps	05
0505	LED lamps (including retrofit LED lamps and household LED luminaires)	Lamps	05
0506	Household luminaires (including household incandescent fittings)	Small equipment	05
0507	Professional luminaires (offices, public space, industry)	Small equipment	05
0601	Household tools (e.g. drills, saws, high pressure cleaners, lawn mowers)	Small equipment	06
0602	Professional tools (e.g. for welding, soldering, milling)	Large equipment	06
0701	Toys (e.g. car racing sets, electric trains, music toys, biking computers)	Small equipment	07
0702	Game consoles	Small IT	07
0703	Leisure (e.g. large exercise, sports equipment)	Large equipment	07
0801	Household medical (e.g. thermometers, blood pressure meters)	Small equipment	08
0802	Professional medical (e.g. hospital, dentist, diagnostics)	Large equipment	08
0901	Household monitoring and control (alarm, heat, smoke, excl. screens)	Small equipment	09
0902	Professional monitoring and control (e.g. laboratory, control panels)	Large equipment	09
1001	Non cooled dispensers (e.g. for vending, hot drinks, tickets, money)	Large equipment	10
1002	Cooled dispensers (e.g. for vending, cold drinks)	Cooling and freezing	10

Source: Balde, C.P. et al., E-waste statistics: Guidelines on classifications, reporting and indicators, United Nations University, IAS – SCYCLE, Bonn, Germany, 2015.

techniques to minimize environmental contamination and potential harmful effects on human health (Iqbal et al., 2015). In view of the environmental problems involved in the management of e-waste, many countries and organizations have drafted national legislation to improve the reuse, recycling and other forms of recovery of such wastes so as to reduce disposal (Cui and Forssberg, 2003). However, the components of e-waste are highly integrated into each other, which make their recycling technologically more complicated than ordinary recyclables such as glass or paper.

Indeed, as already discussed, recycling of e-waste is an important subject not only from the point of waste treatment, but also from the recovery aspect of valuable materials

(Cui and Forssberg, 2003). The factors of interest with regard to e-waste recycling systems include appliance collection, reuse, recycling, disposal, monitoring, financing of recycling systems and, social and environmental impacts associated with the e-waste recycling practices (Xakalashe et al., 2012a). Numerous studies have described various e-waste recycling techniques and/or management procedures. In fact, technical solutions exist for recovering valuables from e-waste as secondary resources with minimal environmental impact (Xakalashe et al., 2012a). Of particular interest in this section are the following three steps: collection and transportation, pretreatment and recovery. The dominant and most valuable materials in e-waste both in quality and value are metals (Xakalashe et al., 2012b); therefore, each step is critical for the recovery of metals (Khaliq et al., 2014).

9.4.1.2.1 Collection and Transportation

Collection and transportation are vital tasks in e-waste recycling, and these tasks can be very costly (Xakalashe et al., 2012a). E-waste can be collected on a voluntary basis or to fulfil legislative regulations (Tanskanen, 2012). However, it is important that e-waste collection is facilitated by appropriate government policies, by effective advertisement for public awareness, and by installing separate collection facilities at public places (Khaliq et al., 2014). Essentially, an efficient collection system largely depends on adequate and accessible collection facilities, as well as consistent information to users (Xakalashe et al., 2012a). Common collection methods of e-waste include (1) curbside collection, (2) special drop-off events, (3) permanent drop-off, (4) talk back and (5) point of purchase (Kang and Schoenung, 2005; Xakalashe et al., 2012a). Table 9.19 gives a summary of the advantages and disadvantages associated with each of the collection methods, as well as the transport responsibilities associated with each method (Kang and Schoenung, 2005; Xakalashe et al., 2012a).

TABLE 9.19
E-Waste Collection Methods and Transportation Responsibilities

	Responsible for Transport			
Collection Method	**To Collection Site**	**To Recycling Site**	**Advantages**	**Disadvantages**
Curbside	–	Recycler or local government	Residence participation, due to convenience	High transportation costs; potential theft and abandonment; need extra sorting
Special drop-off event	Consumer	Recycler or local government	Recycling awareness; ideal for rural areas	Irregular collection amount; storage needed
Permanent drop-off	Consumer	Recycler or local government	High sorting rate; low transportation cost; most cost effective	Regular inspections needed; not effective for all communities
Take-back	–	Original manufacturer, contractor	No collection site needed	High shipment costs; special packaging needed
Point-of-purchase	Consumer	Retailer	Low cost; high visibility if marketed properly by retailer	Retailer commitment; need storage space

Source: Kang, H.Y. and Schoenung, J.M., *Resour. Conserv. Recycl.*, 45(4), 368, 2005.

9.4.1.2.2 Pretreatment

Pretreatment of e-waste is one of the most important steps in the recycling chain (Khaliq et al., 2014). A basic flowsheet diagram of the pretreatment steps is shown in Figure 9.5. The first step in the pretreatment process is manual dismantling, which allows the recovery of whole homogenous parts that may be reusable, or recyclable, for example, whole components, metal, plastic or glass parts, and hazardous components that require further special treatment, for example, mercury-containing components, batteries, cathode ray tubes and liquid crystal displays (LCDs) (Cui and Forssberg, 2003; UNEP, 2009; SEPA, 2011). In summary, at the initial stage, housing, wiring boards and drives, and other components are manually dismantled, liberated, tested and isolated from e-wastes (Khaliq et al., 2014).

Mechanical or physical processing is also an integral part of the pretreatment stage where e-waste scrap is shredded or crushed into pieces using hammer mills (Cui and Forssberg, 2003; Khaliq et al., 2014), thus liberating and reducing the size of the recyclable materials (SEPA, 2011). After the size reduction, the materials are sorted into defined fractions based on their specific physical characteristics, such as weight, size, shape, density and electrical and magnetic characteristics (SEPA, 2011). Typical sorting processes used are screening, magnetic separation of ferrous parts, eddy current separation (electric conductivity) of non-ferrous metals (e.g. copper and aluminium) and density or gravity separation (water or airflow tables, heavy media floating, sifting) of plastics (SEPA, 2011; Khaliq et al., 2014). Alternatively, or in addition, manual sorting or new optical sorting

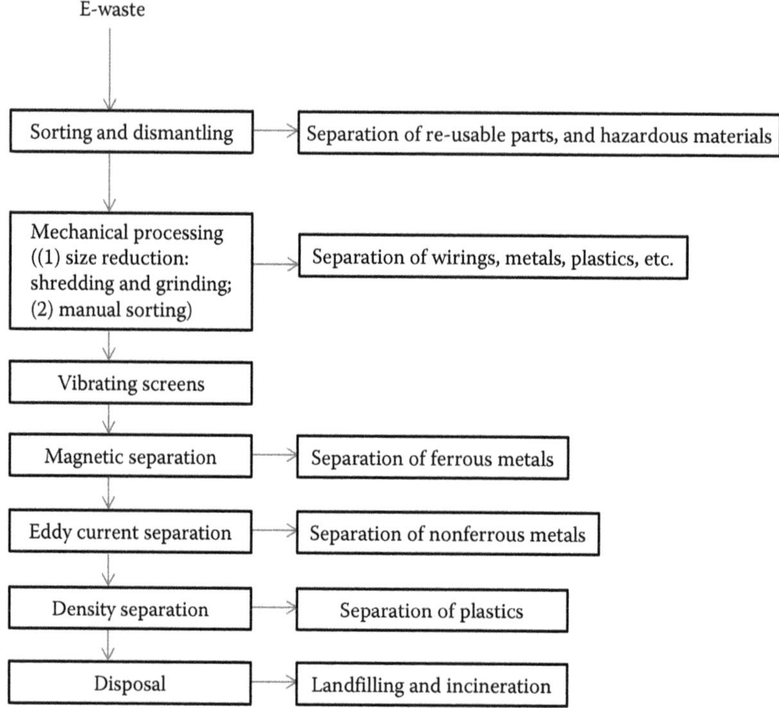

FIGURE 9.5
Schematic representation of the pretreatment step in a typical e-waste recycling process. (From Kang, H.Y. and Schoenung, J.M., *Resour. Conserv. Recycl.*, 45(4), 368, 2005; Khaliq, A. et al., *Resources*, 3, 152, 2014.)

techniques are also used. Furthermore, the sorting may sometimes be supported by screening, as well as further size reduction steps (SEPA, 2011). Final output streams are usually components taken out as a whole (for reuse or further treatment), a magnetic fraction (going to steel plants), an aluminium fraction (going to aluminium smelters), a copper fraction (going to copper smelters) and in some cases various plastic fractions. Usually, a waste fraction is also generated at this stage, which among others consists of a mixture of plastics, glass, wood and rubber. This fraction, which is often called 'the shredder light fraction', is sent for further processing, incineration or landfilling (Cui and Forssberg, 2003; Veit et al., 2005; SEPA, 2011).

9.4.1.2.3 Recovery

After the pretreatment stage, metal-containing components are sent to metal recovery facilities for further upgrading and refining. Further upgrading and refining of the metal-containing fractions are performed by metallurgical processes (SEPA, 2011). Both pyrometallurgical processes, in which the metals are melted, and hydrometallurgical processes, in which the metals are dissolved, are used and often in combination.

9.4.1.2.3.1 Pyrometallurgical Processing Pyrometallurgy, which comprises a range of processing equipment such as rotary kilns, blast furnaces and electric arc furnaces, is the traditional high-temperature method for recovering non-ferrous metals, as well as precious metals, from e-waste (Xakalashe et al., 2012b). However, it is mainly used to recover the copper content of electronic scrap and any other 'noble' metals that dissolve in copper upon melting, such as silver, gold, platinum and palladium (Namias, 2013). Components such as iron and aluminium are rarely recovered in the copper smelting process, but are oxidized to slag (Namias, 2013), though this may depend on the process under investigation (Xakalashe et al., 2012b). Although e-waste can be processed in small furnaces, the most common industrial process is to coprocess the material with copper sulphide concentrates in large copper smelting furnaces, such as copper converters, anode copper furnaces and copper smelting and converting furnaces (refer to Chapter 6) (Namias, 2013).

9.4.1.2.3.2 Hydrometallurgical Processing Following pyrometallurgical processes, hydrometallurgical processes are normally used for the final refining of the metals (SEPA, 2011), but may also be used as an alternative to recovery of metals directly from e-waste (SEPA, 2011; Xakalashe et al., 2012b). Hydrometallurgical processing of e-waste has become more popular in the last decade, due to the fact that hydrometallurgical methods are more exact (or have high precision level), predictable and more easily controlled than pyrometallurgical methods (SEPA, 2011; Namias, 2013). Furthermore, hydrometallurgical processes do not require major capital costs as is the case with smelters (Xakalashe et al., 2012b).

The first step of this process is leaching in which metals are converted into soluble salts in aqueous media. Leaching agents that are normally used include cyanide, halides, thiourea, thiosulphate and sodium hydroxide and acids such as aqua regia, sulphuric acid, nitric acid and hydrochloric acid (SEPA, 2011; Xakalashe et al., 2012b). After leaching, the leachate solutions then go through separation and purification processes in order to concentrate the valuable metals and separate impurities (Namias, 2013). Different types of electrochemical techniques that may be used to separate and purify the metals of interest include solvent extraction, precipitation, cementation, adsorption, filtration, distillation, ion exchange and supported liquid membranes (Cui and Zhang, 2008; SEPA, 2011; Xakalashe et al., 2012b).

9.4.1.2.3.3 Biometallurgical Processing In addition to both pyrometallurgical and hydrometallurgical processes, biometallurgical processes are gaining momentum on the research front; those characterized by bacterially assisted reactions are among the most interesting technologies (Xakalashe et al., 2012b). Biometallurgical processes offer many advantages over other conventional methods due to their relative simplicity, mild operating condition requirements, low capital costs, low energy input, less skilled labour requirements, and being environmentally friendly, among others (Acevedo, 2000; Simate, 2009; Simate et al., 2010).

9.4.2 Spent Catalysts

9.4.2.1 Introduction

Catalysts are used in all sectors of the chemical industry (Yoo, 1998; Fornalczyk, 2012): in basic chemistry (synthesis of sulphuric and nitric acid, ammonia, methanol and aromatics); in petrochemistry; in polymerization chemistry; in refining, in reactions of fluid catalytic cracking (FCC), residue fluid catalytic cracking (RFCC), hydrodesulphurization (HDS) and hydrotreatment; in auto industry to reduce pollution, for removal of NO, CO and hydrocarbons in exhaust emissions and in a variety of many other industrial processes. During the use of catalysts, metal sulphides and oxides and metallorganic compounds are deposited on the catalyst surface, causing loss of catalytic activity and specificity (Mishra et al., 2008). Undoubtedly, therefore, significant quantities of spent catalysts continue to be generated in these industries, thus contributing to large volumes of solid waste (Ahmed and Menoufy, 2012; Akcil et al., 2015). However, the storage, transportation, treatment and disposal of these spent catalysts require compliance with stringent environmental regulations because of their hazardous nature (EPA, 2003; Marafi and Stanislaus, 2008). Spent catalysts are regarded hazardous due to significant amounts of various heavy metal ions (EPA, 2003; Pradhan et al., 2010), as well as non-metallic elements such as elemental sulphur, carbon and oils (Jadhav and Hocheng, 2012). As a result, increasing attention has been paid to minimize spent catalyst waste generation by regeneration and reuse, metals recovery and utilization to produce useful materials, and to develop safe and cost-effective methods for disposal (Marafi and Stanislaus, 2008).

9.4.2.2 Post-Consumer Recycling Processes

This section of Chapter 9 gives a general overview of processing techniques that have been studied and/or used for the recovery of metals from spent catalysts. The section will particularly discuss the recovery of metals from spent catalysts used in petrochemical and the automotive industries. The recovery of metals is an important economic aspect as most of the spent catalysts possess significant concentrations (e.g. varying from 2.5% to 20%) of metals such as Ni, Mo, V, Al, Co, Rh, Pt and Pd (Singh, 2009). Moreover, such waste materials containing high metal concentrations can serve as secondary raw materials (Jadhav and Hocheng, 2012).

Many technologies have been suggested, and some of them are being practised for commercial applications, but pyrometallurgy and hydrometallurgy are the two prominent routes of recovering metals from spent catalysts. Nevertheless, a significant amount of chemicals used in chemical leaching are not environmentally friendly, and thus, biotechnological leaching methods have been developed as alternative methods (Akcil et al., 2015).

In all cases, the metals are recovered as mixed solutions and then separated by conventional separation techniques (e.g. solvent extraction, selective precipitation, ion exchange) (Siemens et al., 1986; Park et al., 2007).

9.4.2.2.1 *Hydrometallurgical Processing*

Hydrometallurgical processes have been in use for many years and have several benefits such as low cost requirements, possible recovery of leachants and decrease of air pollution as there are no particles produced (Akcil et al., 2015). To start with, hydrometallurgical recovery routes require the maximum solubilization of the sample in an appropriate medium for the central steps of separation and recovery of solubilized elements (Valverde et al., 2008). For this purpose, it is necessary to pretreat the sample in order to remove coke and other volatile species present. This step 'cleans' the catalyst surface, thus reducing losses of recoverable metals by physical blocking. However, care must be taken to avoid catalyst ignition during pretreatment, thus forming refractory oxides that are difficult to solubilize in the leaching medium.

It is impossible to cover all studies in this area, and thus, only a few examples will be discussed. The first example is the recovery of various metals in a process developed by the Hall Chemical Co. (Ohio) (Parkinson et al., 1987). In this process, spent catalyst was completely dissolved in HCl. Thereafter, molybdenum was precipitated first, as a sulphide, by adding H_2S. This was followed by the removal of vanadium through solvent extraction, which was reclaimed by adding caustic soda and precipitated as vanadium pentoxide. Nickel was also removed by solvent extraction and then recovered as a sulphate by adding sulphuric acid. Cobalt was extracted in a similar manner to nickel (Marafi et al., 2010). The study also stated that tungsten recovery would be similar to that of molybdenum, but tungsten-bearing catalysts would not be mixed with other metals because the precipitation conditions are almost identical to those of molybdenum.

Valverde et al. (2008) studied the recovery of metals and support components from spent commercial catalysts (CoMo and NiMo/Al_2O_3). Initially, the spent catalysts were pre-oxidized in order to eliminate coke and other volatile species. Pre-oxidized catalysts were then dissolved in H_2SO_4, and the remaining residues separated from the solution. Molybdenum was recovered by solvent extraction using tertiary amines, with Alamine 304 giving the best performance at a pH of around 1.8. Thereafter, cobalt (or nickel) was separated by adding aqueous ammonium oxalate. Before aluminium recovery, by adding NaOH to the acidic solution, phosphorus was removed by passing the solution through a strong anion exchange column.

In the patented process developed by Hyatt (1987), the spent CoMo/Al_2O_3 catalyst was treated with sulphuric acid in the presence of H_2S under pressure in an autoclave. The presence of H_2S resulted in the precipitation of Mo and Co as sulphides while the Al_2O_3 was converted to soluble $Al_2(SO_4)_3$. The metal sulphides were separated from the $Al_2(SO_4)_3$ solution and subjected to oxidation under pressure in an autoclave so as to convert MoS_2 into a solid molybdic acid and the CoS to $CoSO_4$. The molybdic acid was separated by filtration from the $CoSO_4$ solution. The cobalt was recovered by ion exchange. The extraction and recovery of valuable metals, such as Mo and Co together with the alumina as $Al_2(SO_4)_3$, could also be achieved in this process (Akcil et al., 2015).

Readers are referred to Marafi et al. (2010) for more in depth studies pertaining to the recovery of metals from spent catalysts hydrometallurgically, and Table 9.20 is a summary of the experimental studies on recovery of metals from hydrotreating catalyst as reported by Akcil et al. (2015).

TABLE 9.20

Summary of Experimental Studies Carried Out on Spent Hydrotreating Catalysts for Metal Recovery

		Metal Recoveries (%)				
Catalyst	**Leaching Reagent**	**Mo**	**V**	**Ni**	**Co**	**Al**
Mo, V, Ni, Co, Al_2O_3	H_2SO_4	99.7	92.5	–	–	–
Mo, Ni, Al_2O_3	H_2SO_4	93	–	98	–	96
Co, Mo	SO_2 (aqueous)	100	–	–	100	–
Co, Mo	H_2SO_4 + HNO_3 (3:1)	96	–	–	96	–
Co, Mo	H_2SO_4 + $Al_2(SO_4)_3$ or $(NH_4)_2SO_4$	92	–	–	97	–
Mo, V, Ni, Co	H_2SO_4	>90	–	–	>90	–
Mo, V, Ni, Al	Oxalic acid + H_2O_2	90	94	65	–	33
Mo, V, Ni	Tartaric acid	93	94	83	–	–
Mo, V, Ni, Co	Citric acid	94	94	85	–	–
Co–Mo	Leaching with 9 mol/L H_2SO_4 (90°C, 2 h)	99	–	98	98	–
Co–Mo/Ni–Mo	Fusion with $KHSO_4$ (350°C–600°C, 0.5–7 h)	87–90	–	96–99	91–94	–
Spent HDS catalyst	Na_2CO_3 (40 g/L) + H_2O_2 (6 vol%)	–	–	99.4MoO_3	–	
Spent HDS catalyst	Na_2CO_3 (30 g/L)	93	–	98	90	–
Spent hydro-refining catalyst	600°C air roasting + Na_2CO_3 (12 wt.%) for 30 min	–	–	92Na_2MoO_4	–	
Spent hydro-refining catalyst	900°C air roasting + NaCl (20 wt.%)	–	–	90Na_2MoO_4	–	
Co–Mo/Ni–Mo	Two steps alkali-acid procedure	80	–	97	80	–
Spent HDS catalyst	Acid leaching, 70°C (HNO_3/H_2SO_4/HCl = 2:1:1)	–	–	90	99	–
Spent HDS catalyst	Acid leaching, 50°C (0.5 mol/L $H_2C_2O_4$ + 3 mol/L H_2O_2)	–	–	90	65	–
Spent HDS catalyst	Acid leaching, 80°C (H_2SO_4 + H_2O_2)	81	–	72	–	–
Ammonia leaching residue	Water leaching (80°C, 15 min, 50% S/L)	–	–	91	–	–

Source: Akcil, A. et al., *Waste Manage.*, 45, 420, 2015.

9.4.2.2.2 Pyrometallurgical Processing

Pyrometallurgical techniques such as direct smelting, calcination and smelting, chlorination and salt roasting (Parkinson et al., 1987; Howard and Barnes, 1991; Gaballah et al., 1994; Kar et al., 2005; Busnardo et al., 2007; Medvedev and Malochkina, 2007; Jadhav and Hocheng, 2012) employ thermal treatment to effect physical and chemical changes in spent catalysts so as to enable recovery of valuable metals (Akcil et al., 2015). In other words, pyrometallurgical routes involve traditional smelting and subsequent refining methods (Louie, 2005). However, pyrometallurgical processes require a high amount of energy and cause emission of SO_2 gas, which is a severe environmental pollutant (Ojeda et al., 2009; Havlik et al., 2010). Therefore, this route is not economical unless large quantities of spent catalysts are being processed and the metal prices are high (Louie, 2005). Nevertheless, pyrometallurgical and hydrometallurgical methods may also be used together, such that spent catalysts are roasted before chemical leaching (Lee et al., 1992; Park et al., 2007).

9.4.2.2.3 Biometallurgical Processing

For sustainable development, biological processes are often recommended for metal recovery from difficult-to-process metallic materials. As discussed earlier, these techniques are poised to play a significant role not only in the future metallurgical and chemical sector, but also for the treatment of metal-containing wastes and by-products (Lee and Pandey, 2012). Bacteria from the genus *Acidithiobacillus* are the most important microorganisms applied to metal solubilization, particularly because they are autotrophic and tolerate high concentrations of heavy metals (Akcil et al., 2015). Ideally, in aqueous solutions at 25°C, pH < 2.0 and Eh > 500 mV, the metals commonly found in spent catalyst such as Ni, V and Mo have stable soluble ions (Pourbaix, 1974; Mishra et al., 2007; Zeng and Cheng, 2009). Therefore, these solution conditions are consistent with the parameters required for the activity of organisms from the chemoautotrophic, acidophilic, sulphur-oxidizing genus *Acidithiobacillus* (Mishra et al., 2007; Zeng and Cheng, 2009). In such an environment, sulphuric acid, which is generated through the biooxidation of sulphur and/or reduced-sulphur compounds by acidophiles, is the major leaching agent (Ross, 1990).

Mishra et al. (2007) used chemolithotrophic sulphur-oxidizing genus *Acidithiobacillus* bacteria to recover base metals such as Ni, V and Mo from vanadium-rich spent refinery catalysts. Prior to the bioleaching process, the spent refinery catalyst was pretreated with acetone as the solvent. The bioleaching process was carried out in one-step and two-step methods, and the leaching efficiencies in both cases were compared. The recovery for the two-step leaching method was found to be higher than the one-step leaching method. In a similar study, Mishra et al. (2008) extracted Ni, V and Mo from hazardous spent refinery catalysts using bacterially produced sulphuric acid. The thermodynamic parameters such as free energy (ΔG), enthalpy (ΔH) and entropy (ΔS) calculated from the data showed that the bioleaching process was exothermic and feasible. Leaching kinetics of the metals was represented by different reaction kinetic equations; however, only diffusion controlled model showed the best correlation.

A study by Beolchini et al. (2010) used an environmental sample of iron-/sulphur-oxidizing bacteria (i.e. *Acidithiobacillus ferrooxidans*, *Acidithiobacillus thiooxidans* and *Leptospirillum ferrooxidans*) to extract metals from refinery spent catalysts containing naphtha. The bacterial consortium was able to extract 83% Ni, 90% V and 40% Mo from the spent catalyst.

Pradhan et al. (2010) recovered Ni, V and Mo from spent petroleum catalyst. Different leaching parameters such as Fe(II) concentration, pulp density, pH, temperature and spent catalysts particle sizes were studied to evaluate their effects on the leaching efficiency. All the three metal ions followed dual kinetics (i.e. an initial faster kinetic rate that was followed by a slower kinetic rate). Both the leaching efficiencies of Ni and V were found to be higher than that of Mo. The study found that the leaching process followed a diffusion controlled model, and the product layer was observed to be impervious due to formation of ammonium jarosite $(NH_4)Fe_3(SO_4)2(OH)_6$. In addition, the lower leaching efficiency of Mo was observed due to a hydrophobic coating of elemental sulphur over Mo matrix in the spent catalyst. Mishra et al. (2008) attributed the lower solubility of Mo in the reaction process due to its sparingly soluble nature in acidic regime as supported by Pourbaix equilibrium theory. Nevertheless, a successful molybdenite bioleaching requires Mo-resistant organisms or a leach solution that decreases the toxicity of Mo (Akcil et al., 2015).

Several other studies also used biotechnology to recover metals from spent catalysts (Briand et al., 1996; Bosio et al., 2008). A review by Zeng and Cheng (2009) also contains examples of applications of the bioleaching process to the recovery of metals from spent petroleum catalysts. Another review on bioleaching of spent refinery catalysts by Asghari et al. (2013) is also available.

The use of *Aspergillus niger* for the bioleaching of spent refinery processing catalysts has also been reported (e.g. Aung and Ting, 2005; Santhiya and Ting, 2005; Anjum et al., 2010), and a comprehensive review on the fungal leaching of spent catalysts was carried out by Asghari et al. (2013). *Aspergillus niger*, like other fungi, secretes hydroxycarboxylic acids that dissolve metals by lowering the pH and complexion/chelating into soluble organic–metallic complexes (Tzeferis, 1994; Simate, 2009; Simate et al., 2010), and acidolysis is the principal mechanism in bioleaching of metals by *Aspergillus niger* (Johnson, 2006).

9.4.3 Automobile

9.4.3.1 Introduction

Automobiles are one of the most highly recycled commercial products in the world, with more than 95% of the vehicles that have reached their end of life entering the recycling process (Sendijarevic et al., 1997; Bandivadekar et al., 2004; Jody et al., 2010; Sivakumar et al., 2014). At the moment, approximately 86% of the material of a vehicle that enters the automotive recovery infrastructure is recycled, reused or used for energy recovery (Kumar and Sutherland, 2008; Desnica, 2014); and it must be noted that the single largest source of recycled ferrous scrap for iron and steel industries is end-of-life automobiles (Dutta, 1998). The material that is not recovered (also known as automotive shredder residue [ASR] or 'fluff'), mainly non-metallic, is typically landfilled as municipal solid waste (Dutta, 1998; Kumar and Sutherland, 2008, 2009). In general, the infrastructure associated with the recovery and recycling of automobile is mature and operates profitably in many countries (Gupta and Isaacs, 1997; Boon et al., 2001; Kumar and Sutherland, 2008). More specifically, the values of the components and materials that are recovered and recycled in each stage of the infrastructure provide a profit to the operator (Jody et al., 2010). In fact, all the business entities within the automobile recovery infrastructure (i.e. dismantlers, shredders and non-ferrous operators) must be profitable in order to stay in business (Kumar and Sutherland, 2008).

9.4.3.2 Post-Consumer Recycling Processes

The primary goals of the automotive recycling industry are to harvest automobile components for reuse and to recycle the remaining valuable materials into specification-grade commodities that can be used in the manufacture of new basic materials such as steel, aluminium, plastic, copper and brass (Desnica, 2014). Essentially, the process of recycling automobiles can be divided into five major activities, (1) dismantling, (2) remanufacturing, (3) shredding, (4) post-shredder material separation and processing and (5) landfill disposal of ASR (Keoleian et al., 1997; Staudinger and Keoleian, 2001). A generalized flow diagram of automobile recycling processes and associated materials streams is shown in Figure 9.6.

9.4.3.2.1 Dismantling

At its end of life, a vehicle is first taken to either a dismantling facility or a salvage yard/scrapyard where some parts and fluids are removed (Dantec, 2005). Ideally, dismantlers consist of two distinct types: (1) high-value parts dismantlers (high volume, quick turnover operations targeting late model vehicles) and (2) salvage yard/scrapyards (low volume, slow turnover operations accepting most vehicles) (Staudinger and Keoleian, 2001).

At the dismantling facility or a salvage yard/scrapyard, a variety of parts, all vehicle fluids and tyres are removed for either (1) direct reuse (e.g. body panels used to repair collision-damaged vehicles), (2) remanufacture (e.g. clutches, starters, engines), (3) recycle

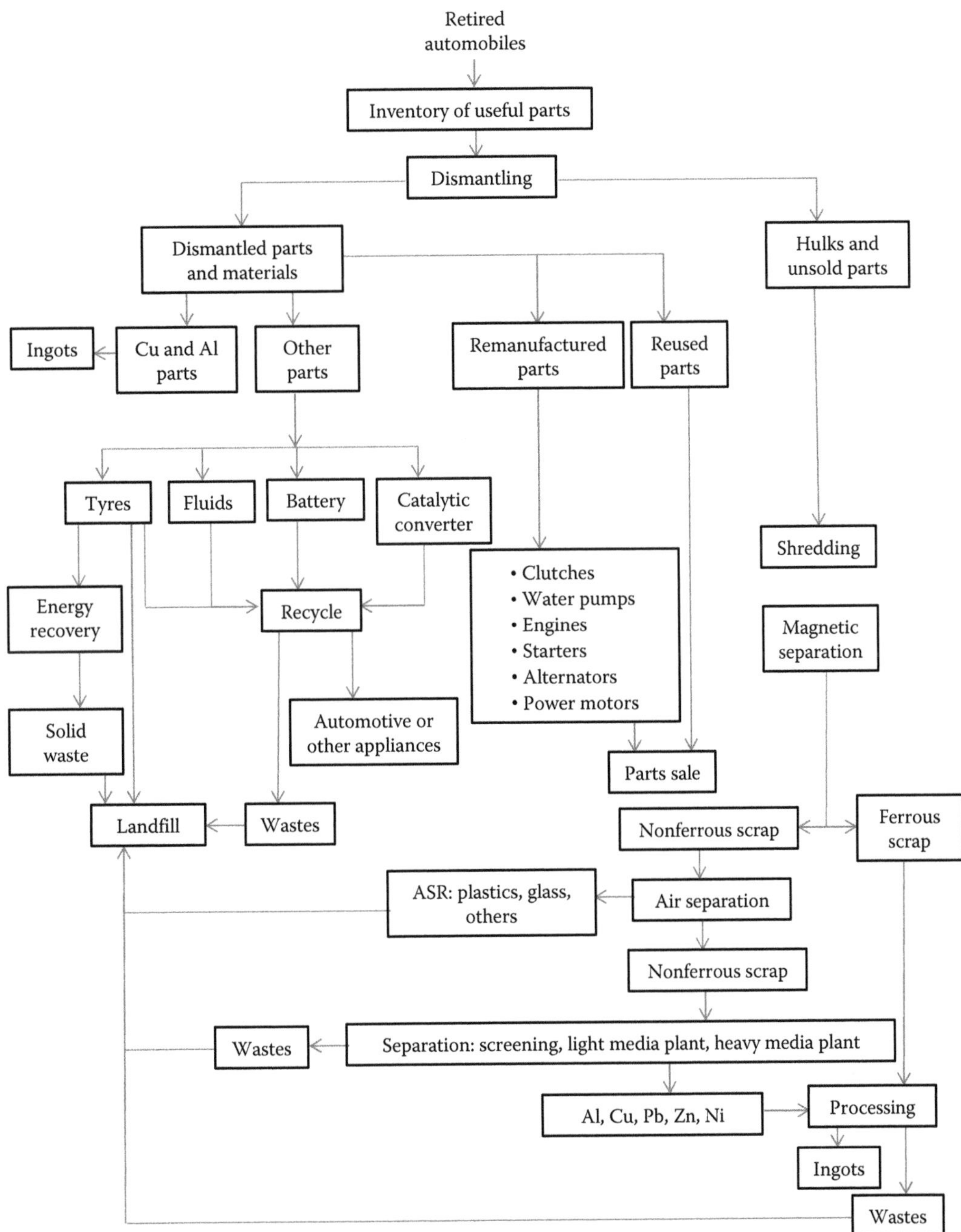

FIGURE 9.6
The automobile recycling process. (From Jody, J.B. et al., *J. Met.*, 46(2), 40, 1994; Keoleian, G.A. et al., Industrial ecology of the automobile: A life cycle perspective, Society of Automotive Engineers, Warrendale, PA, 1997; Jody, B.J. and Daniels, E.J., End-of-life vehicle recycling: The State of the art of resource recovery from shredder residue, 2006, Available at http://www.ipd.anl.gov/anlpubs/2007/02/58559.pdf, Accessed October 2016; Jody, B.J. et al., End-of-life vehicle recycling: State of the art of resource recovery from shredder residue, 2010, Available at http://www.ipd.anl.gov/anlpubs/2011/02/69114.pdf, Accessed January 2016)

(e.g. fluids, batteries, catalytic converters, steel fuel tanks), (4) energy recovery (e.g. tyres) or (5) disposal (e.g. plastic fuel tanks) (Staudinger and Keoleian, 2001).

After reusable and remanufacturable parts (including all fluids) have been removed, the remaining auto 'hulk' is typically crushed and flattened (Staudinger and Keoleian, 2001; Dantec, 2005), which makes it easier to handle and transport to the shredding facility (Dantec, 2005).

9.4.3.2.2 *Remanufacturing*

Remanufacturing and rebuilding of certain components of dismantled vehicles by dismantlers or by repair shops is another important aspect of the recycling process (Jody et al., 2010). There are five steps in the remanufacturing process: (1) dismantling; (2) complete cleaning of all components; (3) component inspection and sorting; (4) repair, refurbishment or replacement of damaged or worn parts and (5) assembly and test (Statham, 2006). Therefore, from the stated steps, it shows that remanufacturing entails the tear down, inspection, repair and/or replacement of subcomponents, as well as testing of the remanufactured component to ensure that the performance specifications of the remanufactured parts are met (Jody et al., 2010).

Automotive components are remanufactured so as to supply lower-cost repair and replacement parts to the automobile consumers. For example, more than 90% of the replacement starters and alternators are remanufactured. Other automotive parts that are typically remanufactured include engines, transmissions, brake systems, and water pumps.

9.4.3.2.3 *Shredding*

Following the dismantling process, auto hulks are sent (typically flattened) to a shredder. At shredder facilities, hulks are inspected prior to shredding to ensure that potentially hazardous components such as batteries, gas tanks and fluids have been removed (Keoleian et al., 1997; Staudinger and Keoleian, 2001). Thereafter, the auto hulk and other materials, including consumer durables (such as home appliances) and other scrap iron and steel, are shredded into fist-sized pieces using hammer mills (Jody et al., 2010).

9.4.3.2.4 *Post-Shredder Material Separation and Processing*

After the auto hulk and other materials have been shredded, two basic separations are made: (1) an initial separation of the combined material stream into ferrous and non-ferrous fractions using a magnetic separation process and (2) separation of the non-ferrous material stream into metal (heavy fraction) and non-metal fractions (light fraction) using a variety of techniques such as air separation (Keoleian et al., 1997; Staudinger and Keoleian, 2001). Essentially, three basic streams are generated during the separation and processing of the shredded materials, (1) ferrous metals (iron and steel), 65%–70% by weight; (2) non-ferrous metals (aluminium, stainless steel, copper, brass, lead, magnesium, zinc and nickel), 5%–10% by weight and (3) ASR (consisting of other materials – plastics, glass, rubber, foam, carpeting, textiles, etc.), 20%–25% by weight.

Recycling of ferrous metal fraction: The separated ferrous metal fraction (containing iron and steel) is sent for recycling to steel smelters. End-of-life automobile scrap is almost exclusively handled by electric arc furnaces, which utilize electric energy to melt and refine scrap in a batch process to make steel products. It must be noted that steel made from scrap is chemically and metallurgically equivalent to steel made from virgin iron ore (Keoleian et al., 1997).

Recycling of non-ferrous metal fraction: The separated non-ferrous metal fraction (e.g. containing aluminium, brass, bronze, copper, lead, magnesium, nickel, stainless steel and zinc) is sent to another specialized facility to separate the stream into its individual metals by a variety of separation techniques. Aluminium and steel are separated by both 'light-media' and 'heavy-media' methods. Copper and brass require additional separation, which is accomplished mainly by image processing. Separated non-ferrous scrap is typically further processed into ingots, for ultimate sale to the non-ferrous scrap market.

ASR fraction: As already discussed, after the ferrous and non-ferrous metals have been recovered for recycling, a large quantity of non-metallic residue called ASR is generated. Two types of ASR streams generated include (1) light ASR, generated at the shredder facility when the non-ferrous fraction is separated into metal and non-metallic streams using air classification processes (the non-metallic fraction being ASR), and (2) heavy ASR, generated at the non-ferrous metal processing facility during separation of the various metal steams (the heavy ASR representing rejected contaminants extracted during processing) (Staudinger and Keoleian, 2001). Therefore, the composition of ASR varies among facilities, because of different types of source materials, models and years of obsolete automobiles and separation processes (Dutta, 1998). Generally, ASR contains a large variety of plastics, rubber, glass, foam, carpeting, textiles and entrained metallic fines, dirt and moisture (Dutta, 1998; Staudinger and Keoleian, 2001).

Landfill disposal of automobile shredder residue: In recent past, due to a variety of reasons (principally low disposal costs), most of the ASR has been sent to landfill for disposal (Staudinger and Keoleian, 2001; Saman, 2006). Nevertheless, reduction of this waste stream through the recovery and recycling of useful metals, organics and inorganics, such as copper, zinc, lead and rubber, plastics, sponge and glass, is the focus of several research works (Daniels et al., 1990; Jody et al., 1994; Shibayama et al., 2006; Nourreddine, 2007; Ciacci et al., 2010; Vermeulen et al., 2011).

9.4.4 Household Equipment

For the purpose of this section, household (or home) equipment (or appliances) may be defined as 'electrical and/or mechanical machines customarily used in a residential home to accomplish some household functions' (Clark et al., 1984; Cooper and Mayers, 2000; Darby and Obara, 2005; Rahimifard et al., 2009; Manomaivibool and Vassanadumrongdee, 2012; Hotta et al., 2014). Examples of household equipment are listed in Table 9.21. Upon comparing Tables 9.18 and 9.21, it is clear that household appliances are a subset of e-waste. In fact according to SEPA (2016), the definition of WEEE (e-waste) includes household appliances such as washing machines and cookers, IT and telecommunications equipment, electrical and electronic tools, toys and leisure equipment and certain medical devices.

9.4.4.1 Post-Consumer Recycling Processes

Since household equipments fall under the category of e-waste, the techniques and/or management procedures discussed in Section 9.4.1 are, therefore, also applicable to waste electrical and electronic household equipments, that is, collection and transportation, pretreatment and recovery. These recycling steps will not be discussed further in this section. However, what is important to note is that the recycling and disposal of appliances is more

TABLE 9.21
Household Appliances

Household Products	Reference
Electric cookers; microwave ovens; refrigerators and freezers; washing machines, dishwashers and tumble dryers; vacuum cleaners and carpet cleaners; small work or personal care appliances; Hi-fi and stereo; radio and personal radio, stereo and CD; televisions (TVs); video equipment; telephones, faxes and answer machines; mobile phones and pagers; computers and peripherals; toys and home and garden tools	Cooper and Mayers (2000)
Big electrical and electronic items (e.g. fridges, TVs), small electrical and electronic items (e.g. hairdryers, shavers)	Darby and Obara (2005)
Microwave oven, electric kettle and desktop computer	Rahimifard et al. (2009)
TVs, digital camera, portable media player, desktop printer, mobile phone, personal computer, refrigerator, air conditioner, fluorescent lamp and dry-cell battery	Manomaivibool and Vassanadumrongdee (2012)
Air conditioners, TVs, electric refrigerators and freezers, electric washing machines and clothes dryers	Hotta et al. (2014)
Air conditioner, microwave oven, dishwasher, dryer, refrigerator, vacuum cleaner and clothes washer	Clark et al. (1984)

complex than waste from 'consumables' such as newspapers, packaging and food or other organic material (Cooper and Mayers, 2000, 2002). Household appliances tend to pass in and out of use, following a 'cascade of use' through which they become financially, functionally and materially degraded, and many items are not disposed of by the original owners, as they are redistributed through reuse (Cooper and Mayers, 2002).

9.5 Concluding Remarks

At every stage of the consumption cycle of consumer metals, whether intermediate or the finished products, waste metals are invariably generated. Therefore, there is a large amount of metals that are discarded as waste after the consumption stage. However, by using appropriate techniques, post-consumer metal wastes can be reprocessed and/or recycled. Therefore, recycling is clearly a waste management strategy, but it can also be seen as an example for implementing the concept of industrial ecology, whereas in a natural ecosystem, there are no wastes, but only products (Frosch and Gallopoulos, 1989; McDonough and Braungart, 2002). Undoubtedly, recycling of metals from post-consumer and/or secondary sources is a growing industry that has major economic and environmental benefits. Indeed, with the steady depletion of natural ores, it is imperative that post-consumer metal wastes are recycled from 'scrap' so as to maintain the demands of metals. In this context, the words 'scrap metals' should be eliminated from the vocabulary. This is even truer in the case of metals (e.g. critical metals) that do not occur in abundant concentration and whose extraction from the scarce primary resources is often expensive.

Some of the benefits of recycling post-consumer metals include (1) conservation of the mineral resources, recycling reduces the need for primary raw materials; (2) saving energy, a large amount of energy is consumed by processing raw materials at the time of production (Hamuth, 2016). Recycling helps to minimize energy consumption, for example,

95% of the energy used to make aluminium from bauxite ore is saved by using recycled materials (EPA, 1990). This also makes the production process very cost effective and beneficial to manufacturers and consumers; (3) decreasing emissions, using recycled materials reduces the need to extract, process, refine and transport mineral ores, etc. into virgin-based metals; therefore, recycling lessens toxic air emissions, effluents and solid wastes that these manufacturing processes create and (4) reduces air and water pollution/emissions associated with landfilling and incineration.

However, recycling of post-consumer waste is multifaceted and has many challenges. In general, waste management of post-consumer wastes faces different kinds of challenges compared to the waste management of the office or factory waste. To start with, it must be noted that not all the recycling challenges are technical (Tanskanen, 2012). The biggest obstacle in recycling is the lack of consumer awareness about collection and recycling possibilities; therefore, there is normally low waste metal collection from the consumers. The diverse and/or complex composition and structure of certain post-consumer wastes (e.g. mobile phones) makes it difficult to develop recycling technologies that are suitable to all product types. However, with the use of proper recycling technologies, most of the metals in post-consumer waste can be recycled (Tanskanen, 2012). Some of the metals such as aluminium tend to degrade after each reuse cycle. Ideally, some recycled products are always not of durable quality. While recycling is good socially and environmentally, economically it is not always cost effective (Hamuth, 2016). For example, setting up new recycling facilities involves high cost. The huge cost can come up as part of erecting a processing facility, setting up segregated collection system, etc. Recycling may also create more pollution as the materials go under the process of cleaning, storage and transportation. Furthermore, a recycling activity might consume more energy and water, thus defeating the purpose of recycling.

In view of the merits of recycling post-consumer waste metals coupled with its challenges, it is, therefore, imperative that a balance is made between the pros and cons. Understanding the impact of recycling is essential because if it is done effectively, it can usher in massive positive results that are beneficial to the mutual existence of human beings and the environment.

References

Abadin, H., Ashizawa, A., Stevens, Y.W., Llados, F., Diamond, G., Sage, G., Citra, M., Quinones, A., Bosch, S.J. and Swarts, S.G. (2007). Toxicological profile for lead. Agency for Toxic Substances and Disease Registry, Atlanta, GA.

Acevedo, F. (2000). The use of reactors in biomining processes. *Electronic Journal of Biotechnology* 3(3): 184–194.

Ahmed, H.S. and Menoufy, M. (2012). New trends in hydroprocessing spent catalysts utilization. In: Patel, V. (ed.), *Petrochemicals*. In-Tech, Rijeca, Croatia.

Akcil, A., Vegliò, F., Ferella, F., Okudan, M.D. and Tuncuk, A. (2015). A review of metal recovery from spent petroleum catalysts and ash. *Waste Management* 45: 420–433.

Alavudeen, A., Venkateshwaran, N. and Winowlin, J.J.T. (2006). *A Textbook of Engineering Materials and Metallurgy*. Firewall Media, New Delhi, India.

Anand, S.H., Balasubramaniyam, R., Thamizharasu, P. and Jayabalan, A. (2016). Welding and analysis of aluminium 5083 alloy with mild steel by friction welding process. *International Journal of Emerging Technology in Computer Science & Electronics* 21(3): 91–94.

Anjum, F., Bhatti, H., Asgher, M. and Shahid, M. (2010). Leaching of metal ions from black shale by organic acids produced by Aspergillus niger. *Applied Clay Science* 47: 356–361.

Antonio, L.C. (2010). Study on recyclables collection trends and best practices in the Philippines. Available at http://www.eria.org/publications/research_project_reports/images/pdf/y2009/no10/Ch03_3R.pdf. Accessed February 2016.

Asghari, I., Mousavi, S.M., Amiri, F. and Tavassoli, S. (2013). Bioleaching of spent refinery catalysts: A review. *Journal of Industrial and Engineering Chemistry* 19: 1069–1081.

Assakkaf, I. (2002). Introduction to structural steel design. Available at http://www.assakkaf.com/courses/ence355/lectures/part2/chapter1.pdf. Accessed December 2015.

Aung, K.M.M. and Ting, Y.P. (2005). Bioleaching of spent fluid catalytic cracking catalysts using Aspergillus niger. *Biotechnology* 116: 159–170.

Authier-Martin, M., Forte, G., Ostap, S. and See, J. (2001). The mineralogy of bauxite for producing smelter grade alumina. *Journal of the Minerals, Metals and Materials Society* 53(12): 36–40.

Ayres, R.U. and Peiró, L.T. (2013). Materials efficiency: rare and critical metals. Available at http://rsta.royalsocietypublishing.org/content/roypta/371/1986/20110563.full.pdf. Accessed 21 February 2017.

Baeyens, J., Brems, A. and Dewil, R. (2010). Recovery and recycling of post-consumer waste materials. Part 1. Generalities and target wastes (paper, cardboard and aluminium cans). *International Journal of Sustainable Engineering* 3(3): 148–158.

Balde, C.P., Kuehr, R., Blumenthal, K., Gill, S.F., Kern, M., Micheli, P., Magpantay, E. and Huisman, J. (2015). E-waste statistics: Guidelines on classifications, reporting and indicators. United Nations University, IAS – SCYCLE, Bonn, Germany.

Bandivadekar, A.P., Kumar, V., Gunter, K.L. and Sutherland, J.W. (2004). A model for material flows and economic exchanges within the U.S. automotive life cycle chain. *Journal of Manufacturing Systems* 23(1): 22–29.

BCS Inc. (2002). Mining industry of the future: Lead and zinc. Available at http://energy.gov/sites/prod/files/2013/11/f4/lead_zinc.pdf. Accessed December 2015.

Beliles, R.P. (1994). Tin. In: Clayton, G.D. and Clayton, F.E. (eds.), *Patty's Industrial Hygiene and Toxicology*, Vol. 2, 4th edn. John Wiley & Sons, New York.

Bell, S., Davis, B., Javaid, A. and Essadiqi, E. (2003). Final report on scrap management, sorting and classification of aluminum. https://www.nrcan.gc.ca/sites/www.nrcan.gc.ca/files/minerals-metals/pdf/mms-smm/busi-indu/rad-rad/pdf/2003-22(cf)cc-eng.pdf. Accessed January 2016.

Bell, T. (2015). Steel grades. Available at http://metals.about.com/od/properties/a/Steel-Types-And-Properties.htm. Accessed December 2015.

Beolchini, F., Fonti, V., Dell'Anno, A., Rocchetti, L. and Veglio, F. (2010). Metal recovery from spent refinery catalysts by means of biotechnological strategies. *Journal of Hazardous Materials* 78: 529–534.

Bo, B. and Yamamoto, K. (2010). Characteristics of E-waste recycling systems in Japan and China. *World Academy of Science, Engineering and Technology* 38: 500–506.

Boni, M. (2003). Non-sulfide zinc deposits: A new–(old) type of economic mineralization. *Society for Geology Applied to Mineral Deposits (SGA News)* 15: 6–11.

Boni, M. and Large, D. (2003). Non-sulphide zinc mineralisation in Europe: An overview. *Economic Geology* 98(4): 715–729.

Bonnin, M., Azzaro-Pantel, C., Domenech, S. and Villeneuve, J. (2015). Multicriteria optimization of copper scrap management strategy. *Resources, Conservation and Recycling* 99: 48–62.

Boon, J.E., Isaacs, J.A. and Gupta, S.M. (2001). Economic impact of aluminium-intensive vehicles on the U.S. automotive recycling infrastructure. *Journal of Industrial Ecology* 4(2): 117–134.

Bosio, V., Viera, M. and Donati, E. (2008). Integrated bacterial process for the treatment of a spent catalyst. *Journal of Hazardous Materials* 154: 804–810.

Bramfitt, B.L. (1998). Structure/property relationships in irons and steels. In: Davis, J.R. (ed.), *Metals Handbook Desk Edition*, 2nd edn. ASM International, Cleveland, OH, pp. 153–173.

Bray, E.L. (2012). Bauxite and alumina. Available at http://minerals.usgs.gov/minerals/pubs/commodity/bauxite/mcs-2012-bauxi.pdf. Accessed 16 November 2015.

Briand, L., Thomas, H. and Donati, E. (1996). Vanadium (V) reduction in *Thiobacillus thiooxidans* cultures on elemental sulfur. *Biotechnology Letters* 18: 505–508.

Brimer, L. (2011). *Chemical Food Safety*. CB International, Oxford, UK.

Buchert, M., Schüler, D. and Bleher, D. (2009). Critical metals for future sustainable technologies and their recycling potential. Available at http://www.unep.fr/shared/publications/pdf/DTIx1202xPA-Critical%20Metals%20and%20their%20Recycling%20Potential.pdf. Accessed 25 January 2016.

Bulten, E.J. and Meinema, H.A. (1991). Tin. In: Merian, E. (eds.), *Metals and Their Compounds in the Environment: Occurrence, Analysis, and Biological Relevance*. Willey-VCH, Weinheim, Germany.

Busnardo, R.G., Busnardo, N.G., Salvato, G.N. and Afonso, J.C. (2007). Processing of spent NiMo and CoMo/Al_2O_3 catalysts via fusion with $KHSO_4$. *Journal of Hazardous Materials* B139: 391–398.

Cabrera, C., Moron, C. and Garcia, A. (2014). Learning process of the steel use in building engineering students. *American Journal of Educational Research* 2(6): 366–371.

Capudean, B. (2003). Metallurgy matters: Carbon content, steel classifications, and alloy steels. Available at http://www.thefabricator.com/article/metalsmaterials/carbon-content-steel-classifications-and-alloy-steels. Accessed 12 December 2015.

Capus, J.M. (2010). *Metal Powders: Global Survey of Production, Applications and Markets 2001–2010*. Elsevier, Oxford, UK.

Chambers, A. (2003). Childhood lead poisoning prevention and intervention programs in Connecticut: A historical analysis. Masters theses. University of Connecticut Health Center Graduate School, Storrs, Connecticut.

Chandler, H. (2006). *Metallurgy for the Non-Metallurgist*. AMS International, Cleveland, OH.

Chapman, D. (1998). High conductivity copper for electrical engineering. Available at http://copperalliance.org.uk/docs/librariesprovider5/pub-122---high-conductivity-copper-for-electrical-engineering/pub-122-hicon-copper-for-electrical-engineering.pdf?sfvrsn=2. Accessed 13 October 2016.

Chatterjee, K.K. (2007). *Uses of Metals and Metallic Minerals*. New Age International, New Delhi, India.

Christopher, S. (2011). Sustainability and value of steel recycling in Uganda. *Journal of Civil Engineering and Construction Technology* 2(10): 212–217.

Ciacci, L., Morselli, L., Passarini, F., Santini, A. and Vassura, I. (2010). A comparison among different automotive shredder residue treatment processes. *The International Journal of Life Cycle Assessment* 15: 896–906.

Clark, W.A.V., Freeman, H.E. and Hanssens, D.M. (1984). Opportunities for revitalizing stagnant markets: An analysis of household appliances. *Journal of Product Innovation Management* 4: 242–254.

Constellium (2014). Aluminium properties and applications. Available at http://www.constellium.com/technology-center/aluminium-alloy-properties. Accessed November 2015.

Cooper, T. and Mayers, K. (2000). Prospects for household appliances: E-Scope Project. Available at http://irep.ntu.ac.uk/6671/1/201121_7265%20Cooper%20Publisher%20rescanned.pdf. Accessed 27 January 2016.

Cooper, T. and Mayers, K. (2002). Discarded household appliances – What destiny? *Proceedings of the 10th International Conference of the Greening of Industry Network*, Göteborg, Sweden, 23–26 June. Available at http://citeseerx.ist.psu.edu/viewdoc/download?doi=10.1.1.548.5572&rep=rep1&type=pdf. Accessed 13 January 2016.

Copper Development Association (CDA) (2016). Copper: Properties and Applications. Available at http://copperalliance.org.uk/education-and-careers/education-resources/copper-properties-and-applications. Accessed October 2016.

Cui, J. and Forssberg, E. (2003). Mechanical recycling of waste electric and electronic equipment: A review. *Journal of Hazardous Materials* B99: 243–263.

Cui, J. and Zhang, L. (2008). Metallurgical recovery of metals from electronic waste: A review. *Journal of Hazardous Materials* 158: 228–256.

CWIT (2014). Recommendation paper to actors on e-waste classifications. Available at http://www.cwitproject.eu/wp-content/uploads/2013/10/Recommendation-paper-to-actors_revisedv2.pdf. Accessed 22 November 2015.

Daniels, E.J., Jody, B.J., Bonsignore, P.V. and Shoemaker, E.L. (1990). Automobile shredder residue: Process developments for recovery of recyclables. Available at http://www.osti.gov/scitech/servlets/purl/6294535. Accessed 16 January 2016.

Dantec, D. (2005). Analysis of the cost of recycling compliance for the automobile industry. MSc thesis. Massachusetts Institute of Technology, Cambridge, MA.

Darby, L. and Obara, L. (2005). Household recycling behaviour and attitudes towards the disposal of small electrical and electronic equipment. *Resources, Conservation and Recycling* 44: 17–35.

Davenport, W.G., King, M., Schlesinger, M. and Biswas, A. (2002). *Extractive Metallurgy of Copper*. Elsevier Science Limited, Oxford, UK.

Davies, N. (2014). The race for bauxite – China's depleting resources add fuel to the fire. Available at http://www.mining-technology.com/features/featurethe-race-for-bauxite-chinas-depleting-resources-add-fuel-to-the-fire-4171637/. Accessed November 2015.

Davis, J.R. (1999). *Corrosion of Aluminum and Aluminum Alloys*. ASM International, Cleveland, OH.

Desnica, E. (2014). The importance of sustainable development in engineering and education. *Technical Diagnostics* 13(1): 13–16.

Dongping, Y. (2013). *Chinese Research Perspectives on the Environment: Urban Challenges, Public Participation, and Natural Disasters*, Vol. 1. Koninklijk Brill NV, Leiden, the Netherlands.

Dossett, J.L. and Boyer, H.E. (2006). *Practical Heat Treating*. ASM International, Cleveland, OH.

Dutta, U. (1998). Compatibility and rheological study of ASR asphalt binder. *Journal of Materials in Civil Engineering* 10(1): 40–44.

Earnshaw, A. and Greenwood, N.N. (1997). *Chemistry of the Elements*, 2nd edn. Butterworth-Heinemann, Oxford, UK.

Ecclestone, C. (2014). Tin sector review. Available at http://ucore.com/Tin_sector_review_Feb2014.pdf. Accessed December 2015.

ECMA (2012). Guidelines for the management of spent catalysts. Available at http://www.cefic.org/Documents/Industry%20sectors/ECMA/ECMA%20GUIDELINES%20FOR%20THE%20MANAGEMENT%20OF%20SPENT%20CATALYSTS-201202.pdf. Accessed January 2016.

Ecofys (2009). Methodology for the free allocation of emission allowances in the EU ETS post 2012: Sector report for the non-ferrous metals industry. Available at http://ec.europa.eu/clima/policies/ets/cap/allocation/docs/bm_study-non_ferrous_metals_en.pdf. Accessed January 2016.

EIC Corporation (1979). A survey of metallurgical recycling processes. Available at http://www.osti.gov/scitech/servlets/purl/5899240. Accessed January 2016.

EPA (1990). Recycling cost analysis and energy balance. Available at http://epa.wa.gov.au/EPADocLib/373_B409.pdf. Accessed January 2016.

EPA (1995). Compilation of air pollutant emission factors. Vol. I: Stationary point and area sources. Available at http://www3.epa.gov/ttn/chief/ap42/toc_kwrd.pdf. Accessed January 2016.

EPA (1998). Sources of lead in soil: A literature review. Available at http://www2.epa.gov/sites/production/files/documents/r98-001a.pdf. Accessed December 2015.

EPA (2003). Hazardous waste management system. *Federal Register* 68(202): 59935–59940.

EPA (2015). Rules and regulations. *Federal Register* 80(8), Tuesday, January 13, 2015. Available at https://www.gpo.gov/fdsys/pkg/FR-2015-01-13/pdf/FR-2015-01-13.pdf. Accessed January 2016.

EU (2005). Directive 2000/53/EC of the European parliament and the council. Available at http://eur-lex.europa.eu/LexUriServ/LexUriServ.do?uri=CONSLEG:2000L0053:20050701:EN:PDF. Accessed January 2016.

European Aluminium Association (2002). The aluminium automotive manual. Available at http://www.european-aluminium.eu/wp-content/uploads/2012/01/AAM-Materials-1-Resources.pdf. Accessed January 2016.

European Commission (2009). Study on the calculation of recycling efficiencies and implementation of export article (Art. 15) of the Batteries Directive 2006/66/EC. Available at http://ec.europa.eu/environment/waste/batteries/pdf/batteries090528_fin.pdf. Accessed January 2016.

European Commission (2010). Critical raw materials for the EU. Available at http://www.euromines.org/files/what-we-do/sustainable-development-issues/2010-report-critical-raw-materials-eu.pdf. Accessed January 2016.

European Union (2008). Directive 2008/98/EC of the European Parliament and of the Council of 19 November 2008 on waste and repealing certain Directives. Available at http://cerrec.eu/files/2008_EU-wasteframeworkdirective_EN.pdf. Accessed January 2016.

European Union (2012). Directive 2012/19/EU of the European Parliament and of the Council of 4 July 2012 on waste electrical and electronic equipment (WEEE) (recast). Available at http://eur-lex.europa.eu/legal-content/EN/TXT/PDF/?uri=CELEX:32012L0019&from=EN. Accessed December 2015.

Faieza, A.A., Hazwan, A.R.M. and Lilehkoohi, A.H. (2015). Crashworthiness studies on various materials, forming process and the effects on vehicle body. *International Journal of Advance Research in Science and Engineering* 4(1): 14–23.

Field, F.R., Ehrenfeld, J.R., Roos, D. and Clark, J.P. (1994a). Automobile recycling policy: Background materials. Available at http://msl1.mit.edu/TPP12399/field-1b.pdf. Accessed January 2016.

Field, F.R., Ehrenfeld, J.R., Roos, D. and Clark, J.P. (1994b). Automobile recycling policy: Findings and recommendations. Available at http://dspace.mit.edu/bitstream/handle/1721.1/1686/field-1.pdf?sequence=2. Accessed January 2016.

Florida Artist Black Smith (FABA) (1989). The clinker breaker. Available at http://blacksmithing.org/CB-Archive/1989/1989-02-cb.pdf. Accessed October 2016.

Fornalczyk, A. (2012). Industrial catalysts as a source of valuable metals. *Journal of Achievements in Materials and Manufacturing Engineering* 55(2): 864–869.

Frosch, R.A. and Gallopoulos, N.E. (1989). Strategies for manufacturing. *Scientific American* 261(3): 144–152.

Gaballah, I., Djona, M., Mugica, J.C. and Solozobal, R. (1994). Valuable metals recovery from spent catalysts by selective chlorination. *Resources, Conservation and Recycling* 10(1–2): 87–96.

Gaines, L. (2014). The future of automotive lithium-ion battery recycling: Charting a sustainable course. *Sustainable Materials and Technologies* 1–2: 2–7.

GEC (2012). Laws and support systems for promoting waste recycling in Japan. Available: http://nett21.gec.jp/Ecotowns/LawSupportSystems.pdf. Accessed January 2016.

Gitlitz, M.H. and Moran, M.K. (2006). Tin compounds. In: Seidel, A. and Bickford, M. (eds.), *Kirk-Othmer Encyclopedia of Chemical Technology*, Vol. 24, 5th edn. John Willey & Sons, New York.

Giurco, D. (2005). Towards sustainable metal cycles: The case of copper. PhD thesis. University of Sydney, Sydney, New South Wales, Australia.

Goodwin, F.E. (1998). Zinc compounds. In: Kroschwitz, J. and Howe-Grant, M. (eds.), *Kirk-Othmer Encyclopedia of Chemical Technology*. John Wiley & Sons, Inc., New York, pp. 840–853.

Goonan, T.G. (2004). Copper recycling in the United States in 2004. Available: http://pubs.usgs.gov/circ/circ1196x/pdf/circ1196X.pdf. Accessed January 2016.

Greenpeace (2015). The e-waste problem. Available: http://www.greenpeace.org/international/en/campaigns/toxics/electronics/the-e-waste-problem/. Accessed December 2015.

Guang-hui, B., Wei, T., Xiang-gang, W., Jin-guo, Q., Peng, X. and Peng-cheng, L. (2010). Alkali desilicated coal fly ash as substitute of bauxite in lime-soda sintering process for aluminum production. *Transactions of Nonferrous Metals Society of China* 20: s169–s175.

Gupta, S.M. and Isaacs, J.A. (1997). Value analysis of disposal strategies for automobiles. *Computers & Industrial Engineering* 33(1–2): 325–328.

Hamuth, P.S. (2016). Boosting recycling for double objectives. Available at http://defimedia.info/boosting-recycling-for-double-objectives-14110/. Accessed February 2016.

Harper, C., Llados, F., Diamond, G. and Chappell, L.L. (2005). Toxicological profile for tin and tin compounds. Agency for Toxic Substances and Disease Registry, Atlanta, GA.

Hatch, J.E. (1984). *Aluminum: Properties and Physical Metallurgy*. ASM International, Cleveland, OH.

Havlik, T., Orac, D., Petranikova, M., Miskufova, A., Kukurugya, F. and Takacova, Z. (2010). Leaching of copper and tin from used printed circuit boards after thermal treatment. *Journal of Hazardous Materials* 183: 866–873.

Helsel, L. and Liu, P. (2008). *Industrial Materials*. Goodheart-Willcox Publisher, Chicago, IL.

Hogan, C.M. (2007). *The Encyclopedia of Earth*. Available at http://www.eoearth.org/view/article/157224/. Accessed December 2015.

Hotta, Y., Santo, A. and Tasaki, T. (2014). EPR-based electronic home appliance recycling system under home appliance recycling act of Japan. Available at http://www.oecd.org/environment/waste/EPR_Japan_HomeAppliance.pdf. Accessed January 2016.

Howard, R.A. and Barnes, W.R. (1991). Process for recovering valuable metals from spent catalysts. US Patent No. 5013533 A. Available at http://www.google.com/patents/US5013533. Accessed 17 January 2016.

Howe, H.E. (1981). Lead. In: *Kirk-Othmer Encyclopedia of Chemical Technology*, Vol. 14, 3rd edn. Wiley & Sons, New York, pp. 98–139.

Howe, P. and Watts, P. (2005). Tin and inorganic tin compounds. World Health Organization, Geneva, Switzerland.

Huyett, G.L. (2004). *Engineering Handbook: Technical Information*. Huyett Expressway, Minneapolis, MN.

Hwang, J.Y., Huang, X. and Xu, Z. (2006). Recovery of metals from aluminum dross and salt cake. *Journal of Minerals & Materials Characterization & Engineering* 5(1): 47–62.

Hyatt, D.E. (1987). Value recovery from spent alumina-base catalyst. US Patent No. 4657745 A. Available at http://www.google.com/patents/US4657745. Accessed 23 January 2016.

International Lead Association (2014). Lead recycling: Sustainability in action. Available at http://www.ila-lead.org/UserFiles/File/ILA9927%20FS_Recycling_V06.pdf. Accessed 15 January 2016.

Iqbal, M., Breivik, K., Syed, J.H., Malik, R.N., Li, J., Zhang, G. and Jones, K.C. (2015). Emerging issue of e-waste in Pakistan: A review of status, research needs and data gaps. *Environmental Pollution* 207: 308–318.

Jadhav, U.U. and Hocheng, H. (2012). A review of recovery of metals from industrial waste. *Journal of Achievements in Materials and Manufacturing Engineering* 54(2): 159–167.

Jameson, C.W., Lunn, R.M., Jeter, S. and Sabella, A.L. (2003). Report on carcinogens background document for lead and lead compounds. Available at http://ntp.niehs.nih.gov/ntp/newhomeroc/roc11/lead-public.pdf. Accessed 11 December 2015.

Janke, D., Savov, L. and Vogel, M.E. (2006). Secondary materials in steel production and recycling. In: von Gleich, A., Ayres, R.U. and Göβling-Reisemann, S. (eds.), *Sustainable Metals Management. Securing Our Future: Steps towards a Closed Loop Economy*. Springer, Dordrecht, the Netherlands, pp. 313–334.

Janke, D., Savov, L., Weddige, H.J. and Schultz, E. (2000). Scrap-based steel production and recycling of steel. *Materials Technology* 34(6): 387–399.

Javaid, A. and Essadiqi, E. (2003). Final report on scrap management, sorting and classification of steel https://www.nrcan.gc.ca/sites/www.nrcan.gc.ca/files/mineralsmetals/pdf/mms-smm/busi-indu/rad-rad/pdf/2003-23(cf)cc-eng.pdf. Accessed 17 January 2016.

Jha, M.K., Kumar, V. and Singh, R.J. (2001). Review of hydrometallurgical recovery of zinc from industrial wastes. *Resources, Conservation and Recycling* 33: 1–22.

Jianjun, W., Shangxing, G., Li, Z. and Qiang, L. (2009). Slag for decopperization and sulphur control in molten steel. *Journal of Iron and Steel Research* 16(2): 17–21.

Jody, B.J. and Daniels, E.J. (2006). End-of-life vehicle recycling: The state of the art of resource recovery from shredder residue. Available at http://www.ipd.anl.gov/anlpubs/2007/02/58559.pdf. Accessed 29 October 2016.

Jody, B.J., Daniels, E.J., Duranceau, C.M., Pomykala, J.A. and Spangenberger, J.S. (2010). End-of-life vehicle recycling: State of the art of resource recovery from shredder residue. Available at http://www.ipd.anl.gov/anlpubs/2011/02/69114.pdf. Accessed 17 January 2016.

Jody, J.B., Daniels, J.E., Bonsignore, V.P. and Brockmeier, F.N. (1994). Recovering recyclable materials from shredder residue. *Journal of Metals* 46(2): 40–43.

Johnson, D.B. (2006). Biohydrometallurgy and the environment: Intimate and important interplay. *Hydrometallurgy* 83: 153–166.

Kang, H.Y. and Schoenung, J.M. (2005). Electronic waste recycling: A review of U.S. infrastructure and technology options. *Resources, Conservation and Recycling* 45(4): 368–400.

Kar, B.B., Murthy, B.V.R. and Misra, V.N. (2005). Extraction of molybdenum from spent catalyst by salt-roasting. *International Journal of Mineral Processing* 76: 143–147.

Kaur, K., Gupta, R., Saraf, S.A. and Sara, S.K. (2014). Zinc: The metal of life. *Comprehensive Reviews Food Science and Food Safety* 13: 358–376.

Keerthiprasad, K., Babu, N. and Chandrashekara, K. (2014). Regression analysis and analysis of variance for EN353 and 20MnCr5 alloyed steels for drilling cutting forces. *International Journal of Engineering Research and Applications* 4(6): 136–146.

Kékesi, T., Török, T.I. and Kabelik, G. (2000). Extraction of tin from scrap by chemical and electrochemical methods in alkaline media. *Hydrometallurgy* 55: 213–222.

Keoleian, G.A., Kar, K., Manion, M.M. and Bulkley, J.W. (1997). Industrial ecology of the automobile: A life cycle perspective. Society of Automotive Engineers, Warrendale, PA.

Khaliq, A., Rhamdhani, M.A., Brooks, G. and Masood, S. (2014). Metal extraction processes for electronic waste and existing industrial routes: A review and Australian perspective. *Resources* 3: 152–179.

Kimbrough, R.D. and Krouskas, C.A. (2012). Contribution of lead in soil to children's lead burden, an update. *Chemical Speciation and Bioavailability* 24(2): 105–112.

Klazema, A. (2014). The chemical properties of copper. Available at https://blog.udemy.com/chemical-properties-of-copper/. Accessed December 2015.

Knežević, M., Korać, M., Kamberović, Z. and Rustic, M. (2010). Possibility of secondary lead slag stabilization in concrete with presence of selected additives. *Metallurgical and Materials Engineering* 16(3): 195–204.

Koyama, K., Tanaka, M. and Lee, J.C. (2006). Copper leaching behaviour from waste printed circuit board in ammonniacal alkaline solution. *Materials Transactions* 47(7): 1788–1792.

Krauss, G. (2015). *Steels: Processing, Structure, and Performance*. ASM International, Cleveland, OH.

Kumar, U. and Singh, D.N. (2013). E-waste management through regulations. *International Journal of Engineering Inventions* 3(2): 6–14.

Kumar, V. and Sutherland, J.W. (2008). Sustainability of the automotive recycling infrastructure: Review of current research and identification of future challenges. *International Journal of Sustainable Manufacturing* 1(1–2): 145–167.

Kumar, V. and Sutherland, J.W. (2009). Development and assessment of strategies to ensure economic sustainability of the U.S. automotive recovery infrastructure. *Resources, Conservation and Recycling* 53: 470–477.

Lee, F.M., Knudsen, R.D. and Kidd, D.R. (1992). Reforming catalyst made from the metals recovered from spent atmospheric residue of desulphurization catalyst. *Industrial and Engineering Chemistry Research* 31: 487–490.

Lee, J.C. and Pandey, B.D. (2012). Bio-processing of solid wastes and secondary resources for metal extraction – A review. *Waste Management* 32: 3–18.

Leon, G. (2006). *The Elements: Zinc*. Marshall Cavendish Benchmark, New York.

Leones, K. (2009). Green rust: How copper defends itself from corrosion. Available at http://ezinearticles.com/?Green-Rust---How-Copper-Defends-Itself-From-Corrosion&id=3465911. Accessed December 2015.

Leppo, H.W. (2009). *What You Really Need to Know to Pass the LEED Green Associate Exam*. Professional Publications, Inc., Belmont, CA.

Lewis, R.J. (1997). *Hawley's Condensed Chemical Dictionary*. John Wiley & Sons, Inc., New York.

Li, J., Yu, K. and Gao, P. (2014). Recycling and pollution control of the end of life vehicles in China. *Journal of Material Cycles and Waste Management* 16: 31–38.

Li, M. and Zinkle, S.J. (2012). Physical and mechanical properties of copper and copper alloys. In: Konings, R.J.M. (ed.), *Comprehensive Nuclear Materials*, Vol. 4. Elsevier, Amsterdam, the Netherlands, pp. 667–690.

Lide, D.R. (2003). *CRC Handbook of Chemistry and Physics*. CRC Press, Boca Raton, FL.

Lloyd, T.B. (1984). Zinc compounds. In: Grayson M. (ed.), *Kirk-Othmer Encyclopedia of Chemical Technology*, Vol. 24, 3rd edn. John Wiley & Sons, New York, pp. 851–863.

Lloyd, T.B. and Showak, W. (1984). Zinc and zinc alloys. In: Grayson, M. (ed.), *Kirk-Othmer Encyclopedia of Chemical Technology*, Vol. 24, 3rd edn. John Wiley & Sons, New York, pp. 835–836.

Louie, D.K. (2005). *Handbook of Sulphuric Acid Manufacturing*. DKL Engineering, Richmond Hill, Ontario, Canada.

Lundgren, K. (2012). *The Global Impact of E-Waste: Addressing the Challenge.* International Labour Organization, Geneva, Switzerland.

Lusty, P.A.J. and Hannis, S.D. (2009). Commodity profile: Copper. Available at http://nora.nerc.ac.uk/7977/1/OR09041.pdf. Accessed October 2015.

Majima, H., Hirato, T. and Awakura, Y. (1993). Dissolution of copper with aqueous cupric amine solution. I. Studies on the selective recovery of copper and iron from motor scrap. *Mining and Material Processes Institute Japan* 109(3): 191–194.

Manomaivibool, P. and Vassanadumrongdee, S. (2012). Buying back household waste electrical and electronic equipment: Assessing Thailand's proposed policy in light of past disposal behavior and future preferences. *Resources, Conservation and Recycling* 68: 117–125.

Marafi, M. and Stanislaus, A. (2008). Spent catalyst waste management: A review. Part I. Developments in hydroprocessing catalyst waste reduction and use. *Resources, Conservation and Recycling* 52: 859–873.

Marafi, M., Stanislaus, A. and Furimsky, E. (2010). *Handbook of Spent Hydroprocessing Catalysts: Regeneration, Rejuvenation and Reclamation.* Elsevier, Amsterdam, the Netherlands.

Master Recycler Program (MRP) (2015). Recycling processes. Available at http://fa.oregonstate.edu/sites/fa.oregonstate.edu/files/recycling/resources/MR_Class/chapter_iv_recycling_processes_2015.pdf. Accessed 5 November 2015.

Matjie, R.H., Bunt, J.R. and van Heerden, J.H.P. (2005). Extraction of alumina from coal fly ash generated from a selected low rank bituminous South African coal. *Minerals Engineering* 18: 299–310.

MatWeb (2015). How alloying elements affect the properties of copper alloys. Available at http://www.matweb.com/reference/copper-alloys.aspx. Accessed December 2015.

McDonough, W. and Braungart, M. (2002). *Cradle to Cradle: Remaking the Way We Make Things.* North Point Press, New York.

McElwee, C.R. (2011). *Environmental Law in China: Managing Risk and Ensuring Compliance.* Oxford University Press, Oxford, UK.

Medvedev, A.S. and Malochkina, N.V. (2007). Sublimation of molybdenum trioxide from exhausted catalyst employed for the purification of oil products. *Russian Journal of Non-Ferrous Metals* 48(2): 114–117.

Menzie, W.D., Barry, J.J., Bleiwas, D.I., Bray, E.L., Goonan, T.G. and Matos, G. (2010). The global flow of aluminum from 2006 through 2025: U.S. Geological Survey Open-File Report 2010-1256. Available at http://pubs.usgs.gov/ofr/2010/1256/. Accessed 5 November 2015.

METI (2007). Towards a 3R-oriented, sustainable society: Legislation and trends 2007. Available at http://www.meti.go.jp/policy/recycle/main/data/pamphlet/pdf/handbook2007_eng.pdf. Accessed 22 January 2016.

Meyrick, G. and Wagoner, R.H. (2001). Physical metallurgy of steel. Available at http://www.tech.plymouth.ac.uk/sme/interactive_resources/tutorials/FailureAnalysis/Undercarriage_Leg/Steel_Metallurgy_Ohio-State.pdf. Accessed December 2015.

Mishra, D., Kim, D., Ralph, D., Ahn, J. and Rhee, Y. (2008). Bioleaching of spent hydro-processing catalyst using acidophilic bacteria and its kinetics aspect. *Journal of Hazardous Materials* 152: 1082–1091.

Mishra, D., Kim, D.J., Ralph, D.E., Ahn, J.G. and Rhee, Y.H. (2007). Bioleaching of vanadium rich spent refinery catalysts using sulfur oxidizing lithotrophs. *Hydrometallurgy* 88: 202–209.

Moore, J.W. (1991). *Inorganic Contaminants of Surface Water: Research and Monitoring Priorities.* Springer-Verlag, New York.

Morici, P., Price, A.H. and Danjczek, T.A. (2013). Melt it here: The benefits of expanding steel production in the United States. Available at http://www.steelnet.org/new/20131219.pdf. Accessed 22 January 2016.

Muchová, L. and Eder, P. (2010a). End-of-waste criteria for aluminium and aluminium alloy scrap: Technical proposals. Publications Office of the European Union, Luxembourg, Luxembourg.

Muchová, L. and Eder, P. (2010b). End-of-waste criteria for iron and steel scrap: Technical proposals. Publications Office of the European Union, Luxembourg, Luxembourg.

Namias, J. (2013). The future of electronic waste recycling in the United States: Obstacles and domestic solutions. MSc dissertation. Columbia University, New York.
Natural Environment Research Council (NERC) (2007). Copper. Available file:///C:/Users/a0009328/Downloads/comm_profile_copper%20(2).pdf. Accessed 19 December 2015.
Nayak, N. and Chitta, R.P. (2009). Aluminium extraction and leaching characteristics of Talcher Thermal Power Station fly ash with sulphuric acid. *Fuel* 89(1): 53–58.
Nielsen, W.D. (2015). Metallurgy of copper-base alloys. Available at http://www.copper.org/resources/properties/703_5//. Accessed 11 December 2015.
Nigo, S., Hirato, T., Awakura, Y. and Iwai, M. (1993). Recovery of copper from motor scrap utilizing ammonia leaching technique. II. Studies on the selective recovery of copper and iron from motor scrap. *Journal of the Mining and Materials Processing Institute of Japan* 109(5): 337–340.
Nourreddine, M. (2007). Recycling of auto shredder residue. *Journal of Hazardous Materials* A139: 481–490.
OECD (1995). Recycling of copper, lead and zinc bearing wastes. Available at http://www.oecd.org/officialdocuments/publicdisplaydocumentpdf/?cote=OCDE/GD(95)78&docLanguage=En. Accessed 16 January 2016.
Ojeda, M.W., Perino, E. and Ruiz, M. (2009). Gold extraction by chlorination using a pyrometallurgical process. *Minerals Engineering* 22: 409–411.
O'Leary, D. (2000). Copper. Available at http://www.ucc.ie/academic/chem/dolchem/html/elem029.html. Accessed 16 December 2015.
O'Neil, M.J., Smith, A. and Heckelman, P.E. (2001). *The Merck Index: An Encyclopedia of Chemicals, Drugs, and Biologicals*, 13th edn. Whitehouse Station, Rahway, NJ.
Outokumpu (2013). *Handbook of Stainless Steel*. Outokumpu, Helsinki, Finland.
Park, K.H., Mohapatra, D. and Nam, C.W. (2007). Two stage leaching of activated spent HDS catalyst and solvent extraction of aluminium using organo-phosphinic extractant, Cyanex 272. *Journal of Hazardous Materials* 148: 287–295.
Parkinson, G., Ushio, S., Hibbs, M. and Hunter, D. (1987). Recyclers try new ways to process spent catalysts. *Chemical Engineering* 94: 25–31.
Pourbaix, M. (1974). *Atlas of Electrochemical Equilibria in Aqueous Solutions*. National Association of Corrosion Engineers, Houston, TX.
Poweleit, D. and Monroe, R. (2006). Steel alloys. Steel Founders' Society of America, Crystal Lake, IL.
Pradhan, D., Kim, D.J., Ahn, J.G. and Lee, S.W. (2010). Microbial leaching process to recover valuable metals from spent petroleum catalyst using iron oxidizing bacteria. *International Journal of Chemical, Molecular, Nuclear, Materials and Metallurgical Engineering* 4(2): 232–236.
Qasioun (2015). Biggest illegal waste dump in Europe. Available at http://qasioun.net/en/article/Biggest-illegal-waste-dump-in-Europe/1816/. Accessed 21 November 2015.
Rahimifard, S., Abubakar, M.S. and Williams, D.J. (2009). Recycling process planning for the End-of-Life management of waste from electrical and electronic equipment. *CIRP Annals – Manufacturing Technology* 58: 5–8.
Rao, S.R. (2011). *Resource Recovery and Recycling from Metallurgical Wastes*. Elsevier, Oxford, UK.
Reddy, Y.P., Kumar, B.J., Raju, G. and Rao, C.S. (2015). Fabrication and thermal analysis of composite pin-fin. *International Journal of Core Engineering and Management* 2(5): 77–89.
Reilly, C. (2002). *Metal Contamination of Food: Its Significance for Food Quality and Human Health*. Blackwell Science Limited, Oxford, UK.
Roney, N., Smith, C.V., Williams, M., Osier, M. and Paikoff, S.J. (2005). Toxicological profile for zinc. Agency for Toxic Substances and Disease Registry, Atlanta, GA.
Ross, G. (1990). *Biohydrometallurgy*. McGraw-Hill Book Company, New York.
Sakai, S. et al. (2014). An international comparative study of end-of-life vehicle (ELV) recycling systems. *Journal of Material Cycles and Waste Management* 16: 1–20.
Saman, M.Z.M. (2006). End of life vehicles recovery: Process description, its impact and direction of research. *Jurnal Mekanikal* 21: 40–52.
Santhiya, D. and Ting, Y.P. (2005). Bioleaching of spent refinery processing catalyst using Aspergillus niger with high-yield oxalic acid. *Journal of Biotechnology* 116: 171–184.

Saravanan, R. and Rani, M.P. (2012). *Metal and Alloy Bonding: An Experimental Analysis*. Springer-Verlag, London, UK.

Savov, L., Volkova, E. and Janke, D. (2003). Copper and tin in steel scrap recycling. *Materials and Geoenvironment* 50(3): 627–640.

Schmidt, M. (2012). Resource efficiency – What are critical metals and how scarce are they? Available at http://umwelt.hs-pforzheim.de/fileadmin/dokumente/2012/Electronic_Displays_2012_Schmidt.pdf. Accessed January 2016.

Séblin, B., Jahazeeah, Y., Sujeebun, S., Manohar and Wong, K.B. (2014). Material Science: MECH 2004. Available at http://www.uom.ac.mu/Faculties/foe/MPED/Students_Corner/notes/EnggMaterials/steelbklet.pdf. Accessed December 2015.

Sendijarevic, V., Pokorski, B., Klempner, D. and Frisch, K.C. (1997). Recent developments in shredder downstream separation processes and recycling options for automotive shredder residue. *SAE Special Publications (New Plastics Application for the Automotive Industry)* 1253: 163–169.

SEPA (2011). Recycling and disposal of electronic waste: Health hazards and environmental impacts. Naturvårdsverket, Stockholm, Sweden.

SEPA (2016). Waste electrical and electronic equipment (WEEE). Available at https://www.sepa.org.uk/regulations/waste/waste-electrical-and-electronic-equipment-weee. Accessed January 2016.

Sepúlveda, A., Schluep, M., Renaud, F.G., Streicher, M., Kuehr, R., Hagelüken, C. and Gerecke, A.C. (2010). A review of the environmental fate and effects of hazardous substances released from electrical and electronic equipments during recycling: Examples from China and India. *Environmental Impact Assessment Review* 30(1): 28–41.

Shamsuddin, M. (1986). Metal recovery from scrap and waste. *Journal of Metals* 38(2): 24–31.

Shantgiri, S.S. (2015). The environmental law. *Indian Streams Research Journal* 5(9): 1–5.

Sharif, A. (2008). Metallurgical aspects of fatigue failure of steel – Part I. Available at http://www.bsrm.info/resources/articles/Fatigue_Seminar_Book-Dr_Ahmed_Sharif_Part-I.pdf. Accessed December 2015.

Sharma, K.D. and Row, B.R.L. (1985). An electrolytic process for recovery of zinc dust from melting furnace slag. *Hydrometallurgy* 13: 377–383.

Shcherban, S., Raizman, V. and Pevzner, I. (1995). Technologies of coal fly ash processing into metallurgical and silicate chemical products. *Proceedings of 210th ACS National Meeting*, Chicago, IL, 20–25 August 1995, Vol. 40(4), pp. 863–867.

Shemi, A. (2013). Extraction of aluminium from coal fly ash using two-step acid leach process. MSc (Eng) dissertation. University of the Witwatersrand, Johannesburg, South Africa.

Shemi, A., Mpana, R.N., Ndlovu, S., van Dyk, L.D., Sibanda, V. and Seepe, L. (2012). Alternative techniques for extracting alumina from coal fly ash. *Minerals Engineering* 34: 30–37.

Shibayama, A., Otomo, T., Takasaki, Y., Cao, Y., Murakami, T., Watanabe, K. and Inoue, H. (2006). Separation and recovery of valuable metals from automobile shredder residue (ASR) fly ash by wet processing. *International Journal of the Society of Materials Engineering for Resources* 13(2): 54–59.

Siemens, R.E., Jong, B.W. and Russel, J.H. (1986). Potential of spent catalysts as a source of critical metals. *Conservation and Recycling* 9: 189–196.

Simate, G.S. (2009). The bacterial leaching of nickel laterites using chemolithotrophic microorganisms. MSc (Eng) dissertation. University of the Witwatersrand, Johannesburg, South Africa.

Simate, G.S., Ndlovu, S. and Walubita, L.F. (2010). The fungal and chemolithotrophic leaching of nickel laterites – Challenges and opportunities. *Hydrometallurgy* 103: 150–157.

Singh, B. (2009). Treatment of spent catalyst from the nitrogenous fertilizer industry – A review of the available methods of regeneration, recovery and disposal. *Journal of Hazardous Materials* 167: 24–37.

Sivakumar, G.D., Barnabas, S.G. and Anatharam, S. (2014). Indian automobile material recycling management. *International Journal of Innovative Research in Science, Engineering and Technology* 3(3): 2754–2758.

Sokic, M., Ilic, I., Vičković, N. and Markovic, B. (2006). Procedures for primary pretreatment and processing of waste tin plates and metallic packages. *Acta Metallurgica Slovaca* 12: 354–361.

Soni, S. and Suar, M. (2015). *Handbook of Research on Diverse Applications of Nanotechnology in Biomedicine, Chemistry, and Engineering*. IGI Global, Hershey, PA.

Statham, S. (2006). Remanufacturing towards a more sustainable future. Available at http://www.lboro.ac.uk/microsites/mechman/research/ipm-ktn/pdf/Technology_review/remanufacturing-towards-a-more-sustainable-future.pdf. Accessed 22 January2016.

Statista (2015). Lead reserves worldwide as of 2014, by country (in million metric tons). Available at http://www.statista.com/statistics/273652/global-lead-reserves-by-selected-countries/. Accessed 4 December 2015.

Staudinger, J. and Keoleian, G.A. (2001). Management of end-of life vehicles (ELVs) in the US. Center for Sustainable Systems, Report No. CSS01-01. University of Michigan, Ann Arbor, MI.

Talbert, R. and Keeton, R.D. (1996). *Steelworker*, Vol. 1. NETPDTC N331, Pensacola, FL.

Tanskanen, P. (2012). Electronics waste: Recycling of mobile phones. In: Damanhuri, E. (ed.), *Post-Consumer Waste Recycling and Optimal Production*. In-Tech, Rijeca, Croatia.

Thompson, S.A. (2011). Properties of lead. Available at https://www.deq.state.ok.us/factsheets/air/lead.pdf. Accessed 4 December 2015.

Thornton, I., Rautiu, R. and Brush, S. (2001). Lead: The facts. IC Consultants Limited, London, UK.

Torma, A.E. (1983). Extraction of aluminium from fly ash. *Metals Berlin* 37(6): 589–592.

Tzeferis, P.G. (1994). Leaching of a low grade hematitic laterite ore using fungi and biologically produced acid metabolites. *International Journal of Mineral Processing* 42: 267–283.

UNEP (2003). Technical guidelines for the environmentally sound management of waste lead-acid batteries. Available at http://www.basel.int/Portals/4/Basel%20Convention/docs/pub/techguid/tech-wasteacid.pdf. Accessed 22 January 2016.

UNEP (2007). E-waste management manual. Available at http://www.unep.or.jp/Ietc/Publications/spc/EWasteManual_Vol2.pdf. Accessed 6 December 2015.

UNEP (2009). Guideline on material recovery and recycling of end-of-life mobile phones. Available at http://www.basel.int/industry/mppi/documents.html. Accessed January 2016.

UNEP (2010). Final review of scientific information on lead. Available at http://www.ucc.ie/academic/chem/dolchem/html/elem029.html. Accessed 11 December 2015.

Uzorh, A.C. (2013). Corrosion properties of plain carbon steels. *International Journal of Engineering and Science* 2(11): 18–24.

Valverde, I.M., Paulino, J.F. and Afonso, J.C. (2008). Hydrometallurgical route to recover molybdenum, nickel, cobalt and aluminum from spent hydrotreating catalysts in sulphuric acid medium. *Journal of Hazardous Materials* 160: 310–317.

van Nostrand (2006). *Tin: Van Nostrand's Scientific Encyclopedia*. John Wiley & Sons, Inc., New York.

Veit, H.M. and Bernardes, A.M. (2015). Electronic waste: Generation and management. In: Veit, H.M. and Bernardes, A.M. (eds.), *Electronic Waste, Topics in Mining, Metallurgy and Materials Engineering*. Springer International Publishing, Basel, Switzerland.

Veit, H.M., Diehl, T.R., Salami, A.P., Rodrigues, J.S., Bernardes, A.M. and Tenorio, J.A.S. (2005). Utilization of magnetic and electrostatic separation in the recycling of printed circuit boards scrap. *Waste Management* 25: 67–74.

Verhoef, E., Dijkema, G. and Reuter, M. (2004). Process knowledge, system dynamics and metal ecology. *Journal of Industrial Ecology* 8(1–2): 23–43.

Vermeulen, I., van Caneghem, J., Block, C., Baeyens, J. and Vandecasteele, C. (2011). Automotive shredder residue (ASR): Reviewing its production from end-of-life vehicles (ELVs) and its recycling, energy or chemicals' valorization. *Journal of Hazardous Materials* 190: 8–27.

Vest, H. (2002). Fundamentals of the recycling of lead-acid batteries. Available at http://www.gate-international.org/documents/techbriefs/webdocs/pdfs/e017e_2002.pdf. Accessed January 2016.

Villanueva, A., Delgado, L., Luo, Z., Eder, P., Catarino, A.S. and Litten, D. (2010). Study on the selection of waste streams for end of waste assessment. Publications Office of the European Union, Luxembourg, Luxembourg.

Virolainen, S. (2013). Hydrometallurgical recovery of valuable metals from secondary raw materials. Doctor of Science (Technology) thesis. Lappeenranta University of Technology, Lappeenranta, Finland.

Wang, F., Huisman, J., Balde, K. and Stevels, A. (2012). A systematic and compatible classification of WEEE. Available at http://ieeexplore.ieee.org/stamp/stamp.jsp?tp=&arnumber=6360480. Accessed December 2015.

Wang, L. and Muhammed, M. (1999). Synthesis of zinc oxide nanoparticles with controlled morphology. *Journal of Materials Chemistry* 9: 2871–2878.

Westrum, B. and Thomassen, Y. (2002). The Nordic Expert group for criteria documentation of health risks from chemicals and the Dutch expert committee on occupational standards. 130. Tin and inorganic tin compounds. *Arbete och Halsa* 10: 1–48.

Xakalashe, B.S., Seongjun, K. and Cui, J. (2012a). An overview of recycling of electronic waste: Part 1. *Chemical Technology* June: 9–12.

Xakalashe, B.S., Seongjun, K. and Cui, J. (2012b). An overview of recycling of electronic waste: Part 2. *Chemical Technology* July: 23–26.

Yang, Z.G., Paxton, D.M., Weil, K.S., Stevenson, J.W. and Singh, P. (2002). Materials properties database for selection of high-temperature alloys and concepts of alloy design for SOFC applications. Available at http://www.pnl.gov/main/publications/external/technical_reports/PNNL-14116.pdf. Accessed October 2016.

Yglesias, C. (2014). *The Innovative Use of Materials in Architecture and Landscape Architecture: History, Theory and Performance*. McFarland & Company Inc., Jefferson, MO.

Yoo, J.S. (1998). Metal recovery and rejuvenation of metal-loaded spent catalyst. *Catalyst Today* 44: 27–46.

Zeng, L. and Cheng, C.Y. (2009). A literature review of the recovery of molybdenum and vanadium from spent hydrodesulphurization catalysts. Part: I. *Metallurgical Processes. Hydrometallurgy* 98: 1–9.

Zhou, K., Shinme, K. and Anezaki, S. (1995a). Dissolution rate of copper in aqueous ammonia solution: Removal of copper from ferrous scrap with ammonia leaching method. *Journal of the Mining and Materials Processing Institute of Japan* 111(1): 49–53.

Zhou, K., Shnme, K. and Anezaki, S. (1995b). Removal of copper from automobile scrap: Removal of copper from ferrous scrap with ammonia leaching method. *Journal of the Mining and Materials Processing Institute of Japan* 111(1): 55–58.

10

Conclusions and Future Outlook

10.1 Introduction

The mining industry has been an important driver of economic growth and has helped to facilitate the rapid development of many countries for decades. However, the ever-increasing demand for mineral and metal resources in order to meet the highly dynamic technological landscape has led to most of the richest mines being rapidly depleted. Reports by the United Nations Environment Programme (UNEP) on 'Critical Metals for Future Sustainable Technologies and Their Recycling Potential' (Buchert et al., 2009) and the U.S. Environmental Protection Agency (EPA) about 'Rare Earth Elements: A Review of Production, Processing, Recycling and Associated Environmental Issues' (Weber and Reisman, 2012) and a number of international initiatives show that global governments have concerns on the security of supply of certain metals. As a result, many mining companies are re-evaluating the current sources of metals, turning to deposits with lower-grade ores. For example, in the nineteenth century, copper was being mined from deposits with 10% ore grade (Gordon, 2002), whereas, today, the world average copper grade is about 0.6%. This trend applies to a number of various other metals such as gold, silver and cobalt. This means the extraction of the metals from such lower-grade ores generates more mine waste per unit extracted as more tonnes of inert materials must be physically removed, crushed, screened and later dumped to yield a reasonable concentrated fraction. This adds to the already extensive quantities of waste currently in existence and, thus, further impacts negatively on the environment and also affects the communities living in the locality of these mines.

Therefore, the production of the large quantity of waste streams together with strict environmental legislature and the associated health and safety impacts on local communities have forced many companies to intelligently design and optimize the metal extraction processes in order to reduce waste production. Furthermore, in a challenging environment with low London Metal Exchange (LME) prices, companies are also looking for better opportunities to extract profitable products from waste produced during mainstream production. Therefore, there has been significant research efforts driven towards the smart utilization of mining and metallurgical waste material through the development of cost-effective reuse and recycling options. The alternative and beneficial use of the waste products can act as a low-cost means to boost profitability and contribute to a sustainable future by allowing companies to sustain their activities during times of low commodity prices.

It is, therefore, imperative to develop a sustainable system loop that can properly recycle and utilize all the valuable resources, which are landfilled as waste materials by converting them into useful products or designing the direct recycling of elements through intelligently designed disassembling techniques at their end of life.

Lottermoser (2011) defined the reuse of wastes as the process that involves the new use or application of the total waste in its original form for a specific purpose directly without any reprocessing. Recycling on the other hand is defined as the extraction of new valuable resource ingredients or use of waste as feedstock with the conversion of the entire waste into a new valuable product or application with some reprocessing. The most desirable benefits of the above-stated approaches are (Lottermoser, 2011):

- Savings in basic raw materials,
- Efficiency and reduction in the intensive use of virgin resources,
- Energy conservation and reduction on ecological impact,
- Increased profitability through increased income and reduced disposal costs; the value of the recovered by-product can offset the cost of the waste management and disposal,
- Improved sustainability for the local communities.

A water treatment plant in eMalahleni, South Africa, owned by Anglo Coal, is a typical example that highlights some benefits that can be realized from waste material to both private and public sectors (Gunther and Mey, 2008). This plant, commissioned in 2007, is the first in the country to successfully treat AMD arising from mining operations. The plant generates water meeting the local municipality quality and also some gypsum as a waste by-product. Some of the water utilized in the eMalahleni local municipality thus, helps to meet 20% of the council's daily water requirements. The rest of the water is reutilized in the coal plant. The company then uses the gypsum by-product to produce bricks which are subsequently used to build houses for the workers at Anglo Coal. The gypsum is used in place of river sand, which means only half of the amount of the river sand is used every time a brick is manufactured. This project has highlighted the significant benefits that can be realised from the treatment of waste and also points to economic resilience (Bansal and McKnight, 2009; Christopherson et al., 2010) that results from the strong competitive advantages created by such an approach, such as development of infrastructure and technology, abundant raw materials and water resources, cooperation, reliable local suppliers and highly trained workforce and community support.

According to Zheng and Kozinski (1997), four basic principles need to be applied in the remediation of solid wastes, although the same principles can also be applied to waste stream solution: reduction of waste generated, recycling of wastes, generation of valuable coproducts

TABLE 10.1

Remediation Principles for Solid Metallurgical Waste

Principles	Procedures
Reduce generation	Substitute less toxic and better-quality raw materials for those currently used. Institute methods that reduce waste by improving the efficiency of existing processing operations. Develop new production methods that limit harm to the environment.
Recycle waste	Recover valuable elements from waste. Reclaim metal scrap. Use wastes to treat wastes. Use combustible wastes as fuel substitutes to generate energy.
Generate co-products	Reuse metallurgical wastes in other industries. Increase the extraction of multiple metals from wastes.
Treat before landfill	Segregate wastes according to their characteristics (e.g. hazardous or nonhazardous). 'Delist' wastes are nontoxic wastes. Treat wastes to reduce their volume, mass and toxicity.

Source: Zheng, G.-H. and Kozinsk, J.A., *Environ. Prog. Sust. Energy*, 15(4), 283, 1997.

and treatment before landfill/disposal. These principles are shown in Table 10.1. However, there are a number of challenges that are associated with the approach proposed in Table 10.1. Some of these challenges are highlighted and discussed in the subsequent section.

10.2 Challenges

According to EPA (2016) waste products can be considered to be of beneficial use if they can

- Be reused in a product that provides a functional benefit,
- Replace a product made from virgin raw material; with the new product meeting product specifications and industry standards,
- Conserve natural resources and reduce greenhouse emissions.

A number of challenges have been identified in view of the potential applications of mineral and metal wastes in alternative industries or sectors and in a beneficial manner as outlined by EPA (2016). These challenges have led most of the companies to continue viewing the material as an inconvenience and not a potential valuable resource. In fact, it is only the advent of stringent regulations, legislations and policies defining standards for protection of the environment that has forced companies to start focusing on reuse and recycling of waste. In addition, the trend towards internalizing the environmental costs of primary metallurgical operations has thus, driven companies towards improving the relative economic attractiveness of secondary application of waste materials. Furthermore, public demand for conservation of natural resources and their most suitable utilization, as well as for generation of the least possible waste, its reuse wherever possible and proper disposal, is also growing (Zheng and Kozinski, 1997). Furthermore, increased knowledge of the harm caused and the potential risks associated with such waste materials being produced have also provided powerful incentives for considering alternative use of the waste materials (IAEA, 2004).

The following sections highlight some of the challenges that may be encountered when trying to gain value from mining and metallurgical processing waste.

10.2.1 Lack of Adequate Regulatory Frameworks and Policies

As highlighted in Chapter 2, most of the current waste management regulations and policies are not linked to resource conservation or ecosystem protection. Furthermore, the existing laws on waste management are not well aligned and vary from one region to the other. There is also a lack of synergy or lack of understanding on the roles that can be played by the public and private sectors.

Too much emphasis has also been paid to the traditional type of waste management systems such as neutralization, landfilling and stabilization. However, it is important to note that landfills can be a major source of methane, and land costs have been steadily increasing in the past few years as competition with the world's ever-increasing population becomes more prominent.

It has already been pointed out that waste is, in most cases, considered valueless, and, as such, most governments spend a significant amount of money on waste collection and disposal without adequate consideration of resource-saving measures and their economic return or input. In other cases, there is also limited effort on reducing waste at source. A major problem also, as noted in Chapter 2, is that, in some countries, the costs associated with breaking the law are sometimes lower than the costs of following the law. For example, paying fines may be cheaper than conducting high-quality and timely environmental impact assessments, paying for expensive waste treatment facilities or installing and using emissions reduction technologies.

Sound policies of repair and reuse can be key strategies for minimizing waste especially from the end user or end-of-life products. For example, automobiles are often repaired to allow further use. In a similar way, Japan has seen a roaring trend in the sale of second-hand cars for reuse in most African countries. Unfortunately, modern technology and sophisticated composite materials often make repair more difficult than it was in the past. In such cases, it becomes very difficult to economically justify the repair or reuse of the end of life products. This is very common in most developed countries where consumers, in many cases, find it cheaper and much easier to buy a new product than to replace and/or repair a worn-out component. This is partly because the easily reparable breakdowns are gradually being eliminated by improved design, and as such, manufacturers prefer replacing the complete unit rather than sell parts. In addition, the rapid pace of change in technology and improved designs in manufactured products has encouraged consumers to replace their products (e.g. cell phones, computers, televisions and refrigerators) at a more rapid rate in such a way that the product is never used to its full end of life. This further adds to the quantity of waste being generated and disposed of to the environment.

10.2.2 Complexity of the Residue Material versus Available Technology

The work presented in this book has further shown that it can be difficult to assign a universal method to the reuse and recycling of all kinds of mining and metallurgical waste. The distribution and quantification of elements in waste is not always the same or simple because each kind of waste has its own physical and chemical properties, and thus, appropriate ways for reuse or recycling can vary according to local environmental conditions (Bian et al., 2012). Ideally, the nature of recycling or reprocessing should be specific to different environments. This is an aspect that needs to be taken into consideration when developing mineral and metal extraction processes. Therefore, the chemical and physical properties of the waste produced need to be critically evaluated to the same level and detail as any raw material deposit if it is to be considered for use in an alternative secondary industry.

The quantity and complexity of the residue from mining and metallurgical processes and post-consumer waste products further create a challenging task for the current recycling technologies. These wastes, as seen in most of the chapters in this book, are generated in such large quantities that metallurgical operations need to be optimized and new techniques developed in order to substantially reduce their volume, mass and toxicity. As already stated, the assemblage of mineral phases and metals in these secondary resources can also be very complex, diverse in physical and chemical composition and structure and highly integrated. This is a particularly serious problem in the case of complex materials such as electronic waste products. In addition, most of the metallic wastes are being disposed into the environment at very low concentrations that make recovery for recycling impracticable in most cases.

While current techniques are successfully treating some of the metallic elements individually, no integrated methods have yet been industrially applied, whereby the various associations of these substances which typically occur in solid wastes can be treated in a single and, therefore, cost-effective process (Zheng and Kozinski, 1997).

However, the complexity of the waste material may need a process route that requires the application of a series of technologies, thus making the overall reprocessing very challenging and increasing the production costs. In addition, when implementing such an approach, the amount of infrastructure, materials, energy and manpower necessary to implement the process may become costly and prohibitive. Currently, not many companies are willing to invest time and money into such processes. Therefore, there is need for technologies that require less infrastructure, materials, energy and manpower in long-term operation and maintenance, which, unfortunately, are not always possible.

10.2.3 Economic Viability

The information presented in this book highlighted that there are currently very few commercial applications in terms of waste utilization. The overall lack of progress on the commercialization front in terms of mineral and metal extraction from waste material also further suggests that there is currently no strong economic case that has been established for the reprocessing of wastes.

Although various methods have been suggested for recycling, their economic viability has not been established, and in most cases, there is a lack of well-documented cost and performance data of the alternative material or uses under a variety of operating conditions. Therefore, there is need for cost and performance evaluation to be carried out. The cost-benefit analysis is the most widely accepted concept for economic appraisal and for designing regulatory framework/policy options and would, therefore, be the ultimate driver in terms of the feasibility of a specific reuse or recycling technology or route. Typically, recycling flow sheets need to be designed for each of the material and economic analyses undertaken to assess their recycling potential (Edraki et al., 2014). Thereafter, overall environmental costs can be determined by various approaches such as ecological risk assessment, life cycle assessment, sustainability operations assessment and an estimate of the ecological footprint (Bian et al., 2012). Such cost-benefit analysis can provide adequate incentives for resource conservation and efficient resource allocation.

Commercialization can only be viable now or in the future if emerging technologies and techniques are sustainable and optimized so as to enable the production of high-quality secondary commodities for the various markets at a cost-efficient level. If the costs of the final products are economically prohibitive, then even the most eco-friendly processing methods will be difficult to implement without regulation or government subsidies (Bian et al., 2012). One approach for minimizing cost is to improve waste processing efficiency, which depends on the optimization of the resource allocation, that is, maximizing the quantity of wastes processed and the associated benefits while minimizing the environmental effects (Bian et al., 2012).

The work presented in this book also suggests that for the waste material to be successfully used in an alternative market, it should be provided in bulk, and with minimal changes to its physical or chemical characteristics or properties (Scott et al., 2005). Modification of the material tends to add to process costs which can have an impact on the overall revenue derived from the waste product. This is true for the waste streams that contain small amounts of valuable mineral or metal that would require a significant level of complex processes to recover. Therefore, interests tend to be shown to the type of waste that is of large volume, low value and requiring minimal processing before application in an alternative industry. This can be seen in the successful utilization of mineral processing tailings as aggregate in the construction industries. These types of wastes are more attractive for most mineral processing companies due to the simplicity of alternative applications and lower costs involved in getting value from them. Thus, if the material can be used in its as-is-state in the downstream applications, there will be higher probability of gaining interest from the waste stream producers for a potential alternative use at an industrial scale.

The other alternative is that the waste should be a source of high-value rare material for which there is a high value demand on an international scale (Mitchell et al., 2003). A typical example is the reprocessing of metallurgical waste and scrap for the recovery of rare earths, titanium, cobalt and other strategic metals.

Efforts have been made towards assessing the economic viability of resource recovery from mining and mineral waste in the project known as the 'Minerals from Waste' (Mitchell et al., 2003). This project was set up and funded by the UK Government's Department for International Development and was run from 1999 to 2002. It was aimed at improving the sustainability of current and former mining and quarrying activities through the investigation of the utilization of the mineral waste as a source of construction and industrial minerals (Mitchell et al., 2003). A series of case studies were also implemented at closed (Namibia) and operational mining sites (Costa Rica) to assess the technical, economic, environmental and societal issues related to utilization of mineral waste from such operations.

In Costa Rica, silica sand waste was evaluated for utilization in ceramic and brick clay, whilst in Namibia, pegmatite mine waste was evaluated for production of mica and feldspar. Both projects were deemed to be economically sustainable. However, significant changes needed to be made in the overall process approach, for example, in terms of the technology applied or the level of quality of the product in order for the projects to be viable. These projects, therefore, indicated that there must be compelling evidence that financial, environmental and community benefits outweigh the project costs and risks for the project to be deemed viable.

10.2.4 Market Availability and Acceptance of the Secondary Products

It is important to note that there is value from the waste only if there is an available market. The market conditions prevailing should not only be considered at the time of reprocessing or reutilization, but a long-term viability of the project should also be undertaken to hedge against the uncertainty of long-term economic factors. For instance, when the waste material is earmarked for the building and construction industry, periodic downturns in the heavy construction industry may diminish the need for large volumes of construction materials at certain time periods. Similarly, the volatility of metals' prices can quickly make a reprocessing strategy impractical and costly. A typical example is the current mining downturn cycle which has significantly impacted the market value of most metal commodities. A few years ago, the price of gold and PGMs skyrocketed, and a lot of small mining companies mushroomed with the focus being the reprocessing of tailings in order to take advantage of the margins in profit. However, the last couple of years have seen a massive slump in the price of most of the commodities. As a result, most of the waste reprocessing companies have closed down, leaving behind unfinished projects, massive waste dumps and the environment in a much worse condition than before the reprocessing or reutilization project.

The location of the market for the valuable product from waste is also an important factor to consider. For the alternative reuse of the waste in a secondary industry to be viable, the waste material should be located as close as possible to the potential market. This is because transportation costs are generally the single factor mainly responsible for limiting the use of mining and metallurgical wastes (Collins and Miller, 1979). If an established market is lacking, the cost of shipping products from the site of the waste material is often substantially higher than comparable shipping costs for conventional materials (Collins and Miller, 1979). Thus, if, for instance, the waste has to be used as construction material or aggregate, the waste must typically be located within a reasonable proximity of towns or construction projects requiring the material.

In addition to market and its location, the reuse of waste material in alternative products should be environmentally safe and acceptable to regulators, stakeholders and the public or consumers. Their properties must be determined in order to assess the potential environmental, health and safety impact that they may have if reused and/or recycled into large-scale applications or consumer products. In such cases, the consumer will not only need to compare the quality of the product from waste with that of an existing product generated from a primary or virgin resource, but also be aware of any potential future impacts. Thus, information in terms of quality, costs and risk would be very important to consider in relation to the products from virgin material. For example, consider the performance of a fertilizer or soil additive applied in brick making and both derived from mining and metallurgical wastes. These would need to be quantified and product quality controlled so that its application could be assessed in relation to other possible fertilizers or commercial bricks and thus, enable the customer to have confidence in the product. These comparative data are in most cases not readily available, thus, further limiting the marketability of the waste-based product (Klauber et al., 2011). Incentives may also need to be supplied to the consumer in order to boost the sales of the waste-generated product over the one generated from the traditional primary raw materials. These could be provided through direct government support, collaborative arrangements between industry, community and government or any combination thereof (Klauber et al., 2011). Even then, the consumer would have to weigh this immediate incentivizing benefit against any perceived associated long-term potential future risks.

In addition, investors and technology vendors may also be reluctant to enter into projects that could lead to long-term harm and subsequent financial liability. Thus, the potential future environmental impact, public health and safety liabilities may inhibit interest in reuse and reprocessing of mining and metallurgical waste material. This suggests a need to develop predictive tools and reliable, field-tested modelling of long-term waste behaviour in alternative applications.

10.3 Future Outlook

Despite the challenges mentioned earlier, it is clear that a lot of countries are focusing on the research and development of alternative metal sources, resource recycling and residue utilization in other industries for sustainability. This is because an income generated from a product from the waste can reduce the financial burden of those left with the responsibility for ongoing monitoring of waste after mine closure, for example. In addition, the production of a saleable product from waste even on a small scale can provide continuous employment, sustaining at least some of the population in the former mining community as long as the market for the product exists and the waste resource remains. Definitely, this approach would help to minimize the formation of what is commonly known as 'white elephant' or 'dead' mining towns. 'White elephant' is a largely common occurrence, especially in developing countries, and can significantly curtail any strides in development that had been made during the mining activities and also accelerate poverty in the community.

Nevertheless, it has been noted that a number of processes or technologies in terms of resource recovery from waste have been proposed, but not implemented. It must be noted also that to successfully convert the waste into a new valuable and marketable material or, to suit the needs of the alternative secondary application, requires much more than the development, refining and optimization of current available technologies.

10.3.1 Waste Management and Utilization Options

The previous sections on challenges have highlighted the need for product market development, techno-economic environmental impact risks analysis and the development of standards and regulations. Putting all these information together suggests that there is a need to have a more strategic and synergic approach in dealing with the waste material for maximum benefit. The following sections highlight some of the potential approaches in attaining value from mining and metallurgical waste material.

10.3.1.1 Zero Waste Process Approach

The practice of mineral processing plants in the past has been to focus on optimization in order to selectively target and recover the valuable mineral species in the most economical way possible, with minor consideration of the environmental outcomes occurring further downstream. However, as seen from various chapters in this book, it is clear that there are more benefits to be gained by avoiding the generation of waste in the first place. This is because this approach has the least environmental impact and possibly involves the least energy and costs spent on waste disposal. Figure 10.1 shows a well-established

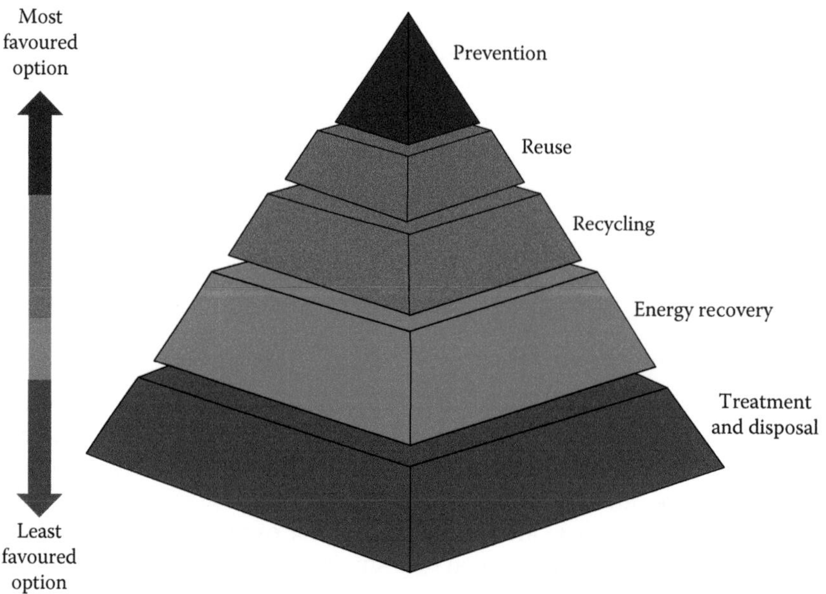

FIGURE 10.1
The mine waste hierarchy, from prevention, through reuse, recycling and energy recovery, to treatment and disposal. (From Lottermoser, B.G., *Elements*, 7, 405, 2011.)

guide for prioritizing waste management practices, with waste prevention being the most preferred option (Hudson-Edwards et al., 2011; Lottermoser, 2011; Mohanty, 2011). Although recycling and reuse have an economic benefit through resource and value recovery, there is an associated cost since a viable and economic process has to be developed and applied for such benefits to be realized.

Although waste prevention has more benefits, it is, however, the most difficult to accomplish. Nevertheless, the minimization and management of waste in metal extraction operations should also focus on waste prevention or minimization through eco-friendly designs and products which include the incorporation of plans for the rehabilitation of mining and metallurgical wastes. This should also include optimization of existing processes for increased efficiency and process intensification in order to increase metal recovery. Most commonly used preventive environmental management practices include waste minimization, pollution prevention, cleaner production, eco-efficiency and industrial ecology (Van Berkel, 2007). These practices tend to promote integrated preventative environmental strategies to processes, products and services throughout in order to reduce or eliminate the creation of pollutants or wastes at their sources (Hilson, 2000). Although eco-efficiency (EE) is not yet widely embraced and used as a framework for enhanced operation and design of processing plants, it has, however, been implemented in existing Australian minerals processing and metals production operations (Van Berkel, 2007).

10.3.1.2 Synergy between the Residue Producers and Residue End Users: Industrial Symbiosis

It is worth noting that there is a significant form of commonality in terms of the potential utilization of some of the waste generated by the extraction processes for different metals. For example, it has been noted that waste material such as jarosite generated during the hydrometallurgical production of zinc and the red mud produced during the production of alumina can both be used as raw material for the construction and building industries such as bricks and additives in cement making, catalysts, glass and ceramics industries. The same can be said for waste produced in the mineral beneficiation and pyrometallurgical processes. These wastes tend to contain a lot of silica and iron which make them good feed material for construction, ceramic and tile industries. Unfortunately, most of the mining and metallurgical wastes have one common disadvantage, which is lack of cohesion, even in very fine fractions. This lack of cohesion or cementitious bond between particles indicates a high potential for slope erosion when used as fill material, thus, there needs to be more research and information on the most effective use of these materials if long-term application is to be maintained. An important consideration in this case would be the incorporation of other industrial waste streams in order to make up for the properties lacking for the ultimate application or conversion into another useful product.

The information given earlier suggests a great need for a more integrated approach, with more cross sharing of ideas and technologies in the research and development sector in order to develop sustainable waste utilization processes. This is achievable only by overcoming the 'artificial boundaries' that tend to be created by researchers and industry players who focus only on one particular area of product development. The idea of industrial waste material synergies and complementary chemistries needs more consideration as it would allow for the development of technologies and possibly lead to more comprehensive exploitation of the different characteristics of the waste material.

There is, therefore, a need to have a more cohesive, synergic and strategic approach to waste utilization in order to realize sustainability in the minerals industry sector. The production and waste generation processes should be adequately oriented towards a structured production route that promotes the waste of one industry as a valued input to another, what is known as industrial symbiosis. Industrial symbiosis is described as the examination of traditionally separate industries located in close geographic proximity as an integrated industrial complex so as to achieve competitive advantages by physical exchanges of previously unused by-products (materials, energy, water) and cooperative usage of infrastructural facilities (energy plants, water treatment plants, treatment facilities, etc.) (Lowe, 1997; Chertow, 2000).

In the context of the material presented in this book, this in essence means a producer of waste material should be considered as 'a supplier of raw material' to the end user (an alternative producer) which may have value for the waste material or use it as a substitute to another feed stream. In contrast, with the philosophy of cleaner production, the main aim is not to minimize the amount of generated waste, but to produce 'valuable waste' that can be reused as a raw material at another industry (Golev and Corder, 2012).

However, opportunities for this approach could be limited by mismatches between waste stream composition and downstream process specifications. Therefore, downstream activities are only viable with the support of the upstream supplier, and a collaborative approach between the two proponents through a reprocessing or recovery-oriented waste management system would be a more effective approach. What would then be essential in this case is that the properties required for the waste material to meet the requirements as feed to the next production plant could be re-engineered in the primary processing plant so that a suitable stock material is generated. This would require deliberate development planning and continuous improvement of cooperation between industries. Clearly, if the producers of the waste were to consider themselves as suppliers of a resource material for application in an alternative neighbouring process, there would be more effort placed in the design of the mainstream production plant and management of the generated waste, that is, design for the future concept.

Once a potential technical synergy match has been identified, both the by-product generating and receiving operations need to identify the business case in developing this synergy (Van Beers et al., 2007). This approach can provide a sustainable business model that generates revenue from waste, reduces corporate liability and delivers an overall improvement to the environment and the communities at large. Proven and viable process technologies, legislation that encourage environmentally beneficial initiatives, enhanced information sharing and closer collaboration between industries are, however, key in providing significant improvements towards such sustainable developments.

The readers are referred to works by Golev and Corder (2012) on regional resource synergies and Van Beers et al. (2007) for case studies and benefits of industrial synergies with particular reference to the Australian Minerals Industry.

10.3.2 Process Integration and Intensification

10.3.2.1 An Integrated Approach to Mineral Waste Utilization

According to Edraki et al. (2014), an integrated approach for mining and metallurgical waste research linked to better processing options will be beneficial. For instance, in order to improve the environmental outcomes, appropriate processes can be included that may not necessarily have short-term cost benefits, but will result in reduced environmental problems and long-term value. These can include, for example, removing pyrite and other sulphides from ores by flotation prior to disposal as well as separating out appropriate clay minerals that can be used for capping purposes (Bruckard and McCallum, 2007). The isolated sulphide component can be used as a source of sulphur for the production of sulphuric acid.

Another approach would be to increase vertical integration of the manufacturing process, either by means of direct ownership or by forcing the firm at the end of the manufacturing value-added chain to take responsibility for the use and disposal of the product (Ayres, 1997). The immediate incentive is that firms will realize that they are more and more likely to be held liable for future environmental harm or health problems associated with their products (or wastes), even if those products were made by another firm or if the potential for harm was not known when the product was originally designed and produced (Ayres, 1997). This would mostly likely drive the firms to think of their products as assets to be conserved, resulting in more emphasis on reuse and recycling.

10.3.2.2 Process Intensification

It is seen from various chapters of this book that the mining and metallurgical industry must develop innovative recovery technologies that essentially turn waste into a resource. In other words, new and more efficient technologies (physical, chemical and thermal) must be developed for recycling and recovery of metals from waste and end-of-life products. Process intensification is one of the most promising development paths for these requirements. Ideally, process intensification consists of the development of novel equipment and techniques that, compared to those commonly used today, are expected to bring dramatic improvements in manufacturing and processing, substantially decreasing equipment-size/production-capacity ratio, energy consumption or waste production and ultimately resulting in cheaper, sustainable technologies (Stankiewicz and Moulijn, 2000). In view of this definition, process intensification can generally be divided into two areas as follows:

1. Process-intensifying equipment, such as novel reactors, and intensive mixing, heat-transfer and mass-transfer devices.
2. Process-intensifying methods, such as new or hybrid separations, integration of reaction and separation, heat exchange or phase transition (in so-called multifunctional reactors), techniques using alternative energy sources (light, ultrasound, etc.) and new process-control methods (like intentional unsteady-state operation).

Obviously, there can be some overlap. For example, new methods may require novel types of equipment to be developed and vice versa, while novel equipment already developed sometimes make use of new, unconventional processing methods.

From the mining and metallurgical point of view, by combining different alternative sources of metals (municipal waste, aqueous wastewaters, electronic goods, low-grade ore and soils) with novel and benign technologies, a closed-loop recycling can be achieved (Górak and Stankiewicz, 2012). Some of the new, efficient, clean and cheap methods needed for the recovery of metals which ensure complete closed-loop recycling include the use of hyperaccumulating plants, biosorption using biomass and bacterial leaching (Dodson et al., 2012). Integrating novel methods with existing metallurgical recovery processes may also be beneficial and necessary, particularly for the purification, concentration and separation of the valuable/targeted metals from impurities (Dodson et al., 2012). The readers are referred to Schonewile et al. (2012) for detailed examples of the applications of clean and novel metallurgical processes.

10.4 Concluding Remarks

The increasing global population's insatiable demand for energy and consumer goods, advances in technology and the desire of individual countries to protect their national security suggests that the demand for metals will be on a steep upward trajectory for a long time to come. The material presented in this book also indicates that there is a significant large amount of waste material being generated in the metal extraction and processing industry. For example, the quantity of solid mining and milling wastes produced by mining for metal ores is typically much greater than the quantities of processed metals. There is, therefore, little doubt that the most important factors guiding the direction of the mining and metal extraction industry in the upcoming years will be depletion of virgin ore material and the environment.

Therefore, there is a need to reduce the intake of virgin materials in the mainstream production processes in order for the industry to remain sustainable. The most economic and efficient option is through recycling and reuse. Instead of adopting the traditional approach of viewing ores from mines in the ground and in increasingly remote areas as the source for elements, recovery from the mining and metallurgical industry waste can be considered, as the new 'mining frontier'. The processing of secondary materials allows for not only reducing the ultimate quantities of waste, but also the reduction of primary metals and energy consumption. For example, every ton of metal that is reused or recycled replaces a ton that would otherwise have to be mined and processed, with all of the intermediate energy and material requirements associated with such activities. Most importantly, recycling would open up a previously untapped source of metals in countries where deposits for those particular metals do not or no longer exist. These measures should limit the demand for new supplies of elements and, thus increase the lifetime of metal reserves infinitesimally. Thus, recycling cannot be ignored in a world attuned to sustainability and environmental and energy-related issues. Increasing the recycling rate and reuse of waste has the potential of providing an improvement in resource efficiency, offering a competitive alternative to an increasingly expensive landfill, incineration and other treatment options, and further offering potential socio-economic benefits to the local communities in close proximity to these operations.

The work presented in this book has also highlighted some of the challenges associated with the value recovery from waste which is evidenced by the lack of significant commercial implementation of some of the proposed recycling and reuse technologies. The current outlook to overcoming some of these barriers is centred on process integration, optimization and through exploitation of synergies existing within the different industries, that is, industry symbioses. It is only through such an approach that the concept of waste as a resource can be realized, thus helping to safeguard the uninterrupted supply of raw materials and products crucial for the global economy and securing the future for the human race in a sustainable, economically viable and environmentally friendly way.

References

Ayres, R.A. (1997). Metals recycling: Economic and environmental implications. *Resources, Conservation and Recycling* 21: 145–173.

Bansal, P. and McKnight, B. (2009). Looking forward, pushing back and peering sideways: Analyzing the sustainability of industrial symbiosis. *Journal of Supply Chain Management* 45(4): 26–37.

Bian, Z., Miao, X., Lei, S., Chen, S., Wang, W. and Struthers, S. (2012). The challenges of reusing mining and mineral-processing wastes. *Science* 337: 702–703.

Bruckard, W.J. and McCallum, D.A. (2007). Treatment of sulphide tailings from base metal and gold operations – A source of saleable by-products and sustainable waste management. *World Gold Congress*, Cairns, Queensland, Australia, 22–24 October, pp. 85–91.

Buchert, M., Schüler, D. and Bleher, D. (2009). Critical metals for future sustainable technologies and their recycling potential. Available at http://www.unep.fr/shared/publications/pdf/DTIx1202xPACritical%20Metals%20and%20their%20Recycling%20Potential.pdf. Accessed 15 June 2016.

Chertow, M.R. (2000). Industrial symbiosis: Literature and taxonomy. *Annual Review of Energy and the Environment* 25: 313–337.

Christopherson, S., Michie, J. and Tyler, P. (2010). Regional resilience: Theoretical and empirical perspectives. *Cambridge Journal of Regions, Economy and Society* 3(1): 3–10.

Collins, R.J. and Miller, R.H. (1979). Utilization of mining and mineral processing wastes in The United States. *Minerals and Environment* 1(1): 8–19.

Dodson, J.R., Hunt, A.J., Parker, H.L., Yang, Y. and Clark, J.H. (2012). Elemental sustainability: Towards the total recovery of scarce metals. *Chemical Engineering and Processing* 51: 69–78.

Edraki, M., Baumgartl, T., Manlapig, E., Bradshaw, D., Franks, D.M. and Moran, C.J. (2014). Designing mine tailings for better environmental, social and economic outcomes: A review of alternative approaches. *Journal of Cleaner Production* 84: 411–420.

Environmental Protection Agency (EPA) (2016). Sustainable management of industrial non-hazardous secondary materials. Available at https://www.epa.gov/smm/sustainable-management-industrial-non-hazardous-secondary-materials. Accessed 5 June 2016.

Golev, A. and Corder, G.D. (2012). Developing a classification system for regional resource synergies. *Minerals Engineering* 29: 58–64.

Górak, A. and Stankiewicz, A.I. (2012). Research agenda for process intensification towards a sustainable world of 2050. Available at http://www.ispt.eu/media/DSD_Research_Agenda.pdf. Accessed May 2016.

Gordon, R.B. (2002). Production residues in copper technological cycles. *Resources, Conservation and Recycling* 36: 87–106.

Gunther, P. and Mey, W. (2008). Selection of mine water treatment technology for the Emalahleni (Witbank). *Water Reclamation WIsA Biennial Conference*, Sun City, South Africa, 18–22 May 2008.

Hilson, G. (2000). Pollution prevention and cleaner production in the mining industry: An analysis of current issues. *Journal of Cleaner Production* 8: 119–126.

Hudson-Edwards, K.A., Jamieson, H.E. and Lottermoser, B.G. (2011). Mine wastes: Past, present, future. *Elements* 7: 375–380.

IAEA (International Atomic Energy Agency) (2004). The long term stabilization of uranium mill tailings. Available at http://www-pub.iaea.org/MTCD/publications/PDF/te_1403_web.pdf. Accessed May 2016.

Klauber, C., Gräfe, M. and Power, G. (2011). Bauxite residue issues: II. Options for residue utilization. *Hydrometallurgy* 108: 11–32.

Lottermoser, B.G. (2011). Recycling, reuse and rehabilitation of mine wastes. *Elements* 7: 405–410.

Lowe, E.A. (1997). Creating by-product resource exchanges: Strategies for eco-industrial parks. *Journal of Cleaner Production* 5(1–2): 57–65.

Mitchell, C.J., Harrison, D.J., Robinson, H.L. and Ghazireh, N. (2003). Minerals from Waste. Recent BGS and Tarmac experience in finding uses for mine and quarry waste. Available at https://www.bgs.ac.uk/downloads/start.cfm?id=1393. Accessed May 2016.

Mohanty, C.R.C. (2011). Reduce, reuse and recycle (the 3Rs) and resource efficiency as the basis for sustainable waste management. Available at http://www.uncrd.or.jp/content/documents/02_Mohanty_CSD19-Learning-Centre.pdf. Accessed May 2016.

Schonewile, R.H., Rioux, D., Kashani-Nejad, S., Keuh, M. and Muinonen, M.E.S. (2012). Towards clean metallurgical processing for profit, social and environmental stewardship. Available at http://web.cim.org/com2015/conference/Pyrometallurgy_Towards_Clean_Metallurgical_Processing_2012.pdf. Accessed May 2016.

Scott, P.W., Eyre, J.M., Harrison, D.J. and Bloodworth, A.J. (2005). Markets for industrial mineral products from mining waste. In: Marker, B.R., Petterson, M.G., McEvoy, F. and Stephenson, M.H. (eds.), *Sustainable Minerals Operations in the Developing World*. Geological Society, London, UK, Special Publication 250, pp. 47–60.

Stankiewicz, A.I. and Moulijn, J.A. (2000). Process intensification: Transforming chemical engineering. *Chemical Engineering Progress* 96: 22–34.

Van Beers, D., Corder, G.D., Bossilkov, A. and Van Berkel, K. (2007). Regional synergies in the Australian minerals industry: Case-studies and enabling tools. *Minerals Engineering* 20: 830–841.

Van Berkel, R. (2007). Eco-efficiency in the Australian minerals processing sector. *Journal of Cleaner Production* 15: 772–781.

Weber, R.J. and Reisman, D.J. (2012). Rare earth elements: A review of production, processing, recycling, and associated environmental issues. Available at https://clu-in.org/download/issues/mining/Hard_Rock/Wednesday_April_4/05_Rare_Earth_Elements/01_Weber_R.pdf. Accessed June 2016.

Zheng, G.-H. and Kozinsk, J.A. (1997). Solid waste remediation in the metallurgical industry: Application and environmental impact. *Environmental Progress and Sustainable Energy* 15(4): 283–292.

Index

B

C

D

E

R

S

Y

Z

Milton Keynes UK
Ingram Content Group UK Ltd.
UKHW050457071024
449327UK00015B/414